ISBN 978-0-282-48105-6
PIBN 10853182

English
Français
Deutsche
Italiano
Español
Português

www.forgottenbooks.com

Mythology Photography **Fiction**
Fishing Christianity **Art** Cooking
Essays Buddhism Freemasonry
Medicine **Biology** Music **Ancient
Egypt** Evolution Carpentry Physics
Dance Geology **Mathematics** Fitness
Shakespeare **Folklore** Yoga Marketing
Confidence Immortality Biographies
Poetry **Psychology** Witchcraft
Electronics Chemistry History **Law**
Accounting **Philosophy** Anthropology
Alchemy Drama Quantum Mechanics
Atheism Sexual Health **Ancient History**
Entrepreneurship Languages Sport
Paleontology Needlework Islam
Metaphysics Investment Archaeology
Parenting Statistics Criminology
Motivational

THE

OPERATIVE MECHANIC,

AND

BRITISH MACHINIST;

BEING A

𝔓𝔯𝔞𝔠𝔱𝔦𝔠𝔞𝔩 𝔇𝔦𝔰𝔭𝔩𝔞𝔶

OF THE

MANUFACTORIES AND MECHANICAL ARTS

OF THE

UNITED KINGDOM.

By JOHN NICHOLSON, Esq.
CIVIL ENGINEER.

LONDON:

PRINTED FOR KNIGHT AND LACEY,

PATERNOSTER-ROW;

AND WILDER AND CAMPBELL, 142, BROAD-WAY, NEW YORK.

1825.

Printed by D. SIDNEY, and Co.
Northumberland-street, Strand.

TO

GEORGE BIRKBECK, Esq. M.D.

PRESIDENT OF THE LONDON MECHANICS' INSTITUTION,

&c. &c. &c.

Sir,

In an age like the present, when the rich and the powerful identify their interests with the welfare of the poor and uninformed, when the wise and the good combine in furthering the diffusion of sound principles and useful knowledge among those who constitute the most important, though hitherto the most neglected, portion of the community, there is not one who can view the future in the past but must anticipate with such data before him, a change as brilliant in its effects, as it is honourable to those who are engaged in promoting it.

The advanced state of science, and the comprehensive views of a just and liberal philosophy, animate those who for many years have compared theory with practice to come forward in the hope of being able to offer something in aid of the common cause.

a 2

Such feelings, Sir, have encouraged me to publish the following pages, which, as an earnest of their future success, I am permitted to dedicate to yourself.

A Work of this kind, combining in the most condensed form the acknowledged principles and recent improvements in Mechanical Science, and professing to be adapted in every possible way to the use of the Mechanic and Machinist, could not well find a Patron more congenial to its Spirit than one, who, during a long series of years, has laboured with no common devotion in promoting their benefit.

I am,

Sir,

Your most obedient, and

much obliged humble Servant,

JOHN NICHOLSON.

at all times important in the construction of Mill-work, more particularly in those parts which have to sustain the greatest force, or put the whole of the other parts of the machinery in motion, we have, next to the Moving Powers, inserted a letter from Mr. RENNIE, jun. to Dr. YOUNG, describing a series of very satisfactory experiments made on this subject.

A description of HYDRAULIC ENGINES next follows; and these are succeeded by certain SIMPLE MACHINES acting as accessories to our manufactures. So that, by the time the reader has advanced thus far, he will have become so thoroughly intimate with machinery, as easily to comprehend and appreciate the several excellencies of our STAPLE MANUFACTURES, which are next unfolded to his view.

The whole was intended to be concluded with an examination of those arts termed MANUAL, in a Treatise on the Art of BUILDING; except, indeed, with the addition of an Appendix, containing a short and concise treatise on PRACTICAL GEOMETRY and MENSURATION, with a Collection of approved RECEIPTS, and a GLOSSARY; but the interest which has lately been excited respecting RAILWAYS and LOCOMOTIVE ENGINES has led to the extension of the Work, about thirty pages, with an article on those constructions.

Although there are several very excellent treatises on Mechanics and Mill-work now extant, yet, presuming on an arrangement widely different to that of others, by which the least erudite and most inexperienced may acquire something more than a mere Superficial Knowledge of Machinery, the Author trusts that the following pages will meet with a favourable reception.

In the course of his labours he has derived material assistance from many of his scientific friends, to whom he thus publicly expresses his acknowledgments; and more particularly to that Gentleman to whom the volume is dedicated.

∴ In a Work of such a nature it is generally understood that extracts are justified, as the description of many things not new are requisite, and the language could not in general be improved. In such cases, however, the authority has, in general, been acknowledged, and in a way calculated to advance the honour and interest of every improver and discoverer.

The volume in its design and execution is offered as a companion to THE WORKSHOP, consequently abstract and theoretical principles have been allowed to mingle no further than has been indispensably necessary to the perfect illustration of the use and application of the object described. The Work has, therefore, no similarity to the Mathematical Illustrations of WOOD, GREGORY, or EMERSON, each of which, and more particularly that of Dr. OLINTHUS GREGORY, deserves to be spoken of with great respect.

'A Book comprehensive and practical, embracing the whole subject as living and contemporaneous, and as connected with private profit and public glory, instructive to individuals and illustrative of the genius of the age in its best direction, has been the object of the Author, and *he* hopes he has not laboured in vain.

LONDON,
March, 1825.

CONTENTS.

ON MATTER.

MECHANICAL POWERS.

MILL GEERING.

CONTENTS.

ANIMAL STRENGTH.

WATER.

WIND.

SIMPLE MACHINES ACTING AS ACCESSORS TO MANU-
FACTURES.

DESCRIPTION OF THE FRONTISPIECE.

—————

This Plate represents a front view of a Steam-Engine connected with a Sugar-Mill, as constructed by Messrs. Taylor and Martineau, who have kindly permitted our draftsman to make a drawing of it.

This Engine, being only twelve times larger than the drawing, is, from its compactness and simplicity of construction, peculiarly applicable to most of the manufactures round the metropolis, where power of a moderate amount is in general required.

It works horizontally, at from 30 to 40lbs. pressure per square inch, without condenser, having metallic pistons and slide-valves, and only requires eight screw-bolts to fasten it to oak sleepers, or frame-work of moderate scantling.

A is a crank connected with the piston-rod, which, as it works in the cylinder horizontally, cannot be seen. B is the cylinder, into which steam is admitted from the boiler, by means of the pipe C C C. The amount of steam flowing into the cylinder is regulated by the throttle valve at D, which is opened and shut at proper intervals by the rod E E E. F F is the governor, or regulator, consisting of two heavy balls, with the sliding collar a, suspended from the top of a vertical spindle b b. at the axis c. This spindle is connected with the main shaft, by a strap passing over the sheeves or pullies, G G G, which cause it to revolve; and as its speed varies with that of the main shaft, the governors F F, according as its speed increases or decreases, have a tendency either to fly from, or approach to, the spindle. This rise or depression of the governor affects the rod E E E, to which it is connected, and regulates the quantity of steam flowing from the boiler into the cylinder.

H is a piece to connect the top part of the piston-rod with the rod I, so that by the motion of the crank the rod I is also moved, which rod moves the slide valves in the cylinder K. By the action of these valves, steam is alternately admitted on the opposite sides of the piston; and as the engine does not condense its steam, there are two pipes, placed one at each end of the cylinder, to carry it off. One of these pipes is seen at N. When the piston has been driven by the force of the steam to the other extremity of the cylinder, the steam, by the action of the slide valves, is shut off from this end, and allowed to flow into the opposite end of the cylinder; the orifice of the pipe N being at the same time opened, the steam at this end is, by the returning action of the piston, driven through the pipe N, and conveyed away under ground, leaving this end of the cylinder ready for a fresh supply.

The power generated by this simple arrangement is made to effect the required purpose by means of the shafting O O O. On this shafting, at a little distance from the engine, is an eccentric, L, to raise the rod M, to pump water into the boiler when required: and at nearly the further end of the shafting is another eccentric, W, which imparts motion to the rod V, for the purpose which we shall hereafter describe.

The rotatory motion which the crank has received from the engine is imparted to the shafting, to the eccentric L, the coupling-box d, the fly-wheel P, the eccentric W, and the pinion Q, which plays in the large cog-wheel, R, on the shaft S, and thence is imparted to the rollers of a sugar-mill, which rollers are moved at equal speeds by the pinions U U.

In this, and most other Sugar-mills, there are three rollers, two at the bottom, and one lying between the other two at the top. Through these rollers sugar canes are passed, and the compressed juice falls into a receiver, from whence it is pumped, by the movement of the rod V, into a copper, or other receiver. At that part of the shafting marked e e, sufficient space is left to allow of play when the canes are passed through the rollers, otherwise the shafting would be very apt to snap and be destroyed.

OPERATIVE

MECHANIC AND MACHINIST.

OF THE ACTION OF FORCES.

ALL matter is continually under the operation of forces, which, if acting upon it equally, and in opposite directions, maintain it in a state of rest. But if a newly created force act upon a body in such a direction, and to such an extent, as to overcome the forces under whose action that body, in common with all other matter, exists, the result will be motion communicated to that body; and in an exact proportion as that newly created force exceeded the amount of forces that were previously acting upon it in the opposite direction. For example, if a man lift a pound weight three feet from the ground, the amount of motion created by that action is exactly equal to the amount his newly created force exceeded the force of gravity or weight which acted on the matter: for if his force had not exceeded the force of gravity, it is manifest that that motion could not have been created; and if the force of gravity had not existed, it is again manifest that the amount of motion would be exactly in proportion to the amount of the whole of the force he had applied. Again, if his force only exactly equalled, and did not exceed the force of gravitation, motion could not have taken place, and the body would have remained at rest.

This state of rest, maintained by the contrary action of two equal forces upon a body, is called *equilibrium*. But the term equilibrium is most commonly applied when one or more bodies are, by the mere force of gravitation, maintained in a state of quiescence, or rest: thus, if a bar of iron, A B, fig. 1, is supported at its centre C, it will balance, or remain horizontal, as the quantity of matter in C A is exactly equal to that in C B, and the amount of the gravitating force proportional to the quantity of matter that is in each arm of the bar: likewise, if a ball, A, fig. 2, be acted upon by a force at B, and by another force, exactly equal, at C, the ball A will be maintained in a state of quiescence, termed equilibrious.

In the common operations of mechanics, the former state of equilibrium frequently occurs; the latter rarely, and never with any permanent duration; by the term equilibrium, therefore, in general, is understood, the position first cited.

Upon duly considering that matter, when between forces acting in opposite directions, is in a state of equilibrious quiescence, it will be manifest, that motion cannot be obtained without destroying the equilibrium. It must not therefore be supposed, that the forces of gravitation or adhesive attraction can produce motion, as has been erroneously urged by some, but rather that all the motion these powers are capable of producing was primarily exerted to bring matter into that state of equilibrium in which we find it. Wherever that equilibrium is disturbed by extraneous causes, the resultant motion, attainable by such disturbance of the general equilibrium, has long since been known, and applied to useful purposes. We may with propriety, therefore, deduce from these considerations, the perfect fallacy of that most ruinous and speculative notion of a perpetually moving force. Many who have wasted their time in attempts to attain that object, have either supposed that the force of gravitation could obtain motion, or that motion once obtained could of itself increase its force; which was about as rational as to suppose that any substance could of itself increase its own bulk. The powers with which nature has supplied us, have, as far as we are aware of, been already applied; and should there be others existing of which we are ignorant, or which we have not reduced to our command, the search for, and developement of such objects, are praiseworthy and valuable: but let us with confidence hope, that the labours of ingenuity will no longer be drawn aside from the paths of prolific study, by this destructive phantasy.

Returning from this digression, when a body is operated upon by a force, and acquires motion, that motion, taking into account the amount of space through which the body passes in a given time, is called *the velocity of the body;* and according as the extent of distance increases or decreases in a greater or less period of time, the velocity is said to . increase or decrease.

If a force acting upon any body, and causing motion, shall continue to act upon it in the same direction, so as to continue to increase that motion, the body, under such circumstances, is said to attain *accelerated velocity.* And if a body be put in motion by a certain force, and another force operate upon it in a contrary direction, so as to tend to bring it to a state of equilibrium; such motion is called retarded motion.*

* The simplest example of accelerated motion, is exhibited in the action of the force of gravitation upon a falling body, where the force continues to act during its descent, and regularly increases in velocity; so that if a body A, fig. 3, be allowed to fall from that position towards the earth, it will pass through sixteen feet during the first second of time, forty-eight feet during the next, and eighty feet during the third. Had its motion been regular during these three seconds of time, it would have passed through only three times sixteen, or forty-eight feet, whereas it has passed through one hundred and forty-four feet, by reason of the force which first caused its motion continuing to act upon it. Now as its velocity increases regularly, we may conclude, that during the performance of the first half of the sixteen feet, it was not proceeding at the rate of sixteen feet per second; and if we suppose it was proceeding only at half that velocity, then it must have travelled through the second half at the rate of thirty-two feet per second; or, if the first eight feet took three quarters of the second, the second eight feet must have been performed in the remaining quarter, therefore, when the body arrived at B, it would be proceeding at the rate of thirty-two feet per second; to which, if we add the force that continues to urge it at the rate of sixteen feet per second, it will exhibit, for the second space, a velocity of forty-eight feet per second: and if for the third space we double its increasing velocity of thirty-two feet, and add that created by the continued force, we shall have twice thirty-two, and sixteen, are eighty, which is the result of experiment. The velocity of bodies under the continuous action of any given force, will, it appears, increase as the odd numbers 1, 3, 5, 7, 9, &c., that is, sixteen feet during the first second, thrice sixteen feet during the next second, five times sixteen during the third second, and so on; or, as the relative portions of the superficial space under equal parts of the perpendicular in a right angled triangle, as represented at fig. 3: where 0 to 1 represents the first second of time, 1 to 2 the second, and 2 to 3 the third. It will be perceived, that under each of these portions, the space contained in the triangle will be as 1, 3, 5; such is uniform accelerated motion. But if the continuous force, which has been shown to increase the velocity, vary in its action upon the body, it is plain the increase will be no longer uniform.

From a clear comprehension of the acceleration of motion in bodies the retardation of motion will be easily conceived: for example, if a body be cast perpendicularly from the earth, as in the firing of a shot from a cannon upwards, the force of the powder, overcoming the force of gravitation, will cause the ball to rise with a certain velocity, whilst that attraction continuing to operate in the opposite direction, checks by regular gradations the created force, and eventually destroys it. Thus the distance which the shot would have accomplished during the first second of time, is reduced by sixteen feet; that which it would have accomplished during the next second, by forty-eight; and so on until the created power is counterbalanced by the force of gravity, and the ball arrives at a state of rest; when the force of gravity acting upon it solely, will cause it to move in the opposite direction, till it descends to the earth

When a ball, attached to a centre by a flexible cord, is put in motion by any one force, which, in common with all other forces, acts in a right line, the motion will be circular. The tendency which such body has to fly from the centre, is called the centrifugal force; and that exerted by the cord to draw it towards the centre, the centripetal force.

When a body is set in motion by any force, it is enabled, to a certain extent, to act on other bodies, and create motion in them; and, as the velocity it obtained was as the power expended to create that motion, so is the power of transmitting that motion to its velocity. This power of communicating motion, or, in other words, this force possessed by matter in motion, is termed *momentum*, or the moving force; and the mode of transmitting it, *impact*; as this force is proportional to the velocity possessed by every particle of matter composing any body, the momentum must be represented by the quantity of matter multiplied by its velocity. For instance, suppose one hundred particles of matter were moving at the rate of one foot per second, the power requisite to overcome their force is exactly the same as that which would be necessary to arrest the motion of one particle moving at the rate of one hundred feet per second: for the velocity of the hundred particles being one foot per second each, their total force would be the force existing in one of them multiplied by one hundred: and again, as the force is in proportion to the velocity, one particle moving at the rate of one foot per second, multipled by one hundred in regard to velocity, will produce a similar result. Also, if a body of one pound weight be moving at the rate of one foot per second, it will possess a certain *momentum*, and if either its weight or its velocity be doubled, its momentum will be likewise doubled: if both be doubled, the momentum will be quadrupled.

Having now considered the action of one and two forces acting together in opposite and similar directions, we will proceed to examine the action of two forces upon a body, acting neither in the same, nor in contrary directions. Thus, if the line A B, fig. 4, represent a force sufficient to carry the body A to the point B, and A C represent another force sufficient to carry the body A to the point C, then A C and A B being equal to C D and B D, and those two forces act upon the body subsequently to each other, we may conceive that the body would, by passing over the lines A B and B D, or A C and C D, be carried to the point D. Now, if they act upon the body at the same instant, the result will be the same, and the total expenditure of the forces will place the body, passing by the line A D, at the point D. Likewise, if the forces A B and A C be not at right angles, as in fig. 5, still as C D and B D are equal, and in similar directions to A B and A C, the motion received from them by A will be represented in amount and direction by the line A D. But supposing A B shall be twice or thrice the power of A C, then the effect will be the same as is shown in fig. 6, where the line A B represents thrice the power of A C. The separate actions of A B and A C will be represented as before by B D and C D, which would place the body A at the point D; therefore their combined force will cause it to pass by the diagonal line A D, as in the former instance. This proves that any number of forces acting upon a body in however many lines, not directly opposite to each other, will be *compounded* into one force : for suppose three forces, A B, A C, and A F, fig. 7, to operate in their several directions at the same instant, on the body A, they will be compounded into the force represented by A I; for if we describe a parallelogram as before by the lines A B and A C, those two forces will be compounded into a force represented by A D; and again, if we do the same with the two forces A C and A F, we shall have the force A H composed of them. We have therefore two forces A D and A H compounded of the three original forces. If we proceed with these two in the same manner, they will be compounded into the force represented by A I; D I and H I completing the parallelogram of which A I is the diagonal: so that any number of forces acting in any number of directions, excepting in opposite ones, may be compounded into one, which is termed their *composant*, and which is always represented by the diagonal of a parallelogram, like that already shown.

The resolution of forces is exhibited by reversing this problem; for as any number of forces may be combined into one force, so may one force be resolved into any number. If a single force be represented by a ball moving with a certain velocity in the direction of the line A B, fig. 8, when it shall come in contact with and act upon the balls C and D, these two balls will each of them move with one half of the velocity with which B was impelled, and in the direction of the lines C H and D I, drawn from the centre of B through each of their centres : so that if the force of B be divided into two equal portions, each of those portions may, by a similar process, be again divided, *resolving* the original force to infinity.

almost universal nature of that resistance called friction; for although the irregularities upon the surfaces of bodies are by no means so manifest as those here represented, still, upon minute examination, we are enabled to discover that the smoothest surfaces contain them; and as the amount of resistance increases in direct proportion as the irregularities present themselves, we are warranted in concluding that all resistance arising from friction owes its origin solely to this cause.

OF THE MECHANICAL POWERS.

THE mechanical powers are six in number, the LEVER, the WHEEL and AXLE, the PULLEY, the INCLINED PLANE, the WEDGE, and the SCREW. A perfect knowledge and thorough appreciation of which should be clearly understood by those who purpose to examine into the effects of mechanical combinations; the whole of which, however intricate, originate from, and are reducible to, one or more of the laws which govern these simple machines.

In demonstrating the mechanical powers, that which is not strictly true must be admitted: the force of gravitation, the retardation of friction, the resistance of the atmosphere, and the irregularity arising from the partial elasticity of the substances of which they are formed, must be excluded, and supposed not to exist.

The first-mentioned power is the *lever*, which is divided into three classes. In fig. 12, A B is a lever, and C the fulcrum, or immovable point on which it rests: now, if a force be applied at B, and the resistance, or the force or weight to be overcome, is at A, then, with the fulcrum so situate between the forces, it is called a lever of the first class; and the operation of the force at B to overcome the resistance at A, will be in proportion as the distance A C is to the distance B C; that is to say, if B C be four times the distance of A C, the force applied at B will be exactly equal to four times the same amount of force at A; or one pound weight at B will counterbalance four pounds weight at A; but to whatever height (suppose one foot) the weight at A be raised, B must descend four times that space, and consequently, to place B in its original position, the force applied must be equal to the raising of four single pounds one foot each, which is the same as the raising of four pounds one foot, as was effected at A.

An actual gain of power does not exist, but the gain in convenience is great; for, by the operation of one pound, four pounds is moved, which, but for the invention of the

lever, could not have been effected. A man whose utmost
strength could lift no more than one hundred and fifty
pounds is by this means rendered capable of giving motion
to four times that weight, although he is obliged to exert his
strength through four times the distance. A lever of the
second class may be represented by supposing A to be the
fulcrum, B the force applied, and C the weight, or resistance
to be overcome. The effect of this lever must be estimated
by comparing the distances C B to A B; the power will increase or diminish exactly in proportion as A B exceeds C B,
and the distance that B moves through will increase exactly
in the same proportion.

Suppose, in reference to the same figure, C to be the force
applied, A the fulcrum, and B the resistance, it will then
represent a lever of the third class. The effect of levers
of this class is to lose power for the purpose of gaining either
motion or distance. For if, in the last mode, the power applied at B increased as the length of A B became greater
than C B, it is plain that in the present case, the resistance
at B is placed in a position to gain by the same law: therefore, the nearer the force is placed to B the greater will be
the effect; and when applied at B, the greatest; but when
the force is at B, it is applied direct to the resistance, and
the lever is abandoned; consequently C, in every position
between A and B, loses power to a greater or less extent.
As the movement of C, in the last case, was one half that
of B, so in the present case will the movement of B be twice
that of C.

In particular operations, levers of each of these classes
have their particular uses. The simplest application of the
first sort may be seen in scissors, shears, forceps, &c. the
pin in the joint is the fulcrum, the hand is the force applied,
and the substance to be cut or pinched is the resistance to
be overcome; the second sort of lever is presented to us in
the cutting knives used by last-makers, where the hand is
the power, the ring into which the other end of the knife
is hooked is the fulcrum, and the object to be cut is the
resistance. Common fire-tongs are levers of the third class,
as they possess a capability of being extended at the extremities: in using them the motion of the hand produces,
perhaps, six times its own motion in the extremities, and a
loss of power exists in a similar proportion; but as they
have to be used only for a short period, the loss of power
is of less importance than the convenience gained. This
last class of lever is frequently introduced in machinery,

for the purpose of obtaining a rapid motion; and as the same object has been aimed at in the construction of almost all animals, we find that nature has introduced it most frequently.

We have considered the operations of the lever by the different dispositions of the acting and resisting forces and the fulcrum, under the supposition that the lines of the direction of the forces were at right angles to the arms on which they operated, or formed tangents to the arcs which the movements of those arms described; but if we alter the form of a lever from that of a right line, and the two forces still maintain their parallel direction, the action on their respective arms will be no longer at right angles, and their effects will in consequence be varied; the mode of estimating those effects must be also changed. Fig. 13, A B C represents a bent lever resting upon its fulcrum B, and having appended to each of its arms the weights D and E, which are equal to each other and in equilibrium, notwithstanding the arm B A is longer than the arm B C. Draw the horizontal line G H, passing through the fulcrum B, then the weights D and E acting in perpendicular lines, we may imagine that D is suspended at the point I, and E at the point K, at which points they will operate similarly. Suppose K I to be the lever, the arms I B and B K will then be equal, and if their distances are multiplied into E and D, which are equal forces, their effects will be equal. The action of parallel forces upon levers, that do not receive the action at right angles to their respective arms, should be estimated by the force multiplied into a line drawn from the fulcrum, perpendicular to the line of direction of each respective force; and whatever may be the form of the lever, it is apparent, that, if the lines of direction vary from being tangents to the arcs described by its arms, their effects must be estimated by the length of perpendiculars let fall upon the lines of direction in a similar manner. Recurring to fig. 13, it will be seen, that if the arm B A rise to the position B L, the perpendicular B I, if allowed to fall upon the line of direction, will be increased from B I to B M, and that the effect of the force D over E will be augmented. This peculiarity is brought into operation in a balance that has a graduated scale on an arc, as A G, the divisions of which arc decreases as it rises in such a manner as to exhibit, by the movement of A, equal portions of force acting upon E. There is however a case of common occurrence, namely, that of drawing a nail with the fang of a hammer, wherein the effects of the applied and resisting forces, being the hand and the nail, although acting with a lever bent at right angles, still operate as though the lever were straight; for the direction of the forces being changed to the exact amount of the same angle as that to which the lever is bent, they continue to act at right angles to their respective arms, which arms will consequently represent perpendiculars let fall from the fulcrum on their respective lines of direction.

The principle of the bent lever is not unfrequently introduced into machinery in order to gain a greater degree of power. Suppose A B C, fig. 14, to represent a bent lever, moving on its fulcrum B, the operating force at A acting in the direction of A D, and the resistance at C in the direction C E : now, as the line of direction of the force C falls upon the fulcrum, it is evident, that no perpendicular can be let fall from the fulcrum upon it, and consequently, the power of C can be nothing in comparison to that of A, whose perpendicular upon its line of direction is B A; for the instant the lever begins to move, suppose to A 1, C 1, then the perpendicular on the line of direction of the force C, assumes a mensurable form, as B, B 1, whilst the power of A has only decreased from B A to B F. From this it will be seen that at the commencement of the action of A, the power over C was indefinite, but instantly after the commencement of that action by the movement of C out of the perpendicular E B, resistance likewise commenced, as the perpendicular from the fulcrum then assumed a mensurable amount.

THE WHEEL AND AXLE.

THE next simple machine classed as a mechanical power, is termed the *wheel* and *axle*, and is represented at fig. 15. A the wheel, B a circular bar called the axle, both turning upon one centre, at C. In general, the force is applied by fixing a rope to the outer rim of the wheel, as represented by D, whilst the resistance, or the weight, or force to be overcome, is represented by E, attached by a rope to the axle. By a simple analysis, this machine will be found to be merely a method of obtaining a continual action of levers of the first class; for if we suppose the radius of the wheel to be the longer arm of the lever, the radius of the axle the shorter arm, and the centre on which they turn the fulcrum, we have a lever of the first class; but from these two members being circular, their radii are an indefinite number of levers, and, by the revolving of the wheel, a

█████ of this class are continually brought into
██. The effective power of the wheel and axle must
████████ by the same mode as a lever of the
████ far as the radius of the wheel exceeds that of
██ ██ increases the power, and so increases the distance
█████████ force has to pass through.

████ and axle is applied in the apparatus for raising
████████, and is introduced in many machines, which
█████ as we proceed.

THE PULLEY.

████████, represented at fig. 16, is the third mechanic
██ is of a circular form, fixed upon a pin that runs
█████ centre at C, and round which it revolves. The
████████ the pulley is by placing a rope over its
███ to the extremities of which, at A and B, the force
████████, and the weight or resistance to be overcome,
█████████ attached, the centre C being supported
█████ D. The operations of this instrument are re-
█████ to the action of a lever of the first class; the
█████ it revolves is the fulcrum, and the radii of the
██ ██ the two arms, which, being equal, no augmenta-
█ █████████ of power can arise.

██ used in this manner, the pulley is but a method
█████ the direction of the applied force. But if in-
████████ in fig. 17, where the end of the line A is
███ to a fixed point, the weight or resistance being
███ the applied force acting upwards, the line from
██ permanently fixed, it will become a fulcrum; and
█████████ radii of the circle assume the position of
█ a lever of the second class; which gains in power as
███████ forces of application and resistance are distant
███ fulcrum; as B, for instance, is twice as far from
██ the weight or force applied at B will raise twice its
███ C.

█ combined action of several pullies is called a tackle,
██ ██ where A and B are two pullies fixed in the position
██████ and C D two others, capable of being either
or lowered; the rope E passes over A, under D, over
d under C, and is permanently fixed at F. It is there-
apparent that if the weight G be suspended from the
s of C and D, (both of which are in the position de-
d at fig. 17,) that each of them will divide its force by
and that one quarter of the weight G, placed at E,

THE WEDGE.

THE fifth mechanical power is the *wedge;* one form of which is shown at fig. 21. It operates in a similar manner to the inclined plane ; but instead of the resistance or force to be overcome being moved along its surface, the plane itself, which is now called the wedge, is forced beneath the object to be raised. Thus, if the wedge A B move upon a level plane to the position A 1, the weight D will be raised from its position to the height D 1; and, consequently, will pass over the whole upper plane of the wedge A B, and ultimately attain the perpendicular height B C. If A B be divided by B C, the quotient, as in the inclined plane, will represent the power which the wedge is capable of exerting ; or if A B is four times B C, the power forcing forward the wedge to A 1 is capable of raising the body D four times its own amount to the position D 1. The wedge represented in fig. 22, is most generally applied to the purpose of dividing wood, where the resisting force to be overcome acts on both sides of it. To estimate the amount of power gained by this form of the instrument, we must consider it as two inclined planes, A B C and C B D conjoined ; and as the forces operating at E and F are equal, we shall have, as A C is to C B, so is the resistance F to the force necessary to overcome it ; and as the force E and the other portion of the wedge are similarly opposed, the total A D is to C B, as the total resistance F and E is to the power necessary to be exerted to counterbalance that resistance ; or, as many times as A D will go into C B, so many times may the resistance contain the amount of the applied force.

THE SCREW.

THE *screw* is the sixth and last of the mechanical powers. In the manner of its construction it is in general said to bear reference to an inclined plane wound about a cylinder ; but as the power of the inclined plane corresponds with that of the wedge, and the mode of applying the facilities they possess, alone forms their difference, and as the screw is almost universally moved to effect the same purposes as the wedge, it would, with greater propriety, as regards its action, bear reference to that instrument.

Fig. 23 represents a cylinder E E, upon which we will suppose the wedge-shaped piece, A B C, is capable of being wound ; when wrapped round such cylinder, it will, by its upper edge B C, represent spiral lines, similar to B D and F G. Now, as the piece A B C is in the shape of a wedge or inclined plane, it should have its power estimated by the line A C compared to the height A B; and if the line B C, wound in its spiral direction, shall just circumscribe the cylinder, the point C will be found directly beneath B, and the distance between C and B, when thus lapped on the cylinder, will represent the line A B, on the perpendicular of the inclined plane or wedge; which, when compared with A C, now represented by the circumference of the cylinder, will give the same data from which the power of the screw, so formed, should be calculated. Consequently the comparison of the circumference of the screw, and the distance between one thread and another, measured on a line parallel to the axis of the screw, is that from which its power should be calculated, or as the distance between the two threads is to the circumference, so the power to be applied is to the resistance to be overcome; or if the circumference be three, and the power one, the force equal to one shall overcome a force equal to three.

Fig. 24 represents a screw of more perfect formation; but the general construction of the screw is so familiar to every one, that we conceive it to be almost needless to enter upon a more minute description. A B represents the acclivity of the plane from which such screw is formed, and the distance between B and C represents what should be compared to the circumference in order to discover the power it possesses.

The screw is applied to mechanical purposes chiefly to obtain great pressures in small distances; and upon examination it will be seen, that they afford a method of using a wedge of an extremely small inclination, and by consequence of great power. The screw is sometimes used for raising exceedingly heavy weights. The hollow screw, or the counterpart in which a screw operates, when in the form of a small movable piece, is called a nut, and the cavity is termed a female screw, the properties of which are, as respects power, exactly similar to the screw.

We have now duly considered the nature and properties of the mechanical powers when in a state of uncombined action; and shall, in the next place, previously to representing them in some of the simplest forms in which they are combined, examine into one more attribute of matter, resulting from gravity.

cogs, or teeth, which being acted upon by any applied force, cause the wheel to revolve; and the axle being similarly furnished with teeth, or cogs, is termed a pinion. The wheel and pinion, therefore, bear a similar relationship to each other as the wheel and axle, and their power must be calculated in the same manner. Suppose A B, fig. 29, to be a shaft on which the handle A C, of twelve inches radius, and the pinion D, of one inch radius, are fixed; and the teeth of the wheel E, of twelve inches radius, acting in those of the pinion D, and upon the shaft of E is fixed the pinion F, of one inch radius, communicating with the wheel G, of twelve inches radius, upon the shaft of which the pulley H, of one inch radius, is fastened; we shall then have the handle A C representing the radius of a wheel, and the pinion D in the situation of the axle; so that there will be a gain of twelve to one: and the wheel E, bearing the same proportion to the pinion F, will also gain in a similar ratio, and G being to H, as E to F, the gain will again be augmented to the same extent; so a force equal to one at C will operate as twelve at D; and twelve at D will operate as a hundred and forty-four at F; and at H as seventeen hundred and twenty-eight. Thus one pound at C will raise seventeen hundred and twenty-eight pounds at H, and the handle C will have to pass through seventeen hundred and twenty-eight times the distance through which the weight I will move. By this form and disposition of wheels and pinions, an accession of power is obtained; but if velocity be required at the expense of power, this train should be inverted. For, if we suppose the pulley H to be turned by a force so as to cause the weight I to pass through one foot, the periphery of the wheel G will have passed through twelve feet, and the periphery of the pinion will have gone through the same distance; but the wheel E being twelve times the diameter of F, it will have passed through twelve times that distance, or a hundred and forty-four feet; and the pinion D, in like manner, will cause C to pass through twelve times that amount of space, or seventeen hundred and twenty-eight feet; whilst the force required at H to cause this motion, must be seventeen hundred and twenty-eight times the resistance at C.

As the circumferences of wheels are proportionate to the circumferences of the pinions they have to act upon, or be acted upon by, so must the number of teeth in the one be to those in the other, otherwise the size of the teeth would not be similar; thus, a wheel that is twelve inches diameter, and a pinion one inch, the circumferences of circles being in proportion to their diameters, the wheel should have

c

the internal face of the circle E, acting with its face F, and causing by its movement the obstacle I to approach nearer to the centre G; this is called the snail movement, and might with propriety be termed a concentric.

Another method of placing a wedge, so as to apply its effects to a revolving motion, is represented in a side and top view at fig. 32, where the wedge A B is placed upon a circular plate C D, turning upon the axis E, and consequently creating motion in the obstacle upon which it acts to the amount of the line G A.

Another movement of considerable accuracy is obtained by the turning of a cone, the principle of whose action is referable to the wedge. Fig. 33 represents a cone fixed upon its axis $i k$. If an obstacle be presented at a, and the cone be caused to pass forward in the direction $k i$, the surface $a c$ will operate as a wedge at $a b c$, raising the obstacle to c; but if during that direct motion the cone is likewise caused to revolve on its axis, the obstacle, instead of passing over $a c$, will pass over the spiral line $a e g d$, to the point d; by this means the operation of a wedge, whose line of inclination is equal to the spiral line $a e g d$, and whose height is equal to $b c$, is brought into action; and if the number of revolutions of the cone be increased during its direct motion, it is plain that the effect of a wedge of infinite elongation may be produced.

The screw is introduced both singly and in a state of combination in many parts of machinery. The combined action of two screws, which avoid the necessity of using a screw of greater fineness, in which the threads would be weakened, is represented at fig. 34, where they are applied to a press. Suppose A A to be a screw fitted in a female screw in the rail B C; and D, a screw that works in the inside of A, having its lower end joined to the upper board of the press H, so that it shall not turn round : now if the screw A A, and the screw D, contain exactly the same number of threads in the inch, by turning A A one revolution, it will proceed downwards exactly the same amount that the screw D will, by the same action, proceed upwards, and the board H will not be moved. But we will suppose that the screw A A contains four threads in the inch, and the screw D six, then, by one revolution, A A will move downwards one quarter of an inch, and D will at the same time, and by the same action, be raised one-sixth of an inch, therefore the board H will move downwards the difference between one quarter and one-sixth, or one-twelfth part of an inch, by every single revolution : which effect is similar to that which would be produced by using a screw of twelve threads to the inch.

· For further elucidation we shall refer the action of each screw to that of a wedge from which the screw has been shown to be derived   Fig. 35 represents two wedges, abh and ecd, each of which may be supposed to represent one lap of a screw of the respective fineness which their heights bh and ec denote. If the wedge abh be caused to pass to the situation $a^1 a h^1$, and is supposed to operate upon the level surface ef, the line ae will be compressed to the line $h^1 c$, by that movement; but if, whilst this action takes place, the wedge ecd be moved to the position $e^1 c^1 c$, and the effect takes place upon its upper surface ed, the line ae will only be reduced to the line $g^1 e$, equal to hd, and will consequently only be compressed to the amount $g^1 a$, which is in effect equal to what a wedge of the fineness of abg would have produced, whose height or line gb is just equal to the difference between ec and hb, as was the case with the screws.

As a gain of power is attainable by two screws or wedges of unequal fineness, performing equal numbers of revolutions, so is the same effect attainable by the unequal revolutions of two screws or wedges of equal fineness.

MILL GEERING.

UNDER this head we purpose to treat of the best formation of the teeth of wheels, of the connection of shafts, termed' couplings, of the disengaging and reengaging of the moving parts, and of the equalization of motion ; and to them we shall annex some further observations upon the general construction of Machinery. To avoid unnecessary repetition, we shall, previously to entering upon the formation of the teeth of wheels, give a general definition of the terms most commonly in use.

Cog-wheel is the general name of any wheel which has a number of teeth or cogs placed round its circumference.

Pinion is a small cog-wheel that has not in general more than twelve teeth; though, when two-toothed wheels act upon one another, the smallest is not unfrequently distinguished by this term ; as is also the trundle, lantern, or wallower, when talking of the action of two wheels.

Trundle, lantern, or *wallower,* is sometimes used in lieu of a pinion. It is represented at fig. 36.

When the teeth of a wheel are made of the same material, and formed of one piece with the body of the wheel, they are called *teeth;* when of wood, or some other material, and affixed to the outer rim of the wheel, *cogs;* in a pinion they are called *leaves;* in a trundle *staves.*

When speaking of the action of wheel-work in general, the wheel which acts as a mover is called the *leader,* and the one upon which it acts the *follower.*

If a wheel and pinion are to be so constructed that the one shall give, and the other receive, impulse, so that the pinion shall perform four revolutions in the time that the wheel is performing one, they must be represented by two circles, which are in proportion to each other as four is to one. When these two circles are so placed that their outer rims shall touch each other, a line drawn from the centre of the one to the centre of the other is termed the *line of centres;* and the radii of the two circles the *proportional radii.* These circles are sometimes called *proportional circles,* but by mill-wrights in general *pitch lines.*

The teeth which are to communicate motion must be formed upon these two circles. The distance from the centres of two circles to the extremities of their respective teeth, is called the *real radii;* and, in practice, the distance between the centres of two contiguous teeth, that is, the distance from the centres of two teeth measured upon their pitch line, is called the *pitch of the wheel.* The straight part of a tooth which receives the impulse is called the *flank,* and the curved part that imparts the impulse, the *face.*

Two wheels acting upon one another in the same plane, having their axes parallel to each other, are called *spur geer;* when their axes are at right, or other angles, *bevelled geer.*

TO DESCRIBE THE CYCLOID AND EPICYCLOID.

FIG. 37. If the circle 1, having a point *a* marked on its circumference, moves along the straight line A C, and at the same time revolves on its axis, the curved line which the point *a* describes is called the cycloid. The point *a* in circle 1 is at its starting place, at B it has reached its greatest height, and at C its lowest depth ; and the curved line A B C described by that point, is the *cycloid.*

Fig. 38. If the circle 1 rolls on another circle, as on the circumference of circle 2, the point *a* describes, in a similar manner to the preceding, the curve *a g h d e,* and the circles 3, 4, 5, 6, exhibit the point *a* in the several positions of *a¹, a²,* a^3, a^4 ; *c a¹* the portion of circle 3 being equal to *c a, c' a'* to $c^2 a$, $c^3 a^3$ to $c^3 a$, and $c^4 a^4$ to $c^4 a$: the line which is thus described is called an *exterior epicycloid*. But if the circle rolls within another circle, as the circle 1, fig. 39, rolls in the inside of circle 2, the line described by the point *a* is then called an *interior epicycloid*.

In fig. 38, the circle *a m n* is called the *generating circle of the epicycloid*, and that portion of the larger circle over which the generating circle rolls in one revolution the *base of the epicycloid*. In the interior epicycloid the generating circle of the epicycloid rolls within the circle of its base.

An epicycloid, either internal or external, may be conceived to be formed of numerous small portions of circles, whose radii are lines drawn from the several points of contact, as c, c^2, c^3, c^4, c being the centre of one, c^2 of another, and c^3 the centre of another, so that these lines are, as respects those several positions, radii of each circle, and perpendiculars to the epicycloid; if, therefore, a line be drawn from any point where the generating circle is in contact with the base to the point which traces the epicycloid, it will fall perpendicular to the epicycloid.

As the several lines drawn from the points of contact of the generating circle are, in all cases, the varying radii for generating the epicycloid, it is plain that when the generating circle shall have passed over half of its base, and consequently have performed half of a revolution, the diameter of the generating circle shall be a line drawn from the point of contact to the generating point, and which line shall, if prolonged, pass through the centre of the circle of the base, so that the tracing point in that part of the epicycloidal line shall be farther from, and in all other points nearer to, the base, as the perpendiculars that fall upon the epicycloid from the points of contact shall in every other position be shorter. Suppose the circle 1, fig. 40, to be a generating circle, and circle 2 to be the circle of the base, if the diameter of circle 1 be equal to the radius of circle 2, the point *a* shall trace the line *a b c* as an interior epicycloid; for if the diameter of circle 1 be equal to half of the diameter of circle 2, so will the circumference of circle 1 be equal to half of the circumference of circle 2, and consequently, when the generating circle 1 shall have performed one revolution upon the circle 2, as its base, the point *a* shall be exactly opposite to the place from where it started : now the diameter of circle 1 *is* equal to the radius of circle 2 when half way, and the tracing point is exactly in the centre of circle 2, which proves, that the epicycloid traced by the circle 1 is a straight line, and the diameter of circle 2,

ON THE TEETH OF WHEELS.

If two cylinders be placed in close contact, motion cannot be communicated to the one without that motion, by means of the irregularities of their surfaces, (of which we have spoken under the article *Friction*,) being communicated to the other, and the smaller cylinder shall perform exactly as many revolutions to one revolution of the larger cylinder, as the larger cylinder contains upon its circumference so many measured circumferences of the smaller cylinder.

Wheels, however, which act by their surfaces only, are ill-calculated to transmit motion to any considerable extent, as the motion which the follower has acquired is not of sufficient power to overcome the great resistance which would, in such case, be opposed to it; consequently it becomes necessary to have projections or teeth, and that form of the teeth will be the best which causes the wheel to act as though the motion were communicated by contact of the pitch lines.

Spur geer, fig. 39*. If the three circles 1, 2, 3, in contact at the point *a*, be made to revolve about their centres, so that they shall continually touch at the point *a*, their motions will be similar to what would have been generated by one communicating motion to the other two by contact; and circle 3 will move as though rolling on the external surface of circle 1, and internal surface of circle 2, and consequently become the generating circle of the exterior epicycloid on circle 1, and the generating circle of the interior epicycloid on circle 2. As the diameter of circle 3 is equal to the radius of circle 2, the interior epicycloid will be a straight line passing through B the centre of circle 2; and, supposing the point *a* to have performed that portion of a revolution which places it at K, a portion of the exterior epicycloid will be represented by the line E K, and a portion of the interior epicycloid by D K. Therefore, as the epicycloids D K and E K are both generated by one motion of the same point on the same circle, they will continually touch at the generating point, and the total surface of E K will pass over the total surface of D K; and if the epicycloid E K be affixed to the external surface of circle 1, and act upon the portion of the epicycloid D K, it will transmit motion to circle 2, as though that motion were communicated by contact of the pitch lines; which proves that E K presents us with the best form of tooth, and which tooth would, when acting upon the radii of the wheel to be driven, move it as though the motion were communicated by contact.

flanks of the teeth. The spaces must be of sufficient depth to allow for the action of the curved part of the teeth and leaves.

Then with the generating circle 1, whose diameter is equal to the proportional radius of the pinion, describe upon the extremities of the sides of each tooth, and upon the circumference of the proportional circle of the wheel as a base, the epicycloids *a b, b n;* and with the generating circle 2, describe upon the proportional circle of the pinion as a base, the epicycloid *q* D, which will give the required form of the teeth and leaves.

For if the projecting epicycloid *a b* push against the radius *f r* of the proportional pinion, the wheel and pinion will move with equal velocity ; and a similar effect will be produced by the epicycloid *p* D being pushed by the radius *o m* of the wheel towards the line of centres.

Fig. 43. When one wheel is to conduct another, it is not necessary that the wheel to be conducted should have teeth of an epicycloidal form ; and were the teeth not subject to wear by friction, there would be no occasion to extend the teeth of the conducted wheel beyond the pitch line; but such being the case, it becomes necessary to form the teeth of the conducted wheel in the manner represented in the figure by the dotted lines.

Mr. Buchanan, in his " Essay on the Teeth of Wheels," objects to this mode of forming the teeth of the conducted wheel, and recommends that a trundle or wheel with cylindrical staves should be adopted, as it will be less acted upon in approaching the line of centres, and consequently have less friction than a pinion or wheel, the sides of whose teeth tend to the centre.

" This will appear," says he, " by fig. 44, which represents a staff, *a*, of a trundle, and a leaf, *b*, of a pinion, turning round on the same centre A, and a tooth adapted to each, turning on a common centre B. The thickness of each of the teeth, and the proportional circle of both wheels, are the same, and the proportional circles of the pinions are also equal, and teeth are each made of the greatest length which the intersection of the curves will admit, which turns out considerably greater in the tooth adapted to the staff. The shaded parts represent the tooth adapted to, and acting upon, the staff; and the dotted lines represent the tooth adapted to, and acting upon, the leaf. The teeth, in both cases, are represented as just at the point where they would cease to move the leaves or staves uniformly; and it appears the staff is conducted considerably further beyond the line of centres than the leaf; hence the staff will be less acted upon in approaching the line of centres."

As the trundle in common use is very weak and imperfect, Mr. Buchanan conceived, that a wheel might be made, which would combine the advantages of both the pinion and trundle, and accordingly had some wheels made, which appeared to answer every expectation.

" These wheels," says he, "were made of cast iron. They were each cast of one solid mass. Fig. 46, No. 1, represents the edge view, and No. 2, a section of one of them; whereby is shown the manner in which the teeth are supported, like the staves of a trundle at each end, and like the leaves of a pinion at the roots, but so very thin there, as to run no risk of having the common fault of pinions, just now noticed. They were difficult to mould : but were they to come more into use, I have no doubt ingenious workmen would soon get over this obstacle." *　" I mentioned," he continues, " in cases where the pinion had few teeth, that in the conducted, whether wheel or pinion, staves should be preferred; but it is obvious, that the method just described, of making a small trundle of cast iron, would not apply to a wheel of a great number of staves. Nor is it in that case so necessary, as the greater the number of teeth are, the longer they will be in losing their proper figure. In such cases, therefore, staves, strictly speaking, should not be used, but teeth made so as to produce the same effect—that is, having their acting parts of the figure of a staff. What is meant will be better understood by inspecting fig. 46, where the lines show the alteration necessary on the tooth A, in order to make it produce the effect of a staff; which staff is represented by the faint dots. The dotted lines on d represent the alteration requisite to adapt it to the staff, it being necessary, as formerly proved, to have it a different epicycloid from what is required to adapt it to a tooth whose acting part is a straight line, tending to the centre of its proportional circle."

" Teeth," says Mr. Tredgold, in the second edition of Mr. Buchanan's work, " seem to be very well adapted for va-

* By casting separate plates with indents to fix the teeth, and bolting them together, the pinion might be made sufficiently strong : such a method indeed is used frequently in crane-work, where it has the important advantage of preventing the wheels getting out of geer.

N. B. This note is by Mr. Tredgold, editor of the second edition of Buchanan's " Practical Essays on Mill-work."

rious purposes, when formed on the principle recommended in the preceding article. I therefore will endeavour to show a simple method of describing such teeth.

"It must be observed, that the teeth to resemble staves are to be always on the conducted wheel or pinion; thus affording the peculiar advantage of the wheel and trundle in either increasing or diminishing velocity.

"Fig. 38*. Let the teeth be divided as usual on the pitch lines, E E, F F; and on the conducted wheel C describe circles, as though there were to be staves. Conceive the centre of one of these staff teeth to be in the line of centres at A, and draw the line A B joining the centres of the staff teeth. Then the radius A b, from the centre A, will describe the curved side b c of the tooth of the conductor, and the curved part b a of the conducted wheel. And since this radius is equal to the pitch diminished by half the diameter of the circle of the staff teeth, and the centres will always be in the pitch lines of the wheels, all the other teeth may be easily described."

The editor then enters into some calculations, which the limits of our work will not permit us to pursue, we therefore refer our readers to the work itself, which embraces much useful information.

Fig. 47. When a pinion is required to have but a slow motion, an internal pinion, which has less friction than the external one, may, in many cases, be adopted with advantage.

To illustrate this, let A, fig. 48, be the proportional circle or pitch line of a wheel, B that of an external pinion, and C that of an internal pinion, all at contact at the point a: now, if motion be communicated to the wheels, so that they move uniformly, it will be seen, that when the point a has arrived at b c d, each of the wheels having travelled over an equal distance from the line of centres D, the space from b to c is much less than that from c to d, and consequently had the wheels moved by means of teeth, the tooth of the internal pinion C would have slid over a smaller part of a tooth of the wheel A, than a tooth of the external pinion B, which proves it would have had less velocity and less friction.

Fig. 49 represents a rack and pinion, recommended by Mr. Tredgold. A B the pitch line of the rack, B C the pitch line of the pinion, and the form of the tooth C D is the involute of a circle; but when the rack impels the pinion, the curved face of each of the teeth of the rack should be a portion of a cycloid, (as A, a, fig. 37,) and the leaves of the pinion straight lines radiating from the centre

of the pinion; the diameter of the generating circle for
describing the cycloidal teeth should be half the proportional
diameter of the pinion. See *Buchanan's Practical Essays
on Mill-work*. Tredgold's edition.

Bevel geer.—We have already stated, that when the axes
of wheels are angular to each other, they are called
bevel geer, in order to distinguish them from spur geer,
whose axes are parallel; it therefore now remains for us
to describe in what manner the teeth of bevel geer differ
from the teeth of spur geer.

Bevel geer is represented by the two cones at fig. 50,
where A B and B C are the axes, and D E and E F their
proportional diameters or pitch lines.

If these two cones are placed in close contact, and motion
is communicated to the one, that motion will, as is already
stated, be communicated to the other, and the motion of
both, as we have shown, when speaking of spur geer,
will be equal.

The epicycloid for forming the teeth of bevel geer, is
generated by one cone rolling upon the surface of another,
while their summits coincide: for example, if a cone C,
fig. 51, having a point *a*, move upon the surface of the
cone D, the point *a* will, in its revolutions, describe the
line A E F, A being the place from where it starts, E its
greatest height, and F its lowest depth; therefore a curved
line drawn from A to E, and continued from E to F,
gives what is called a *spherical epicycloid;* and the base
of the cone C is the *generating circle of the spherical
epicycloid*. The method of using the spherical epicycloid
for forming the teeth of bevel geer is, in every respect,
similar to the method of using the exterior and interior
epicycloid for forming the teeth of spur geer, consequently it
will be needless to repeat it.

Fig. 52. To construct bevel geer we must calculate the
proportional diameters or pitch lines of the wheel and
pinion that are to act upon each other, and then draw
their axes A B and B C. Draw parallel to the axis A B
of the wheel the line D E, and the line F D parallel to
the axis of the pinion, and from the point D, where these
two lines intersect, draw the line D G perpendicular to
A B, and D H perpendicular to B C, and make I G equal to
D I, and K H equal to D K; then D G gives, what is called
the *principal diameter*, or *diameter of the pitch line of the
wheel*, and D H that of the pinion.

Proceed to draw the teeth of the wheel, by fixing one

foot of the compasses in the point at A, and, having extended the other foot to the distance G, sweep the small arc G a, then set off the length of the tooth from G to h, draw the line $b\,c$, tending to a, and sweep the arc $c\,e$, concentric to $b\,n$. Set off from G to f part of the required length of the tooth, from the principal diameter to the root; and draw the line $f\,g$ tending to A, which gives the root of the tooth. Parallel to $f\,g$, draw $a\,e$, and $a\,f\,g\,e$ will represent a section of the solid ring of the wheel.

In an excellent article on mill-work, in Dr. Rees's *Cyclopedia*, the author states, " that the manner of setting out the teeth of cog-wheels, in such a form that they shall act in the most equable manner upon each other, and with the least friction, has been a subject of much investigation among mathematicians and theoretic mechanics ; but the practice and observation of the mill-wrights have produced a method of forming cog-wheels, which answers nearly, if not fully, as well in practice, as the geometrical curves which theory has pointed out to be the most proper. This they have effected by making the teeth of the modern wheels extremely small and numerous. In this case, the time of action in each pair of teeth is so small, that the form of them becomes comparatively of slight importance ; and the practical methods of the mill-wrights (using arcs of circles for the curves) approximates so nearly to the truth, that the difference is of no consequence : and this method is the best, because it so easily gives the means of forming all the cogs exactly alike, and precisely the same distance asunder, which, by the application of any other curve than the circle, is not so easy. The method, which is extremely simple, is explained in fig. 53. The wheel being made, and the cogs fixed in much larger than they are intended to be, a circle, $a\,a$, is described round the face of the rough cogs upon its *pitch diameter*, that is, the geometrical diameter, or acting line of the cogs ; so that when the two wheels are at work together, the pitch circles, $a\,a$, of the two are in contact. Another circle, $b\,b$, is described within the pitch circle for the bottom of the teeth, and a third, $d\,d$, without it, for the extremities. After these preparations, the pitch circle is accurately divided into the number which the wheel is intended to have : a pair of compasses are then opened out to the extent of one and a quarter of these divisions, and with this radius arcs are struck on each side of every division, from the pitch line a, to the outer circle $d\,d$. Thus the point of the compasses being set in the division e, the curve $f\,g$, on one side of the

of the square A, entering into a round hole in the centre of
the square B.

Clutches or glands may be used with much advantage as
a coupling for double bearings. Fig. 57 represents a coupling of this kind; it consists of two crosses, A A and B B,
one fixed to each shaft: B B has its ends bended forward,
and lays hold of A A, which turns that shaft round.*

In boring-mills two kinds of clutches are used. The one
for the smaller kinds of work is represented in fig. 58.
A B is a round plate of cast iron fixed firmly on the shaft C;
D E a lever fixed to the shaft H by the bolt F, and capable
of being moved in the direction of the plate A B, so that it
can lay hold of the projections G G G G, which will admit the
boring shaft H to be thrown in and out of geer at pleasure.

The second kind of boring-mill clutch, or the one that is
used to bore the largest cylinders, is represented in fig. 59.
The only difference between this clutch and the one just
described, consists in having the lever D E to turn on a bolt
at F in a cast iron plate I K L, instead of hanging from the
shaft H. Three spare sets of ears, which are cast on the
plate, to be used in case of those in action breaking, support
the lever near the point of pressure, and take the stress
entirely off the bolt F.

When an engine is started, it frequently happens that the
crank is on the wrong side of the axis of the fly-wheel, so that
both that and the shaft make one or two, and, if the attendant is negligent, several, revolutions in the wrong direction.
To prevent the mischief that would accrue from such an occurrence, a coupling, as is represented in fig. 60, is introduced. A and B are two vertical shafts, maintained in the
same line by a small circular pin, which passes from the shaft
B into a cavity on the shaft A, which cavity is large enough
to admit the pin to lay in it without communicating motion
to the shaft A. The shaft B, which is connected with the
moving power, has a coupling piece with prominences or
teeth, perpendicular on the one side, and inclined on the
other, fixed on its upper end. The coupling or catch box
C, which is capable of sliding freely up and down the square
part of the shaft A, has a correspondent set of teeth; by
which it is evident, that when the shaft B turns the right
way, the perpendicular sides of the teeth of the respective
coupling pieces will act together, and carry round the upper

* For a method of constructing glands we must refer our readers to
Buchanan's Essays on Mill-work.

shaft A ; but when B turns in a contrary direction, the inclined sides of the teeth of the catch-box will slide over the
inclined sides of the teeth of the piece on the shaft B, and
cause the catch-box C to move up and down without communicating motion to the shaft A.

Fig. 61 represents the coupling link used by Messrs. Boulton and Watt in their portable steam-engines. A, a strong
iron pin, projecting from one of the arms of the fly-wheel B ;
D a crank connected with the shaft C ; and E a link to couple
the pin A and the crank D together, so that motion may be
communicated to the shaft C.

Hook's universal joints are sometimes used to communicate motion obliquely instead of conical wheels. Fig. 62
represents a *single* universal joint, which may be employed
where the angle does not exceed forty degrees, and when the
shafts are to move with equal velocity. The shafts A and B,
being both connected with a cross, move on the rounds at
the points C E and D F, and thus, if the shaft A is turned
round, the shaft B will likewise turn with a similar motion in
its respective position.

The *double* universal joint, fig. 63, conveys motion in
different directions when the angle is between 50 and 90
degrees. It is at liberty to move on the points G, H, I, K,
connected with the shaft B ; also on the points L, M, N, I,
connected with the shaft A : thus the two shafts are so connected, that the one cannot turn without causing the other to
turn likewise. These joints may be constructed by a cross
of iron, or with four pins fastened at right angles upon the
circumference of a hoop or of a solid ball : they are of great
use in cotton mills, where the tumbling shafts are continued
to a great distance from the moving power ; for by applying
a universal joint, the shafts may be cut into convenient lengths,
and so be enabled to overcome a greater resistance.

OF DISENGAGING AND REENGAGING MACHINERY.

A knowledge of the best methods of disengaging and reengaging machinery, or, as the workmen call it, throwing in
and out of geer, is found to be highly necessary in most manufactories; and yet it frequently happens that the workmen are
either very ignorant of, or very inattentive to, this important
subject.

Matter possesses a certain property termed *inertia*, which
has a tendency to maintain it in the state in which it actually
is ; that is to say, if a body is set in motion, this property has
a tendency to maintain it for ever in that state, and certainly

, which is out of geer, is suddenly connected with it, or
wn in geer, that the shock proceeding from inertia
s the teeth of the wheels, or causes destruction to some
r part of the machinery. To obviate this as much as
ible, such means should be resorted to, as have been
d in practice to answer best. The risk of breaking
teeth may be considerably lessened by first setting
wheel, that is to be thrown in geer, in motion by the
l

e methods that have been adopted for throwing mary in and out of geer are various ; some of the principal
hich we shall now proceed to notice.

g. 64 represents the sliding pulley. P a pulley, having
low cylindrical bush made so that it can revolve easily
the axle and slide backward and forward upon it; B a
of the bush projecting on one side of the pulley, having
ove sufficiently large to admit the lever L to lay in it
out impeding its motion ; C G a cross or gland fixed
to the axle; and I, one or more teeth, projecting from
pulley on the side opposite to the bush. When the axle
is required to be put in motion, the lever L must be
ed towards the cross or gland C G, so that the teeth
the pulley may catch hold of and carry it round with it.
he fast and loose pulley is represented in fig. 65. B is a
y firmly fixed on the axle A, and C a pulley with a bush,
at it can revolve upon the axle A without communicating
on to it. This contrivance is remarkable for its beauti-
implicity, as the axle A can be thrown in and out of
at pleasure, without the least shock, by simply passing
ap from the one pulley to the other.

backward and forward upon the horizontal shaft by pr
the handle H H ; so that when the shaft B C is rec
to be put in motion, the attendant has only to pus
bayonet into the pulley D E, which will immediately ca
round.

Fig. 67 represents one of the simplest ways of
and reengaging wheels. A B, the bridge of
No. 1, acts as a lever, having its fulcrum at A ; the
end of the bridge B is capable of being lifted by th
K K.   When the wheel, No. 2, is required to be throw
of geer, the key K K is pressed downwards, and the c
the bridge B rests upon the extreme end of the key, as s
by the dotted lines.

The tightening roller is represented in fig. 68.   A
are two pullies, the one to receive, and the other to trai
motion, by means of the strap C : D is the tightening r
fastened to a movable arm E, and connected with a
G F.   When the moving pulley (suppose A) is requir
give motion to the other pulley B, the lever G F mu
pushed downwards, which will tighten the strap by pl
the tightening roller in the position represented by the d
lines, and cause the pulley A to carry the pulley B i
with it.

The friction clutch, represented in fig. 69, is used t
engage and reengage machinery, when the velocity o
moving parts is very great.   A is a pulley, having a
and revolving freely on the shaft S S : B is another pe
having a similar bush, and also capable of revolving ou
shaft : C C is a dish-spring, secured in its place by the
p p, and forcing the pulley B against the collar D, whi
fixed permanently to the shaft.   When motion is req
to be communicated to the shaft S S, the pulley A is m
towards the pulley B, and the teeth projecting from the
of the pulley A, clasps those of the pulley B, and carri
round with it ; and the friction of the pulley B against
collar D, gradually overcomes the inertia, and carries
shaft and connecting machinery also round.

The friction clutch, represented in fig. 70, is a very exce
contrivance, as it prevents all those injurious shocks w
the machinery is apt to receive upon being thrown into
C C is a cross fixed firm on the moving shaft A ;
E is a pulley or drum fixed firm on the shaft to be move
When the shaft B is required to be moved, the clutc
bayonet K is made to pass through the arms of the cross

) the screw-hoop I I, which is by that means carried
ith the shaft A, and the friction caused by the screw-
, turning upon the drum or pulley E, causes the drum
shaft B, to which it is attached, to turn likewise.
riction cone is very similar in its effects to the fric-
.ch. On the moving shaft A, fig. 71, is fixed a cone
on the shaft B is another cone D, made to fit in the
The cone D is movable on a square part of the
and may, by a lever, be moved in and out of geer.
he cone D is moved forward, the cone C receives
by its internal surface.
;. 72 is represented the self-disengaging coupling.
fts, A and B, have each of them a cast iron wheel, with
ique wrought iron teeth; but the wheel on the shaft
ovable, on A it is fixed. When the coupling is en-
the teeth of the wheel C lay hold of the teeth of the
, and carry it, and the shaft A, round with the shaft
G is a bent lever, having its fulcrum at F, which,
he ordinary stress on B, keeps forward the bayonet C,
reight of the part F G; but when a more than usual
omes on the shaft B, the pressure on the oblique
rces the bayonet back, and disengages the coupling,
lever is held by a catch until the coupling is re-
by the hand of the workman.

ON EQUALIZING THE MOTION OF MACHINERY.

egulation of the velocity of a mill is a matter of very
portance to preserve an uniformity of motion, either
c force of the first mover is fluctuating, or when the
re or work of the mill varies in its degree: either or both
causes will occasion the mill to accelerate or diminish
ity; and in many instances it will have a very injurious
ion the operations of the mill. Thus, in a mill for
r cotton, wool, flax, &c., driven by a water-wheel, are
licity of movements, many of which are occasionally
ged, in different parts of the mill, for various purposes.
ids to diminish the resistance to the first mover, and
le mill accelerates. Or, on the other hand, the head
r, which drives the wheel, may be liable to rise and
denly, from many causes, which great and rapid rivers
ect to, and cause similar irregularities in the speed of
el. For such cases judicious mechanics have adopted
inces, or regulators, which counteract all these causes
ularity; and a large mill, so regulated, will move like
, with regard to its regularity of velocity. These

regulators are usually called called *governors,* and are made on different principles. Those most generally used are called flying-balls, operating by the centrifugal force of two heavy balls, which are connected and revolve with a vertical axis. Fig. 189, steam-engine, is the simplest form of this ingenious apparatus: A A is the vertical axis, which is constantly revolving by the machinery; at *a a* two arms or pendulums, *a b, a b,* are jointed, and carry at their extremities a heavy metal ball each, as *b, b;* from the pendulum two chains or iron rods, *d d,* proceed, and suspend a collar *e,* which slides freely up and down the axis, and has a groove formed all round it, in which the end of a forked lever, D, is received; and thus the rising and the falling of the collar *e,* produces a corresponding motion of the end of the lever D; but the collar is always at liberty to turn round with the axis freely within the fork, at the extremity of the lever. The operation of the governor is this: when the vertical axis is put in motion, the centrifugal force of the balls *b, b,* causes them to recede from the centre; and as this is done both together, they cause the collar *e,* and the end of the lever, to rise up: the balls fly out to a certain height, and there they continue as long as the axis preserves the same velocity; as it is the property of a pendulous ball, like *b,* to make a greater effort to return to the perpendicular, in proportion as it is removed farther from it, in consequence of the suspending rod being more inclined, and bearing less of its weight. The weight of the balls to return to the axis may be considered as a constantly increasing quantity; while the quantity of the centrifugal force, causing them to recede from the axis, depends exactly upon the velocity given them. But this velocity increases as they open out, independently of any increased velocity of the axis, in consequence of their describing a larger circle. The combination of these oppositely acting forces causes the governor to be a most sensible and delicate regulator. Thus: suppose the balls hanging perpendicular put the axis in motion with a certain velocity, the centrifugal force will cause the balls to fly out; and this increasing their velocity, (by putting them farther from the centre, and causing them to revolve in a larger circle,) gives them a greater centrifugal force, which would carry them still farther from the centre, but for the counteracting force, viz. the weight of the balls tending to return. This is, as before stated, an increasing quantity, and consequently these opposite forces come to a point where they balance each other; that is, the balls fly out till their weight to return balances the centrifugal force. But if the slightest alteration takes place in the velocity of the axis, the equilibrium is destroyed by the increase or diminution of the centrifugal force, and the balls alter their distance from the centre accordingly, and, by elevating or depressing the end of the lever, operate upon some part of the mill to rectify the cause of the irregularity. In a steam-engine, the lever acts upon a vane or door situated in the passage of the steam from the boiler to the cylinder ; and if the mill loses in velocity, from an increase of resistance, the balls fall together a little, and the consequent fall of the lever opens the door or throttle-valve a little wider, and gives a stronger supply of steam to restore the mill to its original velocity. On the other hand, if the mill accelerates, the balls open out and then close the vane, so as to moderate the supply of steam.

A water-wheel is not so easily regulated by the governor, because the shuttle of a large wheel requires a much greater force to raise or lower it, when the water is pressing against it, than the lever D, can at any time possess; it therefore becomes requisite to introduce some additional machinery, which has sufficient power to move the shuttle, and this is thrown in or out of action by the flying balls. The simplest contrivance, and that which we believe was the regulator first used for a water-wheel, was erected at a cotton-mill at Belper, in Derbyshire, belonging to Mr. Strutt. A square well, or large cistern, was situated close by the water-wheel: it had a pipe leading from the mill-dam into it, to admit water; and another pipe from it to the mill-tail, to take the water away: both were closed at pleasure by cocks or sluices. Within the well was a large floating chest, very nearly filling up the space: it of course rose and fell with the water in the cistern, and had a communication by rack and wheel-work with the machinery for drawing the shuttle, so that the rise and fall of the floating chest elevated and depressed the shuttle of the wheel. The lever of the governor was connected with the cocks in the two pipes in such a manner, that when the mill was going at its intended velocity, both of the cocks were shut; but if the water-wheel went too slowly, the falling of the balls and descent of their lever D, opened the cock in the pipe of supply, and, by letting water into the well, raised the float, and, with it, the shuttle, to let more water upon the wheel, till it acquired such a velocity that the balls began to open out again, and thus shut the cock: or. the other hand, if the mill went too fast, the balls opened the pipe of exit from the well, and then the sinking of the float closed the shuttle till the true velocity was restored.

counteracting the irregularity. If the mill moves too slowly, the balls tend to diminish the feed, and at the same time they raise the upper stone, to set them at a greater distance asunder, that they may require less power to drive them, and consequently suffer the mill, as nearly as it can, to retain its full velocity, though the motive force is greatly diminished. This application of the governor was, we believe, first made by the ingenious captain Hooper of Margate, who invented the horizontal wind-mill. It is a very great advantage, and no wind-mill should be without them. Many wind-mills are provided with flying-balls, which, by very ingenious mechanism, clothe and unclothe the sails just in proportion to the strength of the wind.

In many mills it is of consequence to be able to detect small variations in the velocity, and to ascertain the quantity of them; for the governor only corrects the irregularities, without showing any scale of them. In cases where this is required, it may be done by a very ingenious instrument, invented by Mr. Bryan Donkin, of Fort-place, Bermondsey. He received a gold medal from the Society of Arts, Manufactures, and Commerce, in 1810, for this instrument, which he calls a tachometer.

A front view of Mr. Donkin's tachometer, or instrument for indicating the velocity of Machinery, is represented in fig. 76, and a side view in fig. 77. X Y Z, fig. 76, is the vertical section of a wooden cup, made of box, which is drawn in elevation at X, fig. 77. The whiter parts of the section, in fig. 76, represent what is solid, and the dark parts what is hollow. This cup is filled with mercury up to the level L L, fig. 76. Into the mercury is immersed the lower part of the upright glass tube A B, which is filled with coloured spirits of wine, and open at both ends, so that some of the mercury in the cup enters at the lower orifice, and when every thing is at rest, supports a long column of spirits, as represented in the figure. The bottom of the cup is fastened by a screw to a short vertical spindle D, so that when the spindle is whirled round, the cup (whose figure is a solid of revolution) revolves at the same time round its axis, which coincides with that of the spindle.

In consequence of this rotation, the mercury in the cup acquires a centrifugal force, by which its particles are thrown outwards, and that with the greater intensity, according as they are more distant from the axis, and according as the angular velocity is greater. Hence, on account of its fluidity, the mercury rises higher and higher as it recedes from the axis, and consequently sinks in the middle of the cup; elevation at the sides and consequent depression in middle increasing always with the velocity of rotation. the mercury in the tube, though it does not revolve witl cup, cannot continue higher than the mercury immedi surrounding it, nor indeed so high, on account of the su incumbent column of spirits. Thus the mercury in the will sink, and consequently the spirits also; but as that of the tube which is within the cup is much wider than part above it, the depression of the spirits will be n greater than that of the mercury, being in the same pro tion in which the square of the larger diameter exceeds square of the smaller.

Let us now suppose, that by means of a cord passing r a small pulley F, and the wheel G or H, or in any c convenient way, the spindle D is connected with the mac whose velocity is to be ascertained. In forming this nection, we must be careful to arrange matters so, that the machine is moving at its quickest rate, the ang velocity of the cup shall not be so great as to depres spirits below C into the wider part of the tube. W also, as in the figure, to have a scale of inches and te applied to A C, the upper and narrower part of the tube numeration being carried downwards from zero, which i be placed at the point to which the column of spirits when the cup is at rest.

Then the instrument will be adjusted, if we mark on scale the point to which the column of spirits is depres when the machine is moving with the velocity requi But, as in many cases, and particularly in steam-engi there is a continued oscillation of velocity, in those case have to note the two points between which the col oscillates during the most advantageous movement of machine.

Here it is proper to observe, that the height of the col of spirits will vary with the temperature, when other cumstances are the same. On this account the scale o to be movable, so that, by slipping it upwards or downwa the zero may be placed at the point to which the col reaches when the cup is at rest; and thus the instrum may be adjusted to the particular temperature with the utn facility, and with sufficient precision. The essential part the tachometer have now been mentioned, as well as method of adjustment; but certain circumstances remai be stated.

The form of the cup is adapted to render a smaller quantity of mercury sufficient, than what must have been employed either with a cylindrical or hemispherical vessel. In every case two precautions are necessary to be observed:—first, that when the cup is revolving with its greatest velocity, the mercury in the middle shall not sink so low as to allow any of the spirits in the tube to escape from the lower orifice, and that the mercury, when most distant from the axis, shall not be thrown out of the cup. Secondly, that when the cup is at rest, the mercury shall rise so high above the lower end of the tube, that it may support a column of spirits of the proper length.

Now in order that the quantity of mercury, consistent with these conditions, may be reduced to its minimum, it is necessary—first, that if M M, fig. 76, is the level of the mercury at the axis, when the cup is revolving with the greatest velocity, the upper part M M X Y of the cup should be of such a form as to have the sides covered only with a thin film of the fluid; and, secondly, that for the purpose of raising the small quantity of mercury to the level L L, which may support a proper height of spirits when the cup is at rest; the cavity of the cup should be in a great measure occupied by the block K K, having a cylindrical perforation in the middle of it for the immersion of the tube, and leaving sufficient room within and around it for the mercury to move freely both along the sides of the tube and of the vessel.

The block K K is preserved in its proper position in the cup or vessel X Y Z, by means of three narrow projecting slips or ribs placed at equal distances round it, and is kept from rising or floating upon the mercury by two or three small iron or steel pins inserted into the under side of the cover, near the aperture through which the tube passes.

· It would be extremely difficult, however, nor is it by any means important, to give to the cup the exact form which would reduce the quantity of mercury to its minimum; but we shall have a sufficient approximation, which may be executed with great precision, if the part of the cup above M M is made a parabolic nonoid, the vertex of the generating parabola being at that point of the axis to which the mercury sinks at its lowest depression, and the dimensions of the parabola being determined in the following manner. Let V G, fig. 78, represent the axis of the cup, and V the point to which the mercury sinks at its lowest depression; at any point G above V, draw G H perpendicular to V G; let n be

the number of revolutions which the cup is to perform in
1″ at its quickest motion; let v be the number of inches
which a body would describe uniformly in 1″, with the velocity acquired in falling from rest, through a height = to
G V, and make $G H = \dfrac{v}{3\ 14\ n}$. Then, the parabola to be
determined is that which has v for its vertex, V G for its
axis, and G H for its ordinate at G. The cup has a lid to
prevent the mercury from being thrown out of it, an event
which would take place with a very moderate velocity of
rotation, unless the sides were raised to an inconvenient
height; but the lid, by obstructing the elevation at the sides
of the cup, will diminish the depression in the middle, and
consequently the depression of spirits in the tube: on this
account a cavity is formed in the block immediately above
the level L L, where the mercury stands when the cup is at
rest; and thus a receptacle is given to the fluid, which would
otherwise disturb the centrifugal force and impair the sensibility of the instrument.

It will be observed, that the lower orifice of the tube is
turned upwards. By this means, after the tube has been
filled with spirits by suction, and its upper orifice stopped
with the finger, it may easily be conveyed to the cup and
immersed in the quicksilver without any danger of the spirits
escaping; a circumstance which otherwise it would be extremely difficult to prevent, since no part of the tube can be
made capillary, consistently with that free passage to the
fluids, which is essentially necessary to the operation of the
instrument.

We have next to attend to the method of putting the
tachometer in motion whenever we wish to examine the velocity of the machine. The pulley F, which is continually
whirling during the motion of the machine, has no connection whatever with the cup, so long as the lever Q R is left
to itself. But when this lever is raised, the hollow cone T,
which is attached to the pulley and whirls along with it, is
also raised, and embracing a solid cone on the spindle of the
cup, communicates the rotation by friction. When our observation is made, we have only to allow the lever to drop by
its own weight, and the two cones will be disengaged, and
the cup remain at rest.

The lever Q R is connected by a vertical rod to another
lever S, having, at the extremity S a valve, which, when the
lever Q R is raised, and the tachometer is in motion, is
lifted up from the top of the tube, so as to admit the external

air upon the depression of the spirits; on the other hand, when the lever Q R falls, and the cup is at rest, the valve at S closes the tube, and prevents the spirits from being wasted by evaporation.

It is lastly to be remarked, that both the sensibility and the range of the instrument may be infinitely increased; for, on the one hand, by enlarging the proportion between the diameters of the wide and narrow parts of the tube, we enlarge in a much higher proportion the extent of scale corresponding to any given variation of velocity; and on the other hand, by deepening the cup so as to admit when it is at rest a greater height of mercury above the lower end of the tube, we lengthen the column of spirits which the mercury can support, and consequently enlarge the velocity, which, with any given sensibility of the instrument, is requisite to depress the spirits to the bottom of the scale. Hence the tachometer is capable of being employed in very delicate philosophical experiments, more especially as a scale might be applied to it, indicating equal increments of velocity. But in the present account it is merely intended to state how it may be adapted to detect in machinery every deviation from the most advantageous movement.

General Observations.—In setting out the geering of a mill, it should be the object of the engineer to place the heaviest machinery nearest the moving power, as, in transmitting motion to a great distance, not only the weight of shafting is to be taken into consideration, but the friction which exists in all the different bearings, and which is greatly increased by a small obstacle placed beyond those bearings.

Care likewise should be taken to make as few bearings as possible, still keeping in view that the shafts must not be allowed to swag. Rules might be given for the distances of the bearings of the shafting, if the shafting had only to move itself, but having to carry various sized pullies, both their weight and the weight of the machinery they turn must be taken into consideration, which compel us to forego the attempt; it is, however, necessary to state, that it is better to have a bearing too many than to allow a shaft to bend, as it cannot then run true in its steps or journals.

In forming couplings, great care should be taken to make them fit, so that the coupled shaft may move as though of the same piece with the driving shaft : nor can simplicity be too strongly recommended, that the coupled shaft may, in case of an accident, be instantaneously disengaged, for the loss of time arising from any accident is of serious importance to the manufacturer. Couplings should be placed near the bearings, as there is there the least swag, and the shaft is of course the weakest at the couplings. The same observation is applicable to the disposing of wheels and pullies.

Pullies have been sometimes formed in two halves for putting upon the shaft without taking the shaft down, but their adoption is by no means general, as there is some difficulty in fixing them true whilst the shaft is in its place.

Straps to drive geering should be avoided whenever wheels can be substituted, as they are very liable to stretch and break, and do not transmit regular motion. In fixing the wheels and pullies upon a shaft, which is mostly done by driving wedges in the bush of the wheel or pulley, called *staking them on*, great pains should be taken to have them true, which can only be done by driving the wedges regularly on each side to the same degree of tightness. It most generally happens when one wedge is over-driven, the workmen, rather than take the trouble to alter it, will let it remain; but this is of more importance than is generally imagined, for if a wheel is not true, it cannot work in the pitch line, all round, and where it is out it will shake, or have, what is called, *back-lash*, which, happening always in the same place, will wear the wheels irregularly. If a pulley is not true, it will communicate irregular motion by its strap, and likewise cause an irregular stress upon the shaft on which it works, much to the detriment of the bearing.

Chains have been beneficially introduced as substitutes for straps in driving heavy geer.

Shafts should be circular, as they are less likely to catch any thing, and have a much neater appearance. The same may be said of couplings. The wheels of the geering should be always enclosed in a casing of wood, called *boxing off*, to prevent any thing falling in between them, or accidents occurring to the people who may be working near them. The wheels should be furnished with brushes resting upon their faces, to distribute the grease equally and to keep it between the teeth: and on starting a new pair of wheels, a little emery may be put on with the grease, to bring them to a smooth face.

The following general observations on the construction of MACHINES, and on the regulating of their motions, appear to be highly worthy of the Mill-wright's attention; we have, therefore, extracted them from Dr. Robison's article on *Machinery*, inserted in the Supplement to the *Encyclopædia Britannica*.

When heavy stampers are to be raised, in order to drop on the matters to be pounded, the wipers by which they are lifted should be made of such a form, that the stamper may be used by a uniform pressure, or with a motion almost perfectly uniform. If this is not attended to, and the wiper is only a pin sticking out from the axis, the stamper is forced into motion at once. This occasions violent jolts to the machines, and great strains on its moving parts and their points of support; whereas, when they are gradually lifted, the inequality of desultory motion is never felt at the impelled point of the machine. We have seen pistons moved by means of a double rack on the piston rod. A half wheel takes hold of one rack, and raises it to the required height. The moment the half wheel has quitted that side of the rack, it lays hold of the other side, and forces the piston down again. This is proposed as a great improvement; connecting the unequable motion of the piston moved in the common way by a crank. But it is far inferior to the crank motion. It occasions such abrupt changes of motion, that the machine is shaken by jolts. Indeed, if the movement were actually executed, the machine would be shaken to pieces, if the parts did not give way by bending and yielding. Accordingly, we have always observed that this motion soon failed, and was changed for one that was more smooth. A judicious engineer will avoid all such sudden changes of motion, especially in any ponderous part of a machine.

When several stampers, pistons, or other reciprocal movers, are to be raised and depressed, common sense teaches us to distribute their times of action in a uniform manner, so that the machine may always be equally loaded with work. When this is done, and the observations in the preceding paragraph attended to, the machine may be made to move almost as smoothly as if there were no reciprocations in it. Nothing shows the ingenuity more than the artful yet simple and effectual contrivances for obviating those difficulties that unavoidably arise from the very nature of the work that must be performed by the machine, and of the power employed.

There is also great room for ingenuity and good choice in the management of the moving power, when it is such as cannot immediately produce the kind of motion required for effecting the purpose. We mentioned the conversion of the continued rotation of an axis into the reciprocating motion of a piston, and the improvement which was thought to have been made on the common and obvious contrivance of a crank, by substituting a double rack on the piston-rod, and the inconvenience arising from the jolts occasioned by this change. We have seen a great forge, where the engineer, in order to avoid the same inconvenience arising from the abrupt motion given to the great sledge hammer of seven hundred weight, resisting with a five-fold momentum, formed the wipers into spirals, which communicated motion to the hammer almost without any jolt whatever; but the result was, that the hammer rose no higher than it had been raised in contact with the wiper, and then fell on the iron bloom with very little effect. The cause of its inefficiency was not guessed at; but it was removed, and wipers of the common form were put in place of the spirals. In this operation, the rapid motion of the hammer is absolutely necessary. It is not enough to *lift* it up; it must be *tossed* up, so as to fly higher than the wiper lifts it, and to strike with great force the strong oaken spring which is placed in its way. It compresses this spring, and is reflected by it with a considerable velocity, so as to hit the iron as if it had fallen from a great height. Had it been allowed to fly to that height, it would have fallen upon the iron with somewhat more force, (because no oaken spring is perfectly elastic,) but this would have required more than twice the time.

In employing a power which of necessity reciprocates, to drive machinery which requires a continuous motion (as in applying the steam-engine to a cotton or grist mill,) there also occur great difficulties. The necessity of reciprocation in the first mover wastes much power; because the instrument which communicates such an enormous force must be extremely strong, and be well supported. The impelling power is wasted in imparting, and afterwards destroying, a vast quantity of motion in the working beam. The skilful engineer will attend to this, and do his utmost to procure the necessary strength of this first mover, without making it a vast load of inert matter. He will also remark, that all the strains on it, and on its supports, are changing their directions in every stroke. This requires particular attention to the manner of supporting it. If we observe the steam-engines which have been long erected, we see that they have uniformly shaken the building to pieces. This has been owing to the ignorance or inattention of the engineer in this particular. They are much more judiciously erected now, experience having taught the most ignorant that no building can withstand their desultory and opposite jolts, and that the great movements must be supported by a frame-work independent of the building of masonry which contains it.*

The engineer will also remark, that when a single-stroke steam-engine is made to turn a mill, all the communications of motion change the direction of their pressure twice every stroke. During the working stroke of the beam, one side of the teeth of the intervening wheels is pressing the machinery forward; but during the returning stroke, the machinery, already in motion, is dragging the beam, and the wheels are acting with the other side of the teeth. This occasions a rattling at every change, and makes it proper to fashion both sides of the teeth with the same care.

It will frequently conduce to the good performance of an engine, to make the action of the resisting work unequable, accommodated to the inequalities of the impelling power. This will produce a more uniform motion in machines in which the momentum of inertia is inconsiderable. There are some beautiful specimens of this kind of adjustment in the mechanism of animal bodies.

It is very customary to add what is called a *fly* to machines. This is a heavy disk or hoop, or other mass of matter *balanced on its axis*, and so connected with the machinery as to turn briskly round with it. This may be done with the view of rendering the motion of the whole more regular, notwithstanding unavoidable inequalities of the accelerating forces, or of the resistances occasioned by the work. It becomes a *regulator*. Suppose the resistance extremely unequal, and the impelling power perfectly constant; as when a bucket-wheel is employed to *one* pump. When the piston has ended its working stroke, and while it is going down the barrel, the power of the wheel being scarcely opposed, it accelerates the whole machine, and the piston arrives at the bottom of the barrel with a considerable velocity. But in the rising again, the wheel is opposed by the column of water now pressing on the piston. This immediately retards the wheel; and when the piston has reached

* The gudgeons of a water-wheel should never rest on the wall of the building. It shakes it; and if set up soon after the building has been erected, it prevents the mortar from taking firm bond; perhaps by shuttering the calcareous crystals as they form. When the engineer is obliged to rest the gudgeons in this way, they should be supported by a block of oak laid a little hollow. This softens all tremors, like springs of a wheel carriage. This practice would be very serviceable in many other parts of the construction.

please. It is much better to enlarge the diameter. This preserves the friction more moderate, and the pivot wears less. For these reasons a fly is in general a considerable improvement in machinery, by equalizing many exertions that are naturally very irregular. Thus, a man working at a common windlass exerts a very irregular pressure on the winch. In one of his positions, in each turn he can exert a force of near 70 lbs. without fatigue, but in another he cannot exert above 25 lbs; nor must he be loaded with much above this in general. But if a large fly be connected properly with the windlass, he will act with equal ease and speed against 30 lbs.

This regulating power of the fly is without bounds, and may be used to render uniform a motion produced by the most desultory and irregular power. It is thus that the most regular motion is given to mills that are driven by a single-stroke steam-engine, where, for two, or even three seconds, there is no force pressing round the mill. The communication is made through a massive fly of very great diameter, whirling with great rapidity. As soon as the impulse ceases, the fly, continuing its motion, urges round the whole machinery with almost unabated speed. At this instant all the teeth, and all the joints, between the fly and the first mover, are heard to catch in the opposite direction.

If any permanent change should happen in the impelling power, or in the resistance, the fly makes no obstacle to its producing its full effect on the machine; and it will be observed to accelerate or retard uniformly, till a new general speed is acquired exactly corresponding with this new power and resistance.

Many machines include, in their construction, movements which are equivalent with this intentional regulator. A flour mill, for example, cannot be better regulated than by its mill-stone; but in the Albion Mills, a heavy fly was added with great propriety; for if the mills had been regulated by their mill-stones only, then, at every change of stroke in the steam-engine, the whole train of communications between the beam, which is the first mover, and the regulating mill-stone, which is the very last mover, would take in the opposite direction. Although each drop in the teeth and joints be but a trifle, the whole, added together, would make a considerable jolt. This is avoided by a regulator immediately adjoining the beam. This continually presses the working machinery in one direction. So judiciously were the movements of that noble machine contrived, and so nicely were they executed, that not the least noise was heard, nor the slightest tremor felt in the building.

Mr. Valoué's beautiful pile engine, employed at Westminster bridge, is another remarkable instance of the regulating power of a fly. When the ram is dropped, and its follower disengaged immediately after it, the horses would instantly tumble down, because the load, against which they had been straining hard, is at once taken off; but the gin is connected with a very large fly, which checks any remarkable acceleration, allowing the horses to lean on it during the descent of the load; after which their draught recommences immediately. The spindles, cards, and bobbins, of a cotton mill, are also a sort of flies. Indeed, all bulky machines of the rotative kind tend to preserve their motion with some degree of steadiness, and their great momentum of inertia is as useful in this respect as it is prejudicial to the acceleration or any reciprocation when wanted. There is another kind of regulating fly, consisting of wings whirled briskly round, till the resistance of the air prevents any great acceleration. This is a very bad one for a *working* machine, for it produces its effect by *really wasting* a part of the moving powers. Frequently it employs a very great and unknown part of it, and robs the proprietor of much work. It should never be introduced into any machine employed in manufactures.

Some rare cases occur where a very different regulator is required: where a certain determined velocity is found necessary. In this case the machine is furnished, at its extreme mover, with a conical pendulum, consisting of two heavy balls hanging by rods, which move in very nice and steady joints at the top of a vertical axis. It is well known, that when this axis turns round, with an angular velocity suited to the length of those pendulums, the time of a revolution is determined. Thus, if the length of each pendulum be $39\frac{1}{4}$ inches, the axis will make a revolution in two seconds very nearly. If we attempt to force it more swiftly round, the balls will recede a little from the axis, but it employs as long time for a revolution as before; and we cannot make it turn swifter, unless the impelling power be increased beyond all probability; in which case the pendulum will fly out from the centre till the rods are horizontal, after which every increase of power will accelerate the machine very sensibly. Watt and Boulton have applied this contrivance with great ingenuity to their steam-engines, when they are employed for driving machinery for manufactures which have a very changeable resistance, and where a certain speed cannot be much

without great inconvenience. They have con-
cess of the balls from the axis (which gives im-
tion of an increase of power or a diminution
with. the cock which admits the steam to the
der. The balls, flying out, cause the cock to
nd diminish the supply of steam. The impel-
ninishes the next moment, and the balls again
axis, and the rotation goes on as before,
e may have occurred a very great excess of
ower.

etimes employed for a very different purpose
regulator of motion—it is employed as a *col-
er*. Suppose all resistance removed from the
t of a machine furnished with a very large or
mediately connected with the working point.
ll force is applied to the impelled point of this
ion will begin in the machine and the fly begin
tinue to press, uniformly, and the machine will
his may be continued till the fly has acquired a
tion. If at this moment a resisting body be
working point, it will be acted on with very
or the fly has now accumulated in its circum-
y great momentum. If a body were exposed
o the action of this circumference, it would be
k. Much more will it be so, if the body be
c action of the working point, which, perhaps,
rn while the fly makes a hundred. It will exert
nes more force there (very nearly) than at its
rence. All the motion which has been accu-
ie fly during the whole progress of its accumu-
ted in an instant at the working point; multiplied
ntum depending on the proportion of the parts
ie. It is thus that the coining press performs
y, it is thus that the blacksmith forges a bar of
ing the great sledge hammer round his head,
with force the whole way, this accumulated
once extinguished by impact on the iron. It
ve drive a nail, &c. This accumulating power
occasioned many to imagine that a fly really
r mechanical force to an engine; and, not under-
what its efficacy depends, they often place the
ion where it only adds a useless burden to the
should always be made to move with rapidity.
or a mere regulator, it should be near the first
if it be intended to accumulate force in the

working point, it should not be far separated from it. In a
certain sense, a fly may be said to add power to a machine,
because by accumulating into the exertion of one moment the
exertions of many, we can sometimes overcome an obstacle
that we never could have balanced by the same machine,
unaided by the fly. And it is this accumulation of force
which gives such an appearance of power to some of our
first movers.

ANIMAL STRENGTH.

ANIMAL STRENGTH has been very differently estimated by
different authors ; but this is not to be wondered at when we
consider the many difficulties that ever must attend any
attempt to subject it to an estimate. Physical causes must
sensibly affect the extent and duration of animal exertion,
either in man or beast ; and the only way of coming to any
thing like an accurate result, is to compare the experiments
of the different philosophers who have attended to the subject, This has been already done by Dr. Young, in the
second volume of his *Philosophy*, whose valuable tables we
here present to our readers.

Comparative table of mechanical forces.

In order to compare the different estimates of the force of
moving powers, it will be convenient to take a unit which
may be considered as the mean effect of the labour of an
active man, working to the greatest possible advantage, and
without impediment. This will be found, on a moderate
estimation, sufficient to raise 10 pounds, 10 feet in a second,
for ten hours in a day ; or to raise a 100 pounds, which is the
weight of twelve wine gallons of water, one foot in a second,
or 36,000 feet in a day ; or 3,600,000 pounds, or 432,000
gallons, one foot in a day. This we may call a force of one
continued 36,000"

Immediate force of men, without deduction for friction.

	Force.	Conti- nuance.	Day's work.
A man, weighing 133 pounds, Fr. ascended 62 feet, Fr. by steps, in 34", but was completely exhausted.— Amontons	2,8	34"	
A sawyer made 200 strokes of 18 inches, Fr. each, in 115', with a force of 25 pounds, Fr. He could not have gone on above three minutes.—Amontons..............	6	145	
A man can raise 60 pounds, Fr. one foot, Fr. in 1", for eight hours a day.--Bernouilli	69	8ᵇ	,552
A man of ordinary strength can turn a winch with a force of 30 pounds, and with a velocity of 3½ feet in 1", for 10 hours a day.—Desaguliers	1,05	10ᵇ	1,05

	Force.	Conti-nuance.	Day's work.
men working at a windlass with handles at right can raise 70 pounds more easily than one can ...—Desaguliers	1,22		1,22
can exert a force of 40 pounds for a whole day, assistance of a fly, when the motion is pretty quick, four or five feet in 1″.—Desaguliers, Lect. iv. the annotation it appears to be doubtful whether is 40 pounds or 20	,2		,2
short time a man may exert a force of 80 pounds fly, " when the motion is pretty quick."—Desa-	,3	1″	
going up stairs ascends 14 metres in 1′.—Cou-	1,182	1′	
going up stairs for a day raises 205 chiliogrammes right of a chiliometre.—Coulomb			,412
a spade a man does $\frac{13}{14}$ as much as in ascending -Coulomb			,391
a winch a man does $\frac{4}{9}$ as much as in ascending -Coulomb			,258
carrying wood up stairs raises, together with weight, 109 chiliogrammes to one chiliometre.— b..			,219
weighing 150 pounds, Fr. can ascend by stairs t, Fr. in 1″ for 15″ or 20″.—Coulomb	5,22	20″	
all an hour, 100 pounds, Fr. may be raised one . in 1″.—Coulomb	1,152	30	
ding to Mr. Buchanan's comparison, the force in turning a winch being made equal to the unit, e in pumping will be	,61		
ging	1,36		
ring	1,43		
ing the accuracy of Euler's formula, confirmed by , supposing a man's action to be a maximum when s 2¼ miles an hour, we have 7½ for his greatest , ,04 ′7½ − v)² for the force exerted with any other , and ,0160(7½−v)² for the action in each case : en the velocity is one mile an hour, the action is	,676		
two miles..	,964		
three	,972		
four	,784		
hen five	,5		

d the force in a state of rest becomes 2¼, or about
unds; with a velocity of two miles, 36 pounds; with
, 24 pounds; and with four, 15.

is obvious that in the extreme cases this formula is in-
ate, but for moderate velocities it is probably a tolerable
ximation.

ulomb makes the maximum of effect when a man,
ing 70 chiliogrammes, carries a weight of 53 up stairs,
his appears to be too great a load; he considers 145
grammes as the greatest weight that can be raised.
bserves that in Martinique, where the thermometer is
m below 68°, the labour of Europeans is reduced to one

A horse can in general draw no more up a steep hill than three men can carry; that is, from 450 to 750 pounds; but a strong horse can draw 2000 pounds up a steep hill, which is but short. The worst way of applying the force of a horse, is to make him carry or draw up hill; for if the hill be steep, three men will do more than a horse, each man climbing up faster with a burden of 100 pounds weight, than a horse that is loaded with 300 pounds: a difference which is owing to the position of the parts of the human body being better adapted to climb than those of a horse.

On the other hand, the best way of applying the force of a horse, is an horizontal direction, wherein a man can exert least force; thus a man, weighing 140 pounds, and drawing a boat along, by means of a rope coming over his shoulders, cannot draw above 27 pounds, or exert above one-seventh part of the force of a horse employed to the same purpose.

The very best and most effectual posture in a man is that of rowing; wherein he not only acts with more muscles at once for overcoming the resistance than in any other position, but as he pulls backwards, the weight of his body assists by way of lever.—Desaguliers.

The diameter of a walk for a horse-mill ought to be at least 25 or 30 feet.—Desaguliers.

Some horses have carried 650 or 700 pounds, seven or eight miles, without resting, as their ordinary work; and a horse at Stourbridge carried eleven hundred weight of iron, or 1232 pounds, for eight miles.—Desaguliers, *Experimental Philosophy*, vol. i.

Work of mules.

	Force.	Conti- nuance.	Day's work.
Camusel says, that a mule works in the West Indies two hours out of about 18, with a force of about 150 pounds, walking three feet in a second.—Dr. Young's *Philosophy*	4,5	2ʰ 40ʳ	1,2

These examples exhibit the great advantages which may be gained by directing the exertion of animals in a proper course, their effects being plainly reducible to the operations of mechanical powers. To describe the various modes of applying animal strength, as a first mover of mechanical engines, would greatly exceed our limits; we shall therefore merely state, that the most common machine for receiving the force of animals is the horse-walk, which affords the means of applying the action of that animal to create rotative motion. The horse-walk is formed of an horizontal lever or arm, attached to an upright spindle. The lever should not be less

than twelve feet, as the labour of the animal is great
creased by a small curve, which causes an unequal resis
upon his two shoulders. The machine should be so regu
that the horse may not be required to deviate from his
pace of two miles and a half, with a burthen, an hour.
gives, in which the horse works, should not be immo
fixed to the arms, but hung by a swivel joint, so th
may place himself in the most comfortable position.
work should be supplied to the machinery as regular
possible.

Having, in the preceding account, stated the mean re
of human exertion when applied to regular and un
labour, we shall in the next place proceed to
extraordinary feats of strength, as well as some that ha
appearance of being such, but which were, in reality
mere effects of contrivance and skill, and which might
been performed by almost any men who were poss
of that knowledge of their construction as would enable
similarly to exert their strength to the best advantage,

M. de la Hire, in an *Examination of the Force of*
(vide Memoirs of the Academy of Sciences for 1699,)
"There are men whose spirits flow so abundantly into
muscles, that they exert three or four times more stre
than others do; and this seems to be the natural reas
the surprising strength that we see in some men who
and raise weights which two or three ordinary men
hardly sustain, though these men be sometimes but
moderate stature, and rather appear weak than st
There was a man in this country a little while ago,
would carry a very large anvil, and of whom was rep
several wonderful feats of strength. But I saw anoth
Venice, who was but a lad, and did not seem able to
above forty or fifty pounds, with all possible advantages
this young fellow, standing upon a table, raised from
earth, and sustained off the ground, an ass, by mean.
broad girth, which, going under the creature's belly, was
upon two hooks that were fastened to a plat of small
coming down in tresses from the hair on each side o
lad's head, which were in no great quantity. And al
great force *depended only upon the muscles of the shou
and those of the loins:* for he stooped at first whils
hooks were fastened to the girth, and then raised himsel
lifted up the ass from the ground, bearing with his t
upon his knees. He raised also in the same manner
weights that seemed heavier, and used to say he did

ι α his knees bent, so as to bring the two cords with
lifted to be in the same plane with his ancles and
 of his thigh-bones; by which means the line of
of the man and the whole weight came between the
part of his two feet, which are the supports: then
ended his legs he raised himself, without changing
? direction. That this must have been the manner
:y well assured of, by not only observing those that
uch feats, but having often tried it myself. As for
es of the loins, they are incapable of that strain,
re six times weaker than the extensors of the legs;
found them so in myself.
: the year 1716, having the honour of showing a
y experiments to his late majesty King George I.,
ty was desirous to know whether there were any
 those feats of strength that had been shown half
fore by a man, who seemed by his make to be no
than other men: upon this I had a frame of wood
tand in, (and to rest my hands upon,) and with a
l chain lifted an iron cylinder, made use of to roll
u, sustaining it easily when once it was up. Some
 and gentlemen who were present tried the ex-
afterwards, and lifted the roller; some with more
some with more difficulty, than I had done. This
ighed 1,900 pounds, as the gardener told us.
ls I tried to lift 300 pounds with my hands, (viz.,
 with 150 pounds of quicksilver in each,) which
keed raise from the ground, but strained my back
feel it three or four days; which shows that, in the
ton. the muscles of the loins (which exerted their

"Thomas Popham, born in London, and now about one years of age, five feet ten inches high, with muscl hard and prominent, was brought up a carpenter, whicl he practised till within these six or seven years that showed feats of strength; but he is entirely ignorant art to make his strength more surprising. Nay, som he does things which become more difficult by his dis tageous situation; attempting, and often doing, what h other strong men have done, without making use of th advantages.

"About six years ago he pulled against a horse, sittin the ground with his feet against two stumps driven i ground, but without the advantages which might hav attained by placing himself in a proper situation; the however, was not able to move him, and he thought l in the right posture for drawing against a horse; but in the same posture, he attempted to draw against two he was pulled out of his place by being lifted up, a one of his knees struck against the stumps, which sh it so, that, even to this day, the *patella*, or knee-pan loose, that the ligaments of it seem either to be bro quite relaxed, which has taken away most of the strer that leg."

Dr. Desaguliers then relates the exploits which he sa perform.

"1. By the strength of his fingers, (only rubbed ashes to prevent them from slipping,) he rolled up strong and large pewter dish.

"2. He broke seven or eight short and strong pi tobacco-pipe with the force of his middle finger, havi them on the first and third finger.

"3. Having thrust in under his garter the bowl of n tobacco-pipe, his legs being bent, he broke it to pieces l tendons of his hams, without altering the bending of hi

"4. He broke such another bowl between his fi second finger, by pressing his fingers together sideway

"5. He lifted a table six feet long, which had half a h weight hanging at the end of it, with his teeth, and hel an horisontal position for a considerable time. It i the feet of the table rested against his knees; but, length of the table was much greater than its heigh performance required a great strength to be exerted muscles of his loins, those of his neck, the *masseter* an *poral*, (muscles of the jaws,) besides a good set of teet

" 6. He took an iron kitchen poker, about a yard long, and three inches in circumference, and, holding it in his right hand, he struck upon his bare left arm, between the elbow and the wrist, till he bent the poker nearly to a right angle.

" 7. He took such another poker, and holding the ends of it in his hands, and the middle against the back of his neck, he brought both ends of it together before him; and, what was yet more difficult, he pulled it almost straight again: because the muscles which separate the arms horizontally from each other are not so strong as those that bring them together.

" 8. He broke a rope of about two inches in circumference, which was in part wound about a cylinder of four inches diameter, having fastened the other end of it to straps that went over his shoulders. But he exerted more force to do this than any other of his feats, from his awkwardness in going about it; for the rope yielded and stretched as he stood upon the cylinder, so that when the extensors of the legs and thighs had done their office in bringing his legs and thighs straight, he was forced to raise his heels from their bearings, and use other muscles that are weaker. But if the rope had been so fixed that the rope to be broken had been short, it would have been broken with four times less difficulty.

" 9. I have seen him lift a rolling stone of about 800 pounds with his hands only, standing in a frame above it, and taking hold of a chain that was fastened to it. By this, I reckon, he may be almost as strong again as those who are generally reckoned the strongest men, they generally lifting no more than 400 pounds in that manner. The weakest men who are in health, and not too fat, lift about 125 pounds, having about half the strength of the strongest.

" N. B. This sort of comparison is chiefly in relation to the muscles of the loins; because in doing this, one must stoop forwards a little. We must also add the weight of the body to the weight lifted. So that if the weakest man's body weighs 150 pounds, that, added to 125 pounds, makes the whole weight lifted by him to be 275 pounds. Then, if the strongest man's body weighs also 150 pounds, the whole weight lifted by him will be about 550 pounds, that is 400 pounds and the 150 pounds which his body weighs. Topham weighs about 200 pounds, which, added to the 800 pounds that he lifts, makes 1000 pounds. But he ought to lift 900 pounds besides the weight of his body, to be as strong again as a man of 150 pounds weight who can lift 400 pounds.

" About thirty years ago, one *Joyce*, a Kentish man, famous

for his great strength, showed several feats in London ,
country, which so much surprised the spectators, that
by most people called *the second Sampson.* But
the postures which he had learnt to put his body in
found out by practice, without any mechanical theo
such as would make a man of common strength do suc
as would appear surprising to every one who did no:
the advantage of those positions of the body, yet
then attempted to draw against horses, or raise great w
or to do any thing in imitation of him: because, as
very strong in the arms, and grasped those that tr
strength that way so hard that they were obliged imme
to desire him to desist, his other feats (wherein his ma
acting was chiefly owing to the mechanical advantage
by the position of his body,) were entirely attributed
extraordinary strength.

" But when he had been gone out of England, or had
to show his performances, for eight or ten years, r
ordinary strength had found out the way of makin
advantage of the same postures as *Joyce* had put himse
as to pass for men of more than common strength, by
ing against horses, breaking ropes, lifting vast weight
though they could in none of the postures really perf
much as *Joyce*, yet they did enough to amaze and
and get a deal of money, so that every two or three ye
had a *new second* Sampson.

" About fifteen years ago a German of middle size, a
ordinary strength, showed himself at the *Blue Posts,*
Haymarket, and, by the contrivances above-mentioned,
for a man of uncommon strength, and gained considerabl
of money by the daily concourse of spectators. After
seen him once, I guessed at his manner of imposing u
multitude; and being resolved to be fully satisfied
matter, I took four very curious persons with me to s
again, viz., the Lord Marquis of Tullibardin, Dr Ale
Smart, Dr Pringle, and a mechanical workman who
assist me in my courses of experiments. We place
selves in such manner round the operator, as to be
observe nicely all that he did; and found it so pract
that we performed several of his feats that evening b
selves, and afterwards I did the most of the rest, as I
frame to sit in to draw, and another to stand in a
weights, together with a proper girdle and hooks. I I
showed some of the experiments before the Royal S
and ever since, at my experimental lectures, I expl

reason of such performances, and take any person of ordinary strength that has a mind to try, who can easily do all that the German above-mentioned used to do, without any danger or extraordinary straining, by making use of my apparatus for that purpose.

"In order to explain how great feats may be performed by men of no extraordinary strength, I have in fig. 79, drawn the lower part of a skeleton, containing so many of the bones of the human body as are concerned in these operations, making the figure pretty large, to show the better how the girdle is to be applied.

"The bones marked I S A P H I,* which compose the cavity called the pelvis, contain a bony circle or double arch of such strength, that it would require an immense force to break them by an external pressure directed towards the centre of the circle, or the middle of the pelvis. It is also to be observed, that those parts of this bony circumference, which receive the heads of the thigh-bone above, at, and below A, called the ischium or coxendix, are the strongest of all, so that a very great force may push the heads of the thigh-bones upwards, or, which is the same thing, the upper parts of the coxendix downwards, or towards each other in a lateral direction from A to A, without doing any hurt to the human body.

"Now if the girdle above described be put round the body in the manner represented in the figure, and be drawn downwards at G, by a great weight W, it will press on the os sacrum behind, and the illum; then it will, by its pressure on T T, the great trochanters of the thigh-bones, draw the round heads the faster into their sockets, so as to make them less liable to slip out and strain the ligament by a push directed upwards. So that the semicircular part of the girdle, T C S C T, presses together the bony arch denoted by the same letters, which, according to the nature of arches, is the stronger for that pressure. The abutments of the arch cannot come nearer together by reason of the resistance of the strong bones A P A, neither can they fly outwards, because the girdle keeps them together. Then the thighs and legs T D B are two strong columns, capable of sustaining 4000 or 5000

* These bones are thus distinguished by anatomists: S, the os sacrum; I I, the illum; A A, the os ischium; whose strongest part has on each side an hemispherical concave, in which the round head of the thigh-bone is received and turns round, being held by a strong ligament in its middle; those parts of the bone that join together before, betwixt A A and above P, are called the os pubis, or ossa pubis.

pounds at least, provided they stand quite upright. The
muscles here are put to no strain, being no farther concerned
than to balance each other; that is, the antagonist muscles,
extensors, and flexors, only keep the bones in their places,
which makes them resist like one entire bone formed into an
arch.

" This shows how easily the man, fig. 80, may sustain a
cannon of 2000 or 3000 pounds weight. The same solution
will also serve for the resistance of the man, fig. 81, whom
five men, nay ten men, or two horses, cannot pull out of his
situation when he sits so as to have his legs and thighs in
the horizontal line P F, or in a line inclining downwards
towards A, for then, though there is a difference in the sitting
posture from the standing posture before described, yet by
reason of the mobility of the heads of the thigh-bones in the
acetabula or cavities of the coxendix, the arch is the same
and as strong as before, its abutments being equally supported by the legs and thighs. It is only the bending of the
back-bone above the girdle to bring up the body which makes
the difference of position in the man, though not sensibly in
the resisting parts.

" In breaking a rope the muscles must act in extending the
legs; and that we may the better explain that action, we must
consider a man breaking the rope, as represented in fig. 82.
Suppose a rope fastened to a post at P, or any other fixed
point, is brought through an iron eye L, to the hook of the
girdle H, of the man II I, and so fixed to it by a loop, or
otherwise, as to be quite tight, whilst the man's knees are so
bended as to want about an inch of having his legs and
thighs quite upright. Then if the man on a sudden stretches
his legs and sets himself upright, he will with ease break the
very same rope which held two horses exerting their whole
strength when they drew against him; such as a cart rope,
or a rope of near three quarters of an inch diameter, which
may be broken by a man of middling strength, by the action
of the ten muscles * that extend the legs, five belonging to
each leg.

* The four muscles that extend each leg are described by anatomists
thus :—1. The rectus, arising from the anterior inferior spine of the os ilium,
and inserted, through the medium of the patella, into the anterior tuberosity
of the tibia. 2. The cruralis, situated beneath the former, and arising from
the front surface of the os femoris for a considerable extent, and inserted
into the upper edge of the patella, and also, through the medium of that bone,
into the anterior tuberosity of the tibia 3. The vastus externus, arising from
the root of the trochanter major, and outer side of the os femoris, and inserted into the outer edge of the patella, and again, through its medium, into
the anterior tuberosity of the tibia 4 The vastus internus, which arises from

" In breaking the rope one thing is to be observed, which will much facilitate the performance; and that is, to place the iron eye L, through which the rope goes, in such a situation; that a plane going through its ring shall be parallel, or nearly parallel to the two parts of the rope; because then the rope will in a manner be jammed in it, and not slipping through it, the whole force of the man's action will be exerted on that part of the rope which is in the eye, which will make it break more easily than if more parts of the rope were acted upon. So that the eye, though made round and smooth, may be said in some measure to cut the rope. And it is after this manner that one may break a whip-cord, nay, a small jack-line, with one's hand, without hurting it; only by bringing one part of the rope to cut the other; that is, placing it so round one's left hand, that, by a sudden jerk, the whole force exerted shall act upon one point of the rope. See fig. 83, where the cord to be broken at the point L in the left hand, is marked according to its course, by the letters R T S L M N O P Q, folding once about the right hand, then going under the thumb into the middle of the left hand; where crossing under another part, it is brought back under the thumb again to M, then round the back of the hand to N, so through the loop at L to O, and three times round the little finger at P and Q; which last is only that the loop N O may not give way. Before the hands are jerked from one another, the left hand must be shut, but the thumb must be held loose, lest pressing against the fore finger it should hinder the part T L of the rope from carrying the force fully to the point L; but the little finger and that next to it must be held hard, to keep the loop N O firm in its place.

" There are several cases, wherein it would be of singular use to apply the force of one or more men, by means of the girdle or hook and chain, in the manner above-mentioned; as for example, when the resistance is very great, but the bodies that resist are to be removed but a little way: if we lift very heavy goods a small height, to remove any thing from under them; if we would draw a bolt or staple, and find we cannot do it even with an iron crow, the hand pulling it upwards at the end; then the hook of the girdle being applied at the end of the crow, the force exerted by stretching the legs would be tenfold of what the hands were able to do, without more help at the same place.

" There may also be many occasions on board a ship. I will

the root of the trochanter minor, and inner surface of the os femoris, and inserted into the inner edge of the patella, and likewise, through its medium, into the anterior tuberosity of the tibia with the former muscles.

instance but one.   Let F G, fig. 84, be the tackle for raising
or lowering the main-top-mast, part of which is represented
by m 1, m 2; the block G is fixed below, and as the block F
comes down, it pulls along with it the top rope F B C, m 1
running over the block B, fixed at A, and round the block C
in the heel of the top-mast, so as to draw up the lower end
m 1 of the said main-top-mast, which, when hoisted up to its
due height, is made fast by the iron pin or fid I, which is
thrust through it, and then its own weight and the hole D of
the cap will keep it in its place.   We will suppose that the
force required thus to raise the mast must be that of six men
pulling upon deck at the fall of the tackle, that is, at the
running rope F G K at K on the other side of the main-mast
L l.   Now in order to let down this mast on the sudden, as
in case of hard weather, it is necessary the tackle and power
must be made use of, though it be but to lift it a very little
way, that a man may be able to get out the fid I, before the
said mast can be let down and slip to N on the side of the
main-mast.   I say, that if the hands are so employed otherwise, that instead of six men there be only one man at the
rope K; if he has a strong girdle to which he fastens it, or
makes a bow in the rope itself, to fix it round the lower part
of his back, &c., he may exert much more force in the direction G K than the six men in the common way of pulling;
and if he draws to him, sitting on the ground, and pushing
his feet against the first firm obstacle that he finds, as against
O P, only two inches of the rope G K, he will raise up the
main-top-mast the third part of an inch, which will be sufficient for the iron fid I to be drawn out."   Desaguliers'
Philosophy, vol. i.

WATER-MILLS.

WATER-MILL is the name by which all mills are designated
that receive their motion from the impulse of the water.
As each of these mills will come under their respective heads,
we shall, in the present article, confine ourselves to a minute
description of the different kinds of water-wheels, by whose
axis the force with which they have been impressed may be
transmitted to move any species of machinery, however simple
or complex.
    But, notwithstanding the extensive signification of the
term water-mill when applied to the different branches of
manufacture carried on therein, we have another, and still
more simple division, arising from the peculiar construction
of the water-wheel, termed the undershot-mill, the overshot-

mill, and the breast-mill. There is also another called the mill with horizontal wheels; but as this is very disadvantageous in point of practical utility, we shall forbear to describe it. The *undershot-wheel* is used only in streams, and is acted upon by the water striking the float-boards at the lower circumference of the wheel. In the *overshot-wheel* the water is poured over the top of the wheel, and is received in buckets formed all round the wheel for that purpose. And in the *breast-wheel* the water falls down upon the wheel at right angles to the float-boards, or buckets placed round the circumference of the wheel to receive it.

UNDERSHOT-WHEELS.

Mr. JOHN SMEATON has made numerous experiments upon the different kinds of water-wheels, the results of which were laid before the Royal Society. The time that has elapsed since the period when they were first given to the world, has been sufficient to prove their fallacy, if any had existed; and the high estimation in which they still continue to be held by mathematicians and mechanics, is certain evidence of their value and importance.

Mr. Smeaton prefaces a minute description of the machines and models used by him for his experiments, with an observation, that what he has to communicate on the subject was originally deduced from experiments, which he looks upon as the best means of obtaining the outlines in mechanical inquiry. "But in such cases," says he, "it is very necessary to distinguish the circumstances in which a model differs from a machine in large; otherwise a model is more apt to lead us from the truth than towards it: and, indeed, though the utmost circumspection be used in this way, the best structure of machines cannot be fully ascertained but by making trials with them, when made of their proper size. It was for this reason, though the models and experiments referred to were made in the years 1752 and 1753, that I have deferred offering them to the Society until I had an opportunity of *putting the deductions made therefrom in real practice,* in a variety of cases, and for various purposes, so as to be able to assure the Society that *I have found them to answer.*"

Mr. Smeaton then remarks, that the word *power,* as used in practical mechanics, signifies the exertion of strength, gravitation, impulse, or pressure, so as to produce motion: and by means of strength, gravitation, impulse, or pressure, compounded with motion, to be capable of producing an effect:

and that no effect is properly mechanical, but what requires
such a kind of power to produce it.

Having described the models and machines used for making
his experiments, he observes that with regard to *power*, it is
most properly measured by the raising of a weight, the
relative height to which it can be raised in a given time
being the actual extent; or, in other words, if the weight
raised be multiplied by the height to which it can be raised
in a given time, the product is the measure of the power
raising it; and, consequently, all those powers are equal,
whose products, made by such multiplication, are the same:
for if a power can raise twice the weight to the same height,
or the same weight to twice the height, in the same time
that another power can, the first power is double the second;
but if the power can only raise half the weight to double the
height, or double the weight to half the height, in the same
time that another can, those two powers are equal. This,
however, must be understood to be only in cases of slow and
equable motion, where there is no acceleration or retardation.

In comparing the effects produced by water-wheels with
the powers producing them, or, in other words, to know
what part of the original power is necessarily lost in the
application, we must previously know how much of the power
is spent in overcoming the friction of the machinery and the
resistance of the air; also, what is the real velocity of the
water at the instant that it strikes the wheel, and the real
quantity of water expended in a given time.

From the velocity of the water at the instant that it strikes
the wheel, the height of head productive of such velocity
can be deduced, from acknowledged and experimented principles of hydrostatics: so that. by multiplying the quantity
or weight of water really expended in a given time, by the
height of a head so obtained, which must be considered as
the height from which that weight of water had descended
in such given time, we shall have a product equal to the
original power of the water, and clear of all uncertainty that
would arise from the friction of the water, in passing small
apertures, and from all doubts arising from the different
measure of spouting waters, assigned by different writers.

On the other hand, if the sum of the weights raised by the
action of this water, and of the weight required to overcome
the friction and resistance of the machine, be multiplied by the
height to which the weight can be raised in the time given,
the product will be equal to the effect of that power; and
the proportion of the two products will be in proportion of

the *power* to the *effect*: so that by loading the wheel with different weights successively, we shall be able to determine at what particular load, and velocity of the wheel, the effect is a *maximum*.

The experiments made by Mr. Smeaton may thus be reduced. The circumference of the wheel, 75 inches, multiplied by 86 turns, gives 6450 inches for the velocity of the water in a minute; 1/60 of which will be the velocity in a second, equal to 107.5 inches, or 8.96 feet, which is due to a head of 15 inches; and this we call the *virtual* or *effective head*. The area of the head being 105.8 inches, this multiplied by the weight of water of the cubic inch, equal to the decimal 579 of the ounce avoirdupois, gives 61.26 ounces for the weight of as much water as is contained in the head, upon one inch in depth, 1/16 of which is 3.83 pounds; this multiplied by the depth 21 inches, gives 80.43 pounds for the value of 12 strokes; and by proportion, 39½ (the number made in a minute) will give 264.7 pounds, the weight of water expended in a minute.

Now as 264.7 pounds of water may be considered as having descended through a space of 15 inches in a minute, the product of these two numbers 3970 will express the *power* of the water to produce mechanical effects; which were as follows:

The velocity of the wheel at the *maximum*, as appears above, was 30 turns a minute; which multiplied by nine inches, the circumference of the cylinder, makes 270 inches; but as the scale was hung by a pulley and double line, the weight was only raised half of this, viz. 135 inches.

	lb.	*oz.*
The weight in the scale at the maximum ..	8	—
The weight of the scale and pulley	—	10
The counterweight, scale, and pulley	—	12
Sum of the resistance..	9	6
Or pounds..	9.375	

Now as 9.375 pounds is raised 135 inches, these two numbers being multiplied together, the product is 1266, which expresses the effect produced at a maximum; so that the proportion of the *power* to the *effect* is as 3970 : 1266, or as 10 : 3.18.

But though this is the greatest single effect producible from the power mentioned, by the impulse of the water upon

an undershot-wheel; yet, as the whole power of the wati
not exhausted by it, this will not be the tru
the *power* of the water, and the *sum* of all
ducible therefrom; for, as the water must n
the wheel with a velocity equal to the wheel's
it is plain some part of the power of the water must rea
after quitting the wheel.

The velocity of the wheel at the maximum
minute; and consequently its circumference
rate of 3.123 feet a second, which answers to
inches; this being multiplied by the expense
minute, viz. 264.7 pounds produces 481 for the pu
remaining in the water after it has passed the wheel:
being therefore deducted from the original power 3.5
leaves 3.489, which is that *part* of the power
ducing the effect 1266; and, consequently,
power spent in producing the effect, is to
effect that it produces as 3489 : 1266 : : 10
11 to 4.

The *velocity of the water* striking the
determined to be equal to 86 circumferences of the wheel
minute, and *the velocity of the wheel* at the *maximum* is
30; the velocity of the water will therefore be to that of
wheel as 86 to 30, or as 10 to 3.5, or as 20 to 7.

The *load at the maximum* has been shown to be equa
nine pounds six ounces, and the wheel ceased moving
12 pounds in the scale: to which if the weight of the s
be added, viz. 10 ounces, the proportion will be nearl
3 to 4 between the load at the *maximum* and *that*
which the wheel is stopped.

It is somewhat remarkable, that though the velocity of
wheel in relation to the water turns out greater than ⅐ of
velocity of the water, yet the impulse of the
case of a *maximum* is more than double of what is assig
by theory; that is, instead of ⅓ of the column, it is nea
equal to the whole column.

It must be remembered, therefore, that, in the pres
case, the wheel was not placed in an open river, where
natural current, after it has communicated its impulse to
float, has room on all sides to escape, as the theory suppos
but in a conduit or race, to which the float is adap
the water cannot otherwise escape than by moving al
with the wheel. It is observable, that a wheel working
this manner, so soon as the water meets the float, receive
sudden check, and rises up against the float, like a wave agai

a fixed object; insomuch, that when a sheet of water is not a quarter of an inch thick before it meets the float, yet this sheet will act upon the whole surface of a float whose height is three inches; and, consequently, were the float no higher than the thickness of the sheet of water, as the theory also supposes, a great part of the force would have been lost by the water dashing over the float.

Mr. Smeaton next proceeds to give tables of the velocities of wheels with different heights of water; and from the whole deduces the following conclusions.

Maxim 1. That the virtual or effective head of water, and consequently its effluent velocity, being the same, the mechanical effect produced by a wheel actuated by this water will be nearly in proportion to the quantity of water expended.

Note.—The virtual or effective head of any water which is moving with a certain velocity, is that height from which a heavy body must fall, in order to acquire the same velocity.

The height of the virtual head, therefore, may be easily determined from the velocity of the water; for the heights are as the square of the velocities, and the velocities, consequently, as the square roots of the heights. Mr. Smeaton observed the velocity of the effluent water in all his experiments, and thence calculated the virtual head: he states, that the virtual head bears no proportion to the real head or depth of water; but that when either the aperture is greater, or when the velocity of the water issuing therefrom less, they approach nearer to a coincidence; and consequently, in the large openings of mills or sluices, where great quantities of water are discharged from moderate heads, the actual head of water, and the virtual head, as determined by theory from the velocity, will nearly agree.

For example of the application of his first maxim. Suppose a mill driven by a fall of water whose virtual head is five feet, and which discharged 550 cubic feet of water per minute, and that it is capable of grinding four bushels of wheat in an hour. Now another mill, having the same virtual head, but which discharges 1100 cubic feet of water per minute, will grind eight bushels of corn in an hour.

Maxim 2. That the expense of water being the same, the effect produced by an undershot-wheel will be nearly in proportion to the height of the virtual or effective head. This is proved in the preceding example.

Maxim 3. That the quantity of water expended being the same, the effect will be nearly as the square of the velocity of the water; that is, if a mill driven by a certain quantity of water, moving with the velocity of 18 feet per second, is capable of grinding four bushels of corn in an hour, another mill, driven by the same quantity of water, but moving with the velocity of 22¼ feet per second, will grind nearly seven bushels of corn in an hour; because the square of 18 is 324, and the square of 22¼ is 506¼. Now, say, as 324 is to 4 bushels, so is 506¼ to 6¼ bushels; that is as 4 to 6¼.

Maxim 4. The aperture through which the water issues being the same, the effect will be nearly as the cube of the velocity of the water issuing; that is, if a mill driven by water rushing through a certain aperture with the velocity of 18 feet per second will grind four bushels of corn in an hour, another mill, driven by water moving through the same aperture, but with the velocity of 22¼ feet per second, will grind 51 bushels; for the cube of 18 is 5832, and the cube of 22¼ is 11390¼; then as 5832 is to 4, so is 11390¼ to 7¾.

Maxim 5. The proportions between the power of the water expended, and the effect produced by the wheel, were 3 to 1. Upon comparing several experiments, Mr. Smeaton fixed the proportions between them for large works; that is, if the weight of the water which is expended in any given time be multiplied by the height of the fall, and if the weight raised be also multiplied by the height through which it is raised, the first of these two products will be three times that of the second.

Maxim 6. The best general proportions of velocities between the water and the floats of the wheels will be that of 5 to 2; for instance, if the water when it strikes the wheel moves with a velocity of 18 feet per second, the wheel must be so loaded that its float-boards will move with a velocity of 7.2 feet per second, and the wheel will then derive the greatest power from the water, because as 5 to 18, so is 2 to 7.2.

Maxim 7. There is no certain ratio between the load that the wheel will carry when producing its maximum of effect, and the load that will totally stop it; but it approaches nearest to the ratio of 4 to 3, whenever the power exerted by the wheel is greatest, whether it arise from an increase of the velocity, or from an increased quantity of water; and this proportion seems to be the most applicable to large works. But when we know the effect which a wheel ought to produce, and the velocity it ought to move with whilst producing that effect, the exact knowledge of the greatest load it will bear is of very little consequence in practice.

Maxim 8. The load that the wheel ought to have, in order to work to the most advantage, can be always assigned thus: ascertain the power of the whole body of water, by multiplying the weight of the water expended in a minute by the height of the fall; take one-third of the product, and it gives the effect of power which the wheel ought to produce: to find the load, we must divide this product by the velocity which the wheel should have, and that, as we have before settled, should be two-fifths of the velocity with which the water moves when it strikes the wheel.

In the application of these principles the first thing to be done in a situation where an undershot-wheel is intended to be fixed, is to consider whether the water can run off clear from the wheel, 'so as to have no back-water to impede its motion; and whether the fall which can be obtained by constructing a proper dam to pen up the water, and sluice for it to pass through, will cause it to strike the float-boards of the wheel with sufficient velocity to impel them forcibly forwards; and also, whether the quantity of the supply will be sufficient to keep a wheel at work for a certain number of hours each day.

When we have ascertained the height of the fall of water, that is the height of the surface above the centre of the opening of the sluice, we must find what will be the continual velocity of the water issuing from such opening.

In some cases, we have the velocity of the water given when it issues from the opening of the sluice, and we then require to know what height of column will produce that velocity. These two things we may find by a single rule, and an easy arithmetical operation, which is as follows:

1st. The perpendicular height of the fall of water being given in feet and decimals of feet, the velocity that the water will acquire per second, expressed in feet and decimals, may be found by the following rule:

Multiply the constant number 64.2882 by the given height, and the square root of the product is the velocity required.

Example 1. If the height is two feet, the velocity will be found 11.34 feet per second.

Example 2. If the height is 16.0913 feet, the velocity will be 32.1826 feet per second.

Example 3. If the height is 50 feet, the velocity will be 56.68 feet per second.

Note.—The velocities thus obtained will be only the theo-

it. The principle of this improvement is to make the lower paddles recede from the centre of the axle and to the arms to which they are attached, while the upper paddles proceed to the centre of the axle in equal distances as the others recede; and in the rotation of the wheel, every paddle passes through the various evolutions and positions to which every revolution of the wheel subjects each paddle. The lower paddles describe a greater radius of a circle than the upper paddles, and thereby travel at an increased velocity, or rather they pass at their extreme points through a greater space in the same period of time; this effect renders the lower half of the wheel heavier than the upper, by the eccentric position of the paddles, and the flat ring of iron to which they are attached, and it also increases the speed of any navigable body through the water to which such wheels are applied.

Figs. 85, 86, and 87, are views of my improved wheel with one paddle, as in. fig. 85, at its greatest depth in the water; B, B, B, B, is one of the iron arm frames to which one end of the paddles C, C, C, C, C, C, are attached by the joint-pins D, D, D, D, D, D, to the arm frames B, B, B, B. E E is the flat iron ring or eccentric circle, to which the other ends of the paddles are attached by similar joint-pins F, F, F, F, F, F. G G, are the iron guard or guide-rollers, a section of one of which is shown at fig. 88, which may either revolve on fixed axles, or these rollers may be fixed on revolving axles, whichever is most convenient. The object of these rollers is, to keep the iron circle E E in its proper situation, which is an equal distant position from the centre of the wheel-shaft in a longitudinal direction, and eccentric in a vertical position to the shaft A. These rollers must be placed apart from each other, a distance exactly equal to the diameter of the iron circle E E, consequently the rollers G G must be placed in a line through the centre of the circle E E and which will allow this circle to rub and give motion to the rollers G G, at the speed it revolves. The circle E E, forms an eccentric course, while it rubs on every part of its periphery against the rollers G G. This circle E E may be formed with teeth like the rim of a cog-wheel, and in that case the rollers G G may either one or both of them be formed into spur pinions to fit the teeth of the circle E E, which would be a quick and simple mode for my improved water-wheel to work machinery. I sometimes use two flat, hardened steel springs, as shown in fig. 89, instead of the rollers G G, to keep the circle E E in its proper place; and in certain situations they will be found to answer very well. Great care must be taken to make the joint pin-holes in the iron frames B, B, B, B, exactly an equal distance apart from each other; and it must also be observed in piercing or drilling the joint pin-holes in the circle E E, that they correspond with the holes made in the arm frames B, B, B, B. It will be always advisable to drill both the arm frames, B, B, B, B, and the circle E E, together, that the joint pin-holes in all three may correspond exactly with each other, and particularly from the centre of each. The joint pin-holes, in the paddle-plates or floats, should also be made to correspond with each other; and it is the distance of the holes from D to F in the paddles C C, as shown at fig. 90, which determine the eccentricity of the course of the iron ring E E; and it is by connecting these paddles at D to the arm frames B, B, B, B, and at F to the ring E E, which in the rotation of the whole by the axle A, and by keeping the circle E E in its proper situation as before described, either by the rollers G G, or when the springs H H are substituted for the rollers, that the paddles always preserve a vertical position to the surface of the water, and which cause the upper paddles to approach, whilst the lower paddles recede from the centre of the axle A. Fig. 86 represents a view of the wheel combined with all the paddles connected to both frames of the wheel, with the iron ring or circle E E placed in the middle of the frames and between the sides of the paddles, with the joint-pins in their proper places, with the two lower paddles at their most extended distance from the centre of axle A, whilst the two upper paddles are brought to their nearest situation to that axle; the joint-pins must either have nuts and screws, or other proper fastenings, to keep them in their several places, or split keys, the latter of which I decidedly prefer. The axle A must be properly placed and secured in the iron frames B, B, B, B, in any of the ordinary modes which an experienced and skilful workman will adopt. The number of sets of paddles or floats for any one wheel must be determined according to the magnitude and duty of such wheel; it is the general construction and combination as described, which constitute my improved wheel, and not the number of the paddles or floats, or their magnitude. I should, however, never recommend less than six sets of paddles or floats to be combined in any wheel made on the plan of my improved wheel, although I am aware it would act with a less number, but not so advantageously. The same letters in figs. 85, 86, and 87, represent the same parts in either of these figures, and as far as any of the similar parts are shown in figs. 88, 89, and 90, the letters and characters also distinguish them.

THE OVERSHOT-WHEEL.

This wheel consists of a frame of open buckets, placed round the rim of a vertical wheel, to receive the water from a spout placed over it, so that the buckets on the one side shall be always loaded, while those on the opposite side are empty. The loaded side will of course descend, and the wheel in its revolution will bring the empty buckets under the spout, to be in their turn filled with water.

The principal thing to be attended to in the construction of this wheel is to have the buckets of such a form as will retain the water along the greatest circumference of the wheel: and as this is a thing not easily to be accomplished, numerous contrivances have been resorted to by mill-wrights to determine the best possible form.

Fig. 91 is the outline of a wheel having 40 buckets. The ring of board contained between the concentric circles Q D S and P A R, making the ends of the buckets, is called the shrouding, and Q P the depth of shrouding. The inner circle P A R is called the sole of the wheel, and usually consists of boards nailed to strong wooden rings of compass timber of considerable scantling, firmly united with the arms or radii. The partitions, which determine the form of the buckets, consist of three different planes or boards, A B, B C, C D, which are variously named by different artists. We have heard them called the start or shoulder, the arm, and the wrist (probably for wrist, on account of a resemblance of the whole line to the human arm:) B is also called the elbow.

Fig. 92 represents a small portion of the same bucketing on a larger scale, that the proportion of the parts may be more distinctly seen. AG the sole of one bucket is made about $\frac{1}{4}$ more than the depth G H of the shrouding. The start A B is $\frac{1}{4}$ of A I. The plane B C is so inclined to A B that it would pass through H; but it is made to terminate in C, in such a manner that F C is $\frac{4}{5}$ths of G H or A I. Then C D is so placed that H D is about $\frac{1}{4}$th of I H.

By this construction it follows that the area F A B C is very nearly equal to D A B C; so that the water which will fill the space F A B C will all be contained in the bucket when it shall come into such a position that A D is a horizontal line; and the line A B will then make an angle of

35° with the vertical, or the bucket will be 35° from the perpendicular, passing through the axis of motion. If the bucket descend so much lower that one half of the water runs out, the line A B will make an angle of 25° or 24° nearly with the vertical. Therefore the wheel, filled to the degree now mentioned, will *begin* to lose water at about ¼th of the diameter from the bottom, and *half of the water will be discharged* from the lowest bucket, about $\frac{1}{17}$th of the diameter further down. Had a greater proportion of the buckets been filled with water when they were under the spout, the discharge would have begun at a greater height from the bottom, and we should lose a greater portion of the whole fall of water. The loss by the present construction is less than $\frac{1}{10}$th, (supposing the water to be delivered into the wheel at the very top,) and may be estimated at about $\frac{1}{11}$th; for the loss is the versed sine of the angle which the radius of the bucket make with the vertical. The versed sine of 35° is nearly ¼th of the radius, being 0.18085, or $\frac{1}{11}$th of the diameter. It is evident, that if only ½ of this water were supplied to each bucket as it passes the spout, it would have been retained for 10° more of a revolution, and the loss of fall would have only been about $\frac{1}{14}$th.

These observations serve to show in general, that an advantage is gained by having the buckets so capacious that the quantity of water which each can receive as it passes the spout may not nearly fill it. This may be accomplished by making them of a sufficient length, that is, by making the wheel sufficiently broad between the two shroudings.

Mr. Robert Burns, of Cartside, in Renfrewshire, has made what appeared to be a very considerable improvement in the construction of the bucket. The principle of this improvement consisted in dividing the bucket by a partition of such a height, that the inner and outer portions of the bucket on each side were nearly of equal capacity. See fig. 93. The bucket consisted of a start A B, an arm B C, and a wrest C D, concentric with the rim, and was divided by the partition L M, concentric with the sole and rim. If these buckets be filled one-third, they will retain the whole of the water at 18°, and the half at 11°, from the bottom. These advantages however were found to be counterbalanced by disadvantages; and Mr. Burns did never, we believe, put the construction in practice.

The velocity of an overshot-wheel is a matter of very great nicety; and authors, both speculative and practical, have arrived at very different conclusions respecting it. M. Be-

lida very strangely maintains, that there is a certain velocity
related, to that obtainable by the whole fall, which will
secure to an overshot-wheel the greatest performance.
Desaguliers, Smeaton, Lambert, De Parcieux, and others,
maintain, that there is no such relation, and that the performance of an overshot-wheel will be the greater, as it moves
more slowly by an increase of its load of work. Belidor
again states, that the active power of water lying in a bucket-
wheel of any diameter is equal to the impulse of the same
water on the floats of an undershot-wheel, when the water
issues from a sluice in the bottom of the dam. The other
writers whom we have named assert, that the energy of an
undershot-wheel is but one half of that of an overshot, actuated
by the same quantity of water falling from the same height.
The most generally received opinion is, that the overshot-
wheel does the more work, as it moves slower; and the
following is the reasoning adduced to prove it. Suppose
that a wheel has 30 buckets, and that six cubic feet of water
are delivered in a second on the top of the wheel, and discharged, without any loss by the way, at a certain height
from the bottom of the wheel. Let this be the case, whatever
is the rate of the wheel's motion, the buckets being of a
sufficient capacity to hold all the water which falls into them.
Suppose this wheel employed to raise a weight of any kind,
water for instance, in a chain of 30 buckets, to the same
altitude and with the same velocity. Suppose, further, that
when the load on the rising side of the machine is one half
of that on the wheel, the wheel makes four revolutions in a
minute, or one turn in 15 seconds. During this time 90
cubic feet of water will have flowed into the 30 buckets, and
each have received three cubic feet. In that case each of the
rising buckets contains 1½ feet; and 45 cubic feet are delivered into the upper cistern during one turn of the wheel,
and 180 cubic feet in one minute.

Now, suppose the machine so loaded, by making the rising
buckets more capacious, that it makes only two turns in a
minute, or one turn in 30 seconds; then each descending
bucket must contain six cubic feet of water. If each bucket
in the rising side contained three cubic feet, the motion of the
machine would be the same as before. This is a point none
will controvert. When two pounds are suspended to one
end of a string which passes over a pulley, and one pound to
the other end, the velocity of descent of the two pounds will
be the same with that of a four pound weight, which is
employed in the same manner to draw up two pounds. Our

machine would therefore continue to make four turns in a
minute, and would deliver 90 cubic feet during each turn,
and 360 in a minute. But, by supposition, it is making only
two turns in a minute; which *must* proceed from a greater
load than three cubic feet of rising water in each rising
bucket. The machine must, therefore, be raising *more* than
90 feet of water during one turn of the wheel, and more than
180 in a minute.

Thus it appears that if the machine is turning twice as
slow as before, there is *more than twice the former quantity*
in the rising buckets; and more will be raised in a minute by
the same expenditure of power. In like manner, if the
machine go three times as slow, there must be *more than
three times* the former quantity in the rising buckets, and
more work will be done.

But further we may assert, that the *more* we retard the
machine to a certain practical extent, by loading it with more
work of a similar kind, the greater will be its performance;
and the truth of the assertion may be thus demonstrated.
Let us call the first quantity of water in the rising bucket,
Q; the water raised by four turns in a minute will be $4 \times 30
\times Q = 120 \, Q$. The quantity in this bucket, when the ma-
chine goes twice as slow, has been shown to be greater than
$2 \, Q$; call it $2 \, Q + x$; the water raised by two turns in a
minute will then be $2 \times 30 \times (2 \, Q + x) = 120 + 60 \, x$.
Suppose next, the machine to go four times as slow, making
but one turn in a minute; the rising bucket must now
contain more than twice the quantity $2 \, Q + x$, or more than
$4 \, Q + 2 \, x$, call it $4 \, Q + 2 \, x + y$. The work done by one
turn in a minute will now be $30 \times (4 \, Q + 2 \, x + y) = 120
\, Q + 60 \, x + 30 \, y$. By such an induction of the work accom-
plished, with any rates of motion we choose, it is evident
that the performance of the machine increases with every
diminution of its velocity that is produced by the mere addition of a similar load of work, or that it does the more work
the slower it goes. This, however, is abstracting from the
effects of the friction upon the gudgeons of the wheel, a
cause of resistance which increases with the load, though
not in the same ratio.

We have also supposed the machine to be in its state of
permanent uniform motion. If we consider it only in the
beginning of its motion, the result is still more in favour of
slow motion: for, at the first action of the moving power, the
inertia of the machine itself consumes part of it, and it acquires
its permanent velocity by degrees, during which the resist-

ances arising from the work, friction, &c., increase, till they exactly balance the pressure of the water; and after this the machine no longer accelerates. Now, the greater the power and the resistance arising from the work are, in proportion to the inertia of the machine, the sooner, it is obvious, will it arrive at its state of permanent velocity.

The preceding discussion only demonstrates in general the advantage of slow motion; but does not point out in any degree the relation between the rate of motion and the work performed, nor even the principles on which it depends. But this is not necessary for the improvement of practical mechanics. It is, however, manifest, that there is not, in the nature of things, a maximum of performance attached to any particular rate of motion which should, on that account, be preferred. All, therefore, that we have to do, is to load the machine, and thus to diminish its speed, unless other physical circumstances throw obstacles in the way: for there are such obstacles, as in all machines there are certain inequalities of action that are unavoidable. In the action of a wheel and pinion, though made with the utmost judgment and care, there are such inequalities. These increase by the changes of form occasioned by the wearing of the machine; and much greater irregularities arise from the subsultory motions of cranks, stampers, and other parts which move unequally or reciprocally. A machine may be so loaded as just to be in equilibrio with its work, in the favourable position of its parts; and when this changes into one less favourable, the machine may stop, or, at all events, hobble and work unequally. The rubbing parts thus bear long on each other, with enormous pressures, cut deep into each other, and increase friction: therefore such slow motions should be avoided. A little more velocity enables the machine to overcome those increased resistances by its inertia, or the great quantity of motion inherent in it. Great machines possess this advantage in a superior degree, and, consequently, will work steadily with a smaller velocity.

Mr. Smeaton, in his Experimental Inquiry, previous to examining into the power and application of water, when acting by its *gravity on overshot-wheels*, says, " In reasoning without experiment, one might be led to imagine, that however different the mode of application is, yet that whenever the same quantity of water descends through the same perpendicular space, that the natural effective power would be equal, supposing the machinery free from friction, equally calculated to receive the full effect of the power, and to make

the most of it: for if we suppose the height of a column of
water to be 30 inches, and resting upon a base or aperture
one inch square, every cubic inch of water that departs therefrom will acquire the same velocity or *momentum*, from the
uniform pressure of 30 inches above it, that one cubic inch
let fall from the top will acquire in falling down to the level
of the aperture: one would therefore suppose, that a cubic
inch of water, let fall through a space of 30 inches, and there
impinging upon another body, would be capable of producing
an equal effect by collision, as if the same cubic inch had
descended through the same space with a slower motion, and
produced its effects gradually. But however conclusive this
reasoning may seem, it will appear, in the course of the
following deductions, that the effect of the gravity of descending bodies is very different from the effect of stroke of such
as are *non-elastic*, though generated by an equal mechanical
power."

When Mr. Smeaton had finished his experiments on undershot mills, he reduced the number of floats on the wheel,
which were originally 24, to 12; which caused a diminution in
the effect, on account of a greater quantity of water escaping
between the floats and the floor: but a circular sweep being
adapted thereto, of such a length, that one float entered the
curve before the preceding one quitted it, the effect came so
near that of the former, as not to give any hopes of advancing it by increasing the number of floats beyond 24 in this
particular wheel.

In these experiments the head was six inches, and the
height of the wheel 24 inches, so that the whole descent was
30 inches: the quantity of water expended in a minute was
96¼ pounds, which, multiplied by 30 inches, gives the power
= 2900. After making the proper calculations, the effect was
computed at 1914; the ratio therefore of the *power* and *effect*
will be as 2900 : 1914, or as 10 : 6.6, or as 3 to 2 nearly.
But if we compute the power from the height of the wheel
only, we shall have 96¼ pounds, multiplied by 24 inches
= 2320 for the *power*, and this will be to the effect as
2320 : 1914, or as 10 : 82, or as 5 to 4 nearly.

From another set of experiments the following conclusions
were deduced:

1. The effective power of the water must be reckoned
upon the whole descent, because it must be raised that
height in order to produce the same effect a second time
The ratios between the *powers* so estimated, and the effects
at the *maximum*, differ nearly from that of 10 to 7.6, to that

of 10 to 5.2, that is nearly from 4.3 to 4.2. In those experiments where the heads of water and quantities expended are least, the proportion is nearly as 4 to 3; but where the heads and quantities are greatest, it approaches nearer to that of 4 to 2; and by a medium of the whole, the ratio is that of 3 to 2 nearly. Hence it appears, that the effect of overshot-wheels is nearly double to that of the undershot, and, by consequence, that non-elastic bodies, when acting by their impulse or collision, communicate only a part of their original power, the remainder being spent in changing their figure in consequence of the stroke. The ultimate conclusion is, that the effects, as well as the powers, are as the quantities of water and perpendicular heights multiplied together respectively.

2. By increasing the head from 3 to 11 inches, that is, the whole descent from 27 inches to 35, or in the ratio of 7 to 9 nearly, the effect is advanced no more than in the ratio of 8.1 to 8.4, that is, as 7 to 7.26; and consequently the increase of effect is not one-seventh of the increase of perpendicular height. Hence it follows, that the higher the wheel is in proportion to the whole descent, the greater will be the effect; because it depends less upon the impulse of the head, and more upon the gravity of the water in the buckets: and if we consider how obliquely the water issuing from the head must strike the buckets, we shall not be at a loss to account for the little advantage that arises from the impulse thereof, and shall immediately see of how little consequence this impulse is to the effect of an overshot-wheel. This, however, like other things, is subject to limitation, for it is desirable that the water should have somewhat greater velocity than the circumference of the wheel, in coming thereon; otherwise the wheel will not only be retarded by the buckets striking the water, but a portion of the power will be lost by the water dashing over the buckets.

3. To determine the velocity which the circumference of the wheel ought to have in order to produce the greatest effect, Mr. Smeaton observes, that the slower a body descends, the greater will be the portion of the action of gravity applicable to the producing a mechanical effect, and, in consequence, the greater will be the effect. If a stream of water falls into the bucket of an overshot-wheel, it is there retained till the wheel by moving round discharges it, and consequently the slower the wheel moves, the more water each bucket will receive: so that what is lost in speed, is gained by the pressure of a greater quantity of water acting in the buckets

G

at once. From the experiments, however, it appeared, that when the wheel made about 20 turns in a minute, the effect was near upon the greatest. When it made 30 turns, the effect was diminished about ₁₀th part; and that when it made 40, it was diminished about ¼; when it made less than 18¼, its motion was irregular; and when it was loaded so as not to admit its making 18 turns, the wheel was overpowered by its load. It is an advantage in practice, that the velocity of the wheel should not be diminished further than will procure some solid advantage in point of power, because *cæteris paribus*, as the motion is slower, the buckets must be made larger, and the wheel being more loaded with water, the stress upon every part of the work will be increased in proportion. The best velocity for practice, therefore, will be such, as when the wheel made 30 turns in a minute, that is, when the velocity of the circumference is a little more than three feet in a second. Experience confirms that this velocity of three feet in a second is applicable to the highest overshot-wheels, as well as the lowest; and all other parts of the work being properly adapted thereto, will produce very nearly the greatest effect possible; it is also determined by experience, that high wheels may deviate further from this rule, before they will lose their power, by a given aliquot part of the whole, than low ones can be admitted to do. For a wheel of 24 feet high may move at the rate of six feet per second without losing any part of its power; and, on the other hand, the author had seen a wheel of 33 feet high that moved very steadily and well, with a velocity but little exceeding two feet. The reason of the superior velocity of the 24 feet wheel seems to have been owing to the small proportion that the head, requisite to give the water the proper velocity of the wheel, bears to the whole height.

4. The maximum load for an overshot-wheel, is that which reduces the circumferences of the wheel to its proper velocity; which will be known by dividing the effect it ought to produce in a given time, by the space intended to be described by the circumference of the wheel in the same time; the quotient will be the resistance overcome at the circumference of the wheel, and is equal to the load required, the friction and resistance of the machinery included.

5. The greatest velocity of which the circumference of an overshot-wheel is capable, depends jointly upon the diameter of the height of the wheel, and the velocity of falling bodies; for it is plain that the velocity of the circumference can never be greater than to describe a semi-circumference

while a body let fall from the top of the wheel will descend through its diameter; nor even quite so great, as a body descending through the same perpendicular space cannot perform the same in so small a time when passing through a semi-circle as would be done in a perpendicular line. Thus, if a wheel is 16 feet one inch in diameter, a body will fall through it in one second: this wheel therefore can never arrive at a velocity equal to the making one turn in two seconds; but, in reality, an overshot-wheel can never come near this velocity; for when it acquires a certain speed, the greatest part of the water is prevented from entering the buckets, and the rest, at a certain point of its descent, is thrown out again by the centrifugal force. As these circumstances depend chiefly upon the form of the buckets, the utmost velocity of overshot-wheels cannnot be generally determined; and, indeed, it is the less necessary in practice, as it is in this circumstance incapable of producing any mechanical effect.

6. The greatest load an overshot-wheel will overcome, considered abstractedly, is unlimited or infinite; for as the buckets may be of any given capacity, the more the wheel is loaded, the slower it turns, but the slower it turns, the more will the buckets be filled with water; and, consequently, though the diameter of the wheel and quantity of water expended are both limited, yet no resistance can be assigned, which it is not able to overcome; but in practice we always meet with something that prevents our getting into infinitesimals. For when we really go to work to build a wheel, the buckets must necessarily be of some given capacity, and consequently such a resistance will stop the wheel, as it is equal to the effort of all the buckets in one semi-circumference filled with water. The structure of the buckets being given, the quantity of this effort may be assigned, but is not of much consequence in practice, as in this case also the wheel loses its power; for though here is the exertion of gravity upon a given quantity of water, yet being prevented by a counterbalance from moving, is capable of producing no mechanical effect, according to our definition. But, in reality, an overshot-wheel generally ceases to be useful before it is loaded to that pitch; for when it meets with such a resistance as to diminish its velocity to a certain degree, its motion becomes irregular; yet this never happens if the velocity of the circumference is less than two feet per second, where the resistance is equable.

The reader having now become acquainted with the valuable course of experiments made by Mr. Smeaton, we shall next offer to his notice a few remarks upon the best mode of delivering water upon an overshot-wheel.

In wheels of this construction, it has been, and still is, the common practice, to allow the water to flow into the buckets at the highest point of the wheel; but this system is decidedly bad; for the centre of gravity of the upper bucket is direct over the axle of the wheel, and, consequently, any water poured into that bucket will, instead of creating a rotatory motion, cause a greater pressure upon the pivots of the axle. The greatest advantage would be obtained by causing the water to fall upon the wheel, at an angle of $42\frac{1}{2}$ or 45 degrees, as then the power of the wheel will be augmented by the increased leverage. In constructing wheels upon this principle, however, great care must be taken to allow a sufficiency of room in buckets for the escape of air, otherwise the wheel will not act. The same observation is also applicable to breast-wheels; for we were once present, and witnessed an instance of this kind, at the first starting of a breast-wheel, in which the millwright, in order to obtain the greatest possible effect, had made the back-boards to fit so tight that no water or air could escape; the consequence of which was, the necessity of reducing the whole of the back-boards, to allow air enough to escape for the water to act freely upon the floats.

BURN'S OVERSHOT-WHEEL WITHOUT A SHAFT.

This ingenious machine was invented and erected by the late Mr. Burns, whose mechanical ingenuity we have already had occasion to admire. It is represented in two different sections, in figs. 95 and 96, and forms a large hollow cylinder by its buckets and sole, without having any shaft or axle-tree.

This wheel is $12\frac{1}{2}$ feet diameter, and seven feet broad over all, and has 28 buckets. The gudgeon is 6 inches diameter, by 9 inches long. The flaunch is $1\frac{1}{4}$ inch thick at the extreme points. The arms are of redwood fir, 6 inches square; one piece making two arms in length, where they cross one another at the wheel's centre, $1\frac{1}{4}$ inch of the wood remaining in each, connecting the two opposite arms as one piece. The wheels was made by first fitting the gudgeon into a large piece of hard wood, with the flaunch parallel to the horizon, and in that position the arms and rings were trained and bound fast to it. All the grooves for starts or raisers, and buckets, were cut out before it was removed; first one piece was bolted to the flaunch at *a a*, and so of the others, leaving the distant openings for the cross bars that reach between each arm and its opposite arm. These bars, or pieces, were only 4 inches square, and were of good beech wood, turned round in the body. They were 10 inches square at each end, in which was fitted a strong nut for a bolt, 1¼ inch thick, to go through *b*, and connect the two sides together.

After the arms were trained and fixed right upon the gudgeons, the innermost ring was completed; the tenons were trained on the arms first, and the rings 4¼ inches thick and 8 inches deep, put on by keys driven into the mortice. The remaining tenons were then reduced from 1¼ to 1 inch thick, and the outermost ring, only 3 inches thick by 6 inches deep, was firmly wedged thereon, and bound fast at the other ends by three strong wooden pins, as at C C; to the lower ring, the outside of the uppermost and undermost rings are flush, all the additional thickness of the lower ring projecting inside the buckets.

Some difficulty was found in laying the water properly into the buckets of this wheel, owing to the narrowness of the mouths of the buckets, by the high start or raiser, which was remedied by adopting the following plan.

The openings in the bottom of the troughing should be of iron, and so distant from each other that the water from them is thrown into two separate buckets. The iron curved parts should also be movable, to adjust the openings to the quantity of water necessary for the wheel. Unless the head of water is 12 or 14 inches above these openings, it will be difficult to give it the proper direction into the buckets, especially if the openings are pretty wide for them; for then it deviates the more down from the line of direction, and tends to retard the wheel, by striking on the outside of the bucket. The openings from which the buckets are filled, ought to be 10 inches less in length than the buckets, *i. e.* five inches at each side, otherwise the water is apt to jerk over on each side of the wheel, as the edges of the bucket pass by.

The mode of making and finishing the wheel at Cartside requires very little workmanship, compared to the usual method; and any good joiner will do it as well as a mill-wright. The joiner finished Cartside wheel in six or seven weeks. The construction will be better understood from the following reference to the figures.

Fig. 95 represents three distinct transverse views. The part marked A supposes a part of the shrouding in section, showing

the pins; the part marked B is a section of the wheel thro
any part of the buckets, and showing three of the ties, 1,'
in section.   Part D shows the manner in which the extc
ends of the wheel are finished, also the gudgeons, flaunch,

Fig. 96 is a longitudinal section of the wheel through on
the arms, showing the projection of the shrouding, the mai
in which the arms of the wheel are connected together, and l
wise the manner in which the ties are connected to the gudg

CHAIN OF BUCKETS.

THIS is applicable in many situations where there
considerable fall of water.   This sketch was taken from
in Scotland used to give motion to a thrashing mill;
fig. 97 is so obvious as to need little explanation.
buckets C, D, G, H, &c. must be connected by sev
chains to avoid the danger of breaking, and united int
endless chain, which is extended over two wheels A an
the upper one being the axis which is to communicate mc
to the mill-work; E is the spout to supply the water.
principal advantage of this plan is, that no water is los
running out of the buckets before they arrive at the lo
part, as is the case with the wheel.   Another is, that
buckets being suspended over the wheel A of small diam
it may be made to revolve more quickly than a whee
large diameter, and without increasing the velocity ol
descending buckets beyond what is proper for them.
saves wheel-work when the machine is to be employed, a
a thrashing machine, to produce a rapid motion.   On
other hand, the friction of the chain in folding over the w
at the top, and seizing its cogs, will be very consider.
these cogs must enter the spaces in the open links betw
the buckets, to prevent the chain slipping upon the u
wheel.   We think this machine might be much improve
contriving it so that the chain would pass through the c
of gravity of each bucket, whereas in the present form
weight of each bucket tends to give the chain an extra b

The chain-pump reversed, has been proposed as a subst
for a water-wheel when the fall is very great, and we t
it would answer the purpose with some chance of suc
it would have an advantage over the chain-pump when
ployed for raising water, in the facility of applying
leathers to the pistons on the chain, in the same way as o
pumps, which leathers expand themselves to the insid
the barrel, and are kept perfectly tight by the pressur
the water.   In the chain-pump such leathers canno

employed, because the edges of the leather cups would turn
down and stop the motion, when the cups were drawn
upwards into the barrel. It is the defective mode of leather-
ing the pistons of the chain-pump which occasions its great
friction. In the motion of a machine of this kind, the pistons
would descend into the barrel, and might therefore be leathered with cups like other pumps, so as to be quite tight
without immoderate friction. This machine was proposed
by a Mr. Cooper in 1784, who obtained a patent for it, and
Dr. Robison has again proposed it with recommendation.

BREAST-WHEELS.

THE breast-wheel partakes of the nature both of an overshot and an undershot: it is driven partly by impulse, but
chiefly by the weight of water. The lower part of the
wheel is surrounded by a curved wall or sweep of masonry,
which is made concentric with the wheel, and the float-
boards of the wheel are exactly adapted to the masonry,
so as to pass as near as possible thereto without touching it;
and the side walls are in like manner adapted to the end of
the float-boards or sides of the wheel, the intention being
to let the least possible quantity of water pass without causing
the float-boards to move before it. In fig. 98, the water is
poured upon the top of the wheel over the breasting at I,
the efflux from the mill-dam K being regulated by the sluice
or shuttle M, which is placed in the direction of a tangent to
the wheel, and is provided with the rack R, and pinion P, by
which it can be drawn up so as to make any required degree
of opening, and admit more or less water to flow on the wheel.
The water first strikes on the float, and urges it by its
impulse; but when the floats descend into the sweep, they
form as it were close buckets, each of which will contain a
given quantity of water, and the water cannot escape from
these buckets except the wheel moves, at least this is the
intention, and the wheel is fitted as close as it can be to the
race with that view. Each of the portions of water contained
in these spaces bears partly upon the wall of the sweep, and
partly upon the floats of the wheel; and its pressure upon
the floats, if not exceeded by the resistance, will cause the
wheel to move; hence the action upon all the floats which
are within the sweep of the breasting is by the weight of the
water alone; but the water is made to impinge upon the first
float-board with some velocity, because the surface of the
water in the dam K is raised considerably above the orifice
beneath the shuttle where the water issues.

The upper part of the fall at I is rounded off to a of a circle, called the crown of the fall, and the water over it. The lower edge of the shuttle when put down made to fit this curve, so as to make a tight joint; and consequence, when the shuttle is drawn up, the water will run between its lower edge and the crown in a sheet or stream which strikes upon the first float that presents itself, nearly in a direction perpendicular to the plane of the float-board, or of a tangent to the wheel. The float-boards of the wheel are directed to the centre, but there are other boards placed obliquely which extend from one float-board to the rim of the wheel, and nearly fill the space between one float-board and the next. These are called rising-boards, and the use of them is to prevent the water flowing over the float-board into the interior of the wheel; but the edges of these boards are not continued so far as to join to the back of the next float, because that would make all the boards of the wheel close, and prevent the free escape of the air when the water entered into the space between the floats.

As the water strikes with some force, the rising-boards are very necessary to prevent the water from dashing over the float-boards into the interior of the wheel.

This is the form of breast-wheel employed by Mr. Smeaton in the great number of mills which he constructed; but although he speaks of the impulse of the water striking the wheel, he always endeavoured to make the top of the breast-ing, or crown of the fall, as high as possible; so as to attain the greatest fall and the least of the impulsive action. All rivers and streams of water are subject to variation in height from floods or dry seasons, and in some this is very consider-able: it was therefore necessary to make the crown I of the fall, at such a height as that, in the lowest state of the water R, it would run over the crown in a sheet of three or four inches in thickness, and work the wheel. When the water rose higher in the mill-dam, it would then have a pressure to force it through, and in that case would strike the wheel so as to impel it by the velocity.

Mr. Smeaton was well aware that the power communicated by this impulse was very small. In some cases, where the water was very subject to variation, he used a false or mov-able crown, that is, a piece of wood which fitted to the crown I, and raised the surface thereof a foot or more, so as to obtain the greatest fall when the water stood at a mean height; but when the water sunk too low to run over this movable crown, it could be drawn up to admit the water beneath it,

This effect has since been produced in a more perfect manner by making the crown of the fall a movable shuttle, to rise and fall according to the height of the water in the mill-dam, by which means the inconvenience before-mentioned is avoided.

IMPROVED BREAST-WHEEL, IN WHICH THE WATER RUNS OVER THE SHUTTLE.

FIG. 110 is a section of one of this kind. A is the water which is made to flow upon the float-board B, and urges the wheel by its weight only, the water being prevented from escaping or flowing off the float-boards by the breast or sweep D D, and the side-walls which enclose the floats of the wheel. The upper part of the breast D D is made by a cast-iron plate, curved to the proper sweep to line with the stone-work. On the back of the cast-iron plate the moving shuttle e is applied; it fits close to the cast-iron so as to prevent the water from leaking between them, and the water runs over its upper edge. F is an iron groove or channel let into the masonry of the side-walls, and in these, the ends of the sliding shuttle are received; f is an iron rack, which is applied at the back of the shuttle, and ascends above the water-line where the pinion g is applied to it to raise or lower the shuttle. The axis of the pinion is supported in a frame of wood I I, b H is a toothed sector and balance-weight, which bears the shuttle upwards, or it might otherwise fall down by its own weight, and put the mill in motion when not intended. G is a strong planking, which is fixed across between the two side-walls, and retains the water when it rises very high, as in time of floods; but in common times the water rises only a few inches above the lower edge of the planking. When the shuttle is drawn up to touch this lower edge, the water cannot escape; but when the shuttle is lowered down, it opens a space e through which the water flows upon the float-boards of the wheel.

Fig. 111 is a section of the *most improved form* for a breast-wheel, taken from the Royal Armoury Mills at Enfield Lock, erected by Messrs. Lloyd and Ostel. The general description of this, is like the former, but it is constructed in a better manner, and unites strength with durability. The breast of masonry is surmounted by a cast-iron plate A, 2½ feet high, which is let into the masonry of the side-walls at each end, and the lower part is formed with a flanch, by which it is bolted to the stone breast at top. This plate is made straight at the back for the shuttle B to lie against, and it slides up and down. The ends of the gate are guided

by iron groove pieces or channels which are let into
stone-work of the side walls, and being made wedge-li
they fix the ends of the cast-iron breast fast in its pla
The grooves are not upright, but inclined to the perpendicu
so much, that the plane of the gate is at right angles to
radius of the wheel drawn through the point where the wa
falls upon the wheel. D is a strong plank of wood, extend
between the iron grooves just over the shuttle. When t
shuttle is drawn up it comes in contact with the lower si
of this piece of wood, and stops the water; but the piece D
fixed at such a height, that the water will run clear bene.
it, unless its surface rises above its mean height.

The float-boards of the wheel do not point to the centre
the wheel, but are so much inclined thereto that they
exactly horizontal at the point where the water first flo
upon them. In this way, the gravity of the water has its
effect upon the wheel, and the boards rise up out of the
water in a much better position, than if they pointed to
centre of the wheel; this is more particularly observa
when the wheel is flooded by tail-water penned up in
lower part of the race, so that it cannot run freely away fr
the wheel. The dimensions of this wheel are as follov
diameter 18 feet to the points of the floats, and 14 f
wide; the float-boards are 40 in number, each 16 inc
wide, and each rising-board 11 inches wide. The
formed of four cast-iron circles or wheels, each 14 f
8 inches diameter, placed at equal distances upon the cent
axis, which is 14 feet 8 inches long between the necks
bearings, and 9 inches square; the bearing-necks are
inches diameter. The wheel is calculated to make four re
lutions per minute, which gives near $3\frac{1}{4}$ feet per second
the velocity with which the float-boards move. The fall
water is six feet, and the power of the wheel, when
shuttle is drawn down one foot perpendicular, equal
28-horse power.

BREAST-WHEEL WITH TWO SHUTTLES.

In this wheel the piece of wood marked D in the l.
figure, is fitted into the groove of the shuttle, and is provid
with racks and pinions to slide up and down, independently
the lower shuttle. This enables the lower shuttle to rise a
fall, according to the height of the water, so that the water sh
always run over the top of it, in the proper quantity to wo
the mill with its required velocity, whilst the upper shuttle
only used to stop the mill by shutting it down upon t

lower shuttle, and preventing the water from running over it.
This plan is used when the mill is to be regulated by a
governor, or machine to govern its velocity; in that case the
governor is made to operate upon the lower shuttle, and will
raise it up, or lower it down, according as the mill takes too
much or too little water, and this regulates the supply; but
the upper shuttle is used to stop the mill, and by this means
the adjustment of the lower shuttle is not destroyed, but
when set to work again, it will move with its required
velocity. Fig. 101 is a section of one of the water-wheels
at the cotton-mills of Messrs. Strutt, at Belper, in Derbyshire. The width of this wheel is very great, and to render
the shuttles A B firm, a strong grating of cast-iron is
fixed on the top of the breast K, and the shuttles are
applied at the back of the grating E, so as to slide up and
down against it, the strain occasioned by the pressure of the
water being borne by the grating. The lower shuttle is
moved by means of long screws, a, which have bevelled
wheels, b, at the upper ends, to turn them, by a connection
of wheel-work with the wheel-work of the mill. The upper
shuttle, A, is drawn up or down by racks and pinions, c,
which are turned by a winch, or handle. The bars of the
grating E are placed one above the other, like shelves, but
are not horizontal; they are inclined, so that the upper surfaces of all the bars form tangents to an imaginary circle of
one-third the diameter of the wheel described round the
centre thereof. These bars are not above half an inch thick,
and the spaces between them are 2¼ inches. The bars are of
a considerable breadth, the object of them being to lead the
water, with a proper slope from the top of the lower shuttle
B, to flow upon the floats of the wheel. This disposition
allows the shuttles to be placed at such a distance from the
wheel as to admit very strong upright bars of cast-iron to
be placed between the wheel and the shuttles, for the shuttles
to bear against, and prevent them from bending towards the
wheel, as the great weight of water would otherwise occasion
them to do. These upright bars are very firmly fixed to
the stone-work of the breast at their lower ends, and the
upper ends are fastened to a large timber, D, which is supported at its ends in the side walls, and has a truss-framing
applied to the back of it, like the framing of a roof, to prevent
it from bending towards the wheel. The upright bars are
placed at distances of five feet asunder, so as to support the
shuttles in two places in the middle of their length, as well
as at both ends; and large rollers are applied in the shuttle,

where it bears against these bars, to diminish the friction, which would otherwise be very great.

These precautions will not appear unnecessary when the size of the work is known. The wheel is 21¼ feet in diameter, and 15 feet broad; the fall of water is 14 feet, when it is at a mean height; the upper shuttle is 2¼ feet high, and 15 feet long; the lower shuttle is five feet high, and the same length, so that it contains 75 square feet of surface exposed to the pressure of the water; now taking the centre of pressure at two-thirds of the depth, or 3¼ feet, we find the pressure equal to that depth of water acting on the whole surface; that is, the weight of 3¼ cubic feet of water = 208 pounds, bears on every square foot of surface, which is equal to 15,600 pounds, or near seven tons, on the lower shuttle only; but if we take the two shuttles together, the surface is 112 square feet, and the mean pressure 312 pounds upon each, or 16 tons in the whole. The wheel has 40 float-boards pointing to the centre. The wheel is made of cast-iron. There are two wheels of the dimensions above stated, which are placed in a line with each other, and are only separated by a wall which supports the bearings; for they work together as one wheel, and the separation is only to obviate the difficulty of making one wheel of such great breadth as 30 feet, though this is not impossible, for there is a wheel in the same works 40 feet in breadth, but it is of wood and not iron, and is framed in a particular manner.—Dr. Rees's *Cyclopædia.*

DR. BARKER'S MILL.

DR. DESAGULIERS appears to have been the first who published an account of this machine. He ascribes the invention to Dr. Barker, in the following words : "Sir George Savill says, he had a mill in Lincolnshire to grind corn, which took up so much water to work it, that it sunk his ponds visibly, for which reason he could not have constant work; but now, by Dr. Barker's improvement, the waste water only from Sir George's ponds keeps it constantly to work."

Dr. Barker's mill is shown in fig. 102, where C D is a vertical axis, moving on a pivot at D, and carrying the upper millstone *m*, after passing through an opening in the fixed millstone C. Upon this axis is fixed a vertical tube T T, communicating with a horizontal tube A B, at the extremities of which, A, B, are two apertures in opposite directions. When water from the millcourse M N is introduced into the tube T T, it flows out of the apertures A, B, and, by the reaction or counterpressure of the issuing water, the arm A B, and consequently the whole machine, is put in motion. The bridgetree *a b* is elevated or depressed by turning the nut *c* at the end of the lever *c b*. In order to understand how this motion is produced, let us suppose both the apertures shut, and the tube T T filled with water up to T. The apertures A, B, which are shut up, will be pressed outwards by a force equal to the weight of a column of water whose height is T T, and whose area is the area of the apertures. Every part of the tube A B sustains a similar pressure; but as these pressures are balanced by equal and opposite pressures, the arm A B is at rest. By opening the aperture at A, however, the pressure at that place is removed, and consequently the arm is carried round by a pressure equal to that of a column T T, acting upon an area equal to that of the aperture A. The same thing happens on the arm T B; and these two pressures drive the arm A B round in the same direction. This machine may evidently be applied to drive any kind of machinery, by fixing a wheel upon the vertical axis C D.

In the preceding form of Barker's mill, the length of the axis C D must always exceed the height of the fall N D, and therefore when the fall is very high, the difficulty of erecting such a machine would be great. In order to remove this difficulty, M. Mathon de la Cour proposes to introduce the water from the millcourse into the horizontal arms A, B, which are fixed to an upright spindle C T, but without any tube T T. The water will obviously issue from the apertures A, B, in the same manner as if it had been introduced at the top of a tube T T as high as the fall. Hence the spindle C D may be made as short as we please. The practical difficulty which attends this form of the machine, is to give the arms A, B, a motion round the mouth of the feeding pipe, which enters the arm at D, without any great friction, or any considerable loss of water. This form of the mill is shown in fig. 103, where F is the reservoir, K the millstones, K D the vertical axis, F E C the feeding pipe, the mouth of which enters the horizontal arm at C. In a machine of this kind, which M. Mathon de la Cour saw at Bourg Argental, A B was 92 inches, and its diameter three inches; the diameter of each orifice was 1¼ inch, F G was 21 feet; the internal diameter of D was two inches, and it was fitted into clay grinding. This machine made 115 turns in a minute when it was unloaded, and emitted water by one hole only. The machine, when empty, weighed 80 pounds, and it was half supported by the upward pressure of the water.

This improvement, which was first given by M. Mathon de la Cour, in the *Journal de Physique*, 1775, appeared twent years afterwards in the *American Philosophical Transaction* as the invention of a Mr. Ramsey; and Mr. Waring, who in serted the account, contrary to every other philosopher, mak the effect of the machine only equal to that of a good unde shot wheel, moved with the same quantity of water fallin through the same height.

Dr. Gregory, in his *Mechanics*, vol. ii. has given th paper with some corrections, and recommends it as the be theory. The following rules, deduced from his calculu may be of use to those who wish to make experiments of the effect of this interesting machine.

1. Make each arm of the horizontal rotatory tube or a of any convenient length, from the centre of motion to t centre of the apertures, but not less than one-third (or ninth, according to Mr. Gregory) of the perpendicular heig of the water's surface above their centres.

2. Multiply the length of the arm in feet by .6136, a take the square root of the product for the proper time of revolution in seconds, and adapt the other parts of the n chinery to this velocity; or if the required time of a revoluti be given, multiply the square of this time by 1.629 for t proportional length of the arm in feet.

3. Multiply together the breadth, depth, and velocity second, of the race, and divide the last product by 18. times (14.27, according to Mr. Gregory) the square root the height, for the area of either aperture.

4. Multiply the area of either aperture by the height the fall of water, and the product by 41½ pounds (55.7, according to Mr. Gregory) for the moving force, estima at the centres of the apertures in pounds avoirdupois.

5. The power and velocity at the aperture may be cas reduced to any part of the machinery by the simplest n chanical rules.

TIDE-MILLS.

TIDE-MILLS, as their name imports, are such as empl for their first mover the flowing and ebbing tide, either the sea or a river.

Mills of this kind have not often, we believe, been erect in England, though several of our rivers, and particula the Thames, the Humber, and the Severn, in which the ti rises to a great height, furnish a very powerful mover drive any kind of machinery, and would allow of tide-mi being very advantageously constructed upon their banks. The erection of such mills is not to be recommended universally, as they are attended with a considerable original expense; beside that, some of their parts will require frequent repairs: but in some places, where coal is very dear, they may, on the whole, be found less expensive than steam-engines to perform the same work, and may, on that account, be preferred even to them.

We have not been able to ascertain who was the first contriver of a tide-mill in this country, nor at what time one was first erected. The French have not been so negligent respecting the origin of this important invention, as to let it drop into obscurity; but have taken care to inform us that such mills were used in France early in the last century. Belidor mentions the name of the inventor, at the same time that he states some peculiar advantages of this species of machine. "L'on en attribue," says he, "la première invention à un nommé *Perse*, maître charpentier de Dunkerque, que mérite assurément beaucoup d'éloge, n'y ayant point de gloire plus digne d'un bon citoyen, que celle de produire quelqu'invention utile à la société. En effet, combien n'y a-t'-il point de choses essentielles à la vie, dont on ne connoît le prix que quand on en est privé : les moulins en général sont dans ce cas-là. On doit sçavoir bon gré à ceux qui nous ont mis en état d'en construire partout : par exemple, à Calais, comme il n'y serpente point de rivières, on n'y a point fait jusqu'ici de moulins à eau, et ceux qui vont par le vent chômant une partie de l'année, il y a des tems où cette ville se trouve sans farine, et j'ai vu la garnison en 1730, obligé de faire venir du pain de Saint-Omer, au lieu qu'en se servant du flux et reflux de la mer, on pourroit construire autant de moulins à eau que l'on voudroit : il y a d'autres villes dans le voisinage de la mer sujettes au même inconvénient, parcequ'apparemment elles ignorent le moyen d'y remédier."

Mills to be worked by the rising and falling of the tide, admit of great variety in the essential parts of their construction; but this variety may perhaps be reduced to four general heads, according to the manner of action of the water-wheel. 1. The water-wheel may turn one way when the tide rises, and the contrary when it falls. 2. The water-wheel may be made to turn always in one direction. 3. The water-wheel may fall and rise as the tide ebbs and flows. 4. The axle of the water-wheel may be so fixed as that it shall neither rise nor fall, though the rotatory motion shall be given to the wheel, while at one time it is only partly, at another completely, immersed in the fluid. In the mills we have examined, says Dr. Gregory, the first and third of these divisions have been usually exemplified in one machine; and the second and fourth may readily be united in another; we shall, therefore, speak of them under two divisions only.

1. When the water-wheel rises and falls, and turns one way with the rising tide, and the contrary when it ebbs. In order to explain the nature of this species of tide-mills, we shall describe one which has lately been erected on the right bank of the Thames, at East-Greenwich, under the direction of Mr. John Lloyd, an ingenious engineer of Brewer's-green, Westminster.

This mill is intended to grind corn, and works eight pair of stones. The side of the mill-house parallel to the course of the river, measures 40 feet within; and as the whole of this may be opened to the river by sluice-gates, which are carried down to the low water-mark in the river, there is a 40 feet waterway to the mill: through the waterway the water presses during the rising tide into a large reservoir, which occupies about four acres of land; and beyond this reservoir is a smaller one, in which water is kept, for the purpose of being let out occasionally at low water to cleanse the whole works from mud and sediment, which would otherwise, in time, clog the machinery.

The water-wheel has its axle in a position parallel to the side of the river, that is, parallel to the sluice-gates which admit water from the river; the length of this wheel is 26 feet, its diameter 11 feet, and its number of float-boards 32. These boards do not each run on in one plane from one end of the wheel to the other, but the whole length of the wheel is divided into four equal portions, and the parts of the float-boards, belonging to each of these portions, fall gradually one lower than another, each by one-fourth of the distance from one board to another, measuring on the circumference of the wheel.

This contrivance, which will be better understood by referring to fig. 104, is intended to equalize the action of water upon the wheel, and prevent its moving by jerks. The wheel, with its incumbent apparatus, weighs about 20 tons, the whole of which is raised by the impulse of the flowing tide, when admitted through the sluice-gates. It is placed in the middle of the waterway, leaving a passage on each side of about six feet, for the water to flow into the reservoir, besides that which, in its motion, turns the wheel round. Soon after the tide has risen to the highest, (which at this mill is often 20 feet above the low water-mark,) the water is permitted to run back again from the reservoir into the river, and by this means it gives a rotatory motion to the water-wheel, in a contrary direction to that with which it moved when impelled by the rising tide : the contrivance by which the wheel is raised and depressed, and that by which the whole interior motions of the mill are preserved in the same direction, although that in which the water-wheel moves is changed, are so truly ingenious as to deserve a distinct description, illustrated by diagrams. Let, then, A B (fig 105) be a section of the water-wheel, 1, 2, 3, 4, 5, &c. its floats; C D the first cogwheel upon the same axis as the water-wheel; the vertical shaft F E carries the two equal wallower-wheels E and F, which are so situated on the shaft that one or other of them may, as occasion requires, be brought to be driven by the first wheel C D; and thus the first wheel acting upon F and E at points diametrically opposite, will, although its own motion is reversed, communicate the rotatory motion to the vertical shaft always in the same direction. In the figure the wheel E is shown in geer, while F is clear of the cogwheel C D; and at the turn of the tide the wheel F is let into geer, and E is thrown out; this is effected by the lever G, whose fulcrum is at H, the other end being suspended by the rack K, which has hold of the pinion L on the same axle as the wheel M; into this wheel plays the pinion N, the winch O, on the other end of whose axle, furnishes sufficient advantage to enable a man to elevate or depress the wallowerwheels, as required.

The centre of the lever may be shown more clearly by fig. 104, where $a\, b$ is a section of the lever, which is composed of two strong bars of iron, as $a\, b$; there are two steel studs or pins which work in the grooves of the grooved wheel I, this wheel being fixed on the four rods surrounding the shaft, of which three only can be shown in the figures, as $cd\, ef$ the ends of these are screwed fast by bolts to the sockets of the wallower-wheels, and they are nicely fitted on the vertical shaft, so as to slide with little friction; thus the wallowers may be raised or lowered upon the upright shaft, while the gudgeon, on which it turns, retains the same position.

When the top wallower is in geer, it rests on a shoulder that prevents it from going too far down; and when the bottom one is in geer, there is a bolt that goes through the top wheel socket and shaft which takes the weight from the lever G, at the same time that it prevents much friction on the studs or pins of the lever which works in the grooved wheel L.

When the tide is flowing, after the mill has stopped a sufficient time to gain a moderate head of water, the fluid is suffered to enter and fall upon the wheel at the sluice Q, (fig. 106,) and the tail water to run out at the sluice R.

The hydrostatic pressure of the head of water acting against the bottom of the wheel-frame S, and at the same time acting between the folding-gates T W, which are thus converted into very large hydrostatic bellows, buoys up the wheel and frame, (though weighing, as before observed, nearly 20 tons,) and makes them gradually to rise higher and higher, so that the wheel is never, as the workmen express it, drowned in the flowing water; nor can the water escape under the wheel-frame, being prevented by the folding-gates, which pass from one end to the other of the wheel. In this way the wheel and frame are buoyed up by a head of four feet; and the mill works with a head of 5 or 5¼ feet.

When the tide is ebbing, and the water from the reservoir running back again into the river, it might, perhaps, be expected that in consequence of the gradual subsiding of the water, the water-wheel should as gradually lower; but lest any of the water confined between the wheel-frame at S, and the folding-gates T W, should prevent this, there are strong rackworks of cast-iron, by which the wheel-frame can be either suspended at any altitude, or gradually let down so as to give the water returning from the reservoir an advantageous head upon the wheel; then the sluice R is shut, and V opened as well as X, the water entering at X to act upon the wheel, and flowing out at R. The upper surface of the wheel-frame is quadrangular, and at each angle is a strong cast-iron bar, which slides up and down in a proper groove, that admits of the vertical motion, but prevents all such lateral deviation as might be occasioned by the impulsion of the stream.

At each end of the water-wheel there is a vertical shaft, with wallowers and a first cog-wheel, as F E, and C D; and each of these vertical shafts turns a large horizontal wheel at a suitable distance above the wallowers, while each horizontal wheel drives four equal pinions placed at equal or quadrantal distances on its periphery, each pinion having a vertical spindle, on the upper part of which the upper millstone of its respective pair is fixed. Other wheels, driven by one or other of these pinions, giving motion to the bolting and dressing machines, and different subordinate parts of the mill.

Although the vertical shaft at each end of the water-wheel rises and falls with that wheel, yet the large horizontal wheel turning with such shaft does not likewise rise and fall, but remains always in the same horizontal plane, and in contact with the four pinions it drives. The contrivance for this purpose is very simple, but very efficacious; each great horizontal wheel has a nave, which runs upon friction-rollers, and has a square aperture passing through it vertically, just large enough to allow the shaft P to slide freely up and down in it, but not to turn round without communicating its rotatory motion to the wheel; thus the weight of the wheel causes it to press upon the friction-rollers, and retain the same horizontal planes, and the action of the angles of the vertical shaft upon the corresponding parts of the square orifice in the nave causes it to partake of the rotatory motion, such motion being always in one direction, in consequence of the contrivance by which one or other of the wallowers E F is brought into contact with the opposite points of the first cog-wheel C D.

Several of the subordinate parts of this mill are admirably constructed; but we can only notice here the means by which the direction of the motion in the dressing and bolting machines may be varied at pleasure. On a vertical shaft are fixed, at the distance of about 15 or 18 inches, two equal cog-wheels, and another toothed-wheel, attached to a horizontal axle, is made so as to be movable up and down by a screw, and thus brought into contact with either the upper or lower of the two cog-wheels on the vertical shaft; thus, it is manifest, the motion is reversed with great facility by changing the position of the horizontal axle so that the wheel upon it may be driven by the two cog-wheels alternately. A wheel and pinion working at the other end of the horizontal axle will communicate the motion to the dressing machines.

Mr. W. Dryden, Mr. Lloyd's foreman, employed in the erection of this mill, suggests that a nearly similar mode may be advantageously adopted in working dressing machines in wind-mills; three wheels, all of different diameters, may be employed, two of them, as A and C, turning upon a vertical shaft, and the third, B, upon an inclined one. In fig. 106, the wheels A and B are shown in geer, while C is out; and if A be struck out by some such contrivance as is adopted with regard to the first cog-wheel and wallowers, (fig. 104 and 105,) C would come in contact with B, while A would be free, and so communicate a motion to B the reverse

way. By this contrivance it would be easy, when the winds
are strong and give a rapid motion to the vertical axle, to
bring C to drive B, the wheel on the axle of the dressing-
machines; and on the contrary, when the wind is slack,
and the consequent motion of the machinery slow, let C be
thrown out of geer, and the wheel B driven by the larger
wheel A, as shown in the figure.

We should have been glad to see adopted in this well-
constructed mill, a contrivance, recommended and pursued
by the American mill-wrights, for raising the ground corn to
the cooling-boxes or beaches from which it is to be conveyed into the bolting-machine. In this mill, as in all we
have seen, the corn is put into bags at the troughs below the
mill-stones, and thence raised to the top of the mill-house by
a rope folding upon barrels turned by some of the interior
machinery of the mill. In the American method, a large
screw is placed horizontally in the trough which receives
the flour from the mill-stones. The thread or spiral line of
the screw is composed of pieces of wood about two inches
broad and three long, fixed into a wooden cylinder seven or
eight feet in length, which forms the axis of the screw.
When the screw is turned round this axis, it forces the meal
from one end of the trough to the other, where it falls into
another trough, from which it is raised to the top of the mill-
house by means of elevators, a piece of machinery similar to
the chain-pump. These elevators consist of a chain of buckets,
or concave vessels, like large tea-cups, fixed at proper
distances upon a leathern band, which goes round two wheels,
one of which is placed at the top of the mill-house, and the
other at the bottom, in the meal-trough. When the wheels
are put in motion, the band revolves, and the buckets, dipping into the meal-trough, convey the flour to the upper story,
where they discharge their contents. The band of buckets
is enclosed in two square boxes, in order to keep them clean,
and preserve them from injury.

We shall now proceed,

2. To tide-mills, in which the axle of the water-wheel
neither rises nor falls, and in which that wheel is made always
to revolve in the same direction. A water-wheel of this kind
must, manifestly, at the time of high-tide, be almost, if not
entirely, immersed in the fluid; and to construct a wheel to
work under such circumstances is, obviously, a matter which
requires no small skill and ingenuity.

The first person who devised a wheel which might be
turned by the tide, when completely immersed in it, were

Messrs. Gosset and De la Deuille. Their wheel is described by Belidor in nearly the following terms. Suppose G H (fig. 107) to denote the surface of the water at high-tide, the line L M the surface at low-water, and that the current follows the direction of the arrow N; the problem is to construct the wheel so that it may always turn upon its axis I K. The figure just referred to is a profile of an assemblage of carpentry, which must be repeated several times along the arbor, according to the length which it is proposed to give to the float-boards; and the planks or plates which compose these floats, must be hung to the other parts of the frame by as many joints as are necessary, to enable them to sustain the impulse of the water without bending. The sole peculiarity of this wheel consists in hanging upon the transverse beams in the frame-work, by hinges, the planks which are to compose the float-boards; so that they may present themselves in face, as D, D, D, when they are at the bottom of the wheel, to receive the full stroke of the stream; and, on the contrary, they present only their edges, as A, A, A, when they are brought towards the summit of the wheel; hence, the water having a far greater effect upon the lower than the upper parts of the wheel, compels it to revolve in the order of the letters; instead of which, if the float-boards were fixed as in the usual way, the impulse of the fluid upon the wheel would be nearly the same in all its parts, and it would remain immovable.

We see, at once, that the boards D, D, D, having moved towards M, then begin to float, as at E, E, E, and more still at F, F, F, but that it is not till they arrive at A, A, A, that they attain the horizontal position; after that, having arrived at B, B, B, they begin to drop towards the beams to which they are hooked, and as soon as they have passed the level of the axle I K, the stream commences its full action upon them, which it attains completely between C, C, C, and E, E, E, and this, whether the surface of the water be at G H or at L M; for even in the latter case it is manifest that the float-boards are entirely immersed when in the vertical position P Q. Belidor says, he was present at the first trial of such a wheel at Paris, and that it was attended with all the success that could be desired.

A water-wheel has been lately invented by Mr. Dryden, which will work when nearly immersed in the water of a flowing tide. Fig. 108 is an elevation of this wheel, its upper parts being supposed to stand a foot or two higher than the tide ever rises; the axis of this wheel remains always in one place, and the wheel will work at high-water when the head is at B, and the tail-water at the dotted line A; it will also perform nearly the same work when the head is at C, and the tail-water level with the bottom of the wheel. The floats are all set at one and the same angle, with the respective radii of the wheel, as may be seen in the figure, and are made so as to have an opening of at least an inch between each float and the drum-boarding of the wheel. This opening is intended to prevent the wheel from being impeded by the tail-water; for as the bucket rises out of the water, there can be no vacuum formed in it, there being a full supply of air, in consequence of which the water leaves the wheel deliberately. The case is different with regard to wheels made in the common way; for if such are open wheels, the floats are made in such a manner as to throw the tail-water if they are immersed any depth in it; or, if they are close, the wheel wants proper vent for the air to prevent the formation of a vacuum in the rising bucket, or what is called by the miller " sucking up the tail-water." At D is planking made circular to fit the wheel pretty close for rather more than the space of two floats, so as to confine the water nearly close to the wheel. E, F, G, H, are sluices which are all connected together by the iron bar I, and lifted with the assistance of the wheel, two pinions, and a winch, the first pinion working into the rack K; these sluices are merely for stopping the wheel when occasion requires, although one might be sufficient to supply the wheel. The rings of this wheel may be made either of cast-iron or of wood; the floats may be iron plates rivetted together. The flanches on the arms of the wheel, exhibited in the sketch, are intended to facilitate the fixing of the first cog-wheels; the ring of the wheel may be fixed to the flanches at the extremity of the arms, and the large flanch made fast to the axle will receive the middle part of the wheel.

Fig. 109 is a plan of the house in which either of the two latter wheels may be fixed, showing in what manner the water may be conveyed always on one side of the wheel by the assistance of the four gates A, B, C, and D. When the mill is working from the river, A and B are open, the arrows point out the way the water runs from the river to the basin; and the dotted lines on the contrary the course from the basin to the river, when A, B, are shut, and C, D, opened. These gates are made to turn on an axle, which is about six inches from the middle of the gate, and on the top of the axle is a half-wheel; by some crane-work connected to it, the gate can be opened or shut at pleasure; when a head of water presses against the gates they will open great part of the way of themselves, by only letting the catches that keep them shut be lifted out of their place. X, Y, are two knees of cast-iron, to support the posts that the gates are fixed to. The walls of the building are represented at *a*, *b*, *c*, and *d*.

The reader will now be able to form an estimate of the comparative value and ingenuity of the two kinds of tide-mills here described. The simplicity of construction of the wheels of Gosset, De la Deuille, and Dryden, recommend them strongly; but we entertain some doubts of their being completely successful in practice: had the curious wheel, with the folding-gates, &c. fig. 104 and 106, been placed with its axle perpendicular, instead of parallel, to the course of the river, the water might then have always been admitted to act upon the same side of it, and the hydrostatic pressure would have operated as completely in lowering it continually during the time of ebb, as in raising it continually during the rising of the tide; thus, as appears to us, would the labour of a man be saved, who, according to the present construction, must attend the water-wheel; and all the additional apparatus now requisite to shift the spur-wheels, would at the same time be saved, and a consequent diminution of original expense. Dr. Gregory's *Mechanics*, vol. ii.

In selecting a site for the erection of a mill, the engineer must be careful not to make choice of a spot that is liable to be flooded. When the water in the mill-tail will not run off freely, but stands pent up in the wheel-race, so that the wheel must work or row in it, the wheel is said to be tailed, or to be in back-water or tail-water; which greatly impedes the velocity of the wheel, and, if the flood be great, completely stops it.

Every mill that is well and properly constructed, will clear itself of a considerable depth of tail-water, provided there is, at the time, an increase in the height of the water in the mill-dam or head, and an unlimited quantity of water to draw upon the wheel. Common breast-mills will bear two feet of tail-water, when there is an increase of head, and plenty of water to be drawn upon the wheel, without prejudice to their performance; and mills that are well constructed, with slow moving wheels, will bear three and even four feet and upwards of tail-water. Mr. Smeaton mentions having seen an instance of six feet; and it is a common thing in level countries, where tail-water is most annoying, to lay the wheel from six to twelve inches below the water's level of the pond below, in order to increase the fall of water; and, judiciously applied, is attended with good effect, as it creases the diameter of the wheel, and though it m always work in that depth of tail-water, it will perform as well, because the water ought to run off from the butt of the wheel, in the same direction as the wheel turns.

ON THE CONSTRUCTION OF THE WHEEL-RACE AND WATER-COURSE.

THE wheel-race should always be built in a substan manner with masonry, and if the stones are set in Rom cement, it will be much better than common mortar. earth, behind the masonry, should be very solid, and if it not naturally so, it should be hard rammed and prevent percolation of the water. This applies more pa cularly to breast-wheels, in which the water of the dam reservoir is usually immediately behind the wall or breast which the wheel works, a sloping apron of earth being 1 from the wall in the dam to prevent the water leaking. wall of the breast should have pile planking driven benea to prevent the water from getting beneath, because that mij blow up the foundation of the race. The stones of the n are hewn to a mould, and laid in their places with great cau but afterwards, when the side walls are finished, and the a of the wheel placed in its bearings, a gauge is attached t and swept round the curve, and by this the breast is dres smooth, and hewn to an exact arch of a circle; the side wal in like manner, are hewn flat and true at the place where float-boards are to work. It is usual to make the space tween the side walls two inches narrower at each side, in circular part where the float acts, than in the other parts.

In some old mills the breast is made of wood plankir but this method has so little durability that it cannot recommended.

In modern mills, the breast is lined with a cast-iron pla but we do not approve of this, because it is next to imposi to prevent some small leakage of water through the masom and this water, being confined behind the iron breast, can escape, but its hydrostatic pressure to force up the iron enormous; and if the water can ever insinuate itself behi the whole surface of the plate, rarely fails to break it, if to blow it up altogether. This is best guarded against making deep ribs projecting from the back of the plate,... bedding them with great care in the masonry; these only strengthen the plate, but also cut off the communicati of the water, so that it cannot act upon larger surfaces at once than the strength and weight of the plate can resist. Stone is undoubtedly the best materials for a breasting. In overshot-wheels the loss of water, by running out of the buckets as they approach the bottom of the wheel, may be considerably diminished by accurately forming a sweep or casing round the lower portion of the wheel, so as to prevent the immediate escape of the water, and causing it to act in the manner of a breast-wheel. While this improvement remains in good condition, and the wheel works truly, it produces a very sensible effect; but it is frequently objected to, because a stick or a stone falling into the wheel would be liable to tear off part of its shrouding, and damage the buckets; and again, a hard frost frequently binds all fast, and totally prevents the possibility of working during its continuance; but we do not think the latter a great objection, for the water is not more liable to freeze there than in the buckets, or in the shuttle, and may be prevented by the same means, viz. by keeping the wheel always in motion, a very small stream of water left running all night will be sufficient. Mr. Smeaton always used such sweeps, and with very good effect; it is certainly preferable to any intricate work in the form of the buckets.

Mill-courses.—As it is of the highest importance to have the height of the fall as great as possible, the bottom of the canal or dam which conducts the water from the river should have a very small declivity; for the height of the water-fall will diminish in proportion as the declivity of the canal is increased; on this account, it will be sufficient to make A B, fig. 100, slope about one inch in 200 yards, taking care to make the declivity about half an inch for the first 48 yards, in order that the water may have a velocity sufficient to prevent it from flowing back into the river. The inclination of the fall, represented by the angle G C R, should be 25° 50′; or C R, the radius, should be to G R, the tangent of this angle, as 100 to 48, or as 25 to 12; and since the surface of the water S b is bent from a b into a c, before it is precipitated down the fall, it will be necessary to incurvate the upper part B C D of the course into B D, that the water at the bottom may move parallel to the water at the top of the stream. For this purpose, take the points B, D, about 12 inches distant from C, and raise the perpendiculars B E, D E; the point of intersection E will be the centre, from which the arch B D is to be described; the radius being about

Now, in order that the water may act more advantageously upon the float-boards of the wheel W W, it must assume a horizontal direction H K, with the same velocity which it would have acquired when it came to the point G : but, in falling from C to G, the water will dash upon the horizontal part H G, and thus lose a great part of its velocity; it will be proper, therefore, to make it move along F H, an arch of a circle, to which D F and K H are tangents in the points F and H. For this purpose, make G F and G H each equal to three feet, and raise the perpendiculars H l, F l, which will intersect one another in the points I, distant about four feet nine inches and ⁴⁄₅ths from the points F and H, and the centre of the arch F H will be determined. The distance H K, through which the water runs before it acts upon the wheel, should not be less than two or three feet, in order that the different portions of the fluid may have obtained a horizontal direction; and if H K be much larger, the velocity of the stream would be diminished by its friction on the bottom of the course. That no water may escape between the bottom of the course K H and the extremities of the float-boards, K L should be about three inches, and the extremity o of the float-board n o, should be beneath the line H K X, sufficient room being left between o and M for the play of the wheel, or K L M may be formed into the arch of a circle K M, concentric with the wheel. The line L M V, called by M. Fabre the course of impulsion, (le coursier d'impulsion,) should be prolonged, so as to support the water as long as it can act upon the float-boards, and should be about nine inches distant from O P, a horizontal line passing through O, the lowest point of the fall; for if O L were much less than nine inches, the water, having spent the greater part of its force in impelling the float-boards, would accumulate below the wheel and retard its motion. For the same reason, another course, which is called by M. Fabre the course of discharge, (le coursier de décharge,) should be connected with L M V by the curve V N, to preserve the remaining velocity of the water, which would otherwise be destroyed by falling perpendicular from V to N. The course of discharge is represented by V Z, sloping from the point O. It should be about 16 yards long, having an inch of declivity in every two yards. The canal, which reconducts the water from the course of discharge to the river, should slope about four inches in the first 200 yards, three inches in the second 200 yards, decreasing gradually till it terminates in the river. But if the river, to which the water is conveyed, should, when swollen by the rains, force the water back upon the wheel, the canal must have a greater declivity, in order to prevent this from taking place. Hence it will be evident, that very accurate levelling is necessary for the proper formation of the mill-course.

ON SETTING OUT WATER-COURSES AND DAMS.

THE most ancient mills were undershot-wheels placed in the current of an open river, the building containing the mill being set upon piles in the river. It would soon be observed that the power of the mill would be greatly increased if all the water of the river was concentrated to the wheel, by making an obstruction across the river which penned up the water to a required height; and also to form a pool or reservoir of water. A sluice or shuttle would then become necessary to regulate the admission of water to the wheel, and other sluices would be necessary to allow the water to escape in times of floods; for though in ordinary times the water would run over the top of the obstruction or dam, yet a very great body of water running over might carry away the whole work, by washing away the earth at the foot of the dam, and then overturning it into the excavation. This is an accident which frequently happens to mills so situated; and the danger is so obvious, that most water-mills are now removed to the side of the river, and a channel is dug from the river to the mill to supply it with water, and another to return the water from the mill to the river. The difference of level between these two channels is the fall of water to work the mill, and this is kept up by means of a wear or dam entirely across the river, but the water can run freely over this dam in case of floods, without at all affecting the mill, because the entrance to the channel of supply is regulated by sluices and side walls.

The dam should be erected across the river at a broad part, where it will pen up the water so as to form a large pool or reservoir, which is called the mill-pond or dam-head. This reservoir is useful to gather the water which comes down the river in the night, and reserve it for the next day's consumption; or for such mills as do not work incessantly, but which require more water, when they do work, than the ordinary stream of the river can supply in the same time. The larger the surface of the pond is, the more efficient it will be, but depth will not compensate for the want of surface, because, as the surface sinks, when the water is drawn off, the fall or descent of the water, and consequently the power of the water, diminishes.

The dam for a large river should be constructed with the

utmost solidity; wood framing is very commonly used, masonry is preferable. Great care must be taken, by dri pile planking under the dam, to intercept all leakage of water beneath the ground under the dam, as that loosens earth, and destroys the foundation imperceptibly, whe violent flood may overthrow the whole. It is a com practice to place the dam obliquely across the river, w view of obtaining a greater length of wall for the water to over, and consequently prevent its rising to so great a he in order to give vent to the water of a flood. But th very objectionable, because the current of water const running over the dam, always acts upon the shore or ba the river at one point, and will in time wear it away, i prevented by expensive works. This difficulty is obr by making the dam in two lengths which meet in an angl the vertex pointing up the stream. In this way the cur of water, coming from the two opposite parts of the strike together, and spend their force upon each o without injuring any part. A still better form is a seg of a circle, which has the additional advantage of stre because if the abutments at the banks of the rive firm, the whole dam becomes like the arch of a bridg down horizontally. This was the form generally use Mr. Smeaton.

The foot of the dam where the water runs down shou a regular slope with a curve, so as to lead the water regularly; and this part should be evenly paved with or planked, to prevent the water from tearing it up wh moves with a great velocity.

When the fall is considerable, it may be divided into than one dam; and if the lower dam is made to pen the upon the foot of the higher dam, then the water ru over the higher dam, will strike into the water, and lo force. There is nothing can so soon exhaust the for rapid currents of water as to fall into other water, be its mechanical force is expended in changing the figu the water; but when it falls upon stone or wood, its fo not taken away, but only reflected to some other part channel, and may be made to act upon such a great exte surface as to do no very striking injury at any one time by degrees it wears away the banks, and requires con repairs: for it is demonstrable that, as much of the fo the water as is not carried away by the rapid motion which it flows, after passing the dam, must be expe either in changing the figure of the water, or in wa away the banks, or in the friction of the water running over the bottom.

The cotton-works of Messrs. Strutt, at Belper, in Derbyshire, are on a large scale, and the most complete we have ever seen, in their dams and water-works. The mills are turned by the water of the river Derwent, which is very subject to floods. The great wear is a semicircle, built of very substantial masonry, and provided with a pool of water below it, into which the water falls. On one side of the wear are three sluices, each 20 feet wide, which are drawn up in floods, and allow the water to pass sideways into the same pool; and on the opposite side is another such sluice, 22 feet wide. The water is retained in the lower pool by some obstruction which it experiences in running beneath the arches of a bridge; but the principal fall of the water is broken by falling into the water of the pool, beneath the great semicircular wear.

The water which is drawn off from the mill-dam above the wear, passes through three sluices, 20 feet wide each, and is then distributed by different channels to the mills, which are situated at the side of the river, and quite secure from all floods. There are six large water-wheels; one of them, which is 40 feet in breadth, we have mentioned, from the ingenuity of its construction; and another, which is made in two breadths of 15 feet each, we have also described. They are all breast-wheels. The iron-works of Messrs. Walker, at Rotherham, in Yorkshire, are very good specimens of water-works; as also the Carron-works in Scotland.—Dr. Rees's *Cyclopædia* and Dr. Brewster's *Ferguson*.

PENSTOCK.

The following is a description of a pentrough and stock for equalizing the water falling on water-wheels, by Mr. Quayle.

To ensure a regular supply of water on the wheel, and to obviate the inconveniencies arising from the usual mode of delivering it from the bottom of the pentrough, this method is devised of regulating the quantity delivered by a float, and taking the whole of the water from the surface.

Section of the pentrough. Fig. 99. A, the entrance of the water; B, the float, having a circular aperture in the centre, in which is suspended C, a cylinder, running down in the case below the bottom of the pentrough. This is made watertight at the bottom of the pentrough at F, by a leather collar placed between two plates, and screwed down to the bottom. The cylinder is secured to the float so as to follow its rise

and fall; and the water is admitted into it through the op
in its sides, and there, passing through the box or c
rises and issues at G on the wheel. By this means, a un
quantity of water is obtained at G; which quantity c
increased or diminished by the assistance of a small
and pinion attached to the cylinder, which will rai
depress the cylinder above or under the water line o
float; and, by raising it up to the top, it stops the
entirely, and answers the purpose of the common sh
This pinion is turned by the handle H, similar to a w
handle; and is secured from running down by a rat
wheel at the opposite end of the pinion axis.

K and L are two upright rods to preserve the perpendi
rise and sinking of the float, running through the floa1
secured at the top by brackets from the sides.

M, a board let down across the pentrough nearly t
bottom, to prevent the horizontal impulse of the water
disturbing the float.

Fig. 99*. A transverse section, showing the mo
fixing the rack and pinion, and their supports on the
The rack is inserted into a piece of metal running acro1
cylinder near the top. That the water may pass more {
when nearly exhausted, the bottom of the cylinder is
plane, but is cut away so as to leave two feet, as
fig. 99. The float is also kept from lying on the pen1
bottom by four small feet; so that the water gets un
regularly from the first.

Fig. 99**, An enlarged view of the cylinder, showin
rack and ratchet-wheel, with the clink, and one of the ope
on the side of the cylinder; the winch or handle being o
opposite side, and the pinion, by which the rack is r
enclosed in a box between them.

MR. SMEATON'S PENTROUGH.

FIG. 93*. G represents the pentrough through whic
water flows, and F F strong cross-beams on which it is
ported; the wheel is situated very close beneath the bott
the trough, as the figure shows. E E are two arms o
wheel, which are put together, as shown in fig. 110.
the wooden rim of the wheel; the narrow circle beyond
the section of the sole planking, and on the outside of th
bucket-boards are fixed as the figure shows; one of the bo1
boards, b, of the trough at the end is inclined, and an op
is left between that end and the other boards of the botto
let the water pass through; this opening is closed by a sl

shuttle, c, which is fitted to the bottom of the trough, and can be moved backwards and forwards by a rod d, and lever e, which is fixed into a strong axis f; this axis has a long lever on the end, which, being moved by the miller,

water issues. The extreme edge of the shuttle is cut inclined, to make it correspond with the inclined part b, and by this means it opens a parallel passage for the water to run through, and this causes the water to be delivered in a regular and even sheet; and to contribute to this the edges of the aperture where the water quits it are rendered sharp by iron plates; the shuttle is made tight where it lies upon the bottom of the trough, by leather, so as to avoid any leakage when the shuttle is closed. When the wheel is of considerable breadth, the weight of the water might bend down the middle of the trough until it touched the wheel; to prevent this, a strong beam, O, is placed across the trough, and the trough is suspended from this by iron bolts which pass through grooves in the shuttle, so that they do not interfere with the motion of the shuttle.

Mr. Nouaille took out a patent, in October, 1812, for a new method of laying water upon an overshot-wheel, (see fig. 94,) which he thus describes:—" In my new method of applying water to water-wheels, I cause it to commence its action upon a point of the wheel's circumference, which is about 53 degrees distant from the vertex, or the highest point thereof, instead of applying it at the top of the wheel, as heretofore commonly practised for overshot-wheels. By these means I can have the advantages of a large wheel in situations where the fall would only allow of a smaller, if the water was applied at the top; thus, if there be a perpendicular of 12 feet, I cause a wheel of 15 feet diameter to be made, and of course the water must be made to act upon it at a height of 12 feet, which is three feet perpendicular below the top of the wheel, and at about 53 degrees from the top, measured round its circumference as above stated. I make the pentrough which brings the water to the wheel of such a form that it delivers the water from the bottom of it through the floor, and is directed at such an angle as to fall into the buckets nearly in the direction of the wheel's motion, which will be at an angle of 75 degrees with the horizon; the shuttle or gate slides upon the floor of the trough, so as to cover the aperture, and determine the quantity of water to be let out upon the wheel.

" The exact manner of carrying this principle into effc
particularly explained by the annexed draft, which is a ver
section of a water-wheel on my improved plan. In this
dotted line A A, fig. 116, represents the level of the w
at its full head, and B the level of the tail-water; there
A B is the extreme fall, A C is the depth of the water is
pentrough. Now, instead of the common practice of mak
a wheel of the diameter equal to B C, I make the w
D E F G one-fourth larger than B C, then the water wil
delivered upon it at the point E. The floor C of the p
trough C H L, does not come up to meet the end H ther
but leaves a small space through which the water issue
the direction of the dotted line I I, to the ·buckets of
wheel. The breadth of this space is determined by
shuttle K, which lays flat upon the floor of the pentrou
and slides over the aperture. It is regulated by means t
lever N, acted upon by a screw, rack, or other adjustment
M, and the water is thus delivered in a very thin and rc
sheet into the buckets."

Fig. 117 represents a method of laying on water wl
has for several years been in common use in Yorkshire
the north of England. In this the water is not applied q
at the top of the wheel, but nearly in the same position
the last described; but the advantages of this wheel over
others is, that the water can be delivered at a greater or
height, according to the height at which the water stand
the trough; but in all the preceding methods, if the wate
subject to variations of height, as all rivers are, then
wheel must be diminished, so that in the lowest state of
water it will stand a sufficient depth above the orifice in
bottom of the trough to issue with a velocity rather gre
than the motion of the wheel. In this case, when the w
rises to its usual height, or above it, the increase of fall t
obtained is very little advantage to the wheel; the impro
wheel can at all times take the utmost fall of the water,
when its height varies from three to four feet. A A is
pentrough made of cast-iron; the end of it is formed
grating of broad flat iron bars, which are inclined in
proper position to direct the water through them into
buckets of the wheel. The spaces between the bars are
up by a large sheet of leather, which is made fast to
bottom of the iron trough at a, and is applied against
bars; and the pressure of the water keeps it in close con
with the bars, so as to prevent any leakage. This is
real shuttle, and to open it so as to give the required st

the water to the wheel, the upper edge of the leather is wrapped round a smaller roller, b; the pivots at the ends of this roller are received in the lower ends of two racks, which are made to slide up and down by the action of two pinions fixed upon a common axis which extends across the trough; this axis being turned, raises up or lowers down the roller, and the leather shuttle winds upon it as it descends, or unwinds from it as it ascends, so as to open more of the spaces between the bars, or close them, as it is required. In order to make the roller take up the leather, and always draw it tight, a strap of leather is wound round the extreme ends of the rollers, beyond the part where the leather shuttle rolls upon it. These straps are carried above water and applied on wheels, which wind them up with a very considerable tension, by the action of a band and weight wrapped on the circumference of a wheel, which is on the end of the axis of those wheels.

The water runs over the upper side of the roller, and flows through the spaces between the grating into the buckets of the wheel; the descent of the water passing through the bars, and afterwards in falling before it strikes the bottom of the bucket, is found fully sufficient to produce the necessary velocity of the water, for a fall of four inches produces a velocity of more than four feet per second.

We recommend this as the best method of applying the water, as we see in all other forms that a much greater portion of the fall is given up in order to make the water fall into the wheel; not that any such depth as is commonly given is at all necessary, but the aperture in the trough must be placed so low that the water will run through it in the very lowest states of the water, otherwise the wheel must stop at such times.—Dr. Rees's *Cyclopædia*. *Repertory of Arts*, 1813.

SLUICE GOVERNOR FOR REGULATING THE INTRODUCTION OF WATER UPON WATER-WHEELS OF ALL KINDS.

The ingenious Mr. Burns actually constructed for the Catrine Cotton Mills, the sluice governor, represented at fig. 118, 119, 120, and 121, which was considered of such advantage as to produce a saving of more than 100*l.* per annum. The motion of the water-wheel is communicated by a belt or rope going round the pulley I to the axis E F, which carries the balls G H, fig. 118. This motion is conveyed to the upright shaft T, by the wheels and pinions Q, R, S, T, and the wheel N at the bottom of the shaft drives the wheels

1

O, P, figs. 119 and 120, in opposite directions. When
velocity of the wheel is such as is required, the wheels
move loosely about the axis, and carry the motion no fart
But when the velocity of the wheel is too great, the b
G, H, separated by the increase of centrifugal force,
the box *a* upon the shaft E F. An iron cross *b c*, see fig. 1
is fitted into the box *a*. This cross works in the four pro
of the fork *e b c*, fig. 119, at the end of the lever *d q f e*, wh
moves horizontally round *f* as its centre of motion. W
the box *a* is stationary, which is when the wheel has
proper velocity, the iron cross works within two of
prongs so as not to affect the lever *a f c*, but to allow
clutch *q q*, fixed at the end of the lever, to be disenga
from the wheels. When the cross *b c* rises, it strikes
turning round the prong 3, see fig. 121, which drives aside
lever *e f a*, and throws the clutch *q* into the arms of
wheel P, figs. 119, 120. This causes it to drive round the sl
D C in one direction. When the iron cross *b c*, on
contrary, is depressed by any diminution in the velocit
the wheel, it strikes in turning round the prong 4, wl
pushes aside the lever *e f d*, and throws the clutch *q* into
wheel O. This causes the wheel O to drive the shaft it
opposite direction to that in which it was driven by
Now the shaft D C, which is thus put in motion, drives
means of the pinion C and wheel B, the inclined shaft B
which, by an endless screw, X, working in the tool
quadrant Z, elevates or depresses the sluice K L, and adi
a greater or a less quantity of water, according to the mo
given to the shaft by the wheel P or O. This change
the aperture is produced very gradually, as the train of wh
work is made so as to reduce the motion at the sluice,
centre in which the sluice turns should be one-third of
height from the bottom, in order that the pressure of
water on the part above the centre may balance the pres
on the part below the centre.

MR. FERGUSON'S RULES FOR THE CONSTRUCTION OF UNDERSHOT WATER-MILLS.

WHEN the float-boards of the water-wheel move 1
a third part of the velocity of the water that acts upon th
the water has the greatest power to turn the mill: and w
the mill-stone makes about 60 revolutions in a minute,
found to do its work the best. For, when it makes
about 40 or 50 it grinds too slowly, and when it makes
than 70, it heats the meal too much, and cuts the bra

small, that a great part thereof mixes with the meal, and cannot be separated from it by sifting or boulting. Consequently, the utmost perfection of mill-work lies in making the train so, as that the mill-stone shall make about 60 turns in a minute when the water-wheel moves with a third part of the velocity of the water. To have it so, observe the following rules:

1. Measure the perpendicular height of the fall of water, in feet, above the middle of the aperture, where it is let out to act by impulse against the float-boards on the lowest side of the undershot-wheel.

2. Multiply this constant number 64.2882, by the height of the fall in feet, and extract the square root of the product, which shall be the velocity of the water at the bottom of the fall, or the number of feet the water moves per second.

3. Divide the velocity of the water by 3, and the quotient shall be the velocity of the floats of the wheel, in feet, per second.

4. Divide the circumference of the wheel in feet, by the velocity of its floats, and the quotient will be the number of seconds in one turn or revolution. of the great water-wheel on whose axis the cog-wheel that turns the trundle is fixed.

5. Divide 60 by the number of seconds in a turn of the water-wheel, or cog-wheel, and the quotient will be the number of turns of either of these wheels in a minute.

6. By this number of turns divide 60, (the number of turns the mill-stone ought to have in a minute,) and the quotient will be the number of turns the mill-stone ought to have for one turn of the water or cog wheel. Then,

7. As the required number of turns of the mill-stone in a minute is to the number of turns of the cog-wheel in a minute, so must the number of cogs in the wheel be to the number of staves in the trundle on the axis of the mill-stone, in the nearest whole number that can be found. By these rules the following table is calculated; in which the diameter of the water-wheel is supposed to be 18 feet, (and consequently its circumference 56¾ feet,) and the distance of the mill-stone to be five feet.

Perpendicular height of the fall of water in feet.	Velocity of the water, in feet, per second.	Velocity of the wheel, in feet, per second.	Number of turns of the wheel in a minute.	Required number of turns of the mill-stones for each turn of the wheel.	Nearest number of cogs and staves for that purpose.		Number of turns of the mill-stone for one turn of the wheel by these cogs and staves.	Number of turns of the mill-stone in a minute by these cogs and staves.
					Cogs.	Staves.		
1	8·02	2·67	2·63	21·20	127	6	21·17	59·91
2	11·40	3·72	4·00	15·00	105	7	15·00	60·00
3	13·89	4·63	4·91	12·22	98	8	12·25	60·14
4	16·04	5·35	5·67	10·58	95	9	10·56	59·87
5	17·93	5·98	6·34	9·46	85	9	9·44	59·84
6	19·64	6·55	6·94	8·64	78	9	8·66	60·10
7	21·21	7·07	7·50	8·00	72	9	8·00	60·00
8	22·68	7·56	8·02	7·48	67	9	7·44	59·67
9	24·05	8·02	8·51	7·05	70	10	7·00	59·57
10	25·35	8·45	8·97	6·69	67	10	6·70	60·09
11	26·59	8·86	9·40	6·38	64	10	6·40	60·16
12	27·77	9·26	9·82	6·11	61	10	6·10	69·90
13	28·91	9·64	10·22	5·87	59	10	5·90	60·18
14	30·00	10·00	10·60	5·66	56	10	5·60	59·36
15	31·05	10·35	10·99	5·46	55	10	5·40	50·48
16	32·07	10·69	11·34	5·29	53	10	5·30	60·10
17	33·06	11·02	11·70	5·13	51	10	5·10	59·67
18	34·12	11·34	12·02	4·90	50	10	5·00	60·10
19	34·95	11·65	12·37	4·85	49	10	4·80	60·61
20	35·86	11·92	12·68	4·73	47	10	4·70	59·59
1	2	3	4	5	6		7	8

Example.—Suppose an undershot-mill is to be built where the perpendicular height of the fall of water is nine feet; it is required to find how many cogs must be in the wheel, and how many staves in the trundle, to make the mill-stone go about 60 times round in a minute, while water-wheel floats move with a third part of the velocity with which the water spouts against them from the aperture at the bottom of the fall.

Find 9 (the height of the fall) in the first column of the table; then against that number, in the sixth column, is 70 for the number of cogs in the wheel, and 10 for the number of staves in the trundle; and by these numbers we find in the eighth column that the mill-stone will make 59 $\frac{1}{6}\frac{5}{6}$ turns in a minute, which is within half a turn of 60, and near enough for the purpose; as it is not absolutely requisite that there should be just 60 without any fraction: and throughout the whole table the number of turns is not quite one more or less than 60.

The diameter of the wheel being 18 feet, and the fall of water nine feet, the second column shows the velocity of the water at the bottom of the fall to be 24 $\frac{1}{100}$ feet per second; the third column the velocity of the float-boards of the wheel to be 8 $\frac{1}{100}$ feet per second; the fourth column shows that the wheel will make 8 $\frac{5}{100}$ turns in a minute; and the sixth column shows that for the mill-stone to make exactly 60 turns in a minute, it ought to make 7 $\frac{1}{20}$ (or seven turns and one-twentieth part of a turn) for one turn of the wheel.

Dr. Brewster, in the valuable Appendix which he has annexed to his edition of Mr. Ferguson's works, shows, that the principles upon which the above table is calculated, are erroneous, owing to the author having, with Desagulier and Maclaurin, embraced M. Parent's theory, which Mr. Smeaton, by repeated experiments, proved to be incorrect.

The constant number used by Mr. Ferguson for finding the velocity of the water from the height of the fall, 64.2882, appears to be also wrong. For, from some recent experiments made by Mr. Whitehurst on pendulums, it is found, that a heavy body falls 16.087 feet in a second of time: consequently the constant number should be 64.348.

Dr. Brewster then states, that in Mr. Ferguson's table, the velocity of the mill-stone is too small; and Mr. Imison, in correcting this mistake, has made the velocity too great. From this circumstance, the Mill-wrights' Table, as hitherto published, is fundamentally erroneous, and is more calculated to mislead than to direct the practical mechanic. Proceeding, therefore, upon the practical deductions of Smeaton, as confirmed by theory, and employing a more correct constant number, and a more suitable velocity for the mill-stone, we may construct a new Mill-wrights' Table by the following rules:

1. Find the perpendicular height of the fall of water in feet above the bottom of the mill-course, at K, (fig. 100,) and having diminished this number by one-half of the natural depth of the water at K, call that the height of the fall.

2. Since bodies acquire a velocity of 32.174 feet in a second, by falling through 16.087 feet, and since the velocities

DR. BREWSTER'S MILL-WRIGHTS' TABLE.

*...ich the velocity of the wheel is three-sevenths of the velocity of
...e water, and the effects of friction on the velocity of the stream
...uced to computation.*

Velocity of the water per second, friction being considered.		Velocity of the wheel per second, being 3-7th that of the water.		Revolutions of the wheel per minute, its diameter being 15 feet.		Revolutions for mill-stone, for one of the wheel.		Teeth in the wheel, and staves in the trundles.		Revolutions of the mill-stones per minute by these staves and teeth.	
Feet.	100 parts of a foot.	Feet.	100 parts of a foot.	Rev.	100 parts of a rev.	Rev.	100 parts of a rev.	Teeth.	Staves.	Rev.	100 parts of a rev.
7·62		3·27		4·16		21·63		130	6	89·98	
10·77		4·62		5·88		15·31		92	6	90·02	
13·20		5·66		7·20		12·50		100	8	90·00	
15·24		6·53		8·32		10·81		97	9	89·94	
17·04		7·30		9·28		9·70		97	10	90·02	
18 67		8·00		10·19		8·63		97	11	89·98	
20·15		8·64		10 99		8·19		90	11	90·01	
21 56		9·24		11·76		7·65		84	11	89·96	
22·86		9·80		12·47		7·22		72	10	90·03	
24·10		10 33		13 15		6·84		82	12	89·95	
25·27		10·73		13 79		6·53		85	13	90·05	
26·40		11·31		14·40		6·25		72	12	90·00	
27·47		11·77		14·99		6·00		72	12	89·94	
28 51		12 22		15 56		5·75		75	13	89·94	
29 52		12 65		16 13		5·56		67	12	90 01	
30 45		13·06		16·63		5·41		65	12	89·97	
31·42		13·46		17·14		5·25		63	12	89·99	
32·33		13 86		17 65		5·10		61	12	90·01	
33·22		14·24		18 13		4·96		64	13	89·92	
34·17		14·64		18 64		4·83		55	12	89·84	
2		3		4		5		6		7	

TREATISES ON MILL-WORK.

Künstliche abriss Allerhand, Wasser, Wind-ross, und Hand-mühlen, von Jacob. de Strada a Rosberg, 1617.

Georg. Christoph Laerner Machina torentica nova; oder beschreibung neu erfundanen Drehmühlen, 1661.

Theatrum Machinarum Novum; das ist, neu vermehrter Schauplatz Mechanischen Künste, handelt von Allerhand, Wasser, Wind, Ross, Geu und Hand-mühlen, von Geo. And. Bocklern, 1661.

Contenta discursus Mechanici, concerneatis Descriptionem Optimæ f Velorum horizontalium pro usu Molarum, nec non fundamentum inclinat Velorum in Navibus, habita coram Societate Regia, a R. H. translata Collectionibus Philosophicis M. Dec. num. 3, pa. 61, 1681.

Dissertatio Historica de Molis, quam præside Joh. Phil. Treuer def Jo. Tob. Mühlberger Ratisbonens Jenæ, 1695.

Martin Marten's Wiskundige beschouwinge der Wind of Wadermool vergeleken met die van den heer Johann Lulofs Amsterdam 1700.

Vollständige Mühlen-baukunst, von Leonhard Christoph. Sturm, 1718: Jacob Leopold's Theatrum Machinarum Molinarum, folio, 1724, 1725.

Remarques sur les Aubes ou Palettes des Moulins, et autres Machines m par le Courant des Rivières, par M. Pitot, Mem. Acad. Roy. Paris, 1729.

Joh. van Zyl Theatrum Machinarum Universale of Groot Algemeen, Nool bock, &c., Amsterdam, 1734.

Jo. Caral. Totens Disser. de Machinis Molaribus optime construendis, L Batav. 1734.

Kurze, aber Deutliche anweisung zur construction der Wind und Was mühlen, von Gottfr. Kinderling, 1735

Desagulier's Experimental Philosophy, 2 vols. 4to. 1735, 1744.

Architecture Hydraulique, par M. Belidor, 4 vols. 4to. 1737—1752.

Mr. W. Anderson, F. R. S. Description of a Water-wheel for Mills, Trans. vol. 44, 1746.

Leonh. Euleri, De Constructione aptissima Molarum alatarum dia, Com. Acad. Petrop. tom. 4, 1752.

Mémoire dans lequel on démontre que l'Eau d'une Chûte, destinée à mouvoir quelque Moulin ou autre Machine, peut toujours produire beau plus d'effet en agissant par son poids qu'en agissant par son choc, et q roues à pots qui tournent vite, relativement aux chûtes et aux dépenses d' par M. de Parcieux, Acad. Roy. Paris, 1754.

Jo. Alberti Euleri Enodatio Questionis : quo modo vis Aquæ alatæ t cum maximo lucro ad Molas circumagendas, aliave opera perficienda imp possit, præmio à Societate Regia. Sci. Gotting. 1754.

An experimental Inquiry concerning the Natural Powers of Wind and W to turn Mills and other Machines depending on Circular Motion, by Mr Smeaton, F. R. S. Phil. Trans. 1759.

This, and Mr. Smeaton's other papers are republished with his Reports, M in 4to.

Mémoire dans lequel on prouve que les Aubes de Roues mues par les cou de grandes Rivières feroient beaucoup plus d'effet si elles étoient incli aux rayons, qu'elles ne sont étant appliquées contre les rayons mêmes, con elles sont aux Moulins pendans et aux Moulins sur Bateaux qui sont su Rivières de Seine, de Marne, de Loire, &c. par M. de Parcieux. Mem. Acad. I Paris, 1759.

Joh. Albert Euler's Abhandlung von der bewegung obener Flächen, wen vom Winde Getrieben Werden, 1765.

Schauplatz des Mechanischen Mühlenbaues, Darinnen von Verschied Hand, Trett, Ross, Gewicht, Wasser, und Wind-mühlen Gehandelt Wird, du Johan Georg. Scopp. J. C. Iter Theil, 1766.

Theatrum Machinarum Molariam, oder schauplatz der Mühlenbaue als der Neunte theil von des sel hrn Jac. Leopolds, Theatro Machinarum, Joh. Mathias Beyern, 1767, 1788, 1802.

A Memoir concerning the most advantageous Construction of Water-wheels, &c. by Mr. Mallet of Geneva, Phil. Trans. 1767.

Mémoire sur les Roues Hydrauliques, par M. le Chevalier de Borda, Mem. Acad. Roy. Paris, 1767.

Kurzer unterricht, allerley arten von Wind und Wasser-mühlen auf di vortheilhafteste weise zu erbauen, nebst einigen gedanker über die verbesserung der räderwerks, an den Mühlen, von Joh. König, 1767.

C. G. Bischoff's Beyträge zur Mathesis der Mühlen, 1767.

Détermination générale de l'Effet des Roues mues par le Choc de l'Eau, par M. l'Abbé Bossui, Mem. Acad. Roy. Paris, 1769.

Andreas Kaovenhöfer, Deutliche abhandlung von den rädern der Wasser-mühlen, und von dem einrandigen werke der Schocidemühlen, 1770.

Manuel du Meunier et du Charpentier des Moulins, redigé par Edm. Bequillet, 1775.

Remarques sur les Moulins et autres Machines, où l'Eau tombe en dessus de la Roue, par M. Lambert.

Expériences et Remarques sur les Moulins que l'Eau meut par en bas dans une Direction horizontale, par M. Lambert.

Remarques sur les Moulins et autres Machines, dout les Roues prenant l'Eau à une certaine Hauteur, par M. Lambert.

(The three last articles are inserted in Mem. Acad. Roy. Berlin, 1775.

Ausführliche erklärung der Vorschläge für die Längere dauer de Muhlen-werk, nebst ähnlichen gegenstander, in ein gespräch verfasset, von Johann Christian Fullmann Muhlenmeister, 1780.

Tratado de los Granos y Modo de Molelos con Economic de la Conservaçion de Astos y de las Harinas; escr. en Fr. par M. Beguillet y extract. y trad. al Cast. con algun Notas y un Supplem. por Ph. Marescaulchi, Madrid, 1786.

Suite de l'Architecture Hydraulique, par M. Fabre, 1786.

Mémoires sur les Moyens de perfectionner les Moulins, et la Mouture économique, par C. Bucquet, 1786.

Manuel ou Vocabulaire des Moulins à Pot, à Amst., 1786.

Die Nothigsten Kenntnisse zur Anlegung, Beurtheilung, und Berechnung der Wasser-mühlen, and zwar der Mahl, Oehl, und Säge-Muhlen, zur Anfänger und Liebhaber der Mühlenbaukunst, von Joh. Christ. Huth, 1787.

An Essay proving Iron far superior to Stone of any kind for breaking and grinding of Corn, &c. by W. Walton, 1788.

Mühlenpraktik, oder unterricht in dem Mahlen der Brodfrüchte, für Polizey-beamte, Gaverksleute und Hauswirthe, von L. Ph. Hahn, 1790

The Young Mill-wright and Miller's Guide, by Oliver Evans, Philadelphia, 1790.

Manuel du Meunier, et du Constructeur des Moulins à Eau et à Grains, par C. Bucquet, 1791.

Praktische anweisung zum Muhlen bau, von Lr. Clausen, 1792.

Beschreibung zweir Machinen zur Reinigung des Korns, von Lr. Clausen, 1792.

Instructions sur l'Usage des Moulins à Bras, inventés et perfectionnés par les Citoyens Duraud, Père et Fils, Méchaniciens, 1793.

Theoretisch-praktische abhandlung uber die Besserung der Mühlräder, von dem Verfasser der Zweckmässigen, Luftreiniger, &c. 1795.

A Treatise on Mills, in four parts, by John Banks, 1795.

Handbuck der Maschinenlehre, sur prakiker und akademische lehrer, von Karl Christian Langsdorf, 1797, 1799.

On the Power of Machines; including Barker's Mill, Westgarth's Engine, Cooper's Mill, Horizontal Water-wheel, &c. by John Banks, 1803.

The Experienced Mill-wright, by Andrew Gray, Mill-wright, 1804

The Transactions of the Society of Arts and Manufactures; several of the volumes of which contain improvements in Mill-work.

See also the Repertory of Arts, first series 16 vols. and second series 31 vols, Hachette, Traité Elémentaire des Machines, 4to. Paris, 1811.

Buchanan's Essay on Mill-work, 1811, 8vo.

WINDMILLS.

THE windmill derives its name from the motion it receives from the impulse of the wind.

The date of its invention is not precisely known, though authors generally concur in believing it to have taken place at no very distant period of time. Some state it to have been first used in France in the sixth century: others, on the contrary, assert, that at the time of the crusades it was introduced into Europe from the east, where scarcity of water gave the impetus that led to its discovery.

Windmills are of two kinds, horizontal and vertical.

THE VERTICAL WINDMILL

CONSISTS of a strong shaft, or axis, inclining a little upwards from the horizon, with four long yards, or arms, fixed to the highest end, perpendicular to the shaft, and crossing each other at right angles. Into these arms are mortised several small cross-bars, and to them are fastened two, three, or four, long bars, running in a direction parallel with the length of the arms; so that the bars intersect each other, and form a kind of lattice work, on which a cloth is spread to receive the action of wind. These are called the sails, and are in the shape of a trapezium, usually about nine yards long and two wide.

As the direction of the wind is very uncertain, and perpetually changing, it becomes necessary to have some contrivance for bringing the windshaft and sails into a position proper for receiving its impression. To effect this, two methods are in general use: the one called the post-mill; the other the smock-mill.

POST-MILLS.

IN the post-mill it is accomplished by driving perpendicularly into the earth the trunk of a strong tree, that is held securely upright by several oblique braces, which extend from a platform on the ground to the middle of the tree, leaving 10 or 12 feet of the upper part free from the braces. The part thus left free from obstruction is rounded, and made to pass through a circular collar, formed in the flooring of the lower chamber, and to enter into a socket fixed into the flooring of the upper chamber, and to one of the strongest cross-beams, which must sustain the whole weight of the mill-house, so that, by means of a pivot, or gudgeon, fastened on that part of the post which enters into the socket, the whole machine can turn about horizontally to face the wind. A strong framing, united by joints to the back part of the mill-house, descends in a sloping direction till it touches the ground: the bottom of it is very heavy, and is fastened by cords to some short posts that are driven in a circle, at regular intervals round the mill, to prevent the mill from turning about at every sudden squall. This framing is furnished with steps to serve as a ladder of ascent or descent. At the bottom of it a rope is fastened, and carried thence in an inclined position to the top of the mill, where, by a lever or tackle of pullies, it can be shortened so as to raise the framing from the ground, and then by pushing against it, in the manner of a lever, the whole mill may be turned in any required direction. To obtain more force, a small capstan is often provided to draw a rope fixed to the end of the ladder: this capstan is movable, and can be fastened at pleasure to any one of the posts.

The internal mechanism of a post-mill is exhibited in fig. 123. W X Y the upper chamber; X Y Z the lower one; A B the shaft, or axis, with the cog-wheel G, moving round in order of the letters that describe the sails C D E F, giving motion to the lantern H, and its spindle I K; L M is a bridge to support the said spindle; and N and O P are beams to sustain the bridge. The top mill-stone Q is the only one that moves, and is fixed on the spindle I K by a piece of iron, called the rynd, let in at the lower part of the stone; the lower mill-stone R, is somewhat larger than the other. The corn is put into the hopper S, and runs from thence along the spout T; the spindle I K, being square, shakes in its revolutions the spout T, and causes the corn to fall through the hole V between the stones, where it is ground; the flour then passes through the tunnel a b, and is easily deposited in the chest c; d e is a string going round the pin d, and serving to draw the spout T nearer to, or farther from, the spindle I K, that the corn may be made to run out either faster or slower, according to the velocity of the wind; f g and h i are levers, whose centres of motion are f and m; i h s is a cord going about the pins l and n to wind up and raise the stone Q. By bearing down the end r h, g is raised, which raises the perpendicular N O, the perpendicular raises the cross-beam O P, the cross-beam the bridge L M and the spindle I K, together with the upper mill-stone Q, so that the stones can be set at any required distance apart. The corn is drawn up to the top of the mill by means of a rope rolled about the axis A B; q v is a ladder for ascending to the higher part of the mill. A rod or gripe of pliable wood is fixed at one end s, and at the other tied to the lever t u, movable about at w, which being pressed down stops the motion of the mill at pleasure. When the wind is great, the sails are only in part, or on one side covered, and sometimes only one-half of two opposite sails. The same shaft can have another cog-wheel fixed to the end B, with trundle and mill-stones similar to those already described: by which means the shaft can turn two pair of stones at once; and when one pair only is wanted to grind, the lantern H and spindle I K are taken out from the other.

SMOCK-MILL.

THE other method of bringing the windshaft and sails into a position proper for receiving the impression of the wind is,

by what is called the smock-mill. This mill is more expe
sive in the construction, and more decidedly advantageou
as it can be made of any required dimensions. It is built
the form of a round turret, having at the top of it a wood
ring with a groove in it, furnished with a number of l
truckles, kept equi-distant from each other by their cent
pins being fixed into a circular hoop. Into this groove t
framing of the upper or movable part of the mill, which
called the head, or cap, enters, and a very slight power
alone sufficient to turn it about that the sails may receive t!
action of the wind. The head or cap is very ingenious
contrived to turn itself about whenever the wind change
by a small pair of sails, or fans, fixed up in a frame th
projects from the back part of the head.

Fig. 124, *a* the fans, having on its axis a pinion of 10 leaves *b*, whi
gives motion to a cog-wheel of 60 teeth *c*, its axis *d*, and a pinion
12 teeth at the lower end *e*, turning a bevelled wheel of 72 teeth *f*, a certi
iron shaft *g*, having a pinion of 11 teeth *h*, that works in a circle of 1
cogs. Therefore, whenever the wind changes, it acts obliquely upon t
vanes of the fan, and turns it round, which, giving an impulse to
connecting machinery, brings the main shaft of the sails slowly about to f:
the direction of the wind. The method of this operation is as follows: t
fans, having received the action of the wind, turn round, and the pinion *b*
10 leaves, that is upon its axis, gives motion to the cog-wheel of 60 tee
c, fixed on an inclined axis which has at the lower end the pinion
12 leaves *e*, acting upon the bevelled wheel of 72 teeth *f*, fixed on a verti
iron axis, and giving motion to the pinion of 11 teeth *h*, that works in t
circle of 120 cogs. A B two of the sails (the other two being endw
cannot be seen) fixed on an iron shaft or axis C D, by screwing them
an iron cross formed at one end of it. Upon this shaft is the cog-wheel
that acts upon the lantern F, fixed on a strong vertical shaft extending f
the top to the bottom of the mill, and having on the lower end the la
wheel *i i*, giving motion to the two opposite pinions *k k*, which turn
spindles and the mill-stones G H. A wheel is fixed on the main axis at
to give action to the pinion on the horizontal roller *m*, which has a r
wrapped about it to wind up the sacks of corn. The same wheel I tu
another horizontal axis that has several wheels to receive endless ropes
turning the bolting and dressing machines. On the middle part of
vertical shaft K L is the wheel I, which turns the roller *m*. to draw up
sacks of corn from the lower part of the mill, which is used as a storehou
being divided into as many compartments as the miller may require.
the mill-stone spindle is attached a pair of regulating balls, to regulato
velocity of the mill. For the manner of applying this regulator
fig. 125, *l* a spindle, on which is fixed the pinion *k*, playing into the la
wheel that is attached to the vertical shaft; the lower end of the spin
enters into a square formed on the top of the mill-stone axis at *m*; imn
diately beneath the pinion two iron rods are jointed, bending downwar
having a heavy iron ball *o o* fastened to the end of each; to these rods
attached two links at *p p*, to suspend a collar capable of sliding freely
and down upon the spindle *l*; this collar is embraced by a fork, formed
a steelyard, lying horizontal, and suspended by the fulcrum *q*; *r* is an in
rod fixed at the extreme end of the steelyard, and having at the bottom

_____ to connect it with the lever _s_, whose fulcrum is _t_; this, by means
of an iron rod, suspends one end of the bridge on which the lower pivot of
the mill-stone rests, the other end bearing on a fulcrum, or centre.

Whenever the mill acquires velocity, the iron balls, by their centrifugal
force, will fly out, and elevate the collar, which, acting upon the connecting
_____ will let the upper mill-stone down nearer to the lower one, and the
_____ or friction thus caused will counteract the increased velocity of
the wind. On the contrary, if the wind decreases, the balls will fall
towards each other, and let down the sliding collar, which will raise the
_____ mill-stone, and by increasing the distance between it and the lower
_____ the mill to acquire greater velocity. For this purpose a weight _v_
_____ upon the steelyard, sufficient to raise the stone whenever the
_____ of the collar will permit it so to do. Several notches are cut into
the steelyard for different positions of the fulcrum _q_ and rod _r_, to regulate
_____ effectually the motion of the machinery. For instance, if the wind
_____ blow stronger, and the mill go slower, contrary to the effect expected,
_____ that the regulation is too strong: to remedy this, the leverage of the
_____ must be increased by reducing the distance between the fulcrum _q_
_____ rod _r_, by shifting either of them into different notches. On the other
_____ if the velocity of the mill should increase with the velocity of the
_____ shows that the regulation is not strong enough, and that the
_____ _q_ and the rod _r_ must be set a greater distance apart. Sometimes
_____ that the whole limits of the notches on the steelyard is insuffi-
_____ effectuate the desired object; in such case, the acting length of the
_____ _s_ _s_ must be increased or diminished by removing the fulcrum _t_ to a
_____ or less distance from the suspended rod _v_.

_____ fig. 126 is shown the construction of the horizontal shaft or axis that
_____ the sails. It is an octagonal iron shaft, having two cylindrical necks,
_____ where it rests upon its bearings. At the end it has a kind of box
_____ two mortises, _o_ and _f_, through it in perpendicular directions, to
_____ the sails. At the back of one of these mortises, and the front of the
_____ projecting arm is left in the casting to receive screw bolts for
_____ the sails secure in the mortises. The sails are braced to each arm
by a rope stay, proceeding from the end of a pole, fixed at the end of the
cast-iron axis. Each sail is formed of a sail cloth, spread upon a kind of
lattice work, similar to that described under the head of Post-mill. The
plane of this frame is inclined to the plane of the sail's motion, at such an
angle, that the wind blowing in the direction of the axis acts upon the sails
as inclined planes, and turns them about with a power proportionate to the
size of the sails and the force of the wind. The cog-wheel is fixed on the
axis by bolting its arms against the stanch marked C. The mill-stones are
the same as those described under the head of Flour-mill.

Parent, Euler, and other geometricians have written much
upon the nature and construction of windmills; but as we
consider the experiments and researches made by our own
countryman Smeaton to be far superior in point of practical
utility, we shall content ourselves with giving his opinion
as to the shape, magnitude, and position of the sails.

By Mr. Smeaton's experiments it appears, that when the
sails were set at the angle of 55 degrees with the axis,
proposed as the best by M. Parent and others, they were the
most disadvantageous of any that were tried by him.

On increasing the angle of the ...
to 75 degrees, an augmentation of power ...
the ratio of 31 to 45, and this proves to be the ...
commonly in use when the surfaces of the sails are ...

If nothing more were requisite than to make t...
acquire motion from a state of rest, or to prevent i...
passing into rest from a state of motion, the position ...
mended by M. Parent would be the best; but if the ...
intended, with given directions, to produce the ...
effect possible in a given time, we must reject M. ...
position; and, if use be made of planes, confine ou...
within the limits of 72 and 75 degrees with the axis.

The variation of a degree or two in the angle makes ve...
difference in the effect, when the angle is near upon th...

Mr. Smeaton made several experiments upon a large ...
and found the following angles to answer as well ...
The radius is supposed to be divided into six parts, and ...
sixth, reckoning from the centre, is called one, the extr...
being denoted six.

No.	Angle with the axle.	
1	72°	18
2	71	19
3	72	18 middle
4	74	16
5	77½	12½
6	83	7 extremity.

Having thus obtained the best position of the sails ...
manner of weathering, as it is called by the work...
Mr. Smeaton next proceeded to try what advantage coul...
made by an addition of surface upon the same radius. ...
result was, that a broader sail requires a greater angle; ...
that when the sail is broader at the extremity than near th...
centre, this shape is more advantageous than that of...
parallelogram. The figure and proportion of the en...
sails he found to answer best upon a large scale, when ...
extreme bar is one-third of the radius, or whip, and is d...
by the whip in the proportion of 3 to 5. The triangul...
leading sail, is covered with board, from the point dow...
one-third of its height, the rest with cloth as usual. ...
angles mentioned in the preceding article are found to b...
best for the enlarged sails also; for in practice it is ...
that the sails had better have too little than too much ...

Many persons have imagined that the more sail the gre...
the advantage, and have therefore ... to fill ...
whole area, and by making each sail ... of an ...

according to M. Parent, to intercept the whole cylinder of wind, and thereby to produce the greatest effect possible: but from our author's experiments it appears, that when the surface of all the sails together exceeded seven-eighths of the circular area containing them, the effect was rather diminished than augmented; and consequently, he concludes, that when the whole cylinder of wind is intercepted, it does not then produce the greatest effect for want of proper interstices to escape.

"It is certainly desirable," says Mr. Smeaton, "that the sails of windmills should be as short as possible, but at the same time it is equally desirable the quantity of cloth should be the least that may be, to avoid damage by sudden squalls of wind. The best structure, therefore, for large mills, is that where the quantity of cloth is the greatest in a given circle that can be: on this condition, that the effect holds out in proportion to the quantity of cloth; for otherwise the effect can be augmented in a given degree by a lesser increase of cloth upon a larger radius, than would be required if the cloth were increased upon the same radius."

The ratios between the velocities of windmill sails unloaded, and when loaded to their maximum, turned out different in different experiments, but the most general ratio of the whole was as 3 to 2. In general, however, it appeared, where the power was greater, whether by an enlargement of surface, or a greater velocity of the wind, that the second term of the ratio was less.

The ratio between the greatest load that the sails will bear without stopping, or what is nearly the same thing, between the least load that will stop the sails, and the load at the maximum, were confined between that of 10 to 8, and of 10 to 9, and at a medium about 10 to 8.3, or of 6 to 5; though it appeared on the whole, that where the angle of the sails or quantity of cloth was greatest, the second term of the ratio was less.

The following maxims have been deduced by Mr. Smeaton from his experiments:

Maxim 1. The velocity of the windmill sails, whether unloaded or loaded, so as to produce a maximum, is nearly as the velocity of the wind, their shape and motion being the same,

Maxim 2. The load at the maximum is nearly, but somewhat less than, as the square of the velocity of the wind, the shape and position of the sails being the same.

Maxim 3. The effects of the same sails at a maximum are nearly, but somewhat less than, as the cubes of the velocity of the wind.

Maxim 4. The load of the same sails at the maximum is nearly as the squares, and their effects as the cubes of their number of turns in a given time.

Maxim 5. When the sails are loaded so as to produce a maximum given velocity, and the velocity of the wind increases the load contai the same: first, the increase of effect, when the increase of the veloci the wind is smaller, will be nearly as the squares of those veloci secondly, when the velocity of the wind is double, the effects wil nearly as 10 to 27½; but thirdly, when the velocities compared are than double of that where the given load produces a maximum, the e increase nearly in a simple ratio of the velocity of the wind.

Maxim 6. If sails are of a similar figure and position, the numb turns in a given time will be reciprocally as the radius or length of the

Maxim 7. The load at a maximum that sails of a similar figure position will overcome, at a given distance from the centre of motion, be as the cube of the radius.

Maxim 8. The effect of sails of similar figure and position are square of the radius.

Maxim 9. The velocity of the extremity of Dutch sails, as well as enlarged sails, in all their usual positions when unloaded, or even loade maximum, are considerably quicker than the velocity of the wind.

Mr. Ferguson remarks, that it is almost incredibl think with what velocity the tips of the sails move acted upon by a moderate wind. He several times cou the number of revolutions made by the sails in 10 o minutes; and, from the length of the arms from tip to has computed, that if an hoop of the same size were to upon plain ground with equal velocity, it would go upw of 30 miles in an hour.

RULES FOR MODELLING THE SAILS OF WINDMILLS.

FIG. 127 is a front view of one of the four sails of a wind The letters of reference will serve to explain the terms use of in the following description:

1. The length of the arm, or whip A A, reckoned from the centre great shaft B to the outermost bar 19, governs all the rest.

2. The breadth of the face of the whip A, next the centre, is one-thi of the length of the whip, its thickness at the same end is three-four the breadth, and the back-side is made parallel to the face for hal length of the whip, or to the tenth bar; the small end of the whip is sq and at its end is one-sixtieth of the length of the whip, or half the br at the great end.

3. From the centre of the shaft B to the nearest bar 1 of the latti one-seventh of the whip; the remaining space of six-sevenths of the is equally divided into 19 spaces, so as to make 19 bars; one-ninth o of these spaces is equal to the mortises for the bars, the tenons of are made square where they enter and go through the whip, and quently the mortises must be square also.

4. To prepare the whip for mortising, strike a gage-score at about fourths of an inch from the face on each side, and the gage-score, o leading side 4, 5, will give the face of all the bars on that side; b the other side, the faces of all the bars will fall deeper than the gage according to a certain rule. To find the space to be set off for this pu for each bar, construct a scale in the following manner:

5. Extend the compasses to any distance at pleasure, so that six

that extent may be greater than the breadth of the whip at the seventh bar; set these six spaces off upon a straight line for a base, at the end of which raise a perpendicular; set off three spaces upon the perpendicular, and divide the two spaces that are farthest from the base line into six equal parts each, so that this quantity of two spaces may be equally divided into 12 spaces, marked out by 13 points; from each of these points draw a line to the opposite end of the base, as so many rays to a centre, and the scale is finished.

4. To apply this scale to any given case, set off the breadth of the whip at the last bar, (that is, the bar at the extremity of the sail,) from the centre of the scale along the base towards the perpendicular; and at this point raise a perpendicular to cut the ray nearest to the base; also set off the breadth of the whip at the seventh bar in the same manner, and at this point erect another perpendicular to cut the thirteenth radius. From the intersection of the perpendicular (drawn upon the breadth of the last bar) with the first of the thirteen radii, to the intersection of the other perpendicular with the thirteenth radius, draw an oblique line cutting all the rest, and the distances of each of these last-mentioned points of intersection from the base line is the space which the face of each bar is distant from the gage-line on the driving side.

5. These distances give a difference set off for each bar till the seventh, which same must be set off for all the rest to the first.

6. These mortises must be square to the leading side of the whip.

7. When the mortises are cut, let the face of the whip be sloped off so as to agree with the face of the bars in every part.

8. Two-fifths of the whip are the length of the last or longest bar.

11. Five-eighths of the longest bar must be on the driving side of the whip, and three-eighths on the leading side, each being reckoned from the middle of the whip.

12. The proportion of the mortises already given determines the size of the bars at the mortises, but their thickness must be diminished each way, so as to be only one-half at the ends; but the face must be kept of equal breadth all the way.

13. The leading side goes no farther than the fourth bar, and there only projects one-third of the projection of the last bar.

14. All the bars on the driving side are made hollowing in the arch of a circle, which begins to spring one-third of the length of the bars on the driving side from the whip; and the sweep is such, that if a straight line be applied to the face of the bar from the whip to the end, the face of the bar should leave the straight line about the breadth of the bar.

15. There ought to be three uplongs, as 3, 2, 10, ——— to the driving, and two to the leading side, as 5, 4, to strengthen the lattice. Dr. Rees's *Cyclopædia*.

Mr. Richard Hall Gower, a gentleman in the sea-service of the East-India Company, has made some judicious experiments with a view of determining the proper angles of weather which ought to be given to the vanes of a vertical windmill: his general conclusion is, that each vane should be a spiral, generated by the circular motion of a radius, and of a line moving at right angles to the plane of a circular motion. The construction he deduces from his inquiries is simple, being this: The length, breadth, and angle of weather at the extremity of the vane being given; to determine the angles of weather at different distances from the centre.

Let A B, fig. 129, be the length of the vane; B C its breadth BCD the angle of the weather at the extremity of the vane, & 20 degrees. With the length of the vane A B, and breadth B C, the isosceles triangle A BC: from the point B draw B D perpendi C B, then B D is the proper depth of the vane.

Divide the line A B into any number of parts, (five for instance,) divisions draw the lines 1 E, 2 F, 3 G, and 4 H, parallel to the lin also, from the points of division 1, 2, 3, and 4, draw the lines 1 I, 2 and 4 M, perpendicular to 1 E, 2 F, 3 G, &c. all of them equal in to B D. Join E I, F K, G L, and H M: then the angles 1 E I, 3 G L, and H M, are the angles of weather at those divisions of the and if the triangles be conceived to stand perpendicular to the plan paper, the angles I, K, L, M, and D, becoming the vertical ang hypothenuse of these triangles will, as before suggested, give a peri of the weathering of the vane as it recedes from the centre.

METHOD OF CLOTHING AND UNCLOTHING THE SAILS WH MOTION.

Mr. JOHN BYWATER, of Nottingham, took out a pat 1804, for a method of clothing and unclothing the s windmills while in motion, by which the mill may be c either in whole or in part, in an easy and expeditious m by a few revolutions of the sails, whether they be goin or slow, leaving the surface smooth, even, and reg breadth from top to bottom; and in like manner the or any part of it, may be rolled or folded up to the w pleasure, by simple and durable machinery.

Fig. 130, Nos. 1, 2, 3, are front views of the sails as uncloth clothed, and clothed.

Fig. 131, a ring of iron, or other material, about 4 inches wide, : an inch thick, whose diameter must be sufficient to embrace the sl to which it must be well secured by the stays a a.

Fig. 132, a bevelled wheel, without arms, made of iron, stays edge of the ring so as to turn easily.

Fig. 133, a spur wheel of iron, without arms, made to turn easily pins fixed into four ears b b b b, in the back of the ring; which turned up at their ends to keep it steady.

Fig. 134 is one of the four spindles of iron, or other material, spur nut a, and a bevelled nut b; this spindle passes through fig. c c c c, and the nut a works into the spur-wheel as seen in fig. 133. The four bevelled nuts (fig. 134) work into the bevelled wheels of four cylinders i i i i fig. 130, Nos. 1, 2, 3, and so turn them of those spindles must be shorter than the others when the stocks flush. These cylinders are made of wood of about 3 inches d and are to be placed at the outside of the leading edge of each sa which the cloth is rolled (one edge being fastened on for that when the sail is unclothed. A gudgeon from the end of each runs into an iron fastened to the shaft-head, and is kept in its pla nut screwed to its end. The other end has a gudgeon b, which the eye of the cross iron A, at the points of the whips; f f f f four cy similar to i i i i, placed on the inside the whips; one behind each clothe the sails, by means of ropes o o o o, &c. fastened to them and of the cloth. At the end of each of these four cylinders a nut or

said, *oouf*, to work into the bevelled wheel; fig. 132, whose teeth decline from the centre in proportion as these work from it, which inclination must be reversed when the sails turn in the contrary way, and the gudgeons to run into irons either projecting from the ring or fastened to the key-head like the other cylinders. The gudgeons *g* keep these cylinders steady in the cross iron *h* at the point of the whips, and stays of any shape or number will keep them from springing.

Now, suppose the mill fully clothed, as at 3, all the parts of the machinery revolve with it undisturbed until a lever, fig. 136, which is fastened to the braces or fencing, by the centre pin *a*, fig. 137, on which it turns, and whose end *b* is weighted to hang down towards the breast of the mill, is brought into an horizontal direction by pulling a string attached to the end *a* within-side the mill, which end *b* stops the stud *b*, projecting from the inner surface or back-front of the spur-wheel, fig. 135; consequently the four spur-nuts *a*, at the end of the spindle, fig. 134, and seen at *a a a a*, fig. 135, roll round the spur-wheel, and the bevelled nuts *b* at the other end of the spindle work into the bevelled wheels of the outside cylinders *i i i i*, at 1, 2, 3, in a straight direction behind them, and so turning the cylinders roll the cloth on them till it is rolled up to the whip. The lever is then driven sideways (its spring *e* returning it again) from the stud in the back face of the spur-wheel by the following contrivance:

A screw, *b*, fig. 138, is cut on the gudgeon of any one of the cylinders behind the sail, and a piece of iron, *c*, is tapped to fit it. The end of this iron runs into a slot in the iron *d*, made fast to the shaft-head, to prevent the iron *c* from turning with the cylinder, but allows it to slide up and down, so as to press on that on the iron *a*, which has the eye in it, and raise the end *c* just high enough to drive the lever aside when the cloth is all rolled up, the thread of the screw adjusting it to what number of revolutions you choose to employ for that purpose. The point-end of the iron *c*, is that part of it which pushes aside and passes the lever, fig. 136, and moves on its centre *c*, and must be carried under the spur-wheel so as to act behind it for that purpose. By letting go the string the miller may at any time leave the cloth on the sail where he chooses, likewise the sails may be clothed, or any part thereof, by a lever, similar to *a*, stopping the stud *a*, on the edge of the bevelled wheel, fig. 132, and driven off in a manner similar to the spur-wheel.

Fig. 139 is a stay of wood, fixed to the stock or whip at *a a a a*, 1, 2, 3, to prevent the cylinders from springing too much. In the inside there is left room enough for the cloth to be rolled upon the cylinder through its loop in the eye of this stay. In order to keep the strings, which go over the edge of the shrouds *o o o o*, &c. tight in all weathers, a cord, passing over a spring of any sort or shape, placed under the sail, is fastened to and wound about the upper ends of the cylinders, in a direction contrary to the strings and cloth. To prevent the strings from being driven downwards by the centrifugal force, a ring or two are left on to run along the rods in the direction as *p*, Nos. 2, 3.

The width of the cloth, diameter of the cylinders, and number of revolutions you choose to employ to roll up your cloth, must determine the size of the wheels. In order to furl the cloth instead of rolling it, one end of it must be fastened to the whip and lines passed across the outside of it through loops fastened to its edge, and consequently over the edge of the shrouds, and connected with the cylinder or

K 2

roller, of any shape, placed under the sail, or elsewhere,
other ends of the lines must be connected with the
cylinder or roller; and when the cloth is drawn up in f
towards the whip, so much of these lines will be rolled
the cylinder one way, and off from it the other, as wil
sufficient to let out the cloth again when the same cylin
turning the contrary way, draws the cloth on the sail.
this mode the patentee gets rid of four cylinders, with t
appendages, the work being in other respects the same
rolling the cloths; but since folding gives a surface m
inferior in many respects to rolling, and induces inc
veniences and accidents from which the rolled surface is
he advises the rolling, rather than for a small saving to enc
the inconveniences of folding.

If a sudden gust of wind should arise in the absence of
miller, so as to drive the mill faster than a given velocit
pair of centrifugal balls, like the governor of a steam eng
may be so placed as to adjust the lever so that it
immediately unclothe itself.

BAINES'S VERTICAL WINDMILL SAILS.

MR. ROBERT RAINES BAINES, of Myton, Kingston u
Hull, secured to himself in June 1815, by patent, an impro
ment in the construction of vertical windmill sails.

Fig. 140 represents six sails; the stocks or arms marked A are the
as used for common vertical windmills; the sails marked B are ma
canvass, and fastened to the front sides of the said stocks or arms alon
edges marked a, a, and to the rods or bars marked D, at or near the
marked b, and are also extended by the rods or bars marked E, whic
inserted into or fixed to the backs thereof, and by rods or bars mark
which are inserted into or fastened to the edges of the said sails; eac
is also connected by a bar or rod marked F, as hereinafter descr
with the next following sail. The shafts or rods marked C are faste
the stocks or arms marked A, at or near d, d, by loops or otherwise,
to allow them to move as hinges do. The bars or rods marked D are co
them connected with the shafts or rods marked C by a joint, which
allow the said bars or rods marked D to move from the wind indepen
of the shafts or rods marked C, in case it should blow against the back
of the said sails, but will not allow the said bars or rods marked D to
from the wind independent of the said shafts or rods marked C, whe
wind blows against the front sides of the said sails. The bars or
marked F connect the corners marked e of each sail with the corner o
next following sail at or near the point marked b, leading behind
following sail, and which bars or rods are fastened by hooks, or other p
means, at or near their points, bent to such an angle that if the
should blow against the back sides of the said sails and force them for
the said bars or rods will be unhooked and set at liberty   A rim or c
marked G is fixed by screws or otherwise upon the said stocks or .
marked A, for the purpose of supporting the fulcra or props mark
At I is represented the head or end of a rod or bar which passes thro

the centre of the axletree of the mill, and to which weight may be applied, in the manner well known to mill-wrights, to regulate the said sails towards or from the wind. The bars or cranks marked K are fixed to the shafts or rods marked C, at such an angle, and in such a manner as will, when and as they are acted upon by the levers or bars marked L, either suffer the said bars or rods marked D and the sails to recede from the wind until the said sails present only their edges to it, or will force the said bars or rods marked D towards the wind, until they present to it their breadth. The levers or bars marked L are connected at one of their ends with the head of the aforesaid rod marked I, and at the other ends with the bars or cranks marked K, and form levers resting or acting upon the fulcra or props marked H, and are governed or regulated in their action by the said rod, the head of which is shown at I. The said rods, bars, cranks, loops, and rim, may be made of iron, or other suitable material or materials, and connected at their proper places by joints or otherwise, (so as to fix them or allow their action,) by modes well known to mill-wrights.

CUBITT'S METHOD OF EQUALIZING THE MOTION OF THE SAILS OF WINDMILLS.

Mr. WILLIAM CUBITT, of North Walsham, Norfolk, engineer, took out a patent for this invention in May 1807, which the specification thus describes:

"My invention consists in applying to windmills an apparatus or contrivance which shall cause the vanes, constructed or formed in a new and peculiar manner, to regulate themselves, so as to preserve an uniform velocity under those circumstances in which the wind would otherwise irregularly impel them, as is the case with the sails or vanes of mills of the present construction. I accomplish this object by forming the vanes (for the sake of lightness) with fewer cross bars or shrouds than in the common method, and filling up the remaining open space with small flat surfaces, formed either of boards or sheet iron painted, or any other fit substance, (though I prefer and recommend them to be made of a framing of wood, covered over with canvass.) I hang or suspend the same on their ends by gudgeons, pivots, centres, or any other convenient method, so as to open and shut like valves, (for which reason I shall hereafter so call them,) preferring always to have the centre of motion as near the upper longitudinal edge of the valve as possible, as shown in the drawing, by fig. 141, which exhibits a valve detached. I apply these valves to vanes of the present construction, by suspending them to the cross bars or shrouds of the vane by their longitudinal edges, fastened thereto by joints or otherwise, as may be preferred. These vanes, constructed of valves as above described, and which are represented in the drawing, fig. 142, present a greater or less surface to the wind, according as it acts with more or less force on them; and if the wind be very

fig. 146; where, instead of the studs or levers, the valves may be moved by having pinions fixed to them, and working with each in a rack or slider, as at T. V V are rollers to keep the racks P in their geer. The operation of this apparatus will be clearly comprehended by imagining that if the hook upon the rope Ω, be pulled down to 5, the sheave F with the column E will turn at the same time, putting in motion the rack P with the rod B, which will bring the levers M M into the position represented by the dotted lines: the racks P will have turned the pinions Q till the sliders S and T, with the studs or levers, or racks, (according to whichever method may be used,) bring the valves into the position of the dotted lines, in which position they are represented as having all their surfaces to the wind; therefore, if a sufficient weight be hung to the hook 4, the weight will descend to 5, and keep the valves in the situation of the dotted lines; and supposing the wind to blow upon them with too much force in this state, they will turn on their gudgeons, and raise the weights, so that the superfluous wind will pass through or between them, without exerting an irregular force upon the vanes, so as to produce an unequal velocity."

MILL WITH EIGHT QUADRANGULAR SAILS.

This mill, which is the invention of Mr. James Verrier, is represented in fig. 147.

A A A are the three principal posts, 27 feet 7½ inches long, 22 inches broad at their lower extremities, 18 inches at their upper, and 17 inches thick. The column B is 12 feet 2½ inches long, 19 inches in diameter at its lower extremity, and 16 inches at its upper: it is fixed in the centre of the mill, passes through the first floor E, having its upper extremity secured by the bars G G. E E E are the girders of the first floor, one of which only is seen, being eight feet three inches long, 11 inches broad, and nine thick; they are mortised into the posts A A A and the column B, and are about eight feet three inches distance from the ground floor. D D D are three posts, six feet four inches long, nine inches broad, and six inches thick; they are mortised into the girders E F of the first and second floor, at the distance of two feet four inches from the posts A, &c. F F F are the girders of the second floor, six feet long, 11 inches broad, and nine thick: they are mortised into the posts A, &c., and rest upon the upper extremities of the posts D, &c. The three bars G G G are 3 feet 1½ inches long, seven inches broad, and three thick: they are mortised into the posts D and the upper end of the column B, four feet three inches above the floor. P is one of the beams which support the extremities of the bray-trees or levers; its length is two feet four inches, its breadth eight inches, and its thickness six inches. I is one of the bray-trees into which the extremity of one of the bridge-trees K is mortised. Each bray-tree is 4 feet 9½ inches long, nine inches broad, and seven thick, and each bridge-tree is four feet six inches long, nine inches broad, and seven thick; being furnished with a piece of brass on its upper surface to receive the under pivot of the spindle. L L are two iron screw-bolts, which raise or depress the

of Mr. Smeaton, from which it appears that sails weathered
in the Dutch manner produced nearly a maximum effect, but
also from the observations of the celebrated Coulomb. This
philosopher examined above 50 windmills in the neighbourhood of Lisle, and found that each of them performed nearly
the same quantity of work when the wind moved with the
velocity of 18 or 20 feet per second, though there were some
trifling differences in the inclination of their windshafts, and
in the disposition of their sails. From this fact, Coulomb
justly concluded that the parts of the machine must have
been so disposed as to produce nearly a maximum effect.

In the windmills on which Coulomb's experiments were
made, the distance from the extremity of each sail to the
centre of the windshaft or principal axis was 33 feet. The
sails were rectangular, and their width was a little more than
six feet, five of which were formed with cloth stretched upon
a frame, and the remaining foot consisted of a very light board.
The line which joined the board and the cloth formed, on the
side which faced the wind, an angle sensibly concave at the
commencement of the sail, which diminished gradually till it
vanished at its extremity. Though the surface of the cloth was
curved, it may be regarded as composed of right lines perpendicular to the arm or whip which carries the frame, the extremities of these lines corresponding with the concave angle
formed by the junction of the cloth and the board. Upon this
supposition, these right lines at the commencement of the sail,
which was distant about six feet from the centre of the
windshaft, formed an angle of 60 degrees with the axis or
windshaft, and the lines at the extremity of the wing formed
an angle increasing from 78 to 84 degrees, according as the
inclination of the axis of rotation to the horizon increased
from 8 to 15 degrees; or in the mill-wright's terms, the
greatest angle of weather was 30 degrees, and the least
varied from 12 to 6 degrees, as the inclination of the windshaft
varied from 8 to 15 degrees. A pretty distinct idea of the
surface of windmill sails may be conveyed by conceiving a
number of triangles standing perpendicular to the horizon,
in which the angle contained between the hypothenuse and
the base is constantly diminishing; the hypothenuse of each
triangle will then be in the superficies of the vane, and they
would form that superficies if their number were infinite.

ON HORIZONTAL WINDMILLS.

A VARIETY of opinions have been entertained respecting the relative advantages of horizontal and vertical windmills. Mr. Smeaton gives a decided preference to the latter; but, when he asserts that horizontal windmills have only one-eighth or one-tenth of the power of vertical ones, he certainly forms too low an estimate of their power. Mr. Beatson, on the contrary, who has a patent for the construction of a new horizontal windmill, seems to be prejudiced in their favour. From an impartial investigation, it will probably appear, that the truth lies between these two opposite opinions; but before entering on this discussion, we must first consider the nature and form of horizontal windmills; which we shall do by presenting the reader with a description of the horizontal mill erected at Margate by Captain Hooper.

Fig. 149 is an upright section, and fig. 150 a plan of the building, H H are the side walls of an octagonal building which contains the machinery. These walls are surmounted by a strong timber framing G G, of the same form as the building, and connected at top by cross framing to support the roof, and also the upper pivot of the main vertical shaft A A, which has three sets of arms, B B, C C, D D, framed upon it at that part which rises above the height of the walls. The arms are strengthened and supported by diagonal braces, and their extremities are bolted to octagonal wood frames, round which the vanes or floats E E are fixed, as seen in outline in fig. 150, so as to form a large wheel, resembling a water-wheel, which is less than the size of the house by about six inches all round This space is occupied by a number of vertical boards or blinds F F, turning on pivots at top and bottom, and placed obliquely, so as to overlap each other, and completely shut out the wind, and stop the mill, by forming a close case surrounding the wheel; but they can be turned altogether upon their pivots to allow the wind to blow in the direction of a tangent upon the vanes on one side of the wheel, at the time the other side is completely shaded or defended by the boarding. The position of the blinds is clearly shown at F F, fig. 150. At the lower end of the vertical shaft A A, a large spur-wheel a a is fixed, which gives motion to a pinion c, upon a small vertical axis d, whose upper pivot turns in a brass box bolted to a girder of the floor n. Above the pinion c, a spur-wheel e is placed, to give motion to two small pinions f, on the upper ends of the spindles g, of the mill-stone h. Another pinion is situate at the opposite side of the great spur-wheel a a, to give motion to a third pair of mill-stones, which are used when the wind is very strong; and then the wheel turns so quick as not to need the extra wheel e to give the requisite velocity to the stones. The weight of the main vertical shaft is borne by a strong timber b, having a brass box placed on it to receive the lower pivot of the shaft. It is supported at its ends by cross-beams mortised into the upright shaft b b, as shown in the plan, fig. 150. A floor or roof I I is thrown across the top of the brick-building to protect the machinery from the weather, and to prevent the rain blowing down the opening through which the shaft descends, A small circular hoop K is fixed to the floor, and is surrounded by another hoop or rim L, which is fixed to the arms D D of the wheel. This last is of such a

sails was quadruple the power of one horizontal sail, the dimensions of each being the same. Taking this circumstance into the account, we cannot be far wrong in saying that, in theory at least, if not in practice, the power of a horizontal windmill is about one-third or one-fourth of the power of a vertical one, when the quantity of surface and the form of the sails are the same, and when all the parts of the horizontal sails have the same distance from the axis of motion as the corresponding parts of the vertical sails. But if the horizontal sails have the position A I, E G, in fig. 151, instead of the position C A *d m*, C D *o n*, their effect will be greatly increased, though the quantity of surface is the same; because the part C P 3 *m* being transferred to B I 3 *d*, has much more power to turn the sails. Having this method, therefore, of increasing the effect of horizontal sails, which cannot be applied to vertical ones, we would encourage every attempt to improve their construction, as not only laudable in itself, but calculated to be of essential utility in a commercial country.—See Dr. Brewster's valuable Appendix to Ferguson's *Lectures*.

FLOUR-MILLS.

In fig. 152 we have given a section of a double flour-mill, reduced from Gray's Experienced Mill-wright, with the following account :

A A, the water-wheel. B B, its shaft or axle. C C, a wheel fixed upon the same shaft, containing 90 teeth or cogs, to drive the pinion No. 1, having 23 teeth, which is fastened upon the vertical shaft D. No. 2, a wheel fixed upon the shaft D, containing 82 teeth, to turn the two pinions F F, having 15 teeth, which are fastened upon the iron axles or spindles that carry the two upper mill-stones. E E, the beam or sill that supports the frame on which the under mill-stones are laid. G G, the cases or boxes that enclose the upper mill-stones; they should be about two inches distant from the stone all round its circumference. T T, the bearers, called bridges, upon which the under end of the iron spindles turn. These spindles pass upwards through a hole in the middle of the nether mill-stones, in which is fixed a wooden bush that their upper ends turn in. The top part of the spindles, above each wooden bush, is made square, and goes into a square hole in an iron cross, which is admitted into grooves in the middle and under surface of the upper mill-stone. By this means that stone is carried round along with the trundles F F, when turned by the wheel No. 2. One end of the bridges T T is put into mortises in fixed bearers; and the other end into mortises in the bearers that move at one end on iron bolts, their other ends hanging by iron rods having screwed nuts, as U U; so that when turned forward or backward they raise or depress the upper mill-stones, according as the miller finds it necessary. S S, the feeders, in the under end of each of which is a square socket that goes upon the square of the spindles above the iron cross or rind, and having three or four branches that move the spout or shoe, and feed the wheat constantly from the hopper into the hole or eye of the upper mill-stone, where it is introduced between the stones, and by the circular motion of the upper stone acquires a centrifugal force, and proceeding gradually from the eye of the mill-stone towards the circumference, is at length thrown out in flour or meal. B B the sluice, machine, and handle, to raise the sluice, and let the water on the wheel A to drive it round. No. 3 is a wheel fixed upon the shaft D, containing 44 teeth, to turn the pinion No. 4, having 15 teeth, which is fixed upon the horizontal axle H. On this axle is also fixed the wheel E, on which go the two leather belts that turn the wire engine and dusting mill. L, an iron spindle, in the under end of which is a square socket that takes in a square on the top of the gudgeon of the vertical shaft D. There is a pinion M, of nine teeth, fixed on the upper end of the spindle L, to turn the wheel M M, having 48 teeth, which is fastened on the axle round which the rope Z Z rolls, to carry the sacks of flour up to the cooling benches. By pulling the cord O O a little, the wheel M M and its axle are put into motion, in consequence of that wheel and its axle being moved horizontally, until the teeth of the wheel are brought into contact with those of the pinion at the top of the spindle L: and, on the contrary, by pulling the cord P P, the wheel M and its axle are moved in the opposite horizontal direction, till they are thrown out of geer with the pinion, and the rotatory motion of that wheel stops. But when the sack of flour is raised up to the lever Q, it pushes up that end of the lever, and of course the other end down; by which means the pinion M is disengaged, and thus that part of the machine stops of itself. N N are two large hoppers, into which the clean wheat is put to be conveyed down to the hoppers S S, placed on the frame immediately above the mill-stones. W W the side wall of the mill-house. V, the couples or frame of the roof. Windows to lighten the house.

1 3 represents the surface of the under grinding mill-stone; the way laid out the roads or channels; the wooden bush fixed into the hole in the middle, in which the upper end of the iron spindle turns round; and the rim or hoops that surround the upper one, which ought to be two inches above the stone all round its circumference.

1 4, the upper grinding mill-stone, and iron cross or rind in its place, in the centre of which is a square hole that takes in a square on the iron spindle, to carry round the mill-stone. When the working faces of the mill-stones are laid uppermost, the roads (or channels) go in the same direction in both; so that when the upper stone is turned over, and its surface laid upon the under one, then the channels may cross each other, which assists in grinding and throwing out the flour, the edges of the two furrows then cutting against each other like scissars. They are likewise laid out according to the way the upper stone turns. In those represented in the figures, the running mill-stone is to turn "sunway," or as in what is called a right-handed mill: if the stone revolves the other way, the channels must be cut the reverse way, and then the mill is termed a left-handed one.

The mill-stones are of the utmost importance in the construction of flour-mills, as upon them principally depends the quality of the meal; we cannot, therefore, do better than to give Mr. Ferguson's opinion upon them, as also some subsequent remarks by his editor, Dr. Brewster.

The quantity of power required to turn a heavy mill-stone is but very little more than what is sufficient to turn a light one: for as it is supported upon the spindle by the bridge-tree, and the end of the spindle that turns in the brass foot therein being but small, the odds arising from the weight is but very inconsiderable in its action against the power or force of the water; and, besides, a heavy stone has the same advantage as a heavy fly, namely, that it regulates the motion much better than a light one.

In order to cut and grind the corn, both the upper and under mill-stones have channels or furrows cut into them, proceeding obliquely from the centre towards the circumference: and these furrows are cut perpendicularly on one side, and obliquely on the other, into the stone, which gives each furrow a sharp edge, and in the two stones they come, as it were, against one another like the edges of a pair of scissars, and so cut the corn, to make it grind the easier when it falls upon the places between the furrows. These are cut the same way in both stones when they lie upon their backs, which makes them run cross ways to each other when the upper stone is inverted, by turning its furrowed surface toward that of the lower. For, if the furrows of both stones lay the same way, a great deal of the corn would be driven onward in the lower furrows, and so come out from between the stones without being either cut or bruised.

When the furrows become blunt and shallow by wearing, the running stone must be taken up, and both stones new dressed with a chisel and hammer; and every time the stone is taken up, there must be some tallow put round the spindle upon the bush, which will soon be melted by the heat the spindle acquires from its turning and rubbing against the bush, and so will get in between them, otherwise the bush would take fire in a very little time.

The bush must embrace the spindle quite close, to prevent

any shake in the motion, which would make some parts of the stones grate and fire against each other, while other parts of them would be too far asunder, and by that means spoil the meal in grinding.

Whenever the spindle wears the bush so as to begin to shake in it, the stone must be taken up, and a chisel drove into several parts of the bush; and when it is taken out, wooden wedges must be driven into the holes; by which means the bush will be made to embrace the spindle close all round it again. In doing this, great care must be taken to drive equal wedges into the bush on opposite sides of the spindle, otherwise it will be thrown out of the perpendicular, and so hinder the upper stone from being set parallel to the under one, which is absolutely necessary for making good work. When any accident of this kind happens, the perpendicular position of the spindle must be restored by adjusting the bridge-tree by proper wedges put between it and the brayer.

It often happens that the rind is a little wrenched in laying down the upper stone upon it, or is made to sink a little lower upon one side of the spindle than upon the other; and this will cause one edge of the upper stone to drag all round upon the other, while the opposite edge will not touch. But this is easily set to rights, by raising the stone a little with a lever, and putting bits of paper, cards, or thin chips, between the rind and the stone.

The diameter of the upper stone is generally about six feet, the lower stone about an inch more; and the upper stone, when new, contains about $22\frac{1}{2}$ cubic feet, which weighs somewhat more than 19,000 pounds. A stone of this diameter ought never to go more than 60 times round in a minute, for if it turns faster it will heat the meal.

The grinding surface of the under stone is a little convex from the edge to the centre, and that of the upper stone a little more concave: so that they are farthest from one another in the middle, and come gradually nearer towards the edges. By this means, the corn at its first entrance between the stones is only bruised; but as it goes farther on towards the circumference or edge, it is cut smaller and smaller; but at last finely ground just before it comes out from between them.*

---

* The upper mill-stone, when six feet in diameter, is generally hollowed about one inch at the centre; and the under one rises about three-fourths of an inch. The corn that falls from the hopper insinuates itself between them as far as two-thirds of the radius where the grinding begins; the

When the furrows of mill-stones are worn shallow, and consequently new dressed with the chisel, the same quantity of stone must be taken from every part of the grinding surface, that it may have the same convexity or concavity as before. As the upper mill-stone should always have the same weight when its velocity remains unchanged, it will be necessary to add to it as much weight as it lost in the dressing. This will be most conveniently done by covering its top with a layer of plaster, of the same diameter as the layer of stone taken from its grinding surface, and as much thicker than the layer of stone, as the specific gravity of the stone exceeds the specific gravity of the plaster. That the reader may have some idea of the manner in which the furrows, or channels, are arranged, we have represented, fig. 154, the grinding surface of the upper mill-stone, upon the supposition that it moves from east to west, or for what is called a right-handed mill. When the mill-stone moves in the opposite direction, the position of the furrows must be reversed.

In fig. 156, we have a section of the mill-stone, spindle, and lantern. The under mill-stone M P H G, which never moves, may be of any thickness. Its grinding surface must be of a conical form, the point b being about an inch above the horizontal line P R, and M a and P b being straight lines. The upper mill-stone E F P M, which is fixed to the spindle C D, at C, and is carried round with it, should be so hollowed that the angle O M a, formed by the grinding surfaces, may be of such a size that O n, being taken equal to n M, $n s$ may be equal to the thickness of a grain of corn. The diameter O N of the mill-eye m C, should be between 8 and 14 inches; and the weight of the upper mill-stone E P, joined to the weight of the spindle C D, and the trundle x, (the sum of which three numbers is called the *equipage* of the turning mill-stone,) should never be less than 1550 pounds avoirdupois, otherwise the resistance of the grain would bear up the mill-stone, and the meal be ground too coarse.

In order to find the weight of the equipage; divide the third of the radius of the gudgeon by the radius of the water-wheel which it supports, and having taken the quotient from 2,25 multiply the remainder by the expense of the source, by the relative fall, and by the number 19,911, and you will have a first quantity, which may be regarded as pounds. Multiply the square root of the relative fall by the weight of the arbor of the water-wheel, by the radius of its gudgeon, and by the number 1617, and a second quantity will be had, which will also represent pounds. Divide the third part of the radius of the gudgeon by the radius of the water-wheel, and having augmented the quotient by unity, multiply the sum by 1005, and a third quantity will be obtained. Subtract the second quantity from the first, divide the remainder by the third, and the quotient will express the number of pounds in the equipage of the mill-stone.

The weight of the equipage being thus found, extract its

distance between the stones being there about two-thirds or three-fourths of the thickness of a grain of corn. This distance, however, can be altered at pleasure, by raising or sinking the upper stone.—*Dr. Brewster.*

square root, expressed in pounds, and multiply it by 039, and the product will be the radius of the mill-stone in feet.

In order to find the weight and thickness of the upper mill-stone, the following rules must be observed:

1. To find the weight of a quantity of stone equal to the mill-eye; take any quantity which seems most proper for the weight of the spindle CD, and the lantern X, and subtract this quantity from the weight of the mill-stone's equipage, for a first quantity. Find the area of the mill-eye, and multiply it by the weight of a cubic foot of stone of the same kind as the mill-stone, and a second quantity will be had. Multiply the area of the mill-stone by the weight of a cubic foot of the same stone, for a third quantity. Multiply the first quantity by the second, and divide the product by the third, and the quotient will be the weight required.

2. To find the number of cubic feet in the turning mill-stone, supposing it to have no eye; from the weight of the spindle and lantern subtract the quantity found by the preceding rule, for the first number. Subtract this last number from the weight of the equipage, and a second number will be obtained. Divide this second quantity by the weight of a cubic foot of stone of the same quality as the mill-stone, and the quotient will be the number of cubic feet in E M P F, m C being supposed to be filled up.

3. To find the quantities m N and E M, i. e. the thickness of the mill-stone at its centre and circumference; divide the solid content of the mill-stone, as found by the preceding rule, by its area, and you will have the first quantity. Add b R, which is generally about an inch, to twice the diameter of a grain of corn, for a second quantity. Add the first quantity to one-third of the second, and the sum will be the thickness of the mill-stone at the circumference. Subtract the third of the second quantity from the first quantity, and the remainder will be its thickness at the centre.

The size of the mill-stone being thus found, its velocity is next to be determined. M. Fabre observes, that the flour is the best possible when a mill-stone five feet in diameter makes from 48 to 61 revolutions in a minute. Mr. Ferguson allows 60 turns to a mill-stone six feet in diameter; and Mr. Imison 120 to a mill-stone $4\frac{1}{2}$ feet in diameter. In mills upon Mr. Imison's construction, the great heat that must be generated by such a rapid motion of the mill-stone, must render the meal of a very inferior quality: much time, on the contrary, will be lost, when such a slow motion is employed as is recommended by M. Fabre and Mr. Ferguson. In the best corn-mills in this country, a mill-stone five feet in diameter revolves, at an average, 90 times in a minute. The number of revolutions in a minute, therefore, which must be assigned to mill-stones of a different size, may be found by dividing 450 by the diameter of the mill-stone in feet.

The spindle c D, which is commonly six feet long, may be made either of iron or wood. When it is of iron, and the weight of the mill-stone 7558 pounds avoirdupois, it is generally three inches in diameter; and when made of wood, it is 10 or 11 inches in diameter. For mill-stones of a different weight, the thickness of the spindle may be found by proportioning it to the square root of the mill-stone's weight, or which is nearly the same thing, to the weight of the mill-stone's equipage.

The greatest diameter of the pivot D, upon which the mill-stone rests, should be proportional to the square root of the equipage, a pivot half an inch diameter being able to support an equipage of 5398 pounds. In most machines, the diameter of the pivots is by far too large, being capable of supporting a much greater weight than they are obliged to bear. The friction is therefore increased, and the performance of the machine diminished.

The bridge-tree, A B, is generally from 8 to 10 feet long, and should always be elastic, that it may yield to the oscillatory motion of the mill-stone. When its length is nine feet, and the weight of the equipage 5182 pounds, it should be six inches square; and when the length remains unchanged, and the equipage varies, the thickness of the bridge-tree should be proportional to the square root of the equipage.

Although the mechanism of a flour-mill is exceedingly simple, the profitable manufacture of flour requires considerable experience and attention. We shall therefore give a sketch of the points most particularly to be attended to in such manufacture.

The wheats that grow in Essex and Kent make the best flour. In choosing the wheat much attention should be paid to the thinness of the skin and to its cleanliness from weed. Good wheat may be known by its weight, which should be about 62 pounds per Winchester bushel of 32 quarts. The wheat to be manufactured into the best flour should be winnowed.

The miller judges of the quality of the flour by feeling it, and accordingly as it is too fine or too coarse regulates the upper mill-stone, or increases or decreases the supply of grain. The flour in grinding always acquires a slight warmth, and care must be taken that the warmth does not increase, or the flour will be permanently injured.

The dressing of the flour is of great importance, and too much attention cannot be paid to it. The bran should be in large flakes and free of flour. In grinding the best wheats, in the best manner, the bran will amount to about seven pounds per bushel.

In the process of dressing, the bran is examined as a criterion to know whether too much flour be admitted upon the machine. Care should be taken to have the brush screwed close to the end of this machine.

French stones of about four feet in diameter are expected to grind about five bushels per hour.

Mr. Thomas Fenwick, the author of Four Essays on Practical Mechanics, has made numerous experiments on some of the best mills for grinding corn, in order to form, by practical observation, a set of tables illustrative of the effect of a given quantity of water, in a given time, applied to an overshot-wheel of a given size.

The quantity of water expended on the wheel was measured with great exactness: the corn used was in a medium state of dryness; the mills, in all their parts, were in a medium working state; the mill-stones, making from 90 to 100 revolutions per minute, were from 4½ to 5 feet in diameter.

The result of the experiments was, that the power requisite to raise a weight of 300 pounds avoirdupois, with a velocity of 190 feet per minute, would grind one boll of good rye in one hour; but for the sake of making the following tables hold in practice, where imperfection of construction exists in some small degree, he took it at 300 pounds raised with a velocity of 210 feet per minute, (being $\frac{1}{10}$th more,) and for grinding two, three, four, or five bolls per hour, requires a power equal to that which could raise 300 pounds with the velocity of 350, 506, 677, or 865 feet per minute respectively. The difference of the power requisite to grind equal quantities of wheat to that for rye will be very trifling.

To enable the young mechanist to understand the application of his principles, he adds, as an illustration, that that number of horses, or other applied power, which, by means of a rope (considered as without weight) passing over a single pulley placed over the mouth of a pit or well, can raise out of it a load of 300 pounds avoirdupois, at the rate of 210 feet per minute, will be sufficient to grind one boll of corn per hour; and that a power which, in similar circumstances, can raise the same weight of 300 pounds with a velocity of 350 feet per minute, will be able to grind two bolls of corn per hour, and so on.

Having made some experiments to ascertain the friction of a mill, when going with velocity sufficient to grind two bolls of corn per hour, he relates the manner in which he made them, that the reader may be able to judge of the accuracy of his deductions.

The mill was made quite clear of corn, and the upper mill-stone raised so that it would touch as little as possible on the under stone in its revolutions; then such a quantity of water was admitted to flow on the water-wheel, as to give the mill, when empty, the same velocity it had when grinding corn at the rate of two bolls per hour, which quantity of water was sufficient to raise a load of 300 pounds with a velocity of 100 feet per minute, which was therefore considered by him as the measure of the friction. Now as the power requisite to grind two bolls of corn per hour, including the friction of the mill, is equal to that which can raise a weight of 300 pounds with a velocity of 350 feet per minute, and the friction of the moving parts of the mill is equal to a power which would raise 300 pounds with a velocity of 100 feet per minute; therefore the difference of the two, which is 300 pounds raised with the velocity of 250 feet per minute, is equal to the power employed in the actual grinding of the corn, which is about two-thirds of the whole.

The power equal to raise a weight of 300 pounds avoirdupois, with a velocity of 390 feet per minute, will prepare properly one ton of old rope per week, for the purpose of making paper: and, for preparing in like manner, two tons of the same kind of materials per week, requires a power equal to raise 300 pounds with a velocity of 525 feet per minute, the mill working from 10 to 12 hours per day.

Tables, showing the quantity of water (ale measure) requisite to grind different quantities of corn, from one to five bolls (Winchester measure) per hour, applied on overshot water-wheels from 10 to 32 feet diameter; also the size of the cylinder of the common steam-engine to do the same work.

The water-wheel, 10 feet diameter.			The water-wheel, 11 feet diameter.		
Bolls of corn ground per hour.	Quantity of water requisite, in ale gallons, per minute.	Diameter of the cylinder of a steam-engine to do the same work, in inches.	Bolls of corn ground per hour.	Quantity of water requisite, in ale gallons, per minute.	Diameter of the cylinder of a steam-engine to do the same work, in inches.
1	786	12·5	1	705	12·5
1¼	1056	14·6	1¼	945	14·6
2	1341	16·75	2	1188	16·75
2½	1617	18·5	2½	1454	18·5
3	1894	20·2	3	1723	20·2
3½	2220	21·75	3½	2014	21·75
4	2541	23·25	4	2306	23·25
4½	2891	24·75	4½	2626	24·75
5	3242	26·25	5	2944	26·25

The water-wheel, 12 feet diameter.			The water-wheel, 14 feet diameter.		
Bolls of corn ground per hour.	Quantity of water requisite, in ale gallons, per minute	Diameter of the cylinder of a steam-engine to do the same work, in inches.	Bolls of corn ground per hour.	Quantity of water requisite, in ale gallons, per minute.	Diameter of the cylinder of a steam-engine to do the same work, in inches.
1	655	12·5	1	564	12·5
1½	873	14·6	1½	740	14·6
2	1091	16·75	2	927	16·75
2½	1343	18·5	2½	1140	18·5
3	1576	20·2	3	1353	20·2
3½	1840	21·75	3½	1583	21·75
4	2117	23·25	4	1811	23·25
4½	2408	24·75	4½	2060	24·75
5	2700	26·25	5	2306	26·25

The water-wheel, 13 feet diameter			The water-wheel, 15 feet diameter.		
Bolls, per hour.	Water, gallons per minute.	Cylinder, in inches.	Bolls, per hour.	Water, gallons per minute.	Cylinder, in inches.
1	606	12·5	1	535	12·5
1½	806	14·6	1½	710	14·6
2	1009	16·75	2	894	16·75
2½	1234	18·5	2½	1090	18·5
3	1458	20·2	3	1290	20·2
3½	1705	21·75	3½	1503	21·75
4	1952	23·25	4	1717	23·25
4½	2223	24·75	4½	1967	24·75
5	2494	26·25	5	2211	26·25

The water-wheel, 16 feet diameter.			The water-wheel, 18 diameter.		
Bolls of corn ground per hour.	Quantity of water requisite, in ale gallons, per minute.	Diameter of the cylinder of a steam-engine to do the same work, in inches.	Bolls of corn ground per hour.	Quantity of water requisite, in ale gallons, per minute.	Diam the cyl a stean to do t work, i
1	491	12·5	1	410	1:
1½	650	14·6	1½	595	14
2	811	16·75	2	730	1(
2½	993	18·5	2½	860	18
3	1176	20·2	3	1054	2(
3½	1380	21·75	3½	1227	21
4	1582	23·25	4	1400	2:
4½	1802	24·75	4½	1600	2:
5	2023	26·25	5	1800	2(

The water-wheel, 17 feet diameter.			The water-wheel, 19 diameter.		
Bolls, per hour.	Water, gallons per minute.	Cylinder, in inches.	Bolls, per hour.	Water, gallons per minute.	Cylin inc
1	458	12·5	1	411	1:
1½	628	14·6	1½	550	14
2	770	16·75	2	690	1(
2½	943	18·5	2½	845	18
3	1117	20·2	3	1000	2(
3½	1300	21·75	3½	1165	21
4	1482	23·25	4	1330	2:
4½	1695	24·75	4½	1517	2:
5	1906	26·25	5	1707	2(

The water-wheel, 20 feet diameter.

Bolls of corn ground per hour.	Quantity of water requisite, in ale gallons, per minute.	Diameter of the cylinder of a steam-engine to do the same work, in inches.
1	392	12·5
1½	530	14·6
2	675	16·75
2¼	808	18·5
3	945	20·2
3½	1110	21·75
4	1270	23·25
4½	1445	24·75
5	1623	26·25

The water-wheel, 22 feet diameter.

Bolls of corn ground per hour.	Quantity of water requisite, in ale gallons, per minute.	Diameter of the cylinder of a steam-engine to do the same work, in inches.
1	350	12·5
1½	473	14·6
2	594	16·75
2½	722	18·5
3	860	20·2
3½	1007	21·75
4	1153	23·25
4½	1313	24·75
5	1472	26·25

The water-wheel, 21 feet diameter.

Bolls, per hour.	Water, gallons per minute.	Cylinder in inches.
1	370	12·5
1½	500	14·6
2	635	16·75
2¼	767	18·5
3	900	20·2
3½	1060	21·75
4	1212	23·25
4½	1379	24·75
5	1547	26·25

The water-wheel, 23 feet diameter.

Bolls, per hour.	Water, gallons per minute.	Cylinder, in inches.
1	338	12·5
1½	454	14·6
2	570	16·75
2½	707	18·5
3	824	20·2
3½	964	21·75
4	1124	23·25
4½	1258	24·75
5	1412	26·25

The water-wheel, 24 feet diameter.			The water-wheel, 26 feet diameter.		
Bolls of corn ground per hour.	Quantity of water requisite, in ale gallons, per minute.	Diameter of the cylinder of a steam-engine to do the same work, in inches.	Bolls of corn ground per hour.	Quantity of water requisite, in ale gallons, per minute.	Diameter of the cylinder of a steam-engine to do the same work, in inches.
1	327	12·5	1	303	12·5
1½	436	14·6	1½	403	14·6
2	545	16·75	2	504	16·75
2¼	671	18·5	2¼	617	18·5
3	788	20·2	3	730	20·2
3½	920	21 ·75	3½	852	21·75
4	1050	23·25	4	975	23·25
4½	1204	24·75	4½	1111	24·75
5	1350	26.25	5	1247	26·25

The water-wheel, 25 feet diameter.			The water-wheel, 27 feet diameter.		
Bolls, per hour.	Water, gallons per minute.	Cylinder, in inches.	Bolls, per hour.	Water, gallons per minute.	Cylinder, in inches.
1	316	12·5	1	293	12·5
1¼	418	14·6	1½	385	14·6
2	520	16·75	2	482	16·75
2½	635	18·5	2¼	593	18·5
3	752	20·2	3	703	20·2
3½	876	21·75	3½	822	21·75
4	935	23·25	4	940	23·25
4¼	1150	24·75	4½	1070	24·75
5	1300	26·25	5	1200	26·25

The water-wheel, 20 feet diameter.			The water-wheel, 22 feet diameter.		
Bolls of corn ground per hour.	Quantity of water requisite, in ale gallons, per minute.	Diameter of the cylinder of a steam-engine to do the same work, in inches.	Bolls of corn ground per hour.	Quantity of water requisite, in ale gallons, per minute.	Diameter of the cylinder of a steam-engine to do the same work, in inches.
1	392	12·5	1	350	12·5
1½	530	14·6	1½	473	14·6
2	675	16·75	2	594	16·75
2½	808	18·5	2½	722	18·5
3	945	20·2	3	860	20·2
3½	1110	21·75	3½	1007	21·75
4	1270	23·25	4	1153	23·25
4½	1445	24·75	4½	1313	24·75
5	1623	26·25	5	1472	26·25

The water-wheel, 21 feet diameter.			The water-wheel, 23 feet diameter.		
Bolls, per hour.	Water, gallons per minute.	Cylinder in inches.	Bolls, per hour.	Water, gallons per minute.	Cylinder, in inches.
1	370	12·5	1	338	12·5
1½	500	14·6	1½	454	14·6
2	635	16·75	2	570	16·75
2½	767	18·5	2½	707	18·5
3	900	20·2	3	824	20·2
3½	1060	21·75	3½	964	21·75
4	1212	23·25	4	1124	23·25
4½	1379	24·75	4½	1258	24·75
5	1547	26·25	5	1412	26·25

The water-wheel, 32 feet diameter.		
Bolls of corn ground per hour.	Quantity of water requisite, in ale gallons, per minute.	Diameter of the cylinder of a steam-engine to do the same work, in inches.
1	245	12·5
1½	325	14·6
2	406	16·75
2½	496	18·5
3	588	20·2
3½	690	21·75
4	791	23·25
4½	900	24·75
5	1012	26·25

To make the foregoing tables applicable to mills intended to be turned by undershot or breast water-wheels: from Smeaton's experiments it appears that the power required on an undershot, water-wheel, to produce an effect equal to that of an overshot (to which the tables are applicable,) is as 2·4 to one; and also the power required on a breast water-wheel, which receives the water on some point of its circumference, and afterwards descends on the ladle boards, to produce an equal effect with an overshot water-wheel, is as 1·75 to 1.

le, showing the necessary size of the cylinder of the
ton steam-engine to grind different quantities of
, from 1 to 12 bolls (4 to 48 bushels Winchester
ure) per hour.

Bolls of corn ground per hour.	Diameter of the cylinder of a steam-engine to do the same work, in inches.
1	12·5
1½	14·6
2	16·75
2¼	18·5
3	20·2
3½	21·75
4	23·25
4½	24·75
5	26·25
5½	27·25
6	28·1
6½	29·
7	29·8
7½	31·1
8	32·
8½	33·3
9	34·2
9½	35·2
10	36·
10½	37·3
11	38·
11½	38·85
12	39·5

. B. This table will be applicable to any improved
m-engine, as well as that of the common kind, if the ratio
teir efficacies be known.

APPLICATION OF THE TABLES.

AMPLE I.—*If a stream of water, producing 808 gallons, ale measure,*
inute, can be applied to an overshot water-wheel 20 feet diameter,
quantity of corn will it be able to grind per hour?

Look in the tables under a 20 feet water-wheel, and opposite 808 will be found 2⅜ bolls of corn ground per hour.

EXAMPLE II.—*If a stream of water producing 808 gallons, ale a per minute, can be applied to an undershot water-wheel 20 feet di what quantity of corn can it grind per hour ?*

It is found by the tables, that, if applied on an overshot wate 20 feet diameter, the stream will grind 2⅜ bolls per hour; an page 156, the power required by the undershot to that of the o water-wheel, to produce an equal effect, is as 2·4 to 1; there 2·4 : 1 : 2·5 : 1·04 bolls of corn ground per hour by means of the str

EXAMPLE III.—*If a stream of water, producing 808 gallons, ale a per minute, can be applied on a breast water-wheel 20 feet diameter quantity of corn can it grind per hour ?*

It is found by the tables, that, if applied to an overshot water-w equal size, 2⅜ bolls of corn will be ground per hour; and from pa the power of a breast water-wheel to that of an overshot water-wh produce an equal effect, is as 1·75 to 1; therefore, as 1·75 : 1 : : 2·5 bolls of corn ground per hour by the stream.

EXAMPLE IV.—*Of what diameter must the cylinder of a common engine be made, to grind 10 bolls of corn per hour ?*

By looking on the table, page 157, opposite 10 bolls ground pe the diameter of the steam cylinder will be found to be 36 inches.

FAMILY MILL AND BOLTER.

As a family mill and bolter cannot but be highly use many situations, we shall give a description of one or beginning with that invented by Mr. T. Rustall, of Purb heath, near Portsmouth, who received a premium of guineas from the Society of Arts for his invention.

In fig. 157, A is the handle of the mill; B one of the mill which is about 30 inches in diameter, and five inches in thickness, r with its axis C; D is the other mill-stone, which, when in use, is stati but which may be placed near to or at a distance from the movable B, by means of three screws passing through the wooden block I supports one end of the axis C, after it has been put through a l perforation in the-bed stone. The grain likewise passes throug perforation, from the hopper F, into the mill. F represents the b which is agitated by two iron pins on the axis C, that alternately ra vessel containing the grain, which again sinks by its own weigh consequence of this motion the corn is conveyed through a spot passes from such hopper into the centre of the mill behind, and tl the bed-stone D. G, a paddle, regulating the quantity of corn delivered to the mill, and by raising or lowering which, a larger or s proportion of grain may be furnished; H, the receptacle for the flou which it falls from the mill-stones, when ground; I represents one wooden supporters on which the bed-stone, D, rests. These are so to the block E, and likewise mortised into the lower frame-work mill at K, which is connected by means of the pins or wedges, L that admit the whole mill to be easily taken to pieces; M, a fly-placed at the farthest extremity of the axis C, and on which another l may be occasionally fixed; N, a small rail, serving to keep the hoppe place, the furthest part of such hopper resting on a small pin, admits of sufficient motion for that vessel to shake forward the cor

es in breadth, and 18 inches in depth. A is a movable partition, boat four feet backwards or forwards from the centre of the box, o wooden ribs, which are fixed to the back and front of the box, of which is delineated at the letter B; C, the lid of the bolter, ted open; D, a slider, which is movable in a groove made in the leans of two handles in the back of such lid; E, a forked iron, fixed der D, and which, when the lid is shut, takes hold of the sieve F, res it backwards and forwards on the wooden ribs B, according to tion of the slider; G represents a fixed partition in the lower centre ox, which it divides into two parts, in order to separate the fine ε coarse flour; from this partition the slider A moves each way inches, and thus affords room for working the sieve; H, a board arallel to the bottom of the bolter, and forms part of the slider A; d serves to prevent any of the sifted matter from falling into the rtition; I represents two of the back feet which support the

59 is a view of the top or upper part of the lid of the bolter; R that moves the lengthwise of the bolter; L L the handles by he slider is worked; M, a screw, serving to hold the fork, which motion to the sieve.
60 represents the forked iron, E, separately from the lid.

the mill and bolter may be constructed at a moderate e, and they occupy only a small space of ground. rmer may even be worked in a public kitchen, or a room in a farm-house, without occasioning any ncumbrance.
particular excellence of this mill consists in this cir- nce, that, from the vertical position of its stones, it e put in action without the intervention of cogs or . It may be employed in the grinding of malt, the g of oats for horses, and for making flour, or for all purposes: it likewise may be easily altered, so as to either of those articles to a greater or less degree of

rate of one bushel of wheat within the hour. Besid
industrious farmer will thus be enabled to make comp
experiments on the quality of his grain, and may
himself, at a trifling expense, with flour from his own
without apprehending any adulteration, or without
exposed to the impositions or caprice of fraudulen
avaricious millers.

Lastly, though Mr. Rustall's bolter be more parti
calculated for sifting flour, it may also be applied to '
other useful purposes, and especially with a view to (
the inconveniences necessarily attendant on the leviga
noxious substances, and to prevent the waste of the
particles.

In order to reduce the labour in grinding, and to
the power to the acquired force, and also to simplif
making and grinding, and to reduce the expense att
upon them, so as to enable the farmer and houseke
be independent of the miller's system of grinding, Mr. (
Smart, of the Ordnance-wharf, Westminster-bridge, ii
1814, obtained a patent for certain improvements in mac
for grinding corn, and various other articles, by me
which every article required to be broke or ground is exp
the application of rubbers or crushers, resting on their ful
and pressed against the revolving body by means of
weights, or springs. The rubbers, or crushers, each
on a separate axle, will admit of any irregular surfac
a square to a circle, to revolve against them, as ea
be loaded more or less, by moving the weights on the
further from, or nearer to, the fulcrum; or, if with s
by screwing them more or less down. The rubb
crushers may be plain, grooved, circular sided, conca
any other figure best adapted for the substance to be
or ground. The square or octagon are best adapt
breaking cement-stones, bones for manure, chalk, 1
clay, mortar, &c. For breaking malt, beans, &c. one r
only is wanted; but for wheat, oats, barley, rice, (
flour, or meal, the more rubbers or crushers the fir
article will be ground; and the more flats there are (
revolving body, the more crushers can be applied to advs

HAND-MILLS

ARE most commonly used in grinding coffee and s
but they are sometimes made of a larger size, and u
grind wheat, malt, &c.; in such cases the hand is ge

 effort to a winch handle. In Bockler's *Theatrum Machinarum* there is a description of a mill, in which the effort of a man is applied to a lever moving to and fro horizontally, nearly as in the action of rowing: as this is a very advantageous method of applying human strength, the effort being greatly assisted by the heaviness of the man in leaning back, we shall give a brief description of it.

It is represented in fig. 161. The vertical shaft E G carries a toothed wheel C, and a solid wheel F; the latter being intended to operate as a regulating fly. Upon the crank A B hangs one end of an iron I, the other end of which hangs upon the lever H K, the motion being pretty free at both ends of this bar I. One end of the lever H K hangs upon the fixed hook K, about which, as a centre of motion, it turns. Then, while a man, by pulling at the lever H K, moves the extremity H from H to N, the bar I acting upon the crank A B gives to the wheels C and F half a rotation: and the momentum they have acquired will carry them on, the man at the lever suffering it to turn back from N to H, while the other half of the rotation of the wheels is completed. In like manner another sufficient pull at the lever H K gives another rotation to the wheel C, and so on, at pleasure. The wheel C turns by its teeth the trundle D, the spindle of which carries the upper mill-stone, just as the spindle D carries round the upper mill-stone in fig 156. In this mill the nearer the end of the bar I upon the lever H K is to the fixed hook K, the easier, *cæteris paribus*, will the man work the mill. If the number of teeth in the wheel C be six times the number of cogs in the trundle D, then the labourer, by making ten pulls at the lever H in a minute, will give sixty revolutions to the upper mill-stone in the same space of time.

In the *Transactions of the Society of Arts*, may be seen a description of a mill, invented by Mr. Garnett Terry, of the City-road, for grinding hard substances, by means of a wheel turning upon a horizontal axis instead of a vertical one, as in the common construction. See fig. 162.

THE FOOT-MILL

Is used for grinding corn or any other substance, moved by the pressure of the feet of men or oxen. A judicious construction of the foot-mill is given in G. A. Bockler's *Theatrum Machinarum.*

This mill is represented in fig. 163. A is an inclined wheel, which is turned by the weight of a man, and the impulsive force of his feet while he supports himself, or occasionally pushes with his hands at the horizontal bar H. The face of this wheel has thin pieces of wood nailed upon it at proper distances, to keep the feet of the man from slipping while he pushes the wheel round: and the under side has projecting teeth or waves which catch into the cogs of the trundle B, and by that means turn the horizontal shaft G with the wheel C: this latter wheel turns the trundle D, the axle of which carries the upper mill-stone E. This kind of foot-mill will answer extremely well to grind malt, &c. when no very great power is required.

M

with that disgusting practice among bakers of kneading dough with their bare feet. It would be well if the le ture paid some attention to this, which is still carried some parts of the metropolis, and particularly in the kne of dough for biscuits.

In the public baking-houses of Genoa a machine i which produces a great saving of time and labour. first described in the *Atti della Societa Patriotica di M* vol. ii.

A, in fig. 164, is a frame of wood which supports the axis of the m a wall 11 palms high from the ground may be made use of ins this frame. B, a wall, 3½ palms thick, through which the aforesa passes. C, another wall, similar to the former, and facing it at the d of 21 palms. D, the axis, 30 palms in length, and one palm and on in thickness. E, the great wheel, fixed to the said axis, between the and the wall; its diameter is 28 palms; and its breadth, which is cap holding two men occasionally, is five palms. F, are steps, by trea which, the men turn the wheel very smartly; they are two palms from each other, and one-third of a palm in height. G, a small whe cogs, fixed almost at the further extremity of the axis: its diam 12½ palms. H, a beam of wood which extends from one wall to the being 21 palms in length, and one and a third in thickness. A similar not seen in the figure, is on the opposite side of the axis. I, a tra piece of wood, placed near the wall C; it is fixed into the two bean serves to support the further extremity of the axis; its length is 14 and its thickness one and a third : there is likewise a transverse piece cannot be seen in the figure) 14 palms long, and half a palm thick, close to the wall B. K is a strong curved piece of oak, fixed trans in the side beams H, to receive the axis of the trundle: its length palms, and its thickness 1½. L is a trundle of 5½ palms in di and 1½ in height, which is moved by the cog-wheel G. M is a proceeding from the trundle L, and continued through the cross N bottom of the tub P; its centre is made of iron, partly square and round, and it turns in a socket of brass. The first part of this axis h the trundle L and the cross N is of square iron, surrounded by two of wood, held together by iron hoops, which may be removed at p to examine the iron within; its length is three palms, its diameter one palm. The second part of the axis which is within the tube, i like the first part; its length is 1½ palm, its diameter 1½. The v sheath of this part of the axis is fixed to the bottom of the tub, by the of three screws with their nuts. This axis is distant one-third of from the nearest triangular *beater* of the cross. N, the cross, formed bars of wood unequally divided, so that the four arms of the cross different lengths : one of the two pieces of wood of which the cross is is six palms in length, the other five; their thickness is ⅞ of a and their breadth one palm. O, four pieces of wood, called *beaters*, of a triangular shape, fixed vertically into the extremities of, and underneath, the arms of the fore-mentioned cross; they are 1½ palm in length, and half a palm in thickness, and beat or knead the dough, in the tub at unequal distances from the centre. P is a stout wooden tub, about a quarter of a palm thick, well hooped with iron; its diameter is six palms, its height 1½ in the clear. Fig. 165 is a box or trough of wood, four palms long, and two wide, in which the leaven is formed (in about an hour) in a stove, and which it is afterwards carried to the tub P. Fig. 166 exhibits a view of the handle, cross, &c. with a section of the tub. Fig. 167 is a bird's-eye view of the cross and tub, with the upper ends of the triangular beaters. This tub will contain about 18 rubbi (about 19 bushels) of flour, which is carried to it in barrels; the leaven is then carried to it in the box or trough, and when the whole is tempered with a proper quantity of warm water, the men work in the wheel till the dough is properly and completely kneaded. In general a quarter of an hour is sufficient to make very good dough; but the experienced baker, who superintends, determines that the operation shall be continued a few minutes more or less, according to circumstances.

The measures in the preceding description are given in Genoese palms, each of which is very nearly equal to 9.85 of our inches. The machinery may be varied in its construction according to circumstances, and the energy of the first mover much better applied than by men walking in a common wheel.

In November, 1811, a patent was granted to Mr. Joseph Baker, navy contractor, for a method of kneading dough by action of machinery. The invention consists in having an upright shaft, turning on a pivot, fixed in the centre of a circular trough, so that the dough placed in such trough may be kneaded by a stone or iron roller, on its edge, passing round it in a rotatory motion, being fixed at a due distance by a horizontal bar or axle to the shaft, which is to be turned by means of one or more other horizontal bars likewise fixed thereto, and worked like a capstan by a proportionate number of men or quadrupeds, such horizontal bars having small ploughs fixed to them, so as to run in the trough, and, acting like a plough, cause the dough to present fresh surfaces for each successive revolution.

Independent of the methods above quoted, many others could easily be adapted, to do away with the filthy practice to which we have alluded.

THE STEAM-ENGINE.

THE most prominent feature in modern discovery is the *steam-engine*, which has, with much propriety, been denominated " the noblest monument of human ingenuity." The Marquis of Worcester, who lived in the reign of Charles II., is entitled to the honour of having first directed the attention of mankind to the expansive force of steam when used in a close vessel: but in the book published by him in 1663, he is not sufficiently explicit or intelligible, for us to determine what kind of apparatus, or combination of machinery, was used by him for the guidance of its powers. It is however but reasonable to suppose, notwithstanding the vagueness of his expressions, that it was principally owing to the direction his observations gave to the minds of others that steam began to be used as a first mover of machinery.

When water is exposed to the action of heat, it expands and assumes the gaseous state called steam. When it is confined in a close vessel, and heat is applied to it, the quality of expansion is exerted to a powerful extent; and as the space between the top of the water and the top of the vessel is filled with atmospheric air, the first portion of the power, exerted by the expansion of the steam, is directed towards the displacing of that atmosphere from the situation which its weight had assigned it, and, consequently, such portion of the direct action of the expansive force of steam is to be deducted from its disposable power. This portion of power is, however, ultimately available. For as by a reduction of temperature steam reassumes the state of water, leaving the space which it had occupied again void, the atmosphere which it had displaced returns to its former situation, exerting a force, so to do, exactly tantamount to that which the steam had exerted to displace it. This force may be termed the consequent power of steam. The directing and controlling of these powers, so that they may be applied to the purpose of creating equable motion, is the object attained almost to perfection in the steam-engine: and it is the more accurate control, the more advantageous application, and the more economic production of these powers, that have been aimed at in its various modifications.

For more perfectly illustrating the mode in which steam operates we will suppose the vessel, represented at fig. 168, to be filled with water up to the line A, and the space E occupied with air, and having a plug or piston fitting it at C, and an aperture at D; now if the aperture D be closed, and

lied to the water, as at F, steam will be generated, and by its
t force will raise the piston C upwards; then if the heat be with-
d the vessel suddenly cooled, condensation will take place; the
assuming the form of water, will again occupy the space below
A, and the piston C will return to its place. In this experiment
nsive force of the steam compressed the air in the space E, and
e plug C upwards, we will suppose, to H; but C, in travelling to
ced so much of the atmosphere as occupied the tube from C to H;
ntly, the portion so displaced will seek to resume its natural
, and when the force of the steam is withdrawn by condensation,
ht of that portion of the atmosphere will again return the plug C
ace; by which, it is obvious, that the raising of the plug was the
ction of the steam, and the returning its consequent action, or
on of the atmosphere in consequence of its having been displaced
rce of the steam.

, if we suppose the plug to be in its first situation, as at C, and we
t aperture at D, and apply heat, the steam will rise into the space
expel the air through the aperture D, which being closed, and
ation caused, the space E will be left a vacuum, and the atmo-
eeking to occupy that space will force the plug C down to the line
the movement of the plug C was solely caused by the atmosphere
; itself to regain the position whence it had been expelled by the
the steam through D, and this effect is performed by the conse-
ower of steam alone.

been found by experiments, that the pressure of the atmosphere is
o about 14 pounds weight upon every square inch, so that sup-
the superficies of the aperture of the vessel, fig. 168, to contain one
inch, the power exerted by the steam in raising C to H will be
unt to raising 14 pounds weight that height, together with the
necessary to overcome the friction and weight of the piston C, in
ader; and, that the power exerted by the steam in expelling the
here from the space E, and obtaining its consequent pressure to the
of 14 pounds from A to C; and that the disposable power, obtained
return of the piston from H, will, in the first instance, be equal to the
of 14 pounds weight from C to H, less the amount of the friction of the
C; and, in the second, will be equal to the raising of 14 pounds weight
to A, less the amount of the friction as before. In both these instances
ansive or direct force of the steam has only been considered as equal
displacing of the atmosphere, or what will be equal to 14 pounds
re on each superficial inch; but if the piston C be loaded with any
; the steam will, if urged with sufficient heat, raise it, always premising
e vessel is strong enough to resist the increased pressure. Suppose
e loaded with 10 pounds of weight, the steam must be urged until its
ure is equal to 24 pounds, 10 pounds more, 14 pounds the pressure
atmosphere on each square inch, and the resulting disposable force
e equal to 24 pounds more, the weight of C, less its friction return-
the place from where C was raised; so that, in this case, the pressure
e internal sides of the vessel, tending to burst it, will be equal to
unds per square inch of the internal superficies, the remaining
ads being counteracted by the pressure of the atmosphere on the
al surface, which is equal to 14 pounds of the internal pressure. By
t is evident, that the direct force of steam may be increased without
, whereas the resulting force or pressure of the atmosphere is mani-
bounded to 14 or 15 pounds on the square inch according as its
y varies.

Trusting from this explanation that even those whɔ
unacquainted with matters of this nature will clearly com
hend the manner in which the expansive force of s
operates, we shall proceed to explain the different mecha
combinations which have been formed to render this p
subservient to our will: premising that what we have be
fore called the consequent power of steam will, in futuro
considered the pressure of the atmosphere, which in
it is, having adopted the other term in the introduc
explanation, in order more clearly to impress on the m
that the expansive force of steam is actually the only origin
of power in this machine.

The first apparatus with which we are acquainted,
structe 1 for the purpose of employing steam as power to
in a close vessel, was invented by a Captain Savary,
which, in the year 1698, he obtained a patent. The forr
it, as invented by him, is represented in fig. 169.

a, a close boiler, placed on a furnace, and of sufficient strength to
considerable pressure; B another vessel strongly constructed; *c c* a pip
a cock in it at *i,* by means of which a communication can be made from
at pleasure; *e* a pipe proceeding downwards into a well or other reserv
water; *ff* another pipe proceeding from B to a reservoir placed al
h h is a pipe communicating from B to the pipe *ff,* and having a cc
it at *k* to allow of, or cut off, such communication; *m* is a valve capa
closing the pipe *e* by pressure from above, and of opening it by pr
from below; *l* a similar valve fitted to the pipe *ff,* and capable of
acted upon in a similar manner. If the boiler *a* be filled with water
dotted line, and heat applied by means of the furnace, the steam wil
in the boiler, and, passing through the pipe *c c,* fill the vessel B, and
up the pipe *ff;* the valve *m* being shut by the expansive force of the
pressing upon it: if the cock *i* be now shut and the steam in the ve
condensed, by throwing cold water upon its outer surface, the atmos
pressing on the valve *m* will close it, and the interior of the vessel B n
a vacuum, and the water in the reservoir to which the pipe *e* passes w
forced by the external pressure of the atmosphere into the vessel B
the dotted line, which is supposed to be about 26 feet from the s
of the water in the reservoir, being the length of a column of water, v
taking into account that the vacuum thus formed is not quite perfe
equal to the pressure that will be exerted by the atmosphere. If the
i be opened again, and the steam allowed to press on the surface
water, in B, it will close the valve *m,* and cause the water to ascer
pipe *ff,* through the valve *l,* to the upper reservoir; and when the
is again shut, and the steam in B again caused to condense, the ope
will be repeated, and the weight of the water in the pipe *ff* will clo
valve *l,* and cause the vessel B to be filled by *e* as before.

Such was the construction and mode of operating with the first app
made by Captain Savary. But he, finding it inconvenient to cause
densation by means of throwing cold water upon the outer surface, intro
the pipe *h h* into B, which, by opening the cock *k,* allowed some port
the water to pass from the pipe *ff,* which was always full after the
cr k, and thereby cause condensation more quickly to ensue.

The dial or gauge cocks o, q, were also contrived by Captain Savary, in order to ascertain the height of the water in the boiler. If the surface of the water is above the lower ends of the cocks, and the cocks be opened, water will issue from them; if below, steam. But if the water is at its given level, that is to say, if the surface of the water be in the intermediate space between the ends of the cocks q and n o, water will issue from the former, and steam from the latter. This knowledge was necessary to be obtained; for should the surface of the water exceed the height of the cocks o, there will not be room for a sufficiency of steam to remain above for continued operation.

The application of this engine was confined to the raising of water to small heights, as it operated only by atmospheric pressure; in deep mines it was found not to be effective. Taking into consideration, however, the then imperfect state of knowledge, so far as regarded steam as a first mover, the inventor is entitled to no inconsiderable degree of praise for this specimen of his ingenuity. The greatest objection to this mode is, the great waste of steam, and consequent unnecessary expenditure of fuel, arising from the condensation being effected by allowing cold water to come into contact with the steam in the vessel B.

At the time when the existence of this engine was first made known to the public, the amazing power of steam, which it so plainly demonstrated, began very deservedly to obtain the attention of ingenious men; and disputes for the honour of the discovery took place, the English ascribing it to the Marquis of Worcester, the French to Papin.

Without entering into the minutiæ of this contest, it will be sufficient for us to trace progressively the grand improvements that have taken place in the steam-engine in this country: though not forgetting to mention the accessary improvements derived from foreign aid. Of this latter is the safety valve; an instrument, though in itself simple, of such vast importance, that to it may be attributed the general introduction, and consequent improvement, of the steam-engine to its present existing state of perfection. It was contrived by Dr. Papin, who, at the time of Captain Savary's invention, was making experiments on the power of steam at high temperatures, for the purpose of dissolving bodies. It consists merely of an aperture of a specific dimension, suppose, for instance, one square inch, in or communicating with any close boiler, and a valve properly fitted in that aperture, such valve being on the outside loaded to any extent considered necessary to resist the force of the steam until it has acquired a certain degree of power, computed to be what the boiler is perfectly capable of sustaining without the chance of bursting. Now it is manifest, until the pressure of the steam

overcoming the pressure of the atmosphere again raise it to its original position; and *g* being again closed, and *e* opened, the performance of a similar operation will be effected.

The advantage sought in the construction of this sort of engine was obtained, namely, that of not being required to use steam of greater pressure than the atmosphere, which was the case in Savary's engine, when the height the water was to be raised by the pipe *f f*, fig. 169, exceeded more than about 32 feet. In Newcomen's engine, the weight *o o* balances the piston *c*, and the exertion of the steam is never required to be more than about 14 pounds to the superficial inch; whilst the introduction of the beam affords a movement applicable to the working of pumps, by which water can be raised to any required height.

The cocks *g* and *e* were opened by the hand of an attendant until a boy of the name of Potter, who was intrusted with the management of them, to save himself the labour and attention which they required, ingeniously contrived to attach a piece of string to the levers of the cocks, and to the beam L L, in such a manner, as to procure by its movement their being opened and shut at the proper periods. It was this gave the idea for the construction of that part of the present engine called hand-geer.

The next person who made any considerable improvement in the steam-engine was Mr. Henry Beighton, of Newcastle, who invented the part called the plug-tree, for opening and shutting the valves, which we shall describe in a more advanced part of the work: he also adopted a force-pump, to supply the deficiency of water in the boiler caused by the expenditure in the production of steam. This engine, called *Beighton's fire-engine*, was much used for nearly half a century; and the attention of engineers was directed more towards the economizing of fuel, than the further improvement of the engine.

At length, however, it began to be perceived, that the attainment of a rotatory motion would open an extensive field for the application of its powers to various mechanical purposes; and, accordingly, we find the attention of engineers most actively engaged in endeavouring to effect so desirable an object, which was eventually accomplished by Mr. Matthew Washbrough, of Bristol, who, in the year 1778, obtained a patent for the application of the crank. Though this is by far the best application that has been yet applied, as is evident from its now almost universal adoption, it was for some time superseded by the sun and planet wheels introduced by Messrs. Boulton and Watt.

The sun and planet wheels are represented in fig. 171. A represents the end of the beam, to which is affixed, by a movable centre at F, the rod B, called the connecting-rod, upon the lower end of which is immovably fixed the wheel C; D is a cog-wheel, fixed upon the axis of the fly-wheel G G G G, and capable of revolving with it. When the beam A passes downwards, the wheel C changes from its position at C to C¹, and its cogs, acting upon those of the wheel D, cause that, and the fly-wheel to whirl

it is attached, to perform that portion of a revolution round its axis. The wheel C, from the manner in which it is suspended, tends always to press against D in its downward and upward motions. When C is in the position of C^2, the wheel D, from the velocity it has acquired by its connection with the fly-wheel, causes it to pass under the centre of D, and the beam beginning immediately to ascend, operates by C on the other side of D, giving it another portion of impulse towards a continuous motion whilst passing from C^3 to C^4, and when it arrives at C^5 it will be impelled over the centre D D by the velocity which its own action had given the fly-wheel.

The crank and fly-wheel invented by Mr. Matthew Washbrough is shown in fig. 172. A and B represent the same parts as in fig. 171, but the lower end of B is, in this case, attached to the crank C at c^1, and is capable of revolving round the centre E; the other end of the crank is attached to the fly-wheel D D D D, so that both that and the fly-wheel are capable of revolving round the centre E. When the beam A is depressed, it communicates motion to the crank and fly-wheel to which it is attached, and when it arrives at H, the velocity of the fly-wheel causes it to pass under the centre E, and the beam beginning immediately to ascend, again communicates motion to the fly-wheel.

The length of the connecting rod B should be such that when the beam A is in an horizontal position, the crank will likewise be the same; and when the crank is in the position as at G, its length, together with the length of the crank, such as to permit the other end of the beam, A, to descend until the piston which is attached to it is at the bottom of the cylinder; and the length of these and the cylinder such, that when the crank is in the position H, the piston shall be at the top of the cylinder. This method not only affords a rotative motion, but determines the length of the stroke which the piston may be required to perform, to an exactness that is of great importance; for prior to its introduction, the forcible striking of the piston against the top and bottom of the cylinder, which was attended with injurious effects, was an occurrence by no means unfrequent, and which the sun and planet wheels were in no way calculated to remedy.

Mr. Watt, a native of Glasgow, having his attention accidentally directed to the construction of the steam-engine, discovered that water, when confined in a close vessel, and heated considerably beyond the boiling temperature, would, when the steam was permitted to escape, cool rapidly down to the boiling point, which suggested an idea, that the amount of steam issuing from any vessel was simply in proportion to the amount of heat applied, and that the economizing of fuel could only be obtained by the economizing of steam. Mr. Watt also noticed the great change which took place in the temperature of the cylinder when the cold water was injected to condense the steam; and concluded, that as the coldness of the cylinder would remain after the necessary condenotion had been effected, a wasteful condensation of the newly introduced steam must take place. By experiment he found, that the quantity of steam thus wasted was no less than twice the contents of the cylinder, or three times the quantity which was required for producing the effect sought. The modes to which he had recourse to remedy this defect were first, the substitution of a wooden cylinder, which, upon repeated trials, he was compelled to abandon, on account of the roughness produced by wet and the changes of temperature; secondly, the enclosing of the cylinder with wood, and filling the intermediate space with powdered charcoal; which afterwards was superseded by the introduction of an extra cylinder, that enclosed the working cylinder, and permitted steam to flow round it, which maintained it at a regular temperature. The outer cylinder is termed a jacket, and is now used with advantage.

In the year 1765, Mr. Watt made the grand improvement of effecting the condensation in a separate vessel, communicating only by a pipe to the cylinder, termed the condenser.

Fig. 173, A is the lower part of the cylinder; B the condenser; C the eduction pipe, or pipe of communication between the cylinder and condenser, which is being opened or closed by the cock c; and DD a chamber of cold water, in which the condenser B is immersed. When the cock c is opened, the steam, by its elastic force, rushes into B, where it is condensed; and the space it has left in the cylinder allows a new supply of steam to be generated, which, by the opening of the cock c, also passes into B, and is similarly condensed. The condensing of the steam necessarily accumulates water in B, and reduces the capacity of the condenser; to remedy which, Mr. Watt introduced a small pump, worked by the engine, to withdraw the extra water, as also the air, which in some degree hindered the perfectness of the vacuum: he named it the air-pump. He also found it expedient to permit a small stream of water to run from the chamber DD, by means of a cock, which could regulate the supply according to the temperature of the water in D, or the required rapidity of the condensation. The cistern in which the condenser is placed is called the cold-water cistern; it is continually being supplied with fresh water from a pump connected with a well or pit, and the overplus is discharged by a spout into a drain. The hot water drawn from the condenser by the air-pump is delivered into another cistern, termed the hot-water cistern.

These parts are represented in fig. 174, where their distribution is such as to exhibit them conveniently without taking into view their construction in any particular engine, as their form and relative position is of no importance, and is arbitrarily varied as the different builders think fit, or as convenience or circumstances may in many cases require. A is the cylinder; B the condenser; C the air-pump; EE the cold-water cistern; F the cold-water pump for supplying the same; SS the steam pipe from the boiler; H the eduction pipe from the cylinder to the condenser; OO the hot-water cistern; N the injection cock for allowing a small stream of water to flow into B; P the hot-water pump, which forces a sufficiency of water from the hot-water cistern through the pipe q to keep up the supply in the boiler

t t t t are pump rods, fixed to the beam and worked by its motion; *g* is a suction pipe; and *m* the foot-valve. The pipe I and the valve V, opening upwards, are for the purpose of permitting the steam to pass through to expel the atmosphere from the condenser, called a blow-valve, and then by its own condensation to leave a vacuum on the starting of the engine.

A communication is made from the condenser by a pipe to the upper part of a tube standing in a basin of mercury. The amount of mercury in the tube will indicate the perfectness of the vacuum in the condenser, by exhibiting the height at which it is supported by the pressure of the atmosphere. If the vacuum is perfect, the mercury will stand from 28 to 31 inches.

The only difference between the working of this engine and those already described is, that the pipe HH, instead of being an injection pipe for the admission of cold water to the cylinder, merely leads off the steam to the condenser B, therefore, when condensation is to be effected, the cock *h* is opened to admit the steam from the cylinder into B, there to become condensed, whilst the continued action of the pump C maintains a vacuum in B, by drawing off the condensed water and the air. The cold-water pump F keeps up an abundant supply in the cistern E E, and the superplus is discharged at W; thus maintaining a depression of temperature sufficient to procure condensation in B.

Such was the importance of this alteration in the mode of construction, that one half the quantity of fuel consumed by an engine of the former construction was saved. Still, however, the engine was not complete; the piston was required to have water kept upon its upper surface to keep it air-tight, and as this, in the descent, cooled the cylinder considerably, it was, as is evident from what has been already stated, when speaking of the former mode of injecting cold water, productive of a loss, to which Mr. Watt turned his serious attention, and eventually succeeded in remedying the evil.

The next construction of the engine is what is now called the single acting engine, and as the improvement consisted merely in the cylinder, a diagram of that will be sufficient to make it clearly understood. Finding, as has been stated, the disadvantages attending the open-ended cylinder, Mr. Watt conceived the idea of closing the top of the cylinder, and of causing the piston-rod to work through a close collar stuffed with hemp and grease; and instead of making use of water to procure the piston working air-tight, of using oil or fat as a substitute; and instead of causing it to descend by the pressure of the atmosphere, of employing steam of an expansive force equal to that pressure.

In reference to fig. 175, A is the cylinder; B the piston; CC the stuffing-box filled with hemp moistened with tallow, having its cap screwed firmly down by two screws, so as to form a steam-tight joint round the piston-rod. The piston-rod being turned true and polished, is capable of working up and down through the steam-tight joint. D is the steam pipe leading from the boiler; K the eduction pipe leading to the condenser; H and I are two valves upon the rod L, which passes through stuffing-boxes, or steam-tight joints, at m m. While raising and lowering the rod L, the valves H and I open or close their respective apertures D and K. g is a valve which affords, when open, a communication between the top and bottom of the cylinder through the pipe P E, and from thence through O. Its spindle, which is hollow, works in a steam-tight collar at N, and has the rod L passing through it, likewise steam-tight, in such a manner, that the rod L, with its valves, may be moved independent of the valve g, or the valve g independent of the rod L. If the valves are placed as represented, I and H open, and g shut, and the steam be permitted to come from the boiler through D, it will pass through H, and enter the top of the cylinder at P, depressing the piston B by its elastic force, whilst the lower part of the cylinder is open through O and K to the condenser. When the piston is depressed to B^1, the rod L must be moved downwards, and the valves H and I closed, and the valve g opened, that the steam, by its elastic force, may pass through P E and the valve g, and act equally upon the lower as well as the upper surface of the piston; the piston, therefore, being in an unbiassed state, as regards the pressure, will again be raised to its original situation by the counterpoise weights acting at the other end of the beam, and the steam will pass from above to below the piston. When it has arrived at that place, the valves may be again put in their first position, as represented in the drawing, when that at I, being open, affords a communication through O K for the steam, which now occupies the cylinder below the piston, to pass to the condenser, whilst fresh steam forces through D upon the upper surface of the piston tantamount to the pressure of the atmosphere, and produces an effect fully equal to that obtained from an atmospheric engine.

In the engine just described, the beam was drawn down by the pressure of the steam upon the top of the piston, and when the steam had passed into the condenser, was raised by counterpoise weights placed at its further end; by which it is manifest, that the action of an engine of this construction must be somewhat irregular, for whilst the piston is descending the steam must act upon it to an amount capable of raising the counterpoise weights, and of likewise doing the work put upon it; but when the piston is rising, the actual fall of the counterpoise weights is the only amount of power which the engine is supplying, and from which is to be deducted the friction and weight of the piston. Such engines, therefore, when a continuous and equable force is required, are not so advantageous as those where the rising and depressing of the piston are alike performed by the same means and the same extent of force. This desirable end was accomplished by Mr. Watt in the construction of the double-acting engine, which we shall next notice, in contradistinction to the one that last claimed our attention, and which is now seldom used for any other purpose than that of pumping water

The double-acting engine is formed in such a manner that whilst the steam is pressing on one side of the piston, the cylinder on the other side is always open to the condenser, so that a vacuum exists on the opposite side to that on which the pressure is exerted. This effect is produced by several modes of distributing various pipes, or valves, or cocks. The simplest in its construction is the four-way cock.

Fig. 176. A is the cylinder; B the piston; C the pipe communicating with the boiler; f m a communication with the condenser; I K and H g communicate to the top and bottom of the cylinder; C' and D are two different passages through the plug of a cock, (the plug of the cock is shaded for distinction,) capable of being turned by means of its handle m'. When the cock is as now represented, the steam from C passes through D along H and g to beneath the piston, whilst there will be a direct communication with the condenser from above the piston through K I C' F M. When the cock is turned in the position as represented in fig. 177, both these communications will be inverted by the different dispositions of their openings. The opening e in the plug will then afford a communication from the steam pipe C to I, leading to the upper part of the cylinder; and the opening D will permit a communication from H, leading from the lower part of the cylinder to M, the eduction pipe, leading to the condenser; thus perfecting by its motion the different changes of communication, so that the steam is always made to act on the one side of the piston, whilst the vacuum is effected on the other. The effect on the piston in ascending or descending is equal, and the power is much more regular and convenient for application to rotative motion.

In the single and the double engine the pressure of the atmosphere is not used; but the direct force of the steam which, in the old engines, went to expel the atmosphere, and then by its own condensation to leave a vacuum, is first resorted to, and its condensation leaves a vacuum to assist the action of the next supply. Consequently the single-acting engine has not gained more power than the atmospheric engine in proportion to the steam used, otherwise than by what is saved through keeping the cylinder warm; and the double-acting engine is only a mode of using the steam in a more continuous and unbroken manner, taking just twice the quantity of steam, and exerting twice the power of the single-acting engine.

Other modes for attaining this change of communication, so as to produce the double action, are adopted as circumstances may demand. The one last described, called the four-way cock, was found not to answer in large engines, inasmuch as plugs of sufficient dimensions to afford steamway enough caused so much friction in their collars, that it was an expenditure of considerable power to turn them; the method, therefore, most usually adopted in engines of large

dimensions is, a system of valves, opened and shut at the proper periods by levers.

The internal arrangement of such a system is shown at fig. 173, where C is a pipe leading from the boiler for the conveyance of steam ; and D K a pipe leading to L the condenser ; *o p*, *m n*, are two boxes, each divided internally into three compartments, in which are the valves *e f g h*, capable of opening upwards. From the centre compartment in each of the boxes above and below there is a communication, at *a* and *b*, with the top and bottom of the cylinder. The steam is conveyed along the pipe C into the upper chamber of the upper box, and by means of the pipe *i* passes through it and proceeds to the upper chamber of the lower box, and the pipe K D affords a communication from each of the lower chambers to the condenser L.

If the valve *h* be opened, there will be a communication from the lower part of the cylinder through the aperture *b* and pipe K into the condenser L: and if, at the same instant, the valve *e* be opened, the steam will be admitted through C into the centre chamber of the upper box, and through the aperture *a* into the upper part of the cylinder ; consequently the steam will be admitted to the upper part of the cylinder while the lower part has communication with the condenser, and the piston be forced downwards. If these valves be closed, and the valves *g f* opened, the steam will have access through C *i*, the valve *g*, and aperture *b*, to the lower part of the cylinder, whilst there will be a communication from the top of the cylinder through the aperture *a*, the valve *f*, and the pipe K D, to the condenser, so that the piston will then be forced upwards.

This mode of effecting the changes of communication, though by no means so simple as that of the four-way cock, has an advantage over it that gives it a decided preference. For as the movements of the valves are independent of each other, the shutting off of the steam, and of the communication with the condenser, can, if desired, be effected at different periods, so that the steam may be allowed to act upon the piston only during one half of the stroke, which was discovered by Mr. Watt to effect a saving in steam. Besides this advantage, which can, if required, be effected in other engines by simple contrivances, the valve geer is superior from its lightness, when compared to slides or cocks, adapted to large engines.

In small engines, these advantages, not being of so much importance, are more than counterbalanced by the simplicity obtained by other modes.

The first that we shall notice was invented by Mr. Murray, of Leeds, and is represented in fig. 179, where a section of this apparatus, termed the slide valve, is shown. A is the steam pipe from the boiler : B B B B a close chamber, in which the interior chamber, C C, is capable of sliding upwards and downwards by means of a rod, D D, which passes through a steam-tight box at E. G G is a passage from the chamber B B B B to the upper part of the cylinder ; H H to the lower part of the cylinder : and I I to the condenser. Now when the chamber C C is placed as is represented, it is plain, that the steam has access to the top of the cylinder, and that a communication exists by H H and I I to the condenser. By moving C C upwards, the communication would be reversed, that is, G G from the top of the cylinder would communicate by I I to the condenser, and H H admit steam into the lower part of the cylinder. By the movement of C C upwards and downwards, the alternating effect is produced.

Messrs. Boulton and Watt introduced another very neat and useful mode of distributing the steam in small engines, denominated D valves, on account of the shape they present when seen in a top view.

Fig. 180. A B and C D represent the two D valves capable of moving upwards and downwards by means of the spindle E. Their faces at A and C are made to fit steam-tight upon the inside of the box, and the back part at B and D, as shown at D', is semicircular, to admit of its being packed with hemp in the same manner as the piston. G and H are communications to the top and bottom of the cylinder; I is the steam way in the boiler, which, in some engines, comes from the jacket of the cylinder; and K is the eduction pipe, which carries the steam to the condenser.

Now if the rod E be moved upwards until the lower surface of A B is above the aperture II, the lower surface of C D will be above the aperture G, consequently the steam from I will pass into the cylinder at H, and the steam from the cylinder will pass through G and N K into the condenser. Again, if the upper surface, A B, be below H, C D will be below G, and the steam will pass through the aperture O, down a pipe behind M K, through P, and, by means of G, will have access to the bottom of the cylinder, whilst a communication will be effected by M K and through H, from the top of the cylinder with the condenser. The rod E passes through steam-tight joints, at top and bottom, into boxes.

Fig. 181 represents another manner of forming valves upon this principle, in which a flat surface is introduced at the back, and the packing is effected by having the faces, backs, and sides of the valves made of brass, and fitted to plates of the same material, which can be screwed against the valves as they wear. The slides or moving parts are shadowed. The position in which they now are will admit the steam to pass from I through H, and the condensation to be effected by passing from G through N to K; but if the slides be moved downwards until the aperture C is opposite G, the upper slide will be below H, and the steam will pass through P into the cylinder by means of G, and exhaustion will take place from H through M to K.

Fig. 182 is a combination of what are called concentric valves. Their distribution is exactly similar to those already explained. The spindles CC of the lower valves, in both the top and bottom boxes, pass through the spindles of the upper valves, which are pipes. This mode of constructing the valves was invented by Mr. Murray. They are capable of being moved in many ways; but the one most generally adopted is represented in the figure, where the rods d e, attached to the cross-levers that move on the centres G G G G, act upon the valves. Thus the rod e, by moving downwards, will raise the solid or the lower spindle valve of the lower box, which opens to the condenser, and the hollow spindle the upper valve of the upper box, which admits steam to the top of the cylinder, while the rod d acts upon the other two, and changes the course of the steam and exhaustion, as before shown.

Conceiving that we have sufficiently explained the several modes of guiding the immediate power of steam, we shall in the next place proceed to examine into the construction of the piston.

A section of the piston most commonly used in condensing engines is exhibited in fig. 183. The lower face of the piston is fixed to the rod *d d*, and the upper face is capable of being raised up upon the rod *d*. Hemp rubbed with tallow, called packing, is introduced round the interstice at C F, which, when the upper plate D D is screwed down by the screw at B E, is forced outwards against the sides of the cylinder, so as to render it perfectly steam-tight. As it wears by the friction against the cylinder it must be forced out by tightening the screws; and when entirely worn, the plate *d d* must be raised and the piston fresh packed.

This construction of the piston answers perfectly for condensing engines; but in high-pressure engines the hemp is destroyed so rapidly by the heat and friction, that pistons formed entirely of metal have been introduced with advantage.

The top view of one kind of metallic piston is represented at fig. 184. A A A A is a ring of brass divided into four equal disconnected portions, resting upon the plate B B, which is affixed to the piston-rod *d d*, as seen in fig. 185; the portions of the ring are forced outwards against the sides of the cylinder by springs of any convenient form pressing from the piston-rod N. A', in fig. 185, represents a side view of a similar ring, similarly divided. Its portions are placed upon the ring last described, so that the divisions fall upon the centres of the other four pieces, and are forced against the sides of the cylinder in like manner by springs, and the plate C C is placed over the whole. The upper and lower sides of the plate and rings are all carefully ground, so as to fit steam-tight. This form of piston, though it has, in some cases, been used for a considerable time with advantage, is, upon the whole, defective, as the rings near the interstices between the segments, must open and allow the steam to gain access to the interior where the springs lie, and from thence, through similar interstices, to the other side of the piston.

A metallic piston of a better construction is shown at fig. 186, consisting of six pieces of brass of the forms represented in the figure, by A B C D E F. A B C are circular, and are made to fit the inside of the cylinder, against which they are forced by the wedge pieces D E F, which have springs behind them. When A B C wear so as to divide at the angles, the wedge pieces protrude themselves against the cylinder, keeping the space always filled. These pistons have, in some cases, been used for many years without requiring alteration.

Having now duly considered the construction of the cylinder, or seat of motion, and the means of distributing the operative power so as to produce a reverting rectilinear motion, we shall next proceed to exhibit the manner in which this motion is transferred, so as to maintain its continuance.

The movement of a four-way cock may be accomplished by a plug-tree, which is a perpendicular rod attached to the beam of the engine as represented in fig. 187. O P are two studs or pins placed at such a distance from each other, that the upper one, O, shall force the handle N to the position of N', just as the piston shall have reached the lower part of the cylinder, and the pin P is so placed that it shall carry the handle to its original situation at the time when the piston is again to change its direction

each of the curved ends of the beam, and to the pump
ston rods, as represented in fig. 170, which answered
uired purpose; but in double-acting engines, where
ton-rod forces upwards as well as pulls, some other
of converting the action is required. The most perfect
roduced is the parallel motion, the principle of which
e comprehended by referring to fig. 191.

ose A B to represent one end of a beam, vibrating on its centre at
ther and B will perform the arc C C¹, and carry whatever is
to it in the direction of that arc; then suppose another rod, G H,
length to A B, vibrating on its centre G, the point in the connect-
r, L H, will, by the vibrations of A B and H G, move upwards and
nds in a perpendicular line. For so much as the curve from the
A B draws it towards A, so much will the curve of the radius H G
towards G, and the movements, correcting each other, will cause I
nd fall in a perpendicular line.
re simple mode, which answers very well in small engines, is repre-
a fig. 192. The sling piece attached to the end of the beam has a
rollers, one on each side, which press on each side of the guiding
D D D, and carry the piston, attached at C, in a perpendicular

steam-engines are proportioned to go at a settled
nd to make a certain number of strokes of the piston
nute, which, reckoning the number of feet the piston
in its upward and downward motion in the cylinder,
en by general agreement settled at 200 to 220 feet
inute.
btain this regularity of action, it is manifest that a
ity must be obtained in the amount of power that is to
c the movement; or, in other words, a regular amount
tic force must be exerted at each stroke of the piston.
s somewhat difficult to accomplish, and depends much
ceping the fire, which generates the steam, of a uni-
heat; consequently the care of the fire should only be
ted to one who is well skilled in his business. There
owever, contrivances, called governors, which greatly
in maintaining a regular action, and where very great
is not required, they produce it to a sufficient extent.
governor (which we have already described under the
" On the Equalization of Motion," in Mill-geering)
pon the principle of centrifugal force, and is applicable
o engines that have a rotative motion.
mode of applying the governor is by connecting it by
of levers to a throttle valve, shown in fig. 193. A B
ents a section of the pipe that conveys the steam from
iler, having a small circular fan of iron, capable of being

N 2

Fig. 194 represents a boiler fitted up with all the appendages now generally applied, and set in a furnace of a proper construction. Part of the furnace is shown in a sectional view. B B B B the boiler, the operation of which has been described. C, the steam-gauge, represented at large in fig. 195. Its object is to ascertain the pressure acting in the boiler. It is formed of a bent iron tube, at one end A communicating with the boiler, and at the other end B open to the atmosphere. The tube is filled up to C and D with mercury, and has a thin piece of stick, E, placed in the leg B, which floats perpendicularly in the mercury at D. To the leg B is appended a flat piece of brass, divided into inches, and numbered upwards, to form a scale. The stick is made of such length that the top of it shall be even with the first mark on the scale.

If the steam in the boiler presses against the mercury at C, and raises the surface, D, one inch, (which will be indicated by the end of the stick rising to 1 upon the scale,) it proves that there is one half pound pressing per square inch against the internal surface of the boiler, tending to burst it; for if the section of the bore of the pipe was just one superficial inch the pressure would be supporting one cubic inch of mercury, which will be found to weigh near half a pound; therefore for every two inches rise, one pound pressure may be reckoned, and as condensing engines seldom work with more than three or four pounds pressure upon the inch, the scale need not be longer than eight or nine inches.

C is a strong iron plate, covering a circular or oval hole of about 18 inches diameter, to admit a man into the boiler with a view to clean or examine it.

D is the steam pipe, containing the throttle valve E, to which the rod from the governer is connected. F F are gauge cocks. I I is a feed pipe which passes into the boiler and reaches very near to the bottom. H H H H is a cistern, on the top of the feed pipe; i i is a float, formed of stone, and balanced so as to remain always on the surface of the water in the boiler. By the raising and falling of the water the float acts upon the lever K K¹ and the wire I³, which passes through a steam-tight joint at I⁴, and as the float sinks, draws down the end K, which raises K¹, and the valve M attached to it. By this contrivance when the boiler requires a fresh supply of water, the valve M opens and supplies it from the cistern H H H H.

The feed pipe I I is made to contain a column of water equal to the amount of pressure exerted by the steam in the boiler, which we have already stated should not exceed the supporting of eight inches of mercury. An inch of mercury being equal in weight to about 13½ inches of water, the pipe should be about nine feet high from the surface of the water when the boiler is supplied; and the water in the feed pipe should stand about so that when the pressure is six inches of mercury, or three pounds to the square inch of surface.

The feed pipe contains likewise an iron bucket weight O, hung by a chain which passes over two pullies, P P; to the other end of the chain is attached a plate of iron called a damper. When the steam in the boiler is urged to too great an extent, it forces the water in the feed pipe upwards, and raises

the iron bucket weight O, which lowers the damper into the top of
chimney, and checks the force of the fire.

S is a safety valve, loaded with a determinate weight, and of su
dimension in the bore, as will relieve the boiler of its pressure shoul-
arrive beyond a certain temperature. It is enclosed in a case, to prevent
engine tender having access to it, as some engine tenders have been kn
to load the safety valves, to save themselves the trouble of attending to
fire with that diligence which is necessary, and by which they have end
gered their own lives as well as the lives of others.

A pipe proceeds from this case to the chimney, to carry whatever st
may escape into the flue of the chimney.

There is very frequently another safety valve, which is open to the r
of the engine tender, to indicate when the fire is too high.

T T is a flue formed of sheet iron, and passing lengthwise through
centre of the boiler so near to the bottom that it is always covered with w

The flame and smoke from the fire at n n passes first under the bo
and then immediately returns back through this flue, then dividing i
passes through flues which lead it on both sides the boiler to the chimn

V is a cock for the purpose of emptying the boiler when required t
cleaned or repaired.

Such is the construction and general arrangement of
parts of Messrs. Boulton and Watt's engines both single
reciprocating. We shall now proceed in the examination
some other forms of engines which likewise condense tl
steam.

Mr. Hornblower, conceiving he could obtain greater po
from the complicated force of steam acting in two cylind
obtained a patent, in 1781, for that object. His own accor
taken from the specification, we here transcribe.

"First," says Mr. H., "I use two vessels, in which
steam is to act, and which in other engines are cal
cylinders. Secondly, I employ the steam after it has ac
in the first vessel to operate a second time in the other,
permitting it to expand itself, which I do by connecting
vessels together, and forming proper channels and apertu
whereby the steam shall occasionally go in and out of the
vessels. Thirdly, I condense the steam, by causing it
pass in contact with metalline surfaces, while water is appl
to the opposite side. Fourthly, to discharge the engine
the water used to condense the steam, I suspend a colu
of water in a tube or vessel constructed for that purpose,
the principles of the barometer, the upper end having o
communication with the steam vessels, and the lower
being immersed in a vessel of water. Fifthly, to discha
the air which enters the steam vessels with the condens
water or otherwise, I introduce it into a separate ves:
whence it is protruded by the admission of steam. Sixth
that the condensed vapour shall not remain in the etc

which the steam is condensed, I collect it into
vessel, which has open communication with the steam
id the water in the mine, reservoir, or river.

y, in cases where the atmosphere is to be employed
the piston, I use a piston, so constructed as to
am round its periphery, and in contact with the
ie steam vessel, thereby to prevent the external air
sing, in between the piston and the sides of the
sel."

ing is a description of this engine by the inventor: "Let A and
represent two cylinders, of which A is the largest; a piston
ach, having their rods, C and D, moving through collars at E.
iese cylinders may be supplied with steam from the boiler by
square pipe G, which has a flanch to connect it with the rest of
ipe. This square part is represented as branching off to both
c and d are two cocks, which have handles and tumblers as
ed by the plug-beam W. On the fore side of the cylinders
side next the eye) is represented another communicating pipe,
on is also square, or rectangular, having also two cocks, a, b.
', immediately under the cock b, establishes a communication
upper and lower parts of the small cylinder B, by opening the
ere is a similar pipe on the other side of the cylinder A, imme-
r the cock d.

he cocks c and a are open, and the cocks b and d are shut, the
the boiler has free admission into the upper part of the small
and the steam from the lower part of B has free admission
per part of the great cylinder A; but the upper part of each
no communication with its lower part.

ie bottom of the great cylinder proceeds the eduction pipe K,
lve at its opening into the cylinder; it then bends downward,
ected with the conical condenser L. The condenser is fixed on
M, on which stand the pumps N and O, for extracting the air
which last runs along the trough T, into a cistern U, from which
by the pump V, for recruiting the boiler, being already nearly

Immediately under the condenser there is a spigot valve at
ch is a small jet pipe, reaching to the bend of the eduction pipe
hole of the condensing apparatus is contained in a cistern, R,
r; a small pipe, P, comes from the side of the condenser, and
n the bottom of the trough T, and is there covered with a valve,
kept tight by the water that is always running over it.

the pump-rods, H, cause the outer end of the beam to preponderate,
quiescent position of the beam is that represented in the figure,
being at the top of the cylinders.

e all the cocks open, and steam coming in copiously from the
no condensation going on in L, the steam must drive out all
at last follow it through the valve Q. Now shut the cocks b
l open the valve S of the condenser; the condensation will
commence, and draw off the steam from the lower part of the
ler. There is now no pressure on the under side of the piston
cylinder A, and it immediately descends. The communication,
the lower part of the small cylinder, B, and the upper part of
linder, A, being open, the steam will go from the lower part of
pace left by the descent of the piston of A. It must, therefore,

box is as follows: on the top of the cylinder is a box to contain something soft, yet pretty close, to embrace the piston-rod in its motion up and down; and this is usually a sort of plaited rope of white yarn, nicely laid in, and rammed down gently, occupying about a third of its depth; upon that is placed a sort of tripod, having a flat ring of brass for its upper, and another for its lower part ; and these rings are in breadth equal to the space between the piston-rod and the side of the box. This compound ring being put on over the end of the piston-rod, another quantity of this rope is to be put upon it, and gently rammed as before; then there is a hollow space left between these two packings, and that space is to be supplied with strong steam from the boiler. Thus is the packing about the piston-rod kept in such a state as to prevent the air from entering the cylinder when at any time there may be a partial vacuum above the piston.

Mr. Hornblower's description of this engine was followed by a mathematical investigation of the principles of its action, by the ingenious Professor Robison, which demonstrates that it is the same thing in effect as Mr. Watt's expansion engine; but though this is true, there is a considerable difference in the steps by which the effect is attained, which gives an important advantage when it is reduced to practice. We shall give an investigation in a more popular form, using only common arithmetic. Mr. Hornblower assumed, that the power or pressure of steam is inversely as the space into which the steam is expanded; this is the case with air, and for the present we will grant it to be so with steam, and reason from the same data as the ingenious inventor gives us.

To explain clearly what passes in the two cylinders, we must deviate from the precise form of the engine, and divest ourselves of one complication of ideas, by reducing both cylinders to the same stroke ; therefore, suppose the engine to be made like fig. 197, which represents the two cylinders placed one upon the other, the lower one being double the capacity of the upper one, and both pistons being attached to the same rod, which may be applied to the end of the beam, so that the descent of the pistons must draw up the load at the opposite end of the beam.

Then, if we suppose the small piston to be ten inches in diameter, the great piston must be 14,14 inches; and to avoid all difficulties of the ratio of the expansion, and the pressure of steam, we will suppose the engine to be worked by the pressure of atmospheric air instead of steam ; and for the convenience of round numbers in our calculation we will consider the pressure at only ten pounds per circular inch on the surface of the piston.

The area of the small piston will be 100 circular inches, and being assumed to move without friction, the pressure upon it will be 10 × 100 = 1000 pounds. The area of the great piston is twice as much, or 200 circular inches, and the pressure 2000 pounds.

Suppose both pistons to be at the top of their respective cylinders; let the atmospheric air be admitted to press freely upon the upper surface of the small piston; and suppose the space between the two pistons filled with air of the same density, while there is a perfect vacuum made in the lower part of the great cylinder, beneath its piston.

Under these circumstances, the two pistons will begin to descend with something less than 2000 pounds of pressure on the great piston, by the air contained in the space between the two pistons bearing on the 200 inches of surface with a weight of 10 pounds per inch; and beneath this piston there is nothing to counteract the pressure. At the same time, the small piston, having air of equal density above and below it, is in equilibrio.

This force would balance a load of 2000 pounds; but suppose we diminish the load to 1900 pounds, then the pistons will immediately begin to descend; but they will soon stop, because the air between the two pistons must expand itself, to fill the increasing space occasioned by the equal descent of both pistons in the cylinders, one of which is twice the area of the other; and as the air becomes rarer, its pressure on the great piston must diminish. Now as this same diminution occasions the small piston to have a power of descent, we will first consider the pistons separately, and then conjointly, in their power of descent, with which they draw down the beam.

... power of the ... piston.	Descending power of the small piston.	Combined powers of both pistons.
lbs.	*lbs.*	*lbs.*
...wer will be 2000 ...nce of the ... pounds per ...upon its up... ...nd no pres...	At first the power will be 0. Because the piston is in equilibrio, having 1000 pounds pressing upwards, and 1000 pounds downwards.	At first . . . 2000
...h of the de- ... power will ...inished, by ...crements, to 1600 ...the air be- ...two pistons ...three-fourths ...cylinder, and ...of the great ...ich is a space ...se and one- ...the original ...h it filled; ...spaces will ...1½ and if the ...air is as the ...ortion of the ...it occupies, ...on the great ...be as 4 to 5, or ...2000 = 1600.	At one-fourth, the power will be . . . . . 200. Because the equilibrium does not continue, and at one-fourth of the descent the pressure beneath the small piston is reduced by the expansion of the air between the two pistons to four-fifths of 1000 = 800 pounds, while the pressure above the piston continues to be 1000. The power is, therefore, 1000 − 800 = 200.	At one-fourth . 1800
...f of the de- ...e power will ...inished to 1333⅓ ...t this position ...rom the pis... ...is one-half of ...inder and one- ...s great one, ...pace equal to ...e-half of the ...ed originally. ...will therefore ...and the pres- ...ent piston as ...two-thirds of ...	At one-half of the descent, the power will have increased to . 333⅓. Because the pressure beneath is diminished by the increased rarity of the air to two-thirds of 1000 = 666⅔, while the downward pressure continues to be 1000. The power is therefore 1000 − 666⅔ = 333⅓.	At one-half . 1666⅔
...urths of the ...the power will . . . . 1142⁴⁄₇ ...the air must ...one-fourth of ...cylinder, and ...s of the large ...hich is a space ...se and three- ...the original ...as the spaces ...7 to 4, and the ...the great pis- ...venths of 2000	At three-fourths of the descent, the power will be . . . . . 428⁴⁄₇. Because the pressure beneath is reduced by the rarity of the air to four-sevenths of 1000 = 571³⁄₇; therefore the power is 1000 − 571³⁄₇ = 428⁴⁄₇.	At three-fourths 1571³⁄₇
...um of the cy- ...power will be 1000 ...the air must ...whole of the ...der, a space ...vice the small ...hich it at first ...pressure will ...be one-half of	At the bottom, the power will be . . . . . 500. Because the air beneath the piston is reduced to one-half of its pressure, or 500, which, deducted from 1000, leaves 500.	At the bottom . 1500
...powers ex-} ...the great } 7076 ...its descent. }	Sum of the powers of } the small piston . } 1461	Sum of the combined } 8538 powers }

Upon the action of this engine Dr. Rees, in his Cyclopædia, presents us with the following remarks and comparative statement between it and Mr. Watt's principle of expansion.

Now let us consider how Mr. Watt's principle of expansion would operate in the same circumstance; that is, in a cylinder of 14,14 inches diameter; which is to be supplied with air of 10 pounds pressure per circular inch, until it has completed one-half of its descent, and leaving the remainder of the descent to be accomplished by the expansion of the air already contained in the upper half of the cylinder.

	lbs
At the beginning, the power of descent will be	2000
At one-fourth, the power will still be	2000
At one-half, the power will be	2000
At three-fourths of the descent, the power will be diminished to .	1333⅓

Because the air must occupy one-fourth of the length of the cylinder, in addition to that half of the cylinder which it occupied before the expansion began; therefore the space is one and a half times the former, or as 3 to 2, and the pressure will be two-thirds of 2000.

| At the bottom, the pressure will be . . . . | 1000 |

Because the air is expanded to occupy twice the space it filled before.

8333⅓

The sum total is very nearly the same as the former, but both are greater than they should be, from the imperfect manner in which we have been obliged to make our calculation, so as to express it in common arithmetic, without having recourse to fluxions, which is the only method of treating quantities that are constantly increasing or decreasing by any given law.

The source of the inaccuracy is easily explained: at first we set with the pressure at 2000 pounds in Mr. Hornblower's engine, and did not take into the account that it decreases at all, until the piston has descended to one-fourth, but reasoned as though it diminished all at once at that place; whereas it began to diminish from the very first starting. Here then we have taken a small quantity too much. In the same manner, our process takes no notice of the diminution which happens between one-fourth and one-half of the descent, or between the other points at which we have chosen to examine it; the result is, as if the diminution took place suddenly at each of those points. The remedy for this would have been to have taken the account at a greater number of places, as it is by fluxions alone that we can take an infinite number, so as to obtain a true result. Now in the

the whole stroke, as will appear to any person who

r the trouble to read Professor Robison's investiga-

ut if we consider the difference of the manner in

he whole power is expended during the stroke, we

e great reason to prefer Mr. Hornblower's method,

much greater uniformity of the action; it begins at

nd ends at 1500, whilst Mr. Watt's begins at 2000,

ls at 1000; hence the necessity of those ingenious

nces for equalizing the action in Mr. Watt's patent of

Mr. Hornblower's is not uniform, but approaches

ity more nearly, so that he could have carried the

the expansive principle much farther, in employing

steam, than we believe he ever proposed to do.

ave been thus full upon this subject, because the

more power by the expansion of air or steam acting

le cylinders, has been a favourite idea with many,

e are no less than five different patents for it, but

of these have been upon mistaken notions; neither

tt's nor Mr. Hornblower's can have any advantage

tting off the air, or from a double cylinder, when air

o press the piston; nor could they derive any advan-

m the expansion of steam in their engines, if the

of it was inversely as the space it occupies.

dvantage of the expansive principle arises wholly

peculiar property of steam, by which, when suffered

d itself to fill a greater space, it decreases in pressure

force by a certain law, which is not fully laid down;

the relation between its expansive force and the

hich it occupies is not clearly decided: but Mr.

found that, by applying these properties in their

contained in steam of any given elastic force. All we
with certainty is what is stated in our table of expan
viz. that water, being converted into steam, and confin
a close vessel, when heated until the thermometer indi
a certain temperature, will have a certain pressure or el
force. But here we must observe, that the thermo
indicates only the intensity of the heat, without afford
direct measure of its quantity. When steam is suffer
expand itself into any given space, the quantity of ra
water which will be found to be contained in any given
of steam, in its expanded state, must be undoubtedly pr
tioned to the quantity of water contained in the same
of the steam, before the expansion took place, in the in
ratio of the space which it originally occupied, and that
which it fills when expanded; but we cannot say the
is the case with heat; and it is the quantity of heat
which determines the elastic force.

We believe that, in practice, Mr. Hornblower was not
to obtain any greater effect from the application of
expansive action in two cylinders, than Mr. Watt did in
cylinder. In 1791—2, he erected an engine in Con
at Tin-Croft mine, of which the large cylinder was 27 i
diameter, and worked with a stroke of eight feet long
the small cylinder 21 inches diameter, working with
feet stroke. The only account we have been able to
of the performance of this engine, is from a pamphlet
lished by Thomas Wilson, an agent of Messrs. Boulto
Watt, professedly with a view to prevent the introduct
Mr. Hornblower's engines into that country, in whi
makes it appear, that it raised only 14,222,120 pou
water one foot high with each bushel of coals.

In Mr. Hornblower's own account of his engine, in
gory's *Mechanics*, he informs us, " That an engine
erected in the vicinity of Bath, some years since, o
principle, and under very disadvantageous circumsta
the engine had its cylinders 19 inches and 24 inches
meter, with lengths of stroke in each suitable to the occ
viz. six feet and eight feet respectively. The conde
apparatus was very bad, through a fear of infringeme
Mr. Watt's patent, and the greatest degree of vacuum
could be obtained, was no more than 27 inches of me
The engine worked four lifts of pumps to the depths o
feet, 4500 pounds, 14 strokes in a minute, six feet each,
a cylinder six feet long, and 19 inches diameter, w
great deal of inertia and friction in the rods and bu

Wolf's Table of the relative pressures per square inch; the temperature and expansibility of steam at different degrees of heat above the boiling point of water, beginning with the temperature of steam of an elastic force equal to five pounds per square inch, and extending to steam able to sustain forty pounds on the square inch.

Pounds per Square Inch.		Degrees of Heat.		
5		227½		5
6		230¼		6
7		232¼		7
Steam of an elastic force acting over the surface of the atmosphere upon the safety-valve	8	requires to be maintained by a temperature equal to about	235¼	and at these respective degrees of heat, steam can expand itself to about	8	times its volume, and continue equal in elasticity to the pressure of the atmosphere.
9	237½	9
10	239½	10
15	250¼	15
20	259½	20
25	267	25
30	273	30
35	278	35
40	282	40

And so in like manner, by small additions of temperature, an expansive power may be given to steam to enable it to expand to 50, 60, 70, 80, 90, 100, 200, 300, or more times its volume, without any limitation but what is imposed by the frangible nature of every material of which boilers and other parts of steam-engines can be made. Prudence dictates that the expansive force should never be carried to the utmost which the materials can bear, but rather be kept considerably within that limit.

Having thus explained the nature of his discovery, Mr. Wolf proceeds to give a description of his improvements grounded thereon.

If the engine is constructed originally with the intention of adopting these improvements, it ought to have two steam cylinders of different dimensions, and proportioned to each other, according to the temperature or the expansive force determined to be communicated to the steam made use of in working the engine; for the smaller steam-vessel or cylinder must be a guide for the larger. For example; if steam of forty pounds the square inch is fixed on, then the smaller cylinder should be at least one-fortieth part the contents of the larger one. Each cylinder should be furnished with a piston, and the smaller cylinder should have a communication, both at its top and bottom, (top and bottom being here employed merely as relative terms, for the cylinders may be worked in a horizontal, or any other required position, as well as vertical,) with the boiler which supplies the steam; and the

o

communications, by means of cocks or valves of any construction adapted to the use, are to be alternately opened and shut during the working of the engine. The top of the small cylinder should have a communication with the bottom of the larger cylinder, and the bottom of the smaller one with the top of the larger, with proper means to open and shut them alternately by cocks, valves, or any other well-known contrivance. And both the top and bottom of the larger cylinder should, while the engine is at work, communicate alternately with a condensing vessel, into which a jet of water is admitted to hasten the condensation; or the condensing vessel may be cooled by any other means calculated to produce that effect.

This arrangement being made, when the engine is set to work, steam of a high temperature is admitted from the boiler to act by its elastic force on one side of the smaller piston, while the steam which had last moved it has a communication with the larger steam-vessel or cylinder, where it follows the larger piston, now moving towards that end of its cylinder which is open to the condensing vessel. Let both pistons end their stroke at one time; and let us now suppose them both at the top of their respective cylinders, ready to descend; then the steam of forty pounds the square inch, entering above the smaller piston, will carry it downwards; while the steam below it, instead of being allowed to escape into the atmosphere, or applied to any other purpose, will pass into the larger cylinder above its piston, which will make its downward stroke at the same time that the piston of the smaller cylinder is doing the same thing; and while this goes on, the steam which last filled the larger cylinder in the upward stroke of the engine will be passing into the condenser, to be condensed during the downward stroke. When the pistons in the smaller and larger cylinder have thus been made to descend to the bottom of their respective cylinders, then the steam from the boiler is to be shut off from the top, and admitted to the bottom of the smaller cylinder. The communication between the bottom of the smaller and the top of the larger cylinder is also to be cut off; and the communication is to be opened between the top of the smaller and the bottom of the larger cylinder. The communication between the bottom of the larger cylinder and the condenser is to be cut off, and the steam which, in the downward stroke of the engine, filled the upper part of the larger cylinder, suffered to flow off to the condenser. The engine will then make its upward stroke from the pressure of the steam in the top of the small cylinder acting beneath the piston of the great cylinder, and so on alternately, admitting the steam to the different sides of the smaller piston, while the steam last admitted into the smaller cylinder passes alternately to the different sides of the larger piston in the larger cylinders: the top and bottom of which are at the same time made to communicate alternately with the condenser.

In an engine working in the manner just described, while the steam is admitted on one side of the piston into the smaller cylinder, the steam on the other side has room made for its admission into the larger cylinder, on one side of its piston, by the condensation taking place on the other side of the large piston which is open to the condenser; and that waste of steam which takes place in engines worked only by the expansive force of steam, from steam passing the piston, is prevented; for all steam that passes the piston in the smaller cylinder is received into the larger.

In such an engine, where it may be more convenient for any particular purpose, the arrangement may be altered, and the top of the smaller made to communicate with the top of the larger cylinder; in which case the only difference will be, that when the piston in the smaller cylinder descends, that in the larger will ascend, and *vice versâ*; which, on some occasions, may be more convenient than to have the two pistons moving in the same direction.

This engine is exactly the same in its action as Mr. Hornblower's, which we have before described. The novelty consists in the application of steam of a high pressure thereto, and in proportioning the capacities of the two cylinders to the expansibility of the steam, according to his table. But Mr. We goes on to state, that effectual means must be used to keep up the requisite temperature in all parts of the apparatus into which the steam is admitted, and in which it is not intended to be condensed; and here it may be proper to state, that instead of the usual means of accomplishing this, by enclosing them in the boiler, or in a steam-case communicating with the boiler, a separate fire may with advantage be made under the steam-case containing the cylinders, which in that event will become a second boiler, and must be furnished with a safety-valve, to regulate the temperature. By means of the last-mentioned arrangement, the steam from the smaller cylinder, or steam measurer, may be admitted into the larger cylinder, when kept at a higher temperature than the steam in the smaller cylinder, by which its power to expand itself may be increased; and, on the contrary, by keeping the larger cylinder at a lower temperature than the smaller, its

expansibility will be lessened, which, on particular occ.
and for particular purposes, may be desirable. In ever
care must be taken that the boiler, or case in which the
der is enclosed, the steam-pipes, and generally all the
exposed to the action of the expansive force of the stean
have a strength proportioned to the high
they are to be exposed.

It is not advisable that the proportion of the capa
the smaller cylinder, or steam-measurer, to the capacity
larger or working cylinder, should in any case be smalle
the proportion of the expansion of the steam which i
used in it, as we have stated, yet, in the making of it
considerable latitude may be allowed; for example
steam of forty pounds the square inch, a small cylin
measurer, of one-twentieth, or even larger, instead of
fortieth the capacity of the larger or working cylinder,
with steam of any given strength. And in many cases,
be advisable that this should be the case, because
difficulty of preventing some waste of steam,
condensation, which might lessen the rate of
if not allowed for in the size of the small cylinder
measurer.

In all cases when the engine is ready for working, wl
may be the proportion that has been adopted, or inter
be worked with, it should have its power tried by altori
load on the valve that ascertains the force of the ste
order that the strength of steam best adapted for the
may be ascertained, for it may turn out to be advan
that the steam should be employed in particular en
an elastic force somewhat over or under what i
intended.

Mr. Woolf also states, that Mr. Watt's engines i
improved by the application of his discovery in maki
boiler, and the steam-case in which the working cyli
enclosed, much stronger than usual, and by alteri
structure and dimensions of the valves for admitting
from the boiler into the cylinder in such a manner t
steam may be admitted very gradually by a prop
enlargement of the aperture, so as at first to wired
steam and afterwards to admit it more freely. The re
this precaution is this, that steam of such elastic f
Mr Woolf proposes to employ, if admitted suddenly i
cylinder, would strike the piston with a force as wo
danger the safety and durability of the engine. The a
allowed for admitting steam into the cylinder, or cyl

appear obvious, from attending to what takes place in the working such a piston. When the piston is ascending, that is, when the steam is admitted below it, the space on its upper side being open to the condenser, the steam, endeavouring to pass up by the side of the piston, is met, and effectually prevented by the column of metal, equal or superior to it in pressure, and during the down stroke, no steam can possibly pass without first forcing all the metal through.

In working what is called a single engine, a less considerable altitude of metal is required, because the steam always acts on the upper side of the piston; and in this case, oil, or wax, or fat of animals, or similar substances, in sufficient quantities, will answer the purpose. But care must be taken, either in the double or single engine, when working with this piston, that the outlet which conveys the steam to the condenser shall be so situated, and of such a size, that the steam may pass freely, without forcing before it, or carrying with it, any of the metal or other substance employed that may have passed by the piston; and at the same time providing another exit for the metal, or other substance collected at the bottom of the cylinder, to convey the same into a reservoir kept at a proper heat, whence it is to be returned to the upper side of the piston by a small pump, worked by the engine, or by some other contrivance. In order that the fluid metal used with the piston may not be oxydated, some oil or other fluid substance is always to be kept on its surface, to prevent its coming in contact with the steam; and to prevent the necessity of employing a large quantity of fluid metal, although the piston must be as thick as the depth of the column required, the diameter need be only a little less than the steam-vessel or working cylinder, excepting where the packing, or other fitting, is necessary to be applied; so that, in fact, the column of fluid metal forms only a thin body round the piston.

We have seen an engine of an eight-horse power of this kind at work, with a fluid metal on the pistons: it effectually prevented the leakage. But as it required to have the cylinders twice as long as usual, in order to have sufficient room for the long or thick pistons which it required, and as these pistons must be of considerable weight, the method is not at all applicable in practice; and, indeed, the increase of the bulk of the moving parts is such as to counterbalance the advantage, which is confined to the saving of steam by leakage; for the friction must be greater than in another engine, because the piston must be packed as tight as usual, to be able to sustain a column of fluid metal, which must be more than equal in pressure to that of the steam; and when the steam presses upon the piston, the pressure of the fluid metal to leak by the piston must be double that of the steam: also the friction of so great a surface of fluid metal pressing against the inside of the cylinder is very great.

In 1810, Mr. Woolf had a third patent, the object of which is to prevent the waste of steam from leakage by the piston. For this purpose, he does not allow the steam to come to the piston at all, but causes it to act in a different vessel, and transmits the action thereof to the piston by oil or fluid metal: thus, at the side of the cylinder, he places a separate vessel, communicating with the lower part of the cylinder by a large pipe or passage from the bottom of each; then steam, being admitted into this vessel, will press upon the surface of the oil or fluid metal contained in it, and force the same to pass out of that vessel into the cylinder, where it will act beneath the piston to press the same upwards, a vacuum being at the same time made in the upper part of the cylinder to give effect to the pressure.

The steam is then made to press upon the upper surface of the piston, which is always covered with a quantity of the fluid; and at the same time a vacuum is made in the separate vessel, so as to relieve the surface thereof from all pressure; in consequence the piston is made to descend. It is evident that the piston must be packed so tight as to suffer none of the fluid to pass by it; but this is easy, in comparison with the difficulty of making a packing sufficiently tight to resist the passage of steam, particularly when it is so rare as the expanded steam which Mr. Woolf sometimes uses in his engine. The separate vessel of which we have spoken, is in some cases to be the jacket or space which surrounds the cylinder, which is then to be open at bottom.

This contrivance is ingenious, but we think the necessity of an additional cylinder is an objection which will prevent its adoption in large engines; and for small engines the advantages are not so great.

Since his first patent, Mr. Woolf has erected several small engines, which performed well, and with an evident economy of fuel. But these engines being employed to turn mills, of which the operations do not afford so exact an estimate of the power as the operation of pumping water, Mr. Woolf's engines did not come to a direct and indisputable comparison with those on Mr. Watt's principle, until 1815, when two

large engines were set to work in Cornwall, at Wheal Vor
and Wheal Abraham mines, for pumping water; and these
have since been regularly reported in Messrs. T. and J.
Lean's reports, and of which one of the objects was to ascertain the comparative merit of the double and single cylinder
engines,

The report for May, 1815, states the average performance of these two
engines at 49,980,882 lbs. lifted one foot high for each bushel of coals; and
since that time they have done more than 50,000,000 lbs.

The engine at Wheal Vor has a great cylinder of 53 inches diameter, and
nine-feet stroke; and the small cylinder is about one-fifth of the contents
of the great one. The engine works six pumps, which, at every stroke, raise
a load of water of 37,982 lbs. weight 7½ feet high, which is the length of the
stroke in the pumps. This makes a pressure of 14,1 lbs. per square inch on
the surface of the great piston, and it makes 7,6 strokes per minute. With
respect to its consumption of coals, it raised, in March, 1816, 48,432,702lbs.
one foot high with each bushel; April, 1816, 44,000,000 lbs.; May, 1816,
49,500,000 lbs.; and in June, 1816, 43,000,000 lbs.

From the same reports we learn, that the engine at Wheal Abraham mine
has a great cylinder of 45 inches diameter, working with a seven-feet stroke,
at the rate of 8,4 strokes per minute under a load of 24,050 lbs., which it
raises seven feet at each stroke. Its performance during the above four
months was 50,000,000 lbs.; 50,908,000 lbs.; in May, 56,917,312 lbs.,
which, we believe, is the greatest performance ever made by a steam-engine;
and in June, 51,500,000 lbs,

We must observe, that the variation in the performance
of different steam-engines, which are constructed upon the
same principle, and working under the same advantages, is
the same as would be found in the produce of the labour of
so many different horses, or other animals, when compared
with their consumptive food; for the effects of different
steam-engines will vary as much from small differences
in the proportions of their parts, as the strength of animals
from the vigour of their constitution; and, again, there will
be as great differences in the performance of the same
engine, when in bad or good order, from all the parts
being tight and well oiled, so as to move with little friction,
as there is in the labour of an animal, from his being in good
or bad health, or excessively fatigued; but in all cases, there
will be a maximum which cannot be exceeded, and an average
which we ought always expect to attain.

Fig. 198 is a sketch to show the arrangement of the valves and cylinders
of these two engines; A is the large cylinder, and B the little cylinder, each
enclosed in its steam-case. The steam is admitted from the boiler into the
steam-case of the large cylinder A, by a communication at C; and there is
a communication between this steam-case and that of the small cylinder;
so that all the steam for the supply of the engine passes through both of the
steam-cases, which therefore become part of the communication between the
boiler and the little cylinder, into which the steam is first admitted. D

verted. into steam of a temperature and under a pressure uncommonly high, and also to present an extensive portion of convex surface to the current of flame and heated air from a fire; likewise of other large cylindrical receptacles placed above the former cylinders, and properly connected with them, for the purpose of containing some water and the steam.

These cylindrical vessels are set in a furnace so adapted to them, as to cause the greater part of the surface of each of them, or as much of the surface as may be convenient, to receive the direct action of the fire, or heated air or flame.

Figs. 199 and 200 represent one of these boilers in its most simple form. It consists of eight tubes marked *a*, made of cast-iron, or any other fit metal, which are each connected with the larger cylinder A, placed above them, as is shown in the side view, fig. 200, in which the same letters refer to the same parts as in fig. 199. In fig. 200 is also shown the manner in which the fire is made to act. The fuel rests on the grate-bars at B, and the flame and heated air, being reverberated from the part above the two first smaller cylinders, go under the third, over the fourth, under the fifth, over the sixth, under the seventh, and partly over and partly under the eighth small cylindric tube, all which tubes are full of water. The direction of the flame, until it reaches the last-mentioned tube, is shown by the dotted curved lines and arrows. When it has reached that end of the furnace, it is carried by the flue, O, to the other side of a wall, built beneath the main cylinder A, in the direction of its length, and the flame then returns under the opposite end of the seventh smaller cylinder over the sixth, under the fifth, over the fourth, under the third, over the second, and partly over and partly under the first, when it passes into the chimney. The wall before-mentioned, which divides the furnace longitudinally, answers the double purpose of lengthening the course which the flame and heated air have to traverse, giving off heat to the boiler in the passage, and also of securing the flanges, or other joinings, employed to unite the smaller tubes to the main cylinder, from being injured by the fire. The ends of the small cylindric tubes rest on the brickwork which forms the sides of the furnace, and one end of each of them is furnished with a cover, secured in its place by screws and a flanch, but which can be taken off at pleasure, to allow the tubes to be cleared, from time to time, from any incrustation or sediment which may be deposited in them.

To any convenient part of the main cylinder A, a tube is affixed, to convey the steam to the steam-engine. In working with such boilers, the water carried off by evaporation is replaced by water forced in by the usual means for a high pressure boiler, that is, a forcing-pump; and the steam generated is carried to the place intended by means of pipes connected with the upper part of the cylinder A. In the specification, means are pointed out for applying this plan to the boilers of steam-engines already in use, by ranging a row of cylinders beneath the present boiler, and connecting them with each other, and with the boiler. Directions are also given for constructing boilers composed of cylinders disposed vertically. In every case the tubes composing the boiler should be so combined and arranged, and the furnace so constructed, as to make the fire and flame act around and over the tubes, so as to embrace the largest possible quantity of their surface. It must be obvious to any one, that the tubes may be made of any kind of metal; but cast-iron is the most convenient. The size of the tubes may be varied; but in every case, care should be taken not to make the diameter too great; for it must be remembered, that the larger the diameter of any single tube is in such a boiler, the stronger it must be made in proportion, to enable it to bear the same expansive force of steam as the smaller cylinders. It is not essential, however, to the invention, that the tubes should be of different sizes; but the upper cylinders, especially the one which is called the steam cylinder, should be larger than the lower ones, it being the reservoir, as it were, into which the lower ones send the steam, to be thence conveyed away by the steam-pipe. The following general directions are given respecting the quantity of water to be kept in a boiler of this construction; viz. it ought always to fill, not only the whole of the lower tubes, but also the great steam cylinder A, to about half its diameter, that is, as high as the fire is allowed to reach; and in no case should it be allowed to get so low, as not to keep the vertical necks, or branches, which join the smaller cylinders to the great cylinder, full of water, for the fire is only beneficially employed when applied, through the medium of the interposed metal, to water, to convert it into steam; that is, the purpose of the boiler would in some measure be defeated if any of the parts of the tubes which are exposed to the direct action of the fire, should present a surface of steam in their interior, instead of water, to receive the transmitted heat. This must, more or less, be the case, whenever the lower tubes, and even a part of the upper, are not kept filled with the water.

Respecting the furnace for this kind of boiler, it should always be so built as to give a long and waving course to the flame and heated air, forcing them the more effectually to strike against the sides of the tubes which compose the boiler, and so to give out the greatest possible portion of their heat before they reach the chimney. Unless this be attended to, there will be a much greater waste of fuel than necessary, and the heat communicated to the contents of the boiler will be less from a given quantity of fuel.

When very high temperatures are not to be employed, the kind of boiler just described is found to answer very well; but where the utmost force of the fire is desirable for producing the most elastic steam, the parts are combined in a manner somewhat different, though the principle is the same. In the *Philosophical Magazine*, vol. xvii. p. 40, is a description and drawing of a boiler of this kind, two of which were erected, in 1803, at Messrs. Meux's brewery.

In every case Mr. Woolf uses two safety-valves, at least, in his apparatus, to prevent accidents; a precaution which cannot be too strongly enforced, as it may happen, when but one is employed, that by some accident it may get locked, and the engine and people about it be exposed to the danger of an explosion.

In those engines of Mr. Woolf's which we have seen, he employs boilers like the one described, viz. with two small tubes beneath, which are full of water, and exposed to the

immediate action of the flame, communicating by pe
dicular necks or branches with the large cylinder a
which has water in the lower part, and steam in the u
The only difference from what we have above describ
that the lower and upper tubes are placed in the same (
tion, instead of being at right angles to each other; an
flame proceeds in the direction of their length, inste
crossing them; the lower or water tubes are rather inc
upwards. The metal of these tubes is made very 1
with a view to strength and durability.

The idea of making boilers for raising strong steam
number of small tubes, which can be made stronger th.
large vessel, is not original with Mr. Woolf, Mr. Blak
whom we have before spoken, having proposed it in a
tract which he published in French, at the Hague, in
But his tubes were to be placed over each other, i
inclined direction, and the water, being admitted at the 1
end, ran down within the heated inclined tubes, and be
converted into steam.

Woolf's regulating steam-valve.—Besides the co
safety-valves, Mr. Woolf has also introduced a valve of .
construction into the steam-pipe itself, to regulate the
tity that shall pass from the boiler. In fact, it is a self-a
steam-regulator, and extremely ingenious.

A (fig. 201) is a part of the great or steam cylinder of one of Mr. W
boilers; B B, the neck or outlet for the steam, surmounted by a stea
C, which is joined to the neck B B, by the flanges a, a. The top of
of the steam-box C, marked with the letter D, is well secured in its
and has a hole through it for the rod of the valve to pass; and the i
of the hole is formed to a box to hold a stuffing, and make the rod w
and down steam-tight, the stuffing being kept in its place by mean
collar, screwed down in the usual way, as shown in the figure. By
of a pin b, and the two vertical pieces e, e, the sliding-valve rod is mai
to m, which is a close cover to the hollow cylinder n n. The cover
steam-tight into the conical seat, at the upper end of a collar o o, wl
made fast to the flange a a, and descends into the neck of the boiler, so
a barrel, in which the cylinder fits close. The cylinder n n is of
bottom, having a free communication with the steam in the boiler A
it has three vertical slits cut through the sides, one of which, S, is sho
the plate. The sum of the area of all these slits or openings is equal
area of the opening of the seat or collar o o, in which the cylinder n n v

When the steam acquires a sufficient degree of elastic force to rai:
valve, (that is, the cylinder n n, with its cover m, and the rod R,) to,
with whatever weight the rod may be loaded, then the openings S,
above the steam-tight collar or seat o o, allow the steam to pass int
steam-box C, and to flow off to the engine through the pipe N. Il
quantity of steam that passes is proportioned to the elastic force i
acquired, and the weight with which the valve is loaded; because th
of the openings, S, above the collar o o, will be in that proportion.

This valve may be loaded by applying weights in any of the usual methods; but Mr. Woolf prefers the one shown in the drawing, in which the upper part of the rod R is joined, by means of a chain, to a quadrant of a circle Q, for the purpose of carrying a pendulum weight Z, that admits of being moved nearer to or farther from the centre of the quadrant, according as the pressure of the valve is wished to be increased or diminished.

As the valve rises, the weight moves upwards in the arc n n, giving a continually increased resistance to the farther rising of the valve, proportioned to the horizontal distance of the weight from the centre of Q, of which the weight attains a continual increase by its rise in the arc, according to the horizontal distances measured on the line Q p, pressing through the centre of the weight by perpendiculars from the horizontal line.

Thus, if the weight Z presses down the valve m with a force equal to 20 pounds on the square inch of the aperture in o o, in its present position, when it rises to the position at i, it will press with a force equal to 30 pounds, and at p, with a force equal to 40 pounds on the square inch; so that the rod Z may be made to serve at the same time as an index to the person who attends the fire, nothing more being necessary for this purpose than to graduate the arc, described by the end of the rod Q Z, by experimental trials. In the side of the steam-box C there is an opening, N, to allow the steam to pass from it by a pipe to the steam-engine.

It is plain that the adjustment of the positive pressure on this valve can be determined by sliding the weight Z of the pendulum to a greater or less distance from the centre of motion. Again, to adjust the rate of the increasing forces, so as to correspond with the increasing force of the steam, the radius of the quadrant, Q, must be apportioned to the diameter of the valve and the opening of the slits, S, so that the ascent of the weight, Z, in its quadrant will be correspondent to the varying pressure. This adjustment must be made as nearly as it can be done before the valve is fixed; and to bring it afterwards to an exact regulation, the chain is attached to the rod, R, by a nut and screw; by means of which, any part of the arc can be used that is found most correspondent with the varying pressure, because the rate at which the resistance of the lever increases is much rapid when the pendulum is near to the perpendicular, than when it approaches the horizontal position.

The same effect may be produced, by making the slits in the side of the cylinder narrower at the lower part of the cylinder, instead of being equal.

BELL-CRANK ENGINE.

Messrs. Boulton and Watt, soon after the expiration of their patent for effecting condensation in a separate vessel, introduced a form of engine, called the bell-crank engine, of which we shall represent so much as is necessary to exhibit the alteration in the mode of construction.

Fig. 102 is a side view of the engine. A B C is the bell-crank, there being another exactly similar part on the other side, moving upon a fixed centre, C, the end A D is joined to a cross-piece which works the piston-rod in the cylinder. E serves for the air-pump, and G for the cold-water pump, and the hot-water pump may be worked upon the same bar. The connecting rod from B to H is supposed to be attached to the crank of the fly-wheel at H. Engines of this description are mostly constructed with slide or D valves, which are worked by the beam A C. This form of engine does not possess any particular advantages otherwise than those arising from compactness, which are not of sufficient weight to counter-balance the increased friction. It was, in some few instances, at the commencement of steam navigation, applied to boats, but it was found to answer not so well as the double-beam engine.

VIBRATING ENGINE.

WITH a view to do away with the beam of the engine, and to communicate the motion direct from the piston-rod to the fly-wheel crank, a form of engine has been constructed, which, in engines of small dimensions, where the piston-rod can be made of sufficient strength compared with the weight of the cylinder that is to vibrate, have answered tolerably well. We have seen one of about one-horse power, which had been at work four years.

Fig. 203. A is the cylinder, B the piston-rod, C the crank, D the fly-wheel, E a stand supporting the cylinder pivot F, which has a similar one on the opposite side. One of these pivots is formed like the key of a four-way cock, having a communication to the top and bottom of the cylinder. By the movement of the piston, the cylinder is caused to vibrate, to turn the crank and fly-wheel, and the steam passes alternately to the top and bottom of the cylinder, by the two-way axes on which the cylinder vibrates.

When engines of this construction are formed of any considerable size, there is a danger of bending the piston-rod, and in vibrating, the weight of the cylinder loosens its fitting in the stuffing collar of the cylinder-cap.

ROTATORY ENGINE.

ALL steam-engines as yet noticed, have their action produced by the movement of a piston in a cylinder, and act by what is called a reciprocating motion. In engines of this description, a very considerable degree of power is expended in arresting the motion of the different working parts, and putting them into action in a direct contrary course: this has claimed much attention from engineers, and many attempts have been made to construct an engine in which the action of the steam should operate in a continuous manner, without bringing the parts to a state of rest.

The most obvious mode of attaining this object, is the producing a rotatory motion. One of the most simple engines on this construction is represented in fig. 204, where two sections are shown, the one at right angles to the revolving shaft, the other parallel to it, the same letters in both denoting the same parts. U U U U is a circular steam-case, with the ·

... inclosed by the circular plates V V V V, through which the shaft
... passes. To R is attached by four arms, S S S S, the ring P P, in which
... fans or flat pieces, A and B, are fixed on hinges formed steam-tight, but
... of being shut in upon the ring, as A, or opening and closing the
... steam-course O O O O, as B. To each of these four pieces is attached a
... or tripping piece, C and D, which, during their revolution, touch the
... and raises their respective fans into the steam-way, as shown by the
dotted fan at A¹, just after it has passed the steam aperture, I. The passage
to the condenser is represented at N; G is a camb-piece attached to the
other one, and fitting in a steam-tight manner upon P P P P, serving to
... the fans as they come round. The steam entering at I presses upon
G and A¹, which is supposed to have been just raised to that position, and
... it round, together with the ring P P P P, and the centre shaft R, until
... the aperture through which the steam issues to the condenser, prior
... which the other fan, B, passes the steam-way, and obtains a position to
... the action of the steam, and continue the motion.

The steam-way, O O O O, may be considered to be a cylinder bent round,
and the fans, as they obtrude themselves, act the part of a piston, receiving
the impulse of the steam always on the one side, and effecting the condensation always on the other. It being requisite that the steam-way
should have some termination, the obstacle, G, is indispensable, and the
... of the fans upon hinges, or some other mode, to pass such
..., is unavoidable; and therefore, from being thus compelled to move
the piece acting as a piston continually to and from its fittings, it becomes
extremely difficult to maintain those fittings steam-tight. This, together
with the steam-way not being capable of receiving the cylindrical form,
the inconveniences of great moment. It has been found, therefore, that in
... engines of this construction in a working condition, great
..., which hitherto have not been surmounted; and as at present
these engines exist to no useful end, we shall refrain from describing them
further.

HIGH-PRESSURE ENGINES.

If water be urged greatly by fire, steam of greater pressure is obtained; and it has been long known, that the
extent of the pressure increases in a greater ratio than the
expenditure of heat, which has been an inducement to many
to attempt to use steam at excessive pressures. The pressure generally allowed in high-pressure engines, is not more
than 30, 40, and seldom exceeds 50 pounds to the inch.

In engines where the pressure is so great, the weight of
the atmosphere is not taken into account, and the mode of
obtaining the motion of the piston is, by allowing one end of
the cylinder to be open to the air, whilst the steam acts on
the opposite side of the piston. By this mode of operation,
all the parts appertaining to the promotion of condensation,
are dispensed with, and, consequently, the expense of making
these parts, the friction caused by their operations, and the
attention which was necessary to their well-being, is entirely
saved. This gives to the engine a peculiar degree of simplicity, but it is unfortunately attended with some danger.

Steam was applied in this mode so early as the year 1724, and is described by Leupold, in his *Theatrum Machinarum Hydraulicarum*, vol. ii. p. 93. The engine thus described, is formed with two cylinders, having pistons fitted and attached to two separate beams, whose other ends are connected with two force-pumps. Between the two cylinders is a four-way cock, and as the pistons are weighed and brought down to the bottom of each cylinder, it is evident that by means of this cock, the steam can be let on alternately to the bottom of each cylinder, whilst, at the same time, the opposite cylinder to that in which the steam is admitted has a communication with the atmosphere. Thus, by turning the cocks, the two pistons are alternately raised by the steam, and permitted to descend by the loading of weights attached to their other end. This simple construction of high pressure may be placed on a par with Newcomen's condensing engine.

Mr. Watt presents to our notice this mode of using the direct action of steam, in the latter part of his specification, in 1769; but the most common application of high-pressure steam, of late years, is used in a form of engine invented by Mr. Trevitheck, for the purpose of applying this power to locomotion. He obtained a patent for it, in union with Mr. Vivian, in the year 1802. This engine, from its compactness, is peculiarly applicable to this purpose, as it requires no condensing water, which would be an insurmountable bar to its introduction.

Fig. 205 presents a section of this form of engine. A B is the boiler, A¹ a safety-valve, C D the cylinder, E the four-way cock, G the passage from the boiler, H the passage to the chimney, G¹ for the exit of the steam. E is a four-way cock, F the passage to the top, and K the passage to the bottom of the cylinder. M the piston, N the piston-rod, O the connecting rod, joined to the cranks of the fly-wheel. The beam R is worked by the connecting rod, which has the rod of a small force-pump S attached to it, acting on the other side of the boiler, and forcing water along Q U into the boiler by I. The fire-place is behind the chimney, as seen in the view, and is surrounded on all sides by the boiler. Fig. 206 is a section of the cylinder at right angles to the section at fig. 205. The four-way cock is moved by means of a lever on its axis, which is struck by a tapit upon a rod from the cross-piece C². It must be understood that there is another connecting rod and crank on the farther side of the engine, and that the beam C² connects them.

This engine, we conceive, requires little elucidation. The four-way cock permits the steam to pass alternately to the top and bottom of the cylinder by the passages F and K, and affords it egress by G¹: and the cold water coming to supply the boiler, surrounding it on all sides, imbibes its heat, by which means the boiler is fed with water of a much higher temperature, and the steam is condensed in H, by which a more rapid exit is obtained for it.

This kind of engine was expressly intended for working carriages. A locomotive engine was made by Mr. Trevethick, in South Wales, in 1804, and was tried upon the rail-roads at Merthyr Tydvall. It drew after it as many carriages as carried ten tons of bar iron for a distance of nine miles, without any further supply of water than that contained in the boiler at setting out, travelling at the rate of five miles per hour. Since that period they have been tried in many places upon rail-roads, but their introduction had not become general until 1811, when Mr. Blenkinsop, proprietor of the Middleton Coal-works, which supply the town of Leeds, adopted them for conveying the coals on his rail-road. Mr. Blenkinsop, when he adopted the locomotive engine, took up the common rails on one side of the whole length of the road, and replaced them with rails which had cogs on their upper surface. These cogs are cast at the same time with the rails, and are hollow beneath, to be as light as is con-sistent with strength and durability. The pitch of the cogs is six inches, so that each rail of three feet in length has only six cogs. A wheel which is fixed on an axis which would be that of the fly-wheel at one side of the carriage, works in the teeth of these rails; the whole machine is thus caused to advance along the railway. Many fruitless attempts have been made to produce an engine capable of moving carriages upon common roads; but before this can be effected, the numerous parts of the engine must be made more compact, and its weight considerably reduced.

Observations on the work, &c. &c. of steam-engines in Cornwall, from August 1811 *to May* 1815, *inclusive, by Messrs. Lean.*

Messrs. Thomas and John Lean were appointed to the general superintendence; and the different proprietors, as also the regular engineers of the respective mines, engaged to give them every facility and assistance in their power. Their first Monthly Report was for August, 1811, and included eight engines, which had in that month consumed 32,061 bushels of coals, and lifted 126,126,000 pounds of water one foot high, with one bushel of coals for each engine, being an average duty of 15,760,000 pounds lifted one foot high with each bushel of coals. In the months of September and October the engines reported were nine, and in November and December twelve; and it now evidently appea the regular publication of Messrs. Leans' very usefi had already been attended by some improvements in t dition of the engines; for the average duty for De 1811, extracted from these tables, appears to hav 17,075,000 pounds.

In January, 1812, the number of engines reported w teen, and by the end of that year they were increased teen; and the average duty performed by all the en the last-mentioned month had advanced to 18,200,000

In 1813, the number of engines included in the 1 Reports continued to increase, till in December they v and the average work 20,162,000.

During some of the months of 1814, the engines were 32, and the average duty performed during D was 19,784,000 pounds lifted one foot high with eacl of coals.

The table which is subjoined is an abstract from Leans' Reports, and has been formed by first count: many engines are reported, as in January 1815, 32 e then adding up the column containing the quantity consumed by all the engines during the month, and down the amount, 110,824; in like manner adding column of pounds lifted by each engine one foot high bushel of coals, the amount of which was 637,320,9: lastly, dividing the latter quantity by 32, the nu engines at work, to obtain the average duty perforu 19,916,250 pounds.

TABLE.

	Number of Engines reported.	Bushels of coals consumed by all the engines.	Bushels of coals upon which the report is founded.	Pounds of water lifted one foot high by the coals so reported.	p on bu
1811. August	8	23,661	8	126,126,000	1:
September	9	25,237	9	125,164,000	1:
October	9	24,487	9	121,910,000	1:
November	12	30,998	12	189,340,000	1:
December	12	39,545	12	204,907,000	1:
1812. January	14	50,089	14	237,661,409	1(
February	15	54,349	15	260,514,00C	1'
March	16	59,140	16	274,222,000	1'
April	16	62,384	16	276,233,000	1'

TABLE continued.

	Number of Engines reported.	Bushels of coals consumed by all the engines.	Bushels of coals upon which the report is founded.	Pounds of water lifted one foot high by the coals so reported.	Average of pounds lifted one foot high with each bushel of coals
1812. May	16	51,903	16	273,546,000	17,096,000
June	17	50,410	17	288,076,000	16,940,000
July	17	51,574	17	300,441,000	17,677,000
August	17	44,256	17	314,753,000	18,510,000
September	18	46,536	18	348,396,000	19,355,000
October	18	53,941	18	321,900,000	17,883,000
November	21	57,176	21	381,460,000	18,160,000
December	19	55,784	19	341,803,000	18,200,000
1813. January	19	60,400	19	363,906,000	19,153,000
February	22	58,044	22	438,737,000	19,940,000
March	23	73,862	23	440,642,000	19,157,000
April	23	61,739	23	431,032,000	18,700,000
May	24	58,890	24	463,346,000	19,300,000
June	24	53,110	24	470,157,000	19,590,000
July	23	56,709	23	443,462,000	19,281,000
August	21	50,110	21	416,898,000	19,852,000
September	22	58,008	22	427,148,000	19,415,000
October	26	74,796	26	488,671,000	18,795,000
November	28	77,135	28	537,958,000	19,212,000
December	29	86,273	29	584,721,000	20,162,000
1814. January	28	91,753	28	550,751,000	19,670,000
February	26	78,986	26	536,677,000	20,641,000
March	28	109,904	28	565,406,000	20,193,000
April	29	91,607	29	576,617,000	20,325,000
May	28	79,437	28	569,319,000	20,305,000
June	30	75,343	30	626,669,000	20,888,000
July	27	85,224	27	573,208,000	21,229,000
August	26	70,443	26	545,019,000	20,960,000
September	27	78,167	27	560,608,000	20,763,000
October	32	75,080	32	630,704,000	19,709,000
November	32	82,000	32	637,322,000	19,916,000
December	29	84,669	29	573,744,006	19,784,276
1815. January	32	110,824	32	637,320,990	19,916,250
February	33	101,667	33	710,271,250	21,523,370
March	34	117,342	34	706,071,990	20,766,820
April	35	105,701	35	695,212,340	19,863,210
May	34	107,530	34	669,299,140	20,479,350

From the foregoing table it appears that the average duty of the engines reported, exclusive of Woolf's patent engine, is at this time about 20 millions.

We have purposely omitted Woolf's patent engine, because one of the ends intended to be gained by the Monthly Report

numerous set of experiments they decided, that a horse, working eight hours a day, was capable of raising 33,000 lbs. one foot high, in a minute. Therefore, by dividing the number of pounds an engine can lift one foot high in a minute, it will give the amount of horses' power to which that engine is equal.

An entire view of an engine of the construction termed portable, is represented at fig. 207. A is the cylinder, B the air-pump, C the cold-water pump, D the hot-water pump, E the beam, F the connecting rod, G the fly-wheel, H the eccentric shaft, and I the governor.

It would occupy many volumes to describe the various forms of construction of engines which have, since the knowledge of the power of steam, been contrived; and the information such descriptions would convey would be, comparatively speaking, of very little value, as the majority of them have arisen from men ignorant of the principles of the action of the machine, and whose productions should be classed as futile alterations.

In attempting any improvements, the principles of action should first be taken into consideration. In condensing engines the movement is effected by the alternate increase and decrease of temperature, the perfection of both of which is of great importance. The primary point to be aimed at, therefore, is the maintaining of high temperature whilst the steam is forcing, and reducing it suddenly when the condensation is to be effected. This was taken into consideration in the construction of Newcomen's engine, and was most effectually attained by Mr. Watt.

The other parts of the engine may be examined with a view of improvement, by considering their weight and friction, and by the substituting of a rotatory instead of a reciprocating motion.

Simplicity of construction cannot be too strongly recommended in all mechanical combinations; for there are many contrivances which would certainly be deserving of the name of improvements, were they not inapplicable on account of their intricacy.

Attempts have frequently been made to avoid the use of the air-pump, which takes up a considerable portion of the power of an engine. A water barometer, adapted to the condenser, has been sometimes adopted; and a fall of water has been made to pass over the upper edges and down the orifice of a tube, forcing the air before it. The upper end of this tube communicates with the eduction pipe, and is said to support a vacuum of considerable rarity. Exposing the steam which is to be condensed to an increased surface by passing it along tubes surrounded by water, or amongst tubes containing water, has likewise been frequently adopted. Indeed the exposure of considerable surface to receive heat in the generation of steam, and the same to abstract it in condensation, have been subjected to frequent trial. That an advantage is to be procured by the adoption of such plans is undoubted; but to attain such increased surface an intricacy in the parts, we fear, must be adopted, which will more than counterbalance the advantages gained.

The valves, or those parts of an engine which direct the distribution of the steam, have always had the attention of engineers, and, as we have shown, many elegant combinations have resulted from their ingenuity.

In running a steam-engine, attention should be given to the working parts. The cylinder should be packed with clean hemp and the best tallow, and frequently examined to see that the packing is in order. The steps of the fly-wheel, shaft, and of the crank, beam, &c. should be frequently examined, and kept well oiled with sperm oil, which is the best for all machinery. These parts must be kept from dust, and if dry grindstones are driven in the mill, the dust must be carefully boxed off from the engine. The use of sand on the floor of an engine-house should, for the same reason, be dispensed with.

The method of starting an engine is, first to shut the condensing cock, then to open all the valves to let the steam pass into the jacket, into the cylinder, through the eduction pipe into the condenser, and out at the blow-valve, in order to expel the air from all the parts, and get them to a proper temperature, which will be shown by the steam issuing from the blow-valve; for previously to the parts being sufficiently warmed, the steam in its progress becomes condensed.

When all the parts are heated, the injection water may be let on, and a vacuum procured on one side of the piston, which produces instant action.

The lever of the throttle-valve, which is ultimately to be attached to the governor, should, on starting the engine, be held by the hand of the attendant until the work is thrown on, and the engine has acquired regular motion.

it may be repaired at a trifling cost, and with but little delay."

In examining the effects of this engine, we cannot to a certain extent withhold our approbation ; for the patentee has undoubtedly effected and applied a vacuum, produced by ignition, in a manner different and more manageable than any attempts that have hitherto come to our knowledge. The probability of its entering efficiently into competition with the steam-power, is a question that requires the data of experience, which, in this early state of the invention, cannot be procured.

We understand it is the intention of the inventor to apply the effects of the vacuum thus produced to the movement of a piston in a cylinder, which object will, when attained, afford a much greater scope for the application of its powers, and render it peculiarly applicable to locomotion. The obstacle which at present suggests itself to the attainment of this end is, the difficulty of procuring a rapid condensation without allowing cold water to enter the cylinders at each stroke, which in the present form of construction is allowed, and which greatly aids the operation by keeping the chambers entirely cool. Without, however, seeking for obstacles, we wish the ingenious inventor success in surmounting them.

ON THE STRENGTH OF MATERIALS

An accurate knowledge of the following experiments made by Mr. George Rennie, Jun., and communicated by him in a letter to Thomas Young, M. D. For. Sec. R. S., is of so much importance in the construction of machines, that we have extracted it from the *Transactions of the Royal Society;* to which we have annexed some useful notes by Mr. T. Tredgold.

"In presenting the result of the following experiments," says Mr. Rennie, "I trust I shall not be considered as deviating from my subject, in taking a cursory view of the labours of others. The knowledge of the properties of bodies which come more immediately under our observation, is so instrumental to the progress of science, that any approximation to it deserves our serious attention. The Royal Society appears to have instituted. at an early period, some experiments on

whoever has had occasion to investigate the principles upon which any edifice is constructed, where the combination of its parts are more the result of uncertain rules than sound principle, will soon find how scanty is our knowledge on a subject so highly important. The desire of obtaining some approximation, which could only be accomplished by repeated trials on the substances themselves, induced me to undertake the following experiments.

A bar of the best English iron, about ten feet long, was selected and formed into a lever, whose fulcrum is denoted by *f*, fig. 209. The hole was accurately bored, and the pin turned, which suffered it to move freely. The standard A was firmly secured by the nut *c* to a strong bed plate of cast-iron, made firm to the ground. The lever was accurately divided in its lower edge, which was made straight in a line with the fulcrum. A point, or division D, was selected, at five inches from the fulcrum, at which place was let in a piece of hardened steel. The lever was balanced by a weight, and in this state it was ready for operation. But in order to keep it as level as possible, a hole was drilled through a projection on the bed plate, large enough to admit a stout bolt easily through it, which again was prevented from turning in the hole by means of a tongue *t* fitting into a corresponding groove in the hole. So that, in order to preserve the level, we had only to move the nut to elevate or depress the bolt, according to the size of the specimen. But as an inequality of pressure would still arise from the nature of the apparatus, the body to be examined was placed between two pieces of steel, the pressure being communicated through the medium of two pieces of thick leather above and below the steel pieces, by which means a more equal contact of surfaces was attained. The scale was hung on a loop of iron, touching the lever in an edge only. I at first used a rope for the balance weight, which indicated a friction of four pounds, but a chain diminished the friction one half. Every movable centre was well oiled. Of the resistances opposed to the simple strains which may disturb the quiescent state of a body, the principal are the repulsive force, whereby it resists compression, and the force of cohesion, whereby it resists extension. On the former, with the exception of the experiments of Gauthey and Rondelet, on stones, and a few others, on soft substances, there is scarcely any thing on record. In the memoir of M. Lagrange, on the force of springs, published in the year 1760, the moment of elasticity is represented by a constant quantity, without indicating the relation of this value to the size of the spring: but in the memoir of the year 1770, on the forms of columns, where he considers a body whose dimensions and thickness are variable, he makes the moment of elasticity proportional to the fourth power of the radius, in observing the relations of theory and practice to accord with each other. This

of mechanical writers is the only object attended to, it is of very inferior importance.

The laws of flexure constitute the chief guide in the construction of buildings; and the intention of these notes is to call the attention of experimentalists to this part of the subject; and as it is probable the ingenious author of the experiments now before me may be tempted to resume his labours, I feel certain that he will not feel displeased to have his attention called to some interesting points of inquiry, which he has either neglected to notice, or has not given to the public.—T. T.

Averages.   lbs. avoirdupois.

A prism, having a logarithmic curve for its limits, resembling a column; it was ¼ of an inch diameter by one inch long, broke with... 6954

April 28th.   Trials on prisms of different lengths.

9414·5 {	½ × ½ horizontal	9455
	½ × ¼ ditto	9374
	½ × ⅛ ditto, bad trial, 9006 lbs.	
9982·5 {	½ × ½ vertical	9938
	½ × ¼ ditto	10027

April 29th.   Horizontal castings.

¼ × ⅛	9006
¼ ×	8845
¼ × ⅜	8362
¼ × ½	6437
¼ × ⅝ or one inch long.................................	6391

Vertical castings.

¼ × ⅛	9329
¼ × ¼	8385
¼ × ⅜ a small defect in the specimen	7896
¼ × ½	7018
¼ × ⅝ or one inch.................................	6430

Experiments on different metals.

¼ × ¼	cast copper, crumbled with	7318
¼ × ¼	fine yellow brass reduced ¹⁄₁₆ with 3213· ½ with	10304
¼ × ¼	wrought copper ....¹⁄₁₆ .... 3427· ⅛	6440
¼ × ¼	cast tin ....¹⁄₁₆ .... 552· ¼	966
¼ × ¼	cast lead ....·⅛	483

The anomaly between the three first experiments on ⅛ cubes, and the two second of a different length, can only be accounted for, on the difficulty of reducing such small specimens to an equality. The experiments on ½ inch prism of different lengths give no ratio. The experiments on ¼ inch cubes, taking the average of the three first in each, give a proportion between them and the three on ⅛ cubes,

as 1 : 6·096 in the block castings
as 1 : 7·352 in the horizontal ditto
as 1 : 8·035 in the vertical ditto.

In several cases the proportion is as the cubes.

The vertical cube castings are stronger than the horizontal cube castings. The prisms usually assumed a curve similar to a curve of the third order previous to breaking.

The experiments on the different metals give no satisfactory results. The difficulty consists in assigning value to the different degrees of diminution. When compressed beyond a certain thickness, the resistance becomes enormous.

Experiments on the suspension of bars.

The lever was used as in the former case, but the metals were held by nippers. They were made of wrought-iron, and their ends adapted to receive the bar, which, by being tapered at both extremities, and increasing in diameter from the actual section, (if I may so express it,) and the jaws of the nippers being confined by a hoop, confined both. The bars, which were six inches long, and ¼ square, were thus fairly and firmly grasped.

No.				April 30. 1817.
45	¼ inch, cast iron bar, horizontal	1166	} 1193·5 lbs.	
46	¼ do. do. vertical............. ...	1218		
47	¼ do. cast steel previously tilted............	8391		
48	¼ do. blister steel, reduced per hammer	8322		
49	¼ do. shear steel, do. do.	7977		
50	¼ do. Swedish iron, do. do...............	4504		
51	¼ do. English iron, do. do...............	3492		
52	¼ do. hard gun-metal, mean of two trials	2273		
53	¼ do. wrought copper, reduced per hammer ..	2112		
54	¼ do. cast copper	1192		
55	¼ do. fine yellow brass	1123		
56	¼ do. cast tin	296		
57	¼ do. cast lead	114		

Remarks on the last experiments.

The ratio of the repulsion of the horizontal cast cubes to the cohesion of horizontal cast bars, is 8·65 : 1.

The ratio of the vertical cast cubes to the cohesion of the vertical cast bars, is as 9·14 : 1.

The average of the bars, compared with the cube, No. 16, is as 10·611 : 1.

The other metals decrease in strength, from cast steel to cast lead.

The stretching of all the wrought bars indicated heat.

The fracture of the cast bars was attended with very little diminution of section, scarcely sensible.

The experiment made by M. Prony (which asserts, that by making a slight incision with the file, the resistance is diminished one half) was tried on a ¼ inch bar of English iron; the result was 2920 lbs., not a sixth part less.

This single experiment, however, does not sufficiently disprove the authority of that able philosopher, for an incision is but a vague term. The incision I made might be about the fortieth part of an inch.

Experiments on the twist of ¼ inch bars.

To effect the operation of twisting off a bar, another apparatus was prepared: it consisted of a wrought-iron lever two feet long, having an arched head about 1-6th of a circle, of four feet diameter, of which the lever represented the radius; the centre round which it moved had a square hole made to receive the end of the bar to be twisted. The lever was balanced as before, and a scale hung on the arched head; the other end of the bar being fixed in a square hole in a piece of iron, and that again in a vice. The undermentioned weights represent the quantity of weight put into the scale.

May 30, 1817.

On twists close to the bearing, cast horizontal.

No.		lbs. oz.	
58	¼ in bars, twisted as under with	10 14	in the scale.
59	¼ do. bad casting	8 4	
60	¼ do.	10 11	
	Average	9 15	

Cast vertical.

61	¼	10 8
62	¼	10 13
63	¼	10 11
	Average	10 10

On different metals.

No.		lbs.	oz.	
64	Cast steel	17	9	in the scale.
65	Shear steel	17	1	
66	Blister steel	16	11	
67	English iron, wrought	10	2	
68	Swedish iron, wrought	9	8	
69	Hard gun-metal	5	0	
70	Fine yellow brass	4	11	
71	Copper, cast	4	5	
72	Tin	1	7	
73	Lead	1	0	

On twists of different lengths.

Horizontal.			Vertical.		
No.	Weight in scale.		No.		Weight in scale.
74 $\frac{1}{4}$ by $\frac{1}{4}$ long	7	3	77 $\frac{1}{4}$ by $\frac{1}{4}$ do.		10 1
75 $\frac{1}{4}$ by $\frac{1}{2}$ do.	8	1	78 $\frac{1}{4}$ by $\frac{1}{2}$ do.		8 9
76 $\frac{1}{4}$ by 1 inch do.	8	8	79 $\frac{1}{4}$ by 1 inch do.		8 5

Horizontal twists at 6 from the bearing.

80 $\frac{1}{4}$ by 6 inches long	10 9
81 $\frac{1}{4}$ by do. do.	9 4
82 $\frac{1}{4}$ by do. do.	9 7

Twists of $\frac{1}{2}$ inch square bars, cast horizontally.

No.		qrs.	lbs	oz.	
83 $\frac{1}{2}$ close to the bearing		3	9	12	end of the bar hard.
84 $\frac{1}{2}$ do.		2	18	0	middle of the bar.
85 $\frac{1}{2}$ at 10 inches from bearing, lever in the middle		1	24	0	

On twists of different materials.

These experiments were made close to the bearing, and the weights were accumulated in the scale until the substances were wrenched asunder.

No.	Weight in scale.		No.	Weight in scale.	
86 Cast steel	19	9	91 Hard gun-metal	5	0
87 Shear steel	17	1	92 Fine yellow brass	4	11
88 Blister steel	16	11	93 Copper	4	5
89 English iron, No. 1	10	2	94 Tin	1	7
90 Swedish iron	9	8	95 Lead	1	0

Remarks.

Here the strength of the vertical bars still predominates.

The average of the two taken conjointly, and compared with a similar case of $\frac{1}{2}$ inch bars, gives the ratio as the cubes, as was anticipated.

In the horizontal castings of different lengths, the balance is in favour of the increased lengths; but in the vertical castings, it is the reverse. In neither is there any apparent ratio. In the horizontal castings at six inches from the bearing, there is a visible increase, but not so great as when close to the bearing.

June 4, 1817. *Miscellaneous experiments on the crush of one cubic inch.*

No.		lbs. avoirdupoise.
96	Elm	1284
97	American pine	1606
98	White deal	1928
99	English oak, mean of two trials	3860
100	Ditto, of 5 inches long, slipped with	2572

101 English oak, of four inches long, clipped with		5147°
102 A prism of Portland stone 2 inches long		843
103 Ditto, statuary marble		3216
104 Craig Leith		5883

In the following experiments on stones, the pressure was communicated through a kind of pyramid, the base of which rested on the hide leather, and that on the stone.* The lever pressed upon the apex of the pyramid. Cube, of one and a half inch.

		Spec. g. at 78° aver.
105 Chalk		1127
106 Brick of a pale red colour	2·083	1265
107 Roe-stone, Gloucestershire		1449
108 Red brick, mean of two trials	2·168	1817
109 Yellow face baked Hammersmith paviors, 3 times.......		2254
110 Burnt do. mean of two trials		3243
111 Stourbridge or fire brick		3861
112 Derby grit, a red friable sand-stone	2·316	7070
113 Ditto, from another quarry	2·428	9776
114 Killaly white freestone, not stratified	2·423	10263
115 Portland	2·428	10284
116 Craig Leith, white freestone..............................	2·452	12316
June 5, 6, and 7, 1817.		
117 Yorkshire paving with the strata	2·507	12856
118 Ditto do. against the strata	2·507	12856
119 White statuary marble not veined	2·760	13632
120 Bramley Fall sandstone, near Leeds, with strata	2·506	13632
121 Ditto, against the strata	2·506	13632
122 Cornish granite	2·662	14302
123 Dundee sandstone or brescia, two kinds..............	2·530	14918
124 A two-inch cube of Portland	2·423	14918
125 Craig Leith with the strata	2·452	15560
126 Devonshire red marble, variegated	....	16712
127 Compact limestone..	2·581	17354
128 Peterhead granite hard close grained	.....	18636
129 Black compact limestone, Lanerick..................	2·598	19924
130 Purbeck	2·599	20610
131 Black Brabant marble	2·697	20742
132 Very hard freestone	2·528	21254
133 White Italian veined marble.......................	2·726	21763
134 Aberdeen granite, blue kind.......................	2·625	24550

N. B. The specific gravities were taken with a delicate balance, made by Creighton of Glasgow, all with the exception of two specimens, which were by accident omitted.

Remarks.

In observing the results presented by the preceding table, it will be seen that little dependence can be placed on the

---

* The experiments on woods are considerably below those of other writers, and it appears singular that the four-inch specimen should be stronger than the shorter length. According to Rondelet's experiments, to crush a cubic inch of oak it required from 5000 to 6000 lbs. avoirdupoise

of fir - from 6000 to 7000 lbs.

In the former the pieces were compressed one-third of their length; in the latter one-half of their length (Rondelet's *L'Art de Bâtir*, tom. iv. p. 67.) Mr. Rennie has not stated the diminution of length.

† It certainly would have been preferable to have placed a hard and rigid substance next the stone, in order to secure equality of pressure.

Q

specific gravities of stones, so far as regards their repulsive powers, although the increase is certainly in favour of their specific gravities. But there would appear to be some undefined law in the connection of bodies, with which the specific gravity has little to do. Thus, statuary marble has a specific gravity above Aberdeen granite, yet a repulsive power not much above half the latter. Again, hardness is not altogether a characteristic of strength, inasmuch as the limestones, which yield readily to the scratch, have nevertheless a repulsive power approaching to granite itself.

It is a curious fact in the rupture of amorphous stones, that pyramids are formed, having for their base the upper side of the cube next the lever, the action of which displaces the sides of the cubes, precisely as if a wedge had operated between them. I have preserved a number of the specimens, the sides of which, if continued, might cut the cubes in the direction of their diagonals.

Experiments made on the transverse strain of cast bars, the ends loose. June 8th, 1817.[*]

		Weight of the bars, lbs. oz.	dist. of bearings. ft. in.	lbs. avdir.
135	Bar of 1 inch square	12　6	3　0	897
136	Do. of 1 inch do..........................	9　8	2　8	1935
137	Half the above bar	....	1　4	2329
138	Bar of 1 inch square, through the diagonal	2　8	2　8	851
139	Half the above bar	....	1　4	1587
140	Bar of 2 inches deep, by ½ inch thick	9　5	2　8	2185
141	Half the above bar	....	1　4	4508

* A bar of cast-iron, from a Welsh foundry, which did not yield easily to the file, was laid upon supports exactly three feet apart; the bar was an inch square, and when 308 pounds were put into a scale suspended from the middle of its length, the deflexion was found to be 3-16ths of an inch; whence the height of the modulus of elasticity is 6,386,688 feet. The experiment was made by Mr. R. Ebbels, at Garnons, near Hereford. A joist of cast-iron, nine inches deep, resembling in form the letter I, was laid upon supports 19 feet apart, first on its edge, when the deflexion from its own weight was 3-40ths of an inch. It was then laid flatwise, and the deflexion from its own weight was 3¼ inches. The castings were from Messrs. Dowsons' foundry, Edgware-road. The iron yielded easily to the file. The height of the modulus of elasticity according to the experiment on the
joist flatwise is　　5,100,000 feet,
—— on the edge is 5,700,000 feet.
The deflexion being very small when the joist was on its edge, perhaps it was not measured with the necessary degree of accuracy, as a very small error would cause the difference in the result. The following tablet contains the value of the modulus for cast-iron, according to the experiments above stated.

	Height of modulus in feet.	Experimentalists.
Cast-iron (Welsh)	6,386,688	Ebbels.
Cast-iron	3,500,000	Banks.
Cast-iron, grey French	5,095,480	Rondelet.
Cast-iron, soft do.	4,247,000	Rondelet.
Cast-iron	5,700,000	By my trial.

	Weight of the bars, lbs. oz.	dia. of bearings ft. in.	lbs. sethr.
112 Bar 8 inches deep, by ¼ inch thick	9 15	2 8	3560
113 Half the bar............................	....	1 4	6804
Ditto 4 inches by ½ inch thick	9 7	2 8	3979
Equilateral triangles with the angle up and down.			
edge or angle up..........................	9 11	2 8	1437
──── angle down........................	9 7	2 8	840
Half the first bar.........................	....	1 4	3059
Half the second bar.......................	....	1 4	1656
feather-edged or bar was cast whose dimensions were			
inches deep by 8 wide 10 0		edge up 2 8	3105
Half of ditto			

All these bars contained the same area, though differently distributed as to the forms.

Experiments made on the bar of 4 inches deep by ½ inch thick, by giving it different forms, the bearings at 2 feet 8 inches, as before.

	lbs.	lbs.
formed into a semi-ellipse, weighed	7	4000
Ditto, parabolic on its lower edge....................		3860
Ditto, of 4 inches deep by ½ inch thick ....,........		3979

Experiments on the transverse strain of bars, one end made fast, the weight being suspended at the other, at 2 feet 8 inches from the bearing.

		lbs.
An inch square bar bore ....,......		290
A bar 2 inches deep by ½ an inch thick		630
An inch bar, the ends made fast		1173

The paradoxical experiment of Emerson was tried, which states that by cutting off a portion of an equilateral triangle (see page 114 of Emerson's Mechanics) the bar is stronger than before, that is, a part stronger than the whole. The ends were loose at two feet eight inches apart as before. The edge from which the part was intercepted was lowermost, the weight was applied on the part above, it broke with 1129 pounds, whereas in the other case it bore ... pounds.

Remarks on the transverse strain.

This bar from the cupola, when placed on bearings three feet ... the ends loose, to bear .... 864 lbs.
... all my bars were cast from the cupola, the difference was therefore 33 lbs. ... space of two feet eight inches asunder, as being more convenient ... The strength of the different bars, all cases being the same, ... nearly to the theory, which makes the comparative values as the ... multiplied into the squares of the depths. The halves of the bars ... merely to keep up the analogy. The bar of four inches deep, how... short of theory by 368 pounds. It is evident we cannot extend the ... deepening the bar much further, nor does the theory exactly maintain ... case of the equilateral triangle by .... 243 lbs. ... diagonal position of the square bar, is actually worse than when laid on ... contrary to many assertions.
... quantity of metal in the feather-edged bar was not so strong as ... inch bar.
... semi-elliptical bar, exceeded the four-inch bar, although taken out of it, ... nearly has come near it.
... bar made fast at both ends, I suspect must have yielded, although the ... were made fast by iron straps. The experiments from Emerson, on solids ... might be made; but the time and trouble these experiments ... have compelled me to relinquish further pursuits for the ... However, in the absence of better, they are worthy of the indulgence of the Royal Society; it will not only be a consolation to me that my

Q 2

in the most disadvantageous situation possible: for the load being found always towards the extremity of a radius of the wheel, the arm of the effective lever which answers to it, increases through the whole quadrant the water describes in passing from the bottom of the wheel to the altitude of its centre; so that the power must act in like manner as if it were applied to a winch-handle, and, consequently, cannot act uniformly.

M. de la Faye, to remedy this defect, devised a machine, which may here be described, together with the process of reasoning that led to it.

When we develope the circumference of a circle, a curve is described (i. e. the involute) of which all the radii are so many tangents to the circle, and are likewise all respectively perpendicular to the several points of the curve described, which has for its greatest radius a line equal to the periphery of the circle evolved. The truth of which is shown by geometricians when treating of the genesis of evolute and involute curves.

Hence, having an axle whose circumference a little exceeds the height which the water is proposed to be elevated, let the circumference of the axle be evolved, and make a curved canal whose curvature shall coincide throughout exactly with that of the involute just formed: if the further extremity of this canal be made to enter the water that is to be elevated, and the other extremity abut upon the shaft which is turned; then in the course of the rotation the water will rise in a vertical direction, tangential to the shaft, and perpendicular to the canal in whatever position it may be. Thus the action of the weight answering always to the extremity of a horizontal radius will be as though it acted upon the invariable arm of a lever, and the power which raises the weight will be always the same: and if the radius of the wheel, of which this hollow canal serves as a bent spoke, is equal to the height that the water is to be raised, and consequently equal to the circumference of the axle or shaft, the power will be to the load of water reciprocally as the radius of a circle to its circumference, or directly as 1 to or nearly.

In M. de la Faye's opinion, the machine ought to be composed of four of these canals: but it has often been constructed with eight, as represented in the fig. The wheel being turned by the impulsion of the stream upon the float-boards, the orifices F, E, D, C, &c. of the curvilinear canals, dip one after another into the water which runs into them; and as the wheel revolves the fluid rises in the canals, f, e, d, c, &c. and runs out in a stream P from the holes at O; it is received into the trough Q, and conveyed from thence by pipes.

By this construction the weight to be raised offers always the same resistance, and that the least possible, while the power is applied in the most advantageous manner the circumstances will admit of: these conditions both fulfilled at the same time furnish the most desirable perfection in a machine. Further, this machine raises the water by the shortest way, namely, the perpendicular, or vertical; in this respect being preferable to Archimedes' screw, where the water is carried up an inclined path: and besides this, each curved channel in this wheel empties all the water it receives in every revolution, while the screw of Archimedes delivers only a small portion of the fluid it is charged with, being often loaded with twenty times as much water as is discharged in one rotation; and thus requiring an enormous increase of labour when a large quantity is intended to be raised by it.

The nature and advantages of this wheel evince very forcibly how far the speculations of geometers are from being so unfruitful in useful applications, as is often insinuated by practical men.

3. The wheel just described would, we think, be the most perfect of any that could be employed for raising water, had it not the disadvantage attending the tympanum, which is, that it can only raise water to the height of its semidiameter. As in many cases water is to be raised higher than the radius of any wheel can well be made for practice, we shall next describe a machine called the *Noria,* common in Spain, which raises water nearly through a diameter.

This Noria consists of a vertical wheel of 20 feet diameter, on the circumference of which are fixed a number of little boxes or square buckets, for the purpose of raising the water out of the well, communicating with the canal below, and to empty it in a reservoir above, placed by the side of the wheel. The buckets have a lateral orifice, to receive and to discharge the water. The axis of this wheel is embraced by four small beams, crossing each other at right angles, tapering at the extremities, and forming eight little arms. This wheel is near the centre of the horse-walk, contiguous to the vertical axis, into the top of which the horse-beam is fixed; but near the bottom it is embraced by four little beams, forming eight arms similar to those above described, on the axis of the water-wheel. As the mule which they use goes round, these horizontal arms, supplying the place of cogs, take hold, each in succession, of those arms which are fixed on the axis of the water-wheel, and keep it in rotation.

This machine, than which nothing can be cheaper, throws up a great quantity of water; yet undoubtedly it has two defects: the first is, that part of the water runs out of the buckets and falls back into the well after it has been raised nearly to the level of the reservoir: the second is, that a considerable proportion of the water to be discharged is raised higher than the reservoir, and falls into it only at the moment when the bucket is at the highest point of the circle, and ready to descend. These inconveniences are both remedied by the contrivance mentioned in the next paragraph.

4. The *Persian wheel* is a name given to a machine for raising water, which may be turned by means of a stream A B acting upon the wheel C D E according to the order of the letters; (fig. 210.)

The buckets *a, a, a, a,* &c. instead of being firmly fastened, are *hung* upon the wheel by strong pins, *b, b, b, b,* &c fixed in the side of the rim; which must be made as high as the water is intended to be raised above the level of that part of the stream in which the wheel is placed. As the wheel turns, the buckets on the right hand go down into the water, where they are filled, and return up full on the left hand, till they come to the top at K; where they strike against the end n of the fixed trough M, by which they are drawn, and so empty the water into the trough; from whence it is to be conveyed in pipes to any place it is intended for: and as each bucket gets over the trough, it falls into a perpendicular position again, and so goes down empty till it comes to the water at A, where it is filled as before. On each bucket is a spring, r, which going over the top or crown of the bar w (fixed to the trough M) raises the bottom of the bucket above the level of its mouth, and so causes it to empty all its water into the trough.

To determine the due relation of the power and the weight so that this wheel may be capable of producing the greatest effect, the following may be taken as a good approximation. After having fixed the diameter of the wheel, which must be something greater than the altitude to which the water is to be raised; fix also upon an even number of buckets to be hung at equal distances round the periphery of the wheel, and mark the position of their centres of motion in such a manner that they will stand in corresponding positions in every quarter of the circle: conceive vertical lines drawn through the centre of motion of each bucket in the rising part of the wheel; they will intersect the horizontal diameter of the wheel in points at which, if the buckets were hung, they would furnish the same resistance to the moving force as they do when hanging at their respective places on the rim of the wheel. Thus, supposing there were 18 equidistant buckets; then while eight hung on each side a vertical diameter of the wheel there would be eight on the other side, and two would coincide with that diameter: in this case the resistance arising from all the full buckets would be the same as if one bucket hung on the prolongation of the horizontal diameter at the distance of 2 sin. 20° + 2 sin. 40° + 2 sin. 60° + 2 sin. 80°, these being the distances to the common radius of the wheel.

To know the quantity of water that each bucket should contain, take ⅔ of the moving force of the stream, that is, ⅔ of the weight of the prism of water whose base is the surface of one of the float-boards, and whose height is that through which water must fall to acquire the velocity of the current; so have we the power that should be in equilibrio with the weight of water in the buckets of the rising semicircle. Then say, as the sum of the sines mentioned above is to radius, so is the power just found to a fourth term, the half of which will be the weight of water that ought to be contained in one bucket. Lastly, as the velocity of the wheel will be to that of the stream nearly as 1 to 2⅓, the quantity of revolutions it makes in a determinate time becomes known, and, by consequence, the quantity of water the wheel will raise in the same time; since we know the capacity of each bucket, and the number of them emptied in every revolution of the wheel.

10. Another contrivance for raising water similar to the chain-pump, which is described in another part of the work, is an endless rope with stuffed cushions hung upon it, which, by means of two wheels or drums, are caused to rise in succession in the same barrel, and to carry water with them. From the resemblance of this apparatus to a string of beads, it is usually called paternoster-work. But in this, as well as the chain-pump, the magnitude of the friction is a formidable practical objection.

6. Jets and fountains are not now considered as conducive

to picturesque beauty; nor can they be reckoned of much utility, except perhaps in hot climates; we have not therefore described any in this work. But in the fountain of *Hiero* of Syracuse, a principle is introduced which has been found of great utility in larger works; for the head of water is actually lower than the orifice, but the pressure is communicated by the intervention of a column of air: the construction of this fountain is as follows:

It consists of two vessels K L M N (fig. 212) and O P Q R, which are close on all sides. A tube A B, having a funnel at the top, passes through the uppermost vessel without communicating with it, being soldered into its top and bottom. It also passes through the top of the under vessel, where it is likewise soldered, and reaches almost to its bottom. This tube is open at both ends. There is another open tube S T, which is soldered into the top of the under vessel and the bottom of the upper vessel, and reaches almost to its top. These two tubes serve also to support the upper vessel. A third tube G F is soldered into the top of the upper vessel, and reaches almost to its bottom. This tube is open at both ends, but the orifice G is very small. Now suppose the uppermost vessel filled with water to the height E N, E e being its surface a little below T. Stop the orifice G with the finger, and pour in water at A. This will descend through A B, and compress the air in O P Q R into less room. Suppose the water in the under vessel to have acquired the surface C c, the air which formerly occupied the whole of the spaces O P Q R and K L e E will now be contained in the spaces o P c C and K L e E; and its elasticity will be in equilibrio with the weight of the column of water, whose base is the surface E e, and whose height is A c. As this pressure is exerted in every part of the air, it will be exerted on the surface E e of the water of the upper vessel; and if the pipe F G were continued upwards, the water would be supported in it to a height e H above E e, equal to A c. Therefore, if the finger be now taken from off the orifice G, the fluid will spout up through it to the same height as if it had fallen through a tube whose altitude is e H. So long as there is any water in the vessel K L N M there will be a discharge through the orifice: therefore the play of the fountain will continue whilst the water contained in the upper vessel, having spouted out, falls down through the pipe A B: the height of the water measured from the basin V A W to the surface of the water in the lower vessel O P Q R is always equal to the height measured from the top of the jet to the surface of the water in the vessel K L M N. Now, since the surface E e is always falling, and the water in the lower vessel always rising, the height of the jet must continually decrease, till it is shorter by the depth of K L M N, which is empty, added to the depth of O P Q R, which is always filling; and when the jet is fallen so low, it immediately ceases to play.

7. A machine designed to raise water to a great height for the irrigation of land, in such situations as have the advantage of a small fall, is described in Dr. Darwin's *Phytologia:* as it depends on the principle of Hiero's fountain, it may properly be inserted here.

Fig. 211, *a, b*, is the stream of water.
b, c, c, represents the water-fall, supposed to be 10 feet.

d, c, a, b, a ... [illegible] ... vessel ... [illegible] ... quantity of
water, which may be computed to be about four gallons each.

f, g, h, i, k, l, are leaden vessels, each holding about two quarts.

o, p, two cocks, each of which passes through two pipes, opening the one
and closing the other.

q, r, is a *water-balance,* that moves on its centre *s,* and by which the
two cocks *o* and *p* are alternately turned.

t, u, and *w, x,* are two air-pipes of lead, both internally 1¼ inch in
diameter.

y, z ; y, z ; y, z ; are water-pipes, each being one inch in diameter.

The pipe *b, e, c,* is always full from the stream *a, b :* the small cisterns
g, i, l, and the large one *d,* are supposed to have been previously filled with
water. The fluid may then be admitted by turning the cock *o,* through the
pipe *c, e,* into the large cistern *e.* This water will press the air confined in
the cistern *e* up the air-pipe *w, x,* and will force the fluid out of the cisterns
g, i, l, into those marked *h, k,* and C.—At the same time, by opening B, the
water and condensed air, which previously existed in the large cistern *d,*
and in the smaller ones marked *f, h, k,* will be discharged at B. After a
short time, the water-balance, *q, r, s,* will turn the cocks, and exclude the
water, while it opens the opposite ones: the cisterns *f, h, k,* are emptied in
their turns by the condensed air from the cistern *d,* as the water progressively
enters the latter from the pipe *b, c.*

8. A very ingenious application of the same principle has
been made in the celebrated Hungarian machine, at Chemnitz.
The best account we have been able to obtain of this is the
following :

In fig. 213, A represents the source of water elevated 136 feet above the
mouth of the pit. From this there runs down a pipe D of four inches
diameter, which enters the top of a copper cylinder B, 8½ feet high, 5 feet diameter, and 2 inches thick, and reaches to within 4 inches of the bottom :
it has a cock at I.

This cylinder has a cock at Q, and a very large one at N. From its
top proceeds a pipe V E C, two inches in diameter, which goes 96 feet
down the pit, and is inserted into the top of another brass cylinder C, which
is 6½ feet high, four feet diameter, and two inches thick : the latter containing about 83 cubic feet, which is nearly one half of the capacity of the
former, viz. 170 cubic feet. There is another pipe F O of four inches diameter, which rises from within four inches of the bottom of this lower
cylinder, is soldered into its top, and rises to the trough Z which carries off
the water from the mouth of the pit. This lower cylinder communicates at
the bottom with the water O, which collects in the drains of the mines. A
large cock P serves to exclude or admit this water : another cock M at the
top of this cylinder communicates with the external air.

Now, suppose the cock I shut, and all the rest open : the upper cylinder
will contain air, and the lower cylinder will be filled with water, because it
is sunk so deep that its top is below the usual surface of the mine-waters.
Shut the cocks Q, N, M, P, and open the cock I. The water of the source
A must run in by the orifice J, and rise in the upper cylinder, compressing
the air above it and along the pipe V E C, and thus acting on the surface
of the water in the lower cylinder. It will therefore cause it to rise gradually
in the pipe O F, where it will always be of such a height that its weight
balances the elasticity of the compressed air. Suppose no issue given to
the air from the upper cylinder, it would be compressed into one fifth of its
bulk by the column of 136 feet high : for a column of 34 feet nearly

balances the ordinary elasticity of the 'air. Therefore, when there is an issue given to it through the pipe V EC, it will drive the compressed air along this pipe, and it will expel water from the lower cylinder. When the upper cylinder is full of water, there will be 34 cubic feet of water expelled from the lower cylinder. If the pipe O P had been more than 136 feet long, the water would have risen 136 feet, being then in equilibrio with the water in the feeding pipe D by the intervention of the elastic air; but no more water would have been expelled from the lower cylinder than what fills this pipe. But the pipe being only 96 feet high, the water will be thrown out at Z with a considerable velocity. If it were not for the great obstructions which water and air must meet with in their passage along pipes, it would issue at Z with a velocity of more than 50 feet per second. It issues however much more slowly, and at last the upper cylinder is full of water, and the water would enter the pipe V E and enter the lower cylinder, and, without displacing the air in it, would rise through the discharging pipe O P, and run off to waste. To prevent this there hangs in the pipe V E a cork ball or double cone, by a brass wire which is guided by holes in two cross pieces in that pipe. When the upper cylinder is filled with water, this cork plugs up the orifice V, and no water is wasted; the influx at J now stops. But the lower cylinder contains compressed air, which would balance water in a discharging pipe 136 feet high, whereas O P is only 96. Therefore the water will continue to flow at Z till the air has so far expanded as to balance only 96 feet of water, that is, till it occupies one-half of its ordinary bulk, that is, one-fourth of the capacity of the upper cylinder, or 42⅕ cubic feet. Therefore 42⅕ cubic feet will be expelled, and the efflux at Z will cease; and the lower cylinder is about one-half full of water. When the attending workman observes this, he shuts the cock I. He might have done this before, had he known when the orifice V was stopped; but no loss ensues from the delay. At the same time the attendant opens the cock N the water issues with great violence, being pressed by the condensed air from the lower cylinder. It therefore issues with the sum of its own weight and of this compression. These gradually decrease together, by the efflux of the water and the expansion of the air; but this efflux stops before all the water has flowed out; for there is 42⅕ feet of the lower cylinder occupied by air. This quantity of water remains, therefore, in the upper cylinder nearly: the workman knows this, because the discharged water is received first of all into a vessel containing three-fourths of the capacity of the upper cylinder. Whenever this is filled, the attendant opens the cock P by a long rod which goes down the shaft; this allows the water of the mine to fill the lower cylinder, and the air to get into the upper cylinder, which permits the remaining water to run out of it. Thus every thing is brought into its first condition; and when the attendant sees no more water come out at N, he shuts the cocks N and M, and opens the cock I, and the operation is repeated.

There is a very surprising appearance in the working of this engine. When the efflux at Z has stopped, if the cock Q be opened, the water and air rush out together with prodigious violence, and the drops of water are changed into hail or lumps of ice. It is a sight usually shown to strangers, who are desired to hold their hats to receive the blasts of air: the ice comes out with such violence as frequently to pierce the hat like a pistol bullet. This rapid congelation is a remarkable instance of the general fact, that air by suddenly expanding generates cold, its capacity for heat being increased.

The above account of the procedure in working this engine shows that the efflux both at Z and N becomes very slow near the end. It is found convenient therefore not to wait for the complete discharges, but to turn the cocks when about 30 cubic feet of water have been discharged at Z; and work is done in this way. A gentleman of great accuracy and knowledge of these subjects took the trouble of noticing particularly the performance of the machine. He observed that each stroke, as it may be called, took up about three minutes and one-eighth; and that 32 cubic feet of water were discharged at Z, and 66 were expended at N. The expense therefore is 66 feet of water falling 136 feet, and the performance is 32 raised 96, and they are in the proportion of 66 × 136 to 32 × 96, or of 1 to 0.3422, or nearly as 3 to 1. This is superior to the performance of the most perfect undershot mill, even when all friction and irregular obstructions are neglected; and is not much inferior to any overshot pump-mill that has yet been erected. When we reflect on the great obstructions which water meets with in its passage through long pipes, we may be assured, that by doubling the size of the feeder and discharger, the performance of the machine will be greatly improved; we do not hesitate to say, that it will be increased one-third: it is true that it will expend more water; but this will not be nearly in the same proportion, for most of the deficiency of the machine arises from the needless velocity of the first efflux at Z. The discharging pipe ought to be 110 feet high, and not give sensibly less. Then it must be considered how inferior in original expense this machine must be to a mill of any kind which would raise 10 cubic feet 96 feet high in a minute; and how small the repairs on it need be, when compared with a mill. And, lastly, let it be noticed, that such a machine can be used where no mill whatever can be put in motion. A quantity of water, which would not move any kind of wheel, will here raise one-third of its own quantity to the same height, working as fast as it is supplied.

For these reasons, the Hungarian machine eminently deserves the attention of mathematicians and engineers, to bring it to its utmost perfection, and into general use. There are situations where this kind of machine may be very useful. Thus, where the tide rises 17 feet, it may be used for compressing air to seven-eighths of its bulk ; and a pipe leading from a very large vessel inverted in it may be used for raising the water from a vessel of one-eighth of its capacity, 17 feet high ; or if this vessel has only one-tenth of the capacity of the large one set in the tide-way, two pipes may be led from it, one into the small vessel, and the other into an equal vessel 16 feet higher, which receives the water from the first. Thus one-sixteenth of the water may be raised 16 feet, and a smaller quantity to a still greater height ; and this with a kind of power that can hardly be applied any other way. Machines of this kind are described by Schottus, Sturmius, Leupold, and other old writers; and they should not be forgotten, because opportunities may offer of making them highly beneficial.

Mr. John Whitley Boswell has devised an apparatus which, when attached to such a machine as that at Chemnitz will enable it to work itself without attendance. The description of this will be presented to the reader in Mr Boswell's own words.

Fig. 213.   A is the reservoir, or upper level of water.

B, a chamber made of sufficient strength to bear the internal pressure of a column of water the height of A above it, multiplied by its own base.

C, a chamber of the same strength as B, but of a smaller size; it is placed at the bottom of the pit from which the water is to be raised, and under the level of the water.

These chambers would be stronger with the same materials, if of a globular or cylindrical form; but the square shape is used in the drawing merely for the facility of representing the position of the parts.

D, a pipe from the reservoir A which passes through the top of B and ends near its bottom, to convey water from A to B.

E, a pipe from the top of B to the top of C, to convey air from B to C.

F, a pipe from the bottom of C to the level of the ground at the top of the pit, to carry off the water from the pit.

G, a pipe from the bottom of B to carry off the water from it.

H, a vessel to contain the water used in working the cocks; it is only placed on the top of B to save the construction of a stand on purpose for it.

I, a cock, or movable valve, (worked by the lever there represented,) in the large pipe D.

K, a stop-cock in the small pipe which conveys water from D to H.   Its use is to make the engine work faster or slower, by letting water more or less quick into H; or to stop it altogether from working when required.

L, a movable valve, or cock in the small pipe L K.   The lever which works it is connected by a strong wire with the lever which works I, and is balanced by a weight at its opposite extremity, sufficient to open both these cocks and shut N, when not prevented by a counter weight.

N, a cock in the pipe G to open and shut it as wanted.

O, a self-moving valve in the pipe F, which permits the water to pass upwards, but prevents its return.

P, a self-moving valve at the bottom of C, which permits the water to pass into C, but prevents any from passing out of it; it is furnished with a grating, to prevent dirt getting in.

R, a vessel suspended from the levers of I and L, capable of containing a weight of water sufficient to shut them.

S, a vessel suspended from the lever of N : it must contain water enough by its weight to open N : it is connected by a chain to R, to keep it down as long as N is open.

T, a syphon passing from the bottom of H, near its upper edge, and down again to the mouth of R.

V, a self-moving valve of a sufficient levity to rise, when the water in B comes up to it, and close the pipe E : into which no water would else pass from B.   A ball-cock, such as used in common water cisterns, would do here.

X, a syphon from the bottom of R rising within an inch of its top, and passing down again to the mouth of S.

Y, a small pipe at the bottom of S ; this may have a stop-cock to regulate it, which, when stopped, will also stop the engine.

The mode of this engine's working is as follows : suppose the vessels V, H, R, and S empty of water, and the cocks K and Y open, and the vessel C full of water.   The weight on the lever of L will then open the cocks L and I, on which the water from A will flow into B and H.   As

to B, it will force the air through E into C, which strongly
the water in C, will force it up through the pipe P, till the
to the level of V and closes it, at which time H will be full
quantity flowing in being so regulated by the cock K,) and
flow from it through the syphon T into the vessel R, which
the cock I and L, and prevents any more water coming into
When R is full, the water flows through its syphon X, which
it opens N, which empties B of water, and keeps N open as
is any water in H.

empty, B will be so too, (being so regulated by the cock K,)
a moment or two, R and S will also be empty, which will
ks I and L to open, and all things will be again in the state
for a repetition of the operations described.

engine, the cocks at K and Y should be shut, while S is full
to set it working, they should be open; and this is all the
will require. As no one but an engineer should attempt to
h an engine as this, it was useless to represent the manner of
the pipes by flaches or otherwise, or the proper methods of
closing the parts, which are all well known to such as have
their study.

5, of the New Series of Nicholson's Journal, Mr.
made some further improvements in the applica-
Hungarian machine.

spiral pump is a very curious hydraulic engine,
tes on nearly the same principle as the Hungarian
The first engine of this kind, of which we have
ccount, was invented and erected by H. Andreas
plate-worker of Zurich, at a dye-house in Limmat,
ity of that city. It consists of a hollow cylinder,
large grindstone, turning on a horizontal axis, and
ged in a cistern of water. The axis is hollow at
nd communicates with a vertical pipe. This cylin-
m is formed into a spiral canal, by a plate coiled
it like the main spring of a watch in its box; only
at a distance from each other, so as to form a
r the water of uniform width. This spiral partition
ned to the two ends of the cylinder, and no water
etween them. The outermost turn of the spiral
widen about three-fourths of a circumference from
nd this gradual enlargement continues nearly a
, this part being called the horn : it then widens
forming a scoop or shovel. The cylinder is so
that this shovel may, in the course of a rotation,
al inches into the water. As the cylinder turns
xis, the scoop dips and takes up a certain quantity
efore it emerges again. This quantity is sufficient
horn ; and this again is nearly equal in capacity to
most uniform spiral round.

After the scoop is emerged, the water passes al
spiral by the motion of it round the axis, and drives
before it into the rising pipe, where it escapes. In t1
time, air comes into the mouth of the scoop; and w
scoop again dips into the water, it again takes in
that fluid. Thus there becomes a part filled with wa
a part filled with air. Continuing this motion, a secon
of water will be received, and another of air. The
any turn of the spiral will have its two ends on a lev
the air between the successive columns of water will b
natural state; for since the passage into the rising
main is open, there is nothing to force the water and
any other position. But since the spires gradually d
in their length, it is plain that the column of wat
gradually occupy more and more of the
each. At last it will occupy a complete turn of so
that is near the centre; and when sent further in
continuance of the motion, some of it will run back o
top of the succeeding spire. Thus it will run over i
right-hand side of the third spire, and consequent
push the water of this spire backwards, and raise it
end, so that *it* will likewise run over backwards bef
next rotation be completed. At length this change
position will reach the outermost spire, and some wat
run over into the horn and scoop, and finally into the

But as soon as water gets into the rising pipe, and
little into it, it stops the escape of the air when tl
scoop of water is taken in. Hence there are then two
of water acting against each other by hydrostatic p
and the intervening column of air: they must comp
air between them, and the water and air columns w
be unequal: this will have a general tendency to k
whole water back, and cause it to be higher on the
rising side of each spire than on the right or desc
side: the excess of height being just such as produces th
pression of the air between that and the preceding
of water. This will go on increasing as the water i
in the rising pipe; for the air next to the rising pipe i
pressed at its inner end with the weight of the whole c
in the main: and it must be as much compressed
outer end, which must be done by the water column w
it; and this column exerts this pressure partly by
that *its* outer end is higher than its inner end, and par
the transmission of the pressure on its outer end l
which is similarly compressed from without. Thus

happen that each column of water being higher at its outer than at its inner end, compresses the air on the water column beyond or within it, which transmits this pressure to the air beyond *it*, adding to it the pressure arising from its own want of level at the ends. Consequently, the greatest compression, viz. that of the air next the main, is produced *by the sum of all the transmitted pressures;* and these are the sum of all the differences between the elevations of the inner ends of the water columns above their outer ends: and the height to which the water will rise in the main will be just equal to this sum.

Suppose the left-hand spaces of each spire to be filled with water, and the right-hand spaces filled with air, as is shown, in regard to one spire, in fig. 214. There is a certain gradation of compression which will keep things in this position: for the spaces manifestly decrease in arithmetical progression; and so do the hydrostatic heights and pressures: if, therefore, the air be dense in the same progression all will be in hydrostatical equilibrium. Now this may obviously be produced by the mere motion of the machine; for since the density and compression in each air column is supposed inversely as the magnitude of the column, the quantity of air is the same in all; therefore the column first taken in will pass gradually inwards, and the increasing compression will cause it to occupy precisely the whole right-hand of every spire. The gradual diminution of the water columns will be produced, during the motion, by the water running over backwards at the top from spire to spire, and ultimately coming out by the scoop. Since the hydrostatic height of each water column is now the greatest possible, viz. the diameter of the spire, it is evident that this disposition of the air and water will raise the water to the greatest height. This disposition may be obtained thus: let C B be a vertical radius of the wheel, C being the centre, and B the highest point [the figure may easily be drawn] upon C B, take C L to C B, as the density of the external air to its density in the last column next the rising pipe or main; that is, make C L to C B as 34 feet (the height of the column of water which balances the pressure of the atmosphere) to the sum of 34 feet, and the height of the rising pipe. then divide B L into such a number of turns that the sum of their equal diameters shall be equal to the height of the main; lastly, bring a pipe straight from L to the centre C. Such is the construction of the spiral pump, as originally invented by Wirtz: it certainly indicates very considerable mechanical knowledge and sagacity.

But when the main is very high this construction will require either an enormous diameter of the drum, or many turns of a very narrow pipe. In such cases it will be much better to make the spiral in the form of a corkscrew, than of this flat form like a watch-spring. The pipe which forms the spiral may be wrapped round the frustrum of a cone, whose greatest diameter is to the least (which is next to the rising pipe) in the proportion just assigned to C B and C L. By this construction the water will so stand in every round as to have its upper and lower surfaces tangents to the top and bottom of the spiral, and the water columns will occupy the whole ascending side of the machine while the air occupies the descending side. This form is far preferable to the flat form: it will allow us to employ many turns of a large pipe, and therefore produce a great elevation of a large quantity of water.

The same thing will be still better accomplished by wrapping the pipe on a cylinder, and making it gradually tapering to the end, in such a manner that the contents of each spire may be the same as when it is wrapped round the cone. It will raise the water to a greater height (though certainly with an increase of the impelling power) by the same number of spires, because the vertical or pressing height of each column is greater.

In the preceding description of this machine, that construction has been chosen which made its principle and manner of working most evident, namely, that which contained the same material quantity of air in each turn of the spiral, more and more compressed as it approaches to the rising pipe. But this is not the best construction: for we see that in order to raise water to the height of a column of 34 feet, the air in the last spire is compressed into half its space; and the quantity of water delivered into the main at each turn is but half what was received into the first spire, the rest flowing back from spire to spire, and being discharged at the spout.

But the construction may be such that the quantity of water in each spire may be the same that was received into the first; by which means a greater quantity (double in the instance now given) will be delivered into the main, and raised to the same altitude by very nearly the same force. This may be done by another proportion of the capacity of the spires; either by a change of their calibre, or of the diameters of the solid on which they are folded. Suppose the bore to be uniform throughout, the diameters must so vary that the constant column of water and the column of air, compressed to the proper degree, may occupy the whole circumference. Let A be the column of water which balances the pressure, and H the height to which the water is to be raised. Let A be to A + H as 1 to m. Then it is plain that m will represent the density of the air in the last spire, if its natural density be 1, because it is pressed by the column A + H while the common air is pressed by A. Let 1 represent the constant water column, and consequently it will be nearly equal to the air column in the first spire: then the whole circumference of the last spire must be $1 + \dfrac{1}{m}$, in order to hold the water 1, and to compress the air into the space $\dfrac{1}{m}$ or $\dfrac{A}{A+H}$. The circumference of the first spire is 1 + 1 or 2: and if D and d be the diameters of the first and last spires, we have $2 : 1 + \dfrac{1}{m} :: D : d$, or 2 m : m + 1 :: D : d. If, therefore, a pipe of uniform bore be wrapped round a conic frustrum, of which D and d are the end diameters, the spirals will be very nearly such as will answer the purpose. It will not be *quite* exact, for the intermediate spirals will be rather too large: the conoidal frustrum should in strictness be formed by the revolution of a logarithmic curve. With such a spiral the full quantity of water which was confined in the first spire will find room in the last, and will be sent into the main at every rotation. This is a very great advantage, especially when the water is to be much raised. The saving of power by this change of construction is always proportional to the greatest compression of the air.

The chief difficulty in any of these forms is in determining the form and position of the horn and the scoop; yet on this the performance of the machine greatly depends. The following instructions will render this tolerably easy. Let A B E O (fig. 214) represent the first or outermost spire, of which the axis is C. Suppose the machine immerged up to the axis in the water whose surface is V V': it has been seen that it is most effective when the surfaces K B and O n of the water columns are distant from each

in it in the same manner, leaving the half space B E O filled with
ed air; for it took in and confined only what filled the quadrant
is obvious, therefore, that the quadrant B E must be so shaped
e in and confine a much greater quantity of air; so that when it
to A, the space B E O may contain air sufficiently dense to sup_
olumn A O. But this is not enough: for when the wide mouth
ι *a* rises up to the top, the surface of the water in it rises also,
he part A O *o a'* is more capacious than the part of uniform bore
hat succeeds it, and that cannot contain all the water which it
ʃ held. Since then the water in the spire rises above A, it will
water back from O *n* to some other position *m' n'*, and the pressing
the water column will be diminished by this rising on the other
. Hence it will appear that the horn must begin to widen, not
ut from A, and must occupy the whole semicircle A B E; while
ty must be to the capacity of the opposite side of uniform bore as
of B O and the height of a column of water which balances the
re to the height of that column: for then the air which filled it
he common density will fill the uniform side B E O, when com-
o as to balance the vertical column B O. But even this is not
: for it has not taken water enough. When it dipped into the
. E it carried air down with it, and the pressure of the water in the
ιused that fluid to rise into it a little way; and some water must
ιe over at B from the other side, which was drawing narrower.
ιerefore, the horn is in the position E O A it is not full of water:
ntly, when it comes into the situation O A B it cannot be full, nor
ιlance the air on the opposite side. Hence some will come out
ʃ rise up through the water. The horn must therefore extend at
ι O to B, or occupy half the circumference; and it must contain at
ce as much water as would fill the side B E O. Nay, if it be
rger, there may be no disadvantage; because the surplus of air
takes in at E will be discharged as the end E *e* of the horn rises
o B, and it will leave the precise quantity that is wanted. The
water will be discharged as the horn comes round to dip again
cistern.
rust also secure the proper quantity of water. When the machine
ιch immersed as to be up to its axis in water, the capacity which

11. Desaguliers describes, in the second volume of his *Experimental Philosophy*, a very simple contrivance to raise water, which is this: to one end of a rope is fixed a large bucket, having a valve at its bottom, opening upwards; to the other end is fastened a square frame, and the cord is made to pass over two pulleys, each of about 15 inches diameter, (and fixed in a horizontal plane,) in such a manner that as the bucket descends the frame ascends with equal velocity, and *vice versâ*. The frame is made to run freely upon four vertical iron guide-rods passing through holes at its four corners; and when the bucket is filling with water at the well, the frame stands at the horizontal plane to which the water is to be raised; when the bucket is full, a man steps upon the frame; (his weight, together with that of the frame, exceeding the weight of the vessel and its contained water;) this gives an ascending motion to the bucket, and causes the valve in its bottom to close. When the bucket is raised to the proper height, a hook fixed there catches into a hasp at the side of the bucket, turns it over, and causes it to empty its water into a trough which conveys it where it is required: at this time the man and the descending-frame have arrived at a platform which prevents their further descent, where the man remains till he finds the bucket above is empty; when he steps from the frame, and runs up a flight of stairs to the place from which he descended: the bucket, in the mean while, being somewhat heavier than the frame, descends to the water, and raises the frame to its original position. Thus the work is continued, the man being at rest during his descent, and labouring in the ascent.

Desaguliers employed in this kind of work a "*tavern drawer*," who weighed 160 lbs., whom he desired to go up and down forty steps of 6½ inches each (in all about 22 feet) at the same rate he would go up and down all day. He went up and down twice in a minute; so that, allowing the bucket with a quarter of a hogshead in it to weigh 140 lbs., he is able to raise it up through 22 feet twice in a minute: this Desaguliers estimates as equivalent to a whole hogshead raised 11 feet in a minute, and rather exceeds what he has assigned as a maximum of human exertion.

This machine is in many cases not only the most simple, but the best that can be devised; yet it is one that, without due precautions, is likely to be a very bad one. The frame on which the man steps must be brought up to its place again by a preponderancy in the machine when unloaded; it should arrive precisely at the same time with the man; but it may arrive sooner or later. If sooner, it is of no use, and wastes power in raising a counterpoise which is needlessly heavy, or in fact less water is elevated than the man is able to elevate; if later there is a loss of time. Hence the perfection of this truly simple machine requires the judicious combination of two maximums, each of which varies in a ratio compounded of two other ratios. It will not be difficult, however, to adjust the proportions of the

e bucket and that of the frame: for if B denote the weight of F that of the frame, and φ the force necessary to overcome the 1 the inertia of the pulleys, g denoting 32½ feet, t the time walking up the steps, and s the space ascended or descended, B and F be so adjusted as to satisfy the following equation,

$$s = \frac{B-F}{B+F+\phi} \tfrac{1}{2} g t^2$$

be a spring affording but a small quantity of water, or having fall, it is possible by the loss of some of the water to raise the ly a gentleman's seat, or any place where it is wanted; but in tity than what runs waste, if the place to which the water is to be higher than the spring or reservoir from which the water falls.

ng ago contrived an engine for this purpose: but the first who thing in execution was Gironimo Fanngio, at Rome, in 1618; t in this country was George Gerves, a carpenter, who, in the erected an engine called the Multiplying-wheel Bucket-engine, of Sir John Chester, at Chichley, in Buckinghamshire. This much approved by Sir Isaac Newton, Dr. Desaguliers, and Mr. nd was certainly very ingenious. The water from a spring n a large bucket hanging by a cord from an axle, while a smaller s raised from the same place by a cord hanging from a wheel on xle: a fly and other regulating apparatus were added, to make work itself, which it did for many years without being out of a whole, however, the contrivance is complex; and we are not any other engines of the same kind have been erected. A , with a plate, may be seen in Desaguliers' second volume.

r. H. Sarjeant, of Whitehaven, contrived a very cheap ur raising water, for which the Society for the En- ient of Arts awarded him a silver medal in the year sketch of this simple invention is given in fig. 215.

ine was erected at Irton-hall, which is situated on an ascent of 60 perpendicular height: at the foot of this elevation, about 140 nt from the offices, there runs a small stream of water; and, in cure a constant supply of that necessary fluid, the object was to stream to the house for culinary or domestic uses. With this was formed at a short distance above the current, so as to of about four feet: the water was then conducted through a igh, into which a piece of leaden pipe, two inches in diameter, l, and part of which is delineated at A. n of this pipe is directed in such a manner as to run into the when the latter is elevated; but, as soon as it begins to descend, passes over it, and flows progressively to supply the wooden vell, at the foot of which stands the forcing-pump C, being three iameter. n iron cylinder attached to the pump-rod, which passes t such cylinder is filled with lead, and weighs about 240 is powe works the pump, and forces the water to ascend e through a pipe one inch in diameter, and which is 420 feet in

fixed a cord, which, when the bucket approaches to within four es of its lowest projection, extends, and opens a valve in the he vessel through which the water is discharged. ne in a great degree similar to this was erected some years ago

R 2

by the late James Spedding, Esq. for a lead mine near Keswick, with the addition of a smaller bucket which emptied itself into the larger near the beginning of its descent, without which addition it was found that the beam only acquired a libratory motion, without making a full and effective stroke.

To answer this purpose in a more simple way, Mr. Sarjeant constructed the small engine in such manner as to finish its stroke (speaking of the bucket end) when the beam comes into a horizontal position, or a little below it. By this means the lever is virtually lengthened in its descent in the proportion of the radius to the cosine, of about thirty degrees, or as seven to six nearly, and consequently its power is increased in an equal proportion.

It is evident that the opening of the valve might have been effected, perhaps better, by a projecting pin at the bottom; but Mr. S. chose to give an exact description of the engine as it stands. It has now been some years in use, and completely answers the purpose intended.

The only artificers employed, except the plumber, were a country blacksmith and carpenter; and the whole cost, exclusive of the pump and pipes, did not amount to 5l.

In a letter, dated Whitehaven, April 28, 1801, Mr. Sarjeant observes, that the pump requires about 18 gallons of water in the bucket to raise the counter-weight, and make a fresh stroke in the pump; but it makes three strokes in a minute, and gives about a half-gallon into the cistern at each stroke. He adds, "I speak of what it did in the driest part of last summer; when it supplied a large family, together with work-people, &c. with water for all purposes, in a situation where none was to be had before, except some bad water from a common pump, which has been since removed. But the above supply being more than sufficient, the machine is occasionally stopped to prevent wear, which is done by merely casting off the string of the bucket-valve."

13. Mr. Benjamin Dearborn has contrived an hydraulic engine which may be conveniently added to a common pump, and thereby renders it useful in further elevating water, and particularly in extinguishing fires: the following description of his apparatus is extracted from the *Memoirs of the American Academy*.

Fig. 216. A, B, C, D, represents a pump, the form of which is similar to that of the pumps commonly employed on ship-board.

E, the spout.

F, a stopper.

D, d, a plank-cap, that is fitted to the pump, and provided with leather on its lower surface, being secured by the screws a, b : in the centre is a hole, through which the spear of the pump passes, and round which a leather collar is made, as represented at the letter c.

g, a nut for the screw b.

f, a square piece of wood that is nailed across one end of the plank-cap, through both which the screw a is introduced: a hole is made through such piece and the cap that communicates with the bore of the pump.

G, G, a wooden tube, which may be of any requisite length, and consist of any number of joints; it is made square at the lower extremity, and perforated for the reception of the cock; the upper end being made with a nice shoulder.

en cock, that opens or shuts the communication between the
to tube, being furnished on the opposite side with a handle and
in case it should be found necessary.

two ferules, the object of which is to prevent the tube from

ces, each of which ought to be crossed over another, as nearly
cs as possible.

ons in form of a staple, which surround the tube, and pass
braces; their ends being perforated with holes for fore-locks.

N, is a band made of five pieces of wood; h, l, m, n, a square
lower part of which is a hole for the reception of the extremity
and which piece rests on the shoulder o, p ; to the lower end of
nailed a piece of leather, with a hole in its centre, similar to
the wood. Another piece of leather of the same form is
at top of the tube, and between both is a circle of thin plate-
wo pieces of leather and the brass being pressed between the
the head and the shoulder of the tube. Their edges are deli-

L, M, are the edges of two pieces of plank, of a similar width
d, to which they are closely nailed, each being provided with a
passes through a mortice in the end of the piece O, P: both
holes for a fore-lock at q.

ece of plank of the same width as the sides; the centre of
forated, in order that the tube may pass through ; and in each
s is a mortice for the reception of the tenons.

ap.

wo pieces nailed to the side of the tube; the lower extremity of
nailed with a truck, with a view to lessen the friction of the head
ntal revolution.

event fore-locks, the design of which is to fasten down the head,
the water from escaping at the joint o, p.

a wooden conductor : the extremity marked with the letter Q
while the opposite end, R, is bored with a small auger.

that passes through the conductor and head, and being secured
with a fore-lock or awl : this bolt is rounded near the head, and
e middle.

represents a piece of iron or brass, designed to prevent the head
rom wearing into the wood.

opes for the direction of the conductor.

epresents the head without such conductor.

is a thick brass plate, the centre of which is perforated, so as to
age to impurities, that might otherwise obstruct the conductor :
rpose a piece of leather is nailed under it to the head. The
in the centre is adapted to the size of the bolt, which it prevents
. The conductor has a hollow cut round the bolt on the inside,
size as the circle of holes in the brass; round such cavity is
ne face of the conductor, a piece of leather, that plays on the
e brass plate when the conductor is in motion.

clusion of his memoir, Mr. Dearborn observes, that he has
t of 30 feet on his pump; and, though the severity of the season
ed him from completing it, so that one person only could work
yet he is enabled to throw water on a contiguous building, the
of which is 37 feet from the pump, and between 30 and 40

... it arrives at B, and thence through the engine from B to S.

At the turn B the water enters into a chamber C, the lower ... which terminates in two brass cylinders, four inches in diameter ... pump or pistons of lead, D and E, are capable of moving up ... pieces-work, which pass through a close packing above, ... to the extremities of a chain leading over and properly ... the wheel Q, so that it cannot slip.

The ... pipes D and E are cast in their places, and have no packing ... They move very easily; and if at any time they should become ... be opened ... out by a few blows with a proper instrument, ... them out of their place. On the sides of the two brass ... which D and E move, there are square holes communicating ... and G, which is a horizontal trunk or square pipe, four inches ... inches deep. All the other pipes G, G, and R, are six ... except the principal cylinder wherein the piston H moves; ... cylinder is ten inches in diameter, and admits a nine-foot stroke, ... here delineated as if the stroke were only a three-foot.

... rod works through a stuffing-box above, and is attached to ... the pit-rod, or a perpendicular piece divided into two, so as ... in alternate motion up and down, and leave a space between, ... the fixed apparatus of great cylinder. The pit-rod is ... down into the mine, where it is employed to work the pumps; ... ever applied to mill-work, or any other use, this rod would ... action of the first mover.

... tumbler, or tumbling-bob, capable of being moved on the gud... ... from its present position to another, in which the weight L shall ... inclination on the opposite side of the perpendicular, ... the end K will then be as much elevated as it is now ...

The pipe R S has its lower end immersed in a cistern, by which means it ... without the possibility of the external air introducing ... constitutes a torricellian column, or water barometer; and ... whole column from A to S effectual: as we shall see in our view ...

... suppose the lower bar K V of the tumbler to be horizontal, and ... O so situated, as that the plugs or leaden pistons D and E shall ... to each other, and stop the water-trays G and F. In this state ... though each of these pistons is pressed by a force equivalent ... than 1000 pounds, they will remain motionless, because these ... ... to each other, they are constantly in equilibrio. The ... H being here shown as at the bottom of its cylinder, the tum... ... thrown by hand into the position here delineated. Its action ... and consequently upon the wheel Q, draws up the plug D, and ... so that the water-way G becomes open from A B, and that of ... R; the water consequently descends from A to C; thence to ... into beneath the piston H. This pressure raises the piston, ... there be any water above the piston, it causes it to rise and pass ... into R. During the rise of the piston (which carries the pit-rod ... with it) a sliding block of wood I, fixed to this rod, is brought ... with the sill K of the tumbler, and raises it to the horizontal ... beyond which it oversets by the acquired motion of the wheel L.

The ... rise of the piston, if there were no additional motion in the ... would only bring the two plugs D and E to the position of rest, ... to close G and F, and then the engine would stop; but the fall of ... tumbler carries the plug D downwards quite clear of the hole F, and

Descriptio Machinæ Hydraulicæ curiosæ constructæ Joh. Geor. Faudisi.

...lle Invention de lever l'Eau plus haut que la Source, avec quelques ... au le Moyen de ... par ... de Caus, 166...

...Gravesil ... Monte Sacr. Principia Physico-mechanica diversarum ... ... Pneumaticæ et Hydraulicæ. Venet. 166...

... Machine Hydraulique, par Fruticini. Journ. des Sçav. 166...

Account of this machine is likewise given in the Architecture Hy... of ... tom. 2. and, in the 3d. vol. of Desaguliers' Experi... Philosophy: in both which performances many other hydraulic ... are described.]

Undertaking for raising Water, by Sir Samuel Moreland. Phil. Trans. No. 102.

Hydraulic Engine, by . . . . Phil. Trans. 1675, No. 12...

...Pump, by Mr. Conyers. Phil. Trans. 1677, No. 136.

...Hautefeuille, Réflexions sur quelques Machines à élever les Eaux, ...cription d'une nouvelle Pompe, sans Frottement et sans Piston, ...

...tion des Eaux par toute sorte de Machines, réduite à la mesure, au ...la balance, par le moyen d'un nouveau piston et corps de pompe, et ...eau mouvement cyclo-elliptique, et rejetant l'usage de toute sorte ...velle ordinaires, par le Chevalier Morland. 1685.

...Way of raising Water, enigmatically proposed, by Dr. Papin. ...1685, No. 173. The solutions by Dr. Vincent and Mr. R. A.

...evax, Nouvelles Machines pour épuiser l'Eau des Fondations, ...ou ... simples, font un effet surprenant. 1695. Journ. des Sçav.

...for raising Water by the help of Fire, by Mr. Tho. Savery. Phil. Trans. 1699, No. 253.

...Nouvelle Manière pour lever l'Eau par la Force du Feu : ...

...pour la Construction d'une Pompe qui fournit continuellement ...dans le Réservoir, par M. de la Hire, Mem. Acad. Sci. Paris. 1716. ...tion d'une Machine pour élever des Eaux, par M. de la Faye, ...Sci. Paris. 1717.

...Druckmann's und Joh. Heinr. Weber's Elementar-maschine, ...mittel bey allen wasser-hebungen. Cassel. 1720.

Leupold, Theatri Machinarum Hydraulicarum. 1724, 1725.

...Weidleri Tractatus, de Machinis Hydraulicis toto terrarum ...Marlyensi et Londinensi, &c. 1727. Vide Act. erudit. ...

...ription of the Water-works at London-bridge, by H. Beighton, Phil. Trans. 1731, No. 417.

...count of a new Engine for raising Water, in which horses or other ...ant without any loss of power (which has never yet been prac...and how the strokes of the piston may be made of any length, to ...the loss of water by the too frequent opening of valves, &c. by ...Churchman. Phil. Trans. 1734.

...of the Machine Hydraulique proposée par M. Segner, par ...Euler. Mem. Acad. Sci. Berlin. 1750.

...ication de la Machine Hydraulique de M. Segner à toutes sortes ...et de ses avantages sur les autres Machines Hydrauliques, par ...Euler. Mem. Acad. Sci. Berlin. 1751.

...Segner's machine is no other than the simple yet truly ingenious contrivance known by the name of Barker's mill, which had been described in the 2d volume of Desaguliers' Philosophy, some years before the German professor made any pretensions to the honour of the invention. The theory of it is likewise treated by John Bernoulli at the end of his Hydraulics.]

Recherches sur 'une nouvelle manière d'élever de l'Eau proposée par M. de Moar, par M. L. Euler, Mem. Acad. Berlin. 1751.

Discussion particulière de diverses manières d'élever de l'Eau par le moyen des Pompes, par M. L. Euler, Mem. Acad. Ber. 1752.

Maximes pour arranger le plus avantageusement les Machines destinées à élever de l'Eau par le moyen des Pompes, par M. L. Euler, Mem. Acad. Ber. 1752.

· Réflexions sur les Machines Hydrauliques, par M. le Chevalier D'Arcy, Mem. Acad. Sci. Paris. 1754.

Mémoire sur les Pompes, par M. le Chevalier de Borda, Mem. Acad. Sci. Paris. 1768.

Dan. Bernoulli Expositio Theoretica singularis Machinæ Hydraulicæ. Figuri helvetiorum, exstructæ. Nov. Com. Acad. Petrop. 1772.

Abhandlungen von der Wasserschraube, von D. Scherffer, Priester. Wien. 1774.

Recherches sur les Moyens d'exécuter sous l'Eau toutes sortes de Travaux Hydrauliques, sans employer aucun Epuisement, par M. Coulumb. 1779.

Saemund Magnussen Holm, Efterretning om skye Pumpen. Kiobenhavn. 1779.

Moyen d'augmenter la Vitesse dans le Mouvement de la Vis d'Archimède sur son Axe, tiré des Mémoires Manuscrits de M. Pingeron, sur les Arts utiles et agréables. Journ. d'Agric. Juin, 1780.

The Theory of the Syphon, plainly and methodically illustrated. 1781. (Richardson.)

Memoria sopra la nuova Tromba Funiculare Umiliata, dal Can. Carlo Castelli. Milano. 1782.

Dissertation de M. de Parcieux, sur le moyen d'élever l'Eau par la rotation d'une corde verticale sans fin. Amsterdam et Paris, 1792.

Theorie der Wirzischen Spiral Pumpe, erlautert von Heinr. Nicander. Schwed. Abhandl. 1783.

Jac. Bernoulli, Essai sur une nouvelle Machine Hydraulique propre à élever de l'Eau, et qu'on peut nommer Machine Pitotienne. Nov. Act. Acad. Petrop. 1786.

K. Ch. Langsdorf's Berechnungen über die Vortheilhæftere Benutzung Angelegter Fammelteiche zur Betriebung der Maschinen. Act. Acad. Elect. Mogunt. 1784, 1785.

Nicander's Theorie der Spiral Pumpe. 1789.

Nouvelle Architecture Hydraulique, par M. Prony. 1790, 1796.

A short Account of the Invention, Theory, and Practice of Fire Machinery; or, Introduction to the Art of making Machines vulgarly called Steam-Engines, in order to extract water from mines, convey it to towns, and jets-d'eaux in gardens; to procure water-falls for fulling, hammering, stamping, rolling, and corn-mills; by W. Blakey. 1793.

PUMPS.

1. The construction of pumps is usually explained by glass models, in which the action both of the pistons and valves may be seen.

In order to understand the structure and operation of the common pump, let the model D C B L, fig. 222, be placed upright in the vessel of water K, the water being deep enough to rise at least as high as from A to L. The

be less than 32 feet; otherwise the water will never get above tl
But when the height is less, the pressure of the atmosphere will
than the weight of the water in the pump, and will therefore rais
the bucket; and when the water has once got above the bucket,
lifted to any height, if the rod D be made long enough, and a
degree of strength be employed, to raise it with the weight of
above the bucket without ever lengthening the stroke.

The force required to work a pump will be as the h
which the water is raised, and as the square of the
of the pump-bore in that part where the piston wor
that if two pumps be made of equal heights, and one
be twice as wide in the bore as the other, the widest v
four times as much water as the narrowest, and wil
therefore four times as much strength to work it.

The wideness or narrowness of the pump in any ot
besides that in which the piston works, does not n
pump either more or less difficult to work, except wh:
ence may arise from the friction of the water in
which is always greater in a narrow bore than in a
because of the great velocity of the water.

The pump-rod is never raised directly by such a
E at the top, but by means of a lever, whose longer
the end of which the power is applied) generally exc
length of the shorter arm five or six times, and by th:
gives five or six times as much advantage to the powe
these principles, it will be easy to find the dimensi
pump that shall work with a given force, and dra
from any given depth.

The quantity of water raised by each stroke of th
handle is just as much as fills that part of the bore
the piston works, be the size of the rest of the bore al
below the piston what it will. The pressure of th
sphere will raise the water 32 feet in a pipe exhauste
but it is advisable never to have the piston more th
24 feet above the level of the surface of the water i
the lower end of the pump is placed; and the power
to work the pump will be the same, whether the pis
down to lie on a level with the surface of the well, or
it works 30 feet above that surface, because the weig
column of air that the piston lifts is equal to the w
pressure of the column of water raised by the pressu
air to the piston. And although the pressure of tl
the surface of the well will not raise or force up the
the pump-bore more than 32 feet, yet when the pis
down into the column so raised, the water gets abov

may then be raised to any height whatever above the piston, according to the quantity of power applied to the handle of the pump for that purpose.

Pumps ought to be made so (says Mr. Ferguson) as to work with equal ease in raising the water to any given height above the surface of the well. And this may be done by duly proportioning the diameter of the bore (in that part where the piston works) to the height the water is to be raised, as that the column of water may be no heavier in a long pump than in a short one; or indeed equally heavy in all pumps from the shortest to the longest, on a supposition that the diameter of the bore is the same size from top to bottom; and whatever size the bore be, above or below that part in which the piston works, the power required to work the pump will be just the same as if the bore was of the same size throughout.

In order that a man of common strength may raise water in pumps with the same ease, to any height not less than 10 feet, or more than 100 feet, above the surface of the well, Mr. Ferguson has calculated the annexed table, in which the diameter of the bore is duly proportioned to the height; and in these calculations he supposes the pump-handle to be a lever increasing the power five times.

Height of the water in feet, above the surface of the well.	Diameter of the bore.		Water discharged in a minute, in wine measure.	
	Inches.	100 parts of an inch.	Gallons.	Pints.
10	6	·93	81	6
15	5	·66	54	4
20	4	·90	40	7
25	4	·38	32	6
30	4	·00	27	2
35	3	·70	23	3
40	3	·46	20	3
45	3	·27	18	1
50	3	·10	16	3
55	2	·95	14	7
60	2	·84	13	5
65	2	·72	12	4
70	2	·62	11	5
75	2	·53	10	7
80	2	·45	10	2
85	2	·38	9	5
90	2	·31	9	1
95	2	·25	8	5
100	2	·19	8	1

In the first column look for the number of feet the water is to be raised; then, in the second column, you have the diameter of that part of the bore in which the piston or bucket works; and in the third column, the quantity of water which a man of ordinary strength can raise in a minute by the pump to the given height.

The quantity of water contained in a pipe of either of those heights in the table, supposing the diameter of the bore to be the same from top to bottom, is 4523·2 cubic inches, or 19·58 gallons in wine-measure, as near as the hundredth part of an inch in the diameter of the bore can make it.

Mr. Ferguson has calculated the following table, by which the quantity and weight of water in a cylindrical bore of any given diameter and perpendicular height may be very readily found.

Diameter of the cylindrical bore 1 inch.

Feet high.	Quantity of water, in cubic inches.	Weight of water, in troy ounces.	In avoirdupoise ounces.
1	9·4247781	4·9712340	5·4541539
2	18·8495562	9·9424680	10·9083078
3	28·2743343	14·9137020	16·3624617
4	37·6991124	19·8849360	21·8166156
5	47·1238905	24·8561700	27·2707695
6	56·5486686	29·8274040	32·7249234
7	65·9734467	34·7986380	38·1790773
8	75·3982248	39·7698720	43·6332312
9	84·8230029	44·7411060	49·0873851

For tens of feet high, remove the decimal points one place towards the right hand ; for hundreds of feet, two places ; for thousands, three places ; and so on. Then multiply each sum by the square of the diameter of the given bore, and the products will be the answer.

EXAMPLE :

Qu. What is the quantity and weight of water in an upright pipe 85 feet high, and 10 inches in diameter of bore ? The square of 10 is 100.

Feet high.	Cubic inches.	Troy ounces.	Avoirdupoise ounces.
80	753· 982248	397· 698720	436· 332312
5	47·1238905	24·8561700	27·2707695
85	801·1061385	422·5548900	463·6030815
Multiply by 100		100	100
Answer . . . 80110·6138500		42255·4890000	46360· 308150

Which number (80110·61) of cubic inches being divided by 231, the number of cubic inches in a wine gallon, gives 342·6 for the number of gallons in the pipe ; and 42255·489 troy ounces being divided by 12, gives 3521·29 for the weight of the water in troy pounds ; and, lastly, 46360·3 avoirdupoise ounces being divided by 16, gives 2897·5 for the weight in avoirdupoise pounds.

the force which must be exerted by the labourer becomes much greater than the weight of the column of water which he is raising. If the pump be laid aslope, which is very usual in these occasional and hasty drawings, it is necessary to make a guide for the piston-rod within the trunk, that the bag may play up and down without rubbing on the sides, which would quickly wear it out.

The experienced reader will see that this pump is very like that of Gosset and De la Deuille, described by Belidor, vol. ii. p. 120, and most writers on hydraulics. It would be still more like it, if the bag were on the under side of the partition E, and a valve placed further down the trunk. But we think that our form is greatly preferable in point of strength. When in the other situation, the column of water lifted by the piston tends to *burst* the bag, and this with a great force, as the intelligent reader well knows. But in the form recommended here, the bag is *compressed*, and the strain on each part may be made much less than that which tends to burst a bag of six inches diameter. The nearer the rings are placed to each other the smaller will the strain be.

The same bag-piston may be employed for a forcing-pump, by placing it below the partition, and inverting the valve; and it will then be equally strong, because the resistance in this case too will act by compression.

3. An ingenious variation in the construction of the sucking-pump, is that with two piston-rods, in the same barrel, invented by Mr. Walter Taylor, of Southampton. A vertical section of this pump is given in fig. 224.

The piston-rods have racks at their upper parts working on the opposite sides of a pinion, and kept to their proper positions by friction-rollers. The valves used in this pump are of three kinds, as shown at *a*, *b*, and *c*. The former is a spheric segment which slides up and down on the piston-rod, and is brought down by its own weight; the second, *b*, is called the pendulum-valve; and the third, *c*, is a globe which is raised by the rising water, and falls again by its own weight. Each of these valves will disengage itself from chips, sand, gravel, &c. brought up by the water. In this kind of pump the pistons may either be put in motion by a handle in the usual way, or a rope may pass round the wheel *d e* in a proper groove, the two ends of which, after crossing at the lower part of the wheel, may be pulled by one man or more on each side. A pump of this kind, with seven inch bore, heaves a ton 24 feet high in a minute, with ten men, five only working at a time on each side.

Another improvement of the common pump has been made by Mr. Todd, of Hull. This invention in some particulars bears a resemblance to the ordinary one, but he has contrived to double its powers by the following means:

...is fixed on inverted piston B D, with its bucket and valve upon... ...Upon the top of the barrel there goes off a part F R, either ...or movable by a ball and socket; but in either case ...In this part, at C, is a fixed valve opening upwards. ...that when the piston frame is thrust down into the water, the ...D descends, and the water below will rush up through the valve D, ...above the piston; and that, when the frame is lifted up, the piston ...the water through the valve C up into the cistern P, there to run ...the spout. The piston of this pump plays below the surface of the ...Mr. Martin has described a mercurial pump, which works by ...silver, invented by Mr. Hoskins, and perfected by Mr. Desaguliers; ...other pump of the lifting sort, invented by Messrs. Gosset and De ...and set up in the king of France's garden at Paris, the piston ...works without friction.—*Phil. Brit.* vol. ii. p. 57, &c. ed. 3.

Ctesibius's pump, the first of all the kinds, acts both by ...and pulsion.

...structure and action are as follows: A brass cylinder A B C D, ...with a valve in L, is placed in the water. In this is ...embolus M K, made of green wood, which will not swell in the ...adjusted to the aperture of the cylinder with a covering of ...without any valve. In H is fitted on another tube N H, with ...opens upwards in I.

...the embolus M K being raised, the water opens the valve in L, and ...the cavity of the cylinder; and when the same embolus is again ...the valve I is opened, and the water driven up through the ...

...the pump used among the ancients, and that from ...the others are deduced. Sir S. Morland has ...to increase its force by lessening the friction, ...has done to good effect, insomuch as to make it ...without almost any friction at all.

7. In 1812, the Society for the Encouragement of Arts ...a silver medal on Mr. John Stevens, for an im- ...in the construction of the forcing-pump, by which ...enabled, at a comparatively trifling expense, to raise ...from a well 66 feet below the surface of the ground. ...whole expense of the pump and apparatus was 25l.

...lower part of the pump-tree is four inches in the bore. ...part of the rod which passes through the stuffing- ...made of brass; the elbow and upper pump-trees are ...two-inch bore, and may be easily made of any kind of ...It may also be made to act as an engine to extin- ...by the addition of an air-tight vessel and pipe to ...part.

...drawing is introduced a cap and screw, in preference ...screwing it to the nozzle of the pump, as it is stronger ...more to be depended upon; and when the water is to be ...a great height, a screw is also recommended to be

s 2

made to fit the nozle, that every thing may be always ready for immediate use. The work of this pump is not liable to be injured by frost; and when the well is of considerable depth a brass or metal barrel for the piston to work in should be adopted.

Fig. 228 is a section of a well, in which a pump of this kind is fixed; A A represents the surface of the ground, and B B the brickwork of the well, in which the water stands at the level C, and is, by the pump, to be raised to the surface A A.

D is the lever or handle of the pump, which has the rod *a* jointed to it, and descending to the pump; the rod is made of wood, in several lengths, which are united by joints of iron, in the manner shown at fig. 229; the wooden rods, *a a*, being capped with iron forks *b*, which include the ends of them and are rivetted fast; the ends of the forks are joined together to connect the several lengths.

E is the working barrel, or chamber, of the pump, in which the bucket works; this part is formed of a tree, bored through and having a projecting branch *e*, which is likewise bored obliquely to the barrel, and forms the forcing pipe; in the bottom of the barrel the suction-valve is situated, being at the top of the suction part of the pump, which is bored with a smaller auger than the working chamber, which is also lined with a brass tube, where the bucket works. The top of the barrel is covered by a metal lid. *g*, (see figs. 230 and 231,) which has a stuffing-box in the centre to receive the metal cylindrical part of the pump-rod *h*; to the lower extremity of this the bucket *a* is fixed. The metal lid consists of a ring, which is screwed to the wooden barrel by five screw-bolts, passing through as many ears, projecting from the circumference of the ring; they have eyes below to hook upon pins, which are fixed in the wood, but project sufficiently for these bolts to hold, and are formed into screws above, so as to hold the ring firmly down, by means of nuts screwed upon them. The movable lid of the pump, which has the stuffing-box *g* formed in the centre of it, is screwed to the ring by five screws, and these can be taken out to remove the lid, and draw up the bucket, when it requires to be leathered.

F is the forcing pipe, formed of as many pieces of wooden pipes as are required to make up the length; they are united together by making the upper ends conical, to enter a similar cavity made in the lower end of the next pipe; the lowest piece fits upon the extremity of the projecting branch *e*, and a valve is proposed to be put in the pipe at this joint, to prevent the return of the water, and bear part of the weight of the column from the lowest valve at *f*; the upper end of the pipe has a spout *i*, at which the water is delivered.

M is a second spout, fixed into the pipe lower than the former; it has a screw by which it can be united to a hole, or leather pipe, to convey the water to a distance, or by means of a jet, or branch-pipe, to throw it in the manner of a fire-engine; in this case the upper spout *i* must be stopped up, by a screw-plug or cap; and there is a copper air-tight vessel H, situated at the top of the pipe F, to equalize the pulsative motion of the water as thrown by the pump.

K is a bracket fixed to the pipe F, and projecting over the centre of the pump, where it has a hole to receive the pump rod *h*, and guide it steadily in its motion up and down, that it may not wear the stuffing-box away on one side. As the wooden tubes of which the forcing-pump F is composed may be made from waste or crooked timber, it makes a great difference

.... the low price of such, and that of the straight trees necessary for .... pumps. A wooden plug may be chained to the pump, between .... or scales M and L, so as be ready to stop that which is not wanted ....

Mr. Stephens is of opinion, that it is better to place the .... above the level of the water in the well.

Mr. William Tyror, of Liverpool, took out a patent in ...., 1819, for certain improvements in the construction .... pumps, and in the machinery for working the same.

This improvement consists in having four brass chambers, marked P P P, ... joined together by means of breech-pieces with screws, and .... across the joints; these breech-pieces, marked Q, Q, being cast .... or any other suitable metal. When these are complete, P P, ... is placed upon the breech-piece Q q, fig. 234, and both of them are .... fixed under a box or frame suitable for the purpose.

This box or frame, fig. 236, is furnished with eight brass grooves, O O O .... fastened to the sides with screws; and a crank, or four cranks in .... one crank out of each side of the same piece of square iron or .... suitable metal, one up, one down, one in front, and one in back. .... of the crank, or cranks, is fixed two tooth and pinion wheels, ... distance apart to allow two wheels of the same diameter and .... in round between them, so that the cranks may go round without .... the other wheels, marked C and E. The wheels D and F are .... on the cranks A, A, by means of a screw or pin, and the wheels C and E .... fixed close together, slide to and fro upon the square end .... by means of the guide or sliding gear V, which is fixed in .... out of the nave of the wheel C, by means of a clip and two .... it underneath, and rests in the notches fixed at the .... box or frame, for the end of the guide or gear to rest.

.... are three in number on each side of the box or frame W. .... from the wheel has the guide V, drawn back, with the .... the small wheel F. By moving the guide into the middle .... the wheels C and E are kept between the wheels D and F; and the .... the wheels guides the wheel C on the large wheel D, so that .... much greater when forcing or drawing water from a great ....

.... wheels C and E are placed in the space between the lower .... handle is moved from the upper axle U, and placed upon the .... the crank A A, and the pump is worked without the assistance of .... occasion may require. The machinery is furnished with four .... marked B B B, for the purpose of fixing to them the spear .... pumping-rods, by means of a joint and pin and bolt, the key-bow .... square across in the inside, so as to give the roller-step a fair ....

.... represents the rolling-step, which is formed of two pieces of .... half round, of a thickness according to the strength or size of .... and the other round, like a wheel or sheave in form, and of .... as the other half. This round sheave or wheel is cut half .... middle edgeways, and the piece is then cut off, and a dovetail .... the width in proportion to the crank. The other half is then .... the place from whence the larger piece has been cut, and both of .... together by means of two screws; and the sheave or wheel is .... of its appearance before it was cut. A hole is now drilled through the centre, and it is fixed upon the before mentioned crank or cranks. Y, in fig. 236, represents the larger half of this step; and X, fig. 237, represents the smaller half, with the dove-tailed standing upon it, which fills up the vacancy or room that is made in the large half for fixing it upon the crank.

The ends of each key-bow is set in the grooves, O O O, fig. 238, and the key-bow rods, B B B B, work through holes in the bottom of the box, for which purpose an iron-plate or base is formed with four holes, S S S, fig. 239, and is fixed at the bottom of the box or frame with screws. The rods, by being fitted into the holes in this plate, keep the stroke of the pump perpendicular, while the step rolls backwards and forwards in the key-bow, as they are forced up and down by the cranks moving or turning alternately round.

When this machinery is applied over a forcing-pump, or placed over a fire-engine, it causes a greater quantity of water to be discharged from the cistern or engine, and as it is very powerful, it is highly necessary that it should have a cock of a superior size, to let more water pass through in the same time than ordinary ; for this purpose Mr. Tyror makes the barrel, or that part of the cock where the key or stop goes in on one side, so that there is but one stop to the plug or key, the stop, resting in that part that overhangs the side, admits of room for a full sized water-way to be completely through it, without causing the water to have any bubble or call as it passes through the plug.

Reference to the figures :

Fig. 240 represents a side view of the cock.

Fig. 241, a top view of the same.

Fig. 242, the plug, with the water-way cut out.

Fig. 243 represents the crank, with the tooth and pinion wheels, and the rolling steps.

Fig. 244, the upper axle, with the improved plan of the sliding geer.

Fig. 245, the spear box and rod in the form of the fastening at the top of the key-bow rod, when applied for slupping.

Fig. 246, a front view of the pump standing upon a ship's deck.

9. Mr. Richard Franklin has been rewarded by the Society of Arts for effecting certain improvements in the lifting and forcing pump, by which water can be conveyed into a cistern at the top of the house, to supply all the dressing-rooms, water-closets, &c.

A section of this pump is given in fig. 247.

A A are two pistons; on the upper face of each is a double valve, v v v v; the upper piston-rod passes through the stuffing-box B, and the lower through the stuffing-box C. S is the suction-pipe, and D the discharging-pipe.

Fig. 248 is an external view of the pump ; e e e the lever or handle; F the fulcrum, on which the handle moves; G G is the pump-cylinder; w w the wheels which revolve between the standards s s s s, and which conduct the piston-rods parallel to the cylinder ; e p the conducting-rod, which conveys the motion of the handle to the lower piston; e o the conducting-rod, which gives motion to the upper piston. It is evident, when the handle or lever is lifted, that the upper piston is pressed down, and the lower piston is at the same time elevated, with its valves shut, which forces the water through the upper piston and the discharging-pipe at the same operation.

And when the handle is pressed down, the upper piston rises with its valves closed, and the water in its ascension is forced through the discharging pipe; at the same time the lower piston descends, by which action its valves are opened, and introduces a supply of water equal to the contents of the cylinder, minus the capacity of both pistons. The peculiar advantages of this pump with double pistons are, that with a six-inch stroke it discharges a quantity of water equal to twelve inches of the cylinder: and so, in this proportion, by always doubling the quantity of the stroke, whatever it may be; and thus furnishing a product just equal to two common pumps of the same stroke and capacity of cylinder, and certainly with less than a proportionable friction and expense.

10. The *pumps* that are usually employed *for draining mines* have many inconveniencies, the principal of which we shall here proceed to describe.

1st. As it is necessary for the pumps, whilst sinking, to keep the water very low in the pit, the engine frequently goes too fast, in consequence of the pump drawing up air, and carries up by the violence of the current small pieces of stone, coal, or other substances, and lodges them above the bucket upon the valves, which must considerably retard the working of the pump, and wear the leather.

2dly. When the engine is set to work, (after having been stopped whilst working upon air, and consequently a quantity of air remaining in the pump-barrel, with the small stones, &c. deposited on the valves of the bucket,) it often happens that the compressure of the air by the descent of the bucket, is not sufficient to overcome the weight of the bucket-valves so loaded with rubbish, and the column of water in the stand-pipes; the pump is thereby prevented from catching its water. The usual remedy for this is to draw the bucket out of the working-barrel, until a quantity of water has escaped by its sides to displace the air; this evil often arises from the unnecessary magnitude of the space between the bucket and the clack.

3dly. The pumps being suspended in the pit by capstan ropes, for the purpose of readily lowering as the pit is sunk, the stretching of the ropes (especially when sinking in soft strata) occasions much trouble, by suffering the pumps to rest on the bottom and choke; but the most serious evil is, that the miners, in shifting the pump from one place to another, that they may dig in all parts of the pit, throw them very far out of the perpendicular, thereby causing immense friction and wearing in all parts, besides endangering the whole apparatus, by breaking the bolts and stays, and straining the joints of the pipes.

These inconveniencies have been obviated by Mr. William Brunton, of Butterley Iron-works, in Derbyshire, who, to avoid the pump drawing air, has introduced a side pipe, connecting the parts of the working-barrel which are above and below the bucket, which pipe has a stop-valve, that the miners can regulate with the greatest ease, so as to keep the engine to its full stroke without drawing air, by letting down the water from the upper part of the barrel into the lower, so that it is working again in its own water. Instead of having the whole weight of the lower lift of pumps standing on the bottom, it is fixed in the pit by cross beams, and the miner has only to lift and move an additional pipe or wind-bore which slides upon the lower end of the pump like a telescope, to lengthen down, and this additional wind-bore is besides crooked, and turned aside like a short crank, which, by the facility with which it turns round in the leather collar about the nose of it, can easily be removed into every fresh hole which is made in the bottom by the miners. The pumps are supported in the pit by beams placed across at proper distances, so as to suit the lengths of the pipes, or lengths of the pump, which are nine feet. Short pieces are laid across these, with half-circular holes in them, which being put round the pump, just beneath the flanches, firmly support its weight, but may quickly be removed, when it is required to lower the pumps in the pit; and as they are not fastened by any bolt they do not prevent the pumps being drawn upwards, if it becomes necessary to take out the pumps when the pit is full of water.

The pumps by these means remain stationary, and the suction-pipe lengthens as the pit is sunk, until it is drawn out to its full extent; the whole column is then lowered to the next stanches, and another pipe is added to the top. The pumps being thus kept stationary till nine feet are sunk, the pipe at the top will of course deliver the water at the same level at all times, and instead of being obliged to lengthen the column at every yard sunk, it will only be necessary every nine feet.

Fig. 249 explains the construction of Mr. Brunton's pump, being a section through the centre of the working-barrel and suction piece. A is the door which unscrews to get at the clack of the pump; B is the working-barrel with the bucket D working in it; E is the clack, also shown in figs. 250 and 251; F is the suction-pipe, and G G a movable lengthening piece: this slides over and includes the other when the pump is first fixed; but as the pit is sunk, it slides down over the pipe F, to reach the bottom. The outside of the inner pipe F is turned truly cylindrical and smooth, and the inside of the outer pipe G, at the upper end for about six inches down, is made to fit it. The junction is made perfect by leathers being placed in the bottom of the cup ss, which holds water and wet clay over them, to keep them wet and pliable, and consequently air-tight. The lower extremity of the suction-pipe G terminates in a nose R, pierced with a number of small holes that it may not take up dirt. This nose is not placed in a line with the pipe, but curved to one side of it like a crank, so as to describe a circle when turned round.

By this means the miners, by turning it round upon the pipe F, can always place the nose R in the deepest part of the pit; and when they dig or blast a deeper part, they turn the nose about into it, the sliding tube lengthening down to reach the bottom of the hole, as shown in the figure. By this means there is never a necessity to set a shot for blasting so near the pump foot as to put it in any danger of being injured by the explosion, as is the case of the common pump, in which this danger can only be

which, as they occasion great resistance to the motion of the water, are a waste of power. It acts on the same principle as the ordinary forcing-pump, only that three barrels are connected together for the advantage of raising a constant stream of water.

A A are the barrels, which are bored out truly cylindrical. If the pump is small, the barrels are usually made of brass; but for larger work, cast-iron is used. From one side, near the bottom of each, proceeds a curved pipe B, turning up, and ending with a flaunch, to screw to the under side of the forcing-chamber L. There is also, near the bottom at the opposite side of the barrel, a projecting neck or short pipe D, covered at the end by a door screwed on, that it may be removed to give access to the valve *m*, in the bottom of the barrel. The barrels have projecting rings or flaunches, by which they are screwed down upon the suction-chamber H, which is common to all three barrels; it has a pipe from each of its rods terminating in a flaunch *k*, to screw on the pipes which bring the water to the pump. The upper flaunch, or top of the suction-chamber H, has three holes in it, one under each barrel, and each is covered by a valve shutting downwards, as is shown in the section, fig. 252. These valves are made of iron, to shut down upon hinges like a door, and are covered with leather at the lower side.

Mr. Smeaton made his valves with the centre-pin of the hinge removed backwards from the hole which the valve covers, and it is also raised above the surface of the under side of the valve, by which means the valve opens in some degree on that side where the hinge is, as well as on the other, and any obstruction getting into the valve will be less liable to be detained, and will not have such a great leverage to break the hinge of the valve when the force of the water shuts it down, as it would if the hinge was on a level, and close to the edge of the hole, because the obstacle will not be so near the centre.

The hinge is fastened to the pump by the screw *w*, passing through the metal, and screwing into the hinge; this being withdrawn, and the door D opened, the valve is quite loose, and may be taken out to renew the leather. To give facility to this, the doors D are made oval, as shown in fig. 253. Another similar valve *n* is fitted at the top of each of the pipes B, to cover their apertures; they are all covered by a common forcing-chamber I, which is exactly similar to the suction-chamber, except that it has nozles R, in the top over each valve, and covered with doors to give access to them. The conducting-pipes are carried away from either end of the forcing-chamber, flaunches being provided to unite them. Each barrel is fitted with a piston or forcer M, which consists of three metallic plates, secured to the rod: the middle plate is turned true, and fitted as accurately as possible to the barrel; the upper and lower plates are somewhat smaller. Two round pieces of leather, larger than the barrel, are placed above and below the middle plate, being held fast between it and the upper and lower plates. When forced into the barrels, these leathers turn up and down round the upper and lower plates, forming two cups of leather, which accurately fit the barrel, and will not permit any fluid to pass by them.

The parts of the pump are fastened together by screws and nuts, as will be understood by inspection of the figures. The whole pump is supported on two ground-sills, and by means of two iron branches of the suction-chamber H, the whole pump is bolted down upon the ground-sills.

The action of this pump is simply this : when the piston or forcer of one barrel is raised, it causes a vacuum in it, and the pressure of the atmosphere forces the water up the suction-pipe H, (if not more than 30 or 33 feet,) opens the valve m, in the bottom of the barrel, and fills it with water ; on the descent of the forcer, the lower valve shuts, and the forcing valve n opens, by the water the barrel contained being driven through it into the forcing-chamber L, and thence to any place whither the forcing-pipe is carried. On the reascent of the forcer, the lower valve m opens, and the shutting of the forcing-valve n prevents the water returning into the barrel. The three forcers work up and down alternately, so that while one barrel is sending water up the force-pipe, the others are lifting it up the suction-pipe, and the third con-tinues the action in the interval, when the change of motion takes place between the two. In this manner the pump will raise a very constant stream of water, if the forcers are worked in a proper manner ; which is best done by means of cranks, placed at such an angle to each other, upon the same axis, that they will act in due succession.

12. English ships of war carry four chain-pumps and three hand-pumps, all being fixed in the same well, which also includes the mainmast.

The chain-pump (fig. 254) is no other than a long chain A, with a suffi-cient number of pistons, a, called buckets or saucers, fixed upon it at proper distances; it passes downwards through a wooden tube B, and returns upwards in the same manner, on the other side D, the ends being united together. The chain is extended over two wheels, E and F, called sprocket-wheels; one is placed over the tubes B and D of the pump, and the other at the bottom in the space between the two tubes through which the chain ascends and descends. By turning the upper wheel E, the chain of buckets is put in motion, and the lower part of the wooden tube, in which the chain ascends, is lined with a brass barrel, in which the saucers are fitted. As they are continually ascending in this tube, they raise a constant stream of water, which runs off from the top of the ascending trunk, and is carried by a trunk through the ship's side into the sea. The pump is worked by a crank or winch G, fixed on the axis of the upper wheel, whereon several men may be employed at once; and thus it discharges in a limited time a much greater quantity of water than the common pump, and that with less incon-venience to the labourers.

13. The chain-pump now in use in the navy is of a very improved construction, compared with original chain-pumps. It was introduced by Mr. Cole, under the direction of Capt. Bentinck. The chain of this machine is simple, and not much exposed to damage. It is exactly similar to that of the

of increasing the quantity of water which the machine will raise, unless it was in a considerable degree; and indeed the very best pumps will not raise a much greater proportion with the same power.

The only alteration which has been made on Mr. Cole's pump, since its first introduction, near thirty years ago, is, that they now omit the lower sprocket-wheel altogether, the ascending and descending pipes being so united by a curved metal tube, that the chain passes better than if a wheel was used. The cranks are made to take off, and apply, when wanted, that they may not be in the way; they are long enough for thirty men to work at once; of late it has been proposed to add fly-wheels to them. This would be attended with but slight advantage, and several inconveniences from occupying that room where the men should stand to work, it being an object to employ as many as possible; but if they are crowded, they only incommode each other instead of assisting.

14. The following simple and ingenious method of working a ship's pump, when the crew are either too few in number, or too much exhausted to attend to that duty when the performance is most necessary, namely, in a heavy gale, was put in practice with great success by Capt. Leslie, of the ship George and Susan, on a late voyage from Stockholm to North America. He fixed a spar aloft, one end of which was ten or twelve feet above the top of his pumps, and the other projected over the stern; to each end he affixed a block or pulley. He then fastened a rope to the spears of the pump, and after passing it through both pullies along the spar, dropped it into the sea astern. To the rope he fastened a cask of 110 gallons measurement, and containing 60 or 70 gallons of water. This cask answered as a balance-weight, and every motion of the ship from the roll of the sea made the machinery work. When the stern descended, or when a sea or any agitation of water raised the cask, the pump-spears descended; and the contrary motion of the ship raised the spears, when the water flowed out. The ship was cleared out in four hours, and the crew were of course greatly relieved.

15. *Hand-pumps* have been constructed in great variety for the use of ships; and as they are of great utility, we shall describe two or three of the best.

The ingenious Benjamin Martin invented a ship's pump with two barrels drawing from one suction-pump, so as to raise a constant stream.

This pump has so much merit, that we have given a section of it, fig. 256. Here A is the suction-pipe, conducting the water from the ship's hold up to the pump, where it is enlarged to communicate with both barrels D D, through the valves C C, in the bottom; E E are the pistons of the barrels, with double valves in them; they are not, like other pistons, fitted to slide in the barrels, but are simply brass rings, in which the valves are fitted, and being smaller than the barrels, have large circular pieces of leather fixed on them, the outside edges of which are attached to the insides of the pump-barrels; hence, when the pistons are moved up and down, the leather folds sufficiently to admit the motion, as is shown in the figure; but being close all round, these pistons can have no leakage or friction, and only a small resistance from the stiffness of the leather.

To fasten the edges of the leather piston to the barrels, they are made in two lengths, an upper and a lower, and the leather is introduced in the joint between them, being half fast, and the pump kept together by bars I I, fixed over the barrels, and bolts to press the upper length of the barrels down upon the lower. Both the barrels are included within a box or cistern B B, fixed upon the ship's deck, with trunks L L, which carry off the water as it runs over the tops of the barrels into the cistern. The pump is worked by piston-rods H H, being united by chains to a wheel K, the axle of which is supported by standards from the sides of the cistern B B, and is put in motion by the double lever M, at the end of which cross handles are fixed for several men to work at once. Mr. Martin's pump acts extremely well; the constant stream raised by the alternate action of two barrels upon one pipe, produces an advantage that was shown by experiment, for the water not only rises while the piston rises, but continues to do so even after the piston begins to descend; and therefore the pump was found to deliver more water than was expected from the calculation of the contents of the barrel, and the number of strokes made.

To account for this, it must be considered, that as this pump has both its large pistons working (alternately ascending and descending) at the same time, there must be produced a constant rising column of water in the pipe, whose velocity through a bore of five inches, to supply the barrels of twelve inches diameter each, must be so great, that it cannot be checked or stopped at once, or upon the first descent of the piston; and therefore a surplus of water is produced. Notwithstanding these advantages of Mr. Martin's pump, it has objections, which are serious obstacles to its use on board ships, though in other situations it is a good machine : these are, the shortness of its stroke, which renders it very fatiguing for men to work for a long time; but another more serious objection is, that the leather would, in general, remain dry, and thus become liable to harden and grow stiff, so as to break into holes when used at first, before they become soaked, and to fill the cistern first with water would be very troublesome.

16. The latest improvements in hand-pumps are by Capt. Jekyl, R. N. This gentleman has invented an addition to the pump of an air-vessel, and stuffing-box for the rod to pass

through, by which it will raise the water to a greater height than the head of the pump; and a hose being attached to the pump-spout, by very simple means, the water is conveyed to any part of the ship, and thrown in a jet through a hose-pipe with great force, to extinguish fire, if such a calamity should befall a ship; and thus the pump is rendered of twofold service. The idea of converting the pump to a fire-engine is not new, having been attempted in many different ways by forcing-pumps; but these having pipes proceeding from the lower parts of the barrels and valves, which are not very accessible, are always liable to choke up by obstructions, and have not succeeded in general use. The air-vessel has always been in the way, if made of a sufficient size to answer the purpose of equalizing the stream. Capt. Jekyl has obviated these objections, and without altering the material parts of the hand-pump, has rendered it as complete a fire-engine as can be wished.

This is explained by figure 257, which is a section of the pump through its whole length. A B C is the iron brake or lever to work it; it is branched to the extreme end, and has a wooden pole C, fixed in it, for several men to hold at once; D is the iron stanchion or fulcrum of the brake; it is fixed to the pump-head by means of strong iron hoops at E E and F F, which at the same time strengthen the work of the pump. The centre-pin is to be at the height of two feet six inches above the ship's deck. H are the slings of the pump, united by a forelock or pin to the end of the brake, and suspending the pump-spear I, by means of the joint-piece g. I K is the pump-spear, made of copper in the upper part I, and the lower length K of iron; the latter has the bucket M attached to it. The valve of the bucket is made in a very simple and effective manner, the valve being merely a round plate of brass, with a hole through the centre, to receive the rod upon which it rises and falls, and covers the aperture in the bucket. The bucket is a ring of brass, with a cross bar to fix the rod in: it is made in two thicknesses, one above the other, and a cup of leather is held in between them, projecting all round the upper part of the bucket and turning up, to make a tight fitting in the barrel. The two rings of the bucket are held together by the piston-rod passing through both, and a cross wedge beneath. L is the brass chamber in which the bucket works; it is well fitted into the wood of the pump-tree, so that the water cannot leak by it, and is bored smooth within side.

N is the lower box, fitted into the lower part of the pump-tree, beneath the chamber; it has a groove round it, into which oakum is placed, and when it is put down, makes a tight joint; its valve is of the same construction as that of the bucket, with the addition of a ring or eye on the top of the pin, on which the valve rises and falls. By this eye the box can be drawn up when it needs repair, by first drawing up the bucket of the pump, and putting an iron down into this eye. O O P is the air-vessel; this is a cylinder of sheet-copper, soldered to a cover of brass, within the centre of it is a tube likewise soldered to the cover, through which the upper pump-spear passes, and is fitted round it up with a collar of leather and stuffing. To prevent the escape of the water, it is packed with hemp and two rings of leather. R shows the place of the main barrel, bored through the head of the pump, and confining the cover O O, of the air-vessel; they are fastened by the wedges *d*; it is by these only that the air-vessel is held down; a circle of leather is first put round the air-vessel, just beneath its lid, and this being pressed upon the recess in the wood, makes the joint tight. T is the pump-nozle, which delivers the water. When it is used as a fire-engine, a hose is fixed on by its link-joints, and keys or wedges; the nozle is fixed to the pump by four screw-bolts going through the thickness of the pump, and it is fixed in such a direction as will most conveniently lead to a receiver, fig. 258, which unites the hose from all three of the ship's pumps.

Fig. 259 is the link-joint of the hose, T representing the pump-spout, made of cast-iron, and screwed to the pump-tree; *e e* is the collar or socket, made of brass, with the hose X bound upon it; this has two trunnions, on which a link *f* is fitted, one on each side; these links pass through grooves in the cast-iron piece T, and a key *g*, put down through the link behind it, draws the joint tight, without any screwing or further trouble. The socket *e e* is fitted into the nozle, and has a leather ring to make it tight. The outside of the pump is to be hooped at every three feet, to prevent it from bursting by the pressure of the water. The disposition of the three hand-pumps in a ship's well, renders their connection with a common receiver very convenient to bring all the water into one stream, which will then be very powerful, and more capable of extinguishing a fire than any movable engine.

Two hand-pumps are always placed on the starboard side of the main-mast, in the well; and one of them being the cistern-pump used for washing decks, its foot stands in a small cistern fixed upon the step of the mainmast, and supplied with water by a pipe through the ship's side, with a cock to admit it at pleasure; there is one pump on the larboard side of the mast. Three separate hoses being united with each of the pumps by a link-joint, like fig. 259, at one end, and with three necks *k k k*, of a receiver, fig. 258, by similar joints at the other, brings all the water into one, and a hose being joined by a link-joint, *l*, to the opposite end of the receiver, conveys the whole water to any part of the ship. The receiver has three nozles, *k k k*, at one end, made in a divergent direction, agreeable to the direction in which the hoses come from the three different pumps, and a valve is placed inside, before each hose, to open inwards, in order that the receiver may be used for one or two pumps, whilst the others are repairing or getting ready, or that if any of the hoses burst, the water may not escape from the receiver at the nozle. There are two handles fixed to the receiver, to lift and carry it, as it is to be movable; and when in use, is proposed to be laid upon the grating of the main hatchway, as the most central situation, from whence the hose may be carried in any direction. Z is a branch-pipe or jet, screwed at the end of the great hose X; and it also unscrews at the extreme end, to fit on jets of different bores, in the same manner as all other fire-engines. In working, the pressure of the water condenses the air contained within the receiver, O O P, into a small space, and its reaction to resume its former bulk equalizes the efflux of the water from the nozle of the pump.

In some experiments which we have witnessed upon this pump, it performed as well as could be desired, a single pump forming a very effective engine; but when the three were combined, it was superior in force to any we have ever seen, and would throw a stream of an inch in diameter over the maintopmast-head of a 74-gun ship.

Besides the length of the handle C admitting several men to work at once, an accession of force is gained by a rope *n*, made fast to the brake A B, and conducted through a single block hooked to the deck at *m*, and thence along the ship's deck; at this any number of men may be employed very advantageously to produce the stroke, leaving those at the handle only to return it by lifting the handle. If the ship proves leaky, and the stuffing-box is thought to be an obstruction to the working of the pump, the air-vessel may be taken out by drawing the wedge *d*, and taking out the bars *R*, which confine it; then after taking out the key which connects the joint-piece *g* with the copper rod, also removing the brake, lift out the air-vessel by the two screws of the stuffing-box, and fix on the joint-piece again, but fix the guide-eye H in the lowest pair of holes, so that it will receive the top of the copper rod, and prevent the pump-spear from having any play in the rings.

In this state it acts as a common hand-pump; but the air-vessel can be restored to its place, and be ready for work, in two minutes.

To prevent any of the work from being neglected from carelessness, the inventor proposes that one of the pumps shall be always used to wash the ship by the hose and jet every morning, which it would do much more effectually than by the present mode of raising the water into buckets; and the force with which the jet of water is thrown would very completely wash into every recess of the gun-carriages, and other places where a brush cannot reach; while by this constant exercise the pumps would be always ready, at a moment's notice, upon an alarm of fire.

17. Mr. Robert Clarke, of Sunderland, has proposed a great improvement in the mode of applying men's force to pumping, which is worthy the consideration of seamen. It is to change the posture of standing to sitting, and making the action the same as that of rowing, which, besides that it is by philosophers considered as the most efficacious application of a man's force, it is to seamen most particularly so from their habitual practice of it. He objects to the ordinary action of pumping with a brake, as the posture is weak, and requires much force to preserve it. It oppresses the man by overstretching his loins on one side, and incommodes respiration by the flexure of the body on the other side. Too much motion of the shoulder-joint is required, as the muscles which act on the arm-bone at this joint are disproportionate to the effort they must make when the arm vibrates on the shoulder-joint as a centre, for the force to be communicated by the hand. Besides this, the arms themselves are at one instant enfeebled, by being thrown over the head, and requiring a pull, and the next instant require a pushing effort, which changes of direction in the exertion and restraining force are too continual and rapid for long continuance; in standing the body is a continued dead weight upon the legs.

T

The action of rowing is powerful to a surprising degree, and so well adapted to a man's ease, that he can continue it a greater length of time without fatigue than any other mode of exertion; for though the motion is large, it is made up of easy motions in several joints, the velocity and resistance of which suit the muscles employed. Very little sustaining force is required, for the body is supported, and returns unloaded to its charge: the breathing is free. The manner of carrying this into effect is very simple, the lever or brake being bent at right angles at a certain pin, so that it hangs straight down when it is at rest, instead of being horizontal; then to the lower extremity a rod is jointed, which is carried rather in an inclined direction upwards to the seaman, who is seated before the pump with a rest for his feet. The rod has a cross handle, to hold by both hands, and in some cases it may be made long enough for two men to sit side by side on the same seat; and by drawing and pushing it in the same manner as rowing, the perpendicular lever is caused to vibrate, and the horizontal arm or bended part, which suspends the pump-spear, partakes of the motion sufficiently for pumping.

18. M. de Bonnard, speaking of the *pistons of pumps*, in the *Journal des Mines*, states, that the leathers with which the external circumference of the pistons of pumps are covered, are quickly worn out by the continual friction which they undergo, and the renewing of them is an object of considerable expense in large mining undertakings.

They have therefore used in Saxony, for some years past, pistons without these external leathers in sucking-pumps, and to render the upper part of the piston elastic, by composing it of pieces of wood, which expand or open when the piston rises, and close when it descends.

To obtain this effect, the part of the piston which forms a bucket, is composed of a system of small movable pieces of wood *a a a*, figures 260 and 261, cut obliquely, and disposed so as to cover each other nearly half their breadth; a leather which covers the upper surface of each of these pieces, serves to sustain them, and yet allow them sufficient play. To the under part of the same, pieces of leather are attached, which afford them all the elasticity that is necessary. These leathers are received into slits that are cut round the piston, and directed obliquely to its edges; they are fixed to the pieces of wood by nails, the extremities of which correspond with the notches *c c c*, and to the edges of the solid part of the piston by the screws *d d d*. By this disposition each piece of wood is movable upon a sort of horizontal hinge, and when the piston is raised, the weight of the water with which it is charged, by opening all these pieces, causes them to press one against the other, and against the sides *r r* of the barrel of the pump, so as not to let any of the water escape, and to produce completely the effect of a piston furnished with leather. The interior edges of each of the joints of the movable pieces are covered two by two with leather as at *e e e*, fig. 261, upon which the weight of the water acts as upon the pieces themselves.

All these leathers last a very long time, as well as those of the suckers, because they are not exposed to any friction, which only acts upon the movable pieces of wood. When the piston descends, the water that raises the suckers finds an easy passage, without filtering between the piston and the interior of the barrel of the pump ; an effect which has this additional convenience, that no dirt can be introduced into the joints, which might afterwards prevent the perfect contact of the different pieces.

In 1808, these pistons were tried in several mines in Saxony, and were found very satisfactory. It was only observed that there was some inconvenience attending the use of them where the wells were much inclined; as the pressure of the water above not being equal upon all the movable pieces of the piston, those that were least pressed upon let some of the water pass between them. These inconveniences however exist only in the ordinary pistons.

In some departments pistons with springs are sometimes used, which are composed of movable rubbing-pieces, that are substituted for the leathers that are ordinarily employed. We know that these pistons are used with advantage in the cylinders of some blowing-engines ; but in these pistons the rubbing-pieces are constantly forced against the interior surface of the cylinder by the springs

In the piston with the flexible crown of wood, which M. Bonnard has described, the movable pieces of wood that compose it do not rub against the interior surface of the barrel of the pump, except when the piston ascends, being then pushed by the weight of the column of water that is raised, and they scarcely rub at all against the surface when the piston descends. This peculiar effect assimilates this piston with those that have a flexible crown of leather, or a bucket, and gives it a decided advantage over pistons with springs and cushions. In other departments pistons with springs are sometimes used, which move in cylinders of cast-iron.

These pistons are composed of four pieces of brass, *a a a a*, figures 262 and 263, which are each about three centimetres in height and thickness, and are pushed horizontally by two springs, *b b b b*. Those pieces, which we shall call quadrants, in order that none of the air may escape when they play under the inequalities of the cylinder in which they rise and descend, are each of them something longer than a quarter of the circumference of the cylinder, and towards the extremities they are reduced to half the thickness.

By this means, these quadrants are perfectly covered at the extremities, and prevent the passage of the air in a horizontal direction, while the springs *b b b b* prevent its passage vertically.

In conclusion we shall observe, that these pistons, perfectly joined, have been proved to be proper for driving air with great force. We shall likewise observe, that the quadrants *a a a a* being made of brass, and rubbing against cast-iron, ought to last a very long time ; consequently, the blowing here mentioned have the advantage of not requiring frequent repairs.

19. The following piston, described and recommended by Belidor, seems as perfect as the nature of things will allow. We shall therefore describe it in the author's own words, as a model which may be adopted with confidence in the greatest works.

" The body of the piston is a truncated metal cone, (fig. 264,) having a small fillet at the greater end. Fig. 265 shows the profile, and fig. 266 the plan of its upper base, where appears a cross-bar D D, pierced with an oblong mortise E for receiving the tail of the piston-rod. A band of thick and uniform leather A A (figs. 265 and 267) is put round this cone, and secured by a brass hoop B B, firmly driven on its smaller end, where it is previously made thinner to give room for the hoop.

" This piston is covered with a leather valve, fortified with metal plates G G (fig. 268.) These plates are wider than the hole of the piston, so as to rest on its rim. There are similar plates below the leathers, of a smaller size, that they may go into the hollow of the piston ; and the leather is firmly held between the metal plates by screws H H, which go through all. This is represented by the dotted circle J K. Thus the pressure of the incumbent column of water is supported by the plates G G, whose circular edges rest on the brim of the water-way, and thus straight edges rest on the cross-bar D D of figs. 265 and 266. This valve is laid on the top of the conical box in such a manner that its middle F P rests on the cross-bar. To bind all together, the end of the piston-rod is formed like a cross, and the arms M N (fig. 269) are made to rest on the diameter F F of the valve, the tail E F going through the hole E in the middle of the leather, and through the mortise E of the cross-bar of the box, as well as through another bar, Q R, (figs. 267 and 268,) which is notched into the lower brim of the box. A key V is then driven into the hole I, in the piston-rod ; and this wedges all fast. The bar Q R is made strong ; and its extremities project a little, so as to support the brass hoop B B, which binds the leather band to the piston-box."

This piston has every advantage of strength, tightness, and large water-way. The form of the valve (which has given it the name of the *butterfly-valve*) is extremely favourable to the passage of the water ; and as it has but half the motion of a complete circular valve, less water goes back while it is shutting.

FIRE-ENGINES.

WHEN fire breaks out in a crowded neighbourhood, it carries with it such devastating effects, that any individual who has seriously turned his attention to the constructing of an engine that is in the least calculated to check its progress, must ever be considered as deserving of our praise. Those who have most beneficially directed their attention this way are Messrs. Newsham and Rowntree, whose engines we shall now proceed to describe.

1. A perspective view of Mr. Newsham's fire-engine, ready for working, is represented in fig. 270.

It consists of a cistern A B, about three times as long as it is broad, made of thick oaken planks, the joints of which are lined with sheet copper, and easily movable by means of a pole and cross-bar C *, the fore part of the engine, which is so contrived as to slip back under the body of the bottom and on four solid wheels, two of which are seen at D and E. The hind axle-tree, to which the wheel E and its opposite are fixed, is fastened across under the bottom of the cistern; but the fore axle-tree, bearing the wheel D, &c. is put on a strong pin or bolt, strongly fastened in a horizontal situation in the middle of the front of the bottom of the cistern, by which contrivance the two fore-wheels and the axle-tree have a circular motion round the bolt, so that the engine may stand as firm on rough or sloping ground as if it were level.

Upon the ground next to the hind part of the engine may be seen a leather pipe F, one end of which may be screwed on and off upon occasion to a brass cock at the lower end of the cistern; the other end is immersed in water, supplied by a pond, fire-plug, &c. and the pipe is a sucking-pipe for furnishing the pump of the engine by its working, without pouring water into the cistern. To the hind part of the cistern is furnished a wooden trough G, with a copper grate for keeping out stones, mud, and dirt, through which the cistern is supplied with water when the sucking-pipe cannot be used. The fore part of the cistern is separated from the rest of its cavity by another copper grate, through which water may be poured into the cistern. Those that work the pumps of this engine move the handles, visible at the long sides, up and down, and are assisted by others who stand on two suspended treadles, throwing their weight alternately upon each of them, and keeping themselves steady, by taking hold of two round horizontal rails H I, framed into four vertical rods which reach the bottom of the cistern, and are well secured to its sides. Over the hind trough there is an iron handle or key K, serving to open and shut a cock placed under it on the bottom of the cistern, the use of which we shall explain in the sequel of this article. L is an inverted pyramidal case which preserves the pumps and air-vessels from damage, and also supports a wooden frame M, on which stands a man, who, by raising or depressing, and turning about the spout N, directs the stream of water as occasion requires. This spout is made of two pieces of brass pipe, each of which has an elbow; the lower is screwed over the upper end F, (fig. 271,) of the pipe that goes through the air-vessel, and the upper part screws on to the lower by a screw of several threads, so truly turned as to be water-tight in every direction. The conic form of the spouting pipe serves for withdrawing the water in its passage through it, which occasions a friction that produces such a velocity of the jet as to render it capable of breaking windows, &c. whilst the valves and leather pipes of the engine have sufficient water-way to supply the jet in its greatest velocity. Leather pipes of considerable length may be screwed at one end of the nozle of the engine, and furnished at one end with a wooden or brass pipe for guiding the water into the inner parts of houses.

Between the pyramid-box L, and the fore end of the engine, there is a strong iron bar O, lying in a horizontal position over the middle of the cistern, and playing in brasses supported by two wooden stands; one of which, P, is placed between the two fore stands of the upper rails, and the other is hid in the enclosure over the hind part. Upon proper squares of this bar are fitted, one near each end, two strong brass bars, which take hold of the long wooden cylindrical handles, by means of which the engine is worked; and the treadles by which they are assisted are suspended at each end by chains in the form of a watch-chain, and receive their motion jointly with the handles, that are on the same side, by means of two circular sectors of iron fastened together, and fixed upon proper squares of the middle horizontal bar; the two fore ones may be seen at Q; the two hind ones, represented upon a large scale in fig. 272, differ from the former only in thickness, for the fore sectors are made to carry only one chain each, fastened by one end to their upper part, and by the lower end to the treadles; whereas the sole of the two hind sectors is wide enough to carry two chains each; one set fastened like those of the fore ones for the motion of the treadles; and the other two chains are fastened by their lower ends to the lower part of these sectors, and by their upper ends to the top of the piston-bars, in order to give them motion. See fig. 271, in which the hind sectors and their apparatus are represented as they would appear to a person standing between the two fore wheels, and looking at the hind part of the engine.

The square over the letter A is the section of the middle bar, on which, right over the two barrels, are placed the two sectors BCA and DEA, forged together. FGHK and $fghk$ are the two piston-rods; and the openings between the letters GH and gh, are the spaces through which the hind parts of the two treadles pass. L and M represent two strong studs, rivetted on the other side of the bars on which they are placed; and to each of these is fastened a chain like a watch-chain, and fixed by their upper ends to the upper extremities D and B of the iron sectors, by which they are drawn up and down alternately. These sectors give also an alternate motion up and down to the piston-rods, by means of two other chains less white in the figure, in order to distinguish them from the others; these are fastened by their lower ends to the lower extremities of the sectors E and C, and their upper ends, terminating in a male screw, are made tight to the piston-rods at F and f, by two nuts.

The shape of the piston-rods, and the size and situation of the chains that give them motion, are so contrived, that the vertical axis of the pistons is exactly in the middle of the breadth of the perpendicular part of the chains, and the upper part of the piston-rod taken together. PQ represents one of the two cross-bars through the ends of which pass the handles to which the men apply their hands when they work the engine; these cross-bars are fitted on the middle bar at some distance from the sectors.

The other parts of this useful engine may be understood by the help of fig. 271, which represents a vertical section taken through the middle line of the hind part of the engine, as also the section of the air-vessel, and that of one of the barrels, and likewise the profiles of the hind sectors, and several other parts. AB is the section of the bottom of the cistern, and C that of the hindmost axle-tree. DE is the vertical section of a strong piece of cast brass or hard metal, so worked as to have a hollow in it, represented by the white part, and fixed to the bottom of the cistern; this reaches from the opening D, through the cock W, and afterwards divides itself into two branches, so as to open under the two barrels; one of these branches is exhibited in the figure, and the other is exactly behind this. Through this channel, which may be called the sucking-piece, water is conveyed to the pumps by the pressure of the atmosphere, either from the cistern itself, or from any place at a distance, by means of the leather pipe P, fig. 278, which screws on to the sucking-piece at D, fig. 271, under the hind trough Z, the grate of which is represented by the horizontal strokes. F G represents the vertical section of another piece of cast brass or hard metal, that may be called the communication-piece, having two hollows for conveying the water from under the pistons to the two openings of the mouth of the air-vessel; one of these hollows appears in the figure; the other lies exactly behind this, though not in a parallel direction. Between the section of the sucking-piece D E, and that of the communication-piece F G, may be observed the section of one of the plates of leather, which makes all tight, and forms one of the two sucking-valves, of which there is another just behind this, under the other barrel. R S T is the section of the copper air-vessel, and T V that of the conduit-pipe; this vessel is screwed on to the hind part of the communication-piece, and at top is fastened by a collar of iron to a cross piece of timber.

Between the flanch of the air-vessel and the communication-piece, may be observed the section of one of the plates of leather, making all tight, and forming one of the two forcing valves, of which there is another just behind this, exactly over the other opening of the communication from the air-vessel. These valves are loaded with a lump of cast-iron or lead, having a tail or feet let through the flap of the valve, and cross pinioned under it: and it is to be observed, that though both the valves are represented open in the figure, they are never both open at the same time; for when the engine is not at work, they are closed down by the weights on their upper surfaces; and when the engine works, two are shut, and the other two are alternately opened by the motion of the pistons and the action of the atmosphere, together with the reaction of the air contained in the air-vessel.

H H is the section of one of the barrels of the two pumps, which are both sucking and forcing; as is evident from the position of the valves and the direction of the pistons, each of which is composed of two iron plates, of two wooden trenchers, and of two flat pieces of leather turning one up and the other down. L K represents one of the piston-rods edgewise, behind which is one of the chains, the top screw of which, K, can only be seen. M is the end of the middle bar, and N a section of the hindmost of the two middle stands which support the middle bar. O represents the end of the point of one of the treadles, passing through the rectangular holes of the piston-rods, as in fig. 272. The weight on these treadles brings them and the piston-rods down alternately, and they are raised up again by the help of the other set of chains, one of which may be seen edgewise in this figure, placed on the sole of one of the sectors, &c. see fig. 272.

P Q is part of the cross-bars which carry the handles, seen edgewise, and X Y represents an iron handle, by the help of which the cock W may be turned into the several situations requisite for the use of the engine. The mechanism may be understood by figs. 275, 276, and 277, which represent the horizontal section of it, in three different situations. It has three holes, which are left white in these figures. In fig. 275, the position of the cock is represented when the handles X Y or K are in a direction parallel to D E, or to the middle bar, as in figs. 270 and 271. In this position the water supplied by the sucking-piece enters at D, and proceeds directly through the cock W to the valve under the two pistons; and there is now no communication from the barrel with the cavity of the cistern.

In fig. 276, we have the position of a cock when the handle X Y is turned one quarter of a revolution towards the eye from the last-mentioned situation, in which case there is no communication from the barrels with the outer extremity of the sucking-piece, but the water poured into the fore and hind trough, and passing from thence into the cavity of the cistern, enters the cock sideways at W, and turning at right angles through the cock towards E, proceeds to the barrels of the pumps. Fig. 277* represents the cock W when the handle is placed diametrically opposite to its last situation, in which case there is no communication from the under side of the barrels with the cavity of the cistern or the outward end of the sucking-piece; but this situation affords a communication from the cavity of the cistern with the outside of the engine, and the water left in the cavity of the cistern may by this means be employed when the engine has done working. These engines are made of five or six different sizes.

The principles on which this engine acts, so as to produce a continued stream, are obvious: the water being driven into the air-vessel, as in the operation of common sucking and forcing pumps, will compress the air contained in it, and proportionably increase its spring, since the force of the air's spring will be always inversely as the space which it possesses; therefore when the air-vessel is half filled with water, the spring of the included air, which in its original state counterbalanced the pressure of the atmosphere, being now compressed into half the space, will be equal to twice the pressure of the atmosphere; and by its action on the subjacent water will cause it to rise through the conduit-pipe, and play a jet of 32 or 33 feet high, abating the effect of friction. When the air-vessel is two-thirds full of water, the space which the air occupies is only one-third of its first space; therefore its spring being three times as great as that of the common air, will project the water with twice the force of the atmosphere, or to the height of 64 or 66 feet. In the same manner when the air-vessel is three-fourths full of water the air will be compressed into one-fourth of its original space, and cause the water to ascend in air with the force of three atmospheres, or to the height of 96 or 99 feet, &c. as in the following table:

Height of the water.	Height of the compressed air.	Proportion of the air's spring.	Height to which the water will rise.
½	½	2	33 feet.
		3	66
		4	99
		5	132
		6	165
		7	198
		8	231
		9	264
1/11	1/12	10	297

gine or pump; figs. 282 and 283 are parts of the engine. The
ters are used as far as they apply in all the figures; A A A A
and 281, is a cast-iron cylinder truly bored, 10 inches diameter
ong, and having a flanch at each end whereon to screw two covers,
fing-boxes *a a*, in their centres, through which the spindle, B B, of
.ne passes, and being tight packed with hemp round the collar,
, tight joint; the piston D is affixed to the spindle within the
, and fits it tight all round by means of leathers; at E, fig. 281, a
, called a saddle, is fixed in the cylinder, and fits against the back
indle tight by a leather.

.ave now a cylinder, divided by the saddle E and piston into
s, whose capacity can be increased and diminished by moving the
with proper passages and valves to bring and convey the water;
form a pump. These passages are cast in one piece with the
: one, *d*, for bringing the water, is square, and extends about
l round the cylinder; it connects at bottom with a pipe *e*; at its
er ends it opens into two large chambers *f g*, extending near the
ngth of the cylinder, and closed by covers, *h h*, screwed on; *i k* are
penings (shown by dotted squares in fig. 280) in the cylinder
icating with the chambers; *l m* in *f g* are two valves closing the
the curved passage *d*, and preventing any water returning down
ige *d*; *n o* are two passages from the top of the cylinder to convey
? water; they come out in the top of the cylinder, which, together
: top of the chambers *f g*, form a flat surface, and are covered by
res, *p q*, to retain the water which has passed through them. A
, K, is screwed over these valves, and has the air-vessel *k*, figs. 278
, screwed into its top; from each side of the chamber a pipe, *w w*,
, to which a hose is screwed, as shown in fig. 280. Levers, *s s*,
l to the spindle at each end, as shown in fig. 279, and carry the
H H, by which men work the engine. When the piston moves,
i by the arrow in fig. 281, it produces a vacuum in the chamber *f*,
. part of the cylinder contiguous to it, the water in the pipe *e* then
.e valve *m*, and fills the cylinder.

same motion forces the water contained in the other part of the
through the valve *q*, into the chamber K, and thence to the bose
the pipe *w*; the piston being turned the other way reverses the
n with respect to the valves, though it continues the same in itself.
te *e* is screwed by a flanch to an upright pipe P, fig. 282, con-
with another square iron pipe, fastened along the bottom of the
the engine; a curved brass tube, G, comes from this pipe through
of the chest, and is cut into a screw to fit on the suction-hose
can be used; at other times a close cap is screwed on, and another
p at H, within the chest, is screwed upwards on its socket, to open
small holes in it, and allow the water to enter into the pipe; in this
. engine-chest must be kept full of water by buckets. The valves
le of brass and turn upon hinges. The principal advantage of
me is the facility with which it is cleared from any sand, gravel,

or other obstructions, which a fire-engine will always gather when it work.

The chambers fg; being so large, allow sufficient room to lodge a greater quantity of dirt than is likely to be accumulated in the use of the engine at any one fire, and if any of it accidentally falls into the cylinder, it is gently lifted out again into the chambers, by the piston, without being any obstruction to its motion; to clear the engine from the dirt, two circular plates of five inches diameter, are unscrewed from the lids kk, of the chambers fg, and when cleaned are screwed on again; these screw-covers fit perfectly tight without leather, and can be taken out, the engine cleared, and enclosed again in a very short time, even when the engine is in use, if found necessary.

The two upper valves pq, and chamber K, can also be cleaned with equal ease, by screwing out the air-vessel kk, fig. 278, which opens an aperture of five inches, and fits air-tight, without leather, when closed. The valve may be repaired through the same openings. The use of the air-vessel, kk, figs. 278 and 279, is to equalize the jet from the engine during the short intermittance of motion at the return of the piston-stroke; this it does by the elasticity of the compressed air within it, which forces the water out continually, though not supplied quite regularly from the engine.

The engine from which the drawing was taken, was constructed for the Sun Fire Insurance Company, in London, and from some experiments made by their agent, Mr. Samuel Hubert, appears to answer every purpose.

JACKS.

THE jacks which we purpose here to describe are simple machines used for raising heavy weights.

Fig. 341 represents the common or simple hand jack ; a block of wood about two feet six inches long, 10 inches broad, and six inches wide, is perforated with a square hole or mortise through it lengthwise for the reception of an iron rack B. This rack is formed with a double claw or horn at its upper end. A small pinion C is made to engage in the teeth of the rack. The axis of the pinion is supported in iron plates bolted to each side of the block, and one end of the axis projects through the side plate, with a square to receive a winch or handle, which, being turned round, the pinion elevates the rack B in the mortise, and raises the claw or horns up to the load to which it is applied. To prevent the weight of the load running the pinion back, the handle is detained by a hook or link a, fastened to the outside of the block.

When a greater power is required than the simple rack and pinion are capable of exerting, a combination of wheel-work is used, as shown in the same figure, where A A is the block of wood, which in this case is made sufficiently wide to contain the cog-wheel F, fixed to the pinion C, which acts in the teeth of the rack B. G is a second pinion of four leaves, working in the wheel F: and the axis of this pinion projects through the side of the block for the winch H to be fixed on it. The block A A is made in two halves, and the recess for the wheel F, and the pinion G, is cut out in one of the halves ; the other, being laid flat against it, supports the front pivots

eel and pinions. The two halves are bound together by strong
i *b b*, driven over the outside. The rack has a claw N, at its lower
:ting out sideways through an opening or slit cut through in the
)f the block. This claw can be introduced beneath a stone which
flat upon the ground, and which consequently could not be acted
ie claw on the top of the rack. To prevent the rack descending
s a load upon it, the small click *a* drops into its teeth, but clears
up; when it is not required to detain the rack, this click can be
of the way sideways.

: is a screw jack. The block of wood A A is perforated nearly
ength with a hole sufficiently large to allow the screw B to move
vn without touching. The screw passes through a nut *a*, fixed into
he block A; and if the screw is turned round, it must rise up through
id elevate the claw F. This claw is fitted on the top of the screw
id collar, which allows the screw to turn round without turning
and the claw N, which projects through a groove or opening
e side of the block, is fitted to the screw with a smaller collar.
i of the block has four short points to prevent the machine slip-
i used upon hard ground. To give motion to the screw, the
)f it is formed into a square, and a worm-wheel C is fitted upon
    The teeth of this wheel are engaged by a worm on the axis of
H, and plates of iron, *a b*, are bolted on each side of the block,
iddle of its height, to carry the ends of the axis of the winch and
n which is concealed by the worm-wheel C. When the winch is
id, it causes the wheel C to revolve by the action of the worm in its
as the wheel is fitted on the square part of the screw, it compels
ith it, but at the same time allows the screw to move up and down.

have been also constructed upon the hydrostatic
discovered by Pascal, and which has been applied
:e by the late Mr. Bramah, in this and various
ful machines.

CRANES.

s are certain simple machines in which either the
d axle, or wheel and pinion, are introduced, to effect
ig of heavy loads, such as the loading or unloading
ig at the quays or wharfs, or the raising or lowering
and from chambers or warehouses.
s modes have been adopted to turn the wheel, or
· of the machine which is applied to the same pur-

could be forcibly pressed on the circumference of the
by a lever, to cause such a friction as would prevˌ
weight from descending too rapidly. By this means,
goods may be raised or let down at pleasure, withɔ
danger of injuring the men in the crane. This contriⅴ
ingenious; but the rapid motion of the circumference
large walking wheels, in most cases, rendered it inapp
unless a smaller cog-wheel was fixed upon the same aⅎ
the walking-wheel.

A crane to be turned by winches, was contrived
late Mr. Ferguson, which has three trundles, with d
numbers of staves. Any one of these may be applieⅾ
cogs of a horizontal wheel, mounted on an upright axle
which is coiled the rope for drawing the weight. Thiʃ
has 96 cogs; the largest trundle 24 staves, the next l
the smallest six; so that the largest revolves four tiⅼ
one revolution of the wheel, the next eight, and the ⅎ
sixteen. The winch is occasionally fixed on the ˌ
either of these trundles for turning it, and is applieⅾ
or the other, according as the weight to be raised iʃ
or larger. There is also a fourth trundle acting in th
of the great wheel, and on its axis is a brake and ⅹ
wheel. While the load is drawing up, the teeth
ratchet-wheel slip round below a catch which falls intɔ
and prevents the crane from turning backward, thus de
the weight in any part of its ascent, if the man who w
the winch should accidentally quit his hold, or wish
himself before the weight is completely raised.]
a due allowance for friction, a man may raise, by such ˌ
from three to twelve times as much in weight as
balance his effort at the winch, viz. from 90 to 3
taking the average labour.

Many other constructions of wheel-work are in cˌ
use for cranes. When they are turned by a wincl
prɔper to apply a fly-wheel to the axis of it, both to e
the efforts of the labourer who turns it, and in case ꞩ
dentally lets go the handle, to prevent the load from ⅼ
down so quickly as to endanger any thing. It is conˑ
to have several different powers to a crane of this ꞣ
adapt it for the different burdens to be raised; this
done by employing a train of several wheels, each tuⅼ
a pinion smaller than itself. Thus, suppose the baˌ
which the rope or chain winds to be 12 inches in diˌ
and has a cog-wheel of 96 teeth fixed on the end of i
is turned by a pinion of 12 leaves; on the same aⅹ

a wheel of 32 teeth, moved by a pinion of eight,
d on a third axis, which should carry the fly-wheel.
h of one foot radius can be applied to any of these
xes in the crane, and will give three different powers.
if it is applied to the gudgeon of the barrel, it will
the power of the balance, because the winch describes
: which is twice as large as the barrel on which the
winds; if the winch is fixed on the end of the axis
carries the pinion of 12, and the wheel of 32, it
ve the labourer a purchase of 16 times; and lastly,
he winch is applied to the pinion of eight, his efforts
multiplied 64 times. This simple mechanism is ren-
very complete by fixing a fly-wheel upon the axis of
ion of eight, to prevent all accidents; for which pur-
is more effective than a ratchet-wheel, and requires no
on. The spindles of all the pinions are made capable
ug endwise, for the purpose of disengaging the wheels
ach other at pleasure, that when the wheels are not
:ed, there may be no unnecessary friction in turning
ound.

gibbet of a crane is a very principal member, as we
efore explained; but in its common construction, it
me defects. The rope by which the burden is raised,
exactly over the gudgeon of the vertical beam of the
l is confined between two small vertical rollers, in order
may always lead fair with the pulley or sheave at the
ity of the jib. According to this construction, when-
he jib turns round its axis, the rope is bent so as to
n angle more or less acute, which causes a great in-
of friction, and produces a continual effort to bring
n of the jib into a parallel position to the inner part of
pes. These inconveniences may appear to be trifling,
i actual practice, they are of no small importance; for
ecessarily require a much greater degree of power in
: goods, and the application of a constant force to keep
b in the position that may be requisite; while the par-
ress which is exerted on only a few strands of the rope,
: bent into an acute angle, destroys it in a very short

imple construction of the jib, invented by Mr. Bramah,
es all these defects, and at the same time possesses the
esirable property of permitting the jib, of what is termed
rf or landing crane, to revolve wholly round its axis,
i land goods at any point of the circle described by the
f the jib.

The simplest form of this contrivance is shown in fig. 343, in which A A represents the jib of a warehouse-crane projecting from a wall. It has, as usual, a pulley at the extremity, from which the goods are suspended. The improvement consists in placing a pulley at S, to conduct the rope down through the axis of motion of the jib, the collars or rings a a, on which it swings, being perforated for the purpose. The rope afterwards passes under a pulley b, which conducts it into the house to the crane or machine by which the weight is elevated. The pulley b may be placed between the collars a a, and then there will be no necessity for a perforation of the lower pivot of the jib. When the jib is required to describe a complete circle, instead of the two brackets at a a, fixed to the wall, a cast-iron pillar is used to support the jib, the collars a a fitting upon it; the pillar is hollow, to admit the rope through it, and is firmly fixed in a vertical position, by a plate cast on the lower end of it, and screwed down on the timber of the wharf. Beneath these beams, there is another pulley in place of b, to conduct the rope to the crane.

Fig. 344 represents a crane mounted on four trucks, to be capable of removal from place to place. It was employed on Ramsgate-pier, for lifting stones used in the building, and is extremely well adapted for such a situation, as it requires no fixture, and will take up a weight of four tons with four men, which is sufficient power for such purposes. It was designed and executed by Mr. Peter Kier, by order of the trustees for the management of the harbour at Ramsgate. Its base consists of a cast-iron frame marked A B, nine feet seven inches square, and two tons weight, supported on four cast-iron wheels b b, one pair of which is fixed on a common axle, which moves round on a centre fixed to one side of the frame. This axle has an arm projecting across beneath the frame to the opposite side, where a rack, or segment of a wheel, is fixed on it, as shown at c, engaging a pinion r, shown before the rack, on the top of whose axis a winch is applied at d. Now, by turning this pinion, it twists the wheels round upon the centre, to steer the crane when moving from place to place. A vertical cast-iron shaft marked D F, weighing 23 cwt., is erected on the centre of the iron frame, and is supported by oak braces E E, stepped into boxes cast out of the iron frame A B, at its angles, so as to form a very strong perpendicular column, round which axis the whole crane traverses. The weight of the framing and wheel-work is supported by a steel pivot, or gudgeon, on the top of the shaft F, and is guided by a collar embracing the shaft at I. The framing of the jib, or movable part of the crane, consists of a long beam G H, bearing the pulley G at the extremity, resting on the pivot of the upright pillar in the middle, and the other end supporting the frame for the wheel-work L M N; into this beam are framed two uprights Q Q, suspending the platform I K, on which men who work the crane stand. It is braced by a diagonal stay J P, and a cross piece R, to prevent its bending.

Mr. Bramah's ingenious hydrostatic principle of gaining a great power is applicable in several ways to the raising of heavy weights, and has been frequently employed in powerful cranes. In these the power is not obtained by wheel-work, pullies, or any other ordinary mechanical powers, but on the principle of the experiment called the hydrostatic paradox, which has been known for ages; but the application of its powers to useful purposes is due to Mr. Bramah.

The simplest form is, for a machine to raise a heavy weight

height. A metallic cylinder, sufficiently strong,
truly cylindrical within, has a solid piston fitted
ich is made perfectly water-tight, by leather pack-
its edge, or other means used in hydraulic engines.
n of the cylinder must be made sufficiently strong
other parts of the surface, to resist the greatest
h can ever be applied to it. In the bottom of the
inserted the end of the small tube, the aperture of
imunicates with the inside of the cylinder, and
water or fluids into it; the other end of the pipe
ites with a small forcing-pump, by which the
be injected into the cylinder beneath its piston;
has of course valves to prevent the return of the
w, suppose the diameter of the cylinder to be six
the diameter of the piston of the small pump, or
ly one-quarter of an inch; the proportions between
rfaces or ends of the said pistons will be as the
heir diameters, which are as 1 to 24; therefore the
be as 1 to 576; and supposing the intermediate
een them to be filled with water, or any other
incompressible fluid, any force applied to the small
operate on the other in the above proportion of
Suppose the small piston, or injector, to be forced
n in the act of forcing or injecting, with a weight
, which can easily be done by means of a long
piston of the great cylinder would then be moved
force equal to 1 ton multiplied by 576.

represents a crane constructed upon the hydro-
nciple, that is, by the injection of water from a
p into a large cylinder, which is fitted with a piston,
ck attached to it for the purpose of turning a pinion
xis of a large drum-wheel or barrel, round which
coiled, and from thence passes to the jib.

A A represents the jib, made of iron, and supported upon two
projecting from the wall of the warehouse in which the crane is
erected. The rope passes over the pulley S, and down through
rackets a a, then turns under the pulley b, and comes to the lower
at drum-wheel B. The pinion C is fixed on the same axis with
edgeons turn in small iron frames d, bolted down to the floor of
. The pinion C is actuated by the teeth of the rack D, and a
bore pivot is shown at e, presses against the back of the rack,
ith up to the pinion. The rack is attached to the piston D
er L, in which the power for working the crane is obtained.
ats through a tight collar of leather on the top of the cylinder
.oes not admit of any leakage by the side of it, and therefore if
forced into the cylinder it must protrude the piston from it.
is supported in a wooden frame F F, and has a small copper

U

pipe *g g*, proceeding from the lower end of it, communicating with a small forcing-pump at *k*; this stands in an iron cistern H, which contains the water, and sustains the standard *i i*, for the centre of the handle G, with which the pump is worked by one or two men. The upper extremity of the standard *i i* guides the piston-rod of the pump, to confine it to a vertical motion; *l* is a weight for counterbalancing the handle G of the pump. From what we have said before, the operation of this machine is evident; the power of the cylinder D is in proportion to its size compared with the size of the pump; but as it only acts through short limits, the pinion and drum B are necessary to raise the weight a sufficient height. The operation of lowering goods by this crane is extremely simple, as it is only necessary to open a cock at *m*, which suffers the water to escape from the cylinder into the cistern H, and the weight descends, but under the most perfect command of the person who regulates the opening of the cock; for by diminishing the aperture, he can increase the resistance at pleasure, or stop it altogether.

Fig. 345 is a side elevation of a crane. The post is immovable, and is fixed on an iron frame, with arms extending in the form of a cross, the extremities of which are bolted down by strong screws to large blocks of stone sufficiently heavy to more than counterpoise the weight to be raised by the crane. In the top of the post is fixed a wrought-iron pivot, by which the weight is supported, and a strong cast-iron cap bears on the pivot, and has attached to it two iron frames, one on each side, that receive the pressure from the stay, as well as support the pull of the jib, which is formed of two bars of wrought-iron; the lateral pressure is borne by the bottom of the post, round which two friction-rollers turn to facilitate its motion. This crane will carry five tons with safety.

PRESSES.

:ss is a machine in most extensive use in the
arts. It is usually made of wood, or iron, and
queeze or compress any body very close.

resses generally consist of six members, or pieces ;
at smooth tables of wood or metal, between which
ice to be pressed is placed ; two screws, or worms,
, the lower plank, and passing through two holes
er ; and two nuts, in form of an S, serving to drive
plank, which is movable, against the lower which
id without motion.

used for expressing liquors, are of various kinds ;
most respects, the same as the common presses,
that the under plank is perforated with a great
holes, to let the juice run through into a tub or
iderneath.

nproved *cider-press*, turned by a windlass, is shown
1.

base or foundation with its supporting parts ; B B the cheeks or
' the cross-piece at top, through which the screw passes, and
quently contains the female screw ; E the screw with its
F F the bridge or cross-piece which acts on the pommage ;
ide plank or vat on which the pulp rests in the hair bags, in
ode of the liquor's passing off is seen.

of press may be advantageously employed for packing cloth,
ther goods ; as also in paper-mills, for flattening and rendering
and in the manufacture of woollen cloth, for glazing and setting
i the article in its last stage.

elevations of a very good *screw-press* for a *paper-*
iven in figs. 285 and 286.

e bed, formed of an immense beam of oak ; and each of the
onsists of a long iron bar *b b*, fig. 286, the ends of which are
her, so that it forms a long sink, one end of which receives the
bed A, and the other the end of a massive cast-iron bar D,
ich the screw E is received, and its nut fixed fast therein.
aces of the long links or cheeks, *b, b*, are filled up by rails of
ich support the weight of parts of the press when it is not in
these bear nothing when the press has any articles under
t ; these articles are laid at H, on the bed, and the follower, G,
pon them by the screw, when it is turned by a long lever put
holes in the screw-head F.

rs employed for paper-presses are generally formed with such
ds, and so rapid a spiral, that the elasticity of the paper is suffi-
e it to run back. To these a ratchet-wheel, *a*, is fixed, and a
287, is applied to its teeth ; to prevent its return, the click is
n a bar *b d*, which moves on a centre at B, but the other end is
a catch or lever *f g*. When the press is to be relieved, the

u 2

end *f*, of the catch, is driven back; this relieves the bar *d b*, and the
no longer detaining the ratchet-wheel, the screw runs back.

3. A very ingenious and useful *packing-press* ... 1
invented by Mr. John Peek. It is represented in fig. 28

A A, the frame of the press; B B, the large screws, which, in this p
contrary to those in common use, is fixed and immovable; C, a cir
iron bar, extending beyond the sides of the press, and having thereon
worms, or endless screws E E, which work in two toothed wheels a
to the nuts, and, by turning the winch D, drive the nuts and bed up
down the screws as may be found necessary; F, a stage, suspended f
the bed, and on which the men stand who work the press; such a s
may, if found necessary, be fixed at the other end of the bar, as show
the square shoulder G. The bed of this press must be formed of
pieces of strong wood, which are held together by screws and nuts, pa
through them, as shown at *h h h h*. The great utility of this press consis
its being capable of packing two sets of bales at once; thus answering
purpose of two presses, with more expedition.

4. The *hydrostatic* or *water-press*, or as it is sometir
called *Bramah's Press*, has, for a great number of purpo
superseded the use of the screw-press, over which it posse
great advantages, in all cases where a strong pressure
required. It is one among the many useful inventions of
late Mr. Joseph Bramah, of Piccadilly; and is ingeniou
contrived for applying the *quaqua versum* pressure of fl
as a powerful agent in many kinds of machinery.

These contrivances consist in the application of water
other dense fluids, to various engines, so as, in some instan
to cause them to act with immense force; in others, to co
municate the motion and powers of one part of a machin
some other part of the same machine; and, lastly, to co
municate the motion and force of one machine to anotl
where their local situations preclude the application of
other methods of connection.

The first and most material part of this invention will be clearly un
stood by an inspection of fig. 289, where A is a cylinder of iron, or o
materials, sufficiently strong, and bored perfectly smooth and cylindri
into which is fitted the piston B, which must be made perfectly water-t
by leather or other materials, as used in pump-making. The bottom o
cylinder must also be made sufficiently strong with the other part of
surface, to be capable of resisting the greatest force or strain that ma
any time be required. In the bottom of the cylinder is inserted the en
the tube C; the aperture of which communicates with the inside o
cylinder, under the piston B, where it is shut with the small valve D,
same as the suction-pipe of a common pump. The other end of the t
C communicates with the small forcing-pump or injector E, by mean
which water or other dense fluids can be forced or injected into the cyli
A, under the piston B. Now, suppose the diameter of the cylinder .
12 inches, and the diameter of the piston of the small pump or injecto
only one quarter of an inch, the proportion between the two surfaces or
of the said piston will be as 1 to 2304; and supposing the intermed

reen them to be filled with water or other dense fluid capable of
esistance, the force of one piston will act on the other just in the
ortion, viz. as 1 is to 2304. Suppose the small piston in the
be forced down when in the act of pumping or injecting water
linder A, with the power of 20 cwt. which could easily be done
r H; the piston B would then be moved up with a force equal
multiplied by 2304.

constructed a hydro-mechanical engine, whereby a weight
to 2304 tons can be raised by a simple lever, through equal
much less time than could be done by any apparatus con-
n the known principles of mechanics; and it may be proper
, that the effect of all other mechanical combinations is coun-
y an accumulated complication of parts, which renders them
of being usefully extended beyond a certain degree; but in
acted upon or constructed on this principle every difficulty of
s obviated, and their power subject to no finite restraint. To
it will be only necessary to remark, that the force of any machine
a this principle can be increased *ad infinitum*, either by extending
rtion· between the diameter of the cylinder A, or by applying
wer to the lever H.

) represents the section of an engine, by which very wonderful
y be produced instantaneously by means of compressed air.
ylinder with the piston B fitting air-tight, in the same manner as
n fig. 289. C is a globular vessel made of copper, iron, or other
terials, capable of resisting immense force, similar to those of
D is a strong tube of small bore, in which is the stop-cock E.
ends of this tube communicates with the cylinder under the
and the other with the globe C. Now, suppose the cylinder A
ime diameter as that in fig. 289, and the tube D equal to one
an inch diameter, which is the same as the injector, fig. 289;
se that air is injected into the globe C (by the common method)
es against the cock E with a force equal to 20 cwt. which can
one; the consequence will be, that when the cock E is opened,
3 will be moved in the cylinder A A with a power or force equal
s; and it is obvious, as in the case fig. 289, that any other unlimited
orce may be acquired by machines or engines thus constructed.

is a section, merely to show how the power and motion of one ma-
by means of fluids, be transferred or communicated to another,
stance and local situation be what they may. A and B are two
l, smooth and cylindrical, in the inside of each of which is a
de water and air tight, as in figs. 288 and 289. C C is a tube
nder ground, or otherwise, from the bottom of one cylinder to the
rm a communication between them, notwithstanding their dis-
ever so great, this tube being filled with water or other fluid,
ich the bottom of the piston; then, by depressing the piston A,
B will be raised. The same effect will be produced *vice versâ :*
nay be rung, wheels turned, or other machinery put invisibly
by a power being applied to either.

is a section, showing another instance of communicating the
force of one machine to another; and how water may be raised
ls of any depth, and at any distance from the place where the
power is applied. A is a cylinder of any required dimensions,
s the working piston B, as in the foregoing examples; into the
this cylinder is inserted the tube C, which may be of less bore
ylinder A. This tube is continued, in any required direction,

down to the pump cylinder D, supposed to be fixed in the deep well E E, and forms a junction therewith above the piston F; which piston has a rod G, working through the stuffing-box, as is usual in the common pump. To this rod G is connected, over a pulley or otherwise, a weight H, sufficient to overbalance the weight of water in the tube C, and to raise the piston F, when the piston B is lifted; thus, suppose the piston B is drawn up by its rod, there will be a vacuum made in the pump cylinder D, below the piston F; the vacuum will be filled with water through the suction pipe, by the pressure of the atmosphere, as in all pumps fixed in air. The return of the piston B, by being pressed downwards in the cylinder A, will make a stroke of the piston in the pump cylinder D, which may be repeated in the usual way by the motion of the piston B, and the action of the water in the tube C. The rod G of the piston F, and the weight H, are not necessary in wells of a depth where the atmosphere will overbalance the water in the suction of the pump cylinder D, and that in the tube C. The small tube and cock in the cistern I, are for the purpose of charging the tube C.

By these means it is obvious that the most commodious machines, of prodigious power, and susceptible of the greatest strength, may readily be formed. If the same multiplication of power be attempted by toothed wheels, pinions, and racks, it is scarcely possible to give strength enough to the teeth of the racks, and the machine becomes very cumbersome and of great expense. But Mr. Bramah's machine may be made to possess great strength in very small compass. It only requires very accurate execution. Mr. Bramah, however, was greatly mistaken when he published it as the discovery of a new mechanic power. The principle on which it depends has been well known for nearly two centuries; and it is matter of surprise that it has never before been applied to any useful practical purpose.

5. The *Stanhope printing-press* is delineated in figs. 293 and 294, being elevations, and fig. 295, a plan.

A A is a massive frame of cast-iron formed in one piece; this is the body of the press, in the upper part of which a nut is fixed for the reception of the screw b, and its point operates upon the upper end of a slider d, which is fitted into a dove-tail groove formed between two vertical bars e e, of the frame. The slider has the platen D D firmly attached to the lower end of it; and being accurately fitted between the guides e e, the platen must rise and fall parallel to itself when the screw b is turned. The weight of the platen and the slider are counterbalanced by a heavy weight E, behind the press, which is suspended from the lever F, and this acts upon the slider to lift it up, and keep it always bearing against the point of the screw.

At G are two projecting pieces, cast all in one with the main frame, to support the carriage when the pull is made; to these the rails H are screwed, and placed truly horizontal, for the carriage I to run upon them, when it is carried under the press to receive the impression, or drawn out to remove the printed sheet. The carriage is moved by the rounce or handle K, with a spit and leather girts very similar to the wooden press. Upon the spit or axle, a wheel, L, is fixed, and round this leather belts are passed, one extending to the back of the carriage to draw it in, and two others, which pass round the wheel in an opposite direction, to draw it out.

..... when the handle is turned one way it draws out the car-
..... by reversing the motion it is carried in. There is likewise a
..... from the wheel down to the wooden base M, of the frames,
..... the motion of the wheel, and consequently the excursion of
.....

..... principal improvement of Earl Stanhope's press con-
..... the manner of giving motion to the screw, b, of it,
..... not done simply by a bar or lever attached to the
..... but by a second lever o, g; the screw, b, has a short
..... fixed upon the upper end of it, and this communi-
..... an iron bar, or link, h, to another lever, i, of rather
..... radius, which is fixed upon the upper end of the
..... spindle l, and to this the bar or handle, k, is fixed.
..... when the workman pulls this handle, he turns round
..... l, and by the connection of the rod, h, the screw, b,
with it, and causes the platen to descend and produce
..... But it is not simply this alone, for the power
lever, k, is transmitted to the screw, in a ratio propor-
..... to the effect required at the different parts of the pull;
..... first, when the pressman takes the bar K, it lies in a
..... parallel to the frame, or across the press, and the
lever i (being nearly perpendicular thereto) is also
at right angles to the connecting rod h; but the lever,
..... screw, makes a considerable angle with the rod,
therefore acts upon a shorter radius to turn the screw;
..... the real power exerted by any action upon a lever,
..... be considered as acting with the full length of the
..... its centres, but with the distance in a perpen-
..... drawn line, in which the action is applied to the
..... of the lever. Therefore when the pressman first takes
..... K, the lever i acts with its full length upon a
..... length of leverage, g, on the screw, which will con-
..... be turned more rapidly than if the bar itself was
..... to it; but on continuing the pull, the situation of
..... change, that of the screw, g, continually increasing
..... length, because it comes nearer to a perpendicular
..... connecting rod, and at the same time the lever i,
..... its acting length, because, by the obliquity of
..... the rod, h, approaches the centre, and the perpen-
..... distance diminishes; the bar or handle also comes to
..... favourable position for the man to pull, because he
..... nearly at right angles to its length.
..... causes combined have the best effect in producing
..... pressure, without loss of time; because in the
..... the lever acts with an increased motion upon
..... and brings the platen down very quickly upon the

bring the handle up to its stop with a concussion that shakes his arm very much, and in consequence most pressmen, after a few hours' work, feel inclined to give up the iron press; but when they have once acquired a new habit of standing more upright, and applying only as much force as it requires, the labour of the pull becomes less than that of running the carriage in and out; and men who are accustomed to the iron presses only, would be scarcely able to go through the work of the old press.

Mr. De la Haine has a patent for a Stanhope press, which answers extremely well; the only material alteration is, that he has substituted a spiral or curved inclined plane in place of the screw, which is fixed to the head of the press; and a cross-arm properly formed, and fixed on the upper end of the spindle, which, standing in place of the screw, acts against the fixed inclined plane. The action is very nearly the same as the screw, except that the surfaces admit of being made of hardened steel, and thus diminish the friction very much. The inventor of this for the common press was Mr. Roworth; but Mr. De la Haine has combined it with the levers and iron frame of the Stanhope press.

A common press, of great simplicity, and possessing the same advantage in point of power as Lord Stanhope gains by the compound levers, has been produced by Mr. Medhurst, of Denmark-street, Soho.

6. In November 1813, Mr. John Ruthven, of Edinburgh, took out a patent for an improvement in the printing-press, which differs from those heretofore used in the following particulars:

First, The types, plates, blocks, or other surfaces from which the impression is to be taken, instead of being situated upon a running carriage, as was formerly the practice, are placed upon a stationary platform or tablet, which is provided with the usual apparatus known to printers by the names of tympan and frisket, with points, &c. to receive the sheet of paper and convey it to its proper situation on the types after they had been inked.

Secondly, The machinery by which the power for the pressure is produced, is situated immediately beneath this platform or tablet; and the platen or surface which is opposed to the face of the types, to press the sheet of paper against them, can be brought over the types, and connected at two opposite sides or ends with the machinery beneath the tablet; by this machinery it is so forcibly pressed or drawn down upon the paper, which lays upon the types, as to give

the impression; which being thus made, the platen can be
dismitted from the machinery, and removed from off the types
by the foot, or otherwise, to take out the paper, and introduce
a fresh sheet.

Thirdly, The said machinery for producing the pressure is
a combination of levers, actuated by a crank, or short lever,
turned by a winch, or handle, to which the pressman applies
by his hand; or the pressure may be produced by the tread
of the foot.

Fig. 299 is a horizontal plane; fig. 300, a vertical section
taken through the middle; and fig. 301, an end view; the
same letters of reference being used in each.

A A represent the tablet or surface upon which the types, &c. are laid, its
surface truly flat, and may be made of wood, stone, or metal, or any other
substance used for the carriage of printing-presses. This tablet is mounted
upon a frame of wood or metal, consisting of legs B B, and cross braces
C C, or any other kind of support may be used which will firmly sustain
the tablet at a proper height from the ground. The tablet has a tympan,
9 and 9, joined to it at the end, 9, in the usual manner, and open into the
position of the dotted lines 10, to take off or put on the sheet of paper,
which is confined by the frisket, 11, in the usual manner; the dotted lines,
12, represent the gallows or support for the tympan and frisket when
opened.

For fastening the types upon the tablet, or what the printers call making
register, quoins or wedges may be introduced at the angles, in the usual
manner; but a preferable method is to have screws 13, 13, fitted through
pieces which are made fast to the sides of the tablet, and between the points
of these screws the chase, or frame of types, is held steady upon the tablet,
and may be adjusted.

Beneath the tablet are the levers marked D E, D E, their fulcrums, or
fixed centre-pins, being at D, and they act upon double hooks or clutches,
F F. When the ends E are depressed by means of the third lever I G,
situated beneath and common to both, the connection being made with the
link a, the fulcrum of the lever is at G; and H is a third point to which
the power to actuate it is applied by a connecting rod K, the opposite end
of which is joined to a crank or short lever L M, situated upon an axis or
spindle L, which extends to the front of the machine, and has a winch or
handle N, fig. 299, upon it, for the pressman to turn it by.

The platen of the press is shown at O O; it may be made of wood or
iron, as usual, but must be exactly true on the lower surface, which applies
to the face of the types b b, upon the tablet A A. On the top of the platen
is a strong metal bar P, which may be either cast in one piece with it, or
united to it by screws at r r; at its extremities it has bolts d d, fixed to it
by screws or otherwise; and at their lower ends they must have heads
which are exactly fitted to the clutches or double hooks F F, before described. By means of these the platen is connected with the lever D E, D E,
so that a pressure may be produced when the handle N is turned in the
direction shown by the arrow in fig. 300. This, by turning the lever M
that upon its centre L, pushes the rod K, which acting upon the point H
of the lever G H I, moves it upon its centre G, and depresses the point I,
which being connected with the extremities E of the levers D E, by the
link a, they are made to partake of its motion, and draw down the platen

upon the types by the clutches F F, and hooks *d d*. By returning
N to its original position, the pressure is relieved, and the plat-
removed from the types thus :—At the end of the bar P, two up
figs. 299 and 301, are fixed; and in the ends of these rollers o
marked *f*, are fitted to revolve freely upon their centre-pins. The
having grooves in their edges, run upon sharp angles, formed
upper edge of the two rails R R, which are extended across the fr
press, and project sufficiently behind, as in figs. 299 and 301, being
by brackets *g*, of fig. 301, if necessary. Upon these bars and w
sliders may be used instead of wheels) the platen will run freely,
it backwards and forwards off the types, but when brought over
bolts *d d* will enter the clutches F F, ready to receive the action of t
and give the pressure upon the tympan.

The springs *e* are so adjusted, that when the platen runs bac
forwards upon the rails R, the under surface of it will be sufficien
above the tympan to run clear of it; but when the hooks *d d* an
united, and the pressure given by turning the handle N, these spri
though they have sufficient strength to raise up the platen etc.
tympan, the instant the pressure is relieved.

To draw the platen forward over the types, a handle *h* is fixed
for the pressman to take hold by; but it may be brought by the fo
following manner: the two foremost wheels, *f f*, have links, *k k*
to their centre-pins, to connect them with the upper ends of the
levers *m m*, which are fixed to one common axis *n*, fig. 300, extendi
the whole machine, near the ground; upon the axis a short lever *o*,
is fixed, and a rod *q* unites it to the end of the bent lever *r i*, the
which is made broad, to serve as a paddle for the foot; by depres
the arm *r* draws the short lever *o*, and the long lever *m m* causes t
to advance truly parallel, and come up to the clutches F F.

To make all the work compact, the centres D D of the great lev
of the lower lever G, as well as the pivots L of the winch N, are
ported in one frame composed of two metal cheeks S S, which are
beneath the table, and united thereto by screws, or otherwise, as s
the dotted line in the plan, fig. 299.

The power of the press will depend upon the proportion of the
levers, and the relation between the space described by the motio
handle N, and the descent of the platen O; but it should be obser
the power of this press increases as the handle descends to the lo
position shown in fig. 300; first, because the handle is then in t
favourable position to receive the workman's body; secondly, the le
comes to a position which gives it a great power to force the rod I
is shown by the dotted line L 2, for when the lever and rod com
straight line, its power to force the rod K may be considered as i
great; thirdly, the lever G H is in the most favourable position, marke
receive the action of the rod K, viz. perpendicular to it; fourthly,
G I is in a position to have greater power on the links *e* and the lev
than when it is in a horizontal position. All these sources combi
the best effect in saving time, and at the same time producing i
pressure; for when the pressure first takes hold of the handle N, it
with little advantage with respect to power on the levers, and t
brings the platen down very quickly upon the tympans, with littl
time or motion, till they have assumed positions in which they ex
powerful action upon each other, as above stated; and this action c
to increase until the lever L M and rod K come nearly into a line, w
power is immensely great, and capable of producing any required

uich the parts of the press will withstand without yielding. The handle N is
ade to come to a stop, or rest, which prevents its moving farther than the
ation of the dotted lines, and therefore regulates the degree of pressure
ren upon the work. But to give the means of increasing or diminishing
t pressure at pleasure for different kinds of work, the centre hole of the
n H is made in a piece, which is fitted in a groove in the rod K; therefore
' sliding it in the groove, it has the same effect as lengthening the rod,
hich produces a greater descent of the platen when the handle is brought to
stop; a screw, s, is fitted into the end of the groove, to screw the packing
ght in the groove, and prevent it getting loose in working.

Another method of producing the same effect is to adjust the nuts which
e fitted on the screws at the top of the bolts d d; or it may be done by
nening the screws at r, and fitting packing between the fitting of the
aten and the bar P; the same may be done to adjust the platen parallel, if
prints more at one part than another.

Springs may be applied to take off all shake or looseness in the joints;
may be done in different ways: a strong spring may be fixed beneath the
blet, and act upon the clutch F, to lift it up, and keep the joint tight; or
se small spring may be fixed on the lever D E, (as shown on the opposite
de,) to lift the clutch F, and another being fixed to the lever beneath, and
sting at the end upon a pin in the frame, will lift up the lever and link a,
keep them all tight for working. If it be thought objectionable for the
d K to push endways on the levers, it may be contrived to draw or pull,
f placing the lever M above the spindle L, instead of beneath it, and also
versing the form of the lever G H I; the points G and H to remain as they
e, but the point I to be on the opposite side of the centre, viz. above it;
d with this alteration the drawing of the rod K will produce the pressure,
stead of pushing it, as in the figure.

Fig. 302 shows another arrangement of the lever for a press. In this
ure the same letters are used to denote the same parts, thus: A is the
blet, D E the levers, F the clutches, O the platen, P the cross-bar; the
ds E of the levers are connected by a link a, with a third lever T W, whose
ntre or fulcrum is at V; the power is applied to the long end by a chain t,
hich is conducted over a pulley or roller v, and wound upon a wheel w,
hich is fixed upon the axle of the handle to work the press. To give
eater power, the wheel may be formed like a spiral, instead of circular,
it the chain may lay upon a shorter radius when the pressure is produced.

7. Within these few years, numerous and great improve-
ents have been made in printing-presses; but the best that
e have seen is the invention of Messrs. Bacon and Donkin,
ho exhibited it before the university of Cambridge, by whom
is now employed in the printing of bibles and prayer-books.

Messrs. Bacon and Donkin's press consists in adapting
ie types to be fitted upon, and form the surface of a prismatic
ller, such as a square, pentagon, hexagon, octagon, or other
gure, and mounting this in a frame, with the means of turn-
g it round upon its centres; a second roller is applied in
ich a manner, that its surface will keep in contact with the
rface of the types, which are inked, and the machine being
t in motion, the paper which is to be printed is passed
rough and receives the impression. The types are inked

by a cylinder which is applied to revolve with its surface in contact with them. By this invention, the advantages of types between rollers are obtained, although the types are imposed upon plain surfaces.

Fig. 303 contains a perspective view of a machine, the prism A of which is square in its section, and has the ordinary types or letter-press applied upon its four sides, and firmly attached to it. The pivots at the end of the axis of this prism are supported in the frame B B, and it is caused to revolve by a connection of wheel-work D E and F G, from the winch and fly-wheel at H. The types upon its surface are caused to print upon the paper by means of a second roller I i, called the platen, placed immediately beneath the former, and its surface being formed to a particular curvature, produced by four segments of cylinders, its circumference, when it turns round, will always apply to the surface of the types, and thus a sheet of paper being introduced between them, will receive the impression. The ink is applied to the types by means of a cylinder K K, placed above the prism; it is composed of a soft elastic substance; and that its surface may always apply to the types, its spindle is fitted in pieces L L, which moving upon an axis n, permit the cylinder to rise and fall, to accommodate itself to the motion of the types. The ink-cylinder receives its ink from a second cylinder M M, which is called the distributing-roller, also composed of a soft substance, and is supplied with ink by a third ink-roller N N, which is made of metal, and extremely true. The ink is lodged in quantity against this roller upon a steel plate O O, the edge of which being placed at a very small distance from the circumference, permits the roller, as it revolves, to carry down a very thin film of ink upon its surface, and this being taken off by the distributing-roller, is applied to the surface of the inking-cylinder, which, as before mentioned, inks the types.

The sheet of paper is introduced, as shown in the figure, by placing it upon a blanket, which is extended upon a feeding-board P P, and drawn into the machine at a proper time, by having a small ruler, 2, fixed to it. The ends of this are taken forward by two studs b, attached to endless chains, which are extended from the wheels e, e, at the end of the platen, to other wheels d, d, which are supported in the frame of the feeding-board. The wheels e, e, having teeth entering the links of the chains, cause them to traverse when the machine is turned round, and at the proper time the pins, b, draw the ruler, 2, and blanket forward, and introduce the paper into the machine, and by passing between the prism and platen it is printed, as before mentioned. This is the general action of the machine; and we shall proceed to detail the structure of the several parts.

The type is composed, and made up into pages, in the usual manner; the pages are then placed in frames or gallies a, a, and fastened by the screws at the ends, the shape and size of the gallies being adapted to the size of the page it is intended to print. These gallies are attached to the four sides of the central axis of the prism by the screw-clamps 1, the edges of the gallies being mitred together. By relieving the clamps, the gallies can quickly be removed, and others put in their places. The platen I i is composed of four segments of cylinders, i i, which are attached to the different sides of the central axis I, by means of screws, and these segments being proportioned to the prism, will be the true figure for the platen to produce the required motion, so that the surface, when it revolves, will, in all positions, preserve an accurate contact with the surface of the types. The two wheels D, E, which cause the prism and platen to accompany each other, are formed to correspond with the two. Thus the upper wheel D is square, with its angles rounded off, and the pitch or geometrical outline is exactly of the same size as the square formed by the surfaces of the types. The lower wheel E is of the same shape as the platen, and its pitch-line the exact size of the surface thereof. These wheels being cut into teeth, as the figure shows, will turn each other round, and make their surfaces at the point of contact exactly correspond in their motions, so as to have no sliding or slipping upon each other. To regulate the pressure upon the paper, the bearings, in which the pivots of the platen are supported, can be elevated by screws, 3, and its surface will press with more force upon the types; but that this may not derange the action of the wheels D and E, universal joints are applied in their axles, at R. The inking-cylinder K is caused to preserve its proper distance from the centre of the prism by wheels S, fixed upon its axis, and resting upon shapes T, fixed upon the axis of the prism. Each of the shapes, like the wheel D, has four flat sides, corresponding in that with the surfaces of the types; the angles are rounded to segments of a circle from the centre; the wheels S are of the same size as the inking-cylinder, therefore, as they rest upon the shapes T, they prevent the inking-cylinder passing upon the types with any more than a sufficient force to communicate the ink without blotting. The inking-cylinder is turned round by a cog-wheel N, upon the extremity of the axis of the prism, which is of the same shape as the wheel D, and engages another wheel, W, upon the end of the spindle of the inking-cylinder; the latter wheel likewise gives motion to the distributing-roller by a pinion f, and this again turns the ink-roller by a third pinion g, fixed upon the end of its axis n, which is supported upon bearings B, B, in the frame. The pieces L, L, which support the pivots of the distributing-roller and inking-cylinder, are fitted upon the axis of the inking-cylinder, so as to rise and fall upon its centre; and the distances of the rollers being thus kept invariably the same, their circumferences are kept accurately in contact, to communicate the ink to each other. The steel plate O, which, as before mentioned, regulates the quantity of ink that the roller N shall take round with it, is supported by a piece extended across the fixed frame B B. There are pieces of metal fixed upon this plate by thumb-nuts, which prevent the ink flowing off at the ends, and they enter into grooves formed round the ink-roller N, near its ends. The machine is put in motion by the handle with the fly-wheel H, and this has a small wheel G, turning a large one F, upon the end of the axis b.

The frame supporting the feeding-board P consists of two rails, X, fitted upon the axis of the platen, and supported at the opposite ends by a brace from the framing; they sustain the pivots of the wheels d, d, for the chains; s are two rulers fixed at each side of the feeding-board, and forming a lodgment for the ends of the ruler 2, which is attached to the blanket, and it slides upon these when it is advanced by the chains. The spaces on the platen between the segments i, i, are all filled up by pieces of wood, except one, and in this space the ruler is received when it passes through the machine. In the interval when the spaces between the types are passing over the sheet, and therefore leave the margin between the pages of printing, the paper is not held between the rollers; but to prevent it from slipping during this interval, the blanket and paper are pressed down upon the pieces of wood which fill up in the platen between the segments i, i, by the weight of small rollers or wires 4, supported by cocks 5, projecting from the axis of the prism, and being fitted into the slits at the end of these cocks. The wires are at liberty to rise and fall by their own weight; thus, when they are at the upper part of the revolution, they fall into the spaces at the angle of the prism, between the pages of the types, and thus escape the ink-cylinder; but when they are at the lower part of their revolution, they fall

upon the paper, and press it with sufficient force upon the piec
in the platen to carry the paper forward at the interval when t
not act upon it, and of course while the space between the p
printing is passing through.

The operation of printing being very delicate, and
great accuracy, the machine is provided with man
ments to make it act correctly, which are as follow
segments *i, i,* upon the platen-roller, are attache
central axis I, by three screws at each end ; the tw
ones of these (represented with square heads) draw
ments down upon the central axis, whilst the other
are turned by a screw-driver) bear them off ; ther
means of these screws, the segments can be accur
justed, till they are found by experiment to apply
to the types, and make an equal impression on all
the sheet. To render the whole impression greate
the screws, 3, beneath the bearings of the platen-re
turned as before mentioned. The degree of press
which the ink-roller bears upon the types is regu
increasing or diminishing the size of the shapes '
support its weight. And to render these capable o
ment, each is composed of four pieces, marked 6, att.
screws, 7, to a central piece, or wheel, which is fix
the axis ; and as the edges of these pieces form th
of the shape, they admit of being adjusted by other s
a greater or less distance from the centre, and of co
be made to bear up the ink-cylinder till the pressu
types is equal throughout the whole surface, and i
to supply the ink properly. The ink-cylinder is a(
as to its pressure against the distributing-roller, and
purpose the bearings *k,* which support the cylinder, s
upon the pieces L, to slide, being capable of regul
means of screws. In a similar manner, the distr
roller can be adjusted to a proper distance from the
cylinder. The plate *o* can be adjusted for the distai
the ink-roller N, by screws *p,* fastened by thumb-nu
regulates the degree of colour the impression will l
permitting the roller N to take more or less ink ; be
inking-cylinder K, a rubber, or scraper, is placed,
very lightly against the cylinder, and to prevent
accumulating in rings round the cylinder, it is fitt
centres, and held up by a lever which is suspend
catch *y,* at the end of the piece L. This catch is wi
when the machine is not at work, and then the scrape
down upon its centre, does not touch the cylinde

necessary that the wheels D and E should be placed upon their axis, in such a position that their curvature will correspond with the curvature of the prism and platen. For this purpose the universal-joint R is fitted upon the axis *l*, of the wheel, with a round part, that it may turn on it. A piece of metal, *r*, is fixed fast upon the spindle *l*, and has a hole in it for the reception of a tooth *s*, which is screwed fast upon the universal-joint; then two screws being tapped through the sides of the piece *r*, press upon the end of *s*, and by forcing it either way, will adjust the wheel with respect to the platen till they exactly correspond; another similar adjustment may be applied to the upper axis.

The manner of forming the ink and distributing rollers with an elastic substance is worthy of particular notice. Leather stuffed in the manner of a cushion was first used, but did not succeed, because it became indented with the types; but after many trials, a composition of glue, mixed with treacle, was found to answer perfectly. The roller is made of a copper tube, covered with canvass, and placed in a mould, which is a cylindrical metal tube, accurately bored, and oiled withinside; the melted composition is then poured out into the space of the mould, and when cold, the whole is drawn out of it, with the glue adhering to the copper tube, and forming an accurate cylinder without any further trouble. The composition will not harden materially by the exposure to air, nor does it dissolve by the oil contained in the ink. This machine is well adapted to print from stereotype plates, which the universities have adopted for their bibles and prayer-books.

BRAMAH'S BANK-NOTE PRESS.

6. It was formerly the custom in the Bank of England to fill up the number and dates of their notes in writing, till the year 1809, when the machine invented by Mr. Bramah was adopted for this purpose. By this contrivance, the numbers and dates were inserted not only in a more uniform and elegant manner, but the labour was diminished to less than one-sixth of what it was before.

The copper-plates from which the words of the notes are printed, are double; that is, they throw off two notes at a time upon one long piece of paper. This piece of paper, containing two notes, is then put into the machine, which prints upon them the number and dates in such a manner, that the types change to the succeeding number, and that the whole operation is performed without any attention on the

vious that any combination of the above figures may be produced,
ging them to the highest point of the circle, which is the situation in
they are to be placed when an impression is to be taken. This will
e easily understood, if we consider that the brass plate which covers
cles is put on its place, as represented in fig. 304, at *a*. This brass
has two apertures through it, to receive the two series of types
project up a little above its surface when at the highest. In fig 305,
te is removed to exhibit the interior mechanism.

circles are made to revolve by means of wheels, H, upon an axis
he back axis, parallel to the axis of the circles. The end of it is seen
f. 305, projecting through the frame, and it carries three of the wheels
) of which are at the same distance apart as the two series of figure
to which they apply; the third wheel is placed at an intermediate
e between the other two, and is acted upon by a catch or pallet *b*,
, attached to the axis of the tympan, by means of a joint, in such a
· that it will strike against the highest tooth of the wheel H, and turn
d one tooth. When the handle is lifted up rather beyond the
licular, a stop *a*, fig. 305, upon the axis, meeting a projection *e*,
, on the cover of the box, prevents it from moving farther; but
he handle is returned down the position of the fig. 304, the pallet,
it again meets the tooth of the wheel. gives way upon its joint, and
by without moving the wheel. In this manner it will be seen, that
me the handle is pressed down to take an impression, in raising it up
) place a fresh paper upon the tympan, the pallet moves the wheels H
th, and as the teeth of these wheels engage the teeth of the figure
a similar motion is communicated to them, bringing a fresh number
t the tympan, ready for printing.

to be observed, that the wheels H are of such a thickness, as to en-
ily one of the five type circles at once, and their distance from each
; such, that they take the same circle in the one series as they do in
er. Now, by moving the back axis a small quantity endwise, it is
; that the wheel H can be brought to act upon any of the five circles,
laced in such a position as to be clear of them all. It is for this pur-
at the head I, fig. 305, comes through the frame of the machine;
means of this the axis can be moved on end, and by proper marks
, it may be set to any of the five circles. In these positions it is con-
y a semicircular clip, which enters grooves turned round on the axis,
prives it of longitudinal motion, unless when the clip is raised.
in be done by a nut coming through the back of the frame at K,
. It has a short lever on the inside of it, which, when the nut is turned
raises up the clip, and releases the axis while it is set to the required
and the clip being let fall into the proper groove, confines it from any
motion.

rder that all the circles may stop at the exact point, when the figure
he highest, and consequently when the surface of the figure will be
ital, an angular notch is made on the inside of the figure circles,
intermediate spaces between each figure; and at the lowest points of
cle *e*, fig. 304, a movable pin is fitted into the fixed axis, with a
which gives it a continual pressure downwards. The end of the pin
ed spherical, and well polished, so that when the circle is turned
it is forced into its hole in the axis; but when another notch in the
presents itself, the pin presses out into it, and retains the circle with
rate force in its proper position, until the raising of the tympan, as
described, overcomes the resistance of the pin, and turns the circle

By this contrivance the types always arrange themselves into a

x 2

straight line, after being turned round, without which the impressi
have a very disagreeable and irregular appearance. The tymp
304, is composed of two parts; a solid brass plate, against wi
folds of cloth are placed and secured by the second part, which
frame covered with parchment, and attached to the former by fou
two of which appear at ff, in fig. 305.

The brass plate of the tympan is fastened to the leaf L, fig. 304
ing from the axis by means of six screws. Two of these, only one
A, can be seen in the figure, tend to throw the tympan from the le
the other four, which are arranged one on each side as the tw
draw the tympan in and leaf together. By means of these screws
ing in opposition, the tympan can be adjusted so as to fall exactl
upon the type, and communicate an equal pressure to all parts of t
which is held against the tympan by means of a frisket of j
stretched on a frame which surrounds the tympan, and is m
joints at $k\,k$, fig. 305. The frisket is cut through as is represent
shaded parts in fig. 305, in order to expose the paper where it is t
the impression of the figures, and the No. before the figures, and als
pression of the date, year, and place. The type for these are f
stereotype, and fastened down upon the surface of the brass co
piece containing the day and month being changed every day. In
find the proper position which the paper should occupy upon the
two fine pins are fixed to project from it, and are received into ho
in the brass cover; two dots are printed upon the note from th
plates, and the pins being put through at these dots, ensure the fi
coming on their proper places.

The manner of using the machine is as follows: suppose the
put so far on end as to be detached from all the circles; the figu
arranged by hand, so that the blanks are all uppermost; and t
stereotypes put in for the date. The back axis is then first set, s
wheels H may take the first five circles towards the right han
moving the handle down almost to touch the type, and return
again, the pallet moves the wheels H, and turns the two right-han
bringing up figure 1. The clerk now inks the type with a prin
opens the frisket sheet L, fig. 305, on its hinges, and places
(already printed on the copper-plate press) against the tympan, t
place being determined by the two pins, and the dots printed on
as before mentioned. He now shuts up the frisket sheet, in ord
fine the paper and keep it clean, except in the places where it
printed; then by pressing down the handle F, the impression is gi
on lifting it up again it moves the circles and brings up figure 2.
is now removed, a fresh one put in, and so on, the figure always
every time.

During this operation the two right-hand circles act as units, an
one each time; when 9 are printed in this manner and 0 comes up, t
is moved twice successively without printing, which brings up a 1
then a 1. The back axis is moved, to act upon the second circle
right hand, which now becomes the units, the first circle represen
by moving the handle a, without printing, figure 1 in the seco
comes up, making 11, the next time 12, and so on to 19. The fi
is now put forwards by hand, bringing up 2 and 0, on the second
moving the handle to pass the blank, produces 21, 22, &c. to 30,
first circle is again advanced, bringing up 4; in this manner the
proceeds to 99. The back axis is now shifted to the third circle,
comes units, the second tens, and the third hundreds; the 9 and

which are advanced to bring up 1, 0 is brought up in the second : and the machine itself brings up 0 in the third ; after printing this it changes to 101.

The process now continues through the successive hundreds in the same manner as before till 999. The back axis is now shifted to the fourth circle, and the three first must be advanced by hand when they require it. At 9999 the back axis is shifted to the fifth circle, and it will serve to 999,999, beyond which it is not required to print.

PILE-ENGINE.

The pile-engine is a machine by which piles are driven into the ground for the foundation of the piers of bridges, and various other structures.

The method of driving a pile consists in drawing up a very heavy weight, called a ram or hammer, and by disengaging it from the machinery by which it was raised, letting it fall, by the force of gravity, upon the head of the pile. In the most simple machines the weight is drawn up by men pulling a cord over a fixed pulley, and when it has attained a sufficient height allowing the cord to slip from their hands, which permits the weight to descend with considerable force. The two best pile-engines that we have seen are those invented by Mr. Vauloué and Mr. S. Bunce,

Mr. Vauloué's pile-engine may be thus described. A, fig. 306, is a great upright shaft or axle, on which are the great wheel B, and the drum C, turned by horses joined to the bars S S. The wheel B turns the trundle X, on the top of whose axis is the fly O, which serves to regulate the motion, as well as to act against the horses, and to keep them from falling, when the heavy ram Q is discharged to drive the pile P down into the mud in the bottom of the river. The drum C is loose upon the shaft A, but is locked to the wheel, B, by the bolt Y. On this drum the great rope, H H, is wound ; one end of the rope being fixed to the drum, and the other to the follower G, to which it is conveyed by the pulleys I and K. In the follower G is contained the tongs F, that take hold of the ram Q, by the staple R, for drawing it up. D is a spiral or fusee fixed to the drum, on which is wound the small rope, T, that goes over the pulley U, under the pulley V, and is fastened to the top of the frame at 7. To the pulley block is hung the counterpoise W, which hinders the follower G from acquiring as it goes down to take hold of the ram ; for, as the follower tends to acquire velocity in its descent, the line T winds downwards upon the spiral upon a larger and larger radius, by which means the counterpoise, W, acts stronger and stronger against it ; and so allows it to come down with only a moderate and uniform velocity. The bolt Y locks the drum to the great wheel, being pushed upward by the small lever 2, which goes through a mortise in the shaft A, turns upon a pin in the bar 3, fixed to the great wheel B, and has a weight 4, which always tends to push up the bolt Y, through the wheel into the drum. L is the great lever turning on the axis m, and resting on the forcing bar 5, 5, which goes through a hollow in the shaft A, and bears up the little lever 2.

By the horses going round, the great rope H is wound about the drum C,

are drawn to a scale. In fig. 314, A A denote two oak ground sills, which are firmly bolted down, parallel to each other, upon sleepers let into the ground. At each end of these a vertical iron frame, B B, is erected, to support the gudgeons at the end of a long cylindrical axis, D D, which is turned round by the mill. The cylinder L L, which is to be bored, is fixed immovable over the bar, and exactly concentric with it. A piece of cast-iron K K, L L, (figs. 310, 312, and 313,) called a cutter-head, slides upon the axis, and has fixed into it the knives or steelings *f f f*, which perform the boring. This cutter-head is moved along the bar by machinery, to be hereafter described; by means of which it is drawn or forced through the cylinder, at the same time that it turns round with the axis D. The steel cutters will necessarily cut away any protuberant metal which projects within the cylinder, in the circle which they describe by their motion, but cannot possibly take any more.

The cylinder is held down upon an adjustable framing, which is ready adapted to receive a cylinder of any size within certain limits. Pieces of iron, E E, are bolted down to the ground sills, having grooves through them to receive bolts, which fasten down two horizontal pieces of cast-iron F F, at right angles to them. These horizontal pieces support four movable upright standards G G, which include the diameter of the cylinder L L, which is supported upon blocks, *b b*, below, and held fast by iron bands *a a*, drawn by screws in the top of the standards G G. The cylinder is adjusted, to be concentric with the axis D D, and held firmly in its place by means of wedges driven under the blocks and the standards.

To explain the mechanism by which the cutters are advanced, we must refer to figs. 311, 312, and 313, by the inspection of which it will be seen that the axis D D is, in fact, a tube of cast-iron, hollow throughout. It is divided by a longitudinal aperture *c c*, fig. 310, on each side. At the ends of it is left a complete tube, to keep the two valves together. The cutter-head K K, L L, consists of two parts; of a tube K fitted upon the axis D with the greatest accuracy, and of a cast-iron ring L L, fixed upon K K by four wedges. On its circumference are eight notches, to receive the cutters or steelings *f f*, which are held in and adjusted by wedges. The slider K is kept from slipping round with the axis, by means of two short iron bars *e e*, which are put through to the axis, and received into notches cut in the ends of the sliders K K. These bars have holes in the middle of them to permit a bolt at the end of the toothed rack L to pass through. A key is put through the end of the bolt, which, at the same time, prevents the rack being drawn back, and holds the cross bars *e e* in their places. The rack is moved by the teeth of a pinion N, and is kept to its place by the roller O; the axis of the pinion and roller being supported in a framing attached to the standard B B, as shown in a perspective view of the machine in fig. 314. The pinion is turned round by a lever, put upon the square end of the axis, and loaded with the weight P, that it may have a constant tendency to draw the cutter through the cylinder. This lever is capable of being put on the square end of the axis either way, so as to force the rack back into the cylinder if necessary.

In some boring machines, another contrivance, superior perhaps to what we have now described, is employed to draw the cutter through the cylinder. It consists of four small wheels, one of which is fixed at the right-hand extremity, D, of the bar D D, fig. 314. Another pinion is fastened on the extremity of an axis, analogous to the rack M, having at its other extremity a small screw, which works in a female screw, fixed to the cutter K K at *e*, fig. 310. Below the second pinion is another, con-

.... the same number of teeth, and fixed on a horizontal axis parallel to D D. At the other end of this axis is a fourth pinion, which is driven by the first pinion at the end of the hollow axis D D. The first pinion has twenty-six teeth, the fourth thirty, and the second and third may have any number, provided they are equal. As the axis D revolves, the first pinion fixed on its extremity drives the fourth, which by means of the third, fixed on the same axis with it, gives motion to the second. The second pinion being fixed to an axis within D D, unscrews the screw at its other extremity, and of course makes the cutter advance along the cylinder. This screw has eight threads in an inch, and sixty turns of the axis are required to cut one inch.

To introduce a cylinder into its place in the machine, it is necessary to remove the upper braces, I l, of the bearings upon the standards B B; and by supporting the axis upon blocks placed under the middle of it, the standard, with the pinion N, and roller frame, is removed by taking up the nuts which fasten it to the ground sills A A, the rack M being supposed previously withdrawn. A cutter-block L, of a proper size to bore out the intended cylinder, is now placed upon the slider K, fig. 313, and wedged fast. The cutter-head is then moved to the farther end of the axis, and the cylinder lifted into its place. The standard B is returned, and the whole machine brought to the state of fig. 314, the cylinder being, by estimation, adjudged concentric with the axis D. Two bars of iron are now wedged into the e e in the axis, and applied to the ends of the cylinder; while the axis is turned round they act as compasses to prove the concentricity of the cylinder. Small iron wedges are drawn round the cylinder to adjust it with the utmost accuracy; and in this state the cylinder is ready for boring.

The next operation is fitting the cutters, which are fastened into the block L by wedges, and adjusted by turning the axis round, to ascertain that they all describe the same circle. The boring now commences by putting the mill and axis in motion, and the machine requires no attention, except that the weight P is lifted up as often as it descends by the motion of the cutters or steelings. When the cutters are drawn down through the cylinder, they are set to a circle a small quantity larger, and returned through the cylinder a second time. For common work these operations are sufficient; but the best cylinders are bored many times, in order to bring them to a proper cylindrical surface. The last operation is turning the flanch n of the cylinder perfectly flat, by wedging a proper cutter into the head. This is of great importance to ensure that the lid will fit perpendicular to the axis of the cylinder. The cylinder is now finished and removed.

The accuracy of this machine depends on the boring bar, D D, being turned upon its own gudgeons; and if it is turned to the same diameter throughout, it will certainly be perfectly straight. While the axis is in the operation of turning, a piece of hard wood should be fitted into the grooves of the cylinder. The slider K is first bored out, and afterwards ground upon the axis with emery to fit ......ble.

The elevation of a mill proper for moving two of these represented in fig. 310. The pinion 30 is supposed to be o. a water-wheel, and turns the two wheels 60, 60, which ha axes, with a cross-cut similar to the head of a screw, as is t figure.

The ends of the boring axes have similar notch.... putting keys in between them, the motion may ....ted or discontinued at pleasure, by the ...... key.

FILE-CUTTING MACHINE.

THERE have been various contrivances for ... but the best we are acquainted with is d..... *Transactions of the American Philosophical* .... as follows :

AAAA, fig. 315, is a bench of seasoned oak, th.. is planed very smooth. BBBB the feet of the ben... be substantial. CCCC the carriage on which the ...... moves along the face of the bench AAAA, parallel .. carries the files gradually under the edge of the cutt.. o while the teeth are cut: this carriage is made to mov.. vance somewhat similar to that which carries the log ... of a saw-mill, as will be more particularly described. ... iron rods inserted into the ends of the carriage CCCC, and the holes in the studs EEE, which are screwed firmly ag... the bench AAAA, for directing the course of the carr... rallel to the sides of the bench. FF two upright pillars; n.. into the bench AAAA, nearly equidistant from each end o. edge, and directly opposite to each other. G the lever t carries the cutter HH, (fixed by the screw I,) and works on two screws KK, which are fixed into the two pillars FF, th right across the bench AAAA. By tightening or loosening the arm which carries the chisel may be made to work more o. L is the regulating screw, by means of which the files may b. or finer; this screw works in a stud M, which is screwed i. top of the stud F; the lower end of the screw L bears ag.... part of the arm G, and limits the height to which it can rise. spring, one end of which is screwed to the other pillar F, and t presses against the pillar O, which is fixed upon the arm G; b. it forces the said arm upwards until it meets with the regulati.. P is an arm with a claw at one end marked 6, the other end h join. into the end of the stud or pillar O, and, by the motio. O, is made to move the ratch-wheel Q. This ratch-wheel is f.. axis, which carries a small trundle-head or pinion R, on the o. this takes into a piece SS, which is indented with teeth, and m. against one side of the carriage CCCC; by means of this i is communicated to the carriage. V is a clamp for fastenin. the file ZZ in the piece or bed on which it is to be cut....

clamp or dog at the opposite end, which works by a joint W, firmly fixed into the carriage C C C C. Y is a bridge, likewise screwed into the carriage, through which the screw X passes, and presses with its lower end against the upper side of the clamp V; under which clamp the other end of the file Z Z is placed, and held firmly in its situation while it is cutting by the pressure of the said clamp V. 7 7 7 7 is a bed of lead, which is let into a cavity formed in the body of the carriage, something broader and longer than the largest size files; the upper face of this bed of lead is formed variously, so as to fit the different kinds of files which may be required. At the figures 2 2 are two catches, which take into the teeth of the ratch-wheel Q, to prevent a recoil of its motion; 3 3 is a bridge to support one end, 4, of the axis of the ratch-wheel Q; 5 a stud to support the other end of the axis of that wheel.

When the file or files are laid in their place, the machine must be regulated to cut them of the due degree of fineness, by means of the regulating screw L; which, by screwing farther through the arm M, will make the film finer, and, *vice versa*, by unscrewing it a little, will make them coarser; for the arm G will, by that means, have liberty to rise the higher, which will occasion the arm P, with the claws, to move further along the periphery of the ratch-wheel, and consequently communicate a more extensive motion to the carriage C C C C, and make the files coarser.

When the machine is thus adjusted, a blind man may cut a file with more exactness than can be done in the usual method by the keenest sight; for by striking with a hammer on the head of the cutter or chisel H H, all the movements are set at work; and by repeating the stroke with the hammer, the files on one side will at length be cut; then they must be turned, and the operation repeated for cutting on the other side. It is needless to enlarge much on the utility or extent of this machine; for, on an examination, it will appear to persons of but indifferent mechanical skill, that it may be made to work by water as well as by hand, to cut coarse or fine, large or small, files, or any number at a time; but it may be more particularly useful for cutting very fine small files for watchmakers; as they may be executed by this machine with the greatest equality and nicety imaginable. As to the materials and dimensions of the several parts of this machine, they are left to the judgment and skill of the artist who may have occasion to make one; only observing, that the whole should be capable of bearing a good deal of violence.

RAMSDEN'S DIVIDING MACHINE.

This valuable instrument is the invention of Mr. Jesse Ramsden, to whom the Commissioners of Longitude gave the sum of 615*l.*, upon his entering into an engagement to instruct a certain number of persons, not exceeding ten, in

the method of making and using this machine, in
of two years, say, from the 28th October, 1775, to
tober, 1777; also binding himself to divide all sec
octants by the same engine, at the rate of three sh
each octant, and six shillings for each brass sect
Nonius's divisions to half minutes, for as long tl
Commissioners should think proper to let the engi
in his possession. Of this sum 300*l.* were given to B
den, as a reward for the usefulness of his invention;
for his giving up the property of it to the Commss

The following is the description of the engine
Mr. Ramsden, upon oath:

This engine consists of a large wheel of bell-m
ported on a mahogany stand, having three legs,
strongly connected together by braces, so as to mal
fectly steady. On each leg of the stand is placed
friction-pulley, whereon the dividing wheel rests; t
the wheel from sliding off the friction-pullies, the l
centre under it turns in a socket on the top of the s

The circumference of the wheel is ratched or
method which will be described hereafter) into 21
in which an endless screw acts. Six revolutions of
will move the wheel a space equal to one degree.

Now a circle of brass being fixed on the scr
having its circumference divided into sixty parts,
sion will, consequently, answer to a motion of the
ten seconds, six of them will be equal to a minute,

Several different arbors of tempered steel
ground into the socket in the centre of the whe
upper parts of the arbors, that stand upon the
turned of various sizes, to suit the centres of differe
of work to be divided.

When any instrument is to be divided, the cent
very exactly fitted on one of these arbors; and th
ment is fixed down to the plane of the dividing
means of screws, which fit into holes made in the
the wheel for that purpose.

The instrument being thus fitted on the plane of tl
the frame which carries the dividing point is com
one end by finger-screws, with the frame which ca
endless-screw; while the other end embraces tha
the steel arbor which stands above the instrume
divided, by an angular notch in a piece of harden
by this means both ends of the frame are kept
steady and free from any shake.

The frame carrying the dividing-point or tracer, is made to slide on the frame which carries the endless-screw to any distance from the centre of the wheel, as the radius of the instrument to be divided may require, and may be there fastened by tightening two clumps; and the dividing-point or tracer, being connected with the clumps by the double-jointed frame, admits a free and easy motion towards or from the centre for cutting the divisions, without any lateral shake.

From what has been said, it appears that an instrument thus fitted on the dividing-wheel, may be moved to any angle by the screw and divided circle on its arbor; and that this angle may be marked on the limb of the instrument with the greatest exactness by the dividing-point or tracer, which can only move in a direct line tending to the centre, and is altogether freed from those inconveniences that attend cutting by means of a straight edge. This method of drawing lines will also prevent any error that might arise from an expansion or contraction of the metal during the time of dividing.

The screw-frame is fixed on the top of a conical pillar, which turns freely round its axis, and also moves freely towards or from the centre of the wheel, so that the screw-frame may be entirely guided by the frame which connects it with the centre: by this means any eccentricity of the wheel and the arbor would not produce any error in the dividing; and by a particular contrivance, (which will be described hereafter,) the screw when pressed against the teeth of the wheel always moves parallel to itself; so that a line joining the centre of the arbor and the tracer continued will always make equal angles with the screw.

Fig. 316 represents a perspective view of the engine.

Fig. 317 is a plan of which fig. 318 represents a section on the line IIΔ.

The large wheel A is 45 inches in diameter, and has 10 radii, each being supported by edge-bars, as represented in fig. 318. These bars and radii are connected by a circular ring B, 24 inches in diameter and 3 inches deep; and, for greater strength, the whole is cast in one piece in bell-metal.

As the whole weight of the wheel A rests on its ring B, the edge-bars are deepest where they join it; and from thence their depth diminishes, both towards the centre and circumference, as represented in fig. 318

The surface of the wheel A was worked very even and flat, and its circumference turned true. The ring C, of fine brass, was fitted very exactly on the circumference of the wheel; and was fastened thereon with screws, which, after being screwed as tight as possible, were well rivetted. The face of a large chuck being turned very true and flat in the lathe, the flattened surface A, fig. 318, of the wheel, was fastened against it with holdfasts; and the two surfaces and circumference of the ring C, a hole through the centre and the plane part round b, and the lower edge of the ring B, were turned at the same time.

D is a piece of hard bell-metal, having a hole, which receives the steel arbor *d*, made very straight and true. This bell-metal was turned very true on an arbor; and the face, which rests on a wheel at *b*, was turned very flat, so that the steel arbor *d* might stand perpendicular to the plane of the wheel; this bell-metal was fastened to the wheel by six steel screws.

A brass socket Z is fastened on the centre of the mahogany stand, and receives the lower part of the bell-metal piece D, being made to touch the bell-metal in a narrow part near the mouth, to prevent any obliquity of the wheel from bending the arbor; good fitting is by no means necessary here; since any shake in this socket will produce no bad effect, as will appear hereafter when we describe the cutting-frame.

The wheel was then put on its stand, the lower edge of the ring B, figs. 316, 317, and 318, resting on the circumference of three conical friction pulleys W, to facilitate its motion round its centre. The axis of one of these pulleys is in a line joining the centre of the wheel and the middle of the endless-screw, and the other two placed so as to be at equal distances from each other.

Fig. 316 is a block of wood strongly fastened to one of the legs of the stand; the piece *g* is screwed to the upper side of the block, and has half-holes, in which the transverse axis *h*, fig. 319, turns; the half-holes are kept together by the screws *i*.

The lower extremity of the conical pillar P, figs. 316 and 319, terminates in a cylindrical steel pin *k*, fig. 319, which passes through and turns in the transverse axis *h*, and is confined by a check and screw.

To the upper end of the conical pillar is fastened the frame G, fig. 319, in which the endless-screw turns; the pivots of the screw are formed in the manner of two frustrums of cones joined by a cylinder, as represented at X, fig. 320. These pivots are confined between half-holes, which press only on the conical parts, and do not touch the cylindric parts; the half-holes are kept together by screws *a*, which may be tightened at any time, to prevent the screw from shaking in the frame.

On the screw-arbor is a small wheel of brass K, figs. 316, 317, 319, and 320, having its outside edge divided into 60 parts, and numbered at every sixth division with 1, 2, &c. to 10. The motion of this wheel is shown by the index *y*, figs. 319 and 320, on the screw-frame G.

H, fig. 316, represents a part of the stand, having a parallel slit in the direction towards the centre of the wheel, large enough to receive the upper part of the conical brass pillar P, which carries the screw and its frame; and as the resistance, when the wheel is moved by the endless-screw, is against the side of the slit H which is towards the left hand, that side of the slit is faced with brass, and the pillar is pressed against it by a steel spring on the opposite side; by this means the pillar is strongly supported laterally, and yet the screws may be easily pressed from or against the circumference of the wheel, and the pillar will turn freely on its axis to take any direction given it by the frame L.

At each corner of the piece I, fig. 319, are screws *n*, of tempered steel, having polished conical points; two of them turn in conical holes in the screw-frame near *o*, and the points of the other two screws turn in the holes in the piece Q; the screws *p* are of steel, which being tightened, prevent the conical pointed screws from unturning when the frame is moved.

L, figs. 316, 317, and 321, is a brass frame, which serves to connect the endless-screw, its frame, &c. with the centre of the wheel; each arm of this frame is terminated by a steel screw, that may be passed through any of the holes *g*, in the piece Q, fig. 319, as the thickness of the work to be divided on the wheel may require, and are fastened by the finger-nuts *r*, figs. 316 and 317.

At the end of this frame is a flat piece of tempered steel *b*, fig. 321, wherein is an angular notch; when the endless-screw is pressed against the teeth of the circumference of the wheel, which may be done by turning the finger-screw S, figs. 316 and 317, to press against the spring *t*, this notch embraces and presses against the steel arbor *d*. This end of the frame may be raised or depressed by moving the prismatic slide *a*, fig. 317, which may be fixed at any height by the four steel screws *v*, figs. 316, 317, and 321.

The bottom of this slide has a notch K, figs. 316 and 321, whose plane is parallel to the endless-screw, and by the point of the arbor *d*, fig. 318, resting in this notch, this end of the frame is prevented from tilting. The screw S, figs. 316 and 317, is prevented from unturning, by tightening the finger-nut *w*.

The teeth on the circumference of the wheel were cut by the following method:

Having considered what number of teeth on the circumference would be most convenient, which in this engine is 2160, or 360 multiplied by 6, I made two screws of the same dimension of tempered steel, in the manner hereafter described, the interval between the threads being such as I knew by calculation would come within the limits of what might be turned of the circumference of the wheel: one of these screws, which was intended for ratching or cutting the teeth, was notched across the threads, so that the screw, when pressed against the edge of the wheel and turned round, cut in the manner of a saw. Then having a segment of a circle a little greater than 60 degrees, of about the same radius with the wheel, and the circumference made true, from a very fine centre, I described an arch near the edge, and set off the chord of 60 degrees on this arch. This segment was put in the place of the wheel, the edge of it was ratched, and the number of revolutions and parts of the screw contained between the interval of the 60 degrees were counted. The radius was corrected in the proportion of 360 revolutions, which ought to have been in 60 degrees, to the number actually found; and the radius, so corrected, was taken in a pair of beam-compasses; while the wheel was on the lathe, one foot of the compasses was put in the centre, and with the other a circle was described on the ring; then half the depth of the threads of the screw being taken in dividers, was set from this circle outwards, and another circle was described cutting this point; a hollow was then turned on the edge of the wheel, of the same curvature as that of the screw at the bottom of the threads, the bottom of this hollow was turned to the same radius or distance from the centre of the wheel, as the outward of the two circles before-mentioned.

The wheel was now taken off the lathe, and the bell-metal

piece D, fig. 318, was screwed on as before dire
after this ought not to be removed.

From a very exact centre a circle was described on the rin
317, and 318, about four-tenths of an inch within where the
teeth would come. This circle was divided with the great
was capable of, first into five parts, and each of these into
parts were then bisected four times, (that is to say,) suppo
circumference of the wheel to contain 2160 teeth, this bein
five parts, each would contain 432 teeth; which being divi
parts, each of them would contain 144; and this space bisec
would give 72, 36, 18, and 9; therefore each of the last d
contain nine teeth. But, as I was apprehensive some err
from quinquesection and trisection, in order to examine the a
divisions, I described another circle on the ring C, (fig. 322
an inch within the former, and divided it by continual bisect
1080, 540, 270, 135, 67½, and 33¾; and as the fixed wire (t
presently) crossed both the circles, I could examine their
every 135 revolutions; (after ratching, could examine it at ev
not finding any sensible difference between the two sets of di
ratching, made choice of the former; and, as the coincidence of
with an intersection could be more exactly determined than
division, I therefore made use of intersections in both circles be

The arms of the frame L, fig. 322, were connected by a thin
of three-fourths of an inch broad, having a hole in the middle
of an inch in diameter; across this hole a silver wire was fixe
line to the centre of the wheel; the coincidence of this wire
sections was examined by a lens seven-tenths of an inch focus,
which was attached to one of the arms L.* Now a handle o
fixed on the end of the screw, the division marked 10, on the
set to its index, and, by means of a clamp and adjusting-scre
pose, the intersection marked 1 on the circle C was set exac
with the fixed wire; the screw was then carefully presse
circumference of the wheel, by turning the finger-screw S, t
the clamp, I turned the screw by its handle nine revolutions,
section marked 240 came nearly to the wire; then, unturni
screw S, I released the screw from the wheel, and turned th
till the intersection marked 2 exactly coincided with the
means of the clamp before-mentioned, the division 10 on th
set to its index, the screw was pressed against the edge of the
finger-screw S; the clamp wire removed, and the screw turne
tions till the intersection marked 1 nearly coincided with the fi
screw was released from the wheel by unturning the finge
before; the wheel was turned back till the intersection 3 c
the fixed wire; the division 10 on the circle being set to
screw was pressed against the wheel as before, and the scre
nine revolutions, till the intersection 2 nearly coincided with
and the screw was released; and I proceeded in this manner
were marked round the whole circumference of the whee
repeated three times round, to make the impression of the
I then ratched the wheel round continually in the same dire
ever disengaging the screw; and, in ratching the wheel abo
round, the teeth were finished.

* The intersections are marked for the sake of illustration; th
invisible, they lying under the brass plate.

It is evident if the circumference of the wheel were even one tooth or ten minutes greater than the screw would require, this error would in the first instance be reduced to one part of a revolution, or two seconds and a half; and these errors or inequalities of the teeth be equally distributed round the wheel at the distance of nine teeth from each other. Now, as the screw in ratching had continually hold of several teeth at the same time, and these constantly changing, the above-mentioned inequalities soon corrected themselves, and the teeth were reduced to a perfect equality. The piece of brass which carries the wire was now taken away, and the cutting-screw was also removed, and a plain one (hereafter described) put in its place; on one end of the screw is a small brass circle, having its edge divided into sixty equal parts, and numbered at every sixth division, as before mentioned.

On the other end of the screw is a ratchet-wheel c, having sixty teeth, covered by the hollowed circle d, fig. 320, which carries two clicks that catch upon the opposite sides of the ratchet when the screw is to be moved forwards. The cylinder S turns on a strong steel arbor F, which passes through and is firmly screwed to the piece Y; this piece, for greater firmness, is attached to the screw-frame G, fig. 319, by the braces v; a spiral groove or thread is cut on the outside of the cylinder S, which serves both for winding the string, and also giving motion to the lever J on its centre, by means of a steel tooth n, that works between the threads of the spiral. To this lever is attached a strong steel pin m, on which a brass socket, r, turns; this socket passes through a slit in the piece p, and may be tightened in any part of the slit by the finger-nut f; this piece serves to regulate the number of revolutions of the screw for each tread of the treadle R.

T, fig. 316, is a brass box containing a spiral spring; a strong gut is fastened and turned three or four times round the circumference of this box; this gut then passes several times round the cylinder S, and from thence passes to the treadle R, fig. 316. Now, when the treadle is pressed down, the string pulls the cylinder S round its axis, and the clicks catching hold of the teeth on the ratchet carry the screw round with it, till, by the tooth n working in the spiral groove, the lever J, fig. 319, is brought near the circle, and the cylinder stopped by the screw-head, a, striking on the top of the lever J; at the same time the spring is wound up by the other end of the gut passing round the box T, fig. 316. Now, when the foot is taken off the treadle, the spring, unbending itself, pulls back the cylinder, the clicks leaving the ratchet and screw at rest till the piece t strikes on the end of the piece p, fig. 316; the number of revolutions of the screw at each tread is limited by the number of revolutions the cylinder is allowed to turn back before the stop strikes on the piece p.

When the endless-screw was moved round its axis with a considerable velocity, it would continue that motion a little after the cylinder, figs. 316 and 319, was stopped; to prevent this, the angular lever v was made, that when the lever J comes near to stop the screw a, it, by a small chamfer, presses down the piece x of the angular lever; this brings the other end, y, of the same lever forwards, and stops the endless-screw by the steel pin, p, striking upon the top of it: the foot of the lever is raised again by a small spring pressing on the brace v

D, two clamps, connected by the piece a, slide one on each arm of the frame L, figs. 316, 317, and 321, and may be fixed at pleasure by the four finger-screws e, which press against the steel springs to avoid spoiling the arms; the piece q is made to turn without shake between two conical pointed screws f, which are prevented from unturning by tightening the finger-nuts N.

The piece m, fig. 321, is made to turn on the piece q, by the conical pointed screws, s, resting in the hollow centres e.

As there is frequent occasion to cut divisions on inclined planes, for that purpose the piece, γ, in which the tracer is fixed, has a conical axis at each end, which turn in half-holes; when the tracer is set to any inclination, it may be fixed there by tightening the steel screws β.

Description of the engine by which the endless-screw of the dividing-engine was cut.

Fig. 324 represents this engine of its full dimensions, seen from one side. Fig. 323, the upper side of the same, as seen from above.

A, represents a triangular bar of steel, to which the triangular holes in the pieces B and C are accurately fitted, and may be fixed on any part of the bar by the screws D.

E is a piece of steel whereon the screw is intended to be cut; which, after being hardened and tempered, has its pivots turned in the form of two frustrums of cones, as represented in the drawings of the dividing-engine, fig. 320. These pivots were exactly fitted to the half-holes F and T, which were kept together by the screws z.

H represents a screw of untempered steel, having a pivot I, which turns in the hole h; at the other end of the screw is a hollow centre, which receives the hardened conical point of the steel pin m. When this point is sufficiently pressed against the screw, to prevent its shaking, the steel pin may be fixed by tightening the screws Y.

N is a cylindric nut movable on the screw H; which, to prevent any shakes, may be tightened by the screws O. This nut is connected with the saddle-piece P, by means of the intermediate universal joint W, through which the arbor of the screw H passes. A front view of this piece, with a section across the screw-arbor, is represented at X. This joint is connected with the nut by means of two steel slips S, which turn on pins between the cheeks T, on the nut N. The other ends of these slips, S, turn in like manner on pins a; one axis of this joint turns in a hole in the cock b, which is fixed to the saddle-piece; and the other turns in a hole d, made for that purpose in the same piece on which the cock b is fixed. By this means, when the screw is turned round, the saddle-piece will slide uniformly along the triangular bar A.

K is a small triangular oar of well-tempered steel, which slides in a groove of the same form on the saddle-piece P. The point of this bar or cutter is formed to the shape of the thread intended to be cut on the endless-screw. When the cutter is set to take proper hold of the intended screw, it may be fixed by tightening the screws e, which press the two pieces of brass upon it.

Having measured the circumference of the dividing-wheel, I found it would require a screw about one thread in a hundred coarser than the guide-screw arbor H, and that on the stub E, on which the screw was to be cut, were proportioned to each other to produce that effect, by giving the wheel L 198 teeth. and the wheel Q 200. These wheels communicated with each other by means of the intermediate wheel R, which also served to give the threads on the two screws the same direction.

The saddle-piece P is confined on the bar A by means of the pieces g, and may be made to slide with a proper degree of tightness by the screws a

LATHES AND TURNING APPARATUS.

THE lathe is a very useful engine for turning wood, ivory, metals, and other materials.

The common lathe is composed of two wooden cheeks or sides, parallel to the horizon, having a groove or opening between; perpendicular to these are two other pieces, called puppets, made to slide between the cheeks, and to be fixed down at any point at pleasure. These have two points, between which the piece to be turned is sustained; the piece is turned round backwards and forwards by means of a string put round it and fastened above to the end of a pliable pole, and underneath to a treadle or board moved with the foot. There is also a rest which bears up the tool, and keeps it steady.

We shall now proceed to give Mr. J. Farey's description of the improved lathes manufactured by *Mr. Henry Maudslay*, of Margaret-street, Cavendish-square.

A, fig. 325, is the great wheel, with four grooves on the rim; it is worked by a crank B, and treadle C, in the common way; the catgut which goes round this wheel passes also round a smaller wheel D, called the mandrel, which has four grooves on its circumference, of different diameters, for giving it different velocities, corresponding with the four grooves on the great wheel A. In order to make the same band suit, when applied to all the different grooves on the mandrel D, the wheel A can be elevated or depressed by a screw, a, and another at the other end of the axle; and the connecting rod, C, can be lengthened or shortened by screwing the hooks at each end of it further out of, or into it. The end M, fig. 326, of the spindle of the mandrel D, is pointed, and works in a hole in the end of a screw, put through the standard E, fig. 325; the other end of the bearing F, fig. 326, is conical, and works in a conical socket in the standard F, fig. 325, so that, by tightening up the screw in E, the conical end, F, may at any time be made to fit its socket; the puppet G has a cylindric hole through its top to receive the polished pointed rod d, which is moved by the screw e, and fixed by the screw f; the whole puppet is fixed on the triangular prismatic bar H, by a clamp, fig. 332, the two ends of which, a, b, are put through holes, b, in the bottom of the puppet under the bar, and the whole is fixed by the screw e pressing against it; by this means the puppet can be taken off the bar without first taking off the standard I, as in the common lathes; and the triangular bar is found to be far preferable to the double rectangular one in common use. The rest j is a similar contrivance: it is in three pieces; see figs. 327, 328, and 329. Fig. 328 is a piece, the opening, a, b, c, in which, is laid upon the bar H, fig. 325; the four legs, d d d d, of fig. 329, are then put up under the bar (into the recesses in fig. 328, which are made to receive them) so that the notches in d d d d may be level with the top of fig. 328; the two beads, e f, in fig. 327, are then slid into the notches in the top of d d d d, fig. 328, to keep the whole together; the groove i is to receive a corresponding piece on e f, fig. 327, to steady it; the whole of fig 327 has a metallic cover, to keep the chips out of the grooves. It is plain that, by tight-

is put upon the bar H, fig. 325, and fixed in the place of the rest j, by the clamp, fig. 232, the distance from the centre a is adjusted by the screw A; this moves the slide, fig. 335, in the grooves i, figs. 331 and 334, with the apparatus upon it; by the screw m, figs. 331 and 335, as before described, the slide, fig. 336, may be moved in a direction perpendicular to the bar H, fig. 325, and its projections o o, acting against the slits p p, figs. 331 and 337, as inclined planes, will raise or lower the plate B, as is required.

The tool, which has been before fixed in the holders b b, can be set at the proper angle by loosening the screw e, as previously described; and, lastly, the tool with the holders and slider a, can be advanced or withdrawn by working the screw e. The nuts of the screws c and b, fig. 331, are not fixed to the sliding plates, but are held by two pins t, fig. 335, which fit the grooves u, fig. 334, in each side of the nut; by these means the sliding plate can at any time be taken out by only unscrewing one of the brass slides from the grooves i, without taking out the screw or nut. In order to make the pieces always fit their slides, the two pieces of brass y y, fig. 331, which are at the sides of the groove, have elliptic holes for their screws v, so as to admit, when the screws are slackened, of being pushed inwards by the screw w which works in a lump of metal cast with the part A A.

The large lathes which Mr. Maudslay uses in his manufactory, instead of being worked by the foot, as represented in fig. 325, are worked by hand; the wheel and fly-wheel which the men turn works by a strap on another wheel fixed in the ceiling directly over it; on the axis of this wheel is a larger one, which turns another small wheel or pulley fixed in the ceiling, directly over the mandrel of the lathe; and this last has on its axis a larger one which works the mandrel H, by a band of catgut. These latter wheels are fixed in a frame of cast-iron, movable on a joint; and this frame has always a strong tendency to rise up, in consequence of the action of a heavy weight, the rope from which, after passing over a pulley, is fastened to the frame; this weight not only operates to keep the mandrel-band tight, when applied to any of the grooves therein, but always makes the strap between the two wheels on the ceiling fit. As it is necessary that the workman should be able to stop his lathe, without the men stopping who are turning the great wheel, there are two pulleys or rollers (on the axis of the wheel over the lathe) for the strap coming from the other wheel on the ceiling; one of these pulleys, called the dead pulley, is fixed to the axis and turns with it, and the other, which slips round it, is called the live pulley; these pulleys are put close to each other, so that by slipping the strap upon the live pulley, it will not turn the axis; but if it is slipped on the other, it will turn with it; this is effected by a horizontal bar, with two upright pins in it, between which the strap passes. This bar is moved in such a direction as will throw the strap into the live pulley, by means of a strong bell-spring; and in a contrary direction it is moved by a cord fastened to it, which passes over a pulley, and hangs down within reach of the workman's hand; to this cord is fastened a weight heavy enough to counteract the bell-spring, and bring the strap upon the dead pulley to turn the lathe; but when the weight is laid upon a little shelf, prepared for the purpose, the spring will act and stop it.

Mr. Maudslay has likewise some additional apparatus for cutting the teeth of wheels, in which the face of the mandrel D, fig. 325, has seventeen concentric circles upon it, each divided into a different number of equal parts, by small holes.

There is a thin stop x, fig. 325, which moves round on a screw fixed in the standard F; this stop is made of thin steel, and is so fixed, that when it is turned up, and its point inserted into any of the divisions of the mandrel, it will have a sufficient spring to keep it there; the wheel to be cut is fastened, by means of a chuck, to the screw *s*, and after it has been turned, and brought to the proper shape, the rest, *j*, is to be taken away, and the slide-tool substituted; a square bar is then put into the two holders *b b*, fig. 331; this bar has two branches for holding the ends of a spindle, near one end of which is a pulley, and at the other are four chisels fixed perpendicularly into the spindle for cutting out the teeth, (instead of the circular saw commonly used:) the pulley is turned (with the intervention of several wheels to augment the velocity) by the same great wheel as the lathe, with 7300 revolutions per minute; the mandrel is then fixed by the stop x, fig. 325, and the cutter advanced towards the wheel, by the screw *c*, fig. 331. When it has cut that tooth, the cutter is withdrawn, and the mandrel turned to another division, and a tooth is cut again as before. At that part of the frame of the cutting-spindle where the bar which is fixed in the holders of the slide-tool connects with the two branches, there is a joint, by which the cutting-spindle can be set in an inclining position for cutting oblique teeth, like those which are to work with an endless-screw. The great velocity with which this spindle turns soon generates by friction and resistance a degree of heat sufficient to expand it very sensibly; but this ingenious mechanist, foreseeing such a circumstance, has judiciously compensated for it in his construction, by making the spindle so short as to play loosely in its sockets at the commencement of the motion; but after a few seconds the expansion is such as to cause the whole to fit together as it ought to do, and the work of cutting to proceed with accuracy and safety.

Mr. *Smart*, of the Ordnance-wharf, Westminster, has made some very useful improvements in the art of turning, and particularly has struck out a simple method of turning cylinders and cones in wood.

His turning machine is illustrated in figs. 339 and 340, where the legs or stiles L, the puppets A B, the cheeks *o o*, the pikes and screws M, N, R, with the handle D, are but slightly varied from the usual construction. Round the mandrel E passes a band F F, which also encompasses a large wheel, not shown in the figure; and when this large wheel is turned round with moderate swiftness, it communicates a rapid velocity to the mandrel F, and the long piece of wood G, which is proposed to be made cylindrical. This piece is previously hewn into an octagonal form. The cutting frame H contains a sharp iron tool, which is to answer the purpose of the common turning gouge, and which is fitted into the frame so as to project a little be-yond its inner part, after the manner of a carpenter's plane-iron for round or ogee work. Then, while the piece G is turning swiftly round by a man working at the great wheel, another man pushes the frame H gently on from L towards M, the lower part of that frame fitting between the cheeks *e*, and sliding along between them. By this process, the piece G is re-duced to a cylinder, moderately smooth; and, in order to render the smooth-ness as complete as need be, a second cutter, and its frame I, adapted to a rather smaller cylinder than the former, is pushed along in like manner from L to M. This operation may be performed with such speed, that a very accurate cylinder of six feet long, and four inches diameter, may be fixed to the lathe and turned in much less than a minute.

Mr. Smart turns a conical end to one of these cylinders with great facility, by means of a cutting-blade fixed in an iron hol-low conical frame K, the smaller end of which admits the pike from the screw S, fig. 340, to which one end of the cy-linder G is attached; as the cylinder turns rapidly round, the cutter K is conducted gently along it by means of the hollow frame, and soon gives the conical shape to the end of the cylinder, as required.

Some important directions for turning screws, ovals, cubes, rose-work, swath-work, &c. may be seen in *Moxon's Mechanic Exercises:* see also, "Tour pour faire sans Arbre toutes Sortes de Vis," *par M. Grandjean,* in "*Recueil des Machines et Inventions approuvées par l'Acad. Roy. des Sciences,*" tom. v.; and Mr. Healy's method of cutting screws in the common turning-lathe.

Previously to entering upon the several branches of our manufactures, where machinery will be found in its most complex state, it may, perhaps, be considered not alto-gether irrelevant, if we take a cursory view of the manner in which we have conducted the reader thus far. In the first place, we have taken up the subject by treating of the Me-chanical Powers, and the attributes of matter, as if he were totally unacquainted with the science; and having given him every necessary information with respect to the fundamental principles, have then proceeded to demonstrate the Moving Powers; thus progressively leading him on to a perfect com-prehension of the invariable laws of mechanics, before we have ventured to introduce to his notice certain simple machines acting, either separately or conjointly, as accessors to our manufactures. These we have now also portrayed, and so amply, that we feel satisfied he will, though totally destitute of the science at the commencement, be able fully to compre-hend and appreciate the several excellencies of the various combinations which will now be unfolded to him.

IRON MANUFACTURE.

WORKS for the manufacture of iron, owing to the sums necessary to be expended in their erection, have within these few years, been confined to a very limited but the spirit of enterprise which has of late, espeially since the French revolution, manifested nearly the whole of our manufactures, conjoined to immense capitals acquired by many individuals, and the of employing them to a better advantage, have given manufacture of this highly valuable metal a more character.

The ores from which the metal is extracted country, found, in general, to consist of iron oxygen and various proportions of earthy matter.

The earthy matter in a state of combination with be divided into two classes ; the one called argillaceous abounding in excess of alumine, or clay ; the other from abounding in lime. The former is by far common ; indeed it is owing to the ore being so met with in an argillaceous state, that iron-making very inattentive to its quality, and that we them use limestone as a flux when the ore already with calcareous ingredients.

Both lime and clay, when separately subjected to heat of the blast-furnace, are infusible ; but on being together in certain proportions, are too fusible even common purposes of brick-making. An alloy of two is also fusible at a temperature much less than th metical mean of the metals themselves.

Such being the case, it is much to be regretted, th masters, in general, are so very ignorant of, and in to, the fusibility of the different combinations of this which causes them so frequently to be at a loss what to the furnace in order to produce the most fusible An analysis of the ore, by which they might learn the proportions of its earthy constituents, and the qu limestone or clay to be added as a flux, would, in prove much to their advantage.

In the usual process of smelting, the coke is always quantity, and the proportions of ore and limestone varied according to the quantity of iron to be made, working order of the furnace. In proportion to the quantity of lime and ore that is added to the standard quantity of the coke, the furnace is said to carry a greater or less burthen. Some furnaces carry so little burthen as not to produce more than 13 or 14 tons per week; whilst others, with the same sized furnace, will yield 60 and even 70 tons in an equal time. The burthen of the last-mentioned furnaces is very great, the ore to the coke being, in some cases, as 13 to 7. The quality of the iron is uniformly inferior.

The burthen of the furnace will vary according as the iron to be made is required to possess more or less carbon; for instance, in making No. 1, or the best iron, which contains the greatest portion of carbon, the burthen must be considerably less than that required to make less carburetted iron, or what is called white-iron, or forge-pig.

To afford a general idea of the proportions of the materials, we shall state the quantities, given by Mr. Mushett, as used at a blast-furnace, making good melting iron, which is of an intermediate quality between No. 1 and the forge-pig. The ore is argillaceous, containing on the average about 27 per cent. of iron; the coal rather soft, but not very bituminous, and contains a large proportion of carbonaceous matter; and the limestone, which is that abounding in shells, from Critch, in Derbyshire, is very good. It works with a bright tuyere, and receives from the blast about 2,500 cubic feet in a minute, through a circular aperture of 2¼ inches in diameter.

It is usual at this, and most other furnaces, to divide the men into two classes, one class to relieve the other every 12 hours; these periods are called shifts. The average charges of coke per shift are 50 (each 2¼ cwt.) or about the same. The quantity of calcined ore for the manufacture of good melting iron is upon a par with the coke; and for forge-pig, or the least carburetted variety, six of coke to seven of ore. The limestone unburnt, under the same circumstances, is to coke as 4 to 11; and for melting metal, retains a similar ratio. With the above charge per day, that is, for twelve hours, this furnace makes on the average about 40 tons melting iron per week.

After the ore is dug, it is drawn from the pit by the power of steam-engines; it is then, in order to extract the arsenic and sulphur, subjected to a process called *roasting*. This process consists in laying the ironstone in strata with refuse pit-coal, called in Staffordshire *slack*, and setting fire to it on the windward side, burning it in large heaps in the open air. When the ore has been roasted, it is taken to the smelting or blast furnace, the lower part of which is filled with either charcoal or coke; the coke is always a fixed quantity, and the proportion of limestone added to the ore is according to the quantity of heterogeneous matter with which the metal is combined.

A section of the blast-furnace is represented in fig 346. A, at the top of the furnace, is an opening for the introduction of the materials; B, the body of the furnace; C, the place where the blast is introduced; and D, a cavity to receive the metal when released from the earthy matter.

The materials in the furnace are, previously to the introduction of the blast, heated simply by the draught of the atmosphere; the coke and limestone to a bright red or white heat, and the iron-ore to a melting heat.

When the blast is introduced, the metal immediately above it is brought into a state of fusion, and penetrates through the fuel into the cavity D. The ore and fuel that were above it sink down to fill up the space left by the ore melted and the fuel consumed. This next comes under the operation of the blast, and is similarly reduced.

The men who attend the furnace keep adding fuel, ore, and limestone, through the opening A, at the top, and the operation of smelting goes on, until the melted iron, in the cavity D, rises nearly to a level with the tuyere-irons, or blast-pipes.

The melted iron is then tapped, by driving a round-pointed bar into a sort of loam, with which the hole is stopped, and run into moulds made in sand; in this state it is called pig or cast-iron.

When the slag, in smelting, has a greenish-grey appearance, it is a certain sign that the furnace is in excellent order; and when the colour changes to black, it denotes that something is going wrong.

In making the best iron, called No. 1, which possesses the greatest quantity of carbon, it frequently happens that a portion of the iron will unite with a great excess of carbon; and as this carburet is less fusible than the iron, it will, previously to the iron entering into the pig-moulds, be seen floating at the top in the form of scales. The appearance of this substance, called by the workmen *kish*, is a sign that the furnace is working the best sort of iron; indeed, it is so common an attendant on the production of the most highly carbonized iron, that the workmen have applied the term *kishy* to that peculiar sort of iron.

The limestone and the earths being much lighter than the metal, float upon its surface, and gradually rising as the metal

.............; is ultimately discharged over a *dam-stone*, which, though at right angles, we have represented by the dotted lines enclosing the letter I. T is called the *tymp-stone*, and forms a bridge over the cavity in which the liquid cinder rises; t is the *tymp-plate*, to give the stone greater ...., as e, which is called the *dam-plate*, does the dam-stone. The cinder, if not taken to mend the roads, is thrown away as useless.

.........as two blasts are introduced, as may be seen in fig. 347. That ...., and that furnace, which can reduce the greatest quantity of fuel in a certain time, will always produce the greatest quantity of iron.

The blast conveyed into the furnace is from 1,000 to 4,000 feet per minute; and it is worthy of remark, that the quantity of metal fused does not agree with the ratio of the blast; for ...., a blast of 1,500 feet per minute will manufacture ..... of smelting iron per week; a blast of 3,000 only 30 ....., and a blast of 6,000, which is double the last, and four ..... the amount of the first blast, only 36½ tons per week; and again, a blast of 2½ lbs. per square inch will manufacture ..... 25 tons per week; while two pipes of the same ..... as the last, with a blast of 3 lbs. per square inch, will never exceed 30 tons.

In the summer months, owing to the increased temperature of the atmosphere, the furnaces yield little better than one .... the quantity that they do in winter, and the iron is of .... inferior quality. In some manufactories, by adding a ..... quantity of fuel, the quality of the iron is preserved; but in others no addition of fuel will compensate for the ..... either in quality or quantity.

In presenting the section of a blast-furnace, represented in ....., we do not pretend to recommend it as the best form of construction. Different iron-masters have variously con-structed their furnaces, and as each of them can boast of .... in some degree successful, it were needless to give a ..... of any particular one; we shall therefore briefly .... that of whatever materials the buildings be made, ... should be taken that they contain no more moisture than ..... absolutely necessary for their proper construction.

In the erection of the wall, a space of about six inches ..... be left from the bottom to the top; in this aperture .... fragments of sand-stone, not exceeding the size of an ..... may be introduced, so that when the expansion, pro-ceeding from the fire-building of the interior, causes the ..... immediately in contact to push outwards, the masses of sand-stone are instantly reduced in size, and filling the

interstices occasioned by their former angular shap
much less space, and present to the flame or fire,
be inclined to penetrate so far, a solid vertical s
sand, after having secured the expansion of the f
the extent of some inches. The effects of the pre
thus diverted from the shell of the building, and l
pulverization of the sand-stone.

The advantages resulting from this plan may
doubled, by using a double lining of fire-bricks, betv
of which and the common building a similar vacancy
left, but filled with sharp sand, containing no more
than serves to compact it in a firm body; as this
becomes gradually expelled in the slow heating or
of the furnace, the sand occupies less bulk, or, whi
same in effect, is then susceptible of a greater degrr
pression when the gradual expansion of the furna
on. It is evident that the force is here also divert
the sand, in place of acting immediately with a te
enlarge the circumference of the building.

Over and above these precautions, the annealing
of the furnace in a progressive and regular manner
be carefully attended to, and continued for two
months at least.

Many methods have been adopted to obtain a re
uniform blast. The first that we shall notice, and
in pretty general use, is, by discharging the air
blowing-cylinder into an intermediate cylinder of l
meter, called the regulator; in this vessel is a loo
which is forced up by the air from the blowing-cyli
being weighted, it descends during the returning st
continues to press the air into the furnace, by whi
a more steady and uniform blast is kept up than
effected by the first cylinder alone.

As this method of regulating the blast has been
be far from perfect, other means have been resorte
a view of obtaining the desired end. The one c
water-regulator consists of a large cistern, in which
of less area and capacity is inverted. Through th
of the smaller cylinder, which is, from its being
uppermost, a pipe communicates with the blowing-
This inner cistern is filled with water, as is also t
between the inner and outer cistern to the same level
supposing the air to be forced from the blowing
through the above-mentioned pipe into the inner cis
water, being displaced by the air, will descend in

_____, and rise up between the two vessels till the column
of water on the outside be equal to the required force
of the blast; this column would be about 4 lbs. upon a square
inch, and about nine feet. Another pipe proceeds from the
same cavity in the inner vessel to the furnace, and communicates nearly a uniform blast, varying only with the outer
column of water, which will be less as the outer surface of the
_____ is greater.

This contrivance, though for some time considered an
important discovery, has, in many instances, been abandoned,
owing to its carrying water, both in a state of spray, produced
by the agitation, and in a state of vapour, into the furnace;
by which both the quality and quantity of the iron was
materially affected.

Another mode has been attempted to equalize the blast,
called the air-vault. The first experiment of this nature was
tried at the Clyde iron-works, by excavating a large cavity
in a rock, into which the air was forced by the blowing
machine; but the trial was unattended by success, partly from
the vault not being air-tight, and partly from the moisture
which exuded from the rock mixing with the air.

A more successful experiment was made at the Carron
iron-works. An air-vault of wrought-iron plate has been
_____ in one of the furnaces at Bradley, in Staffordshire,
and it appears to answer very well. Its form is a cylinder
about 10 or 12 feet diameter, and 50 or 60 feet long.

_____ an average deduced from a series of experiments made by
_____[*] it appears, that when the outer air was from 63° to 66°,
the air immediately after its escape from the blowing cylinder, into a
_____, was increased from 63° to 90°, and from 66° to 96°. In
_____ of thirty experiments the air in the act of condensing was raised
_____. This would have the effect of increasing its volume not less than 1/12
of the whole, and the increased pressure of the blast by this cause alone,
would be nearly half a pound upon an inch. Or, in other words, if the
_____ introduced into the furnace at 60°, the same quantity would be
_____ with half a pound less pressure upon an inch than if it were
_____ any means of cooling the air after its condensation, in all
_____ of the year, must be attended with beneficial consequences. If the
_____ were made of wrought-iron, and its surface constantly kept wet,
the _____ from so great a surface, if freely exposed on all sides to the
_____ cool the air very considerably. Indeed, without the aid of the
_____ the effect would be such as to recommend its adoption. It was
_____ that in the summer season there would be some advantage in
_____ the air under ground for a considerable distance before it entered
the blowing machine; but the resistance arising from the friction on the

* *Edinburgh Encyclopædia*, Dr. Brewster, 4to.

sides of the channels through which it must pass has been found obstacle that overrules it.

The pig-iron which has the smallest portion of c. the best adapted for conversion into malleable iron : a proof that the pig-iron has only to lose its carbon to malleable, we shall state the fact, that we have in this (at this present time, many manufactories upon a larg for the express purpose of converting articles made iron, such as nails, cutlery, &c. into iron perfectly m. without altering in the slightest degree the figure g them in the casting. We have even seen nails made way welded together, and when cold bent at right a a vice.

The method of releasing the pig-iron of its carbo converting it into what is called wrought, or malleal. is, by placing it in an open furnace, termed a refinc by some a run-out furnace, heated by cokes, and su to the operation of a very powerful blast. The pig laid upon the cokes, and is soon melted, leaving muc impurity behind. This is termed refining it. The when melted is run into plates, about four inches thi as soon as it becomes set, is thrown into water, which it more frangible, and easier to be broken.

The refining furnace is represented in figs. 348 and 349. A is or trough, made of cast metal, having a bottom of fire-stone This recess is surrounded on three sides by a cavity, through whi is constantly passing from the cistern C; p p are two pipes c with the blowing machine, and entering into conical opening refining furnace. These pipes are kept cool by water from the which runs off at the pipe b e c. B is a shallow recess, about fu deep, to receive the melted mass.

When the cake of metal is broken into lumps of venient size, it is taken to the puddling furnace, wh heated with coals, without the aid of an artificial bl. soon as the metal becomes heated, and begins t or has a frosty appearance, the furnaceman throw small quantity of water to keep it at a proper tempe and keeps stirring and moving it about, so that tho makes its escape. The water that is thrown in to p the temperature also assists in some degree the decar tion. The quality of iron depends much upon the at that is paid to it during this process.

When the iron is deprived of the carbon, or fusil perty it before possessed, the furnaceman rolls it balls of one half or three quarters of a cwt. each. It brought out of the furnace and placed under a tilt-hammer, or passed through the rolls, or rollers, which consolidates it, and forces out more of the impure parts. A considerable loss in weight is sustained in this process, not only from the iron losing its impurities, but also from the surface of the bloom or bar oxydizing and falling off in scales whilst being worked. The loss which is thus sustained in weight is generally estimated at one-sixth or one-seventh of the whole.

A section and elevation of the puddling furnace is represented in fig. 350. A is the door for the admission of metal, having a small square hole h, for the introduction of the rake and other tools used by the furnaceman. B is the chimney; C the ash-pit; and D the grate. At E is a circular cavity, where the prepared metal is laid, and the flame passes over it up the chimney B. The heat of the furnace is so intense that without having the door for a guard, and the small hole h for the introduction of the imple-ments, the furnaceman could not approach it; nor indeed can he as it is without suffering great inconvenience. The hole is also of use for him to look into the furnace to observe how the work is going on. At first the light is too intense to be borne, but by practice the eye at length becomes accustomed to it, and is able perfectly to distinguish the different masses as they lay in the furnace.

The iron having undergone this process is taken to the shears and cut into lengths of about one or two feet, and in order to impart closeness and solidity is piled into pieces of seven or eight together, and heated in another furnace, very similar to the one just described. There is no occasion this time to remove them about, for the iron having lost its carbon is infusible. When it is of a sufficient heat, which the furnaceman from practice can easily tell by his eye, it is again brought to either the hammer or the rollers, and is worked into a bar. This is called No. 2 iron. Again, further to improve the quality, it is cut up, piled, and worked over again; and is then called No. 3, or best iron. The more the iron is worked the purer it becomes, and the grain becomes more closely united; but of course it becomes more expensive.

Two kinds of hammer, moved by machinery, are used in iron-works. The one called the *forge-hammer* is represented in fig. 353. The first mover gives motion to the shaft A A, by means of a cog-wheel acting upon the pinion B. The shaft is regulated by a fly-wheel C, and has at the further end a number of cogs, which by passing under the shaft, or helve, D, lift the hammer E. F is a strong horizontal beam, inserted in the post G, and loaded with heavy pieces of metal, at H, to prevent it receiving motion from the hammer. Another large beam of wood, made of either oak, or ash, but most frequently the latter, is inserted in the posts I K. The hammer in its ascent strikes against this beam, called the rabbit, which by its elasticity reacts upon the hammer, and causes it to descend with greater velocity than would be produced by gravity alone.

The construction of a *tilt-hammer* differs from that of the

forge, by being poised on a centre of motion, about th
or two-thirds of the length of the helve from the h
from receiving its motion from cogs acting upon th
the helve. In some few cases the ash spring is pl
the head of the hammer similarly to that above de
but, in general, the tail of the helve is made to strik
a fixed floor, and the hammer from the force it has
continuing to rise after the tail strikes the floor,
bends, and by its elasticity causes the hammer to
with greater force upon the anvil.

The *tilt-hammer* is represented in fig. 364. It is taken from
made at the Carron iron-works in Scotland, after designs of the
Mr. Smeaton. It is adapted for forging iron into bars. The
is extracted from Dr. Rees's *Cyclopædia.*

Having described the manner in which the tilt-hammer *is*
with the first-mover, (drawings of which may be seen in the
author proceeds to explain the figure above referred to; *e* the ir
the hammer, *f* its centre of motion, and *d* the tail or extreme
which the cogs of the wheel act, and which is plated with iron on
side, to prevent it from wearing.

P is the anvil-block, which must be placed on a very firm fou
resist the incessant shocks to which it is subjected: the centre
of the hammer, is supported in a cast-iron frame *g h,* called
When the cogs of the wheel strike the tail of the hammer sudd
and raise the head, the lower side of the tail of the hammer stri
support *n,* which acts to stop the ascent of the head of the ham
it arrives at the desired height; but as the hammer is thrown
considerable velocity as well as force, the effort of the head to
motion, after the tail strikes the stop *n,* acts to bend the helve
hammer, and the elasticity of the helve recoils the hammer dow
anvil with a redoubled force and velocity to that which it wo
from the action of gravity alone.

To obtain this action of recoil, the hirst *g h* must be held do
as possible; and for this purpose, four strong iron bolts are ca
from the four angles of the bottom plate *h,* and made fast to the
of stone R R, upon which the whole rests; upon this basis are p
layers of timber, *i k l m,* which are laid one upon another, and t
of each layer are laid cross-ways over the others. Each layer
several pieces laid side by side, and they are slightly trenailed
form a platform. Each platform is rather less than that upon wh
so as to form a pillar of solid timber; on the top of which the h
g h, is placed, and firmly held down by the four bolts, which desce
all the platforms, and have secure fastenings in the solid mason

The stop *n* is supported by a similar pillar, but smaller, and c
three layers: the upper piece *n,* which is seen cross-ways, is
feet long, and the under side is hollowed, so that the piece bear
the two ends, leaving a vacancy beneath it, which occasions it
spring every time the tail *d* of the hammer strikes upon it, and t
recoiling action very much.

The axis on which the hammer moves is formed by a ring o
through which the helve of the hammer is put, and held fast
round it. The ring has a projecting trunnion on each side, on

..... conical point, which is received in a socket firmly fixed in the beam .... by screws and wedges, one of which is seen at r. These two ...... are thus capable of adjustment, so as to make the hammer face fall ...... upon the anvil.

... the Carron iron-works, these hammers are worked from the same ..... In such case it is necessary to have the three wheels that com...... motion to their respective hammers of different sizes and numbers .... to produce that velocity in each hammer which is best adapted for .... work it is to perform; thus the wheel for the hammer, which is repre...... in fig. 322, has eight cogs, and therefore produces eight blows of the .... for each revolution of the fly-wheel; the wheel for the middle ...... has 12 cogs; and the wheel for the smaller hammer 16; the ...... will therefore make two strokes for every one of the great hammers. .. fixing the three wheels upon the great shaft, care is taken that they ...... produce the blows of the different hammers in regular succession, ... equalize as much as possible the force which the water-wheel must ..... The wheels are fixed on the shaft by means of a wedging of hard ......, driven in all round; the wood being capable of yielding a little to ...... shocks occasioned by the cogs meeting the tails of the hammers, ... the concussions less violent.

The following are the principal dimensions:

The head of the great hammer weighs 3½ cwt. and it is intended to make 155 blows per minute; it is lifted 17 inches from the anvil at every blow.

The middle hammer is 2 cwt. and makes 225 blows per minute; it is lifted 14 inches each time.

The small hammer weighs 1½ cwt. and makes 300 blows per minute; it is lifted only 12 inches.

.. produce these velocities, the great axis upon which the cog-wheels ...... must make 18¾ turns per minute; and the pinion upon this axis ..... in proportion with the cog-wheel upon the shaft of the water-wheel ...... the water-wheel must make 6¼ revolutions per minute; the ...... being 18 feet diameter, its circumference will be 18×3.1416 ...... or 56¼ feet; this multiplied by 6·25 is about 353 feet motion .... or divided by 60 = 5·9 feet motion per second for the circum...... of the water-wheel.

... tilt-mills employed in the manufacture of steel, do .... have the great hammer, but the largest they use is about .... size of the middle one, and is adapted for welding faggots .. steel, to make sheer steel: the other two hammers are ..... the size of the smallest just described, and are made .. work much quicker, viz. from 350 to 400 blows per ...... This is very easily accomplished by making the ...... upon the fly-wheel shaft in proportion to the cog...... that acts upon it, and is fixed to the water-wheel, as .. in to 4.

... highly valuable metal, having undergone these pro...... is now sold, and is used by smiths for an innumerable ...... of purposes. Indeed, when we reflect upon the many ...... of men, women, and children, who are daily em...... in the manufacture and working of this metal; when we consider the immense number of families of miners, melters, refiners, smiths, and other handicraftsmen, who, in all the civilized parts of the world, look up to this particular branch of manufacture for their maintenance and support; when we consider, that the once obscure and inconsiderable village of Merthyr Tydvil, though wild, barren, and sterile, and too poor to produce even the common necessaries of life, has been peopled in the teeth of every obstacle, and, within the space of seventy years, has, through the manufacture of this metal, become by far the largest and most populous town in Wales; we cannot but rejoice that this metal is one of the staple manufactures of Great Britain.

When this metal has become too much worn to answer longer the purpose for which the smith designed it, it is sold to the 'dealers in marine stores,' who assort it into three parcels; one called *coach-tyre*, consisting of the old tyre of coach and other wheels; another *bushel iron*, being remnants of old hoops, and different pieces of iron of similar nature; and another *scrap* or *nut-iron*, consisting of old nails, screws, nuts, and pieces of that description.

These are sold to the manufacturer to be remanufactured. The process of *remanufacturing* is as follows:

Two pieces of iron, each forming three sides of a square, are fixed to a wooden bench, about 10 or 12 inches apart. In the space between these two pieces are placed two rods of iron, about three-eighths of an inch square, one rod being placed close to each of the pieces. On these rods are laid pieces of old hoop, previously straightened, and cut to the proper lengths of 12 or 14 inches, according to the intended length of the faggot. The ends of the hoop rest upon the bottom of each of the pieces of iron first described, and similar pieces of hoop are ranged upon each side, while the interior is filled with bushel or scrap iron. The top is then covered with hoop, and the whole pressed tightly down, and bound, by bringing the ends of the three-eighths rod together, and screwing them round. This is termed a *faggot*, being about 12 or 14 inches long, and six inches square.

The faggot is then carried to a furnace not much unlike the puddling furnace, and when sufficiently heated is brought out, and passed through the rollers, and made into what are called *blooms*. These blooms are generally about two feet long, by three or four inches wide, and two thick.

The blooms are again exposed to the heat in the furnace, and when at a proper temperature are taken out and passed through the rollers, either those represented in fig. 351, or those in fig. 352, accordingly as they are to be made into hoops, or bars. The hoop-rollers are represented in fig. 351; the bar-rollers in fig. 352.

Tables of the average weight of bars, squares, and bolts, 10 feet in length.

BARS.

Inches.	C. qr. lb.	Inches.	C. qr. lb.	Inches.	C. qr. lb.
6 × ¾	1 1 15	3¾ × ¾	— 3 12	2¼ × ⅝	— 1 23
⅝	1 0 13	⅝	— 2 24	½	— 1 10
½	— 3 19	½	— 2 8	⅜	— 1 1
5½ × ¾	1 1 1	⅜	— 1 20	2⅛ × ¾	— 2 2
⅝	1 0 6	3½ × ¾	— 3 5	⅝	— 1 18
½	— 3 10	⅝	— 2 18	9/16	— 1 14
5 × ¾	1 0 13	½	— 2 4	½	— 1 9
⅝	— 3 23	⅜	— 1 16	⅜	— 1 0
½	— 3 2	3¼ × ¾	— 2 27	2 × ¾	— 1 24
4¾ × ¾	1 0 10	⅝	— 2 14	⅝	— 1 15
⅝	— 3 19	½	— 1 27	9/16	— 1 11
½	— 2 25	⅜	— 1 14	½	— 1 6
⅜	— 2 5	3 × ¾	— 2 22	⅜	— 0 26
4½ × ¾	1 0 4	⅝	— 2 8	1⅞ × ¾	— 1 20
⅝	— 3 13	½	— 1 23	⅝	— 1 12
½	— 2 21	⅜	— 1 10	9/16	— 1 9
⅜	— 2 11	2¾ × ¾	— 2 14	½	— 1 5
4¼ × ¾	— 3 25	⅝	— 2 2	⅜	— 0 24
⅝	— 3 7	½	— 1 20	1¾ × ¾	— 1 17
½	— 2 17	⅜	— 1 7	⅝	— 1 10
⅜	— 2 0	2½ × ¾	— 2 8	9/16	— 1 5
4 × ¾	— 3 19	⅝	— 1 25	½	— 1 2
⅝	— 3 1	½	— 1 15	⅜	— 0 23
½	— 2 12	⅜	— 1 4	1⅝ × ¾	— 1 11
⅜	— 1 24	2¼ × ¾	— 2 5	⅝	— 1 3
				9/16	— 1 0

SQUARES.		BOLTS.	
Inches.	C. qr. lb.	Inches.	C. qr. lb.
3	2 3 0	3	2 0 18
2⅞	2 2 3	2⅞	1 3 22
2¾	2 1 8	2¾	1 3 6
2⅝	2 0 11	2⅝	1 2 17
2½	1 3 18	2½	1 1 23
2⅜	1 2 24	2⅜	1 1 11
2¼	1 2 5	2¼	1 0 24
2⅛	1 1 14	2⅛	1 0 9
2	1 0 25	2	— 3 24
1⅞	1 0 8	1⅞	— 3 9
1¾	— 3 21	1¾	— 2 26
1⅝	— 3 2	1⅝	— 2 16
1½	— 2 21	1½	— 2 3
1⅜	— 2 11	1⅜	— 1 24
1¼	— 1 25	1¼	— 1 14
1⅛	— 1 15	1⅛	— 1 5
1	— 1 6	1	— 0 27
⅞	— 0 26	⅞	— 0 20
¾	— 0 19	¾	— 0 15
⅝	— 0 13	⅝	— 0 10
½	— 0 8	½	— 0 17

STEEL MANUFACTURE.

WHEN iron has lost all its carbon, and has become malleable, it can be reimpregnated with carbon, to a certain extent, without materially injuring its malleable properties.

The compound of iron and carbon thus produced is called *steel.*

To reimpregnate the iron with carbon, it must be put into a close vessel, called a *cementing pot,* and stratified with powdered charcoal.

The pots are made with a peculiar kind of stone, termed *fire-stone,* which is found abundantly in the neighbourhood of Sheffield. It possesses the properties of not being liable to crack by the heat, or of entering into fusion. These pots in the interior dimensions are from 10 to 15 feet long, and from 24 to 30 inches square. Each bar of iron is completely covered with powdered charcoal, and the last stratum of it is usually made much thicker than the rest, and kept close with a mixture of sand and clay, to prevent the charcoal from entering into combustion with the outer air. Two of these pots only are contained in a furnace at a time, and fire is gradually employed till the heat is little short of what would be required to fuse the steel.

A vertical section, and horizontal plan, of the converting furnace is shown in figs. 355 and 356. In both figures the same letters denote the same parts.

C C is the external cone, built in a substantial manner of stone or brick-work. Its height from the ground to its vertex, in order to procure a good draught of air, should not be less than 40 or 50 feet; and to procure a still stronger heat a cylindric chimney of several feet in length is most generally fixed on the top of the cone. The lower part of the cone, which may be made of any dimensions, is built either square or octangular. The sides are carried up until they meet the cone, giving the furnace the appearance of a cone cut to a square or octangular prism at its base, and exhibiting the parabola where every side intersects the cone.

Inside the conical building is a smaller furnace, called the vault, built of fire-brick or stone, which will withstand the action of the most intense heat. D D, in the section, is the dome of the vault, and E E are its upright sides, the space between which, and the wall of the external build-ing, is filled with sand and rubbish. A B represent the two pots that contain the iron to be converted into steel. The space between them is about one foot in width, and the fire-grate is directly beneath it. The pots are supported by a number of detached courses of fire-brick, as shown at e e, in fig. 355, which leave spaces between them, called flues, to conduct the flame under the pots; in the same manner, the sides of the pots are supported from the vertical walls of the vault, and from each other, by a few detached stones, represented by f, placed so that they may intercept as little as possible of the heat from the contents of the pots. The adjacent sides of the pot are supported from one another by small piers of stone-work, which are also perforated to give passage to the flame. The bottoms of the pots are built of a double course of brick-work, about six inches thick; the sides nearest together are built of a single course of stone, about five inches in thickness; and the other parts of the pot are single courses about three inches, the sides not requiring so much strength, because they have less heat and pressure to resist.

The vault has ten flues, or short chimneys, F F, rising from it, two on each side, to carry off the smoke into the great cone, shown in fig. 356, communicating with each side, and two at each end. In the front of the furnace an aperture is made through the external building, and another corresponding in the wall of the vault; these openings form the door, at which a man enters the vault to put in or take out the iron; but when the furnace is lighted, these doors are closed by fire-bricks luted with fire-clay. Each pot has also small openings in its end, through which the ends of two or three of the bars are left projecting in such a manner, that by only removing one loose brick from the external building, the bars can be drawn out without disturbing the process, to examine the progress of the conver-sion from time to time; these are called the tap-holes; they should be placed in the centre of the pots, that a fair and equable judgment may be formed from their result of the rest of its contents.

a b, in the elevation, is the fire-grate, formed of bars laid over the ash-pit I, which must have a free communication with the open air, that it may convey a current of fresh air to supply the combustion. The ash-pit

of days and nights is allowed for the gradual cooling of the furnace.

The steel when taken from the converting furnace is found on its surface to be covered with blisters; and on being broken is found to be full of cavities within, for this reason it is called *blistered steel.*

To make it sound and tenacious, it is put into a furnace, and moderately heated, and is then exposed to the action of the tilt-hammer, which we have already described. This is called *sheer steel.*

The steel is made of different degrees of hardness, by giving it more or less carbon, according to the different degrees and duration of the heat applied.

The steel used in the manufacture of coach-springs contain the smallest portion of carbon; a somewhat greater quantity is used in the different branches of cutlery, and in the make of agricultural implements; and the greatest dose of all is required for files, which cannot be too hard, provided the steel be sufficiently malleable to be worked.

Cast steel, which is entirely free from the defects of blistered steel, and is, in some degree, preferable to sheer-steel, is made, by placing small portions of the bars of blistered steel into a crucible, capable of containing about 30 pounds weight.

These crucibles are made of Stourbridge clay, mixed with a small portion of powdered charcoal, which makes them much less liable to crack in the heating or cooling. They are furnished with covers, which are more fusible than the body of the vessel, and, on that account, soon enter into a state of partial vitrification; by which means they become closely luted at the time the steel is at a temperature sufficiently high to be destroyed by the oxygen of the atmosphere.

The fuel employed for melting steel should consist of the hardest cokes, which will give a great heat for a longer continuance than the soft cokes.

When the metal is fused it is taken from the furnace, and poured into iron-moulds, which form it into ingots of an octagonal shape, about 30 inches long.

These ingots, like the bars of blistered and sheer steel, are again heated, and drawn into bars by the operation of the tilt-mill. By means of this machinery the ingots of cast-steel can be drawn into bars one-third of an inch square; and by the hands it can be drawn into rods of a much smaller size.

The manufacture of steel has been greatly improved within

a short period, and it can now be fused with so small
tion of carbon, as will admit of its being welded cith
iron or another piece of steel.

The most singular property belonging to steel is the
hardening by being heated red-hot, and suddenly
and the hotter the steel be made, and the colder th
into which it is plunged, the harder will be the steel.
is generally employed for this purpose; and spring v
considered to be the best. File-makers state, that t
which is inevitable in their hardening water, makes th
harder, and they sometimes put sulphuric acid into it
same purpose.

In hardening steel in thin plates, such as saws, parti
when of cast-steel, quenching in water would cause t
crack, and make them so hard as not to be useful.
have, in consequence, recourse to some substance wl
not so good a conductor of heat. Oil, with tallow and
wax, and resin dissolved in it, is generally employed fo
articles. If the steel be heated red-hot, it mostly ret
its original state. This, however, is sometimes not tl
with thin plates of cast-steel. In giving various de
heat from the hard state, it becomes more soft and less

In the year 1789, Mr. David Hartley took out a pat
a method of tempering steel by the aid of a pyrome
thermometer, applied near to the surface of the articl
at the same time recommended the use of heated
which (he says) many dozens of razors or other tools
be tempered at once with the utmost facility, and the v
degrees of heat necessary for different purposes might sp
be determined by experiment. (See Nicholson's Jo
vol. i, quarto.) An improvement of this principle ha
since suggested by Mr. Parkes, by providing a bath of
of some kind of fusible metal for the tempering of
species of edged tool, which contrivance would, in his op
give to this operation a greater degree of certainty, tha
ever been experienced by those who have conducted
manufactories.

WIRE MANUFACTURE.

WIRE is made of various ductile metals; but as the
facture of the whole is very similar, we shall confini
selves principally to a description of the manufacture o
wire, which is by far the most extensive article of comn

The process of wire-drawing consists in drawing a piece of metal through a hole in a steel plate, which forms it into a regular and even thread, of great length, according to the quantity of metal supplied.

The first part of the process in the manufacture of iron wire is, to subject the iron to the action of a tilt-hammer till it be reduced to a size that will admit of its being drawn through the plate. The tilt-hammer used is similar to that which we have described in the article " Iron Works." It weighs about 100 pounds, and makes 130 strokes per minute. A smaller tilt-hammer, weighing about 50 pounds, and making 20 strokes per minute, is also used for the wire-work.

To prepare the iron for the draw-plate, the workman heats six or eight inches of the end of a large bar, and works it under the small tilt-hammer until it is drawn out into a small and regular round rod, of about six feet in length: Before it has time to cool another workman straightens it, and cuts off with a hammer upon an anvil the rod thus formed, and puts the remainder of the bar into the forge to be again heated.

In manufacturing common wire, the bars may be advantageously run through a pair of rollers, instead of exposing them to the action of the tilt-mill; but as the iron in rolling does not acquire so much tenacity as in the hammering, this process should not be attempted in the manufacture of the best wire.

The rod being thus prepared by one of these methods, is next drawn through a hole in the draw-plate, either by a strong machine with a chain, or else by a lever-machine.

The machines used in the process of wire-drawing are, first;

The common draw-bench, which consists of a strong plank of wood fixed on legs, like a stool or bench. It is represented in fig. 357. A is an axis fixed in a horizontal position, so that it can be easily turned round by means of the four levers B B, fixed like radii on the end of the axis. C is a strong strap or chain, capable of being wound about the axis or roller, and connected by means of a link with the pincers D. E is a draw-plate, perforated with holes of different sizes, lodged against two strong iron pins a a, which are fixed in the bench, and left standing up perpendicularly, so that the plate can rest against them. The wire is passed through the draw-plate E, and is seized by the pincers D, which, by turning the arms or levers B B, winds about the roller, and draws the wire through the plate.

Fig. 358 represents another kind of draw-bench, where a rack and pinion are used, instead of a roller and strap or chain, as above-mentioned. If this machine be turned by a winch the motion is more uniform, which is of importance for some purposes. For instance, if a piece of metal be drawn rapidly through the draw-plate, it will in passing through be greatly compressed, and on emerging will expand a little; but if it be drawn through the plate slowly, it will lose its expansive property. Now in the common draw-bench, or one we first described, the motion communicated by means of the arms B B is very irregular, and the wire is consequently sometimes drawn through the plate with a fast, and sometimes with a slow motion, which causes it to be of different degrees of quality; but in using the rack and pinion, by means of a winch, the motion is regular, and the quality uniform.

In France the roller or windlass is not employed, but the pincers are attached to a lever, which alternately draws them backwards and forwards by the power of the water-wheel.

The pincers are so constructed, that they open and release themselves from the wire when they move towards the draw-plate; but when drawn from the draw-plate close and bite the wire with a force that will draw it through the plate.

A machine of this kind is represented in fig. 359. A B is a wooden lever, which moves round an iron bolt or pin p, as a centre of motion; C is an iron link, connected with the upright part of the lever A B, and having its lower end formed like a ring to seize the ends of the pincers. The pincers are supported upon an inclined plate of iron i, which has a groove to receive the head of the pincers, to direct them in their motion to and from the draw-plate.

The end B of the lever is depressed by cogs, affixed to the axis of the water-wheel, which draws the wire through the plate; but when the cogs quit the end of the lever, it is returned to its former position, by means of a rope fastened to the end of B, and to a strong wooden pole, fixed to the top of the roof of the building, which acts as a spring. As the lever returns to its place, the pincers, by their own weight, slide down the inclined plane, and in their descent open sufficiently to allow the wire to slide through them, without extricating itself from their jaws; and on the next descent of the lever, they close upon the wire, and draw another portion through the plate.

Three of these machines, of different sizes, are, in general, employed in a wire-mill; the largest draws two inches of the wire at each stroke, and makes about forty-eight strokes per minute; the next four inches; and the third five inches. This last makes about sixty-four strokes per minute. This mode of drawing wire is very simple, but defective; for much time is lost in the returning of the pincers; they sometimes fail to take hold; and wherever they bite they make deep marks upon the wire, which are not more than two inches apart in the great wire, and five inches in the smaller.

Fine wire is always made from the large wire, by reducing it and lengthening it out by repeated drawings. The large wire is usually manufactured at the wire-mills in the country, and sometimes is reduced to small wire at the same establishments, but those who have occasion to use much wire usually purchase the large sort, and reduce it themselves.

A hand machine, represented in fig. 360, is used for this purpose. A is a roller, or cylinder, turning upon a vertical pin, fixed in the bench B; C a handle turned by manual labour; E the draw-plate; and *a a* the pins against which it rests. The wire to be drawn is placed upon a reel D, which turns upon a vertical pin. This reel is sometimes placed on the table, and sometimes in a tub containing starch-water, or beer that has become acid. This last is to loose the oxyd from the surface of the wire, which it has acquired in the process of annealing. Fig. 361 represents a very simple and complete wire-drawing machine, capable of drawing three wires at once. A R are two rollers or barrels with cog-wheels, T V, on the ends of their axis. S is a pinion which is turned round by means of a handle B, and communicates motion to the cog-wheels T V. Both these wheels are fitted upon round parts of the axis of their respective rollers, so as to slip or turn freely round with the same; but a square is formed on the axis outside of the wheel, and a clutch or catch, *t* or *r*, is fitted on this square part, so as to turn always round with the axis. The catch is at liberty to slide upon the axis in the direction of its length, by means of a lever W, which operates upon both catches at once. When either of them is pushed back in contact with the wheel, it intercepts two studs which project from the face of the wheel, and then compels the axis or roller to turn round with the wheel; but when the catch is drawn away from the wheel, then the wheel will slip round upon its ms without communicating any motion. By means of the lever W, only one wheel can be engaged at once, and the other must be free. The draw-plate is firmly fixed between the two rollers, and it has a great many holes; the rollers are long enough to receive three wires at the same time. Each roller has a groove in it parallel to the axis, into which a bar of metal is fitted, and will exactly fill it up.

When the wires are introduced through the holes in the plate, the ends are laid across this groove; the bar is then put in and fastened by a simple contrivance, and it fastens the ends of the wires beneath it, so that they become attached to the roller; then by turning the handle, B, round, the two wheels are put in motion in contrary directions; and that wheel which is connected with its axle by its catch, will turn its barrel round, and wind up the wires so as to draw them through the plate E. The other roller being at the same time detached, its wheel is at liberty to turn round in a contrary direction to the wheel, as fast as the wires are drawn off from it. When the whole length of the wires has been drawn through the plate, they are detached from the roller, the ends introduced through smaller holes in the plate, and fastened again to the roller; then the lever W is shifted, to disengage that wheel which operated before, and engage the other. This being done, the rollers will be turned in an opposite direction, and will wind back the wires, although the handle B is turned the same way round.

After the wire has been drawn three or four times, the metal becomes so hard and fibrous that it would not draw any more without breaking; it therefore requires to be heated in the fire to restore its ductility; for this purpose it must be taken off the barrels. A roller, M, is provided to wind the wire upon and draw it off from the barrel; this roller is turned round by a handle, *m*, fixed on the extremity of its axis; and the wire which is wound upon it in a coil is slipped off sideways. This machine is well adapted to be worked by a mill, because the handle may always be turned in the same way.

Fig 362 represents a machine that is used for reducing the wire to be employed in the manufacture of musical instruments, or in making cards for wool and cotton. A A A A are conical rollers, called blocks, each having a bush, through which passes a vertical spindle. These spindles are connected with wheel-work, situated beneath the bench, and being round are capable of revolving without communicating motion to the rollers. When the rollers are required to be engaged, they are lifted up from the bench, till two knobs, fixed in the hollow part of each, come in contact with a cross-bar fixed on the top of each spindle, which immediately carries them round. So long as any wires are supplied by the reels E E E E, the stress of the wires passing through the draw-plates will hold the rollers and spindles clasped together; but as soon as the whole of the wires have passed through the draw-plates, the rollers will become disengaged, and fall upon the bench. The tubs in which the reels are placed contain stale-beer grounds, or starch-water, for the purpose which we have already noticed.

The French draw-plates are the most esteemed, and, in time of war, a good French draw-plate has been sold for its weight in silver. M. Du Hamel, in *Les Arts et Métiers*, vol. xv. gives the following account of the process of making the draw-plates for the large iron-wire.

A band of iron is forged of two inches broad and one inch thick. This is prepared at the great forge. About a foot in length is cut off, and heated to redness in a fire of charcoal. It is then beaten on one side with a hammer, so as to work all the surface into furrows or grooves, in order that it may retain the substance called the potin, which is to be welded upon one side of the iron, to form the hard matter on which the holes are to be pierced. This potin is nothing but fragments of old cast-iron pots; but those pots which have been worn out by the continued action of the fire are not good; the fragments of a new pot which has not been in the fire are better.

The workman breaks these pieces of pots on his anvil, and mixes the pieces with charcoal of white wood. He puts this in the forge, and heats it till it is melted into a sort of paste; and to purify it he repeats the fusion ten or twelve times, and each time he takes it with the tongs to dip it in water. M. Du Hamel says, this is to render the matter more easy to break into pieces.

By these repeated fusions with charcoal, the cast-iron is changed, and its qualities approach those of steel, but far from becoming brittle, it will yield to the blows of the hammer and to the punch, which is used to enlarge the holes. The bar of iron which is to make the draw-plate is covered with a layer of pieces of the potin, or cast-iron thus prepared. It is applied on the side which is furrowed, and should occupy about half an inch in thickness. The whole is then wrapped up in a coarse cloth, which has been dipped in clay and water, mixed up as thick as cream, and is put into the forge. The potin is more fusible than the forged iron, so that it will melt. The plate is withdrawn from the fire occasionally and hammered very gently upon the potin, to weld and in some measure amalgamate it with the iron, which cannot be done at

They employ the iron manufactured in the departments of L'Orne and La Haute Soane, as being of the best quality. The first produces the best wire for making screws, nails, and pins, as much on account of its hardness as its fine polish, which resembles steel-wire. In this respect, it is superior to the iron of Haute Soane; but from its ductility the latter can now be made extremely fine, and it appears to be most free from heterogeneous particles.

The smelted iron, prepared and hammered, being in a state nearly fit for their purpose, is transported at a small expense to L'Aigle, by the rivers and canals. They have a forge to reduce the steel and iron of Normandy, which arrives in large pieces, into small and regular bars.

When the iron is formed into an irregular bar of about a centimetre, near four-tenths of an inch in diameter, they begin to draw it into wire. Although it be already much extended by hammering, it is in the first place passed four times through the drawing-plate; then its molecules become disposed lengthways, and exhibit fibres at their utmost extension. The fibres must be removed by means of heat, which disperses and divides them; and after that the wire may again be reduced three numbers. The fibres which are reproduced by this operation are again removed by heat. The whole process is five times repeated, consequently the wire is passed through fifteen numbers; after which, a single exposure to the fire is sufficient to fit it for passing six others, whereby it is reduced to the thickness of a knitting-needle.

The steel-wire, being much harder, requires to be passed through forty-four numbers, and to be annealed every other time.

The machine which draws the steel-wire, must go slower than that which draws the iron; for the first being very hard, and offering more resistance to the drawing-plate, should be pulled out with more care, since the quickness ought to be proportioned to the resistance, and reciprocally; and if they depart from this principle the results will vary. Thus, for example, the iron of the department of L'Orne, which is more compact than that produced at Haute Soane, if drawn by the same machines, augments to hardness, and is weakened when it is brought to too great a degree of fineness. But this iron, which is very hard, and capable of receiving a very high polish, is to be preferred for certain uses.

In order to anneal the wire, they formerly employed a large and elevated furnace, with bars of cast-iron to support the wire in the middle of the flames. It contains 7000 pounds weight, so contrived as to contain equal portions of each number. They are so arranged that the thickest wires receive the strongest heat; therefore, the whole is equally heated in the same space of time.

The operation lasts three hours with a fire well kept up, and it might be imagined that this apparatus was completely adapted to the purpose; but there are imperfections in this method, because it leaves the wire exposed to the contact of the atmospheric air, the oxygen of which it seizes with extreme avidity; whence a considerable quantity of oxyd is occasioned, and also an operation to free it from the scales, which consists of beating the bundles of wire with a wooden hammer wetted with water.

Notwithstanding this precaution, there often remains a portion of oxyd adhering to the surface of the metal, which streaks the draw-plate, or fixes on the wire, and gives it a tarnished appearance, and causes it to break when it is brought to a great degree of fineness. This furnace is only used for the steel-wire, or the iron from L'Orne, which is less liable to change; and besides, being harder, is not easily attacked by the oxygen.

In order to diminish the waste that the fire occasions, they have contrived another process, which consists in dipping the bundles of wire into a basin of wet clay before they put them into the furnace; and they are left in the furnace to dry before the fire is lighted, without which precaution the clay would peel off from the iron.

For making wire for cards, M. Mouchel invented another furnace. It is round, and about one metre six decimetres in diameter, and one metre eight decimetres in height, without including its parabolic arch, and the chimney above it. The interior is divided by horizontal grates into three stories; the lowest receives the cinders, the second is the fire-place, and into the third, or upper place, they slide a roleau of wire, weighing 150 kilogrammes, which is enclosed in a space comprised between two cast-iron cylinders, being luted to prevent the admission of air between them. The flames circulate about the outside of the first, and within the interior of the second, which defends the wire from atmospheric air. The diameter of the largest cylinder is about one metre four decimetres; that of the second one metre; thus the space comprised between them is two decimetres, on an elevation of five decimetres. There must be several pair of cylinders provided, because whilst one pair is in the furnace, another must be prepared to receive a fresh roleau of wire; they are changed every hour by means of a long iron lever, with which a single man can easily push them in, and draw them out again, as the cylinder slides on cast-iron rails.

They are very careful not to open the cylinders immediately on their being drawn out of the fire; for the rolapus of wire contained in them, being still red, would oxydate quite as much as if they had been heated in the midst of the flames without the least precaution.

The opening contrived for the passage is on the side, and has a door of cast-iron, with a groove which winds round the furnace; the fire-place has one something similar to it; that of the ash-hole is vertical, in order that it may be raised to increase the fire at will.

When the iron-wire is reduced to the thickness of a knitting-needle, it is made up into bundles of 125 kilogrammes (275 lbs.) each, into a large iron vessel, in order to anneal it sufficiently to be reduced for the last time. This vessel is placed upside-down in the middle of a round furnace, which is so con-structed as to sustain burning coals all round it, and of which it consumes 35 kilogrammes (77 lbs.) before the operation is completed. The cover must be carefully luted, as the slightest admission of air is sufficient to burn the external surfaces of the wire to an oxyd, which cannot afterwards be reduced.

When one of these vessels is sufficiently heated, it is filled with water containing three kilogrammes (six pounds and a half) of tartar, and suspended over the flames of the furnace to make it boil; this solution, without attacking the metal, frees it from the grease and the little oxyd that adheres to it. This is the last operation in which the wire is exposed to the fire, and it is then in the proper state for being reduced to the utmost degree of fineness it is capable of sustaining, and will preserve enough of the effect of the annealing to require it no more; but when the natural hardness of the iron varies, this last exposure to the fire should take place in proportion to its thickness. As steel loses its capacity of extension much sooner than iron, it is annealed until it is no thicker than a sewing-needle. The space which is left in the vessel is filled up with charcoal-dust, which prevents it from losing the quality of steel, and preserves the heat long enough to give it the proper degree of pliancy.

As Messrs. Mouchel always use iron and steel at the same manufactory, they have been able to reduce their operations to a general system; and to attain this end, have determined a graduated scale, by which the wire will not be more stretched in the drawing-plate in one number or size than another.

The following is the method they contrived, in order to form this scale for the iron-wire :—They take a certain quantity of various thicknesses, which has been drawn as fine as the iron will bear; the smallest size is 100,000 metres (109,333 yards) in length to the kilogramme, 2·2 pounds avoirdupois; they note the weight that each might be capable of supporting without breaking; this being expressed by figures, it is easy, by a few interpolations, to express them in a progressive scale. This kind of scale has been partly formed by comparing the weights of the different sizes with equal lengths, from which gauges or calibres may be made for the use of the workmen. These gauges are certain guides, which they cannot mistake, except through great carelessness. If they had not these gauges, they would often pass the wire through holes in the drawing-plates that are too large for it, whence it does not acquire the strength it should have in proportion to its thickness, and loses its hardness; they might also pass it through holes that were too small, which would weaken it, and render it very brittle. In the latter case, it frequently happens that the steel of the drawing-plate, being unable to sustain the force to which it is exposed, will give way, as if the plate were too soft; and the wire will be brittle at the beginning, and soft and too thick at the other extremity.

The greatest part of the fine wire at Messrs. Mouchel's manufactory is drawn by workmen who are dispersed about the country; but they have also a machine which moves twenty-four bobbins in a horizontal direction, which only requires the workmen to look after it. It is upon the bobbins that the wire is reduced to the different degrees of thinness desired; therefore this is the last operation in the art of drawing iron and steel wire, although it has all requisite qualities given to it in the workshop of the wire-drawer.

The wire is still incapable of being made into needles and fishing-hooks until it has undergone another operation for dressing and straightening the wire, by which it is made to lose the bend or curve that it acquires on the bobbins.

This work consists in drawing the wire between pins fixed on a piece of wood, and which act to bend the wire, first in one direction and then in the opposite, in a waving line, of which the waves are at first larger, but decrease gradually, and the last bend of which tends to force the wire into a straight line. The dresser is obliged constantly to adjust the pins by inclining or raising them with strokes of the hammer. Also, for each number of wires, the pins must be at different and calculated distances. This requires a workman of intelligence, diligence, and address.

An ingenious instrument is now appropriated to this operation, and removes all difficulty. Six little puppets of very hard steel are substituted for the nails of the ordinary instrument, and are fixed on parallel bars of metal, so jointed together that the movement of them all will be parallel, and the puppets are widened or brought nearer together by screws; the wire is drawn between these puppets in a zigzag, or waving line, and the repeated flexures break the sinuosities of the wire. There is a conductor of the wire to the puppets, and another conductor which serves to prevent the wire from being shaken. There are slight grooves at the extremity of the puppets, to give a passage to the wire. A scale sustained by a screw indicates the distance at which the puppets should be placed from each other, to straighten each size of wire; this forms nearly an invariable rule, and the dresser saves a third of the time which is employed in regulating the plan of the instrument formerly used. There is nothing more to be done but to draw out the wire by means of a wheel, on which he reels it, and then form it into bundles to be delivered to the consumers.

The steel wire of France is proper for many purposes. It is brought from Messrs. Mouchel for making knitting needles in the English fashion, shoemakers' needles, and other similar articles; it may be also used for needles of all sizes, and even for cards for wool-combing; but as this steel is much more expensive than the iron-wire, it is very seldom used for the latter purpose.

The method of preparing the draw-plates is described by Messrs. Mouchel, and is different from that before described.

For making wire for cards, two sorts of drawing-plates are used, large and small ones; the first, for the sort of wire that we have been describing, is drawn with the pincers, as fig. 359, and with the bobbin or reel, which is a cylinder, adapted to the axis turned by the water-mill, and is used in preference, to avoid the marks made on the wire by the pincers; the small drawing-plates are used for such wire as may be drawn by hand. The steel which they employ for these drawing-plates should never vary in quality, except that the smallest pieces are made of the finest steel. Several bars of iron are disposed in the furnace in the form of a box without a lid, the weight being according to the use for which they are intended to serve. The workman fills each of these boxes with cast-steel, and having covered it over with a luting of clay, it is exposed to a fierce fire until the steel is melted. His art consists in seizing the proper moment to withdraw the plate from the fire; he raises the luting, and blows on it through a tube, in order to drive off all heterogeneous parts, and then amalgamates the iron by light blows; after it is cool, he replaces it at the fire, and fusion again takes place, but to a less degree than before; as soon as

works the steel with light blows of the hammer, to purify and solder it with the iron. This operation is repeated from seven to ten times, according to its quality, which renders it more or less difficult to manage. During this process, a crust forms on the steel, which is detached from it the fifth time of its exposure to the fire; because this crust is composed of an oxydated steel of an inferior quality. It sometimes happens that two, and even three, of these crusts are formed of about two millimetres, or one-sixteenth of an inch, in thickness, which must also be removed.

After all these different fusions, the plate is beaten by a hammer wetted with water, and the proper length, breadth, and thickness, are given to it. When thus prepared, the plates are heated again, in order to be pierced with holes by punches of a conical form; the operation is repeated five or six times, and the punches used each time are progressively smaller. It is of importance that the plate never be heated beyond a cherry-red, because if it receives a higher degree of heat, the steel undergoes an unfavourable change. The plates, when finished, present a very hard material, which nevertheless will yield to the strokes of the punches and hammer, which they require when the holes become too much enlarged by the frequent passing of the wire through them.

When the plates have been repaired several times, they acquire a degree of hardness which renders it necessary to anneal them, especially when they pass from one size to another; sometimes they do not acquire the proper quality until they have been annealed several times. Notwithstanding all the precautions which are taken in preparing the plates, the steel still varies a little in hardness, and according to this variation they should be employed for drawing either steel or iron wire; and if the workman who proves them finds that they are too soft for either the steel or iron, they are put aside, to be used by the brass-wire drawers.

A plate that is best adapted for drawing of steel-wire is often unfit for the iron; for the long pieces of this latter metal will become smaller at the extremity than at the beginning, because the wire, as it is drawn through the plate, is insensibly heated, and the adhering parts are swelled, consequently pressed and reduced in size towards the latter end. The plates that are fit for brass are often too soft for iron, and the effect resulting is the reverse of that produced by a plate that is too hard.

The smallest plates which Messrs. Mouchel use are at the least two centimetres, or eight-tenths of an inch, in thickness, so that the holes can be made sufficiently deep; for when they are of a less thickness, they will seize the wire too suddenly, and injure it.

2 A 2

This inconvenience is much felt in manufactories where they continue to use the plates for too long a time; as they become exceedingly thin after frequent repairs. One of Messrs. Mouchel's large plates reduces 1,400 kilogrammes (3,000 lbs. avoirdupois) from the largest size of wire to that which is of the thickness of a knitting-needle; 400 kilogrammes (880 lbs.) of this number are afterwards reduced to one single small plate to N° 24, which is carding-wire; and to finish them, they are passed through twelve times successively.

Wires are frequently drawn so fine as to be wrought along with other threads of silk, wool, or hemp; and thus they become a considerable article in the manufactures.

Dr. Wollaston, in 1813, communicated to the Royal Society the result of his experiments in drawing wire. Having required some fine wire for telescopes, and remembering that Muschenbroeck mentioned wire 500 feet of which weighed only a single grain, he determined to try this experiment, although no method of making such fine wire had ever yet been published. With this view, he took a rod of silver, drilled a hole through it only one-tenth its diameter, filled the hole with gold, and succeeded in drawing it into wire till it did not exceed the three or four thousandth part of an inch, and could have thus drawn it to the greatest fineness perceptible by the senses. Drilling the silver he found very troublesome, and determined to try to draw platina wire, as that metal would bear the silver to be cast round it. In this he succeeded with greater ease, drew the platina to any fineness, and plunged the silver in heated nitric-acid, which dissolved it, and left the gold or platina wire perfect.

LEAD MANUFACTURE.

Lead ore is found in most parts of the world. In Britain the principal lead-mines are situated in Cornwall, Devonshire, and Somersetshire; in Derbyshire, Durham, Lancashire, Cumberland, and Westmoreland; in Shropshire, Flintshire, Denbighshire, Merionethshire, and Montgomeryshire; at the lead-hills in Scotland, on the borders of Dumfriesshire and Lanarkshire, in Ayrshire, and at Strontian in Argyleshire.

The smelting of the ore is performed by either a low furnace, called an ore-hearth, or a reverberatory furnace. In the former method, the ore and fuel are mixed together,

metals, are combined with various kinds of earthy matter, which require them to be well pounded before they are introduced into the reverberatory or smelting furnace. The pounding is sometimes performed by women using hammers, and sometimes the ores are pounded or crushed by causing them to pass through iron rollers loaded with great weights. After the ores have been pounded or crushed, the earthy matter is separated by washing.

The powder to be washed is put into a riddle or sieve, and placed in a large tub full of water; when, by a certain motion, the lighter or earthy parts are separated and thrown over the edge of the riddle, while the metal, which, as we have before stated, is always considerably heavier than its accompanying ingredients, is retained. There are some impurities, however, which cannot be separated by this process, consisting principally of *blind*, or *black-jack*, called *mock* ore, and pyrites, or sulphuret of iron, named *Brazil*.

In the process of smelting, the ore is spread upon the concave hearth, so that the flame may act upon it, and release the sulphur. When the sulphur has escaped, the lead combines with oxygen, and the oxyd of lead, thus formed, combines with and reduces the earthy matter to a liquid, which floats upon the surface of the metal, and for the remainder of the operation, protects it from the action of the oxygen. The temperature of the furnace is now considerably raised, to separate as quickly as possible the lead from the liquid scoria; after which a considerable portion of the scoria is tapped off, leaving only so much behind as is necessary to protect the metal from the action of the oxygen. The fire is now slackened, and a quantity of slack, or refuse pit-coal, thrown into the furnace, which serves to diminish the heat, and to concrete the melted scoria; though this last part of the process is not well done unless powdered lime be also added. The scoria being now hardened, is broken to pieces by a rake, and thrust to the opposite side of the furnace, where it is taken out through the apertures already mentioned.

The lead is now tapped, in a manner similar to that described in the manufacture of iron, and is allowed to run into a large iron pan, from whence it is laded into moulds to cast into pigs. When the ores abound with blind, or black-jack, or sulphate of iron, it becomes necessary to add the fluat of lime, as a flux.

The scoria is still found to contain some lead, independent of that in the state of oxyd, and chemically combined with it, and is consequently exposed to the heat of another furnace, being a species of blast, and called a slag-hearth, which fuses the teeth, and causes the metal to penetrate through it, and fall into a cavity, where it is protected from the agency of the blast, and from whence it is taken and cast into pigs.

As all lead ores contain more or less of silver, we shall extract from Dr. Rees's Cyclopædia the method by which the silver, by the oxydation of the lead, is extracted.

"A shallow vessel, or cupel, is filled with prepared fern-ashes well rammed down, and a concavity cut out for the reception of the lead, with an opening on one side for the mouth of the bellows, through which the air is forcibly driven during the process. The French smelters cover the surface of the ashes with hay, and arrange symmetrically the pieces of lead upon it. When the fire is lighted, and the lead is in a state of fusion from the reverberation of the flame, the blast from the bellows is made to play forcibly on the surface, and in a short time a crust of yellow oxyd of lead, or litharge, is formed, and driven to the side of the cupel opposite to the mouth of the bellows, where a shallow side or aperture is made for it to pass over; another crust of litharge is formed and driven off, and this is repeated in succession till nearly all the lead has been converted into litharge and driven off. The operation continues about forty hours, when the complete separation of the lead is indicated by a brilliant lustre on the convex surface of the melted mass in the cupel, which is occasioned by the removal of the last crust of litharge that covered the silver. The French introduce water through a tube into the cupel, to cool the silver rapidly, and prevent its spitting out, which it does when the refrigeration is gradual, owing probably to its tendency to crystallize. In England the silver is left to cool in the cupel, and some inconvenience is caused by the spirting, which might be avoided by the former mode,

"The silver thus extracted is not sufficiently pure; it is again refined in a reverberatory-furnace, being placed in a cupel lined with bone ashes, and exposed to greater heat; the lead which has escaped oxydation by the first process, is converted into litharge, and absorbed by the ashes of the cupel.

"The last portions of litharge in the first process are again refined for silver, of which it contains a part which was driven off with it. The litharge is converted into lead again by heating it with charcoal; part is sometimes sold for pigment, or converted into red-lead. The loss of lead by this process differs considerably, according to the quality of the lead,

The litharge commonly obtained from three tons of lead
amounts to 58 hundred weight; but when it is again reduced
to a metallic state, it seldom contains more than 52 hundred
weight of lead, the loss on three tons being eight hundred
weight. The Dutch are said to extract the silver from the
same quantity of lead with only the loss of six hundred
weight."

Having explained the process by which pig-lead is extracted from the ores, it now remains for us to show the
methods by which pig-lead is manufactured into sheet-lead,
or into the tubes called lead-pipes.

In the manufacture of sheet-lead, the ingots or pigs are put
into a large caldron or furnace built with free-stone and
earth. Near this furnace is the table or mould on which the
sheet is to be cast; it is made of large pieces of wood, well
jointed, and bound at the ends with bars of iron, and has
a ledge or border of wood, about two or three inches thick,
and one or two high, called the *sharps*. The tables are
usually about three or four feet wide, and from eighteen to
twenty feet long. The table is covered with very fine sand,
which is prepared for the casting by moistening it with clean
water, working it together with a stick, beating it flat with
a mallet, and smoothing it with a piece of brass or wood.

A long narrow piece of wood, with notches cut in each end,
so as to fit the ledges, is placed over the table, and is so
arranged, that the space between it and the sand shall be
proportionate to the intended thickness of the plate. The
workman gradually slides the strike from one end of the
table to the other, by which means he obtains a sheet of the
requisite, and in all parts of equal, thickness.

At the top of the table is a large triangular iron peel or
shovel, with its fore part bearing upon the edge of the table,
and the hinder part on a tressel, somewhat lower than the
table; the design of which is, to prevent the liquid metal
running off at the fore side, where there is no ledge. The
metal being sufficiently fused, is taken out of the furnace or
caldron with a large iron ladle, and is put into the peel,
where it is cleansed of its impurities by using another large
iron ladle pierced like a scummer. The handle of the peel is
now raised, which causes the liquid metal to run into the
mould, while the workman, with the strike, regulates the
thickness. When the sheet is of the required thickness,
the handle of the peel is lowered, and the sheet is allowed to
cool. When set, the edges on both sides   planished in
order to render them smooth and straight.

The method above described is only used in casting large sheets of lead; in casting sheets of smaller dimensions, the table or mould, which is placed in an inclined position, is, in lieu of sand, covered with a piece of woollen stuff, nailed down at both ends, and over that is placed a very fine linen cloth.

In this process great attention must be paid to the heat of the liquid metal, and a piece of paper is used as a test; if the paper take fire, the lead is too hot, and would destroy the linen; if it be not shrunk and scorched, it is not hot enough.

When the sheets are required to be very thin, it is necessary to make the peel and strike of one piece. It is a kind of wooden box without a bottom, being closed only on three sides; the back of it is about seven or eight inches high, and the two sides, like two acute angles, diminish to the top; the width of the middle makes that of the strike, which again makes that of the sheet to be cast.

The strike is so placed, that the highest part is towards the lower, and the two sloping sides towards the upper end of the table. The top part of the table, where the metal is poured in, is covered with a pasteboard, which serves as a bottom to the case, and prevents the linen from being burnt while the metal is pouring in.

The strike or peel being filled with lead, according to the intended size of the sheet, two men, one at each side, seize hold of it, and with greater or less velocity, as the sheet is to be more or less thick, force it down the inclined table; for the thickness of the sheet always depends upon the velocity with which the strike slides down the table. The sheet-lead, after casting, is frequently reduced by rollers.

As this particular department is so intimately connected with the business of a plumber, we shall not be considered as departing from the subject by inserting the following tables, from *Hutton's Mensuration*.

Plumber's work is commonly estimated by the pound or hundred weight; but the weight may be discovered by the measure of it, in the manner below stated. Sheet-lead used in roofing, guttering, &c. is commonly between seven and twelve pounds weight to the square foot; but the following table shows by inspection the particular weight of a square foot, for each of several thicknesses.

Thickness	Weight to a square foot	Thickness	Weight to a square foot
·10	4·899	·15	
·11	6·489	·16	
	6·554	·17	9·831
·12	7·078	·17	10·023
	7·373	·18	10·618
·13	7·668	·19	11·207
·15	8·226	·2¼	11·797
¼	8·427	·21	12·387

In this table the thickness is set down in tenth hundredths, &c. of an inch; and the annexed corresponding numbers are the weights in avoirdupois pounds and thousandth parts of a pound. So the weight of a square foot to ·10 or ·11 of an inch thick is 5 pounds and ·899 of a pound; and the weight of a square foot to one-ninth of an inch thickness is 6 pounds and ·489 of a pound. Lead pipe of an inch bore is commonly 13 or 14 pounds to the yard in length.

EXAMPLES:

1. How much weighs the lead which is 39 feet 6 inches long, and 3 feet 3 inches broad, at 8¼ lbs. to the square foot?

Decimals.		Duodecimals.
39·6		39 6
3¼		3 3
118·5		
9·876		
		129 4 6
128·375		8¼
8¼		
		1027
1027·000		64
64·1875		2
1091·1875		

Answer...... 1091 ¼ lbs.

2. What cost the covering and guttering of a roof with lead, at 18s. per cwt.; the length of the roof being 43 feet, and the breadth of it being 32 feet, the guttering 57 feet long, and 2 feet wide; the former 8·427 lbs. and the latter 7·373 lbs. to the square foot?—Answer, 115l. 9s. 7½d.

It is now time to direct our attention to the manufacture of *lead pipes*, which are universally employed for water-pipes, from the facility of bending them in any direction, and soldering their joints.

Lead pipes are sometimes cast in an iron mould, made in two halves, forming, when put together, a hollow cylinder of the size of the intended pipe; in this cylinder, or mould, is put an iron rod or core, extending from the top to the

bottom, and leaving all round a space between it and the
cylinder of the intended thickness of the pipe. The lead is
poured in at a spout, formed by two corresponding notches
cut in each half of the mould; and a similar hole is made at
another place for the escape of air. The mould is fastened
down upon a bench, upon which, at one end, and in a line
with its centre, is a rack, moved by toothed-wheels and
pinions.

When the pipe is cast, a hook at the end of the rack is put
into an eye at the end of the iron core, which, by the action
of the cog-wheels and pinions, is drawn so far out, that about
two inches of it only remain in the end of the pipe; the two
halves of the mould, which fasten together by wedges or
screws, are now separated from the pipes, and are fastened
upon the iron core, and the two inches of lead pipe attached
to it; melted lead is again poured into the mould, which,
uniting with the end of the first piece, forms the pipe of considerable length; and the operation is repeated till it be of
the length required.

Another and a much better method is, to cast the lead in an
iron mould upon a cylindrical iron pipe, of a size proportioned
to the bore of the pipe to be made, and leaving a space between the core and the mould three or four times the thickness of the intended pipe, and in short lengths, which are
afterwards drawn through holes in pieces of steel, similar to
the process of wire-drawing, till the pipe is reduced to the
required thickness.

Another method is that for which the celebrated iron-
manufacturer, Mr. John Wilkinson, of Brosely, took out a
patent in 1790, and which, since the expiration of his patent,
has been successfully practised by many other manufacturers.
This method consists in casting a circular piece of lead, about
eighteen inches long, with a core or hole longitudinally
through its centre. This piece is of considerably larger
diameter than that of the pipe intended to be made. The
core or hole at one extremity suddenly decreases, so as to
form on the internal surface of the piece of lead a stop or
shoulder, against which a polished iron triblet or mandrel,
which has been passed thus far along the core, rests. This
triblet or mandrel is of somewhat greater length than the
length required of the pipe to be manufactured, which,
generally speaking, is from seven to nine feet. An iron screw,
having a loop at the opposite end, is then passed down the
other end of the core, and is screwed into that part of the
mandrel which rests against the shoulder. In this state

the mandrel, with the circular piece of lead fixed fast on it, is taken to the drawing-table.

The drawing-table in the principle of its operation resembles that the block described in the Wire Manufacture in every respect, though it is far more powerful. The table generally used is about thirty feet long, by two feet wide; having at one end a powerful cylinder with a chain attached to it. This cylinder receives motion from a steam-engine, or other power, and can be thrown in and out of geer by any adaptation of any one of the appropriate modes described under the article " Mill-geering." About two-thirds the length of the bench from the cylinder, or roller, are two pins or stops which hold a steel plate, which has a gradation of conical holes. Through the largest of these holes, which is somewhat less than the diameter of the circular piece of lead, the loop that is screwed on to the end of the mandrel is passed, and attached to a hook at the extremity of the chain, which chain is affixed to the cylinder or roller. The cylinder being now thrown into geer, the piece of lead is drawn through the hole in the steel plate, which diminishes it in diameter, and increases it in length; and this operation is carried successively through the series of gradually decreasing holes in the draw-plate, until the pipe is reduced to the required diameter. The cylinder is now struck out of geer, and the mandrel liberated from the chain, which is immediately attached to the other end of it. The steel draw-plate being now removed, the stops against which it rested allow the mandrel to pass between them, but detain the lead pipe, which, consequently, by striking the cylinder into geer, allows the mandrel to be extricated from it. A small portion of pipe being cut off at both ends, the pipe is considered finished. Through the whole of the operation, great care is taken to keep the mandrel and steel plate well oiled.

As no acid can pass through lead pipe without being more or less affected by its deleterious qualities, it is necessary in cases where acids are used, to have pipes made of iron, or of lead lined with tin. To line lead pipe with tin, the lead pipe must be cast in a vertical mould, which has a core of somewhat larger diameter than the intended bore, the pipe passing down its centre. When the pipe is cast, and the metal is set, this mandrel is drawn out of the mould, and another of smaller diameter is substituted. About as much coarse resin as will lay on a shilling is now thrown into the space between the pipe and the core or mandrel now passed down the mould. This resin by the heat of the lead

, and runs to the bottom of the mould. The melted
now poured in, the resin will float on its surface,
sequently, as the tin rises, anoint the tin in every
l act as a flux, and unite the two vessels. As soon
is set, the last-mentioned mandrel is drawn out,
external mould being removed, the lead now lined
s, when quite cold, ready to be submitted to the
f drawing. Various other equally simple processes
ed to this purpose.

PAPER MANUFACTURE.

that highly valuable substance, which enables us
nicate our thoughts to persons situate at the most
arters of the civilized globe, is manufactured from
he aid of machinery.
formerly necessary to assort with great care the
h were intended to be manufactured into paper;
but the whitest and best, and which, consequently,
most expensive, could be made into paper of the
lity; but since the introduction of chlorine (which
vered by Scheele) into our bleaching establishments,
sity of this assortment has been greatly obviated;
soon conceived that that chemical agency, which
ble of bleaching linen, was also applicable to the
of the rag during the process of paper-making.
period of the process, when the rag is coming into
pulp, chlorate of lime, which was first manufactured
nuant, of Glasgow, is introduced into the vat; and,
mical action on the fibre, whitens or bleaches the
ss; thus enabling the manufacturer to produce a
d much finer quality of paper from rags of a se-
quality, than he had heretofore done from rags of
expensive description. It must, however, be ad-
at, as in all bleaching processes where the fibre is
ess deteriorated by the action of chlorine, the paper
ured and whitened by this agent is not so strong as
erly produced; as may be observed in some thick
ifully white papers frequently offered to the public

at astonishing low prices, which arc manufactured from coloured and inferior rags, with a superabundance of the chlorate of lime introduced in the process of the manufacture. By this, therefore, it is evident that the chlorate of lime, when used too abundantly, will rot or destroy the fibre of the whole; but when judiciously applied, it produces a paper of superior colour, and of adequate strength for all practical purposes.

The paper-mill consists of a water-wheel, or other first mover, connected with a combination of toothed and other wheels, so arranged as to cause the cylinder in the *washer*, and the one in the *beating* engine, which will be hereafter described, to make from 120 to 150 revolutions per minute. On the same shaft, and of the same size as the water-wheel, is a toothed or cogged wheel, which plays in a pinion; the spindle of this pinion is furnished with a crank, which, by means of a connecting rod, gives a reciprocating motion to a lever, for the purpose of working two pumps, which raise a constant stream of water from the mill-dam. This stream of water is kept running through the rags in the washing-engine, to carry away the dirt separated from them by the operation. The structure of an engine is more minutely explained by figs. 371, 372, 373, 374, &c.; fig. 371 being a section through the length of the engines, and fig. 372 a horizontal plan.

The large vat or cistern, A A, is of an oblong figure on the outside, the angles being cut off; but the inside, which is lined with lead, has straight sides and circular ends. It is divided by a partition, B B, also covered with lead. The cylinder C is fixed fast upon the spindle D, which extends across the engine, and is put in motion, as before described, by the pinion E, placed on the extremity of it. The cylinder is made of wood, and furnished with a number of teeth, or cutters, fixed fast on its circumference, parallel to the axis, and projecting about an inch, as is shown on a larger scale at fig. 375.

Immediately beneath the cylinder, a block of wood, H, is placed, and provided with similar cutters to those of the cylinder, which, when they revolve, pass very near the teeth of the block, but do not touch; the distance between them being capable of regulation, by elevating or depressing the bearings on which the necks D, D, of the spindle are supported. These bearings are made on two levers, F, F, which have tenons at their ends, fitted into upright mortises, made in short beams, G, G, bolted to the sides of the engine. (See also fig. 373.) The levers, F, F, are movable at one end of each, the other ends being fitted to rise and fall on bolts, in the beams G, as centres.

The front one of these levers, or that nearest to the cylinder C, is capable of being elevated or depressed, by turning the handle of the screw *b*, which, as shown in fig. 373, acts in a nut *a*, fixed to the tenon of F, and comes up through the top of the beam G, upon which the head of the screw

... its bearing. Two brasses are let into the middle of the levers P, P, and form the bearings for the spindle of the engine to work upon. Th... it is used to raise or lower the cylinder, and cause it to cut finer or ... by enlarging or diminishing the space between the cutters in the ... those of the cylinder.

... K, figs. 371 and 372, is a circular breasting made of boards, and ... with sheet-lead; it is curved to fit the cylinder very truly, and ... but very little space between the teeth and breasting. An inclined ... K leads regularly from the bottom of the engine-vat to the top of ... breasting; and at the bottom of it the block, H, is fixed.

... engine is supplied with water by a pipe, Q, bringing it from the ... this pipe delivers it into a small cistern, M, adjoining, and com... ...ting with the engine. The pipe has a cock, P, to stop the entrance ... when required, or to regulate the quantity of its discharge. ... cistern has a grating fixed across it, covered with a hair-strainer, ... catch any extraneous matter which may come in with the water, or a ... bag is sometimes tied over the orifice of the cock, P, through ... all the water must be filtered. When the engine is filled with water, ... quantity of rags put in, they are, by the revolution of the cylinder, ... between its cutters and the teeth of the block H. This cuts them ... them, by the rapid motion of the cylinder, the rags and water are ... over the top of the breasting, upon the inclined plane; in a short ... this raises more rags and water into that part of the engine-vat; and ... ...dency to restore the equilibrium puts the whole contents of the vat ... motion, down the inclined plane K, and round the partition BB, ... they come to the cylinder again in about the space of 20 minutes; ... the rags are repeatedly cut and chopped in every direction, till they ... ...ed to a pulp.

... circulation is of advantage, in turning the rags over in the engine, ... cause them to present themselves to the cutters in a different direction ... time; for as the cylinder cuts or clips in straight lines, in the same ... as a pair of shears, it is requisite to cut the rags across in different ... ...tions, to reduce them to a pulp.

... manner of the cutting is this: the teeth of the block are placed ... inclined to the axis of the cylinder, as shown by fig. 374, but the ... of the cylinder are parallel to its axis; therefore, the cutting edges, ... they meet, are at a small angle, and come in contact first at one ... and then successively the contacts proceed along to the other end, so ... ...ny rags interspersed between them are cut in the same manner as ... would be between the blades of a pair of shears. Sometimes the ... cutters, h, in the block, are bent to an angle in the middle, instead ... being straight, and inclined to the cylinder; in this case, they are ... elbow plates, and of course the two ends are both inclined to the ... of the cylinder in opposite directions. In either case, the edges of ... ...te of the block cannot be straight lines, but must be curved, to ... ...emselves to the curve which a line traced on the cylinder will of ...

... plates or cutters of the block are united, by screwing them altogether, ...ing them into a cavity cut out in the wooden block H; their edges ... ...ed away on one side only; as shown at h in the section, fig. 374. ... block is fixed in its place by being made dove-tailed, and truly fitted ... the bottom of the cistern, so that the water will not leak by it. The ... ...ft comes through the wood-work of the chest, and projects a small ... on the outside of it, being kept up to its place by a wedge, so that ... ...withdrawing this wedge, the block becomes loose, and can be removed, to sharpen the cutters, as occasion requires. This is done on a grindstone, the plates being first separated from each other.

The cutters of the cylinder are fixed into grooves, cut in the wood of the cylinder, at equal distances from each other round its circumference, in a direction parallel to its axis; the number of these grooves is twenty; and for the washer, each groove has two cutters or bars put into it; then a fillet of wood is driven fast in between them, to hold them firm; and the fillets are kept fast by spikes driven into the solid wood·of the cylinder. The beater is made in the same manner, except that each groove contains three bars and two fillets, as shown in fig. 375.

In the operation of the cylinder, it is necessary that it should be enclosed in a case, or its great velocity would throw all the water and rags out of the engine. The case is a wooden box, L L, enclosed on all sides except the bottom; one side of it rests upon the edge of the vat, and the other upon the edge of the partition B B. The lines, c c, represent the edges of wooden frames, which are covered with hair or wire-cloth; and immediately behind these, the box is made with a bottom, and a ledge towards the cylinder, which makes a complete trough.

The dark spaces, e e, in fig. 371, show the situation of two openings, or spouts, through the side of the case, which lead to flat lead-pipes, b, b, fig. 372, which are placed by the side of the vat; the beam, F, being cut away for them. There are waste pipes, to convey away the foul water from the engine; for the cylinder, as it turns, throws a great quantity of water and rags against the sieves; the water goes through them, and runs down into the trough at e e, and from thence into the ends of the lead pipes, b, b, fig. 372, by which it is conveyed away; d, d, fig. 371, are grooves for two boards, which, when put down in their places, cover the hair-sieves, and stop the water from going through them, if it is required to retain the water in the engine. This is always the case in the beating-engines, and therefore they are seldom provided with these waste-pipes, or at most on one side only; the other side of the cover being curved, to conform to the cylinder. Except this, the only difference between the washing-engine and the beater is that the teeth of the latter are finer, having 60 instead of 40 bars on its circumference; and it revolves quicker than the washer, so that it will cut and divide those particles which pass through the teeth of the washer.

The rags being now reduced to a state of pulp, we shall in the next place proceed to show the method of forming it into sheets of paper.

It was formerly the custom to allow a small, but sufficient portion of the pulp to flow on a sieve furnished with two handles, which sieve was agitated by a workman until the pulp had subsided or settled regularly throughout the surface. This, when it had passed through the usual processes of pressing, drying, &c. constituted a sheet of paper; and its texture was indicated by the fineness of the quality of the wire out of which the sieve was constructed.

This unmechanical and desultory mode of operation has been obviated by improvements effected by many ingenious persons; but the machines which are now almost universally employed, and which have most decidedly superseded all

same object, was the invention of the
The action and arrangement of this
hanism consists in having a horizontal
d length, furnished with a roller or
)ver which is stretched an endless web
requisite texture or fineness for the
nfactured. At one end of the frame,
ediately over, one of the cylinders, is a
ito which the pulp is received, whence
; slit or opening, which is regulated by
ie surface of the web beneath. At this
. the cylinders are set in motion, and
rly forward with a tremulous motion,
isperses the pulp regularly over the
web. This tremulous motion is im-
the machinery by an eccentric move-

' arrives in this crude and wet state at
eb of the further cylinder, it is wiped
:r cylinder, covered with felt or flannel,
ı series of similar cylinders, and finally
wound off in a coil or hank so long as
Thus, paper, by the action of this
y be manufactured to an unlimited
th that is compatible with the manu-
Γhe reel or winder being now with-
r is cut on both sides, forming sheets
th of the machine and reel on which it

the different speeds of the various
ie web, and afterwards pressing the
e action of the reel, and the tremulous
to the whole of the machinery by the
lso the regular supply of the pulp for
the paper to be manufactured, form a
n of ingenuity and mechanical know-
be lamented that the inventors and
s great source of national industry,
ed a reward adequate to the benefit
ed on their country.
r which a paper-mill can command to
rally limits the extent of its trade;
s should attend to every improvement
can increase their effect.

2 B

internal part large enough to permit the passage of water, but _____
intercept fibres of rags; thirdly, it must be so contrived that the surface do
not yield from its perfectly cylindrical form, notwithstanding a very considerable degree of pressure upon it; fourthly, it must be furnished with small
flat rings for the purpose of covering part of its surface; at the same time
may be several pairs of these rings of different widths, in order to vary the
proportion of the surface which is left uncovered, provided the same
cylinder is extended for making different sized papers; fifthly, it must be
hung upon an axis in a horizontal position, and firmly fixed in bearings, so
that it may be turned by any convenient power; sixthly, the number of
small apertures on the external surface must open into a less number of
large ones, communicating with the internal surface, with solid intervals
between them; seventhly, it ought not to be made of wood because it
would be liable to warp, nor of iron because it would rust, and injure the
paper; brass or any other strong metal would be found most convenient.
To construct a cylinder possessing the requisites above-mentioned, of
which the dimensions must be according to the size and thickness of the
paper it is intended for making, the patentee takes a brass cylinder, perfectly
smooth inside and outside, excepting a small portion at each end which is
left plain, and turns the outside so as to resemble a screw, the threads of
which are about a quarter of an inch apart, and the twenty-fifth part of an
inch in depth, with a round edge. He then drills holes between the threads,
which are cut in a taper form, the diameter at top being the width of the
interval between the threads, and at the bottom reduced to one half that
size; the space on the outer surface of the cylinder, left between these
holes on each side, is equal to the breadth of the thread; notches are cut in
the threads for the purpose of letting in cross wires, the diameter of which is
equal to that of the threads, so that when they are laid into the notches and
soldered, or otherwise fastened down, the surface of the cylinder will resemble network, with openings of an oblong shape, and having the surface
of all the interstices plane with each other, and wound to an equal curve.
It is then covered with an endless web of woven wire, which is drawn tight
over it. The ends of the cylinder are cut down, or rabbeted, so that a ring
may be made to slide on each end; and the ends of the wire are fastened to
this ring by means of small plates, which are put over the wire, and screwed
down upon the rings by means of screws which pass through the wire.
These rings are also furnished with other screws for the purpose of extending them out from the cylinder, and the wire being fastened to them. It is
by that means stretched and drawn tight down upon the surface of the
cylinder.

In the annexed drawing, fig. 380, a to b represents a transverse section
of a segment of the cylinder C C C, being the holes; d d d, the cross wire;
e e e e, the thread of the screw, which is shaded.

Fig. 381 is a plan of a portion of the external surface of the cylinder,
wherein A to B shows it without the cross wires, or the external wove
wire; C C C are the holes, e e e the thread of the screw with the notches
out, a a a, for the reception of the cross wires; B to C shows it with the
cross wires let in, d d d, which are soldered or otherwise fastened at their
ends into the ends of the cylinder. C to D shows it with the woven wire
laid over it, through which the surface of the cylinder is seen underneath.

Fig. 382 is a section of a part of the cylinder at one end, where the holes
are marked c c c, the cross wires d d d, the threads e e e; the external wove
wire f, is represented by a red line, it is carried under the plates g, and
fastened, by means of a number of screws, h, down upon the ring i, which
is shaded, and after it is fixed in that manner at each end of the cylinder,

passes through the upper part of the surface of the cylinder, and off through the orifice as in the same section. The cylinder is yellow, the caps red, and the parts which answer to the rings are

Fig. 386 is a front view of one of the caps.

Fig. 387 is a sectional elevation of the machinery, in a state of tion for the manufacture of paper.

Fig. 388 is a plan of the same. Each part in the elevation, fig on a line with the same part in the plan, fig. 388; and every part is with the same letter in both. A is a circular stuff-chest; into w stuff is admitted from the engine. B is an agitator, consisting of a of arms, connected with the spindle C, which passes up through a in the centre of the chest, and this being turned by the bevelled cog-n keeps the stuff in motion in the chest, and also, by means of riggers F F, gives motion to another small agitator, in the smaller v which is for the purpose of receiving the stuff from the first chest, . conveyed through the pipe H, the aperture of which is enlarged or co by means of a conical valve, which is acted upon by some apparat the principle of a ball-cock, so that as the vessel fills with stuff it g closes the orifice; by this means the stuff in the smaller vessel G kept at a uniform height, and the head being uniformly tho sa discharge through the pipe J at the bottom will be always equ large chest A may be of any shape or dimensions, and agitated convenient manner; the smaller chest G ought to be circular, as 18 inches diameter, and the same depth. The use of it is to uniform discharge, which would not take place if the stuff were to p the large chest without any intermediate vessel, because its passage the pipe would be more or less rapid, according to the height of t which would be continually varying in proportion to the consum accumulation of stuff. In the pipe J there is a cock K, by means the quantity of stuff that is permitted to pass may be regulated greatest degree of nicety; and when it is once ascertained what p of stuff is required, no variation in the supply can take place. sort of cock or valve for the purpose will be such a one as leaves a ing for the stuff, which is nearly round or square, because if it were the stuff might lodge. The pipe J descends into the pipe K, throug there is a constant and rapid flow of water, and it carries away th of pulp from the pipe J, and they pass together into the vessel L, i there are two agitators M M, kept in pretty quick motion by mean riggers N N. In this vessel and in the pipe K the water and stuff intimately mixed, and formed into pulp, of a proper consistency for v but it is to be observed, that in making paper by this method al times as much water should be introduced into the pulp as is mad in the ordinary modes of paper-making. From the vessel L the pu through the pipes O O into the vessel P, which has been before d in fig. 384, and called a back. Q Q are waste pipes, for adjusting th of the head, or, in other words, the level of the pulp. In the back hollow cylinder, described in figs. 380, 381, 382, 383, 384, and 3 the cylinder being in motion in the direction described in the d the water is constantly flowing through the surface of it from the to the point T, that is to say, through every part which is covered pulp, and, as the water passes through, the fibres of rag are left surface, so that they are generally accumulating on any given par surface of the cylinder during the whole of its passage from the poi the point T: and when it emerges from the pulp at the point quantity requisite for the composition of a sheet of paper is c

It is to be observed, that the principle here developed would admit
less eligible modifications, such as confining a body of the pulp
surface of an endless well of woven wire, carried round cylinders, a
outline section, fig. 389, or supporting it on a cylinder of a large si
the outline section, fig. 390, without applying the pneumatic pres

In fig. 389 the cylinder *a b c* should be hollow, and have pervious
In fig. 390 the cylinder might be of a more simple construction th
described in figs. 380, 381, 382, 384, and 385, but unless of a ve
size indeed, it could only be made use of for making very thin pap
cause the water requires so long time to run off before the paper wil
of any mechanical pressure.

It is to be observed, that in making paper by this method, after a
quantity of fibres of rag are deposited on the surface of the cylinder,
ders the passage of the water and the accumulation of more fibres t
cult, that without a considerable height of pulp the pressure will
sufficient to force the water through the cylinder, and the fibres of ra
upon it, the consequence of which would be, that the fibres of rag a
lated on the surface of the cylinder would be washed off by the pulp,
much disturbed before they arrived at the point T, which is the level
pulp in the back; to obviate this, it will be necessary to add the pre
the atmosphere to the weight of the water in making thick papers
may be done by extending one side of the trough V below the level
pulp, so as to cause a suction under that part of the cylinder w
covered by the pulp, as well as under that part which has emerged f
For this purpose a wider trough would be necessary; but at all eve
exact proportion of the cylinder, covered by the trough, is not mate
cause it will be found by experience what width is sufficient for dryi
paper, so as to enable it to have the pressure of the roller *a*. The r
ought to press on the cylinder R about the point which is over one
the trough V, and, according as the trough is shifted, the roller sh
shifted also; but this pressure ought to be not less than forty-five
above the level of the axis, because, otherwise, part of the water
out of the paper will be absorbed by it again, whereas, from the pea
acts in, in the drawing, fig. 387, the water will be sucked into the
The roller *a* should not be fixed in bearings, but confined down up
cylinder by weights, suspended upon each end of the axis, which
adjusted according to circumstances, and in all cases the principal p
should be upon the roller *b*. The water which runs through the cylin
figs. 387 and 388, and out at the end, falls in the first instance into t
tern C, from whence it passes through the pipe *d* into the cistern
from thence is, by means of a pair of double-acting pumps, *f f*, forced t
the pipe K into the vessel L, so that it continually returns for the same p
of conducting the pulp to the cylinder from the pipe J. The pipe G i
of gauge, by means of which, after the pulp rises to a proper height
vessel L, the remainder of the water is carried off into the cistern G
there may be a waste pipe for conveying off the superfluous quantity.
water drawn from the cylinder R, through the trough V, by means
air and water pumps X X, may run to waste. The size of the cylin
and of the trough V, must be regulated according to the substance a
mensions of the paper it is intended for making. Fifteen inches w
sufficient for the diameter of a cylinder intended for making paper eq
substance to a paper twenty-two inches by seventeen inches and
weighing twenty pounds per ream: the length of the cylinder is a
arbitrary. The thickness of the paper made by a cylinder may be ad

a various ways : first, by using cylinders of various diameters; secondly, by accelerating or retarding the motion of the cylinder; thirdly, by varying the proportion of the surface of the cylinder, which is covered with pulp; fourthly, by varying the consistency of the pulp. The periphery of the cylinder ought to move at the rate of about thirty-six feet per minute; the pulp ought at all events to be very thin, and therefore the most eligible mode of adjusting the thickness of the paper would be by varying the proportion of the surface of the cylinder, which is covered with pulp; consequently for thicker papers a larger cylinder would be necessary, or a back may be made use of, extending higher up towards the point Z, so as to cover a larger proportion of the surface of the same cylinder : and for thinner paper a back might be made use of covering less of the cylinder, as in fig. 390, by means of the cock in pipe J. The quantity of pulp supplied to the cylinder can be adjusted with the greatest accuracy, consequently the thickness of the paper may be preserved uniform, or varied at discretion, provided the thickness of the pulp in the chest A, and· the motion of the cylinder R be continued uniform. By means of the gauge pipes Q the level of the pulp in the back P can be varied till the most eligible point for the cylinder to emerge from the pulp is ascertained, and the supply of water through the pipe r must be adjusted accordingly. It may be laid down as a general rule, that the thicker the paper the higher should be the level of the pulp in the back. In order to close the trough V tight upon the cylinder R, the patentee proposes packing it all round the top, where it comes in contact withinside of the cylinder, as at the points nn, in the section fig. 380.

The mode of packing is so well known, that it is unnecessary to give any description, except the representation in the drawing. The friction of the back P upon the cylinder may be taken off by strips of woollen cloth or leather, particularly at the line across from the point S.

Figs. 391 and 392 are for the purpose of explaining a more simple mode of construction. A is a hollow cylinder, with a pervious surface, which may be used in cases when the pneumatic pressure is not applied; a a a is the thread of a screw; b b b represent cross-bars, carried across the internal surface, parallel with the axis. The best mode of constructing it will be to cast a cylinder with the bars in the inside, and to cut the screw deep enough to form an opening between every bar. It should be furnished with cross wires, c c c, and covered with wove wire, in the same way as the cylinder L. It might be made on a larger or smaller scale, according to the purpose for which it is required. The roller b, in figs. 387 and 388, may be made in this manner, but stronger, as a great degree of pressure is intended to take place upon it.

When the machinery is to work, the agitator and pumps should be set in motion; first by turning the shaft K, and then the cylinder R, by means of the cog-wheel P, which gives motion to the rollers a and b, by cog-wheels q and r. The mode of giving motion, and the situation of the pumps and stuff-chest, may be arranged according to convenience, but the motion ought to be perfectly regular.

and on that account is much avoided. An objection has been started to the applying of this power, under a supposition, that the animal by changing his speed would injure the cotton; it is almost superfluous to add that many simple contrivances may be adapted to equalize the motion, and prevent these dreaded effects.

When the cotton has undergone either of these processes, it is packed, and exported to the European markets.

When it arrives in this country, it is again submitted to the action of machinery for the further separation of the extraneous matter, unless it is to be spun into coarse yarn, when the preceding process is considered sufficient.

The first machine that we shall describe as used in this country for the further cleaning of the particles is called a *picker*, and is represented in fig. 393. A and B are two rollers, having an endless-cloth, C D, stretched over them. This cloth is called the feeding-cloth, and its upper surface is, by the revolution of the rollers, always carried towards D. E and F are two fluted rollers, which nearly touch each other, and revolve, so that their touching surfaces pass towards G H. G H I K are cylinders, covered on their outer surfaces with long blunt pins, making about 250 revolutions, in the direction of the letters, per minute. L L is a grating of wires for the seeds to fall through, when the cotton carried by the feeding-cloth is delivered by the small rollers upon the face of G H. By the rapid revolution of G H, the cotton is thrown against the top O P, and is carried forward and delivered upon the cylinder I K, which in like manner carries it rapidly round, draws it over the grating, and delivers it back upon the lower face of G H, which after having drawn it over the remainder of the grating, and divested it of the remainder of the seeds and particles of dust, deposits it in the box R R.

This machine is liable to injure the staple of the cotton, and is therefore superseded by another called a *batter*, represented in fig. 401. In this machine, the feeding-cloth upon the rollers A and B carries forward the cotton to the rollers c and d, which deliver it upon the curved rack or grating d e, while a scotcher, g h, revolving rapidly upon its axis, strikes the cotton with its two edges g and h, and divides it; at the same time a draught of air, created by the revolution of the fan I, blows the cotton forward over the grating K K, divests it of the superfluous parts, and ultimately deposits it in a box at the end.

The cotton is now considered in a state fit for the operation of spinning; which is differently performed according to the purposes to which the yarn is to be applied. The different sorts of spinning may be classed under the respective heads of Jenny, mule, and water spinning.

Mule-spinning, which is by far the most perfect process, and by which the finest yarn is produced, shall first have our attention.

In this process, when the finest yarn is to be produced, the cotton, instead of being submitted to the operation of either of the machines before described, is cleansed entirely by the hand. The mode of effecting this is, by spreading the cotton

upon a strong netting of cords stretched on a frame,
ing it with osier wands till divested of its impurities
undergoes the elementary operations of carding,
stretching and plying, and twisting; the whole of w
essential in the manufacture of mule yarn.

Carding is performed by two kinds of engines, one
called the breaker, operates upon the cotton prepara
being submitted to the operation of the other, c
finisher.

A card is a kind of brush, formed by making wires into
staples, as represented in fig. 394. The two legs of the staples
through holes in a flexible piece of leather, and present to the s
form similar to that shown in the figure, where A B is the leathe
the wires forced through it. Cards are formed in two ways; the
sheet-card, is made about four inches wide, and 18 inches lon
length corresponding with the width of the main cylinder, which
to cover; the other, called fillet-card, is made in one continuo
fillet, and is used for covering the doffer cylinder. The teeth of th
are placed pointing in the direction of the length of the fillet, and
cover the cylinder to which they are applied; whereas in sheet-c
is left between every sheet, as may be seen on the main cylinder,

Fig. 395 represents a sectional view of the immediate working
breaker carding-engine. A is the main cylinder, covered with s.
B the doffer cylinder, covered with fillet-card; C C C are the tops
feeding-cloth supplied with cotton, which has been previousl
moving forward over the roller, f, by means of the roller g, and
the cotton between the feeding-rollers H H, which carry it to the
linder. The main cylinder revolves rapidly in the direction of th
carries the cotton upward between itself and the tops, which a
with sheet-cards, about 1½ inches to 2 inches wide, so that the
nearly as possible, follow the curve of the main cylinder. I is th
cylinder, having a wooden roller I¹, laying upon its upper surfa
is the doffer or taker-off, having affixed to it the steel comb
doffing-plate.

The doffing-plate may be seen more at large in fig. 396, which
a front view of the doffer cylinder on a larger scale. On insp
figure, it will be seen, that the doffing-plate L L, whose low
formed like a comb, is fastened across the whole of the doffer cyl
is supported by the two uprights m m, fastened on two cranks on
n³ n². The upper parts of these uprights, m m, are fastened to corr
cranks at n n, so that the doffing-plate, by the revolution of th
made to move downwards while in contact with the doffer cylinde
wards while away from it. The cotton is taken in by the feedi
and is carried up by the main cylinder and passed between it and
or flats, whose teeth lie in an opposite direction to these of
cylinder, and by whose united action the cotton is combed, di
cleansed, and its fibres placed in a direction more parallel to each

The main cylinder, by its revolving motion, is soon
with cotton, and is divested of it by the doffer cylinde
is placed so as nearly to touch it, and which moves a
slower speed, in the direction of the dart. The effec
engine would therefore be to distribute the cotton equ

cylinder, the top cards, and the doffer cylinder; but
g-plate, by the action already described, is continually
the doffer cylinder; whose points are consequently
to receive a fresh supply from the main cylinder.
ng-plate continually strips the doffer cylinder of the
tton, which it delivers upon the lapping cylinder in
nuous web of about 18 inches wide, which is the usual
the engines for fine work.

the top cards are covered with cotton, an attendant
ted to take them off and to divest them of the cotton
of a card nailed on a board, which he carries in his
that purpose.

antity of work delivered to the engine is ruled by the
he cylinders and quality of the cotton. When it has
rough the engine, and is wound upon the lapping
(which is so adjusted as to contain about 20 laps,)
ant lifts up the roller l¹, makes a division in the
eb, and takes it off the roller.

operation we are presented with the first act of ply-
ubling, which is introduced in the process of spin-
der to obtain equality in the strength and thickness

on is in this state called a lap, and is immediately taken to a
ine, which, in general, is disposed back to front, immediately after
engine, as may be seen in fig. 397. The construction of the
ine is exactly similar to that of the breaker-engine, except that
aving a lapping cylinder, the cotton, when it leaves the doffer, is
gh a mouth-piece, R, formed like the end of a trumpet, by means
s and t, and is delivered into the can W. The rollers s and t
in section in this figure, and in a front view in fig. 396. Pre-
wever, to leaving this process, we shall make a few remarks, as
much propriety, considered the very foundation of all good

eaker-engine for spinning fine cotton is generally
ith cards of a fineness that will admit 225 teeth, or
s, in a square inch; and the finisher 275, or 550.
iers are much divided on this subject, and in some
same work is performed with cards one-fifth coarser
in others. The top cards are in general one-tenth
nd those of the doffer cylinder one-tenth finer, than
the main cylinder: and in some manufactories, at the
of the engines, where the cotton first arrives, coarser
have been introduced, with a view of divesting the
the largest particles of extraneous matter, and in
tances have been again laid aside as superfluous.
t be set easy in the leather, which should be thin
ng. The card-engine is driven by a strap passing

from a drum over a fast and loose pulley, fixed on the shaft of the main cylinder. The fast and loose pulley is represented in fig. 65 ; and its utility has been explained in the article MILL-GEERING.

To return to the manufacture, the cotton, which is now in the can from the card-engines, in the form of a sliver, is next submitted to the process of drawing, represented in fig. 398. In this process three or four card-ends are brought in tin cans, and passed between the rollers A B and C D, which revolve with different velocities ; that is, the rollers C and D revolve much quicker than A and B, and the top rollers A and C are made to press upon B and D by means of the weight e. Now, supposing four slivers be placed together, and passed through the rollers A B and C D, and that C D revolve so much quicker than A B, that the sliver will become four times its original length, the cotton will, by such elongation, be reduced in thickness three-fourths, that is, to the same thickness of a single sliver when first brought to the rollers. By this process the fibres of the cotton are laid more parallel to each other, in the direction of the length of the sliver, and the operation is repeated, by plying the slivers which have passed the rollers, and passing them through a similar set. The sliver, when thus plied and reduced, is drawn through the mouth-piece G, by the rollers E and F, and delivered into another can.

After the cotton has been plied and drawn as many times as the spinner, from the quality of the cotton, and the intended quality of the yarn, considers necessary, it is carried to the roving-frame.

The *roving-frame*, which is much used in mills where mule-spinning is carried on, is represented in fig. 402, and is termed the *can roving-frame*. A B, two rollers, moving at a slower speed than C D ; A and C are pressed upon the rollers B and D by the weight E, as may be seen in a front view, fig. 402, and section, fig. 403. The cans (fig. 402) are represented, the one shut, and the other open ; the latter opens by means of hinges, after raising the ring g. The cans are capable of revolving upon their spindles h h, and are supported in an upright position by the collars i i, and have at their upper extremities funnel-shaped pieces, k k.

If two slivers of cotton are brought from the drawing-frame, and passed between the rollers A B and C D, the processes of plying and drawing will again take place ; and the rollers C D will feed the end thus formed into the can through the mouth-piece at k, which, by revolving rapidly upon its axis, will impart to the end, or sliver, a slight degree of twist. When the can is filled, the rollers are thrown out of geer, and the motion ceases; the can is then opened, and the cotton, or as it is now called, the roving, is taken out and wound upon a bobbin, and in that state is carried to a machine called a stretcher.

Some objections exist against this species of roving ; first, from the necessity of taking the roving out of the can for the purpose of winding it upon a bobbin, during which it is liable to sustain much damage from the fibres being in a very slight state of adhesion ; and secondly, from the roving receiving its twist solely from the revolution of the can in which it rests, and by which the twist is not equally diffused over the whole length of the roving. The first objection was attempted to

The yarn, delivered from the stretching-frame in the form of a cop, is taken to the mule, which is, though much lighter, both in the form and action of the parts, very similar to the stretching-frame. The spindles also are of a smaller size, and are situated nearer to each other.

The *mule spinning-frame* differs from that of the stretching-frame insomuch as the act of stretching is added to the other operations ; for when the frame E E E has receded a certain distance, generally about one yard, the rollers C C C cease to move, and the frame still continuing to recede, stretches the yarn. During this process, the spindles on the frame E E E move considerably quicker, in order to save time. The stretching is performed with a view to elongate and reduce those places in the yarn which have a greater diameter, and are less twisted than the other parts, so that the size and twist of the yarn may be more uniform throughout. When the cops are full, they are taken from the moving spindles, and placed on stationary parts of other mules, as at A, and the yarn is again submitted to the same process, until it is reduced and spun to the proper fineness, both as respects the diameter and the twist ; during the whole of which process, the yarn can be continually joined, so that the cops, which are in separate pieces, can be added to each other in parts, or otherwise, as the continual elongation of the yarn in the course of the different operations of each mule may require. The pieces are joined by children, called *piecers*, who are in attendance on each mule, to join any yarn that may be broken in the act of stretching or twisting.

The drums, which drive the spindles in those parts of the mule that recede, receive their motions from bands communicating with the moving power ; but the advancement and recession of the carriage, for the purposes of receiving and stretching the yarn, as before described, is performed by means of a wheel moved by hand-labour. A spinner is enabled by experience to judge of and regulate both these operations, as also the building of the cop, which is a matter of very great nicety ; for if the cop is not well built, the yarn will not run off even when it is to be used. The number of spindles on a mule amount frequently to 300. The yarn produced by mule-spinning, being by far the most perfect, is employed in the fabrication of the finest articles, such as lace and hosiery ; and when it is twisted in two, four, or six plies, is used for sewing-thread.

Jenny-spinning is of earlier date, and a much less perfect process than mule-spinning ; consequently it is but little used, except in the manufacture of yarn for coarse goods. In this spinning, the cotton, after having been cleansed by some of the processes already described, is, preparatory to being exposed to the action of the jenny, immersed in a solution of soap and water, to divest it of the glutinous matter generally found on the surface of this and other vegetable fibres ; it is then, after the soap and water has been pressed from it, put into a warm stove, and when dry, is considered to be in a fit state to be exposed to the operation of the carding-engine.

The carding-engine used in jenny-spinning is different in its construction to the one before described ; for in mule and water spinning there is a breaker and a finisher engine ; but the engine used in this process is called the double-engine ; the first part, or breaker, is in the same frame with the second part, or finisher, and the doffer from the first part delivers the cotton upon the main cylinder of the second part, which, in like manner, delivers it upon the second doffer. The second doffer, instead of being covered with fillet-cards, as the doffer of the single engines, is covered with sheet-cards, like the main cylinder, but being of smaller dimensions, has generally only twelve cards upon it ; therefore the web of cotton combed from such doffer by the doffing-plate is not in one continuous piece, but in several pieces or portions, equal to the quantity attached to each sheet-card upon the doffing-cylinder.

As the several small portions are delivered by the comb, they fall into the concave part of a smooth arc that is equal to one-third of a circle. In this arc a cylinder of smooth mahogany slowly revolves in such direction that the lower surface in the arc passes from the engine. This cylinder has small cavities or flutes on its surface, in a parallel direction to its axis ; the angles on the projections between the flutes are taken off, so that the several portions of web which fall from the doffer into the arc are seized by the flutes, and carried forward on the concave face of the arc, and formed into a sliver, about half an inch in diameter, and of a length corresponding with the breadth of the carding-engines, which is about from 24 to 34 inches. The portions thus rolled are called rows, rolls, or rowans.

In this state, the cotton may be considered in the same relative state of progress as a card-end in mule or water spinning ; but it is evident that this mode of spinning is very deficient for the purposes of fine yarn, insomuch as in the rowans the fibres of the cotton are laid across the longitudinal direction in which they are to be spun, so that the advantage derived in the other process of carding, from the fibres being

2 c

placed in a direction parallel to the intended length of the yarn, is entirely lost. In this process, also, the advantage of plying, which we have noticed as taking place on the lapping cylinder, is omitted.

When the rowans are perfected by the mahogany cylinder, they are taken up by children, and placed upon the feeding-cloth of a machine called the *billy*, or *roving-billy*, the operation of which is called roving or slubbing; but the latter expression is now but seldom used, except in the manufacture of woollen. This machine is in its construction and action very similar to the mule, as is the feeding-cloth to that described in the machine called the picker and batter.

The feeding-cloth lays in a slanting position, and the rowans are placed upon it so that they can pass lengthwise in the direction of its action, and be delivered over the upper roller between two pieces of board which possess a capability of clasping and again relieving them. The rowans are then attached to revolving spindles, which have an advancing and receding motion similar to the mule or drawing-frame. By this revolution and recession the spindles perform the operation of spinning and stretching; and at such intervals as the spindles are stretching and twisting, the feeding-cloth stops, and the clasps seize hold of the roving, and detain it till sufficiently spun and twisted, when it relieves it in order to allow a further portion of the rowan to be fed. The roving having by this means received a certain degree of twist, is built on a spindle in the form of a cop, as in mule-spinning, and is then taken to the machine called the jenny.

The operation of the *jenny* is nearly the same as the roving-billy; the only material difference is, that the cops of roving to be spun are fixed upon a moving carriage, which has clasps to hold the roving while in the act of being stretched and spun into yarn.

Having now concluded the process of jenny-spinning, it will be seen, that drawing and plying, the two essential requisites for producing fine yarn, by placing the fibres parallel to the length of the twist, are wanting, and that fine yarn, in consequence, cannot be produced; but the fibres in this process being placed in a direction more across the length of the twist, give to the yarn a rich fulness which renders it preferable for the weft of heavy goods, for which it is esteemed.

Water-spinning differs both from the mule and jenny spinning; but the carding and drawing machines are the same as those used in the process of mule-spinning. When the cotton has passed through the carding and drawing machines, it is

When the reel has made 80 revolutions, a small b
connected with the machinery rings, and warns the
to stop the motion of the reel. The portion thus
called a lay, and seven of these lays wound upon
reel constitute a hank, which is taken from the
causing one of the horizontal bars, supplied with a
fall inwards. The circumference of the reel is a ya
half; consequently the hank measures 840 yards.
of the twist is expressed by stating how many han
the pound weight: thus, the yarn called N° 100 is th
takes 100 hanks of 840 yards each to weigh an av
pound. Yarn can be spun upon mules as fine as 2
to the pound; but in water-twist and jenny-spinning
exceeds 60 or 70.

The plan of the buildings in which the cotton-
machinery is placed, is generally in the form of a
ogram, of a length proportionate to the extent of th
facture carried on therein, and about thirty feet wide.
best constructed mills; the carding and other pre
machines are placed on the lowest floor; the mo
stretching frames on the next; and so on progres
the machines improve the fineness of the yarn. Th
jennies, and water-frames are placed with their line
dles across the building; and the card-engines have
of their cylinders parallel to the long wall of the
Four or six rows, breakers and finishers, are
nately.

The steam-engine, or first mover, is placed at on
the building, and the motion is communicated by a b
shaft running the whole length of the building, whic
mits the motion to vertical shafts with bevel-wheel
wheels transmit the motion to horizontal shafts in t
floors.

———————

WOOL MANUFACTURE.

This well-known staple is in the process of the m
ture divided into two distinct classes, *long wool,* or
spinning; and *short wool,* or the *spinning of woolle*

ON WORSTED SPINNING.

Having by means of machinery accomplished the
tion of a thread of cotton, the application of the prin

other fibres would naturally follow; and although some difficulty might be expected to occur in adapting the rollers to different staples, yet this was soon overcome. The methods of forming threads from long wool and from flax, by the hand, were very different, yet each was spun from the middle, not from the end, of the respective fibre. In hand-spinning, the pluck, that is, the portion plucked from the sliver or combed wool, was placed across the fingers of the left hand and from the thick part of it, the fibres were drawn, and twisted, as the hand was withdrawn from the end of the spindle, to which it had been previously attached. The revolution of the wheel, effected by the right hand, conveyed by a band to the whirl, or pulley on the spindle, produced the requisite twist to give firmness to the thread; and by a very gentle motion of the same wheel, the thread being brought nearly perpendicular to the spindle, it was wound upon the spindle to form the cop. From this it was transferred to the reel, and became a hank, of a definite length, but varying in weight with the thickness of the thread. In this state it was transferred to the manufacturer to be converted into the different fabrics of shalloon, calimanco, bombasin, &c.

A few years after the introduction of cotton machinery, an obscure individual of the name of Hargraves, previously unknown as a mechanic, who had been long employed by Messrs. William Birkbeck and Co. at Settle, in Yorkshire, in the management of a branch of the worsted manufactory, attempted to spin long wool by means of rollers. He constructed working models of the necessary preparing machinery, and of a spinning-frame, by the assistance of persons accustomed to the construction of cotton machinery, and succeeded so completely, as soon to induce his employers to build a large mill for its application. By degrees his plans became known to the trade, and many large manufactories have subsequently been erected for this purpose. Contrary to the earlier anticipations on this subject, it has been found, that mill-spun yarn answers better for the coarse as well as the finer fabrics, than that produced by the hand, which it has entirely superseded.

The first process after the wool of the fleece has been properly sorted, as it is termed, and washed, is combing. This is either done by the hand or by machinery, invented for that purpose some years since by the ingenious Dr. Cartwright. The object of each mode is to arrange the fibres as much as possible parallel to each other, which, as they have

a somewhat tortuous form, and are of considerable
requires them to be frequently drawn from each other
exertion of the strength of the wool-comber or the m:
In this state they form a bundle of fibres about six
length, called a sliver, and this being laid upon the s
ing or drawing frame, constitutes the commencement
preparing process. The wool passes through several [
rollers of which the first and last are of course the es
ones, the intermediate moving with equal velocitie
consequently serving merely to conduct the skein:
received in cylindrical cans; and three such skeins
passed through another drawing-frame, and stretched i
progress, become fitted for roving, the last step in th
paratory processes. Allowing for the difference in dis
of rollers and weights, which on account of the leng
adhesiveness of the fibres of wool, are both nece
greater than with cotton, the description of the bobbin
machine already introduced, will be sufficiently explan

Spinning, the concluding process, is effected by m
two pairs of rollers moving with unequal velocities, anc
mediate auxiliaries.

The loosely twisted thread from the roving bobbin, E, fig. 403, i
carried forwards by the holding rollers A, a, and supported as it
by the two pairs, C, c, and D, d. It is then drawn between the rol
and having been thus brought to a proper thickness, is twisted
flier L, fixed on the top of the spindle, through which at K it pas
then taken up by the bobbin M, which moves round with the sp
axis, although not equally quick. The ultimate thickness or siz
thread is determined by the difference of velocity in the holding and
pairs of rollers; that is of A, a, and B, b, which in their operation e
imitate a pair of hands. The celerity of the three pairs of rollers n
the back of the frame is equal; consequently no stretching tak
amongst them. The upper rollers of the first and last pair are
down upon the lower, by weights, F, G, much heavier than H, I, w
supported by the axes of C, D; these being only required steadily
forward the skein, and prevent the remote ends of the fibres of t
from starting, whilst B, b, are pulling their other extremities.
rollers belonging to one division, or box, as it is commonly ter
represented in fig. 409, where the drum, which moves the spindles
a bevelled pinion at the top of its axis conveys motion to the rolle
shown. The pinion on the right extremity of the roller, acting
train of wheels properly adjusted, imparts the required relative m
succession, to the rollers beyond.

SHORT WOOL.

Short wool is wrought into the finest cloths for p
wear, and is spun in a manner similar to cotton, as de.
in jenny-spinning.

SILK MANUFACTURE.

SILK is a very fine and delicate thread, the produce of a small insect, called *bombyx*, or the silk-worm; which is not less curious on account of the changes it undergoes in its existence, than valuable for the beautiful fibre which it spins,

The egg, requiring not the care of parental incubation, is by the solar heat brought into existence, and the bombyx or silk-worm thus produced lives upon the leaves of the mulberry-tree until it has arrived at maturity, when, spinning itself up in a small bag, about the size of a pigeon's egg, it is changed into an aurelia. In this state it continues till about the fifteenth day when it is changed into a butterfly, and, if not prevented, eats its way through the silken prison, to expand its newly acquired wings in the sun.

The ball or cocoon, which the ingenious little insect has been at so much pains to spin, to secure itself from its enemies and the effects of the weather, is the substance we call silk; and many who have examined it with attention are of opinion that it will extend to the distance of six English miles.

In order to secure the silk for the purposes of the manufacturer, it becomes necessary to destroy the insect so soon as the cocoon is completed, which is on or about the tenth day. The cocoon is of various colours; but the most predominant are flesh colour, orange, and yellow. The whole of them, however, are lost in the process of scouring and dying, and therefore it is not necessary to wind them on separate reels.

The balls, preparatory to being wound off into skeins or hanks, are immersed in hot water, which dissolves a natural gum, by which the fibres are united together, so that a single thread taken from the reel will be found to be composed of numerous small fibres or threads in the state produced by the worm.

The silk is imported into this country thus wound off into skeins, and in order to undergo the processes of the manufacturer is wound upon bobbins; and each thread being, as we before have stated, composed of several fibres, receives a certain degree of twist, that the constituent parts may be united more firmly together than they can possibly be by the gum alone. When they have been subjected to thus much of the manufacture, they are wound upon fresh bobbins, and two or three threads twisted together, to form a strong thread for the weaver, who warps and finally weaves the silk into various beautiful and useful articles, by a process very similar to that used in the weaving of cotton and linen.

In Piedmont, where very excellent silk is produced, the manufacture is carried on by aid of the silk reel represented in fig. 424.

The balls or cocoons are thrown into hot water contained in a copper basin or boiler, A, about 18 inches in length, and six deep, set in brick-work, so as to admit of a small charcoal fire beneath it. B B is a wood frame sustaining several parts of the reel ; D is the reel upon which the silk is wound ; C is a guide which directs the thread upon it ; and E F the wheel-work which gives motion to the guide. The reel D is merely a wooden spindle, having four arms mortised into it to support the four batteus or rails on which the silk is wound.

Upon the end of the wooden spindle of the reel, and within the frame B, is a wheel of 22 teeth, which gives motion to another wheel C, fixed upon the end of the inclined axis E F, and having twice the number of teeth ; at the end of this inclined axis is another wheel G, of 22 teeth, playing in a horizontal cog-wheel with 35 teeth. This wheel turns upon a pivot fixed in the frame, and has a pin fixed in it at a distance from the centre, to form an eccentric pin or crank, and give a backward and forward motion to the slight wooden rail or layer C, which guides the threads upon the reel ; for this purpose, the threads are passed through wire loops or eyes, C, fixed into the layer, and the end thereof opposite the wheel and crank F is supported in a mortise or an opening made in the frame B, so that the revolution of the crank will cause the layer to move, and carry the threads alternately towards the right or left. There is likewise an iron bar H, fixed over the boiler at H, and pierced with two holes, through which the threads pass to guide them.

In the operation of reeling, it is well known, that if the thread be wound separately it will be totally unfit for the purposes of the manufacturer ; consequently the ends of the threads of several balls or cocoons are joined and wound together, and when any one of them breaks or comes to an end, its place is supplied by a new one, and thus by continually keeping up the same number the united threads may be wound to any required length.

The reeling is conducted by a woman, who, when the balls or cocoons have remained a sufficient time in the hot-water contained in the boiler A, to soften the gum, takes a whisk of birch or rice-straw, about six inches long, cut stumpy like a worn-out broom, and brushes the cocoons with it, which causes the loose threads to adhere to it ; these she disengages from the whisk, and by drawing them through her fingers cleans them from the loose silk, which always surrounds the cocoon, till they come off clean, which operation is called *la battue*. When the silk has been perfectly cleansed, she passes four or more of the threads, if she intends to wind fine silk, through each of the holes in the thin iron bar H, and afterwards twists the two compound threads, consisting of four cocoons each, about 20, or 25 times round each other, that the four ends in each thread may the better join together by crossing each other, and that the thread of the silk may be round which otherwise would be flat.

The threads when thus twisted together are passed through the eyes of the loops, C, of the layer, and thence are conducted and made fast to one of the rails of the reel. As it is of consequence in the production of good silk, that the thread should have lost part of its heat and gumminess before it touches the bars of the reel, the Piedmontese are by law obliged to have 38 French inches between the guides, C, and the centre of the reel; and the layer must also, under a penalty, be moved by cog-wheels instead of an endless-cord, which, if suffered to grow slack, will cause the layer to stop and not lay the threads distinctly, and that part of the skein will be glued together, whereas the cog-wheel cannot fail: when the skeins are quite dry, the reel is removed from the frame, and by the folding of two of its arms, by means of hinges, the skeins are taken off, and with some of the refuse silk are tied into hanks.

Although from the foregoing description the operation must appear very simple, it is a matter of very great nicety to wind an even thread, and the difficulty of keeping the thread always even is so great that, except when using a thread of two cocoons, they do not say a silk of three, four, or six cocoons; but a silk of three or four, four or five, five or six cocoons. In a coarser silk it cannot be calculated even so nearly as to four cocoons, and consequently they say, from 12 to 15, from 15 to 20, and so on.

It is also necessary that the water in the boiler be kept at a certain temperature; for if the water is too hot, the thread is dead and has no body; if too cold, the ends of the threads do not join well, and form a harsh silk. The threads themselves indicate when the water is not at the proper degree of temperature, by frequent breaking when it is too hot; and coming off entangled, and in a woolly state, when too cold.

In the process of winding the woman has always a bowl of cold water by her, into which she occasionally dips her fingers, and frequently sprinkles it upon the iron bar H, that the threads may not be burnt by the heat of the basin; it also serves to lessen the temperature of the water in the boiler when approaching the boiling point.

All kinds of silk which are simply drawn from the cocoons by the process of reeling are called raw silk, and is denominated coarse or fine according to the number of fibres of which the thread is composed. In preparing the raw silk for

the thread is slightly twisted, in order to enable it to
he action of the hot liquor without the fibres separating
'ng up. The silk-yarn employed by the weavers for
of or weft of the stuffs which they fabricate, is composed
or more threads of the raw silk, slightly twisted by the
machinery; and the thread employed by the stocking-
r is of the same quality, but composed of a greater
er of threads, according to the thickness required.

nzino silk consists in combining together two or more
s of silk, each of which has in the first instance been
d by itself, and afterwards the whole are twisted toge-
This operation, with the exception of the elongation
cotton, closely resembles roving in the Cotton Manu-
. The process consists of six different operations.
c silk is wound from the skein upon bobbins in the
ig-machines. 2. It is then sorted into different quali-
3. It is spun or twisted on a mill in the single thread,
ist being in the direction of from right to left, and more
tight, as the purposes to which the silk is to be applied
equire. 4. Two or more threads thus spun are doubled
wn together through the fingers of a woman, who at
ie time cleans them, by taking out the slubs which
are been left in the silk by the negligence of the foreign
. 5. It is then thrown by a mill, that is, two or more
s are twisted together, either slack or hard, as the
acturer may require; but the twist is in an opposite
ion to the first twist, and it is wound at the same time
ins upon a reel. 6. The skeins are sorted according
ir different degrees of fineness, and then the process is
etc.

first operation which the raw silk undergoes is winding,
, drawing it off from the skeins in which it is imported,
'nding it upon wooden bobbins, in which state it can
the other machines.

h of the skeins is extended upon a slight reel called
t; it is composed of four small rods, fixed into an axis,
all bands of string are stretched between the arms to
: the skein, but at the same time the bands admit of
' to a greater or less distance from the centre, so as to
sc the effective diameter of the reel, according to the
! the skein, because the skeins, which comes from dif-
vary in size, being generally an exact yard,
measure, of the country where the silks are
swifts are supported upon wire pivots, upon
freely when the silk is drawn off from them;

but in order to cause the thread to draw with a gentle force, a looped piece of string or wire is hung upon the axis within-side the reel, and a small leaden weight is attached to it, to procure friction. The bobbins which draw off the threads are received in the upper part of the frame, and are turned by means of a wheel beneath each, the bobbin having a small roller upon the end of it, which bears by its weight upon the circumference of the wheel, and the bobbin is thereby put in motion to draw off the silk from the swift. A small light rod of wood, called a layer, which has a wire eye fixed into it, is placed at a little distance from, and opposite to, each bobbin, so as to conduct the thread thereupon ; and as the layer moves constantly backwards and forwards, the thread is regularly spread upon the length of the bobbin. The motion of the layer is produced by a crank fixed upon the end of a cross-spindle, which is turned by means of a pair of bevelled wheels from the end of the horizontal axle, upon which the wheels for turning all the bobbins are fixed.

These winding-machines are usually double, to contain a row of bobbins and swifts at the back as well as in front. Two of these double frames are put in motion by cog-wheels from a vertical shaft, which ascends from the lower apart-ments of the mill, where the twisting-machines are placed. The winding-machines require a constant attendance of children to mend the ends of threads which are broken ; or when they are exhausted, they replace them by putting new skeins upon the swifts. When the bobbins are filled, they are taken away, by only lifting them up out of their frame ; and fresh ones are put in their places.

A patent has been lately taken out by Messrs. Gent and Clarke, for a new construction of the swifts for winding-machines : they are made with six single arms, instead of four double ones ; and the arms are small flat tubes made to con-tain the stems of wire forks, which receive the skein, instead of the bands of string in the common swifts. These forks admit of drawing out from the tubes until the swift be suffi-ciently enlarged to extend it ; but as they extend the skein at six points instead of four, as in the common ones, the motion is more regular. Instead of the weight which causes the friction, a spring is used to press upon the end pivot of the axis, and make the requisite resistance.

The twisting of the silk is always performed by a spindle and bobbin, with a flyer, but the construction of the machine is frequently varied. The limits of our plate do not admit a representation of the great machines, or throwsting-mills, at their full extent; but the principle is the same as fig. 426, which we have extracted from Dr. Rees's Cyclopædia, varying the description a little, to agree with the present improved state of the manufacture.

In fig. 426, we have given a drawing of a small machine, which is similar in the parts which act upon the silk; and indeed many mills employ such machines constructed on a large scale. The one in our plate contains only thirteen spindles, and is intended to be turned by hand, a method which is too expensive for this country, but is common in the south of France, where many artisans purchase their silk in the raw state, and employ their wives or children to prepare it by these machines, which they call *ovales*, because the spindles, *b b*, are arranged in an oval frame, G H.

B is the handle by which the motion is given; it is fixed on the end of a spindle R, which carries a wheel D, to give motion to a pinion upon the upper end of a vertical axle E; this, at the lower end, has a drum or wheel F, to receive an endless strap or band, *a a*, which encompasses the frame G, and gives motion to all the spindles at once. The spindles *b b* are placed perpendicularly in the frame G H, their points resting in small holes in pieces of metal, which are let into the oval plank G; and the spindles are also received in collars affixed to an oval frame H, which is supported from the plank G, by blocks of wood; *d* and *a* are small rollers supported in the frame G H, in a similar manner to the spindles; their use is to confine the strap *a*, to press against the rollers of the spindles with sufficient force to keep them all in motion.

The thread is taken up, as fast as it is twisted, by a reel K, which is turned by a wheel A, and a pinion *i*, upon the end of the principal spindle R. The threads are guided by passing through wire eyes, fixed in an oval frame L, which is supported in the frame of the machine, by a single bar or rail *l l*, and this has a regular traversing motion backwards and forwards, by means of a crank or eccentric pin R, fixed in a small cog-wheel, which is turned by a pinion upon the vertical axis E; the opposite end of the rail *l* is supported upon a roller, to make it move easily. By this means the guiders are in constant motion, and lay the threads regularly upon the reel K, when it turns round, and gathers up the silk upon it as shown in the figure.

One of the spindles is shown at *r* without a bobbin, but all the others are represented as being mounted and in action. A bobbin, *o*, is fitted upon each spindle, by the hole through it being adapted to the conical form of the spindle, but in such manner that the bobbin is at liberty to turn freely round upon the spindle; a piece of hard wood is stuck fast upon each spindle, just above the bobbin, and has a small pin entering into a hole in the top of the spindle, so as to oblige it to revolve with the spindle, this piece of wood has the wire-flyer, *b*, fixed to it; the flyer is formed into eyes at the two extremities; one is turned down, so as to stand opposite to the middle of the bobbin *o*; and the other arm, *b*, is bent upwards, so that the eye is exactly over the centre of the spindle, and at a height of some inches above the top of the spindle. The thread from the bobbin, *o*, is passed through both the eyes of this wire, and must evidently receive a twist when the spindle is turned; and at the same time, by drawing up the thread through the upper eye, *b*, of the flyer, it will turn the bobbin round, and unwind therefrom. The rate at which the thread is drawn off from the bobbin, compared with the number of revolutions which the flyers make in the same time, determine the twist to be more or less hard. This circumstance is regulated by the proportion of the wheel A to the pinion *i*, from which it receives motion; and these can be changed when it is required to spin different kinds of silk.

The operation of the machine is very simple. The bobbins filled with silk in the winding-machine, fig. 425, are put loose upon the spindles at *c*, and the flyers are stuck fast upon the top of the spindles; the threads are conducted through the eyes of the flyers *b*, and of the layers L, and are then made fast to the reel K, upon which it will be seen that there are double the number of skeins to that of the spindles represented, because one half of the number of spindles is on the opposite side of the frame, so that they are hidden. With this preparation the machine is put in motion, and continues to spin the threads by the motion of the flyers, and to draw them off gradually from the bobbins, until the skeins upon the reel are made up to the requisite lengths. This is sometimes known by a train of wheel-work at *n o p*, consisting of a pinion, *n*, fixed upon the principal spindle R, turning a wheel, *o*, which has a pinion fixed to it, and turning a larger wheel *p*; this has another wheel upon its spindle, with a pin fixed in it, and at every revolution raises a hammer and strikes upon a bell, *s*, to inform the attendant that the skeins are made up to a proper length.

In the silk-mills they employ two different machines, one for the first operation on organzine, and the other for the second operation.

Thus, after the silk is twisted it must be wound on fresh bobbins, with two or three threads together, preparatory to twisting them into one thread. In the original machines at Derby this was done by women, who, with hand-wheels, wound the threads from two or three of the large bobbins upon which the silk is gathered instead of the reels, and assembled them two or three together upon another bobbin, of a proper size to be returned to the twisting-mill.

In 1800, Mr. John Sharrar Ward, of Bruton, obtained a patent for a new method of doubling silk, worsted, cotton, or flax, which we intend to describe here; for though various modes are adopted for this purpose, one will be sufficient to give an idea of the whole. Whatever number of threads may be required to be doubled together, they may by means of this invention be doubled to the greatest certainty; for if at any time any one of the threads, or union of threads, to be so doubled, should break, it will immediately stop the other thread or threads until the broken thread shall be repieced, which secures a constant double thread, or union of threads; and the manner in which the same is to be performed will, we trust, be clearly understood by the subjoined description.

Fig. 429. A is a roller carrying round the bobbin B, which draws the threads C C from the bobbins D D; consequently the balls E E, and the thread-wires F F, move round on the pins G G. H H are two wood or iron standards, at the tops of which are hung two regulating thread-wires, I I. When either of the threads C C break, the thread-wire through which it passes falls down, and the tail part K rises up to a level with the ball E, and stops the other thread-wire from going round, and consequently the thread that passes through it, and prevents the bobbin B from taking it up; but the roller A continues its motion. L L are guide-wires for the threads to pass over; M is a slide, moved by a short wheel or crank, to lay the threads level on the bobbins.

Fig. 430 is another doubling-machine, the form varied, but the principle the same as fig. 429. A is a roller, whereon lies a smaller one, marked B, the axis of which goes through the bobbin; C is a slide, for the same purpose as M, in fig. 429; D D two bobbins, with spindles through them, on each of which is fixed a wheel E E; F F are two thread-wires hung at G G. When either of the threads break, the wires drop between the teeth of the wheel, and stop the other thread, the bobbin and roller B stopping at the same time; but the roller A continues moving, as A in fig. 429.

The bobbins being thus filled with double or triple threads, are carried back to the throwsting-machine, and are there spun or twisted together in a manner similar to that before described. At this period, the silk is a marketable article, and is passed from hand to hand.

The silk being now spun, is put into a boiler filled with hot water, into which is put a small quantity of soap, in order to divest the silk of its gum. In the earlier processes, the gum was necessary for the purposes of the manufacture, for the silk, had it been divested of it, would have assumed a fine downy appearance similar to that of cotton, and must have undergone similar operations before it could have been formed into a thread; this, indeed, is necessary for that portion of waste silk which is drawn from the cocoons in the first operation of reeling; also for those cocoons which have been reserved for breed, or which is made in the operations of twisting just mentioned, through which the moth or butterfly has eaten a hole, and rendered them impracticable to be wound off into silk.

The silk is now taken to the warping-mill, which, being a precursor to the act of weaving, will be noticed under that head.

At this present moment several improvements are in progress for winding and throwing silk upon a new principle; indeed, the silk manufacture now may be compared with what the cotton manufacture was about thirty years ago. There appears to be taking place in every department the same great and rapid improvements; and it is much the opinion of practical men, that the machinery now in use will, in the course of a very few years, be entirely superseded, and that this branch of our manufactures will ultimately be almost, if not quite, as great a source of national prosperity as the cotton manufacture.

The art of throwing silk was first introduced into this country by Mr. John Lombe, who, with considerable ingenuity, and at the risk of his life, took a plan of one of these complicated machines in the king of Sardinia's dominions from which, on his return, he, in conjunction with Mr. Thomas Lombe, established a similar set of mills at Derby. Parliament granted them a patent for fourteen years; and, on being petitioned at the end of that term for a renewal, granted them 14,000*l.* instead, on condition that they should allow a perfect model to be made, and placed in the Tower for public inspection.

FLAX MANUFACTURE.

FLAX undergoes various processes before it can be worked into cloth or other articles; these processes are very different, and require different sorts of implements and machinery, in order to their being properly performed. Flax, for the purpose of being formed into cambric, fine lawn, thread, and lace, is dressed in rather a different manner to that which is employed for other purposes; it is not scotched so thoroughly as common flax, which from the scotch proceeds to the heckle, and from that to the spinner; whereas this fine flax, after a rough scotching, is scraped and cleansed with a blunt knife upon the workman's knee, covered with his leather apron; from the knife it proceeds to the spinner, who, with a brush, made for the purpose, straightens and dresses each parcel before she begins to spin it.

In the *Swedish Transactions* for the year 1747, a method is given for preparing flax in such a manner as to resemble cotton in whiteness and softness, as well as in coherence; for this purpose, a little sea-water is directed to be put into an iron pot, or an untinned copper kettle, and a mixture of equal parts of birch-ashes and quick-lime strewed upon it, a small bundle of flax is to be then opened and spread upon the surface, and covered with more of the mixture, and the stratification continued till the vessel is sufficiently filled. The whole is then to be boiled with sea-water for ten hours, fresh quantities of water being occasionally supplied in proportion to the evaporation, that the flaxy matter may never become dry. The boiled flax is to be immediately washed in the sea, by a little at a time, in a basket, with a smooth stick at first, while hot; and when grown cold enough to be borne by the hands, it must be well rubbed, washed with soap, laid to bleach, and turned and watered every day for some time. Repetitions of the washing with soap expedite the bleaching; after which, the flax is to be beat, and again well washed; when dry, it is to be worked and carded in the same manner as common cotton, and pressed between two boards for forty-eight hours. It is now fully prepared and fit for use. It loses in this process nearly one-half its weight, which, however, is abundantly compensated by the improvement made in its quality, and its fitness for the finest purposes.

The *flax-brake* is a hand instrument, or machine, which was originally, and for many ages, chiefly employed in breaking and separating the boon or core from the flax, which is the cuticle or bark of the plant. In performing this business, the flax being held in the left hand, across the three under teeth, or swords of the brake, shown at A, figs. 432 and 433, the upper teeth or swords B, fig. 432, and *b*, fig. 433, are then with the right hand quickly and often forced down upon the flax, which is artfully shifted and turned with the left hand, in order that it may be fully and completely broken in its whole length.

The *flax foot-brake* is an implement, or machine, of the brake kind, invented in Scotland, by which flax is broken and scotched with much greater expedition than by the hand instrument just described, and in a more gentle and safe manner than by the flax-mill. By this contrivance, the boon or stem is well broken, and the sloping stroke given as with the scotcher, while the machine is moved by the foot. The treadle is of considerable length, on which account it is put in motion with great facility, and assisted in it by means of a fly. The scotchers are fixed upon the rim of a fly-wheel. But though these machines may be highly useful where mills turned by water cannot be established, they are probably much inferior in point of expedition, and the economy of labour. A brake of this kind is represented in different views, in figs. 434 and 435, in which is shown, by A, the three under brake-teeth, or swords, seventeen inches long, three inches deep, one inch and a quarter thick at the back, and a quarter of an inch at the fore-part or edge.

B the edges, two inches and three-quarters asunder at the end next the guide B, and two inches asunder at the other end.

C displays the two upper teeth, about an inch shorter than the under teeth; and

D represents the brake-mallet, about thirty-three pounds English weight.

F is a compound foot-treadle, which is eight feet four inches between the fulcra F, raised at F eight inches above the ground, or rather five inches higher than the lance of the workman; E is two feet four inches between the fulcra G, and is raised at G eighteen inches above the ground; that is, fifteen inches higher than the lance of the workman.

H the sword, or upright timber rod, which turns the wheel by the treadle-crank.

I the treadle-crank, of seven inches and a half radius.

K the fly-wheel, four feet and a half diameter, above sixty pounds English weight. As here represented, it is bent or cast iron; but it may also be made of timber.

L brass cods or bushes.

m M the lifting-crank; M is fixed firm upon the axle of the fly, while the crank m, about eight inches radius, plays freely round the axle. In position first, M begins to take round the crank (which by the lever R pulls up the mallet); when it comes to position second, the mallet is again at liberty, and by its weight pulls up the crank (faster than the fixed pieces move) into position third.

R may be observed that the treadle-crank is advanced about one-eighth part of the circle before the lifting-crank.

a a small pulley, which turns easily round on the end of the crank, and to which a rope is fixed.

O a piece of timber which prevents the roller from falling in upon the axle, but which should not rub against the rope in its coming down.

P shows where the rope passes between two friction-rollers, which are so placed that it comes down three or four inches, or half the radius of the lifting-crank on the side of the plummet-line, crossing the centre of the wheel; that is, to the side on which the crank turns when it pulls down the rope.

Q a pillar, which serves only to support the guard for the rope O, and the friction-rollers at P.

R the lever.

S the lever-pillar.

T part of the mallet-frame.

U two pillars which guide the brake-mallet.

V an iron spring which receives the leap of the mallet, and throws it the quicker down.

W the pillars which support the fly.

X U the pillars which bear the brake-teeth and mallet.

Y Y the spur and cross that support the pillars.

Z Z the bottom frame-piece.

a the broad stool upon which the workman stands, three inches above the ground.

The lifting crank and pulley are shown separately, in different views, at M m n, and m n.

The brake-teeth are made of good beech or plane-tree; the brake-mallet of plane-tree, ash; elm, birch, or oak; and the sword, or upright timber-rod, between the treadle and the treadle-crank, of beech, ash, or oak. The fly-wheel, if timber, should be made of oak, ash, beech, elm, or plane-tree. All the other parts of timber worth mentioning may be made of fir-wood.

At fig. 436 is shown the ground plan of the whole.

This brake may at any time be converted to a beater of flax and hemp, by removing the brake-teeth, and putting in their place flat boards. In the upper of these boards may be driven 32 nails, the heads about three-quarters of an inch long, and the points of the heads about a quarter of an inch in diameter; the points of the nail-heads may be placed one inch clear asunder, and at equal distances, as in this way any of the nails may most easily be drawn out in repairing the mallet. An iron hoop put about the mallet will prevent its bursting with the driving in of the nails. In the time of beating, the narrow end of the mallet is placed towards the workman, and where there is much work in that way, the mallet and fly may be made heavier, and then two or more workmen can work together upon the foot-treadles, which may also be made equally long.

The *flax-hackle* is an instrument or tool constructed for the purpose of hackling or straightening the fibres of the flax, which is seen at figs. 437 and 438. It has many teeth, fixed in a square flat piece of wood, as seen at A and B. When used, it is firmly fixed to a bench before the workman, who strikes the flax upon the teeth of the hackle, and draws it quickly

: teeth. To persons unacquainted with this kind of work, this
a very simple operation; but in fact it requires as much practice
the method of hackling well, and without wasting the flax, as any
ation in the whole manufacture of linen. The workmen use
arser and wider-teethed hackles, according to the quality of the
ally putting the flax through two hackles, a coarser one at first,
finer one in finishing it.

rippling-comb is an instrument or tool which is formed by letting
or more long square teeth nearly upright, in a long narrow piece of
that their different angles shall come nearly to touch each other.
g the flax through between these teeth, the balls or pods in which
. contained are forced off. It is seen at A and B, fig. 439. If
:o be regarded more than the seed, it should, after polling, be
lie some hours upon the ground to dry a little, and so gain some
o prevent the skin or harl, which is the flax, from rubbing off
ling; an operation which ought by no means to be neglected, as
: put into the water along with the flax, breed vermin, and other-
the water; the balls also prove very inconvenient in the grassing
ng. In Lincolnshire and Ireland they think that rippling hurts
nd therefore, in place of it, they strike the balls against a stone.
ils for rippling should not be great, as that endangers the lint in
g-comb. After rippling, the flax-raiser will perceive that he is
ort each size and quality of the flax by itself more exactly than
:fore have done it.

and and foot methods of breaking and scotching the
however, too tedious in their operation to give satis-
) the manufacturers, in the present advanced state of
cal science; consequently mills have been con-
by which these preparatory operations are much
d. .

mills are constructed in great variety; but one of
with which we are acquainted is described in *Gray's
iced Millwright*, in nearly the following terms:—

i is the *plan*. A A, the water-wheel; C C, the shaft or axle
:h it is fixed; B B, a wheel fastened upon the same shaft,
102 teeth, to drive the pinion D, having 25 teeth, which is
i the middle bruising-roller; E, a pinion in which are 10 teeth,
the wheel B, which is fastened upon the under end of the
ilar shaft that carries the scotchers; M M, the large frame that
me end of the shaft C, and the perpendicular axle; N N are
which the rollers turn that break or bruise the rough flax; I A
: machine and handle to raise the sluice when the water is to be
wheel A A, to turn it round; G G, doors in the side walls of the
; I K, windows to lighten the house; H H, stairs leading up to

l is the *elevation*. A A, the water-wheel upon its shaft C C, on
.ft the wheel B B is also fixed; this latter wheel containing 102
turn the wheel E, having 25 teeth, which is fastened upon the
:uising-roller. F F is a vertical shaft, upon the lower end of
fixed a pinion having 10 teeth, which is driven by the wheel B.
two arms that pass through the shaft F; and upon these arms are
with screwed iron bolts, the scotches that clear the refuse off the

flax. D D, the frames which support one end of the axle C, the vertical shaft, and the breaking-rollers; L is a weight suspended by a rope, the other end of which is fastened to a bearer, as is seen in fig. 442; S S, a lever, the short arm of which is attached to the frame that the gudgeons of the upper roller turn in; and by pushing down the long arm, the upper roller is, when necessary, so raised as to be clear of the middle one. N N, the end walls of the mill-house; R R, the couples or frame of the roof; H, a door in the side wall; I K, windows.

Fig. 442 is a *section*. A A, the great water-wheel fixed upon its shaft, and containing 40 aws, or float-boards, to receive the water which communicates motion to the whole machinery. B B, a wheel fastened upon the same axle, having, as before mentioned, 102 cogs, to drive the wheel C, of 25 teeth, which is fixed upon the middle roller, No. 1. The thick part of this roller is fluted, or rather has teeth all round its circumference; these teeth are of an angular form, being broad at their base, and thinner towards their ontward extremities, which are a little rounded, to prevent them from cutting the flax as it passes through betwixt the rollers. The other two rollers, Nos. 2 and 3, have teeth in them of the same form and size as those in the middle roller, whose teeth, by taking into those of these two rollers, turns them both round. The rough flax is made up into small parcels, which being introduced betwixt the middle and upper rollers, pass round the middle one; and this either having rollers placed on its off-side, or being enclosed by a curved board that turns the flax out betwixt the middle and under rollers, when it is again put in betwixt the middle and upper one, round the same course, until it be sufficiently broken or softened, and prepared for the scotching-machine. The bearer in which the gudgeon of the roller No. 1 turns, is fixed in the frame at C; and the gudgeons of the rollers Nos. 2 and 3 turn in shders that move up or down in grooves in the frames S S: The under roller is kept up to the middle one by the weights D D, suspended by two ropes going over two sheeves in the frames S S; their other ends being fastened to a transverse bearer below the sliders in which the gudgeons of the roller No. 3 turn. The weights D D must be considerably heavier than the under roller and sliders, in order that its teeth may be pressed in betwixt the teeth of No. 1, to bruise the flax when passing between the rollers. The whole weight of the roller No. 2 presses on the flax which passes between it and No. 1. There is also a box fixed on the upper edge of its two sliders to contain a parcel of stones, or lumps of any heavy metal, so that more or less weight can be added to the roller, as is found necessary. O O is the large frame that supports one end of the shaft which carries the two wheels A B, and vertical axle F F; on the lower end of which is fixed the pinion turned by the wheel B, and having 10 teeth. In the axle F are arms upon which the scotches are fastened with screwed bolts, as seen at G G, fig. 441. These scotches are enclosed in the cylindrical box E E, having in its curved surface holes or porches at which the handful of flax are held in, that they may be cleaned by the revolving scotchers. H H, the fall or course of the water; I I, the sluice, machine, and handle for raising the sluice to let the water on the great wheel. The gudgeons of the axles should all turn in cods or bushes of brass. K K, the side walls of the mill-house; G G, doors; L L, windows.

Having proceeded thus far, the reader will have become acquainted with the various modes of preparing flax for the operation of spinning, which operation, from the copious manner in which we have treated of it under the article COTTON MANUFACTURE, requires but little elucidation.

About the year 1787, Messrs. Kendrew and Porthouse, of Darlington, obtained a patent for spinning a flaxen thread by means of machinery; prior to that time, we believe, the rock and wheel, variously modified, occasionally for superior spinners to form two threads at once, were universally employed. In Ireland especially, even at the present day, this method is much practised. The flax, rendered straight and smooth by hackling, is wrapped loosely round the rock, from which it is gradually drawn by the left hand, whilst the thumb and fore-finger of the right, moistened with water, are employed in adjusting the fibres, and directing the thread. A bobbin and flyer, placed upon a horizontal spindle, serve to give the twist, and to take up the finished thread; their motion is derived from a wheel, impelled by the foot through a treadle and crank, by means of an endless-band passing round a pulley of much smaller diameter, which is fixed upon the spindle.

The straightness and smoothness of the fibres of flax, so different from the corrugation and adhesiveness of cotton and wool, with their extraordinary length, seemed to demand an arrangement in machine-spinning very different from what has been already delineated.

In the patent alluded to, the hackled flax was extended upon a horizontal frame, at fig. 410°, to be carried between the rollers B b, and afterwards to pass along with the cylinder C, (revolving with a velocity equal to that of any point of the circumference of B,) under several successive rollers, until it arrived at the drawing-rollers D d; the twist and removal of the thread then taking place by the flyer and bobbin, as before described. The rollers F., F, G, H, I, if of equal weights, will, on account of their respective positions, press with unequal force; the one resting upon the vertex of the cylinder being evidently the most efficacious, and with the surface beneath acting probably the part of a pair of holding-rollers to fibres of the length of nearly one-fourth of the circumference; whilst for fibres which are longer or shorter, the other rollers, according to their place, will answer the same purpose. In this, however, there is no new principle; and although modified, it amounts merely to the operation of holding and drawing rollers. From some impediments thrown in the way of the Scotch flax-spinners by the patentees before mentioned, they began, we believe, in no long time, to place their rollers in a straight line, at distances suitable to the length of the fibres. Of the excellence of this arrangement a working model made for the Andersonian Institution in Glasgow, in the year 1803, afforded sufficient evidence.

We shall now proceed to give a description of a patent, obtained in the year 1806, by Messrs. Clarke and Bugby, for effecting certain improvements in a machine, intended to be worked by hand-labour, for the spinning of hemp, flax, tow, and wool.

Fig. 445 represents an oblique view of the front of a frame containing six spindles, (but frames may contain an indefinite number of spindles.). A, the spindle or a bow passing through the whole frame, having ten bosses of brass or cast-iron thereon, each about four inches diameter, each boss supplying one spindle; B, a pinion of twelve leaves upon the end of the spindle A, connected with the wheel C, of eighty teeth, fixed upon the end of a small iron spindle F, covered with wood and extending through the whole frame; D, a slack or intermediate pinion of any size at discretion, connected with another similar pinion, the latter connected with a wheel of 120 teeth, which is fixed upon an iron spindle G, of about 1⅜ inch diameter, and extending through the whole frame; but the wheels B C D and E may be varied in their numbers, to increase or diminish the draught of the substance operated upon, as may best suit its quality or the ideas of the workman. The pinion B is so contrived as to slip off the end of the spindle A, to make room for a smaller or larger one; by means whereof a larger or shorter thread may be spun from the same sized rovings; *a a a a a a a a a* represent ten roved slivers of hemp, flax, tow, or wool, passing between the iron spindle G and rollers in pairs pressed against them by springs or weights; these springs or weights must be of sufficient force to hold back the slivers or rovings so securely, that they may only pass on with the movement of the spindle; these pairs of pressing rollers are placed behind the spindle. The use of the small iron spindle F, covered with wood, and left rather larger than the spindle G, is, with pressure of the small wood roller, made up in pairs *b b b b*, and so contrived that each pair may roll upon two slivers, to bring them down straight, and preserve the twist which they receive in the roving-machine till the slivers leave them. The bosses on the spindle A have likewise wooden rollers in pairs pressed against them by springs or weights, between which the drawn, lengthened, or extended slivers pass to the spindle, the rollers having each a tin conductor *c c c c c c c c c*, to bring the material under operation as centrically as possible between the wood rollers and the bosses; but all the above-mentioned parts of the machine is so similar to the common upright frames for spinning flax, that a person conversant with them will not be at a loss to make it all. H is a wheel of wood four feet in diameter, having its rim about two inches thick, with a groove in its periphery for a small cord or band. In its centre is a rule or stock of wood through which the spindle I passes, and extends into its frame about one-fourth of its length. To enable the person that turns the winch to reach all the spindles at work, with the hand that is not engaged in turning, to remove any obstacle that may arise to the spindles, the arbor or spindle of the wheel I has its bearing on the sides of the frame that contains it, marked L L L L; this frame, with the wheel H, the arbor I, and the winch K, is similar to that part of a machine called a mule-jenny, used for spinning cotton; this frame is supported in a horizontal position at the outer end by two legs marked M M, and a screw pin which passes through K, the front upright, *a* A, fig. 444, and made tight with the thumb-screw *a*; the screw passes through a groove or mortise at the end of the wheel frame, to enable the workman to adjust the wheels N and O, as it will be found necessary to change the wheel N, to make such alteration in the twist as the size of the yarn may require, or as the workman may think proper. P and Q are bevel wheels of equal size, the former fixed upon the rule or stock of the wheel H, and connected with Q upon the spindle R, taking round with it the wheel N, which is connected with the wheel O. Upon the embossed spindle or arbor A, *a a a a a a a a a a*, are spindles standing on a carriage with four wheels, similar to the carriages used in mule-jennies for spinning cotton, having at each of them, at *d d d d d d d d d*, a convex seat of wood of any convenient size, not less than the bottom of the bobbins or quills *e e e e e e e e e*; these bobbins or quills are about six inches long and 1½ inch diameter at the bottom, and three-quarters of an inch diameter at the top; but the sizes must be varied according to the size of the yarn. Perhaps four or five variations will be sufficient to spin yarn for tarpaulins or sail-cloth, up to fine yarns, fit for good sheeting and fine weavings. T a pulley, over which a band, from S, runs and returns, to draw out the carriage upon the four wheels described; W, the cylinder which drives the spindles.

Fig. 444 exhibits a side view. A, the wheel mentioned above in fig. 445, and there marked H; B, the winch by which it is turned by hand; C C C C, the frame wherein it works; D and E are blocks of wood on each side of the said frame to raise the wheel, so that the winch may be clear of the carriage F F, and apparatus G G; the two end wheels upon the carriage containing the spindles having two more corresponding on the opposite side thereof. H, a groove upon the end of a cylinder, which drives the spindles, and stretches through the carriage frame, for the diameter of which no certain rule can be laid down, as it depends upon the length or size of the yarn, taken into account with the other parts of the machinery. N N N N N N N, a small band passing over the wheels A K, H I L and M, by which the groove wheel H and its cylinder are moved, and the spindles driven. O a treadle shaft, represented by S S, in fig. 445, passing through the frame or part thereof at the option of the workman, connected with a tumbler at the end of the embossed spindle or arbor A, in fig. 445, by a small band, wound five or six times round each of them, and passing over the wood groove wheel Q, and made fast to the back of the carriage F F; this tumbler, by the motion of A, is, at the return of the carriage, locked to the wheel R; and unlocked when the carriage is not to its destined place.

The carriage is drawn on by the weight of S fastened to a cord, which passes over the groove wheel T, and is connected with the front of the carriage; U, the wheel on the arbor containing the holder shown in fig. 445. V, the cylindrical roller on a stirt fixed therein, and rolling at every return of the carriage on the plane W and X, which raises and falls the fallers and holders, so as to distribute the yarn upon the bobbins from top to bottom; the wheels Y Z, A 2, and B 2, are the same wheels shown by B, C, D, and E, in fig. 445; 1, 2, &c. spools containing the rovings.

This machinery is calculated to save the heavy expense of currents of water, erecting spacious buildings, waterworks, steam-engine, &c. and to spin hemp, flax, tow, and wool, at such an easy expense, as to bring it within the reach of small manufactories. This machinery is also constructed upon such safe and simple principles, that no length of experience is necessary to enable even children to work it; and the use of water, steam, &c. being rendered unnecessary, it occupies so little space, that it may be placed in small rooms, out-buildings, or other cheap places. To effect the above purpose, it was necessary to get rid of the lanier or flyer upon the spindle used in the old machinery for spinning hemp and flax, which requires a power in proportion of 5 to 1, and to surmount the difficulty that arose from the want of elasticity in these substances. This want of elasticity in the substance to be operated upon, is compensated and provided for in this machinery; and upon this compensation and provision, effected by the various means hereafter mentioned, the return of the carriage without any assistance from the work-person, and the traverse for distributing the yarn upon the bobbins or quills, lay the stress upon the patent. The most simple mode of compensating the want of elasticity, and which is recommended in preference to the other, is that of having a holder of large wire for every spindle fixed in an arbor or shaft extending from one end of the carriage to the other.

This arbor or shaft, with the holders, may be considered as a large and improved substitute for what is called a faller in the mule-jennies for spinning cotton, fig. 443. Let A represent the arbor or shaft, *b b b b b b b b b b* the holders fixed therein with the elliptical eyes, through each of which a thread passes from the bosses on the arbor A, in fig. 445, to its spindle. B, a spindle, which may be from 10 to 13 inches long. C, the whirl, wherein a small worsted band from the cylinder H, fig. 444, works D, a convex seat upon the spindle, whereon the concave bottom of the bobbin or quill E rests.

F, a piece of buffalo skin or metal screwed or nailed to the rail I, having a hole in it, through which the spindle passes, and by which it is kept steady; G, a wire bent at right angles at *a*, and the bent part driven into the rail A, so that it may be removed to or from the whirl C, and by the other crook, *b*, prevent the spindle from running out of its step H, which is a screw of brass or other metal passing through the rail K. The wire of which the holder is made, after forming the elliptical eye, is left or extended beyond the uppermost part at *e*, that the yarn may be conveniently slipped in when occasion may require it; these holders for each thread are for the purposes of keeping the yarn in a state nearly vertical over the tops of the spindles when the carriage which contains them is coming out, and being released from that situation at the beginning of the carriage's return, and thrown into nearly a horizontal position, so as to bring the yarn below the top of the bobbins or quills upon the spindles; and then being curved and raised again by the wheel U, and its cylindrical roller moving upon the plane W and X, fig. 444, distributes the yarn upon the bobbins or quills, and prevents it from cockling, hinkling, or improperly doubling or twisting together. The seats upon the spindles described by D, are turned convex, and the bottoms of the bobbins and the bottoms of the quills concave, to keep the bobbins or quills in a more central state upon the seats. The concavity of the bobbins or quills exceeding the convexity, throws the weight of the bobbins or quills upon the peripheries or extremities of the seats, and ensures the rotary motions of the bobbins or quills with that of their spindles. We prefer the convex and concave surfaces before described; but other surfaces will have nearly the same effect, if so contrived (as they easily may be) to bear upon the peripheries or extremities of the seats as well as of the bobbins or quills. The hole through the bobbin or quill, fig. 446, is rather larger than the spindle, that it may not be obstructed in its motion round the spindle, which motion takes place at every return of the carriage, and as often as any thing obstructs the coming forward of the sliver of which the yarn is formed. At one end of the arbor whereon the holders are fixed is a counterpoise L, fig. 443, having a socket, and made fast;

the arbor by a thumb-screw m, the round ball at the top being led to, counterbalance the holder. This counterpoise, when the holders are in a vertical state, declines about 10 or 15 degrees towards the horizon, but while the holders are thrown down, and under the government of the cylindrical roller V, upon the wheel U, is in a different situation; but the roller V, arriving at B 3, fig. 444, on the return of the carriage, the holders are precipitated to a height where the counterpoise overbalances them, and locks the wheel M, fig. 442, or U, in fig. 444, in the ratchet a, where it remains until the carriage has reached its destined place, where the tail of the catch O strikes against a pin in the frame CCCC, fig. 444, and releases it, the said roller then resting upon the frame U X. A second method of compensating and providing for the want of elasticity in hemp and flax, which is a part of the discovery, is, to fix a round bar of wood, about 1½ inch in diameter, the whole length of the carriage, about three or four inches above the tops of the spindles, so that the outer surface, or that next the work-person, may be perpendicularly, or nearly so, over the tops of the spindles, the inner side having pieces of wood or metal nailed, or otherwise fixed thereto, leaving only small spaces between each for the yarn to pass through; the use of these pieces is to prevent the threads getting together and entangling, see fig. 447. A A A A represents a bobbin filter used in the mule-jennies for spinning cotton with counterpoise B, wheel C, with its cylindrical roller D, with the plane W and X, before described by figs. 445, 444, and 443. E E, spindles with their whirls, concave seats, bobbins or quills, with their concave bottoms, F F F F F F F F F the pieces of wood or metal, nailed or otherwise fastened to the round piece of wood, to prevent the thread getting together. In this case every thing applied to or used with the arbor, containing the holders above mentioned, may be applied or used.

A third method of compensating the want of elasticity in hemp and flax, which the patentees describe as a part of their contrivance, invention, and discovery, is the fixing each spindle in a small frame A A, fig. 448; b, a stop of brass; C, a common mule spindle, with its whirl D; E and F two skirts of iron fixed one on each side of the frame A A, equally in a line with the groove in the whirl D, and moving in holes in two cheeks g g, fastened into the rail H, on the small frame A A. On the back side thereof, next to the cylinder, is a small roller, moving on two pivots, so planted, that when the spindle is in an upright position, the band from the cylinder which drives it may just run free of it, and as the spindle frame, A A, is kept to the rail I by a tender spring made of wire, wound round a pin about half an inch in diameter, that the spindle may yield to the yarn in all cases when necessary, the said roller is to prevent the band which drives the spindle out of the whirl, when the spindle leaves its vertical position.

Fig. 449, a side view of the little frame in fig. 448; A the frame, B the spindle, C the whirl therein, D the end of the roller and one of its supporters. This apparatus requires the faller last mentioned and described by fig. 447, with its appendages for laying the yarn upon the spindles, no seat on the spindle or bobbin in this case being wanted, nor any more than a thin piece of paper, or something thin, round the spindle, to enable the spinner to take the yarn off with safety and ease. A fourth and last-mentioned mode of compensating and providing for the want of elasticity in hemp and flax, and preventing breakages and other accidents from any tightness in the yarn, occasioned by any obstruction or other circumstance, and which is part of this invention and discovery, is, by driving the common mule-spindle with a slack band, having the yarn to pass over the holders described in fig. 443, or over the round bar described in fig. 447, with all the other apparatus for laying the yarn upon the spindles, &c. This last method cannot be used to advantage in any case, but may be substituted for either of the three methods described above for spinning yarn for sail cloths, sackings, tarpaulins, or other coarse or heavy goods.

WEAVING.

In the preceding articles, cotton, wool, silk, and flax, we have traced the process of forming the four commonest fibrous materials to the state of thread or yarn, and now purpose, in general terms, to treat of their further destination in the formation of the various superficial structures termed web or cloth.

The structures which come under the name of cloth are formed of two distinct layers of yarns, generally crossing each other at right angles, termed the warp, and the woof or weft; and as all cloths, however varied, are constructed of these two distinct portions of threads, the mind will, when made to comprehend the mode and form of loom used in the weaving of one material, be able easily to conceive its application to other sorts, varying only in dimensions and strength, according as the weight of the yarn or size of the cloth may demand.

Prior to commencing the weaving of any material, it is necessary to prepare the yarn for the loom, one process of which preparation is, in fact, the measuring and arranging the threads that are to compose the warp in a parallel direction, termed *warping*.

The warp, or that layer of threads which extends the length of the piece to be woven, requires the most attention in the preparation. To form a warp, it is necessary to be very particular in the number and quality of the yarns; for upon their fineness, length, and breadth, depend the fineness, length, and breadth of the piece to be woven. Though this may appear a very simple operation, yet, the performing of it with expedition and accuracy demands some mechanical skill. The machine for effecting this object is termed the warping-mill, and, though considerably larger, may be compared to the reel described in the cotton manufacture: but the spindle upon which it revolves is placed perpendicularly. The reel cannot easily be made of such dimensions that a thread measured upon its circumference shall be equal to the required length of the warp; and consequently the layer of yarns is placed in a direction parallel to the axis of the reel, and wound upon it in a spiral direction till they arrive at the upper end, when the motions of the mill and the yarns are reversed, and a fresh layer is placed upon the same parts of the reel. By this method of plying the layers of yarn, it is obvious, that a small number of ends may be doubled so as to form the required breadth of the warp. If the twist is spun on cops, it must, prior to warping, be wound on bobbins, which bobbins are placed in a frame to be wound off upon the warping-mill.

The next operation to be effected in the manufacture of cotton goods is that of dressing the warp; that is, impregnating it with certain gummy or gelatinous matter, and coating the surface of the yarns, to enable the warp to sustain the abrasion to which it is subjected in the process of weaving, as will be seen when that process is described. In preparing the wool and silk yarns for the looms, dressing is in general only required for the finest sort, when a little mucilage of gum arabic, or of jelly made from rabbit or other light skins, is used to increase in a slight degree the tenacity.

As it is of considerable importance in the dressing of warps to have the materials dispersed equally over the surfaces, many ingenious mechanics have constructed machines for that purpose; the general principle of which are the placing of the warp on a roller, and immersing it in the mucilage, which allows it to be drawn off covered with mucilaginous matter. The superfluous mucilage is brushed off, and the yarn is put in a frame, and by means of revolving fans is dried and rendered fit to be put in the loom. In cases where the manufacturer operates separately, the weaver dresses the warp, by extending and carefully brushing it over with paste, and drying it in the air, prior to placing it in the loom.

Before we proceed to give a description of the looms used in the manufacture of cloth, it is requisite that the reader should be acquainted with the various sorts of structure arising from the different dispositions of the warp and weft, termed fabric.

The simplest mode of disposing of the warp and weft is called common fabric; and taking into account the quantity of yarn used for a given superficies of cloth, is certainly, so far as respects its strength and durability, the most advantageous mode of distributing it.

Fig. 412 is a section of a piece of cloth wove in the common fabric. The circles represent the warp in section, and the weft is seen passing alternately above and below each succeeding yarn, and the return, or next layer of weft, passing beneath those threads over which it had passed before, and *vice versa*.

Fig. 413 represents a section of a piece of cloth wove to a twilled pattern. In this the yarn of the weft passes alternately over four and under one of the threads of the warp, and *vice versa* in its return.

Fig. 414 represents the section of a dimity or kerseymere, in which the weft passes over four and under four, then over one and under four, and over four and under one, which places it in a position to begin again; when the passage of the weft, as it regards the warp, is exactly reversed.

Fig. 415 shows the construction of a double cloth woven with two warps. This mode of weaving is mostly applied to the construction of carpets, and the transposition of the colours in the pattern arises from it. All the divers modes of passing the weft among the warp may be introduced in this figure, and whatever is effected with one web of warp is alternately effected with the other, as may be seen by the figure. It is therefore easy to conceive, that all the various patterns in woven goods are obtained by differently disposing of the warp, that is, lifting a greater or smaller quantity of it at a time, which places the weft on either surface of the cloth at pleasure.

The common loom, or that which is destined to weave cloth of the common fabric, is the most simple in its construction, as the mode of lifting and depressing the portions of the warp are similar at each throw of the shuttle. A top view of a loom of this description is given in fig. 416.

A is a warp.

B a roller upon which the warp is wound, called the yarn-beam, and applied to maintain the warp in a stretched position by means of a lever passing through one of its ends and tightened with a string, as will be more clearly understood by referring to the perpendicular section at No. 2 of this figure.

C C C are rods placed between the threads of the warp to keep them separate, so that they may pass forward clear of each other, when the warp is fed forward and filled with the weft, these rods are at different periods moved towards the warp-roller B.

At D are the heald or heddles, formed of two rods, one above and the other below the piece, and connected together by numerous strings, through which distinct portions of the warp pass, and by means of the treadles below are lifted and depressed. A detached view of two leaves of a heddle is represented in section in fig. 417.

$a\,a$, $a^1 a^1$, are the top and bottom bars, and the two lines $a^2 a^2$ represent two adjacent yarns of the warp, so that when $a\,a$ rises it carries with it one thread, while the other thread which is passed through the lower loop of this heddle is depressed by the other heddles. The next part, E, fig. 416, is a frame to carry the reed, called the lay. A portion of the reed is shown detached in fig. 418. It is, except when used in cloth of the coarsest description, formed of flatted wires, placed parallel to each other, and governed, as to their thickness and adjacency, by the fineness of the fabric in which they are to be used.

The lay, fig. 416, which carries the reed, is hung from a bar capable of vibrating on gudgeons in the upper frame of the loom. The two thin elastic pieces of wood which suspend the lay are called swords, and may be seen at F^1, F^1, fig. 419. The reed thus hung is just beyond the line of the shuttle-flight, and has one or two threads of the warp passed between each of its wires, which wires are termed dents. Its use is to strike home the thread of the weft immediately after it has been delivered by the flight of the shuttle; it is therefore pushed by the weaver towards the yarn-beam, prior to each flight of the shuttle, and when the weft has been delivered, it is allowed to return and strike home that individual thread.

The next part of the loom is the shuttle-boxes, which are placed at F, F. In weaving narrow goods, the shuttle is passed between the warp by the hands of the weaver, but when the cloth is fine, or of a breadth to preclude this mode, the fly-shuttle, which is much more compact, and has a spindle to carry a cop upon it, is introduced. This form of shuttle is represented in fig. 420. The shuttle with its cop is placed in the shuttle-box, which is of dimensions just sufficient to receive it. In fig. 419 is represented the reed and lay at $F^a F^a$. The shuttle is driven to the opposite boxes by means of a small piece of wood, called a driver, which lies behind the shuttle in each box, and is capable of being swiftly drawn forward by a string attached to it, and connected with a handle, G. The weaver holds the handle in his hand, and by a jerk throws the shuttle across the web into the opposite box, and then, by bringing the lay towards him, strikes home the weft. The flight of the shuttle requires adjustment or skill, as its impetus must be proportioned to the weight of the yarn which it carries, and the freedom with which the cop unwinds.

If two or three colours of weft are to be put into a piece, so as to form a pattern, there are two or three shuttles to be thrown; in such case, the shuttle-boxes are formed in three parts, as represented by the dotted lines. This combination of shuttle-boxes is capable of being moved upwards and downwards upon the lay by the small levers, H, H, fixed upon the swords, and worked by the handle I, so that the shuttle to be thrown may be brought opposite to the division in the warp through which it is to fly.

As the cloth is perfected, it is led over the breast-beam K, fig. 416, and is, by means of a ratchet wheel, wound upon the roller L, which is termed the cloth-beam. At m is a stretching-rod, formed of two pieces, and lashed with a piece of band, in such manner, that the ends are forced outwards, as may be seen in the figure. This rod has small points at each end, which pass through the selvage of the cloth, and serve to keep the cloth stretched, as otherwise the action of the weft would occasion it to pucker and lay in hollows. The weaver sits behind the breast-beam, and in fine work, where the breast-beam is dispensed with, behind the cloth-roller.

Such is the construction of loom used in plain-weaving; and by examining it attentively it will be seen, that by an additional number of heddles any required movements of the warp can be effected, and by varieties of weft other diversifications attained almost to infinity. The greatest skill required in the act of weaving by hand is the directing of the flight of the shuttle, where the impetus given should just suffice to deliver it in the opposite box. The striking home of the weft should be done with a regular force, and the preparatory operations carefully attended to, that the warp may wind off freely and regularly with an equal tension in all its parts.

From an examination of the movements of so simple a machine as the loom, the mechanist will instantly conceive the practicability of applying power to produce the necessary movements; we shall, therefore, present the reader with two combinations of this class which are called power-looms: the first invented by a Mr. Millar.

THE OPERATIVE MECHANIC

Fig. 421 represents a section of a power-loom, in which all the operations are effected by means of treadles, moved by wipers or eccentrics.

A, the main shaft, to which the power is communicated, carrying the wipers, one of which is seen at A. A¹, the yarn-beam; B, three rollers, on the lower of which the cloth is wound after passing above and between the two upper ones; C, C, the heddles; D, D, the treadles, to which the heddles are attached by means of a line passing over a pulley, in such manner, that tne depression of one heddle occasions the raising of the other; E, E, the lay carrying the reed; the motion is given to the lay by a wiper moving a treadle that is attached to the lay by means of the line and crank at F. The return motion for striking home the weft is given by a weight hanging over a pulley, as may be seen in the figure. The flight of the shuttle is occasioned by attaching the strings from the drivers to another treadle, which treadle is worked at the proper periods by another wiper.

Another form of power-loom, called the crank-loom, is used, and varies from the preceding in the mode by which the movement is given to the heddles. In this construction of loom the revolving shaft is placed immediately under the heddles, which are suspended over a pulley similarly to the loom last described; but the motion is given to them by means of their being attached to two opposite cranks on the shaft. The motion is given to the lay by a crank upon another shaft, which is made to revolve twice, while the shaft that moves the heddles revolves once. By this, it is evident, that the warp is opened, and the shuttle thrown twice, during one revolution of the first shaft.

The flight is given to the shuttle by the cords of the drivers being attached to an upright lever, as represented in fig. 422.

The cords, c, c, of the drivers are attached to the lever, e, which, by means of the arms, h, i, is caused to vibrate in opposite directions upon the centre, g, being alternately struck, by two projecting pieces upon the first mentioned crank-shaft, which causes the lever, e, to vibrate in a plane parallel to the crank-shafts, and gives flight to the shuttle, at the period when the warp is opened.

In either of these plans for working looms by power, if the number of heddles is required to be increased, in order to produce any figure, it is easily effected, by varying the number and the position of the cranks or wipers.

But, though great variations in the movements of a warp may be effected by using many heddles, yet when the number of heddles and the number of cranks are increased, great objections arise to their being used; consequently, another construction of loom, called the draw-loom, is introduced when complicated figures are to be woven. In this loom the changes are effected by raising one portion of the warp entirely out of the way, while the other is wrought by the heddles at the time it is being filled with the weft; the part raised is then lowered into work, and other yarns of the warp lifted out of the way.

A loom on this principle is shown in fig. 428. For weaving carpets by this method, every yarn of the warp has a line attached to it, which lines are brought together in one connected piece, according to the portion of warp to be raised at once; and carried over the pulleys as at A, and attached to the fixed beam at B. This portion is called the tail. Below the warp these lines, which are called the simples, are kept in a state of tension by weights, as at C; and in order to keep them distinctly apart, are made to pass through a board perforated with holes at D. Other lines are attached to the tail, capable of being pulled by handles, as at E, by which means, such portions of the warp as are required can be raised. By this contrivance the greatest intricacy of pattern can be attained; but the attaching of the simples to the different parts of the warp, by means of small eyes of metal through which the threads of the warp are made to pass, is a work of considerable labour. Damask table-cloths are produced by this loom.

It would occupy too much room were we to enter with more exactness into the great variety of looms which ingenuity has constructed; what we have said therefore we trust will be sufficient to convey to the reader a perfect knowledge of the principles of forming those various fabrics which are termed cloth. In the weaving of ribands and other ornamental works, many extraneous substances, totally unconnected with the warp or weft, are thrown in, which affords the designers an additional scope for the display of embellishments. These substances are merely held in the fabric by the intersection of the two staple parts, the warp and the weft, and are by the weavers denominated whips.

In the formation of cloth from the yarn of cotton, silk, hemp, and long wool, denominated worsted, the fabric when taken from the loom is, so far as regards the weaving, in a perfect state; the further operations, both mechanical and chemical, which it undergoes, may properly be considered as tending merely to its further embellishment. These operations consist generally of singing the superfluous fibres from the surface of the cloth, by drawing it over hot irons, and after bleaching or dying, submitting the cotton and linen goods to pressure between heavy iron cylinders, for the purpose of giving it a gloss, and the worsted, called camblets or stuffs, between warm copper-plates, called hot-pressing, to give it a smooth and finished appearance.

In the formation of cloth from short wool, of which our wearing apparel is made, the loom cannot be said similarly to have completed the operation. In this branch of manufacture, the yarn is woven in a common loom in the manner we have shown, and called common fabric, but when the piece is taken from the loom the web is too loose and open, and is consequently submitted to another operation, called fulling. The cloth, after it is, by repeated washings, divested of the oil that was put in it in the act of carding the wool, is taken to the fulling-mill, where it is immersed in water, and subjected to repeated compressions by the action of a large beater formed of wood, which repeatedly changes the position of the cloth, and by its continuous action causes the fibres to felt and combine more closely together, so that the beauty and stability of the texture is greatly improved. The cloth is next submitted to the dyer, if so destined ; but in many colours for the best cloths this process is effected in the wool prior to the commencement of the manufacturing.

The cloth then undergoes the operation of the gig-mill, which is formed of a cylinder, somewhat similar to that of a carding-engine, covered with the heads or burs of a large species of thistle, called teasels. This engine is used to raise the fibres or nap, and lay it in a parallel direction, which, in fine cloth, is cut off by shears, and the cloth then undergoes hot-pressing to bring it into a state fit for the market.

Upon considering the various methods of fabricating cloth in general, it may easily be conceived, that great scope is afforded to the practice of dishonest modes of forming fabric apparently valuable, such as the introducing of weft or warps of different qualities and hiding them by the mode of plying the other part. In trying the strength of cloth it should always stand both in the direction of the warp and the weft; and the substances of which all parts are formed should be known by separate examination, and not by mere superficial inspection, as the surface may easily be formed to hide defects, and display apparent value.

ROPE-MAKING.

In rendering the hemp-plant proper for the uses of the rope-maker, it has to undergo a variety of processes.

The first of these is *retting*, that is, exposing it to the action of the dew, or water ; the former termed *dew-retting;* the latter, by which the finest hemp is produced, *water-retting.* In both or either of these processes, the quality of the hemp is said to be influenced by the state of the weather, and the finest to be produced when showers have mostly prevailed.

In *dew-retting*, the hemp-stalks, immediately after being pulled, are spread out, in a thin, even, and regular way, so as to keep exact rows, on a fine level piece of close old sward land, for the space of three, six, and sometimes eight weeks, as circumstances may require; during which, they are turned two or three times in the week, according to the state of the atmosphere.

The motive for thus spreading it out upon the ground is, that the dew, by penetrating into the plant, may render the separation of the rind from the stem or bar easy to be accomplished. When the dew has acted upon it sufficiently for this purpose, it is tied up into large bundles, and carried home and slacked, or otherwise it is put into a covered building, till wanted to be formed into hemp.

This process, called *grassing*, requires great nicety and attention, that the texture of the hemp may not be injured either by too long continuance on the sward, or by being removed before the hempy substance is rendered sufficiently separable.

In *water-retting*, the much more common and speedy method is, to tie the hemp-plant into small bundles, by means of bands at each end, and in general to deposit it bundle upon bundle, in a direct and crossing manner, in a pond of standing water, to form what is called a bed of hemp. This bed, when formed of as great a thickness as the depth of the water will admit, which some think can hardly be too great, though five or six feet are the usual depths, is loaded with large pieces of heavy wood until the whole is immersed in the water. In choosing ponds, those should be preferred that have clayey bottoms.

When the hemp-plant has remained in the water for about five or six days, varied according to the nature of the pond and the state of the weather, it is taken out, and conveyed to a piece of mown grass or other sward land, which is free from all sorts of animals. Here the bundles are untied, and the hemp-stalks spread out thin, stem by stem. While in this state, especially in moist weather, they must be carefully turned every second day, to prevent their being injured by the worm casts. In this way they are kept for about five or six weeks, when they are gathered up, tied in large bundles, and kept perfectly dry in a house or small stack, till wanted for use.

In some of the northern parts of Scotland, the hemp, after it has been pulled, and cleared of its leaves, seeds, and branches, by means of a ripple, is formed into bundles of twelve handfuls each, and steeped in a manner similar to flax, till its reed becomes capable of parting from the bark. In this process, it is favourable to give it rather too much than too little of time; and let it be observed, that the most slender hemp requires the greatest length of time in the water. Where the quantity of hemp is only small, the hempy part may be separated from the reed by hand-labour; but where it is large, drying and breaking it in the manner of flax is strongly recommended.

After the hemp has been taken out of the water, it is not spread out upon the grass-ground in the way of flax; but set up in an inclined position against cords arranged for the purpose, or by any other method by which it can receive the full benefit of the air till it be perfectly dried, which may be known by its rising in blisters from the boon. As soon as it has been reeded, it must be divested of the mucilaginous material which it contains, by pouring water upon and repeatedly squeezing it. In this part of the process great care must be taken to prevent the fibres from getting entangled, as by that means great waste will be incurred.

M. Brealle, on the Continent, has suggested a mode, very different to any of these, for the purpose of steeping hemp, the advantages of which, it is asserted, have been fully proved by numerous trials. The process consists in heating water in a vessel, or vat, to the temperature of from 72 to 75 degrees of Reaumer, and dissolving in it a quantity of green soap, in the proportion of 1 to 48 of the hemp: the body of the water being about forty times the weight of the hemp. When this preparation is made, the hemp is thrown into it, and floats on the surface, and the vessel being immediately covered, the fire is put out. In this state the hemp is allowed to remain for two hours, at the expiration of which time it will be found to be fully steeped.

The principal superiority of this method, besides the great saving of time and expense, is said to consist in the hemp affording a greater proportion of tow. The value of the fuel, as well as the time employed in the process, should, however, be well considered in such cases. Besides these, it is said to promote the cultivation of the hemp crops, by the facility which it affords to the preparation, even in such situations as are not contiguous to rivers, streams, or ponds; it also obviates any ill consequences that might possibly ensue from the putrid effluvia of the atmosphere, and the corruption of the waters, induced by it, which last are well known to destroy the fish contained in them, as also to prove hurtful to the cattle that drink of them.

In consequence of the great trouble and expense attendant upon the process of water-retting, the hemp is frequently left for seed; in which case, it is commonly stacked up and well covered for the winter season, in order that it may be thinly spread out in about January or the following month. Where this can be executed during the period of a snow, the hemp comes much more readily to a good colour, and forms strong coarse cloths; but it is far inferior to that pulled in due season, and which has undergone the water-retting operation.

Various contrivances have been made in the form of ponds and pits, for the steeping of hemp; but the one which seems to possess the most merit is described in the Norfolk Report, as the invention of Mr. Rainbeard; by using of which, the hemp can be deposited in the pit, without the necessity of any person getting wet. The pond is an old marl-pit, with a regular slope from one side, where the hemp is prepared, to the depth of eight feet on the other side. On the slope, above the water, the hemp is built into a square stack upon a frame of timber, of such a height as will float and bear a man without wetting his feet: this is slid down upon the frame into the water, and a person on the opposite bank draws it to the spot where it is to be sunk. Mr. Rainbeard has found by experience, that the hemp does soonest at the bottom, and would not object to 16 feet of water. By means of this very useful contrivance he can put in a waggon-load in an hour. The sheaves are taken out one by one in the usual way; but it is suggested, that some more expeditious and simple contrivance, either upon the principle of the lever, or some other, should be resorted to, for effecting the desired purpose.

In preparing the hemp for the hackle, the work is executed chiefly by the beetle, first using a coarse, and then a finer brake: it may, however, be more expeditiously performed by the rollers of a lint-mill. In either mode, shaking the handfuls frequently with force is necessary. In cases where the plant has not been sufficiently watered to loosen the rind, the operation of peeling must be performed by the hand.

Another method for effecting this purpose is the hemp-mill, which is much used in America. It consists simply of a large heavy stone in the form of a sugar-loaf, having the small end cut off. Thus shaped, it readily moves round in a circle, when passing upon a plane. The motion is given by the impulse of water on a wheel, and the hemp, deposited on the receiving floor of the mill, is, by the weight and revolutions of the stone, perfectly crushed and broken.

Still, however, the fluted rollers of a lint-mill are the best means of performing the work, provided sufficient care be taken to guard against accidents.

When the hemp has been completely broken, it is submitted to another operation, called swingling, or scotching; the intention of which is, to separate the reed from the hemp. This operation is sometimes performed by a labourer, who takes a handful of hemp in his left hand, and while holding it over the sharp edge of a board, strikes it with the fine edge of a long flat straight piece of wood, usually termed a swingle-hand or scotcher; but this way is both laborious and tedious, consequently, mills moved by water, having a number of scotchers fixed upon the same axle-tree, and moving with great velocity, are much more frequently used. The work, in this case, is executed with much greater expedition, and far less fatigue of the workman; but the velocity of the mill occasions a great waste of hemp.

Before the hemp prepared in this manner is subjected to the hackle, it mostly undergoes another process, termed beetling; by which the fibres of the hemp become more loosened and divided. The beetles employed with this intention are moved by the power either of the hand or water, which may be considered best.

The implements used in preparing hemp for the operation of spinning are so very similar to those described in the preparatory processes of the flax-manufacture, that we do not consider it necessary to give more than a general description of the processes; we shall therefore conclude this article with an accurate description of a patent, taken out by Mr. George Duncan, of Liverpool, in March 1813, for his improvements in the different stages of rope-making, and for certain machines adapted for the same.

The first part of the process which he has described, is that for *spinning the yarn for all kinds of cordage, lines, and twine.*

In this part of the invention there are two railways, adjoining and parallel with each other, fixed along the spinning-ground or rope-walk, from one end of it to the other. Upon each of these railways a machine for spinning the yarn is made to travel alternately backwards and forwards, one setting off from the bottom of the ground at the same time that the other sets off from the top, and as they both travel at the same rate, the former arrives at the top of the ground at the same time the latter arrives at the bottom.

These spinning-machines are in every respect similar to

each other, and are respectively furnished with two sets of twisting spindles; one set being placed at one end of the machine, with the hooks facing the top of the spinning-ground; the other at the opposite end, with the hooks facing the bottom. The spinners employed in spinning with them are, accordingly, divided into two companies, and arranged similar to the machines; the one company at the top, the other at the bottom, of the ground. The number of spindles in each set of each machine should be equal, and also equal with, or rather not fewer than, the number of spinners in each company; or, in other words, as there are in each machine two equal sets of spindles, (four sets in all,) the number of separate spindles in the two machines should not be fewer than double the whole number of spinners employed; because one set only in each machine is occupied at the same time in spinning, the other set being in the mean time engaged in retaining the yarns last spun from it, and following them back to the winding-machine.

The manner in which the operation is performed is as follows:—The spinning-machines are placed, as before described, one at each end of the spinning-ground, on its respective railway, ready to set off. Each spinner of the two companies immediately attaches his hemp or flax to the spindle of the machine that is nearest to him; and the motions of both machines, excepting those of that set of twisting-spindles facing the opposite company, are then struck into geer, and each machine recedes from its own company, spinning and leaving the yarn on separate guides or hooks as it proceeds onwards, the one down and the other up the ground, the one arriving and striking itself out of motion at the bottom, when the other arrives and strikes itself out of motion at the top. Each spinner of the two companies then detaches the yarn in his hand from the hemp or flax which he was spinning, and fixes the end of the piece that is spun to a winding-up reel, in a machine stationed behind or near him, while the other end still remains attached to the spindle-hook of the machine on which it was spun, and which is now at the further end of the ground.

The machines have now changed their company; that is, the machine which formerly belonged to the company at the top, now belongs to the company at the bottom; and that which belonged to the company at the bottom, belongs now to the company at the top. Each spinner, therefore, of both companies, immediately attaches his hemp or flax to the spindles left vacant by the opposite company, and the motions of the spindles to which the hemp or flax is attached being struck into geer, the spinning proceeds as before; during which, the respective winding-machines wind up the yarns last spun, regularly as the spinning-machine, to whose spindles they are attached, and which have remained stationary, advances towards it from the other end.

The spinning of the two sets of yarn, and winding up of the two last spun, being finished at one and the same time, the whole machinery again stops, and each spinner, as before, immediately detaches the yarn in his hand from the hemp or flax, and then detaches from one of the spindle-hooks of the spinning-machine just arrived, the yarn which he had on the former occasion spun, and which has just been wound up on one of the reels of the winding-machine, as closely as the short length necessarily intervening between the spinning and winding machines will allow. The two ends of these yarns he splices together, so that the yarn just spun, lying on the guides or hooks along the whole length of the spinning-ground, is now ready to be wound up. The spinners then attach their hemp to the emptied hooks, the machines are again struck into motion, the spinning and winding go on as before, and the same procedure continues to be renewed each time.

The general principle which constitutes this mode of operation, and consequent facilities of the invention, are, that one set of spindles in each machine is always employed in spinning, while the yarns spun by and attached to the hooks of the other set at the opposite end of the machine are winding up, so that every spinner is constantly kept at work in spinning, excepting the short interval when splicing the yarn, and preparing to set on to spin. Throughout the whole of the operation, therefore, whatever one of the spinning and the winding machines may be performing, and one of the company of spinners employed in doing, the other spinning and winding machines, and company of spinners, are, in every respect, similarly engaged.

An endless rope, driven by any external machinery, gives the travelling and twisting motions to both spinning-machines; and the whole of these motions are connected with and bear a given proportion to each other, capable of being regulated to suit the speed required. The two winding-machines may also be driven by the endless rope. All or any of these machines may, nevertheless, be driven by distinct endless ropes, or by any other method or methods in use for driving locomotive machinery, provided the proportionate speed be kept up.

The application of the rack, hereafter described, will be the most accurate for regulating the travelling-movement of the spinning or any other machine, on a rope-walk; but as the resistance in the present case is very trifling, the motion given to the truck-wheels of the spinning-machine, as shown in the engravings, will answer the purpose at less expense.

As a further improvement, Mr. Duncan invented an additional apparatus for giving an after-twist, which either may or may not be adopted. The object of this improvement is, to prevent the yarns losing strength, which otherwise they do, by losing twist by the counter-twist of the strand, and other subsequent operations. This is effected in a simple and convenient way, by continuing the twisting for a sufficient length of time after the travelling motion of the spinning-machine has ceased; by which means, an additional twist is given to the yarns after they are spun to their full length; and this is done while the spinners are splicing them at the other end of the spinning-ground, and preparing again to set on to spin, without occupying any of their time for the purpose. The effect of this part of the invention for the after-twist has been produced in different ways by others; but by modes either so complicated or expensive, as not to be conveniently available in practice.

The principal advantages to be derived from this method of spinning are the following:

First. The spinners are enabled, at much less expense, to spin a greater quantity of hand-spun yarn than they can by any other method in the same space of time; because, except while splicing the threads, they are constantly occupied in spinning; and have neither to hook up the threads, nor to travel up and down the walk, so that their time and attention is solely confined to the delivering of the hemp or flax from their hands. Secondly. The speed of the spinning-machine, besides being uniform, is proportioned to the full extent of work the spinners can conveniently accomplish, which, in some measure, compels them to produce the greatest possible quantity; and as the machine is constructed so as to lift the threads off the hooks, and follow them up to the winding-machine, little or no attendance is required from wheel-boys or followers. Thirdly. The spinners are enabled, partly from their whole skill and attention being confined to the one object, of simply delivering the hemp or flax from their fingers, and partly from the requisite degree of twist being given by machinery, to produce a superior quality of yarn.

And fourthly. The hemp to be spun by this machinery may either be dressed in the usual way, or prepared and dressed on the hackling-machines, which draw out the whole in a long sliver ; and in either case it may be spun from the end of the fibres, by which means the strongest yarn is formed. It may, however, be spun from the spinner's waist, and with more convenience than in the common way, because, by this method, the spinners remain always in a room at each end of the spinning-ground ; consequently, the hemp is not so liable to be discomposed as when they have to travel up and down its whole length ; neither, for the same reason, is it so liable to be wasted.

The expense attendant upon erecting the spinning machinery, with all its connections, and the power requisite to drive it, is comparatively trifling. In making the spinning machinery, various forms or shapes may be adopted, as they may be made to travel on railways on the ground, or on railways suspended from the beams of the rope-walk, or fixed upon or above them, as may be thought proper ; and the disposition of the necessary machinery may be arranged and diversified to suit the situations of the railways, and the construction and conveniences of the rope-walk.

The mode shown in figs. 469, 470, and 471, is most preferred by Mr. Duncan, merely because it occupies least room in proportion to the number of spindles the spinning-machines can employ. The whole width of a rope-walk, required by this mode, need not be more than six feet, except at each end, where, of course, sufficient width convenient for the spinners, and for the winding-machine and dressed hemp, must be allowed. In this narrow space no less than twenty-four threads may be constantly kept spinning at one time, and as many winding up ; so that besides every other advantage, the saving in the original cost of a spinning-ground on this plan, and in the expense of covering it in, will be considerable.

In the explanations to the figs. 478 and 479, is shown how other arrangements may be made by which the competent mechanic will be enabled to adapt or diversify the form of the machines, and form and disposition of the machinery, to suit any situation in a ropery which is most convenient, or of little use for other purposes.

In that part of the drawing, entitled " Rope-spinning," from A to B is supposed to be the spinning-ground, shown as broken off in the middle, for want of room to show it in full length ; C C and C C, on each side of the break, is one railway ; D D and D D the other.

Fig. 469 is a plan of one of the spinning-machines, at the top of the ground, on the railway C C.

Fig. 470 is the other, exactly of the same construction, at the bottom of the ground, on the railway D D ; and

Fig. 471 is a side elevation.

Though both machines are precisely similar, yet some parts of the machinery are omitted in some of the figures, that other parts may be seen more distinctly, and in none of the figures are the whole shown together,

E, wherever it occurs, shows the endless rope which drives both machines, and w the rollers or pullies suspended from the beams L over head, which guide and carry it.

The same characters, in all the above figures, wherever they may be used, denote the same part.

F shows the framing of the machines.

a and b two grooved sheeves, fastened upon the two upright shafts c and d, (as best seen in fig. 471,) driven contrary ways by means of the endless rope.

The manner in which the rope goes round, and grasps the sheeves, and occasions their contrary motion, is best seen in fig. 470.

In fig. 471, e and f are two spur-wheels, fixed upon the shafts c and d, with a view to equalize their motion.

The twisting motions for each set of spindles are driven by their respective shafts, which, at the same time, drive all the travelling motions. But the twisting motions for one end only are driven at a time; for while the shaft that is nearest to the end from which the spinners are spinning, is driving the twisting motions of that set of spindles, together with all the travelling motions, the shaft nearest to the other end is not driving any, though both shafts are then revolving.

g and h in the travelling motions are two pinions, upon loose rounds, alternately driving the wheel i, which is fast upon the short upright shaft h, and shown only in fig. 471. On the lower end of this shaft is the bevel-wheel l, driving, by means of another bevel-wheel, m, the cross shaft n, upon one end of which are fastened the sheeves 1, 2, 3, and upon the other the sheeves 4, 5, 6, seen best in fig. 469. These sheeves are of different sizes, and one only at each end of the shaft is in use at a time; o and p are the two axles of the truck-wheels, having the sheeves 7, 8, 9, fastened upon o, and upon p the sheeves 10, 11, 12. The four truck-wheels q are also fastened upon the shafts o and p, and motion is given to them by two belts; one driven from any one of the sheeves 1, 2, 3, and the other from any one of the sheeves 4, 5, 6, each belt running upon its corresponding sheeve on the axles of the truck-wheel. The turning round of these axles impels, according to the motion given to them, the machine forward; which motion can, by the sheeves being of different sizes, be regulated as occasion may require.

On the lower end of the shafts c and d of the twisting motions are the sheeves r and s, on loose rounds, each in its turn carried about by catch-boxes, (as shall be hereafter explained,) and driving the upright rollers or cylinders G and H, by means of belts going round the sheeves t and v, fastened on the axles of the rollers. These rollers give motion to the twisting spindles by separate bands or belts passing round a whirl on each spindle; in each machine are placed twenty-four spindles, twelve at each end, or six at each corner of each end, the positions of which are seen in figs. 469 and 470. In either of these figures, one spindle at each of the four corners only appears, the other five being ranged in a direct line underneath; but the manner in which they range is seen in fig. 471. In fig. 471, twelve, or one half of the number of spindles, appear on the nearest side: six, or half a set, at each end; the other half appear similarly situated on the opposite side. These two sets are alternately employed, the one in spinning, and the other in holding and following the yarns that are winding up; s s, fig. 471, are ratchet-wheels and catches, placed on the axles of the rollers i and H, to keep the yarns from untwisting when winding up.

Figs. 469 and 470 best show the carriers, projecting from the frame in which the twisting spindles run; the form and use of them, and of the whirls and spindles, are so obvious, that it is not necessary to point them out in any of the figures by a distinguishing character of reference; and, for the same reason, none of the bands or belts are marked.

Having so far described the different motions and appurtenances of the spinning-machine, we shall now proceed more particularly to explain the manner in which they operate.

By an inspection of fig. 471, it will be seen, that the catch of the catch-box 13 is in contact with the catch of the pinion *h*, and the catch of the catch-box 14 with the catch of the sheeves *I*, and that the catches of the boxes 15 and 16 are not in contact with the corresponding pinion *g* and sheeve *r*.

17 and 18 are two separate levers, one at each end of the machine, alternately used for striking into geer the catches above mentioned; the lever 17 serving for the two boxes 13 and 14, and the lever 18 for the two boxes 15 and 16. 19 and 20 act as swingers or levers from the joints 21 and 22, having claw ends to grasp the catch-boxes 13 and 14; and being coupled with the main lever 17, by means of the connecting rod 23, move them either up or down. When in geer they are held firm by the sneck 24; but on running against a fixture at one end of the spinning-ground, are pulled back, which causes all the motions of the machine to stop. The machine, however, may at any time be stopped, as occasion may require, by pulling back the sneck by hand. The two main levers 17 and 18 are so weighted at the handle end, that when disengaged from their snecks the catch-boxes always fly out of geer. The machine is put in motion by raising the main lever into the sneck by hand. All the machinery on each side of the wheel *i*, at each end of the machine, is precisely alike; the description therefore given of one end may answer for the other. The twisting motions at each end are never in geer at the same time; for those at one end are engaged in spinning one set of threads, while those at the other, whose spindles retain and follow up the other set of threads (last spun) to the winding-machine, remain at rest. All the four catch-boxes, 13, 14, 15, and 16, constantly go round with the shafts *c* and *d*, by means of feathers in the shafts acting in grooves in the boxes.

When the catch-box 14 is in contact with the sheeve *s*, it gives motion to the set of twisting-spindles belonging to the roller H, at the same time its accompanying catch-box 13, by being in contact with the pinion *h*, gives a retiring or travelling movement to the whole machine, which is effected by the wheel *i* communicating motion to the four truck-wheels, by the means which have been before described. The wheel *i* is common to both pinions, being turned one way by one pinion, when the machine is retiring from the top of the ground, and the contrary way by the other when it is retiring from the bottom of the ground; consequently, the cross shaft *n*, which derives its motion from the wheel *i*, turns at the same time both the truck-wheel axles, one way when retiring from the top, and the contrary way when retiring from the bottom; the shaft *n* being common to both truck-wheel axles.

Fig. 472 is the plan of a winding-machine, placed at the top of the spinning-ground, containing twelve reels, corresponding with the number of spindles in each spinning-machine.

Fig. 473 is a plan of a similar winding-machine, placed at the bottom of the ground, containing the same number of reels. Both these winding-machines are mounted so high above the ground as to allow the yarn winding on them to pass over head, that the spinners may have room to move underneath. In the engraving they are placed rather nearer the spinning-machines than they ought to be, from want of room in the plate. When all the spindles of the two spinning-machines are employed, one half in spinning and the other half in following up the yarns to the winding-machine, as has already been described, all the reels of both winding-machines are of course at the same time fully employed in winding up.

Fig. 474, at one end of the ground, shows a side view of the reels, placed on their spindles, a description of which is unnecessary, as the movements and construction of such machines are well known and understood. In the figure they are represented as only winding one yarn on each reel, in order to explain the improved or patent method of rope-making; but more than one yarn may be wound on any one of these, or any other kind of reel, that may be used in my method of spinning, for the convenience of the common method of rope-making, or for other purposes. It is neither convenient, nor necessary, that the endless rope should cease motion, when the spinning-machines have arrived and struck themselves out of geer at the top and bottom of the ground; consequently we will suppose, that the endless rope is in motion, and all the other parts of the machinery at rest, excepting the two shafts c and d, and their respective catch-boxes. The catch-boxes of each shaft in both machines are in the position of 15 and 16, as seen in fig. 471. Each spinner in the two opposite companies having now attached the hemp to the spindles, nothing remains to be done but to raise by hand the lever 17, fig. 471, (which in the engraving appears to be already done,) and the corresponding lever in the opposite machine; which will cause the spinning and winding to proceed in the manner already described.

When the machines stop, each spinner splices his thread, and throws it on the nearest guide s, to keep it out of the way, and to conduct it to the winding-machine. The grooved sheeves a and b, on the top of the shafts c and d, to which the endless rope gives the first motion, may be changed when required for sheeves of a larger or smaller diameter, for the purpose of diminishing or increasing all the motions in a proportionate degree. For the same purpose the wheel or sheeve, which gives motion to the endless rope, may also have grooves of different diameters. The sheeves that may be changed for increasing or diminishing the twisting motions, are the four sheeves t r for one end of the machine, and v s for the other, as seen in fig. 471. In order to obtain more or less travelling motion, the belts may be made to run either on the sheeves 1 and 9, and 6 and 10, or on 2 and 8 and 5 and 11, or on 3 and 7 and 4 and 12, as seen in fig. 469.

Fig. 475 (within which are figs. 476, 477, 478, and 479) represents an end view of a rope-ground building, set down as eighteen feet wide inside. It is merely divided into portions, to show some different modes of diversifying spinning-machines upon this principle, the different situations in which they may work, and the proportion of room they may occupy according to the number of spindles.

Figs. 478 and 479 show end views of two forms of spinning-machines, different from each other, and from the one already described; but all upon the same principle. The machine in fig. 478 is represented as moving on a railway M M, underneath the beam L, having spindles both above and below. The parts shown in the figure are as follows: N N two of the truck-wheels. O P the endless rope sheeves. Q one of the rollers for turning the spindles. R a sheeve on the end of the roller, answering the same purpose as t and v in fig. 471. W part of a sheeve on the truck-wheel axle, answering the same purpose as one of those on the axles o and p in figs. 469 and 470. The carriers, the whirls, the spindles, and their bands, in this machine are the same as in the spinning-machine already described; and their situations are so obvious, as not to require particular characters of reference. Such part of the figure as consists of framing will be obvious. One side of the railway is fixed to the post K; the other side is fixed and rests upon the iron fixture S, hanging from the beam L, which also serves the same purpose for the adjoining railway. X one of the guide pullies for the endless rope. T a rail, (which may occasionally be removed,) laid across, and answering the purpose of both railways, from post V to K, having upright pins at proper distances, for the purpose of bearing and keeping separate the yarns of the lower spindles. The hooks fixed to the under side of the beam L are to answer the same purpose for the yarns of the upper spindles. In order to lay the yarns upon these hooks, a separate guide upon each spindle is fixed upright in a slender rail, fastened to, and projecting two or three feet from, and parallel with, each end of the machine. The shape of these guides is the same as the hooks in the beam, with this exception, that each one has an eye at the point, to convey the yarns in a slanting direction from the spindles, and to lay them upon their respective hooks in the beam when spinning, and also to lift them off when winding up. The manner in which the guides pass between the hooks may be seen in fig. 477, where c represents the projecting rail; b^a the guides, two only of which are marked; it must here be understood, that the spindles are not opposite the eyes of their respective guides, but exactly opposite the upright part of them, and on a level with the eyes. There are many other ways by which the hooking up of the yarns may be effected, but the method just described is conceived to be the most simple. The space between the post V and the iron fixture S, is the room to be occupied by the other spinning-machine.

The spinning-machine, fig. 479, is represented as moving on a railway, laid upon the beam L. It will be seen that this machine is nothing more than the lower part of the one last described, having no spindles on the upper part. The guide pins are in this method driven into the beam. The empty space to the right of this machine is the room to be occupied by its fellow. The letters of reference used in fig. 10, and its appurtenances, apply to the same parts whenever they are used in fig. 479, and its appurtenances. Though these machines (the end views of which are shown in figs. 478 and 479) are different from each other in form and arrangement of machinery, and also from the form and arrangement of the one shown in figs. 469, 470, and 471, yet the same principle of the travelling and twisting motions is applicable to all, and therefore it is unnecessary to enter into further explanations respecting them.

Fig. 480 shows the method of giving the after-twist. As the apparatus for this purpose is to be applied to each of the endless rope shafts in each machine, a description of the apparatus as belonging to one of them may be sufficient. This figure is a side view of the apparatus, and is represented as applied to the shaft d, in fig. 471; the same characters of reference there used being retained in the present figure, where they denote the same parts. The apparatus for the after-twist is merely an addition, which on the lower part of the figure consists of a catch on the under side of the sheeve s, a corresponding catch and catch-box 25, carried round by the shaft, and the lever or swinger 26 to act on the catch-box. The rod 23 is lengthened, to connect this swinger with the other two. On the upper part of the figure is the remainder of the apparatus, consisting of a catch on the upper side of the catch-box 13; the worm on a loose round on the shaft 27, with a catch on the under side, to operate with its corresponding catch on the catch-box 13; the screw-wheel 28 to act in the worm; the arm 29, on a loose round on the axis of the screw-wheel, confined near the circumference of the wheel by the staple 30, but having play the width of the staple, the end of the arm farthest from the axis being intended to act on, and press down, the swinger 19; and the spring 31, fixed on the wheel, which presses against the back of the arm.

The whole of the machinery in the figure is represented out of the geer, and is in the position as when ready to set off from one end of the spinning-ground, to follow the yarns to the winding-machine. In that position it continues until it has arrived at the winding-machine, and the yarns are disengaged from the spindles when the main lever 17 is lifted up by hand into the catch 24, for the purpose of putting the spinning and travelling motions into geer, as formerly described. By the raising up of this lever the swinger 19 is pulled down, and the arm 29 is thus disengaged, which having play within the staple, swings forward by its own weight, clear of the swinger, which is hollowed or bent at that place for the purpose. The object of this is, that the arm may not be in the way of the swinger 19 the next time it goes into geer with the worm. When the machine has returned to the other end of the ground, and the yarns consequently are spun to their full length, the catch 24, on which the main lever rests, is thrown back by the machine going against a fixture in the ground, as has been before mentioned, and the lever (being sufficiently weighted at the handle end) drops down, by which means the travelling motion is stopped. The twisting motion would also be stopped, were it not, (as will be seen from the figure) that though the catch-box 14 is thrown out of geer with the sheeve s on the upper side, the catch-box 25 will be at the same instant, and by the same movement, thrown into geer with it on the under side, so that the twisting motion continues. The under side of the catch-box 13, being thrown out of geer with the pinion A, (which stops the travelling motion) the upper side of it will be at the same instant, and by the same movement, thrown into geer with the worm 27, which consequently gives motion to the screw-wheel 28; the arm 29 (which it will be recollected is then hanging down) is also carried round with the wheel; and when it comes in contact, having nearly made one revolution with the swinger 19, it forces it down, and by this means puts the catch-box 25, as well as its own, out of geer, and causes the whole of the machinery to stop. The use of the spring 31, pressing against the back of the arm, is to cause it to force, as soon as the catches 13 and 25 are out of geer, the swinger 19 a little further down, which it will then be enabled to do, in consequence of the resistance against it being decreased; the object of this is to prevent any jarring of the catches when in that situation. The spring is prevented forcing too far by a stop. Another method of forcing the swinger thus much further down may be adopted, by fixing a pin or fang to project from the framing of the machine, so as the end of the spring above mentioned may come in contact with it a little before the time when the arm begins to force down the swinger, in order that the arm may be relieved from the pressure of the spring until the arm has forced the swinger down nearly to the point of sending the catches out of geer, at which time the end of the spring, having got free of the pin, comes with a sudden blow against the back of the arm, and thus sends down the catches clear of those with which they were in geer; the spring in this case also is prevented from forcing too far by means of a stop. The time the screw-wheel is in going round is the time allowed for the after-twist: but should one wheel not allow sufficient time for the purpose the motion may be further decreased, by any usual and well-known means, and change wheels may be applied to suit the different kinds of yarn.

In tempering the strands of all kinds of cordage, whether shroud, hawser, or cable-laid, it is well known that from various causes an inequality of tension between the different strands intended for the same rope takes place, and is most commonly apparent during or after the operation of hardening, some of the strands becoming too slack, others too tight, and

consequently of unequal lengths, though originally they
may have been of equal length, and have received the same
twisting or number of turns by machinery of the most improved
and perfect construction. In cases, therefore, where this inequality appears, the strands require to be rectified, by being
brought to an equal degree of tension, in order that each
may bear its equal portion of strain in the rope when made.
The operation for effecting this object is commonly called
tempering the strands; and the method in general practice is
to give more twist to a slack strand, or to take twist out of a
tight one, or to do both. In some rope-grounds where the machinery is driven by steam, or other considerable power, the
method adopted is to give more twist to the slack strands,
which is done by stopping the twisting of the tightest strand,
by throwing its hook out of geer, and to keep it waiting in
that position until the slack strands have twisted up to the
same tension.

These methods in most cases are defective; because the
strand to which more twist is given is thereby rendered less
pliable, and is of smaller circumference; consequently it cannot top or lay up in the rope evenly and regularly with the
other strands which have less twist; for the harder twisted
one will in the rope sink inwards, and the others stand outwards and form more of a spiral round the harder twisted one,
which will thereby have more than its proportionate strain
in the rope to bear, and will also be least enabled, when
under a strain, to stretch up, so as to avail itself of the assistance of the others, and by consequence must be the first
to break. Should the inequality of tension be occasioned by
any original inequality of thickness in the strands, the smallest
one will, during the process of hardening, become the slackest, and in tempering by twisting it up to the tension of the
tightest, the inequality of size will by that means be increased;
for the more it is twisted, the still smaller in circumference,
as well as shorter in length, will it become. But, supposing
all the strands were originally of equal thickness, and that the
inequality in tension proceeded entirely from an error in the
original lengths; it is plain, that, by tempering according to
the methods in question, (and no other methods, after the
strands are fixed on the hooks, and the work has commenced,
can, by any machinery hitherto in use, conveniently be adopted,)
the same defective principle still applies, which, by causing
one strand to be harder twisted, and consequently to become
of a smaller size, and another to be softer twisted, and to become of a larger size, prevents the whole from jointly forming

a regular and perfect rope, and to stretch equally when under a strain, as already described.

As a more convenient, accurate, and certain method, than any hitherto practised, appeared to be necessary, Mr. Duncan invented and adopted a new mode of tempering the strands of all kinds of cordage, whether shroud, hawser, or cable-laid : the nature and general principle of which is, to cause any one strand-hook of the foreboard, or foreboard-machine, when the strand attached to it requires to be tightened, to recede from its corresponding opposite hook of the sledge or stranding machine, to which the other end of the strand is attached; or when it requires to be slackened, to cause it to advance towards its corresponding opposite hook, thus bringing all the strands to an equal tension, without one strand-hook making more revolutions than another. And, what is of essential importance to this invention, the operation may be performed leisurely, as occasion may appear to require, either before, during, or after hardening the strands, without stopping the twisting, or other motions, or occasioning any interruption to them whatever; and with more ease, minute accuracy, and useful effect, than by any other method yet practised for the purpose.

In order more particularly to exemplify and illustrate this part of the invention, we have annexed engravings of the machinery which Mr. Duncan has contrived and adapted for the purpose.

In fig. 481, A B C represent the upper part of the framing in which the machinery, placed at the foreboard, is fixed; C being the front of it, facing or looking down the rope-walk. D is a toothed wheel, receiving motion from any external machinery. This wheel drives the other toothed wheel E, and either of them can be changed to suit the speed of the motion required. The toothed wheel E is fixed upon the axis of, and gives motion to, the toothed wheel or fluted cylinder F; which cylinder drives the four pinions 1, 2, 3, 4, whose axles, to the hooks of which the rope strands are attached when twisting and tempering, are the four strand hook spindles a, b, c, d. To answer the purpose of the invention, the strand hook spindles, besides having the rotative or twisting motion which we have already described, are so contrived, for the tempering of the strands, that any one or more of them may, while the twisting motion is or is not going on, be made to slide, in a horizontal direction, parallel with the axle of the cylinder F, along any part of its length, either backward or forward, as shall now be explained. The strand hook spindles having to slide, as has been said, in a direction parallel to the axis of the cylinder, are of course placed in that direction, and so as their pinions may pass each other. The positions of these pinions round the cylinder are seen in fig. 482, which represents a front view of the machinery; the same references in each figure being used to denote the same part. As all the four strand hook spindles, with their accompaniments and immediate connections, are precisely the same, a description of one will be sufficient; we will therefore take the spindle b, in fig. 481. G H is a long or male screw, a few inches longer than the cylinder, upon which is fitted the nut or female screw, *e*, having spokes or · arms, to admit of it being turned by hand. Joined fast to this long screw is a head-piece or claw *f*, within which a carrier or step is fitted, and in which the adjoining end of the strand hook spindle revolves. Two collars, *g* and *h*, fitted on the spindle, one on each side the step, cause the spindle to accompany the long screw, either backward or forward, when moved by turning the nut *e*, the rotative motion of the spindle going on at the same time if required; *i* and *k* are two steps or guides, fixed on the cross framing B and C, through which the spindle may pass and repass, and in which it also revolves; *l* is a carrier of the same description, fixed on the cross framing A, through which the long screw may pass and repass, but without revolving. Fast upon, and projecting from the head-piece *f*, and consequently accompanying the long screw and spindle in the sliding movement, (see the side view, fig. 483,) is the tongue *m*, the end or point of which is fitted to pass along during that movement in a slot in the rail *n*, fixed parallel with the long screw and spindle, between the two cross bearers A and B.

The object of this contrivance is to prevent the spindle (one end of which, as has been shown, revolves within the head-piece of the long screw) from carrying round the screw along with it, and to keep the screw and its head-piece at all times steady, and in a direct line with the spindle. For the purpose of keeping the long screw stationary in the situation to which it may have been last set, the pull of the strand on the hook (by pressing and abutting the screw-nut *e* against the back of the carrier *l*) will always be found to be sufficient. The diameter of the cylinder F may be about two feet, and that of each of the fore pinions 1, 2, 3, 4, about one foot, more or less, according to the speed desired, and the discretion of the mechanic. The pitch of their teeth should be the same as that of the teeth of the cylinder. The length of the cylinder should at least be equal to the greatest difference or inequality of length ever likely to take place between the slackest strand and the tightest strand intended for the same rope, previous to, or during, the operation of hardening, when they are both brought to an equal tension by tempering according to this method. The inequality of length, or, in other words, of tension, which takes place in the strands during the process of hardening them, is generally found to be in proportion to their circumference, and is more in a set of the large strands than in the small.   In rope-walks, therefore, where cordage of the largest size is manufactured for the use of his Majesty's navy, the length of this cylinder should not be less than four feet; but Mr. Duncan has found, by experience, in manufacturing cordage for merchantmen of the greatest burthen, that few cases occurred where it was requisite for the length to be more than three feet. In rope-walks where cordage on the common principle is manufactured, some additional length is necessary. Each of the four pinions is fastened upon the middle of the length of its strand hook spindle. Supposing, therefore, that the pinion 2 should be set so as to be exactly at one end of the cylinder next to the cross framing B, it must be enabled to slide along to the opposite end next the cross framing C, and also back again to B; for this purpose the spindle must always be kept in its steps or carriers, *i* and *k*, which support it, and in which it both slides and revolves, and therefore it requires to be double the length of the cylinder, besides an additional length equal to the spaces in its passage occupied by the necessary steps, framing, clearances, &c. The length of the long screw G H, and of the rail *n*, are each the length of the cylinder, and correspond with, or are a few inches longer than the sliding distance, to allow for steps, &c. as above. It has been shown, that the cylinder drives the four strand hook spindles, and that any one of them can be moved, by means of its screw, either backward or forward, without interrupting its own rotatory motion, or the rotatory motion of any of the others; the teeth of the pinions being for this purpose kept in geer with, while at the same time they are made to slide along between, the teeth, or in the flutings of the cylinder.

Suppose, then, that the strands are attached to their respective hooks, and the pinions set so as to be all at an equal distance from each end of the cylinder, and all the strand hook spindles going round, twisting and hardening the strands, the operation of tempering is performed merely by turning round, by hand, as often as required, any one or more of the screw-nuts either way about, as the case or cases may require, according as any one or more of the strands require slackening or tightening for bringing them all to an equal tension. Thus, in order to slacken a tight strand, its hook must be advanced forward further from the front of the framing C; and in order to tighten a slack strand, its hook must be drawn in, towards the framing.

Fig. 484 is a side view, representing some variation in the machinery for effecting the sliding movement of the strand hook spindles on the same principle, and answering the same purposes, as the plan in fig. 481, already described. After the full descriptions and explanations already given, a very short account will be sufficient to make this fully understood : *b* is a strand hook spindle, similar to those in fig. 481, excepting that the pinion 2 is not made fast upon it, because it has to pass or slide through the axle hole of the pinion. In order that the spindle may at the same time revolve with the pinion, the slot 10 is cut upon one side of the spindle, (the length of the slot being the sliding distance,) which slot receives a feather or key in the axle hole of the pinion, through which the slotted part of the spindle is to pass and repass, as occasion may require, the feather always remaining in the slot to carry round and give the rotatory or

2 F

twisting motion to the spindle. The parts *f*, *g*, and *h* are exactly the same as the parts which have the same characters in fig. 481. I is a rack, (to answer the same purpose as the long screw in fig. 481,) which the pinion *o*, by means of the handle *p*, moves either backward or forward. The ratchet-wheel, *q*, and its catch, hold the rack and pinion stationary in the situation to which they may be set; *i i* and *k* are the steps in which the spindle revolves, and through which it also slides; *r r* are two rings or washers, loose upon the spindle, between the steps *i i* and the pinion 2, intended to qualify the friction during both the sliding and the rotatory operation of the spindle; *s* is a guide or step for the rack to slide in, made square at the bottom, which renders the tongue and slot, shown in fig. 483, unnecessary. The wheel K, receiving motion from any external machinery, drives the pinion 2. Change wheels, for varying the motion, may be applied to this method in the same way as in fig. 481.

From what has already been described, it will appear, that the strand hook spindle may, by means of the rack I and pinion *o*, be drawn or slided either backward or forward, through its pinion 2, without interrupting its rotatory motion; the pinion 2 always keeping in geer with the wheel K, by which it is driven, and which latter may receive its motion from any external machinery. Referring, therefore, to the former description, it will be evident, without further explanation, by what means the strands are to be tempered by this variation in the machinery.

The reader will observe that there are two principles by which the strands of cordage may be tempered or brought to an equal tension; the one by causing any one or more of the strand hook spindles either to advance or recede, whereby an equal tension will be effected without one spindle making more revolutions than another; and the other, that of causing any one or more of the strand hook spindles to be at rest while the others are revolving; whereby an equal tension will be effected by an unequal number of revolutions. If one of these two principles only is to be adopted, Mr. Duncan prefers the former, as being generally more appropriate and effectual. As, however, it sometimes occurs in practice, that the application of the one principle, sometimes of the other, and sometimes of both, proves to be the most proper and effectual remedy, Mr. Duncan has invented a still more perfect method, by which either or both of the principles may be practically applied in one set of machinery. This object which had never, we believe, been before accomplished, is effected merely by applying to either of the two varieties of machinery before described, an additional apparatus, so that all kinds of cordage-strands may thereby be tempered, either entirely, by the principle of causing any one or more of the strand hook spindles to advance or

......; or entirely, by causing any one of more of the strand hook spindles to be at rest while the others are revolving; or partly by the one and partly by the other, according as the original cause occasioning the inequality of ...... in the different strands may point out; the whole, or any part, of the operations going on, either together or separately, as may be found convenient, without interruption to each other.

Fig. 485 is a plan showing the additional machinery for tempering, by combining the two principles as adapted for the first described machinery, represented in fig. 481. The difference between the machinery of fig. 481, and that of fig. 485, consists chiefly in the latter having its pinion 2 loose upon the twisting spindle δ, but confined between two collars, which are fixed upon the spindle. The reason of the pinion running on a loose round is that it may be either put in or out of geer with the spindle, by means of the catch-box t and lever u. The catch-box has a slot, fitting a feather of the spindle, in order that it may revolve with it, as well as slide in or out of geer, when moved by the lever. The ratchet-wheels v and w are ...... upon the spindle, one having teeth cut the reverse of the other, that either of the two palls x and y may, when the spindle is thrown out of geer with the pinion, prevent the strands from untwisting, as otherwise the spindle would be at liberty to be acted upon by the force of twist already in the strand. The pall y is flat towards the point for holding against the ratchet-wheel w for a right-hand twist, and the pall x is hooked towards its point for holding the wheel v for a left-hand twist. So far, this apparatus would serve the purpose either of keeping in geer, or stopping the rotatory motion of the spindle, provided it were not also required to perform the sliding movement. In order, therefore, to complete the apparatus for both these purposes, the arm z, fastened to the claw or hand-piece f, and forming one piece with the long screw G H, stretches alongside, parallel with the spindle, so that its other end is nearly opposite to the pinion, where it is furnished with two ears, having each an eye or ring, 7 and 7, fitting easy upon the round iron rod 8; which rod is fixed parallel with the spindle, between the cross framing B and C. The step 9, on the cross framing B, serves as a guide for the arm z. It is necessary that the distance between B and C should be as much longer, than the distance in fig. 481, as the length taken up or occupied by the catch-box and ratchet-wheels. The spindle also will require this additional length. The arm z, during the sliding movement, has to conduct with it, along the rod 8, the lever u, and the two ratchet palls x and y, the rod serving them also as a guide during the sliding movement, and at all times as an axle. Though the pinion 2 is always in geer with, and carried round by, the cylinder F, fig. 481, yet the spindle only goes round when put in geer with the catch-box by the lever; therefore the rotatory motion of the spindle may at any time, and for any space of time, be stopped, for the purpose of causing the twist o ...... to cease, while at the same time the other strands are twisting up. Though only one spindle is here spoken of, it is evident that all or any of them may, from being furnished with the apparatus now described, be made either to give, or cease from giving twist, while any one or more of the spindles either may or may not, as required, be performing the sliding movement.

Fig. 486 is a side view, showing the method adapted for the second described machinery, as represented in fig. 484, the apparatus in this case

applying to the narrow wheel, and that in the former case of fig. 485, applying to the wide wheel or cylinder F. The difference between the one and
the other is, that the spindle and pinion in fig. 485, slide together, as in
fig. 481, whereas in the figure now to be described the spindle slides
through the pinion, as in fig. 484. The spindle *b* in this figure is similar to
that in fig. 484, having a slot, to receive a feather, which is fixed in the
catch-box *t*. The pinion 2, which is always in geer with the wheel K, is
fastened on the bush 11, running loose upon the spindle *b*. This bush,
being furnished with the collar 12, serves, by means of its revolving in the
cavity 13, adjoining the step *i*, to keep the pinion in its proper place,
during the spindle's sliding movement: *i* and *k* are two steps, answering
the same purpose as those of the same characters of reference in fig. 484 :
v and *w* are two ratchet-wheels, fast to each other, but not fast on the
spindle, having a feather, fitting the slot of the spindle, in order that they
may hold it fast when occasion requires, and that it may pass and repass
through them during the sliding movement. These ratchet-wheels are
furnished with their two palls *x* and *y*, altogether answering the same purpose as those described in the former fig. 485. The catch-box *t*, having
also a feather, fitting the slot, is furnished with a lever, (not shown in the
figure,) answering the same purpose as the one marked *u*, in fig. 485 ; but
to suit the present case, it works on a stationary pivot, fixed to the framing
of the machine. The two ratchet palls also work on pins fixed to the
framing ; and their wheels *v* and *w*, being furnished with the rim or
fencing 14, are kept always opposite to the palls, by means of a bracket,
(fixed to the framing, but not seen in the figure,) hollowed out to receive
the rim. It will be evident, from what has here been said, that the operation of striking in and out of geer the rotatory motion of the spindle is
performed exactly in the same manner, and also answers the same purpose,
as that described under fig. 485 ; and that the sliding movement of the
spindle in both cases is performed in the same manner, and answers the
same purpose, as described under figs. 481 and 484, either one or other of
the methods, under figs. 485 and 486, combining the two principles of tempering strands in the manner previously pointed out.

Though in the first-described machinery it has been shown,
that the sliding movement of the strand hooks may be
effected by means of a male and female screw, and in the
second-described machinery by that of a rack and pinion,
yet it will be seen, that either means may with equal propriety be applied to either machinery. And a competent
mechanic, from what has been described, will easily perceive
that any other power, such as that of a lever, weight, or rope
and pulley, or one or more of them combined, may be applied
for the same purpose, though in the first preference be given
to the screw, and in the next to the rack.

The next part of the invention to be described is a
new method of regulating both the backward and forward
travelling movements of any sledge or other locomotive machine that is or may be used in a rope-walk. The backward movement of the stranding-sledge, or the retrograde
movement of that machine towards the bottom of the rope-
walk by which strands are drawn out, in rope-walks where

the improved or patent principle of rope-making is adopted, has hitherto been effected by means of a rope applied in different ways for the purpose. In some cases the rope is made to haul the sledge backwards, by fastening one end of it to the sledge, and the other round the capstan or barrel, at the bottom of the rope-walk; and in other cases the rope is stretched tight along, and made fast at each end of the rope-walk, and two or more doubles or bights of it passing round and grasping the same number of grooved binding sheeves in the sledge, which revolve by connection with the rotative motion of the strand hooks, from which the other motions are derived: thus the sledge works itself backwards along the rope.

The great object to be attained in regulating this backward motion is, to cause it always to preserve a certain speed in a given ratio with that of the rotative motion, in order that the strands may always receive the degree and uniform distribution of twist intended. But in whatever way a rope has hitherto been applied for that purpose, the object has never been effectually attained, nor the operation conveniently performed, because, from the elasticity and specific gravity of the rope itself, extended along the whole length of the rope-walk, it has been found impossible to keep it accurately stretched, and equally tight from one end to the other, so that when the sledge is in motion, particularly when first struck into gear, it pulls up the slack of the rope from the bottom of the rope-walk, and its retrograde motion is thus retarded in proportion as the rope may stretch, slip, or give way.

The retrograde motion loses therefore its relative speed commensurate with that of the rotative motion of the strand hooks, which have in the mean time, without interruption, continued to put twist into the strands. Instances are not unfrequent where they have been twisted to such a degree as nearly to break them asunder before the rope could be tightened sufficiently to cause the sledge to move on at its proper speed; and, on the whole, it is obvious, that by the present method of drawing out the strands, they can neither receive their proportionate twist nor the distribution of it. The labour required in applying the rope is besides extremely inconvenient and troublesome, because it requires to be first fixed to the sledge, or round its binding sheeves, at the top of the rope-walk, then tightened, and afterward disengaged at the bottom, on every single occasion of drawing out a strand or set of strands. The plan also is expensive, because the constant wear and tear is considerable, and requires the rope to be frequently renewed. An iron chain may indeed be applied for the purpose, and though not requiring to be so frequently renewed, it is equally objectionable with the rope in most other respects, and on some accounts more so. The forward movement of the stranding, topping, and dragging sledges, is that slow progressive movement necessarily required towards the top of the rope-walk by the shortening or shrinking up of the strands in twisting, while forming on the common principle; and of the strands and cordage, either common or patent, whilst hardening and topping. It will readily be seen, that this movement should also be uniformly regular, in a given proportion to the twisting motion, and that the travelling distance should be neither more nor less than the length the strand or rope ought to shrink up. According to the usual method, a number of press barrels or weights are placed on the stranding or topping sledge, or on a drag sledge, attached to their tail end, to serve as a resistance against the pull of the strand or rope when shrinking up. But as the quantity of weight to be applied is to be varied and proportioned to the size of the strand or rope, and degree of twist required, and as the friction of the drag on the ground is greater on some parts than on others, the operation, depending on criterions so uncertain, must be attended with a great degree of irregularity, both with regard to the sledge sliding faster or slower, and also with regard to the whole length to which, eventually, it may be dragged; thereby occasioning a proportionate irregularity, both in the distribution and total quantity of twist or lay in the strands or rope, corresponding with their length.

The object, therefore, of this invention, with regard to the *backward* movement, is, to cause the sledge, or any other locomotive or travelling machine used, or that may be used, in a rope-walk, to travel and recede down the walk at one uniform speed, such as shall be predetermined as proportionate with the rotatory speed of the twisting hooks of the machine, so as to cause the twist to be uniformly regular throughout each operation. And the object of this invention, with regard to the *forward* movement, is, to cause the sledge, or other movable machine, to which any kind of strand or rope may be attached, for the purpose of being formed, hardened, or laid, to travel slowly, and advance up the walk, during the operation, at one uniform predetermined motion, and precisely the length or distance assigned to it, equal to that which the strands or rope ought to shrink up in the operation.

: travelling speed of the machine,
ard or forward, becomes at all times
a certain ratio, with the speed of its
whole machinery being composed of
no part of it is liable to slip or yield.
th of travelling and twisting motions,
e wheels, to suit each other in that
any other machine or machines that
and the same operation. The whole
n by an endless rope, receiving its
achinery at the top of the rope-walk,
in use for driving locomotive ma-
f scarcely observe, that it is not
se of producing accurate work, that
as all the others should be uniform,
riginal motion be quicker or slower
er during the operation, the motions
till keep their proportionate speed.
l be in the time in which the work
ave mentioned the particular cases
invention is more essentially useful;
s the application of the rack, in
an invention subservient to every
r process of rope-making, for which
motion to any machine, either
a a rope-walk or elsewhere, may be

g entitled, "Backward and Forward Move-
side view of a travelling sledge, or locomotive
and for the purposes referred to, moving on
side view of the rack-way laid down and
N, or other suitable material, supported in
e bottom of the rope-ground. This machine

is represented as driven by the endless rope O; 13 and 14 are two guide pullies, to conduct the rope in going on and coming off the large sheeve or grooved wheel P, round which that rope (driven by external machinery, and running from top to bottom of the rope-ground) passes, by which the first movement in the sledge is given. This sheeve, giving motion to the spindle or shaft Q, and being coupled with the shaft R, turns the pinion 1, which drives the pinion 2, upon whose shaft, S, is the small bevel-wheel 3, driving the large bevel-wheel 4, upon whose shaft again is the spur-wheel 5, driving the other wheel 6; which last wheel works in the rack-way. This wheel is not fast upon its shaft, being capable of sliding thereon, for the purpose of being put in and out of geer with the rack by means of the lever T. The machine travels on the railway on four truck-wheels: the two shown in this figure are marked 7. The pinions 1 and 2 are change-able, to suit the different travelling speeds required.

So far as has been now described refers only to the backward movement of the machine; which movement, it must be understood, is in the direction along the rack-way, as from A towards B. The contrary, or *forward*, movement is of course in the direction from B towards A, and is effected by giving a reverse turn to the wheel 6, which works in the rack-way. The necessary machinery for this purpose is the small pinion 8, on the shaft Q, driving the wheel 9, on the shaft U; which last shaft, and the one coupled with it, W, lie parallel with, and extend to, the end of the shaft R, in order that the pinion 10, fixed on the end of W, may, when required, work in the pinion 2. The shaft S then becomes common to both the pinions 1 and 10, and may, as required, be driven by either the one or the other, the pinion 1 being for the backward movement, and the pinion 10 for the forward move-ment, one of them therefore must be out of geer while the other is in geer. The figure shows the pinion 10 as out of geer. But supposing it to be in geer with 2, and the pinion 1 out of geer with it, the effect is, that a con-trary motion is given to the wheel 6, which works in the rack-way, by means of the intervening wheels 3, 4, and 5, before described. The twisting mo-tions of this machine are produced by the shaft Q being continued to the front of the machine, where the wheel 11, on the end of the shaft, drives the counter-wheel 12, from whence the required degree of speed is given to the twisting hooks. From what has been before described, it will be seen, that the backward and forward motions of the machine are produced by means of the wheel 6 working in the rack-way either way about as required. As, therefore, any predetermined quantity of twist may be given by means of the change wheels 11 and 12, whilst at the same time the machine may be made to travel at any given predetermined speed, either backward or forward, by means of the change wheels 1, 2, and 10; and as the twisting as well as the travelling motions are driven by one and the same impulse, originating in the machine at the grooved wheel P; they must always preserve a rela-tive speed to each other in such proportion as may be assigned to them. A forked lever, clasping on the catch-box 15, serves either to put in or out of geer all the motions of the machine, excepting that of the grooved wheel P.

Fig. 488 is a view of the back end of the same machine, showing as much of the machinery as is necessary for understanding it. The same characters of reference used in fig. 1 denote the same parts in this.

Fig. 3 is a plan of part of the rack-way. A is the rack, and N N is the wood sleeper upon which it is fastened. The forward motion of the sledge is a remarkably slow movement; the speed of the wheel 6 therefore requires to be considerably reduced. The wheels shown in the figures will not re-duce the motion sufficiently slow to suit every possible occasion; but enough is shown to enable a mechanic readily to produce any degree of motion that may be required.

All or any part of the machinery which we have described may be driven by the power of steam, water, wind, or animals.

In the course of describing the different machines, and their component parts, adapted for the various purposes of the invention, we have seldom taken notice either of their dimensions or of the materials of which they may be made, because no fixed rules can be given: but any competent mechanic, from what we have shown, will be enabled to apply such sizes, and use such materials, as may be suited and proportioned to the nature and design of each machine, and to the power which is to drive it, particularly when we add, that the figures in the plates marked "Tempering, and Backward and Forward Movements," are made out on a scale ⅛ of an inch to a foot, and that the dimensions there given are such as may with effect be applied in practice.

SAW-MILLS.

SAW-MILLS, constructed for the purpose of sawing either timber or stone, are moved by animals, by water, by wind, or by steam. They may be distinguished into two kinds; those in which the motion of the saw is reciprocating, and those in which the saws have a rotatory motion. In either case the researches of theorists have not yet turned to any account: instead therefore of giving any uncertain theory here, we shall proceed to the descriptive part, and refer those who wish to see some curious investigations on this subject to a Memoir on the Action of Saws, by Euler, en Mem. Acad. Roy. Berlin, 1756.

Reciprocating saw-mills, for cutting timber, and moved by water, do not exhibit much variety in their construction. The saw-mill represented in fig. 450 is taken from Gray's Experienced Mill-wright; but it only differs in a few trifling particulars, from some which are described in *Belidor's Architecture Hydraulique*, and in Gallon's Collection of Machines approved by the French Academy.

The plate just referred to shows the elevation of the mill. A A the shaft or axle upon which is fixed the wheel B B, (of 17½ or 18 feet diameter,) containing 48 buckets to receive the water which impels it round C C, a wheel upon the same shaft containing 96 teeth, to drive the pinion No. 2, having 24 teeth, which is fastened upon an iron axle or spindle, having a coupling-box on each end that turns the cranks, as D D, round; one end of the pole E is put on the crank, and its other end moves on a joint or iron bolt at F, in the lower end of the frame G G. The crank D D, being turned round in the pole E, moves the frames G G up and down, and those having saws in them, by this motion cut the wood. The pinion No. 2 may work two, three, or more cranks, and thus move as many frames of saws. No. 3 an iron wheel having angular teeth, which one end of the iron K takes hold of, while its other end rolls on a bolt in the lever H H. One end of this lever moves on a bolt at I, the other end may lay in a notch in the frame G G so as to be pulled up and down by it. Thus the catch K pulls the wheel round, while the catch I falls into the teeth and prevents it from going backwards.

Upon the axle of No. 3 is also fixed the pinion No. 4 taking into the teeth in the under edge of the iron bar, that is fastened upon the frame T T, on which the wood to be cut is laid: by this means the frame T T is moved on its rollers S S, along the fixed frame U U; and of course the wood fastened upon it is brought forward to the saws as they are moved up and down by reason of the turning of the crank D D. V V the machine and handle to raise the sluice, when the water is to be let upon the wheel B B, to give it motion. By pulling the rope at the longer arm of the lever M, the pinion No. 2 is put into the hold or gripe of the wheel C C, which drives it; and by pulling the rope R, this pinion is cleared from the wheel. No. 5, a pinion containing 24 teeth, driven by the wheel C C, and having upon its axle a sheave, on which is the rope P P, passing to the sheave No. 6, to turn it round; and upon its axle is fixed the pinion No. 7, acting on the teeth in an iron bar upon the frame T T, to roll that frame backwards when empty. By pulling the rope at the longer arm of the lever N, the pinion No. 5 is put into the hold of the wheel C C; and by pulling the rope O, it is taken off the hold. No. 8, a wheel fixed upon the axle No. 9, having upon its periphery angular teeth, into which the catch No. 10 takes, and being moved by the lever attached to the upper part of the frame G, it pushes the wheel No. 8 round; and the catch, No. 11, falls into the teeth of the wheel, to prevent it from going backward, while the rope rolls in its axle, and drags the logs or pieces of wood in at the door Y, to be laid upon the movable frames T T, and carried forward to the saws to be cut. The catches Nos. 10 and 11 are easily thrown out of play when they are not wanted. The gudgeons in the shafts, rounds of the cranks, spindles, and pivots, should all turn round in cods or bushes of brass. Z, a door in one end of the mill-house at which the wood is conveyed out when cut. W W, walls of the mill-house. Q Q, the couples or framing of the roof. X X X, &c. windows to admit light to the house.

Saw-mills for cutting blocks of stone are generally, though not always, moved horizontally; the horizontal alternate motion may be communicated to one or more saws, by means of a rotatory motion, either by the use of cranks, &c. or in some such way as the following. Let the horizontal wheel A B D C, fig. 451, drive the pinion O N, this latter carrying a vertical pin P, at the distance of about one-third of the diameter from the centre. This pinion and pin are represented separately in No. 2 of fig. 451. Let the frame W S T V, carrying four saws, marked 1, 2, 3, 4, have wheels, V, T, W, W, each running in a groove or reel, whose direction is parallel to the proposed direction of the saws: and let a transverse groove P R, whose length is double the distance of the pin P from the centre of the pinion, be cut in the saw-frame to receive that pin. Then, as the great wheel revolves, it drives the pinion, and carries round the pin P; and this pin being compelled to slide in the straight groove PR, while by the rotation of the pinion on which it is fixed its distance from the great wheel is constantly varying, it causes the whole saw frame to approach and recede from the great wheel alternately, while the grooves in which the wheels run confine the frame, so as to move in the direction T t, V v. Other blocks may be sawn at the same time by the motion of the great wheel, if other pinions and frames running off in the directions of the respective radii, E B, E A, E C, be

...... by the teeth at the quadrantal points B A and C. And the contrary ...... of these four frames and pinions, will tend to soften down the jolts, ... equalise the whole motion.

The same contrivance, of a pin fixed at a suitable distance from the centre of a wheel, and sliding in a groove, may serve to convert a reciprocating into a rotatory motion; but it will not be preferable to the common contrivance by means of a crank.

When saws are used to cut blocks of stone into pieces having cylindrical ......, a small addition is made to the apparatus. See figs. 452 and ..... The saw, instead of being allowed to fall in a vertical groove, as it ...... the block, is attached to a lever or beam F G, sufficiently strong; this ...... has several holes pierced through it, and so has the vertical piece E D, which is likewise movable towards either side of the frame in grooves in the top and bottom pieces A L, D M. Thus the length E G of the radius can be varied at pleasure, to suit the curvature N O; and as the saw is moved backwards and forwards by proper machinery, in the direction C B, B C, it cuts lower and lower into the block, while, being confined by the beam F G, it cuts the cylindrical portion from the block P, as required.

When a complete cylindrical pillar is to be cut out of one block of stone, the first thing will be to ascertain in the block the position of the axis of the cylinder; then lay the block so that such axis shall be parallel to the horizon, and let a cylindrical hole of from one to three inches diameter be bored entirely through it. Let an iron-bar, whose diameter is rather less than that of the tube, be put through it, having just room to slide freely to and fro as occasion may require. Each end of this bar should terminate in a screw, on which a nut and frame may be fastened; the nut-frame should carry three flat pieces of wood or iron, each having a slit running along the middle nearly from one end to the other, and a screw and handle must be adapted to each slit: by these means the frame work at each end of the bars may readily be so adjusted as to form isosceles or equilateral triangles; the iron-bar will connect two corresponding angles of these triangles; the saw to be used, two other corresponding angles; and another box of iron or of wood, the two remaining angles; to give sufficient strength to the whole frame. This construction, it is obvious, will enable the workman to place the saw at any proposed distance from the hole drilled through the middle of the block; and then, by giving the alternating motion to the saw-frame, the cylinder may at length be cut from the block as required. This method was first described in the Collection of Machines approved by the Paris Academy.

If it were proposed to saw a conic frustrum from such a block, then let two frames of wood or iron be fixed to those parallel ends of the block which are intended to coincide with the bases of the frustrum, circular grooves being previously cut in these frames to correspond with the circumferences of the two ends of the proposed frustrum; the saw being worked in these grooves, will manifestly cut the conic surface from the block. This, we believe, is the contrivance of Sir George Wright.

The best method of drilling the hole through the middle of the proposed cylinder seems to be this: on a carriage running upon four low wheels let two vertical pieces (each having a hole just large enough to admit the borer to play freely) be fixed, two or three feet asunder, and so contrived that the pieces and holes to receive the borer may, by screws, &c. be raised or lowered at pleasure, while the borer is prevented from sliding backwards and forwards by pieces upon its bar, which are larger than the holes in the vertical pieces, and which, as the borer revolves, press against these pieces: let a part of the boring bar between the two vertical pieces be square, and a grooved wheel with a square hole of a suitable size be placed upon this part of the bar; then the rotatory motion may be given to this bar by an endless-band, which shall pass over this grooved wheel and a wheel of much larger diameter in the same plane, the latter wheel being turned by a winch-handle in the usual way. As the boring proceeds, the carriage with the borer may be brought nearer and nearer the block, by levers and weights.

Circular saws, acting not by a reciprocating, but by a rotatory motion, have been long known in Holland, where they are used for cutting wood used for veneering. They were introduced into this country, we believe, by General Bentham, and are now used in the dock-yard at Portsmouth, and in a few other places; but they are not as yet so generally adopted as might be wished, considering how well they are calculated to abridge labour, and to accomplish, with expedition and accuracy, what is very tedious and irksome to perform in the usual way. Circular saws may be made to turn either in horizontal, vertical, or inclined planes; and the timber to be cut may be laid upon the plane in any direction; so that it may be sawed by lines making any angles whatever, or at any proposed distance from each other. When the saw is fixed at a certain angle and at a certain distance from the edge of the frame, all the pieces will be cut of the same size, without marking upon them by a chalked line, merely by causing them to be moved along, and keeping one side in contact with the side of the frame; for then as they are brought one by one to touch the saw revolving on its axle, and are pressed upon it, they are soon cut through.

Mr. Smart, of the Ordnance Wharf, Westminster Bridge, has several circular saws, all worked by a horse, in a moderate

..ed walk; one of these intended for cutting and boring tenons, used in this gentleman's hollow masts, is represented in fig. 454.

HOPQR is a hollow frame, under which is part of the wheel-work of the horse-mill. A B C D E F are pullies, over which pass straps or bands, the parts of which out of sight run upon the rim of a large vertical wheel; by means of this simple apparatus the saws S S are made to revolve upon their axles, with an equal velocity, the same band passing round the pullies D C, upon those axles; and the rotatory motion is given to the borer G by the band passing over the pulley A. The board I is inclined to the horizon in an angle of about 30 degrees; the plane of the saw S is parallel to that of the board I, and about a quarter of an inch distant from it, while the plane of the saw S' is vertical, and its lowest point at the same distance from the board I. Each piece of wood K, out of which the tenon is to be cut, is about four inches long, and an inch and a quarter broad, and ⅜ of an inch thick. One end of such piece is laid so as to slide along the ledge at the lower part of the board I, and as it is pushed on, by means of the handle L, it is first cut by the saw S, and immediately after by the saw S'; after this the other end is put lowest, and the piece is again cut by both saws: then the tenon is applied to the borer G, and as soon as a hole is pierced through it, it is dropped into the box beneath.

By the above process, at least 30 tenons may be completed in a minute, with greater accuracy than a man could make one in a quarter of an hour with the common hand saw and gimlet. Similar contrivances may, by slight alterations, be used for many other purposes, particularly all such as may require the speedy sawing of a great number of pieces into exactly the same size and shape. A very great advantage attending this sort of machinery is, that when once the position of the saws and frame is adjusted, a common labourer may perform the business just as well as the best workman.

BARK-MILL.

This bark-mill is constructed for the purpose of grinding and preparing bark till it is fit for the tanner.

Bark-mills, like most other mills, are worked either by means of horses, by water, or by wind.

One of the best mills we have seen described for these purposes is that invented by Mr. Bagnall, of Worsley, in Lancashire. This machine will serve not only to chop bark, to grind, to riddle, and pound it; but to beam or work green hides and skins out of the mastering or drench, and make them ready for the ouse or bark-liquor; to beam sheep-skins, and other skins, for the skinner's use; and to scour and take off the bloom from tanned leather, when in the currying state.

Fig. 455 is a horizontal plan of the mill; fig. 456 a longitudinal section of it; fig. 457 a transverse section of it.

A, the water-wheel, by which the whole machinery is worked.

B, the shafts.

C, the pit-wheel, which is fixed on the water-wheel shaft B, and turns the upright shaft E, by the wheel F, and works the cutters and hammer by tapets.

D, the spur and bevel wheels at the top of upright shafts.

E, the upright shaft.

F, the crown-wheel, which works in the pit-wheel C.

G, the spur-nut to turn the stones I.

P, the beam, with knives or cutters fixed at the end to chop or cut the oark, which bark is to be put upon the cutters or grating i, on which the beam is to fall.

Q, the tryal that receives the bark from the cutters i, and conveys it into the hopper H, by which it descends through the shoe J to the stones I, where it is ground.

K, the spout, which receives the bark from the stones, and conveys it into the tryal L; which tryal is wired, to shift or dress the bark as it descends from the stones I.

M, the trough, to receive the bark that passes through the tryal L.

R, the hammer, to crush or bruise the bark that falls into the dish S, which said dish is on the incline, so that the hammer keeps forcing it out of the lower side of the said dish, when bruised.

k, a trough, to receive the dust and moss that passes through the tryal Q.

T, the bevel-wheel that works in the wheel D, which works the beam-knife by a crank V, at the end of the shaft u.

W, the penetrating-rod, which leads from the crank V to the start x.

x, the start, which has several holes in it to lengthen or shorten the stroke of the beam-knife.

y, the shaft, to which the slide-rods h h are fixed by the starts n n.

h, the slide rod, on which the knife f is fixed, which knife is to work the hides, &c. On the knife are two springs a a, to let it have a little play as it makes its strokes backwards and forwards, so that it may not scratch or damage the hides, &c.

s, is a catch in the slide rod h, which catches on the arch-head e; and the said arch-head conveys the knife back without touching the hide, and then falls back to receive the catch again.

l, the roller to take up the slide-rod h, while the hides are shifting on the beam b, by pulling at the handle m.

b, the beam to work the hides, &c. on. Each beam has four wheels, p p, working in a trough-road, g g, and removed by the levers c c. When the knife has worked the hides, &c. sufficiently in one part, the beam is then shifted by the lever c as far as is wanted.

d, a press, at the upper end of the beam, to hold the hide fast on the beam while working.

e, an arch-head, on which the slide-rod h catches.

f, the knife fixed on the slide-rod h, to work the hides, &c.

i, cutters or grating to receive the bark for chopping.

The beam P, with knives or cutters, may either be worked by tapets, as described, or by the bevel-wheel T with a crank, as V, to cut the same as shears.

The knife f is fixed at the bottom of the start, which is fixed on the slide-rod h; the bottom of the start is split open to admit the knife, the width of one foot; the knife should have a gudgeon at each end, to fix in the open part of the start; and the two springs a a prevent the knife from giving too much way when working. The knife should be one foot long, and four or five inches broad.

The arch-head c will shift nearer to or further from the beam h, and will be fixed so as to carry the knife back as far as is wanted, or it may be taken away till wanted.

The roller l is taken up by pulling at the handle m, which takes up the slide-rod so high as to give head room under the beam-knife; the handle may be hung upon a hook for that purpose. The slide-rod will keep running upon the roller all the time the hide is shifting; and when the hide is fixed, the knife is put on the beam again by letting it down by the handle m. There may be two or more knives at work on one beam at the same time, by having different slide-rods; there should be two beams, so that the workman could be shifting one hide, &c. while the other was working. The beam must be flat, and a little on the incline; as to the breadth, it does not matter; the broader it is, the less shifting of the hides will be wanted, as the lever c will shift them as far as the width of the hide, if required. Mr. Bagnall has formed a kind of press d, to let down, by a lever, to hold the hide fast on each side of the knife, if required, so that it will suffer the knife to make its back stroke without pulling the hide up as it comes back. The slide-rod may be weighted, to cause the knife to lay stress on the hide, &c. according to the kind and condition of the goods to be worked.

Hides and skins for the skinner's use are worked in the same way as for the tanner's.

Scouring of tanned leather for the currier's use can be done on the beam, the same as working green hides; it is only taking the knife away, and fixing a stone in the same manner as the knife by the said joint, and to have a brush fixed to go either before or after the stone. The leather will be much sooner and better secured this way than by hand.

The whole machinery may be worked by water, wind, steam, or any other power; and that part of the machinery which relates to the beaming part of the hides, may be fixed to any horse bark-mill, or may be worked by a horse or other power separately.

OIL-MILLS.

As these kingdoms do not produce the olive, it would be needless to describe the mills which are employed in the southern parts of Europe; we shall therefore content ourselves with a description of a Dutch oil-mill, employed for grinding and pressing linseed, rapeseed, and other oleaginous grains; and, to accommodate our description still more to our local circumstances, shall employ water as the first mover; thus avoiding the enormous expense and complication of a windmill.

Description of fig. 456.

1 is the elevation of a wheel, over or under shot, as the situation may require.

t, the bell-metal socket, supported by masonry, for receiving the outer gudgeon of the water-wheel.

3, the watercourse. ,

Fig. 459.

1, a spur-wheel upon the same axis, having 52 teeth.

2, the trundle that is driven by No. 1, and has 78 staves.

3, the wallower, or axis for raising the pestles. It is furnished round its circumference with wipers for lifting the pestles, so that each may fall twice during one turn of the water-wheel : that is, three wipers for each pestle.

4, a frame of timber, carrying a concave half cylinder of bell-metal, in which the wallower (cased in that part with iron plates) rests and turns round.

5, masonry supporting the inner gudgeon of the water-wheel and the above-mentioned frame.

6, gudgeon of the wallower, which bears against the bell-metal step fixed in the wall. This double support of the wallower is found to be necessary in all mills which drive a number of heavy stampers.

Fig. 460 is the elevation of the pestle and press-frame, their furniture, the mortars, and the press-pestles.

1, the six pestles.

2, cross-pieces between the two rails of the frame, forming, with these rails, guides for the perpendicular motion of the pestles.

3, the two rails ; the back one is not seen. They are checked and bolted into the standards, No. 12.

4, the tails of the lifts, corresponding with the wipers upon the wallower.

5, another rail in front, for carrying the detents which hold up the pestles when not acting. It is marked 14, in fig. 464.

6, a beam a little way behind the pestles ; to this are fixed the pulleys for the ropes, which lift and stop the pestles. It is represented by 16, in fig. 464.

7, the said pulleys with their ropes.

8, the driver which strikes the wedge that presses the oil.

9, the discharger, a stamper which strikes upon the inverted wedge, and loosens the press.

10, the lower rail with its cross-pieces, forming the lower guides of the pestles.

11, a small cog-wheel upon the wallower for turning the spatula, which stirs about the oil-seed in the chauffer-pan. It has 28 teeth, and is marked No. 6, in fig. 464.

12, the four standards, mortised below into the block, and above into the joists and beams of the building.

13, the six mortars hollowed out of the block itself, and in shape pretty much like a kitchen-pot.

14, the feet of the pestles rounded into cylinders, and shod with a great lump of iron.

15, a board behind the pestles, standing on its edge, but inclining a little backwards. There is such another in front, but not represented here. These form a sort of trough, which prevents the seed from being scattered about by the fall of the pestles, and lost.

16, the first press-box, (also hollowed out of the block,) in which the grain is squeezed, after it has come for the first time from below the mill-stones.

17, the second press-box, at the other end of the block, for squeezing the grain after it has passed a second time under the pestles.

18, frame of timber for supporting the other end of the wallower in the same manner as No. 4, fig. 459.

19, small cog-wheel on the end of the wallower, for giving motion to the mill-stones; it has 28 teeth.

20, gudgeons of the wallower, bearing on a bell-metal socket fixed in the will...

Fig. 461. Elevation and mechanism of the mill-stones.

1, upright shaft, carrying the great cog-wheel above, and the runner mill-stones below in their frame.

2, cog-wheel of 75 cogs, driven by No. 19 of fig. 459.

3, the frame of the runners.

4, the innermost runner, or the one nearest the shaft.

5, outermost ditto, being farther from the shaft,

6, the inner rake, which collects the grain under the outer runner.

7, the outer rake, which collects the grain under the inner runner. In this manner the grain is always turned over and over, and crushed in every direction. The inner rake lays the grain in a slope, of which fig. 465 is a section; the runner flattens it, and the second rake lifts it up, as is marked in fig. 466; so that every side of the grain is presented to the mill-stone, and the rest of the legger or nether mill-stone is so swept... that not a single grain is left on any part of it. The outer rake is also furnished with a rag of cloth, which rubs against the border or hoop that surrounds the nether mill-stone, so as to drag out the few grains which might otherwise remain in the corner.

8, the ends of the iron axle which passes through the upright shaft, and through the two runners. Thus they have two motions: first, a rotation round their own axis; secondly, that by which they are carried round upon the nether mill-stone, on which they roll. The holes in these mill-stones are made wide; and the holes in the ears of the frame, which carry the ends of the iron axes, are made oval up and down. This great freedom of motion is necessary for the runner mill-stones, because frequently more or less of the grain is below them at a time, and they must therefore be at liberty to roll over it without straining, and perhaps breaking, the shaft.

9 and 10, the border or hoop which surrounds the nether mill-stone.

11 and 12, the nether mill-stone and masonry which support it.

Fig. 462, plan of the runner mill-stones, and the frame which carries them round.

1, 1, are the two mill-stones.

3, 3, 3, the outside pieces of the frame.

4, 4, 4, the cross-bars of the frames, which embrace the upright shaft 5, and give motion to the whole.

6, the iron axis upon which the runners turn.

7, the outer rake.

8, the inner ditto.

Fig. 463 represents the nether mill-stone seen from above.

1, the wooden gutter which surrounds the nether mill-stone.

2, the border or hoop, about six inches high all round, to prevent any meal being scattered.

3, an opening or trap-door in the gutter, which can be opened or shut at pleasure; when open, it allows the bruised grain, collected in and above along the gutter by rakes, to pass through into troughs placed below to receive it.

4, portion of the circle described by the outer runner.

5, portion of the circle described by the inner one. By these we see that the two stones have different routes round the axis, and bruise more meal.

6, the outer rake.

7, the inner ditto.

8, the sweep, making part of the inner rake, occasionally let down to

2 G

sweeping off all the seed when it has been sufficiently bruised. The pressure and action of these rakes is adjusted by means of wooden springs, which cannot be easily and distinctly represented by any figure. The oblique position of the rakes (the outer point going foremost) causes them to shove the grain inwards, or toward the centre, and at the same time to turn it over somewhat in the manner as the mould-board of a plough shoves the earth to the right hand, and partly turns it over. Some mills have but one sweeper; and indeed there is great variety in the form and construction of this part of the machinery.

Fig. 464, profile of the pestle-frame.

1, section of the horizontal shaft.

2, three wipers for lifting the pestles.

3, little wheel of 28 teeth for giving motion to the spatula.

4, another wheel which is driven by it, having 20 teeth.

5, horizontal axle of ditto.

6, another wheel on the same axle, having 13 teeth.

7, a wheel upon the upper end of the spindle, having 12 teeth.

8, two gudeu, in which the spindle turns freely, and so that it can be shifted higher and lower.

9, a lever, movable round the piece No. 14, having a hole in it at 9, through which the spindle passes, turning freely. The spindle has in this place a shoulder, which rests on the border of the hole 9, so that by the motion of this lever the spindle may be disengaged from the wheel-work at pleasure; this motion is given to it by means of the lever 10, 10, movable round its middle. The workman employed at the chauffer pulls at the rope 10, 11, and thus disengages the spindle and spatula.

11, a pestle seen sidewise.

12, the left of ditto.

13, the upper rails, marked No 3, in fig. 460.

14, the rail marked No. 5, in fig. 460. To this are fixed the detents, which serve to stop and hold up the pestles.

15, a detent, which is moved by a rope at its outer end.

16, a bracket behind the pestle, having a pulley through which passes the rope going to the detent 15.

17, the said pulley.

18, the rope at the workman's hand, passing through the pulley 17, and fixed to the end of the detent 15.

This detent naturally hangs perpendicular by its own weight. When the workman wants to stop a pestle, he pulls at the rope 18, during the rise of the pestle. When this is at its greatest height, the detent is horizontal, and prevents the pestle from falling, by means of a pin projecting from the side of the pestle, which rests upon the detent, the detent itself being held in that position by hitching the loop of the rope upon a pin at the workman's hand.

19, the two lower rails, marked No. 10, fig. 460.

20, great wooden, and sometimes stone, block, in which the mortars are formed, marked No. 21, fig. 460.

21, vessel placed below the press-boxes for receiving the oil.

22, chauffer, or little furnace, for warming the bruised grain.

23, backet in the front of the chauffer, tapering downwards, and opening below in a narrow slit. The hair-bags on which the grain is to be pressed after it has been warmed in the chauffer, are filled by placing them in this backet. The grain is lifted out of the chauffer with a ladle, and put into these bags; and a good quantity of oil runs from it through the slit at the bottom into a vessel set to receive it.

the spatula attached to the lower end of the spindle, and turning round the grain in the chauffer-pan, and thus preventing it from going to the bottom or sides, and getting too much heat.

the first part of the process is bruising the seed under the er-stones; that this may be more expeditiously done, if the runners is set about two-thirds of its own thickness er the shaft than the other; thus they have different ls, and the grain, which is a little heaped towards the re, is thus bruised by both. The inner rake gathers it nder the outer stone into a ridge, of which the section is sented in fig. 465; the stone passes over it, and flattens it. gathered up again into a ridge, of the form of fig. 466, r the inner stone by the outer rake, which consists of parts; the outer part presses close on the wooden er which surrounds the nether stone, and shoves the seed mely inwards, while the inner part of this rake gathers hat has spread towards the centre. The other rake has at near the middle of its length, by which the outer half can be raised from the nether stone, while the inner continues pressing on it, and thus scrapes off the moist. When the seed is sufficiently bruised, the miller lets the outer end of the rake; this immediately gathers the le paste, and shoves it obliquely outwards to the wooden where it is at last brought to a part that is left unboarded, it falls through into troughs placed to receive it. These that have holes in the bottom, through which the oil drips time of the operation. This part of the oil is directed a particular cistern, being considered as the purest of the e, having been obtained without pressure, by the mere ing of the hull of the seed.

some mills this operation is expedited, and a much er quantity of this best oil is obtained, by having the if masonry which supports the legger formed into a little te, and gently heated; but the utmost care is necessary vent the heat from becoming considerable. This, ling the oil to dissolve more of the fermentable substance e seed, exposes the oil to the risk of growing soon very ld; and in general it is thought a hazardous practice, the oil does not bring so high a price.

hen the paste comes from under the stones, it is put into air-bags, and subjected to the first pressing. The oil obtained is also esteemed of the first quality, scarcely for to the former, and is kept apart (the great oil-cistern divided into several portions by partitions.)

e oil-cakes of this pressing are taken out of the bags, en to pieces, and put into mortars for the first stamping.

Here the paste is again broken down, and the parenchyma of the seed reduced to a fine meal; thus free egress is allowed to the oil from every vesicle in which it is contained. But it is now rendered much more clammy by the forcible mixture of the mucilage, and even of the finer parts of the meal. When sufficiently pounded, the workman stops the pestle of a mortar, when at the top of its lift, and carries the contents of the mortar to the first chauffer-pan, where it is heated to about the temperature of melting bees'-wax, (this, we are told, is the test,) and all the while stirred about by the spatula. From thence it is again put into hair-bags, in the manner already described; and the oil which drops from it during this operation is considered as the best of the second quality, and in some mills is kept apart. The paste is now subjected to the second pressing, and the oil is that of the second quality.

All this operation of pounding and heating is performed by one workman, who has constant employment by taking the four mortars in succession. The putting into the bags, and conducting the pressing, gives equal employment to another workman.

In the mills of Picardy, Alsace, and most of Flanders, the operation ends here; and the produce from the chauffer is increased, by putting a spoonful or two of water into the pan among the paste.

But the Dutch take more pains. They add no water to the paste of this their first stamping; they say that this greatly lowers the quality of the oil. The cakes which result from this pressing, and are then sold as food for cattle, are still fat and soft. The Dutch break them down, and subject them to the pestles for the second stamping; these reduce them to an impalpable paste stiff like clay. It is lifted out, and put into the second chauffer-pan; a few spoonfuls of water are added, and the whole kept for some time as hot as boiling water, and carefully stirred all the time. From thence it is lifted into the hair-bags of the last press, subjected to the press, and a quantity of the lowest quality is obtained, sufficient for giving a satisfactory profit to the miller. The cake is now perfectly dry and hard, like a piece of board, and sold to the farmers. Nay, there are small mills in Holland which have no other employment than extracting oil from the cakes which they purchase from the French and Brabantees: a clear indication of the superiority of the Dutch practice.

The nicety with which that industrious people conduct all their business is remarkable in this manufacture.

In their oil-cisterns the parenchymous part, which unavoidably gets through, in some degree, in every operation, gradually subsides, and the liquor, in any division of the cistern, comes to consist of strata of different degrees of purity. The pumps which lift it out of each division are in pairs; one takes it up from the very bottom, and the other only from one half the depth. The last only is barrelled up for the market, and the other goes into a deep and narrow cistern, where the dreg again subsides, and more pure oil of that quality is obtained. By such careful and judicious practice, the Dutch not only supply themselves with this important article, but annually send considerable quantities into the very provinces of France and Flanders, where they buy the seed from which it is extracted. When we reflect on the high price of labour in Holland, on the want of timber for machinery, on the expense of building in that country, and on the enormous expense of wind-mill machinery, both in the first erection and the subsequent wear and tear, it must be evident that oil-mills erected in England on waterfalls, and after the Dutch manner, cannot fail of being a great national advantage. The chatellenie or seigneurie of Lille alone makes annually between 30,000 and 40,000 barrels, each containing about 26 gallons.

What is here delivered is only a sketch. Every person acquainted with machinery well understands the general movements and operations; but the intelligent mechanic well knows that operations of this kind have many minute circumstances which cannot be described, and which, nevertheless, may have a great influence upon the whole. The rakes in the bruising-mill have an office to perform which resembles that of the hand, directed by a careful eye and unceasing attention. Words cannot convey a clear notion of this; and a mill constructed from the best drawings, by the most skilful workman, may gather the seed so ill, that the half of it shall not be bruised after many rounds of the machinery. This produces a scanty return of the best oil, and the mill gets a bad character; the proprietor loses his money, is discouraged, and gives up the work. There is no security but by procuring a Dutch millwright, and paying him with the liberality of Britons. Such unhoped-for tasks have been performed of late years by machinery, and mechanical knowledge and intuition is now so generally diffused, that it is highly probable we should soon excel our teachers in the branch; but this very diffusion of knowledge, by encouraging speculation among the artists, makes it a still greater risk to erect a Dutch oil-mill, without having a Dutchman, acquainted with its most improved present form, to conduct the work.

COLOUR AND INDIGO MILLS.

THE reducing of earths, vegetable substances, and metallic oxyds to an impalpable powder, is still in a great degree effected by manual labour, by moving a heavy stone with a smooth surface, called a muller, upon a slab of the same material. To effect this work upon a larger scale, and to secure the workman from the ill effects of the poisonous and noxious vapours of the paint, which is not unfrequently ground with litharge of lead, Mr. Rawlinson, of Derby, has invented a machine which we here describe. It is represented in fig. 467.

A, the roller, or cylinder, made of any kind of black marble. Black marble is esteemed the best, because it is hardest, and takes the best polish. B, the concave muller, covering one-third of the roller, and of the same material, fixed in a wooden frame b, which is hung to the frame E at i i. C is a piece of iron, about an inch broad, to keep the muller steady, and is fixed to the frame with a joint at f. The small binding screw with a fly-nut, which passes through the centre of the iron plate at c, is for the purpose of laying more pressure upon the muller, if required, as well as to keep it steady. D is a taker-off, made of a clock-spring, about half an inch broad, and fixed similar to a frame-saw in an iron frame K, in an inclined position to the roller, and turning on pivots at d d. G is a slide-board to draw out occasionally, to clean, &c. if any particles of paint should fall from the roller; it also forms itself for the plate H, to catch the colour as it falls from the taker-off. F is a drawer for the purpose of containing curriers' shavings, which are used for cleaning paint-mills. E is the frame.

Previously to putting the colour in the mill, it must be pulverized in a mortar, covered in the manner of the chemists, when they levigate poisonous drugs, or rather in an improved mill, used at Manchester, by Mr. Charles Taylor, for grinding indigo in a dry state, a drawing and description of which is annexed. After undergoing this process of dry-grinding, which is equally necessary for the marble slab now in use, it is mixed with either oil or water, and is with a spatula, or palette-knife, put on the roller, near to the top of the concave muller. Motion being given to the roller, it, without any difficulty, carries the colour under the muller, and in a few revolutions spreads it equally over the surface. When ground sufficiently, it is taken off, both cleanly and expeditiously, by the taker-off described, which, for that purpose, is held against the roller, while the roller is turned the reverse way. The muller only requires to be cleaned when the workman changes the colour, or ceases from the operation; it is then turned back, being hung on pinions to the frames at i i, and is cleaned with a palette-knife or spatula; afterwards a handful of carriers' shavings is held against the roller, which, in two or three revolutions, cleans it effectually.

The roller of Mr. Rawlinson's machine is sixteen inches and a half in diameter, and four inches and a half in breadth; and the concave muller which it works against covers one-third of the roller. It is therefore evident, that, with this machine, he has seventy-two square inches of the concave marble muller in constant work on the paint, and that he can bring the paint much oftener under the muller in a given space of time than with the common pebble muller, which, being seldom more than four inches in diameter, has scarcely sixteen square inches at work on the paint, whereas the concave muller has seventy-two.

The quantity ground at once in the mill must be regulated by the degree of fineness of which it is required, that which is the finest requiring the smallest quantity to be ground at once. The time requisite for grinding is also dependant upon the state of fineness; but Mr. Rawlinson observes, that his colour-grinder has ground the quantity of colour which used to serve him per day in three hours; the colour also was more to his satisfaction, and attended with less waste.

When the colour is ground, Mr. Rawlinson recommends, instead of drawing the neck of the bladder up close in the act of tying it, to insert a slender cylindrical stick, and bend the bladder close round it; this, when dry, will form a tube or pipe, through which, when the stick is withdrawn, the colour may be squeezed as wanted, and the neck again closed by replacing the stick. This is not only a neater and much more cleanly mode than the one usually adopted, that of perforating the bladder, and stopping the hole with a nail, or, what is more common, leaving it open, to the detriment of the colour; but the bladder, not being injured, may be repeatedly used for fresh quantities of colour. The barrel of a quill may be inserted in the neck of the bladder, as a substitute for the stick, and the end being cut off, may be closed by a small piece of wood.

In order to make the whole of the process of colour-grinding complete, we shall here insert a description of the indigo-mill used by Mr. Charles Taylor, of Manchester, for grinding indigo in a dry state, which may with equal advantage be similarly employed for colours. It is represented in figs. 408 and 409.

L, fig. 408, represents a mortar, made of marble or hard stone; one made in the common way will answer. M. a muller, or grinder, nearly in the form of a pear; in the upper part of which an iron axis is firmly fixed, which axis at the parts N N turns in grooves or slits, cut in two pieces of oak, projecting horizontally from a wall, and when the axis is at work are secured in the grooves by iron pins O O. P, the handle, which forms a part of the axis, and by which the grinder is worked. Q, the wall in which the oak pieces N N are fixed. R, a weight, which may occasionally be added if more power is wanted.

Fig. 468°, shows the muller or grinder, with its axis separate from the other machinery; its bottom should be made to fit the mortar. S is a groove cut through the stone.

On grinding the indigo, or similar substance, in a dry state, in this mill, the muller being placed in the mortar and secured in the oak pieces by the pins, the indigo to be ground is thrown above the muller into the mortar; on turning the handle of the axis, the indigo, in lumps, falls into the groove cut through the muller, and is thence drawn under the action of the muller, and propelled to its outer edge within the mortar, whence the coarser particles again fall into the groove of the muller and are again ground under it; which operation is continued till the whole of it is ground to an impalpable powder; the muller is then easily removed, and the colour taken out.

A wood cover in halves, with a hole for the axis, is usually placed upon the mortar, during the operation, to prevent any loss to the colour, or bad effects to the operator.

POTTERY.

THE clays best adapted for the manufacture of earthen-ware are excavated in Dorsetshire, and the next in quality in Devonshire.

The natural compounds, called clays, consist generally of pure clay, or alumine, combined with either silex or lime, and sometimes magnesia, and the oxyd of iron. The presence of the magnesia may easily be detected by its imparting a soapy feel; and the iron by the clay burning to different shades of red, proportionate to the quantity it contains. The magnesia has obtained the name of soap-rock, and a marked variety of it steatite.

The clay is first put into a trough about five feet long, by three wide, and 2½ deep, with a certain proportion of water, and subjected to the process called *blunging*, which is obviously akin to blending, or mixing. This is performed with a long piece of wood formed in the shape of a blade at one end, and with a cross-handle at the other. The bladed end is put into the trough, and moved backwards and forwards, up and down, with violence, till the clay be broken and well levigated. The coarser particles of the clay sink to the bottom of the trough, while the finer parts remain suspended in the solution; and clay is continued to be added until the solution has acquired the consistence of thick cream. This thick liquid is passed into a large tub, and afterwards through fine hair and silk lawn sieves, and then mixed with certain proportions of a liquid of ground calcined flints and Cornish stone, which, likewise, have been passed through silk lawn sieves.

The china clay, which is used in every kind of earthenware except the cream colour, is sometimes put into the mass, and blunged with it; at other times it is put into another tub, and blunged separately, and is then mixed in proper proportions with the other slip.

The slip is now passed into another large stone or wood cistern, and the parts, which have not been previously, are now added, and the whole is passed through fine lawn into a reservoir, from whence it is pumped upon the slip-kiln.

When a steam-engine is used, the clay is thrown into a vertical cast-iron cone, about two feet wide at top, and six feet deep. Inside of this cone are fixed strong knives, having a spiral arrangement and inclination, and radiating towards the centre. In the centre of these is worked a perpendicular shaft, with similar radiating knives, so that the knives, by the revolution of the shaft, cut in pieces every thing that is thrown into the cone, and force downward, agreeably to the nature of the screw, whatever may be put in till it is discharged through an orifice at the bottom.

The clay, thus reduced to powder, is next subjected to the process of blunging. For this purpose it is thrown into a large circular vat, or cistern, having a strong vertical shaft of wood, with arms formed like a gate as radii, worked by the power of the steam-engine. The vat is nearly filled with proper proportions of water and clay, which, by the rapid motion of the shaft, becomes well levigated and mixed; clay or water being added until the liquid is of the consistence of cream. The liquid is then passed along several troughs, at the end of each of which is fixed a fine hair or lawn sieve. These sieves have a quick horizontal motion communicated to them by crank machinery, which causes the slip to pass through into a large reservoir, where it remains till pumped upon the kiln.

The flint in its crude state is the common flint used for striking fire, which consists principally of pure silex. The

method of calcining it is, by placing it in a small
kiln, about nine feet deep, and altogether not like
that used to burn limestone. When red-hot it is
of the kiln and thrown into cold water, in order to
aggregation, and make it easier to reduce to pow
flint is next broken into pieces, either by manual
machinery. Where manual labour is employed a
found adequate to break per diem enough to fill
flint-pans, 12 feet diameter.

In the other process the flints are put on a
grating, and are struck by large hammers, another
chinery, till they be so reduced as to fall through
into a cavity, from whence they are taken to the

The *flint-mill* consists of a large circular vat
inches deep, with a step fixed in the centre at the
for the axis of a vertical wood or iron shaft. The
end of the shaft is surmounted by a large cog-wheel
to which the moving power is applied. The arms
at right angles, four leaves, or paddles, to each of
which are fixed chert-stones. Large blocks of chert
are also placed in the vat. The flints being put in,
the whole is covered with water, to prevent any
arising, which had formerly a very injurious effect
being communicated to the shaft, the chert-stones
ried round with considerable velocity, and the which
being of a very fragile nature, are, by their reciproca
reduced to an impalpable powder.

This semi-fluid is put into another vat, that has
vertical shaft, and when a large quantity of water
introduced, the power is applied and the whole
levigated. In this process, the weighty particles at
bottom, and the finest remain in suspension; which
passed into a reservoir that has certain apertures for
off the surplus water, till it has subsided to a stat
the potter's use. This is a very important process
attended with some difficulty. It is at present
formed by Mr. Sampson Hanley, of Sandon Mill.

The manufacturer should be very choice in select
stones to be employed in the grinding; the
tain calcareous carbonates, such parts will be abra
by mixing with the silicious matter will, in a sui
process, prove a serious injury.

A few years ago a loss to the amount of
pounds was experienced by some manufacturers,
very injudiciously purchased stones that had been gr

a person ignorant of the art, and who had employed stones for the grinding containing carbonate of lime.

The average weight of an ale pint measure of the pulp of flint is 32 oz.; and of clay 24 oz.

In some manufactories the pulps are mixed together in a large vat, by a process similar to that first described of mixing the clay with the water. But however the mixing be accomplished, great attention must be paid to the relative specific gravity of each fluid, and more of the solution of the flint, or the clay, must be added, till a pint of the mixture weighs the determined number of ounces. It is by the consistence and weight of these materials, that the manufacturer is enabled to ascertain, the proper proportions requisite for each kind of pottery; and it is from these that he can calculate, whether there be a probability of making any improvement that will yield him a profitable return.

When the proper proportions of slop clay and flint have been well blunged together, the liquid is pumped out of the reservoir on the top of the slip-kiln,

The *slip-kiln* is a kind of trough formed of fire-bricks, of various sizes, from 30 to 60 feet in length, by from 4 to 6 in breadth, and about 12 inches in depth. Flues from the fire-places pass under these troughs, and the bricks of which they are formed being bad conductors of heat, a slow and advantageous process of evaporation is carried on, which gives uniform consistence to the mass.

The porcelain clay is never allowed to boil, but is carefully evaporated at a slow heat on a plaster-kiln; the gypsum being run on old moulds pulverised, and thus forming a level surface.

The slip-maker carefully attends to the evaporation, and at proper intervals turns over with a paddle the thickened mass from one end to the other, else the part nearest to the bricks would become hard, while the surface were fluid. To regulate the heat three different thicknesses of bricks are employed, the thickest being placed nearest to the fire-place, where is the greatest excess of heat.

When a sufficient quantity of the moisture is evaporated, which is indicated by the cessation of apparent effervescence, or the absence of air-bubbles on the surface of the mass, the composition, still called clay, is removed to the flags.

If the evaporation were continued longer the clay could not be formed into the required shapes, either on the wheel, or by the vat, but would be, what is called knotty, lumpy.

The clay is cut out of the kilns in square masses, by means

of spades, and is thrown into a heap; where is a
uniform temperature of cold and moisture. The
can lie after coming off the kiln the better it will b
time is arbitrarily varied by the want of room, of
of capital.

When the clay is first taken off the kiln, it is, p
the air-bubbles remaining in it, and partly from
dissipation of the heat requisite for evaporation, t
be worked. On this account it is well incorporated
or tempered, by beating with wooden mallets. I
cut into small pieces with a paddle, not much unlik
and from the paddle each piece is, with all the fo
workman, propelled upon the mass. These two c
are repeated until a proper consistence pervades,
whole is supposed to be well-tempered.

When the clay is required for the thrower the
slapping follows next. This is performed by a st
who places a large mass, about half a hundred-wei
a convenient and strong bench. He then, with a t
wire, cuts the mass through, and taking up the p
cut off, he, with his utmost strength, casts it down
the mass below; and continues the operation as l
considered necessary.

This is a very laborious process, and is absolute
sary to drive out any air-bubbles which may happen
in the mass after it has been beaten: for should an
in the clay the pieces on being fired would blister
owing to the rarefaction of the air by the heat. On
important account, the process is continued until,
wherever cut by the brass wire, exhibits a surface,
smooth, and homogeneous.

In several of the largest manufactories the labou
ping the clay is superseded by mechanical contriv
quantity of the mass from the slip-kiln, when rathe
thrown into a large conical iron vessel, (similar to
played in breaking the clay,) with strong knives fi
with a given inclination, with corresponding knives
from a vertical shaft, moved by the steam-engine wi
and regular motion. By these means, all the clay
the cone is very minutely separated, and pressed
by a screw, so that the mass just cut, and divided, is
squeezed together again, and is then similarly aff
other knives below. At the bottom of the cone on o
a quadrangular aperture, through which the clay is
forced, and is by a thin brass wire cut into bric

pieces of from 50 to 60 pounds weight. Sometimes these masses are for particular purposes returned into the tone, and undergo the process a second time.

Wedging the clay is a similar process, though never omitted by the *presser*, or *squeezer*, however well it may have been beaten by the slip-maker. The presser cuts off, with a thin brass wire, a piece of clay from the mass, which he slaps forcibly between the palms of his hands, and then with great violence throws it on the board; continuing the operation until the commixture is so complete that there is no probability of any air-bubbles remaining. If one of the two first pieces of clay had been white, and the other black, the mass, after undergoing these processes, would present wherever cut a uniform grey colour.

It is owing to the mass being properly wedged that that consistency and tenacity is obtained, which enables the workman to employ it with facility and confidence in the fabrication of the different pieces of pottery which he has to make. The clays for vessels require different degrees of wedging; and some kinds require much more careful and continued wedging than others.

The clay may now be considered ready for the *thrower*. The *throwing-wheel*, or, with greater propriety, the *throwing-engine*, consists of a large vertical wheel, having a winch or handle affixed to it, and a groove on the rim for the introduction of a cord. The whole is fixed upon a strong movable plank, by which the cord can be slackened or tightened at pleasure, and then upon a frame nearly triangular, or half-oval, and about 30 inches in height, with a broad ash hoop placed edgewise on the fore part, about six inches deep.

In the centre of this frame is a vertical spindle, with its lower end fitted and working in a step. A little above this is a pulley, with grooves for three speeds of the propelling power, connected with the throwing-wheel by means of a cord or belt; and a little higher up is a pivot turned to fit and work in a collar-step. On the upper end is a stout wooden circular top, which revolves horizontally, and is in diameter about seven inches; and other tops of different diameters are in readiness to be fixed on, according to the intended size of the vessel to be made.

The engine is set in motion by manual labour, applied at the winch, and another man, called the *baller*, cuts with a thin piece of brass wire a piece of clay from the mass on the bench, and forms it into a ball, which he gives to the thrower. If china is to be made, the baller, previously to forming the

clay into a ball, breaks it in two, and violently slaps it together between the palms of his hands. The thrower forcibly
throws the ball down upon the horizontal revolving top of
the engine, and dipping his hands frequently into water, to
prevent the clay adhering to them, fashions it into a long
thin column, which he again forces down into a lump, and
continues to repeat the operation until he is satisfied that
the air-bubbles, which might have remained in the clay after
the processes of slapping and balling, are dispelled.

The thrower now directs the speed of the engine to be
lessened, and with his fingers, which he frequently dips in
water, he gives the first form to the vessel; then with
different *profiles*, or *ribs*, he forms the inside of the vessel
into whatever shape may be required, and smoothes it by
removing the *slurry*, or inequalities.

If a number of vessels of the same size be required, the
thrower has a peg placed as a gauge, which serves to direct
him in the width and depth; and when the vessel has two
diameters, as the neck and body in a jug, he has two pegs to
guide him.

The thrower forms all circular vessels in this manner; and
he employs different sized ribs to finish the shapes, or swell
of the edge, &c. When he has thus given the first form to
the clay, he cuts the vessel from the head of the engine, by
passing a thin brass wire through the lowest part of the clay,
which separates it, and allows it to be easily lifted off, and
placed by the baller on a long board or shelf, where it is
left to dry a little preparatory to being turned, or properly
smoothed and shaped.

Where large vessels are made, and the power of a steam-
engine applied, according to Mr. J. Wedgwood's method, a
pair of vertical cones is used, the apex of the one being
opposite to the vertex of the other. One of these cones is
driven directly by the steam-engine, and transmits motion
to the other by means of a broad belt or strap of leather,
which is always equally tight in any and every part of the
cones, because they are equal and reversed; but it is plain,
that the speed of the driven cone will vary much according
as the belt is at the top or the bottom of the driving cone.
When the belt is at the bottom or thinnest part of the driving
cone, the driven cone moves very slowly; as the belt is made
to ascend, the speed of the driven cone increases, and ultimately attains its maximum when the belt is at the top. A
strap is attached from the driven cone to the spindle of the
throwing-engine, and the speed is varied at the thrower's

pleasure, by a boy working a directing winch. When the article is finished, the machine is thrown out of geer.

For forming saucers, and other small circular articles, there has been recently introduced a small vertical shaft, called a finger, on the top of which is a turned head, suited to receive the mould on which the saucers, &c. are to be formed.

When the clay is in one peculiar state, called the *green state*, it is the most suitable and proper for performing to the greatest advantage the remaining operations and processes of turning, handling, trimming, &c.

The *turning-lathe* is the same as used by wood-turners. The end of the spindle, outside the headstock, has a screw thread, upon which is screwed *chocks* of wood, of a tapered form, and of different diameters, according to the size of the interior of the articles of pottery to be turned. The turner stands very steady, and receives from an attendant the vessel to be turned, which he fixes upon the chock, and then with a tool presses the edges close down.

The tools are of different sizes, from one quarter of an inch to two inches in breadth, and six inches in length, made of thin iron, like hoop-iron, the end for cutting being turned up about a quarter of an inch, and ground sharp.

Motion being communicated to the lathe, the turner applies his tool or tools to the various parts of the surface that require reduction of substance, either as regards thickness, or the suitable shapes of rims, feet, &c. When this is completed, a contrary motion is communicated to the spindle, during which the turner applies the flat part of his tool to the vessel, and by gentle pressure gives it a smooth surface, and solid texture.

In the turning-lathes moved by steam some particular arrangements are made. A horizontal shaft runs the whole length of the room; and opposite to each lathe is a drum, which communicates motion to a set of pulleys, of various sizes, fixed on an arbor or shaft, by means of a leather belt. Upon this arbor, or shaft, is a loose pulley, connected by a crossed belt with a small pulley fixed on the spindle of the lathe, which evidently will, whenever the strap from the drum is directed upon the loose pulley, receive a retrograde motion. The spindle has pulleys counter to those fitted on the arbor, and as they are ever revolving, the directing of the belt from them to the spindle, by a guide moved by the workman's foot, will increase or diminish the speed during the turning of the vessel under operation; and when it is finished, by moving the drum-strap on another pulley, retrograde motion is given, during which the turner smooths off his article, as before noticed.

The *engine-lathe* is of the kind employed to give unto circular articles of hardware a milled edge; consequently, it differs from the other, or common lathe, in the formation of the end of the spindle, and the appendages to the headstock. Certain thin circular plates of steel, into whose edges are cut, at regular intervals, and of different degrees of breadth, deep incisions, are made to screw very firmly on the end of the spindle above the chock. The collar-step of the spindle is so fitted that it can be effected by a screw pin, which gives it the requisite horizontal shuffling motion. Opposite to the steel-plate is fixed an iron piece that fits into the incisions. The turner's tools are filed to give the particular form to the designed ornament, and the vessel, having been previously turned in the plain way, receives a shuffling motion backwards and forwards as the spindle slowly revolves, and only when the incision admits the piece of iron will the vessel be in contact with the tool of the workman. When the iron is against the rim the surface remains untouched by the tool. Numerous very elegant and curiously indented porcelain articles are formed by the engine-lathe. The black Egyptian circular tea-pots will exemplify every species of engine-lathe turning.

As the vessels as soon as turned are in the best green state, they are, as soon as possible, passed to the *handler*, who fixes the spouts, handles, and all other requisite appendages. Such spouts, handles, or appendages, as are in any way curved, oval-shaped, or ornamented, are formed in moulds of two or more parts, as will be seen hereafter when speaking of squeezing.

For handles, and some other articles of appendage, a press is used, consisting of an iron cylinder, six inches wide, and ten inches deep. This cylinder has a strong bottom, with an aperture in the centre, to which is made to fit differently shaped plugs. It has a piston acting by a screw, that works in a bent iron bow, fastened to the block on which the cylinder rests. The aperture being supplied with a plug of the required form, some clay is put into the cylinder, and the piston forced down, by turning the screw, which causes the clay to protrude through the aperture in the shape required. The workmen cut it into lengths, as wanted, and bend it into the required form, and when sufficiently dry, affix it to the vessel by slip. Slip is likewise used to affix all other appendages. - When a tube is wanted, a pin is fixed in the clay that protrudes through the aperture of the cylinder, a pin is fixed above the centre of the plug. The vessel, being allowed a short time to dry, is cleared of all the superfluous clay by a knife. The vessel is then trimmed with other tools, and the whole of the joints cleaned off with a moist sponge, which, while it carries off all excrescences, gives to the whole uniform moisture.

We shall, previously to mentioning the process of squeezing, take notice of the modeller and the mould-maker, whose occupations are very distinct branches of the art.

The *modeller* has great scope for the exertion of natural and acquired ability, taste, and ingenuity: for on him depends the elegance, size, figure, adaptation, and correct arrangement of suitable ornaments. His business consists in taking a large lump of well-tempered clay, and modelling it, by continued carvings, with a sharp narrow-bladed knife, into the rough figure: he then commences the trimming process, by removing all excrescences, inserting any additions, and finally with a great variety of suitable tools, made of ivory, wood, or metal, gives to the whole the several touchings and retouchings requisite for finishing.

The modellers of the present day have attained much excellence, and as a proof we need only to state, that many who have seen the *Portland* or *Barberini vase* (for modelling of which Mr. Wedgwood is said to have paid Webber the enormous sum of four hundred pounds) declare, that any good modeller would now execute the whole himself in less than a month, and with a proper assistant in a fortnight. The branch of modelling, however, is by far more common now than it was in the time of Mr. Wedgwood; and good workmen obtain fair remuneration for their labour.

The *mould-maker* receives the model, and forms from it the requisite moulds, by employing plaster of Paris.

The gypsum or native sulphate of lime plaster is first ground in a mill, similar to a flour-mill. It is then put in a long trough, under which runs a flue communicating with the fire, to effervesce until all the water is expelled. This process is called both *boiling* and *burning*. The workman has his mouth and nose always well covered, to prevent his inhaling any of the dusty particles, which would, if taken inwardly, be very prejudicial to the lungs.

The mould-maker forms, and secures by a broad strap, a casing of thick clay round the model: he then mixes in a jug, containing a certain quantity of water, the proper portion of the soft impalpable powder or plaster, and stirring it

quickly, that the water may have an opportunity
it thoroughly, pours it upon and around the mod
instances gently or briskly shaking the mass.
immediately given out, and the whole very soo
compact mass. After standing a short time, tl
easily separated from the model, and each part i
stove to be dried.

When the moulds are found to be perfect, tl
dry, by which they retain the property of absorb
with great rapidity, so that the squeezer can o
his work from them readily, and when this is t
mould is said to *deliver* easily.

In some of the principal manufactories lar
plaster are fitted up as shelves, which serve
purpose of holding the newly-formed articles, an
ing the drying, by absorbing a portion of the mc

The workman, called the *dish-maker*, who
for dishes, plates, saucers, wash-bowls, & h
always cuts off from the mass a piece of clay acc
size and strength of the article he has to mak
again cuts asunder, or breaks with his hands, r
operation of forcibly slapping them together, to
air-bubbles from remaining in it. The piece is
a flat surface of board, or plaster, and the wor
heavy lump of clay, with a level under-surface
holding in the hand, beats the clay to the thinne
is intended to form. These pieces of clay are
called *bats*.

For wash-bowls, dishes, or plates, the workma
whirler, uses a vertical spindle, surmounted wit
block, ten inches diameter, and about two inche
this he places his plaster-mould, and with a bat
properly upon it; he then with one hand gives
whole, while with the other, dipped in water, h
clay very close to the plaster-mould : then, wh
tional piece is required, as the ledge, or foot, it
with slip, and firmly squeezed to the other clay.
a suitable thin tool or utensil of pot, of the
inside, is applied, to give the proper shape :
The sponge is now again employed to clean o
cences; the whole is cut to its size, finished with
and set to dry a little, and a horn tool is empl
it off.

The moulds are capable of being used five or
succession each day, because as soon as one has

it is set in a stove to dry, and as the workman proceeds regularly, each is allowed equal time for drying.

When the bowls, dishes, or plates, are taken off the moulds, and have been pared round the edge with a thin bladed knife, they are slightly polished by the hand, and afterwards laid on each other in quantities of four, eight, twelve, or more, according to their size and strength, to dry and harden, preparatory to being placed in saggers for the biscuit-oven.

The *squeezer* generally uses moulds which have two or more parts. The moulds for figures have their parts numbered. He takes a bat of a proper size and thickness, and lays it in one part of the mould, then with a large sponge beats and well forces it into all the cavities; he next takes another part, on which is the bottom, and presses the two parts together; he then rolls a piece of clay, and forces it into those parts of the article where the mould joins together, and afterwards cleans off all the excrescences, and secures the parts by a leather strap, so that they cannot come asunder while the mould is in the stove, or on the shelf, to dry to the green state. When he takes the strap off, the parts of the mould are carefully separated, and the vessel finished, by the joints being pared, cleaned, and sponged. The spouts, handles, covers, ornaments on the outside, and figures, are similarly formed and finished off.

This part of the process was formerly performed by casting; but *casting* is now only employed for the most elegant irregular shapes, where strength is not important.

The very dry mould is well closed together, and strapped for security. Some clay is then mixed with pure water till it be reduced to a pulp of the consistence of cream. This is poured into the mould until it be filled, and the plaster, of which the mould is formed, absorbs the water from the clay that is contiguous to it, and leaves a coating of clay attached to the mould. The pulp is then poured out, and the coating allowed a short time to dry; a second charge of a much thicker consistence is then poured in, and forms a body sufficiently thick for the article intended, and when a coating is again formed, the remainder of the pulp is poured off, and the mould placed a short time near a stove, and when sufficiently dried is separated, and the article left to dry to the green state: the seams of the joints are then smoothed off, and the article is finished by the skill of the workman, and when thoroughly dried is placed in a sagger for the biscuit-oven.

All the articles made in the clay by these various processes,

2 H 2

after being finished in reference to their shapes, figures, sizes, ornaments, &c. are placed on boards, and left to dry by the temperature of the apartment where they were made, or put into a drying-house, green-house, or stove.

The *sagger-maker** is expected to know the exact proportions of marl, old ground saggers, and sand, that are required to form the best saggers for either pottery or porcelain. Saggers are of different sizes, shapes, and depths, formed of a very porous composition, and capable of bearing, without being fused, a most intense heat. The bottom of each sagger has a thin layer of fine white sand, to prevent the pieces of pottery touching and adhering to it.

For porcelain flat ware, as plates, &c. the sagger is also firmly filled with very dry flint, to preserve each piece in its proper shape. When a sagger is filled with clay ware, on its outer edges are placed thick pieces of coarse clay, called *wads* from their being employed to wedge or closely join the interstice between two saggers, as well as to support the edges, and preserve equal pressure.

Each pile of saggers placed in an oven is called a *bung;* and the man who places the ware in the saggers, and the saggers in the oven, the *oven-man.*

The potter's oven, for both biscuit and gloss firings, is very much like that in which bricks and tiles are usually burnt in most parts of the kingdom; that is, a cylindrical form, surmounted by a dome. Around this oven are formed fire-places or mouths, whence the fire passes into horizontal flues in the bottom, and perpendicular flues, called bags, on the inside, and so ascends through all the interstices of the bungs of saggers, until the surplus escapes through the aperture in the dome of the oven.

Most ovens are surrounded by a high conical building, called a *hovel,* large enough to allow the man to wheel coals to the requisite places, and to pass along to supply each mouth with fuel; and at the same time to protect both him and the oven from rain or any other atmospheric inclemency.

The saggers are sometimes placed to dry in the sides of the hovel, and sometimes in a smoke house.

The biscuit-oven is always the largest upon the premises. The workman is called the *biscuit fireman,* and is employed from 48 to 50 hours at a time. The heat is gradually in-

---

* The word "sagger" is by many supposed to be a corruption of safe-guard; but we are disposed to date its origin to the Hebrew, from the word *sagar*, to burn. It is a baked earthen vessel into which others are placed when put into the kiln.

creased throughout the time, but porcelain does not require it so long, as it more readily allows the heat to be raised.

In different parts of the oven, where they can be easily extracted, rings of Egyptian black clay are placed, as trials, by which an experienced fireman can tell how much longer the process must be carried on, not within an hour, as indicated by Wedgewood's pyrometer, but within ten minutes. Hence the pottery district has a very pertinent proverb: "*Nothing beats a trial.*"

The name of the ware thus fired is *biscuit*, because of its being to appearance and feel like ship-bread when well baked; the surface is devoid of any appearance except that of a tobacco-pipe, sometimes tinged by the intense heat. When the saggers are taken out, the articles are carefully sorted, and all injured pieces are rejected.

If pottery were used in the biscuit state, it would, in some cases, be permeable to water; hence wine-coolers, *alcazaras*, are always in the biscuit state. The best size of wine-coolers is that which just admits the bottle, for then the air of the room can very little affect the water in the cooler, which consequently, by passing from the inner to the outer surface, effects the purpose sooner; a humid coating being thus presented to the action of the surrounding atmosphere, the evaporation causes a consequent quicker diminution of heat than could take place with a dry surface

All articles of pottery which have but one colour, and many that have several, are in general ornamented either by the pencil, or by impressions taken from copper-plates. The former is called *blue*, or *biscuit-painting*, the latter *blue-printing*. Both processes take place on the biscuit, prior to the ware being dipped in the glaze. If the ware were not previously fired, and were capable of being handled about for the painting, the water, used to soften the colours, would soften the ware; and the impressions from plates could not be clearly, even if at all, transferred to the ware; water also could not be employed to wash off the paper, and the water which contains the components of the glaze would be absorbed by the clay body, which would by this means be rendered so soft as not to preserve its shape in the oven.

It has been thought that advantage might result from being able to mix some substance with the clay of enamelled ware, which would resist the action of water, as a suitable glaze might then be first employed, and one firing answer for both the biscuit and the gloss, which would save one operation, as well as the time, labour, and expense of fuel.

In *blue-painting*, the colour is mixed with water and gum, and carefully laid on the biscuit ware. As every stroke leaves a mark in the pores of the vessel, great attention must be paid to the pattern, for a stroke once made can never be rubbed out. After the pattern is finished, the ware is allowed to dry by the atmosphere, and is then dipped in the glaze; it is afterwards exposed to heat in the gloss-oven, which fuses the minerals contained in the colours, and gives to each a coating of true gloss: about 4000 young women are employed in this branch of pottery, and by their industry support themselves in a respectable manner.

Blue-printing is the impressions taken from engraved copper-plates by means of a rolling-press. The blue-printer lays the plate upon a stove while the oily colouring substance is rubbed in, and by the heat the metalline particles contained in the oil flow and sink more readily into the engraved lines. The colour is oxide of cobalt, fluxed with different substances, and in suitable proportions, for the pale or dark blues.

The superfluous colour is carefully cleaned off the hot plate, which is laid on the press, and covered with a piece of coarse tissue paper, which has been first brushed over with a strong lie of soft soap, called *sizing*. The whole is now passed through the press, and the heat of the plate dries the paper, renders it more adhesive of colour, and also more easy to be extracted from the plate. The impression when taken off the plate is given to a girl, called a *cutter*, who cuts it into shapes, and hands the parts to a woman, (the *transferrer*,) who puts them on the biscuit, and when she has properly arranged them rubs them till the several pieces are completely affixed to the biscuit article: the article is then left for a short time to imbibe the colouring matter; after which, the paper is well washed off with clean water, and the article is put into a kiln to dissipate the oil. Sometimes the outline of a pattern is printed on the ware, and the colours are afterwards added with a pencil.

The earthenware is now ready to receive the smooth coating called *glaze* or *gloss*. The employing of this glaze, though in general, is not always, with a design to prevent the vessel from imbibing the liquid that may, at any future time, be poured into it; because some bodies of earthenware are, before glazed, impermeable to liquids of any kind; but with a design of accomplishing a more important object, that of hiding the substance of the vessel, which is not always either for fineness of texture or whiteness of colour, of a very prepossessing appearance. A coating of *mere* glaze would, by its transparency, only expose these defects; even if it were sufficiently contractile and expansile, by sudden changes of temperature, to admit of its being used. Hence is employed a vitrifiable composition of oxides of lead, glass, tin, &c. somewhat resembling common flint glass, readily made fusible by a little alkali and hardened flint, which will, when well managed, possess sufficient opacity, and by applying a certain degree of heat, flow and vitrify, and render fusible any flint or clay in contact with it, and thus not only fill up the pores of the biscuit article, but cover the whole with an opaque coating, that may be regarded as of real flint glass.

As the glaze that suits one, composition of ware will not suit another, owing to the difference in kinds as well as proportions of materials, it is ever requisite that the components of the glaze be carefully appropriated to the hardness, density, &c. of the components of the clay; because a good glaze should always possess the property of remaining, after being fired, unaffected by heat or cold, in exactly the same ratio as the clay, else on any sudden change of temperature, there would be a counter action between the body and the glaze.

When the article is short-fired, it is always more susceptible of the components of the glassy surface, and becomes altogether crazed, or full of little cracks, which render it permeable to water, and receptive of oily and greasy, and other heterogeneous substances, and ere long the article will, by constant usage, appear very much like a rotten substance.

Crazing is the technical term for the cracking of the glaze, whatever be the cause: whether it arise from excess of alkali in the materials composing the glaze, the deceptive union of the body and glaze, the unsuitableness of the body to the materials of the glaze, the components of the glaze not being equally fusible at the heat employed, or the heat for the proper fusion of the glaze being too high for the body itself.

Mr. Parkes states, that a little lime mixed in the clay will prevent crazing; but manufacturers are of opinion that the fact is the contrary. Lime will in a slight degree add to the transparency of porcelain, but ever render it liable to craze. If the articles, whether biscuit or gloss, be taken out of the oven before tolerably cool, the temperature of the air will most generally affect them, and especially the glaze, which is not then properly annealed.

The *glaze* is a vitrifiable composition, about the consistence of, and in appearance, very much like new cream. It is essential that it be thin, and when fired, possess a degree of opacity to approach as nearly as possible to, and yet be below the fusibility of the biscuit, that the combination may be more intimate and permanent. Hence its composition varies for each body, consonant with the view and experience of the manufacturer; and it is very seldom that it can be applied to another body without previously altering its composition.

In some instances the cost of glazing is much less than in others; though economy is sought for in all, and each manufacturer regards his own as the best and cheapest of the kind for the purpose to which it is employed. Great care is taken that the *recipes*, which are considered very valuable, be kept as much as possible secret among themselves, to prevent foreign potters availing themselves of them, to the injury of our manufacture.

Raw glazes are employed for the common pottery, such as toys, jugs, tea-ware, &c. They are generally composed of white-lead, Cornish-stone, and flint, ground by a hand-mill. We have seen a few raw glazes for porcelain of a very good quality; but fritt glazes are mostly used, and are of excellent quality.

Fritt is derived from a certain combination of different materials being well mixed together, or *fritted*, and then calcined; which procures a union of all the parts, and a solidity and purity not otherwise attainable. The fritt is generally placed where it can be affected by a sufficient heat to fuse all its ingredients, without volatilizing the uncombined alkali.

Lynn sand is occasionally one of the ingredients employed in the fritt. Some persons use soda, to render the fritt more fluent while being fired. In some instances, common salt is used along with a portion of potash, which decomposes it, and drives off part of its impurity. The remaining impurities are driven away in the process of fritting. Let it be remembered, however, that brilliancy of glaze is formed only by lead; and that the employment of salts ever produces a poor appearance.

The calcined fritt is pounded, picked, sifted, and ground to an impalpable powder, after which it is mixed with certain proportions of white-lead and flint, and again ground in a very powerful mill. The finer it is ground, the more serviceable it is for the purpose; the glaze is every way better, is more level on the ware, more readily and easily fired, of greater brilliancy, and scarcely ever liable to craze.

The lead causes the other components to vitrify at a certain heat; and accordingly as more or less is used, the glaze becomes harder or softer. Many objections have been made to its employment: those in reference to vessels for domestic purposes we have already noticed; and in reference to the dippers being subject to paralysis (which is supposed to result from the lead,) every aid is afforded by preventives, and where attention is paid to personal cleanliness, and the water and towel placed for their use are employed, deleterious effects can seldom be experienced.

The materials being well-ground and in a state of fluidity, are next put into the dipping tub. As the materials are heavy it is requisite to keep the powder suspended, and uniformly dispersed through the mass, which weighs about 32 oz. per pint. By the side of this tub stands the dipper, and a boy, his assistant. The boy is employed in brushing the articles, and delivering them, one by one, to the dipper, who dips them quickly into the liquid, and as soon as he takes them out, turns them rapidly about, that the thickness of the liquid may be equal in all the parts. The water is imbibed by the porosity of the biscuit, and there is left a coating of the substances, sufficiently hard to continue affixed until the article be placed in the sagger. The article is then placed on a board, another is similarly dipped, and thus it proceeds until the quantity be finished, when the whole are put into saggers.

When a flat piece has been dipped, it is placed on a board, in which are a number of nails, about an inch above the surface; the superfluous compound runs off, the remainder quickly dries, and soon admits of being moved; which effects a saving in fuel and materials, and the articles are better glazed.

Hollow pieces and blue-printed ware, are placed on hair sieves, or on four pieces of sheet iron, from two to three feet long, called a *fiddle*; in three minutes the dipped articles are sufficiently dry to be removed to the board, and a few minutes afterwards to be placed in the saggers.

In the inferior earthenware certain metallic oxides, as of copper, &c. are mixed with the glaze. These kinds of glazes are distinguished by the name of *dips*. When the article has been thus dipped, it is finished on a turner's lathe, to mark what is to be white, and when the appendages are affixed it is dried in the oven.

The articles are again put into the saggers to fuse the glaze, and as in this process each would attach itself to the other, were they to come in contact, pieces of clay of different sizes and shapes, called stilts, cockspurs, rings, pins, bats, &c. are put to keep them apart.

The saggers are, as before, piled in the gloss-oven, which seldom holds more than one half the quantity of ware fired in the biscuit-oven. The gloss-fireman raises the temperature as quickly as possible to a height sufficient to fuse the glaze, which is much lower than the heat of the biscuit-oven, and usually keeps it fired from 16 to 19 hours. Trials, made of native red clay, are found very essential in this operation to prevent the ware being more intensely heated than the biscuit body will bear; for as clay contracts by every addition of heat, were the heat of the gloss-oven to exceed the heat for the biscuit, the articles would be further contracted, and would be either crooked in shape, or injured in the glaze. The coating of glaze which adhered to the biscuit is, by this firing, uniformly spread over the surface, the particles are fused altogether, and the ware, when cold, appears to be covered with perfect gloss.

As the gloss-oven is sometimes fired to a greater degree of heat than some colours will bear, another process is employed, called enamelling, because the designs are more elegant in their execution and form, and the colours are burnt into the glaze of the pottery. These designs are of the finest description, and are most delicately executed upon the glossy surface.

The colours used are generally of a mineral or metallic nature. For blacks, oxide of umber and cobalt, and black oxide of copper. The best oxide of iron is produced by causing heated air to act upon iron.

For purples and violets, precipitate of cassius, and calx of gold.

For greens, oxide of copper, and precipitate of copper.

And for blues, oxide of cobalt.

These oxides are all in an impalpable powder, and are mixed with a certain powder as a flux, and are so prepared as never to spread beyond their lines, or injure the drawing while being fired.

Each colour is ground with a muller on a large hard stone, and is incorporated with acid of tar, oil of turpentine, or whatever oil may be deemed suitable, and is evaporated. Camel-hair pencils are used to lay the colours on the pottery.

As both males and females are employed in this branch, the men are called painters, the women paintresses: but in blue-painting, where no men are employed, the women are called blue-painters.

This is the finest and most durable species of painting, and it is capable of being employed for the most elegant and durable embellishments, as neither air, nor wear, can affect either the beauty of the design, or the brilliancy of the colours.

Gilding requires the precipitate of gold from its solution to be properly mixed with oil of turpentine, and great pains must be taken in laying it on the pieces, which is done in a manner similar to the preceding. When the article is heated, the oxygen flies off and leaves on the ware the gold in a metallic state; but the natural brilliancy of the gold is wanting, consequently, a burnisher of agate, blood-stone, or steel, is applied to the gold, first moistened with flint-water, to procure the bright and shining property of the precious metal, which is, by that means, quickly brought in view. This, when the gold is not too much lowered by fluxing, will scarcely ever tarnish.

Black-printing is a very distinct and curious process. The workman boils a quantity of glue to a certain consistence, and pours it on very smooth dishes, to the thickness of an eighth or a quarter of an inch, according to the size of the plate he may have to use. This, when cold, is cut into sizes for the plate, called *papers;* and he makes as many as he can conveniently use in his routine of working.

Then taking a copper-plate, properly engraved, he rubs into it some well-boiled oil, and having properly cleansed the plate, forcibly presses the glue-paper against it; the latter being firmly fastened to a piece of wood to be held in one hand, and the paper being laid on a boss or cushion held in the other. The oil in the plate adheres by the pressure to the glue-paper, and he carefully, but firmly, presses it and the piece of pottery together; then separates them, and with fine cotton slightly sprinkles the colour (which is in an impalpable powder) upon the design left by the oil. After a certain time the oil has evaporated sufficiently to permit all superabundant colour to be wiped off, which is done with much delicacy and attention, by using old silk rags, and the black printed pottery is placed in the enamel-kiln, where the glaze and colour fuse and incorporate.

The enamel-kiln is commonly made in the shape of a chemist's muffle, from about six to ten feet long, and three to five feet wide; having from one to four mouths, according to the size of the kiln, and the purposes to which it is applied; these mouths are made for the admission of fuel. In this kiln the articles are very carefully placed in layers, or thin bats, until the whole be filled; the mouth is then stopped, and the kiln fired for about eight or ten hours.

The articles, when painted, gilded, or black printed, are

subjected to a third firing in the enamel-kiln, which fuses both
the glaze and the colours, and the mineral or metallic particles
flow and become incorporated into the glassy surface.

Lustre ware consists of an inferior quality of the materials
worked into the usual forms, and having the hue of gold,
platina, or copper, &c. fixed on the glaze, whose great
brilliance, when first made, occasioned it to be thus named.

The very easy method of performing the operation, and the
quick sale which the articles obtain, has caused it to become
so common, and of a quality so inferior, as to be little esteemed
by potters.

The pottery to receive lustre is made and glazed for the
purpose. That for *gold lustre* is made of the red clay of the
district, and when fired gloss, has just a sufficient tint left to
give to the articles that peculiar shade of colour observable on
viewing them. A very common article of cream-colour is
commonly used for the *silver lustre*.

The oxide used for lustre, as gold, platinum, &c. is mixed
with some essential oil by the application of heat, and the
fluid is brushed over the surface of the articles. Sometimes
ornaments are formed on the surface. For this purpose, a
thick fluid of soot or lamp-black is laid on the articles, by
brushes, according to the patterns, and the articles are then
heated in a very hot iron oven, and afterwards have the lustre
brushed over them. When dry, they are placed in a kiln,
similar to that for enamel ware; which, being carefully fired,
dissipates the oxygen, loosens the ornamental article, and re-
stores the metallic lustre to a degree almost equal to its primitive brilliance; but in some cases it is of a coppery and
steely brilliance.

In Messrs. Rileys' shining black biscuit porcelain, the ware
is of a jet black jasper, or porcelain body, having undergone
a high degree of vitrification, which elicits a lustre, or bright
vitrified polish on the surface, of the appearance of black
coral, without a glaze, which is of considerable importance in
point of durability, elegance, and usefulness. It is warranted
never to change its elegant quality by time or use, and will
clean with water, equal to a piece of the finest porcelain. It
has a decided advantage over the *dry body*, or *common
Egyptian black*, which is generally scoured and oiled to give
the surface a smooth appearance, by which it imbibes dust
and becomes offensive, and the substance of which it is composed being of a porous nature, it becomes saturated with the
liquids poured into it, which eventually prove unwholesome,
as well as disagreeable to the hands and sight; the whole of

these disadvantages is obviated by Messrs. Rileys' black lustre, which, being perfectly vitrified, allows no liquid to be imbibed.

The direful effects of using lead in the manufacture of pottery are manifest by severe cholics, paralysis of the limbs, and often the untimely death of the workmen; and yet this dangerous mineral forms the glaze of the *common red pottery*, in which much of the food of the lower classes is prepared. Lead is slightly soluble in animal oil, more copiously in the acids of our common fruits, and more especially when their action is aided by the heat required in cookery. It is not improbable that many of the visceral disorders of the poor, who use such pottery, are attributable to this little suspected source; and that it is to procure the temporary removal of the pain occasioned by the action of the lead, that they habituate themselves to the deleterious use of distilled spirits.

It was on this view of the subject that the Society for the Encouragement of Arts, Manufactures, and Commerce, were induced to offer their largest honorary premium for the discovery of a glaze for such red pottery, composed of materials not any way prejudicial to the health, and which from its cheapness, and fusibility at the comparatively low temperature required by red pottery, might supersede the use of lead in that branch of manufacture.

This important object was eventually discovered by J. Meigh, Esq. of Shelton, who was well persuaded of the possibility of its accomplishment, and who, without any other stimulus than a desire to benefit mankind, first fully ascertained what particular objects were contemplated by the Society, and then communicated his successful process; by which any makers of red pottery, who may choose to depart from long-established usage, which is but too often the greatest obstacle to improvement, may easily remove the source of the mischief, and considerably improve the quality of the ware, and effect a saving in materials in fuel.

After this view of the subject, we shall not be required to apologize for giving the process in Mr. Meigh's own language. "The common coarse red pottery, being made of brick-clay, is very porous, and is fired at as low a temperature as possible, to save the expense of fuel, and to avoid fusion, or variation of shape, which would result from highly firing common clay; consequently there is needed a glaze to fill up the pores, that the vessel may contain fluids. This glaze must be very fusible, and cheap; hence, for transparent vessels, litharge, and for black opaque, common lead ore, are used. A glaze of lead, whether altogether or in part, is objectionable, because, first, when quickly raised to the temperature of boiling water, it cracks from different expansibility of the clay and the glaze, so that the liquid permeates the body of the vessel; and secondly, the glass of lead, whether alone or mixed with small proportions of earthy matter, is very soluble in vinegar, in the acid juices of fruits, and in animal fat when boiling."

The injurious effects arising from these have been already stated. Mr. Meigh therefore proposes, that a mixture of red marl, which can be easily ground in water to an impalpable paste, and will remain suspended therein for a long time, be employed to dip the vessel in, so that its pores may be filled with the fine particles of the marl, preparatory to glazing; which is performed with a mixture, of the consistence of cream, of equal parts of black manganese, glass, and Cornish-stone, (chiefly felspar,) well ground and mixed together; in a white glaze the manganese is omitted. After undergoing this process, the ware is well dried and fired, as usual.

Mr. Meigh also proposes a substitute for the materials of the common red pottery, consisting of four parts common marl, one part red marl, and one part brick-clay. The ware made in this way is of a reddish cream-brown, harder, more compact, and less porous, than the red pottery; more economical to the potter, and calculated to contribute in no inconsiderable degree to the health of the lower classes who use the red pottery.

The aim of the principal manufacturers has been to obtain the composition of a clay and glaze for porcelain, which, when fired, should be very fine in its texture, extremely white in colour, possess considerable transparency, and at the same time be able to bear different degrees of heat and cold. That the reader may understand more fully the several peculiarities which are considered by manufacturers as essential to perfect porcelain, we shall state,

That the first and most important quality is a superiority in the *whiteness* of the porcelain; that its appearance be free from any specks, and that it be covered with a rich and very white glaze of almost velvet softness in appearance, and of best flint-glass smoothness to the touch.

That the second important and essential quality is *durability*, or a substance whose components will bear, without being injuriously affected, a sudden and rapid increase of temperature, and particularly to sustain unaltered, the action of boiling water.

That the third essential quality is *transparency*, which is admitted to be, in some measure, requisite, but certainly not entitled to that high degree of preference so frequently given to it, the best porcelain being a shade less transparent than a kind much inferior.

Formerly, connoisseurs very highly estimated porcelain of a fine granular texture; but this criterion of excellence cannot always be relied on.

To ascertain the texture of an article, it must be fractured, and thereby tacitly destroyed; the semi-vitrification and fineness of texture observable in one piece will not be so obvious, but there will be a varied appearance in different pieces, though all be fabricated at the same time, and from the same mass of materials.

Stone-china is formed of a compound of Cornish-stone and clay, blue clay, and flint; and with a glaze, consisting of lead, bottle-glass, Cornish-stone, and flint. It is very dense and durable, but less transparent than bone-china; and is very much used for jugs, and the larger sorts of vessels.

Iron-stone china is not very transparent; but possesses great strength, compactness, density, and durability. It is not much used for tea-ware, but has very suitable properties for dinner and supper services, jugs, and ornaments. It was discovered by Messrs. G. and C. Mason, and has been more productive than any other species of pottery or porcelain.

Felspar china, which has been only very recently introduced, is the most noted of all the porcelains; it results from the introduction of certain proportions of a fresh material into both the clay and glaze.

Cornish-stone is a species of granite in a state of decomposition, and contains much felspar. Cornish-clay is found in situations where this decomposition is in progress. The decomposing granite is broken up with pickaxes, and the fragments are thrown into running water, whose action washes off, and keeps in suspension, the slight argillaceous particles miscible with that fluid. The water is discharged into pans or pits, where the particles subside, and the water is evaporated, formerly by the atmosphere, but now by heated flues passing under the reservoirs. When the water is evaporated, the substance is cut out in square lumps, and placed on shelves to dry, when it becomes extremely white, and in the state of an impalpable powder. It is then packed up in casks, and forwarded to the manufacturers.

The clay of the best felspar porcelain is formed of certain

proportions of china-stone and felspar; the mixing of the proportions requires much attention, for an excess of felspar would cause the vessels to shrink in the biscuit-oven, prior to the fusion of the clayey particles, which causes its transparency; and an excess of china-clay would increase the opaqueness. In both cases, the glaze would expand and contract in a ratio differing from that of the biscuit, and cause the pieces to be crazed. The fusibility of native felspar is owing to its containing about 13 per cent. of potash, which causes it to be one of the best materials for glazing porcelain. Calcined bone is used, and renders the clay very white; but it should be employed with judgment, as its great contractibility causes the articles wherein it is used to excess, to crack on sudden changes of temperature.

Beside the porcelain or china-clay already noticed, the manufactures use *four* other kinds; the two first from Devonshire; the two last from Dorsetshire.

The *black clay* is remarkable for the fact, that the bituminous matter which gives the colour whence it derives its name, flies off by firing; and the blacker the clay when first dug, the whiter will be the pottery.

The *cracking clay* is used because of its extremely beautiful whiteness when fired; but it requires very exact proportions of flint, otherwise the ware will crack during the firing of the biscuit.

The *brown clay* burns very white without cracking, and some manufacturers use much of it; but as the ware does not so readily imbibe the particles of melting glaze, the liability of the ware to craze causes others to reject it altogether. This clay is with difficulty sifted through the lawn, requires much longer *weathering*, or exposure to the action of the atmosphere, for the separation of its particles, and to prevent crazing, different proportions of other materials; but the greatest objection to it is, that some of the kind sent within these few years has always burned inferior in colour to what it formerly did.

The *blue clay* is the best, and the most expensive. It forms a very white and solid body, and requires a much greater proportion of flint, which considerably improves the quality of the ware; but the proportions require very strict attention, and a higher degree of biscuit-fire.

The *cream-coloured pottery* has its name from the tint of its colour being that of new cream. It is, when well made, and properly fired, very sonorous, sufficiently hard to elicit sparks by the application of steel, and will contain liquids without being permeated by them. When it is of good quality, it will resist the action of nitre, glass of lead, and other fluxes, which renders it of great utility in all domestic and chemical processes where great heat is used. Care must be paid to the current of air while the pottery is in contact with fire, otherwise its hardness and density, by preventing its sudden contraction or expansion, renders it very liable to break.

Wedgwood's cream-coloured pottery is allowed to retain

total rejection of lead is not compatible with perfection in pottery.

The *blue-printed pottery* is a very popular kind, and most persons who have seen it placed near the preceding, must have remarked that it is of a finer kind, with a very different tint or colour.

The best species is in considerable demand for dinner, dessert, tea, and supper services; while its cheapness has caused it to supersede almost every kind of ware.

The difference is caused by two peculiarities; one in the clay, arising from the employment of a greater proportion of blue and porcelain clays and flints; the other in the glaze, from certain components being mixed together, and calcined into a frit, which is often picked and sifted, then ground together with glass and white lead, and mixed with certain proportions of Cornish-stone and flint.

One kind of this pottery has its glaze varied to capacitate it for enamelling. The blue printed tea-ware has recently obtained the name of *semi-china*, owing to its being, when well fired, very fine, white and neat, and possessing some degree of transparency.

The *chalky pottery* is a very excellent and beautiful kind, having a delicate white appearance, of fine texture, and glassy smoothness. The nature of the clay and the glaze renders it very proper for enamelling, as smalts are introduced, in accordance with the views of the maker, to effect the tints.

The clay is boiled on a plaster-kiln, and consists of certain proportions of porcelain, blue and Welsh clays, pulverized, calcined, or raw flints, Cornish-stone, white enamel, tinged with smalts; and some persons add calcined bone and plaster of Paris. This ware requires a most ardent fire for the biscuit.

The glaze is composed of a frit of glass, Cornish-stone, flint, borax, nitre, red-lead, potash, Lynn sand, soda, and cobalt calx. After fritting, and being well fired, it is ground and mixed together with white-lead, glass, flint, and Cornish-stone.

The *fine red pottery* is formed of almost equal proportions of yellow brick-clay and the red from Bradwall-wood; an inferior sort is made for lustre-ware.

In the Hall-field colliery, east side of Henley, is found a marl, which, when properly prepared, by levigating and drying, will alone form a very beautiful light red, of four distinct shades, according to the intensity of the firing. This was discovered by Mr. G. Jones, in 1814, who commenced a manufacture of this kind of ornamental pottery for Messrs. Burnett, to be shipped to Holland; but the sudden return of Napoleon from Elba so disconcerted the arrangements, that the elder Mr. Burnett died suddenly, and Jones did not long survive the disappointment he experienced.

The introduction of ochre will change the red to a brown colour.

The *bamboo, or cane-coloured pottery*, is a very beautiful kind, employed chiefly for ornamental articles, and the larger vessels of tea-services. It is never glazed outside, though one kind has the outside vitrified. The insides of tea-ware are well washed with a liquid which forms, when fired, a thin coating of glass. The colour varies from that of a light bamboo to almost a buff: but the prevalent colour is nankeen. The best clay or body is formed of proportions of black marl, brown clay, Cornish-stone, and shavings of cream-coloured pottery.

The *jasper pottery* was invented by Mr. J. Wedgwood. It is extremely beautiful; and is formed of blue and porcelain clay, Cornish-stone, Cork-stone, (sulphate of barytes,) flint, and a little gypsum, tinged with cobalt calx.

The *pearl pottery* is a superb kind for elegant and tasteful ornaments, and is so much valued, that the workmen are usually locked up, and employed only on choice articles. The components of the clay are blue and porcelain clay, Cornish-stone, a little glass, and red-lead. This forms the best body for apothecaries' mortars; but it is more expensive, and more durable, than the common mortar body.

The *black Egyptian pottery* is now so very popular for tea-services, that few persons are ignorant of what is meant by this denomination. Its components are cream-coloured slip, manganese, and ochre; sometimes glazed with lead, Cornish-stone, and flint; and the inside is washed with white-lead, flint, and manganese. It was the custom formerly to grease the outside with butter or suet, to give it a bright appearance.

The ochreous material is obtained from the water that is pumped out of the collieries. This water is carried along channels in which are placed small weirs, to afford an opportunity for the precipitation of the sediment. When a sufficient quantity has accumulated, the water is diverted, the weirs are emptied, and the thick fluid is thrown into small pools, called sun-pans, whence the moisture is evaporated by the solar heat. This substance is afterwards burned with small-coal, which renders it proper for use.

The unpleasantness of the grease, requisite to give bright-ness to the black, having been a subject of general complaint,

2 i 2

Messrs. Riley, of Burslem, were induced to attempt to remedy it; the result of which was, the invention of a new black porcelain, with a bright burnished, vitrescent appearance, superior to any other kind of dry-body pottery. It never imbibes dust, or absorbs moisture; and it can be cleaned with water equally as well as the finest porcelain, and always retains the appearance of a beautiful black coral.

The *drab pottery* is useful for articles which require strength to be united to ornament, as flower-pots, water-jugs, &c. It is formed of blue, porcelain, and Bradwall-wood clays, Cornish-stone, and black marl, mixed with nickel; one kind is made of turners' shavings of cream-coloured ware made into slip, and mixed with nickel. The inside is rendered white by a wash of slip, flint, and porcelain clay.

It has for some time been usual for ladies of taste and acquirements in the fine arts, to purchase porcelain in its glazed state, for the exercise of their talents and ingenuity in ornamenting their own tea-services. This very pleasing amusement is often aided by manufacturers, who readily afford every assistance in their power to facilitate the easy enamelling of such services; they supply proper mineral colours, and the rectified oil of amber, for the best purposes, and the best oil of turpentine for others; and they attend to the proper firing of the enamel, burnish the gold, and dress off the whole for the table.

The different combinations of materials appear to be of less importance in the fabrication of good pottery, than due regard to well-determined proportions. All clays have some proportions, more or less, of metallic matter, which cause great difference in their appearance, and the effects produced on them by fire. All clays vary in colouring according to the ardency of the fire; hence the oven-man's greatest care is, to place the saggars in the most appropriate parts.

The chief ingredients are clay and flint; for no pottery will be perfect unless made of suitable clay, with a definite proportion of flint. The great difficulty is to unite beauty and goodness in the same composition. If too much flint be used, the pottery, after being fired, will crack on exposure to the air; and if too little, the glaze will not be retained on it after firing. Every kind of clay that is dried alone will crack; for if pure argillaceous earth be made sufficiently soft to be wrought on the potter's wheel, it will, while drying, shrink one inch in twelve, which will inevitably cause it to craze.

Pure clay *(alumina)* is always opaque, and the flint *(silica)* always transparent; but both are prepared previously to

HOROLOGY

In the early ages, time was measured either by the sun-dial or clepsydra; in the former, by the shadow of a wire, or of the upper edge of a plane, erected perpendicularly on the dial, falling upon certain lines meant to indicate the hour; in the latter, by the escape of water from a vessel through a small orifice, which vessel had certain marks upon it to show the time the vessel was discharging.

These modes are now superseded by the use of clocks, watches, and chronometers, which indicate time by the movement of machinery.

Under this general head of Horology, therefore, we propose to treat of the structure of the several kinds of machines now used for the exact measurement of time; in doing which, the article will of necessity be divided into three sub-heads, Clocks, Watches, and Chronometers; and to them will be annexed two others, treating of some of the best kinds of pendulums and escapements.

CLOCKS.

Clocks are certain machines, constructed in such a manner, and so regulated by the uniform action of a pendulum, as to measure time, in larger or smaller portions, with great exactness

Fig. 489 represents the profile of a clock. P is a weight suspended by a rope that winds about the cylinder or barrel C, which is fixed upon the axis *a a;* the pivots *b b* go into holes made in the plates T S, T S, in which they turn freely. These plates are made of brass or iron, and are connected by means of four pillars Z Z, and the whole together is called the frame.

The weight P, if not restrained, would necessarily turn the barrel C with an uniformly accelerated motion, in the same manner as if the weight were falling freely from a height; but the barrel is furnished with a ratchet-wheel R K, the right sides of whose teeth strike against the click, which is fixed with a screw to the wheel D D, as represented in fig. 490, so that the action of the weight is communicated to the wheel D D, the teeth of which act upon the teeth of the small wheel *d*, which turns upon the pivots *c c.* The communication or action of one wheel with another is called the *pitching ;* a small wheel like *d* is called a *pinion*, and its teeth the *leaves* of the pinion. Several things are requisite to form a good pitching, the advantages of which are obvious in all machinery where teeth and pinions are employed. The teeth and pinion-leaves should be of a proper shape, and perfectly equal among themselves; the size also of the pinion should be of a just proportion to the wheel acting upon it: and its place must be at a certain distance from the wheel, beyond or within which it will make a bad pitching.

The wheel E E is fixed upon the axis of the pinion *d;* and the motion communicated to the wheel D D by the weight is transmitted to the pinion e, consequently to the wheel E E, as likewise to the pinion e, and wheel F F, which moves the pinion f, upon the axis of which the crown or balance wheel G H is fixed. The pivots of the pinion f play in holes of the plates L M, which are fixed horizontally to the plates T S. In short, the motion begun by the weight is transmitted from the wheel G H to the pallets I K, and by means of the fork U X, rivetted on the pallets, communicates motion to the pendulum A B, which is suspended upon the hook A. The pendulum A B describes, round the point A, an arc of a circle, alternately going and returning; if, therefore, the pendulum be once put in motion by a push of the hand, the weight at B will make it return upon itself, and it will continue to go alternately backward and forward till the resistance of the air upon the pendulum, and the friction at the point of suspension at A, destroys the originally impressed force. But as, at every vibration of the pendulum, the teeth of the balance-wheel G H act so upon the pallets I K, (the pivots upon the axis of these pallets play in two holes of the potence e t,) that after one tooth, H, has communicated motion to the pallet K, that tooth escapes, then the opposite tooth, G, acts upon the pallet I, and escapes in the same manner; and thus each tooth of the wheel escapes the pallets I K, after having communicated their motion to the pallets in such a manner that the pendulum, instead of being stopped, continues to move.

The wheel E E revolves in an hour. The pivot e of the wheel passes through the plates, and is continued to r; upon the pivot is a wheel N N, with a long socket fastened in the centre; upon the extremity of this socket, v, the minute-hand is fixed. The wheel N N acts upon the wheel O, the pinion, p, of which acts upon the wheel g g, fixed upon a socket which turns along with the wheel R. The wheel g g makes its revolutions in twelve hours, upon the socket of which the hour-hand is fixed.

From the foregoing description it is evident, first, that the weight P turns all the wheels, and at the same time continues the motion of the pendulum; secondly, that the quickness of the motion of the wheels is determined by that of the pendulum; and thirdly, that the wheels point out the parts of time divided by the uniform motion of the pendulum.

When the cord from which the weight is suspended is entirely run down from off the barrel, it is wound up again by means of a key, which goes on the square end of the arbor at Q, by turning it in a contrary direction from that in which the weight descends. For this purpose, the inclined side of the teeth of the wheel R, fig. 490, removes the click C, so that the ratchet-wheel, K, turns while the wheel D is at rest; but as soon as the cord is wound up, the click falls in between the teeth of the wheel D, and the right side of the teeth again act upon the end of the click, which obliges the wheel D to turn along with the barrel, and the spring A keeps the click between the teeth of the ratchet-wheel R.

We shall now explain how time is measured by the pendulum; and how the wheel E, upon the axis of which the minute-hand is fixed, makes but one precise revolution in an hour. The vibrations of a pendulum are performed in a shorter or longer time in proportion to the length of the pendulum itself. A pendulum of 3 feet 8¼ French lines in length makes 3,600 vibrations in an hour, that is, each vibration is performed in a second of time, and for that reason it is called a *seconds pendulum ;* but a pendulum of 9 inches 2¼ French lines makes 7,200 vibrations in an hour, or two vibrations in a second of time, and is called a *half-second pendulum.* Hence, in constructing a wheel whose revolution must be performed in a given time, the time of the vibrations of the pendulum, which regulates its motion, must be considered. Supposing, then, that the pendulum A B makes 7,200 vibrations in an hour, let us consider how the wheel E shall take up an hour in making one revolution. This entirely depends on the number of teeth in the wheels and pinions. If the balance-wheel consists of 30 teeth, it will turn once in the time that the pendulum makes 60 vibrations ; for at every turn of the wheel, the same tooth acts once on the pallet I, and once on the pallet K, which occasions two separate vibrations in the pendulum ; and the wheel having 30 teeth, it occasions twice 30 or 60 vibrations. Consequently, this wheel must perform 120 revolutions in an hour, because 60 vibrations, which it occasions at every revolution, are contained 120 times in 7,200, the number of vibrations performed by the pendulum in an hour.

In order to determine the number of teeth for the wheels E F, and the pinions *e f,* it must be remarked, that one revolution of the wheel E must turn the pinion *e* as many times as the number of teeth in the pinions is contained in the number of teeth in the wheel. Thus, if the wheel E contains 72 teeth, and the pinion *e* six, the pinion will make 12 revolutions in the time that the wheel makes one; for each tooth of the wheel drives forward a tooth of the pinion, and when the six teeth of the pinion are moved, a complete revolution is performed; but the wheel E has by that time only advanced six teeth, and has still 66 to advance before its revolution be completed, which will occasion 11 more revolutions of the pinion. For the same reason, the wheel F having 60 teeth, and the pinion *f* six, the pinion will make 10 revolutions while the wheel performs one. Now the wheel F, being turned by the pinion *e,* makes 12 revolutions for one of the wheel E; and the pinion *f* makes 10 revolutions for one of the wheel F ; consequently the pinion *f* performs 10 times 12, or 120 revolutions in the time the wheel E performs one. But the wheel G, which is turned by the pinion *f,* occasions 60 vibrations in the pendulum each time it turns round; consequently, the wheel G occasions 60 times 120, or 7,200 vibrations of the pendulum, while the wheel performs one revolution ; but 7,200 is the number of vibrations made by the pendulum in an hour, and consequently the wheel E performs but one revolution in an hour ; and so of the rest.

From this reasoning, it is easy to discover how long a clock may be made to go for any length of time without winding up : 1. by increasing the number of teeth in the wheels ; 2. by diminishing the number of teeth in the pinions ; 3. by increasing the length of cord that suspends the weight ;

4. by increasing the length of the pendulum; and 5. by adding to the number of the wheels and pinions. But, in proportion as the time is augmented, if the weight continues the same, the force which it communicates to the last wheel, G H, will be diminished.

It only now remains for us to take notice of the number of teeth in the wheels which turn the hour and minute hands. The wheel E performs one revolution in an hour; the wheel N N, which is turned by the axis of the wheel E, must likewise make only one revolution in the same time; and the minute-hand is fixed to the socket of this wheel. The wheel N has 30 teeth, and acts upon the wheel O, which has likewise 30 teeth, and the same diameter; consequently the wheel O takes one hour to a revolution. Now the wheel O carries the pinion p, which has six teeth, and which acts upon the wheel $g g$, of 72 teeth; consequently the pinion p makes 12 revolutions while the wheel $g g$ makes one, and of course the wheel $g g$ takes 12 hours to one revolution; and upon the socket of this wheel the hour-hand is fixed. All that has been here stated concerning revolutions is equally applicable to watches as to clocks.

Clock-work, properly so called, is that part of the movement which strikes the hours, &c. on a bell; in contradistinction to that part of the movement of a clock or watch which is designed to measure and exhibit the time on a dial-plate, and which is termed *watch-work*.

Fig. 491 represents the clock part. H is the first or great wheel, moved by means of the weight or spring at the barrel G. In 16 or 34 hour clocks, this wheel has usually pins, and is called the *pin-wheel*; and in eight-day pieces, the second wheel, I, is commonly the pin-wheel, or striking-wheel, and is moved by the former. Next the striking-wheel is the detent-wheel, or hoop-wheel, K, having a hoop almost round it, wherein is a vacancy at which the clock locks. The next is the third or fourth wheel, according to its distance from the first, called the *warning-wheel*, L. The last is the flying-pinion, Q, with a fly or fan, to gather air, and so bridle the rapidity of the clock's motion. To these must be added the pinion of report, which drives round the locking-wheel, called also the *count-wheel*, which has, in general, eleven notches, placed at unequal distances, to make the clock strike the hours.

Besides the wheels, to the clock part belongs the rash or ratch, which is a kind of wheel with twelve large fangs, running concentrical to the dial-wheel, and serving to lift up the detents every hour, and make the clock strike; the detents, or stops, which being lifted up and let fall, lock and unlock the clock in striking; the hammer, as 8, which strikes the bell R; the hammer-tails, as T, by which the striking-pins draw back the hammers; latches, whereby the work is lifted up and unlocked; and lifting-pieces, as P, which lift up and unlock the detents.

We shall now proceed to give a description of an ingenious

clock, contrived by the late Dr. Franklin, of Philadelphia, that showed the hours, minutes, and seconds, with only three wheels and two pinions in the whole movement.

The dial-plate of this clock is represented by fig. 492. The hours are engraved in spiral places, along two diameters of a circle, containing four times 60 minutes. The index A goes round in four hours, and counts the minutes from any hour it has passed by to the next following hours. The time as appears in the figure is either 32½ minutes past 12, or past 4, or past 8; and so on in each quarter of the circle, pointing to the number of minutes after the hours the index last left in its motion. Now, as one can hardly be four hours mistaken in estimating the time, he can always tell the true hour and minute by looking at the clock, from the time he rises till the time he goes to bed. The small hand B, in the arch at top, goes round once in a minute, and shows the seconds as in a common clock.

Fig. 493 shows the wheel-work of the clock. A is the first or great wheel; it contains 160 teeth, goes round in four hours, and the index A (fig. 492) is put upon its axis, and moved round in the same time. The hole in the index is round; it is put tight upon the round end of the axis, so as to be carried by the motion of the wheel, but may be set at any time to the proper hour and minute, without affecting either the wheel or its axis. This wheel of 160 teeth turns a pinion, B, of ten leaves; and as 10 is but a sixteenth part of 160, the pinion goes round in a quarter of an hour. On the axis of this pinion is the wheel C of 120 teeth; it also goes round in a quarter of an hour, and turns a pinion D, of eight leaves, round in a minute; for there are 15 minutes in a quarter of an hour, and 8 times 15 is 120. On the axis of this pinion is the second-hand B, (fig. 492, and also the common wheel E, fig. 493, of 30 teeth, for moving a pendulum (by pallets) that vibrates seconds, as in a common clock.

This clock is not designed to be wound up by a winch, but to be drawn up like a clock that goes only thirty hours. For this purpose, the line must go over a pulley on the axis of the great wheel, as in a common thirty-hour clock.

One inconvenience attending this clock is, that if a person wake in the night, and look at the clock, he may possibly be mistaken in the four hours, in reckoning the time by it, as the hand cannot be upon any hour, or pass by any hour, without being upon, or passing by, four hours at the same time. In order, therefore, to avoid this inconvenience, the ingenious Mr. Ferguson contrived the following method.

In fig. 494, the dial-plate of such a clock is represented; in which there is an opening, *a b c d*, below the centre. Through this opening, part of a flat plate appears, on which the 12 hours are engraved, and divided into quarters. This plate is contiguous to the back of the dial-plate, and turns round in 12 hours; so that the true hour or part thereof, appears in the middle of the opening, at the point of an index, A, which is engraved on the face of the dial-plate. B is the minute-hand, as in a common clock, going round through all the 60 minutes on the dial in an hour; and in that time, the plate seen through the opening *a b c d* shifts one hour under the fixt, engraven index A. By these means the hour and minute may be always known at whatever time the dial-plate is viewed. In this plate is another opening, *e f g h*, through which the seconds are seen on a flat

pendulum-ball to describe but small arcs in its vibrations. Some men of science think small arcs are best; but wherefore we know not. For whether the ball describes a large or a small arc, if the arc be nearly cycloidal, the vibrations will be performed in equal times; the time therefore will depend entirely on the length of the pendulum-rod, not on the length of the arc the ball describes. The larger the arc is, the greater the momentum of the ball; and the greater the momentum is, the less will the time of the vibrations be affected by any unequal impulse of the pendulum-wheel upon the pallets.

The greatest objection to Mr. Ferguson's clock is, that the weight of the flat ring on which the seconds are engraved, will load the pivots of the axis of the pendulum-wheel with a great deal of friction, which ought by all possible means to be avoided; and yet one of these clocks, recently made, goes very well, notwithstanding the weight of this ring. This objection, however, can easily be remedied by leaving it out; for seconds are of very little use in common clocks not made for astronomical observations; and table clocks never have them.

Having thus described this clock, we shall next proceed to give a description of a clock, by the same ingenious mechanic, for showing the apparent daily motions of the sun and moon, the age and phases of the moon, with the time of her coming to the meridian, and the times of high and low water, by having only two wheels and a pinion added to the common movement.

Mr. Ferguson's clock for exhibiting the apparent daily motions of the sun and moon, and state of the tides, &c.

The dial-plate of this clock is represented by fig. 496. It contains all the twenty-four hours of the day and night. S is the sun, which serves as an hour index, by going round the dial-plate in twenty-four hours; and M is the moon, which goes round in twenty-four hours fifty minutes and a half, from any point in the hour circle to the same point again, which is equal to the time of the moon's going round in the heavens, from the meridian of any place to the same meridian again. The sun is fixed to a circular plate, as fig. 497, and carried round by the motion of the plate, on which the twenty-four hours are engraven, and within them is a circle divided into twenty-nine and a half equal parts for the days of the moon's age, accounted from the time of any new moon to the next after; and each day stands directly under the time (in the twenty-four hour circle) of the moon's coming to the meridian, the twelve under the sun standing for mid-day, and the opposite twelve for mid-night. Thus, when the moon is eight days old, she comes to the meridian at half an hour past six in the afternoon; and when she is sixteen days old, she comes to the meridian at one o'clock in the morning. The moon M, fig. 496, is fixed to another circular plate, of the same diameter with that which carries the sun; and this moon-plate turns round in twenty-four hours fifty minutes and a half. It is cut open, so as to show some of the hours and days of the moon's age; on the plate below it that carries the sun, and across this opening at *a* and *b* are two short pieces of small wire in the moon-plate. The wire *a* shows the day of the moon's age, and time of her coming to the meridian, on the plate below it that carries the sun; and the wire *b* shows the time of high water for that day, on the same plate. These wires must be placed as far from one another, as the time of the moon's coming to the meridian differs from the time of high-water at the place where the clock is intended to serve. At London-bridge it is high water when the moon is two hours and a half past the meridian.

Above this plate that carries the moon, there is a fixed plate N, supported by a wire A, the upper end of which is fixed to that plate, and the lower end is bent to a right angle, and fixed into the dial-plate at the lowermost or midnight twelve. This plate may represent the earth, and the dot at L, London, or any other place at which the clock is designed to show the times of high and low water.

Around this plate is an elliptical shade upon the plate that carries the moon M : the highest points of this shade are marked High Water and the lowest points Low Water : as this plate turns round below the fixed plate N, the high and low water points come successively even with L, and stand just over it at the times when it is high or low water at the given place; which times are pointed out by the sun S, among the twenty-four hours on the dial-plate : and, in the arch of this plate, above twelve at noon, is a line H, that rises and falls as the tide does at the given place. Thus, when it is high water, (suppose at London,) one of the highest points of the elliptical shade stands just over L, and the tide place H is at its greatest height : and when it is low water at London, one of the lowest points of the elliptical shade stands over L, and the tide place H is quite down, so as to disappear beyond the dial-plate. As the sun S goes round the dial-plate in 24 hours, and the moon M goes round it in 24 hours 50½ minutes, the moon goes so much slower than the sun as only to make 28½ revolutions in the time the sun makes 29½; and therefore the moon's distance from the sun is continually changing; so that at whatever time the sun and moon are together, or in conjunction, in 29½ days afterwards they will be in conjunction again. Consequently the plate that carries the moon moves so much slower than the plate that carries the sun, as always to make the wire *b* shift over one day of the moon's age on the sun's plate in 24 hours.

In the plate that carries the moon, there is a round hole *m*, through which the phase or appearance of the moon is seen on the sun's plate, for every day of the moon's age from change to change. When the sun and moon are in conjunction, the whole space seen through the hole *m* is black : when the moon is opposite to the sun (or full) all that space is white; when the moon is in either of her quarters, the same space is half black and half white; and different in all other positions, so as the white part may resemble the visible or enlightened part of the moon for every day of her age.

To show these various appearances of the moon, there is a black shaded space, fig. 497, as N *f* F *l*, on the plate that carries the sun. When the sun and moon are in conjunction, the whole space seen through the round hole is black, as at N ; when the moon is full, opposite to the sun, all the space seen through the round hole is white, as at F; when the moon is in her first quarter, as at *f*, or in her last quarter, as at *l*, the hole is only half shaded ; and more or less accordingly for each position of the moon, with respect to her age ; as is abundantly plain by the figure.

The wheel-work and tide-work of this clock are represented by fig. 498,

th, and the other ¼th of a semicircle, supposing their teeth and
be respectively equal to one another, but if both wheels are cut in
g-engine by the same cutter, the inequality will fall in the teeth
in either cases, the action of one of the teeth must be bad if the
roperly proportioned, and periodic jerks will be the consequence,
a wheel-work going by a clock or watch movement, ought to be
Whether or not Mr. Ferguson had the dial of the clock at
Court in his eye when he contrived the simple mechanism of this
will not undertake to affirm; but we think it extremely probable
d, particularly as he has copied the position of the annular train
of his clocks. Being in the habit of calculating numbers proper
cating given periods of time in clocks, watches, orreries, &c. we
ed our thoughts towards the improvement of this clock, as well
r pieces of mechanism, so far as relates to accuracy; and beg
ay before the reader the alteration that has occurred to us, for
the clock before us more perfect than it is in the state above described.
describing the Hampton Court clock, we endeavoured to prove
the moon's age is indicated by the difference of the velocities of
ands, moving in the same direction, and representing the sun
, the latter ought to pass the xii o'clock point, on each day 50ᵐ
later than on the preceding day; but by Mr. Ferguson's calcu-
see the daily retrogradation is 50ᵐ 526, and the difference
ints to an entire day's motion in a little more than 952 days; or
upwards of 32 lunations, as we have stated. What therefore
in this case, is a couple of divisible numbers that shall be to
very nearly in the ratio of 24ʰ to 24ʰ 50ᵐ 473, which numbers, by
arithmetical process become familiar to us by practice, we have
d to be 2368 : 2451. These are the nearest possible numbers
e got without ascending higher in the scale of continual ratios, and
s capable of reduction into composite numbers thus: 2368 taken
act is equal to 74 ×32 and 2451 = 57 × 43; therefore the train
ill be the wheel-work required; the solar wheel of 74 teeth being
evolve with a tube as an arbor in 24 hours, by the clock move-
st impel the wheel of 43 placed on a stud, or otherwise on the
of the frame, at one side of it, and this wheel of 43 must have
driver, 32, pinned to it, to impel the last wheel, 57, or lunar
iced on a solid arbor, concentrically behind the solar wheel,
to Mr. Ferguson's position, and the dials and other designs of
face may remain precisely as described; so that instead of the
19 impelling two unequal wheels at once, we shall have a pair of
els pinned together, one impelled by, and the other impelling its
ere the motion must be taken from an arbor of twelve hours,
wheel of 37 to actuate the 74 in twenty-four hours, instead of
of eight hours, as Mr. Ferguson proposed; which mode is equally
c.
roof of the accuracy of our calculation, we have by direct propor-
63 : 2451 : : 24ʰ: 24ʰ 50ᵐ 4729729, &c: hence the deviation from
here only 0000271 of a minute in each lunar day, which will
it to an error of an entire day in less than 1,862,472 such days,
ore; may be assumed as no bad substitute of the truth itself; see-
ick will never be expected to go so long without clearing or stop-
some external cause.
d it occur to the reader that 32 lunations constitute a period long
r the clock of Mr. Ferguson to go, before a new rectification, we
to suggest to him, that in the space of a lunar day there are two

tides and two ebbs, consequently an error of three-quarters of an
lunation will place the tide-plate H, three hours wrong in the sp.
four months, and in nearly eight months an high water will be c
low water, and the reverse in the next eight months, which is
indispensable error.

"That the clock-maker may not be at a loss how to apply the
have proposed for the inaccuracy of Mr. Ferguson's solar and l
we shall conclude our description of the clock before us with a
the exact dimensions of the parts proposed to be substituted.
the wheel of communication of 37 teeth at 12 per inch, mea
pitch line, its geometrical diameter will be 98 or $\frac{11}{12}$ of an i
practical diameter, with the addendum for the ends of the teeth
wheel of 74 being double will have its geometrical diameter eq
and its practical one 2·02; the fellow of this last or solar-w
geometrical diameter by the same proportion, 1.14, and its p
1.20; the distance of the stud from the centre of motion of th
lunar wheels, must necessarily be the sum of the geometrical r
two last wheels, namely, $1\cdot96 + 1.14 \div 2$ which is $=1\cdot55$; ag
of the geometrical radii of the remaining two wheels, 32 and .
also equal to 1·55, in order that the centres of motion of the sol
wheels may exactly coincide; but a wheel of a geometrical diame
$1\cdot55 \times 2$ or 3.10 inches and of $32 + 57$ or 89 teeth, will have
nine teeth per inch, and the practical diameters of wheels 32 an
same, will be respectively 1·21 and 2·1. The calliper suitabl
proportions and dimensions is given, of their full size and dim
fig. 498°, which needs no farther explanation, except that the wl
32 are so nearly of a size that one circle represents both, as pinn
and revolving with a contemporary motion round a stud or s
centre, going into the front plate of the clock-frame. The small
acts deeper into the teeth of its fellow than the 43, by reason of t
teeth than the other, though the wheel is of the same size."

In the year 1803, the Society for the Encourageme
&c. presented to Mr. John Prior, of Nessfield, Yo
reward of thirty guineas on account of his contrivanc
striking part of an eight-day clock. As this in
likely to be useful we shall describe it here. It cor
wheel and fly, with six turns of a spiral line, cut
wheel for the purpose of counting the hours. The p
this spiral elevate the hammer, and those above ar
use of the detent. This single wheel serves the p
count-wheel, pin-wheel, detent-wheel, and the fly-i
has six revolutions in striking the twelve hours. If w
a train of wheels and pinions used in other striking p
made without error, and that the wheels and pini
turn each other without shake or play, then, allowing
supposition to be true, (though every mechanic knows
Mr. Prior's striking part would be found six times s
others, in striking the hours 1, 2, 5, 7, 10, 11; twe
superior in striking 4, 6, 8; and eighteen times in s
9, and 12. In striking 2, the inventor purposely

rfection equal to the space of three teeth of the wheel;
n striking 3, an imperfection of nine or ten teeth; and
oth these hours are struck perfectly correct. The flies in
's turn round at a mean, about sixty times for every knock
c hammer, but this turns round only three times for the
purpose: and suppose the pivots were of equal diameters,
influence of oil on them would be as the number of
ations in each. It would be better for clocks if they
no warning at all, but the snail piece to raise a weight
what similar to the model Mr. P. sent for the inspection
at respectable society.

striking part of this clock is represented in fig. 499.

the large wheel, on the face of which are sunk or cut the six turns of a

the single worm screw, which acts on the above wheel, and moves
'C.

the spiral work of the wheel A. The black spots show the grooves
hich the detents drop on striking the hour.

the groove into which the locking piece F drops when it strikes 1,
on which place it proceeds to the outward parts of the spiral in the
ssive hours, being thrown out by a lifting piece H at each hour; the
detent G being pumped off with the locking piece F, from the pins on
eel A.

triking the hour of 12, the locking piece, having arrived at the outer
at II, rises up an inclined plane, and drops by its own weight into
ner circle, in which the hour 1 is to be struck, and proceeds on in a
ssive motion through the different hours till it comes again to 12.

he hammer-work made in the common way, which is worked by thirteen
n the face of the spiral.

. 500, K, the thirteen pins on the face of the spiral, which work the
er-work.

the outer pins which lock the detent.

the pump spring to the detent.

the fourth century, an artist named James Dondi con-
ted a clock for the city of Padua, which was long con-
d as the wonder of the period. Besides indicating the
, it represented the motions of the sun, moon, and
ts, as well as pointed out the different festivals of
car. On this account Dondi obtained the surname of
logio, which became that of his posterity. A short time
William Zelander constructed for the same city a
still more complex; which was repaired in the sixteenth
ry by Janellin Turrianus, the mechanist of Charles V.
t the clocks of the cathedrals of Strasburgh and of Lyons
uch more celebrated. That of Strasburgh was the work
onrad Dayspodius, a mathematician of that city, who
ed it about 1578. The face of the basement of this
exhibits three dial-plates; one of which is round, and
ists of several concentric circles; the two interior ones of

2 K

which perform their revolutions in a year, and serve to mark
the days of the year, the festivals, and other circumstances of
the calendar.   The two lateral dial-plates are square, and
serve to indicate the eclipses both of the sun and the moon.

Above the middle dial-plate, and in the attic space of the
basement, the days of the week are represented by different
divinities, supposed to preside over the planets from which
their common appellations are derived.   The divinity of the
current day appears in a car rolling over the clouds, and at
midnight retires to give place to the succeeding one.   Before
the basement is seen a globe borne on the wings of a pelican,
around which the sun and moon revolved; and which in that
manner represented the motion of these planets, but this
part of the machine, as well as several others, has been deranged for a long time.   The ornamental turret, above this
basement, exhibits chiefly a large dial in the form of an astro-
labe; which shows the annual motion of the sun and moon
through the ecliptic, the hours of the day, &c.   The phases
of the moon are seen also marked out on a particular dial-
plate above.   This work is remarkable also for a considerable
assemblage of bells and figures, which perform different motions.   Above the dial-plate last mentioned, for example, the
four ages of man are represented by symbolical figures : one
passes every quarter of an hour, and marks the quarter by
striking on small bells ; these figures are followed by Death,
who is expelled by Jesus Christ risen from the grave : who,
however, permits it to sound the hour, in order to warn man
that time is on the wing.   Two small angels perform movements also; one striking a bell with a sceptre, whilst the
other turns an hour-glass at the expiration of an hour.   In
the last place, this work is decorated with various animals,
which emitted sounds similar to their natural voices; but
none of them remain, except the cock, which crows immediately before the hour strikes, first stretching out its neck
and clapping its wings.   Indeed it is to be regretted that a
great part of this curious machine is now entirely deranged.

The clock of the cathedral of Lyons is of less size than that
of Strasburgh, but is not inferior to it in the variety of its
movements ; it has the advantage also of being in a good condition.   It is the work of Lippius de Basle, and was exceedingly well repaired in the last century by an ingenious clock-
maker of Lyons, named Nourisson.   Like that of Strasburgh,
it exhibits, on different dial-plates, the annual and diurnal
progress of the sun and moon, the days of the year, their
length, and the whole calendar, civil as well as ecclesiastic.

The days of the week are indicated by symbols more analogous to the place where the clock is erected; the hours are announced by the crowing of a cock, three times repeated, after it has clapped its wings, and made various other movements. When the cock has done crowing, angels appear, who by striking various bells, perform the air of a hymn; the annunciation of the virgin is represented also by moving figures, and by the descent of a dove from the clouds; and after this mechanical exhibition the hour strikes. On one of the sides of the clock is seen an oval dial-plate, where the hours and minutes are indicated by means of an index, which lengthens or contracts itself, according to the length of the semidiameter of the ellipsis over which it moves.

A very curious clock, the work of Martinot, a celebrated clock-maker of the seventeenth century, was formerly to be seen in the royal apartments at Versailles. Before it struck the hour, two cocks on the corner of a small edifice crowed alternately, clapping their wings; soon after, two lateral doors of the edifice opened, at which appeared two figures bearing cymbals, beat upon by a kind of guards with clubs. When these figures had retired, the centre door was thrown open, and a pedestal, supporting an equestrian statue of Louis XIV., issued from it, while a group of clouds separating, gave a passage to a figure of Fame, which came and hovered over the statue. An air was then performed by bells; after which the two figures reentered, the two guards raised up their clubs, which they had lowered as if out of respect to the presence of the king, and the hour was then struck.

While, however, we have thought it right to describe these ingenious performances of foreign artists, we must not neglect to mention the equally ingenious workmanship of some of our own countrymen. We now refer to two clocks made by English artists, as a present from the East India Company to the Emperor of China. These two clocks are in the form of chariots, in each of which a lady is placed in a fine attitude, leaning her right hand upon a part of the chariot, under which appears a clock of curious workmanship, little larger than a shilling, which strikes and repeats, and goes for eight days. Upon the lady's finger sits a bird, finely modelled and set with diamonds and rubies, with its wings expanded in a flying posture, and which actually flutters for a considerable time by touching a diamond button below it; the body of the bird, in which are contained the wheels that animate it as it were, is less than the 16th part of an inch. The lady holds in her left hand a golden tube, little thicker than a large pin, on the top of which is a small round box, to which is fixed a circular ornament not larger than a sixpence, set with dia- monds, which goes round in or near three hours in constant regular motion. Over the lady's head is a double umbrella, supported by a small fluted pillar the size of a quill, and under the larger of which a bell is fixed, at a consider- able distance from the clock, with which it seems to have no connection, but from which a communication is secretly conveyed to a hammer that regularly strikes the hour, and repeats the same at pleasure, by touching a diamond button fixed to the clock below. At the feet of the lady is a golden dog.

In a work like the present, however we may wish to pursue this interesting subject through its progressive steps of im- provement, and to do justice to the numerous scientific and ingenious men who have from time to time effected those improvements, we are compelled to confine ourselves within certain limits, which preclude us from entering more fully into detail in this article; we therefore refer such of our readers, who wish to pursue the subject, to the catalogue of writings in Dr. Young's Natural Philosophy.

We shall next proceed to give a description of the mecha- nism of an ordinary watch, and to annex thereto a useful set of tables, published originally by Mr. W. Stirt.

WATCHES.

Figure 501 represents the interior works of an ordinary watch with the crown-wheel escapement, as they remain on the pillar-plate when the upper part of the frame, shown by fig. 505, is unpinned and removed; and fig. 502, which is a section of the whole frame and its contents, shows the connection of all the parts, as though the calliper were in one right line. These two figures, by having the same letters of reference, mutually explain each other. The mainspring which actuates all the wheels and pinions, that are called, in one general term, the movement, is contained in the circular box a, seen in the different views in the separate figs. 501, 502, and 508, in the last of which its parts are given in a detached state, viz. the box; the relaxed spring immediately above lying in a spiral form; the arbor with its pin, on which the interior end of the spring is hooked, and the lid through which the pivot of the arbor penetrates; this spring is forced into the box by a tool on purpose when it is strong; and then the exterior end is hooked to a pin in the circular edge of the box, so that if the box is made to turn round while the arbor is held fast, the spring begins to coil at the centre, and is thereby said to be wound up. The same effect would be produced if the box were held fast, and the arbor only were turned; but in the latter case the chain, which requires to be uncoiled from the spring-box as this spring is wound up, would remain unmoved; it is necessary therefore that the box be turned while the arbor is at rest, which is thus effected: one end of the chain is made fast to the side of the spring-box, and the other to the fusee b, after being coiled several times round the circumference of the box ; then as the square end of the spring-box arbor is held by the small ratchet and click c, seen on the reversed face of the pillar-plate in fig. 507, so that it cannot revolve, it is obvious that inserting a key on the square of the fusee arbor, and turning it in a proper direction, will wind the chain upon the spiral groove of the fusee, while it is unwound from the box ; and during this operation, the spring will be coiled up to the centre of the box, or put into its state of greatest tension for pulling the fusee back again. ·The rapid motion which the fusee would have in a retrograde direction when pulled by the whole force of the coiled spring, is prevented by the train of wheel-work and balance, thus : the great wheel d is not fast to the thick end of the fusee, as appears in the drawings, but carries a click and click-spring s, as seen in fig. 503, while the ratchet-wheel, seen in fig. 504, is made fast to the fusee ; the consequence of which contrivance is, that while a key applied to the fusee arbor winds up the watch and fills the fusee groove with the chain, until the guard driven by it catches the beak at the small end of the fusee, the click, in fig. 503, slides over the teeth of the ratchet in fig. 504, without acting on them, and thus leaves the great wheel d at rest, in connection with the pinion e on the centre or minute wheel arbor ; but when the spring acts on the fusee in a contrary direction, the click attached to the great wheel is laid hold of by the teeth of the ratchet, which thus makes it fast to the end of the fusee ; or in other words, until the spring wants winding up again, which usually happens once in 28 or 30 hours ; but it is commonly wound up once in 24 hours more or less. The action of the great wheel d, on the pinion e, is that of a long lever driving a short one; or this wheel may be said to act under a mechanical disadvantage, when an increase of velocity, but a loss of power, is experienced by the pinion ; again, on the same central arbor of this pinion e is rivetted the centre wheel f, which revolves in an exact hour, as we shall see presently, and this wheel drives the pinion g, on the arbor of the third wheel h, also with a mechanical disadvantage, for the force it imparts to the pinion i, on the arbor of the contrate wheel, is again diminished in the ratio of the diameter of the wheel to that of its pinion ; thus the force of the mainspring is continually diminishing, as it is transmitted through the train, and when the contrate wheel comes to be actuated, it has just force enough to drive the horizontal pinion on the balance wheel l, so that the alternate impulse given by its teeth to the pallets of the balance verge are just sufficient to perpetuate the oscillation from right to left, under all the obstacles of friction, dirt, wear, and the air's resistance. It is a curious fact that the crown-wheel escapement, though the oldest that we know of, is still the most in use in common watches, probably from the facility with which it is constructed ; for certainly it is more under the influence of the irregularities of the mainspring's force than any other escapement. The properties and action of this escapement have been minutely explained in page 516 of the article Escapement, with reference to fig. 523, to which explanation and figure we request our reader's attention.

In order that the force applied to pallets of the verge at each oscillation may not sensibly vary, it was found necessary to equalize, as much as possible, the variable forces of the mainspring in its different states of tension ; and the most practical way of doing this has been found, to convert the cylinder on the arbor of the great wheel, which would have been proper for a gravitating body, used as a maintaining power, into a figure of a parabolic form, that is, into a solid, generated by the revolution of a parabola, in order that, as the force of the spring becomes greater by increased tension, its action on the great wheel might be lessened in a similar proportion, by a gradual decrease of the radius of the fusee, round which the chain is wound, to impart the force thus modified. Every separate spring, therefore, has not only its average force proportioned to the balance it is destined to actuate, when diminished by transmission through a given train, but requires its scale of varying forces to be nicely counteracted in every degree of tension by the shape of the fusee ; and this is done by means of a tool, called a fusee adjusting tool, which is nothing more than a lever with a sliding weight, attached to the square end of the fusee arbor, as represented in fig. 509; for when the weight on the lever is an exact counterpoise to the force of the mainspring in every part of the successive revolutions of the fusee, as the spring is wound up by the lever instead of a key, then the shape of the fusee is proper, but not otherwise. Hence, whenever a new mainspring is put to a watch, the fusee ought to be adjusted in the fusee engine according as the adjusting tool determines.

The comparative forces of the spring at the extreme ends of the fusee may be adjusted by the small ratchet c, on the back of the pillar-plate in fig. 507, but when the spring is put to a suitable degree of tension to act well at both extremities of the fusee, it must not be altered by the ratches' click, but the intermediate forces must be equalized by a due shape given to the fusee. We have insisted the more on this part of the mechanism being attended to, because, as the *primum mobile*, it is the basis of all other motions. The number of rounds that the spiral of the parabolic fusee may be cut into depends on the length of the pillars of the frame, or, which is the same thing, the shallowness of the watch. The French frequently leave out the fusee, and attempt to equalize the forces of the mainspring by tapering it ; and with detached escapements, this mode may sometimes answer tolerably, but with the crown-wheel escapement a fusee is indispensable. Again, the number of teeth in the great wheel, and in the centre pinion, depends on the number of rounds in the spiral of the fusee.

In a thirty-hours' watch, with six turns of the fusee, the great wheel must have Ψ or five times as many teeth as the centre pinion; so that if this has six leaves, the wheel must have $5 \times 6 = 30$ teeth ; but if eight, then $5 \times 8 = 40$; if the spiral has seven turns, the great wheel 48, and the pinion 12, then the time of going will be $\frac{4}{5} \times 7 = 28$ hours; also, if there be $5\frac{1}{4}$ turns on the fusee, 50 teeth in the wheel, and 10 leaves in the pinion, the period of going will be $27\frac{1}{2}$ hours, or $\frac{4}{5} \times 5\frac{1}{4} = 5 \times 5\frac{1}{4} = 27\frac{1}{2}$; but if 24 hours only were required as the period, with six turns and a pinion of 12, the great wheel would be required to have 48.

Thus when an alteration is made in either the pinion, the wheel, or the turns in the fusee, a corresponding variation may be made in the others, to produce the same period of going, but still the centre wheel revolves once in an hour. In the commonest watches the pinions have only six leaves each, which do not act so well as pinions of higher numbers; but in the best watches, and in all chronometers, the leaves and teeth are more numerous. The pivot-holes, particularly of the verge and escapement arbor, have jewels for the purpose of diminishing the friction, in the best watches; but detached and remontoire escapements are the best correctives of the unequal impulses given through the medium of the train in the different states of its foulness.

The potence m, and small or counter potence n, that hold the pivots of the balance-wheel, are small cocks seen in fig. 502, both in their attached and detached states, and are screwed to the top or upper plate within the frame; but the springs, buttons, and joints of the case, are not exhibited, as forming no part of the movement. Fig. 505 represents the outer face of the upper plate, with the balance p, the cock o, and balance-spring s, called the pendulum-spring, from its having the properties of the pendulum; by means of this spring, not only is the regulation made steady, but the adjustment for time is effected. In every balance-spring there is a certain length, to be taken as the effective length, by which the going of the watch to which it is applied is limited to exact performance; and when this length is determined by experiment, a pin is put in the stud that holds the exterior ends, as at 4, in fig. 505, to prevent its being altered; but as the variation of temperature will alter the momentum of the moving-balance, the effect thereby produced is a loss of time in the rate, in hot weather, and a gain in cold weather, by an alternate increase and decrease in the dimensions of the balance itself, as well as by some alteration in the spring. To remedy this defect, in an ordinary watch, the contrivance shown in fig. 506 is introduced; the wheel t is placed under the graduated circle r, seen in fig. 505, and a circular rack u, fig. 506, that holds the curb or slit-piece 5, seen in both figures, is moved by a sliding motion given to it, when a key is applied to the squared arbor of the figure circle, and thus the effective length of the spiral spring is limited by the position of the curb 5; and according as the key is turned forwards or backwards, towards the words 'fast' or 'slow' engraved on the cock, the shortened or lengthened spring alters the rate of going, till the proper length is found that suits the season in question.

In Harrison's time-piece the curb was moved by an expansion-lever of two metals, that acted by means of the change of temperature; but in the best chronometers of more recent times, the compensating levers constitute the three portions into which the rim of the balance is divided, and the adjustment for time, as well as compensation for temperature, are by means of heavy screws, which form a part of the moving balance. In these more perfect machines, the length of the spring, which is now made helical or cylindrical, is first determined such, that the long and short vibrations are performed in the same time, and this is called the isochronal lengths, which is not afterwards altered by subsequent adjustments.

The last portion of the watch which demands our explanation is the dial-work, for producing the hours and minutes; this will be easily understood by reference to figs. 502 and 507. When the pinion called the cannon-pinion, seen near the minute-hand in fig 502, is inserted on the arbor of the hour or centre wheel, to which it fits rather tight by friction, it revolves therewith in an hour, and receives the minute or hour hand on its protruding squared end; then this pinion drives the wheel x round a stud on the pillar-plate, and with it a pinion w made fast to its centre; which pinion again drives a second wheel, v, round the tube of the cannon-pinion in twelve hours; and to this the hour-hand is attached. This diminution of twelve revolutions from the cannon-pinion to the hour-wheel might be effected by one pinion driving a single wheel of twelve times its number of teeth; but as the motion must be brought back to the centre of the dial again, two more wheels, or a wheel and pinion, are necessary to be introduced, and these are therefore made a part of the train, and no large wheel or small pinion is wanted, for the ratio 12; 1 may be more conveniently obtained by two factors, viz. 4 : 1 and 3 : 1; thus, suppose the cannon-pinion to have 15 leaves, its wheel may have $4 \times 15 = 60$ teeth for wheel x, and if wheel v be the same, its pinion will be $\frac{60}{3} = 20$, and the train $\frac{60}{15} \times \frac{60}{20} = \frac{360}{30} = \frac{72}{6}$ or $\frac{60}{5} = \frac{12}{1}$ or 12; so that when the pinions are fixed upon for the dial-work, the wheels are readily determined, and vice versâ.

The following Tables, somewhat differently arranged, were published by W. Stirt, an ingenious balance-wheel and fusee cutter.

A TABLE OF TRAINS FOR WATCHES;

Showing the Number of Turns on the Fusee and Teeth in the Balance-wheel, with the Beats in an Hour, and the number of Seconds in which the Contrate or Fourth Wheel revolves; for the easy Timing of Watches by the Vibrations of the Pendulum.

9 Teeth in the Balance-wheel

Second wheel 58 6 Third wheel pin. 60 8	60 6	60 6	60 6	60 6	64 6	64
Third wheel 56 6 Contrate pin... 56 7	58 6	58 6	60 6	60 6	60 6	60
Contrate wheel 54 6 Balance pin. .. 80 6	52 6	56 6	51 6	60 6	54 6	80
Beats 14,616 in an hour........ 14,400	15,080	16,210	16,200	18,000	17,280	14,4
Seconds 39⁹⁄₁₀, in which the 4th wheel revolves ....} 60	37½	37¼	36	36	33¼	6(

A Table of Trains for Watches *continued.*

11 *Teeth in the Balance-wheel.*

Second wheel 48 6 Third wheel pinion....	54 6	54 6	56 7	56 6	56 6
Third wheel 45 6 Contrate pinion	45 6	50 6	45 6	54 6	56 6
Contrate wheel 70 6 Balance pinion	65 6	60 6	78 6	54 6	55 6
Beats 15,400 in an hour................	16,087	16,500	17,160	16,632	17,567
Seconds 60, in which the 4th wheel revolves	53⅓	48	60	42⅔	41¼

58 6	58 6	58 6	58 6	58 7	60 6	60 6	60 6	60 6	60 6	60 6
52 6	54 6	54 6	56 6	56 6	50 6	52 6	54 6	54 6	54 6	55 6
52 6	52 6	54 6	54 6	56 6	52 6	52 6	50 6	52 6	54 6	52 6
15,973	16,588	17,226	17,817	15,879	15,888	16,520	16,500	17,160	17,820	17,477
42⅔	41⅓	41⅓	39½	54⅓	43	41½	40	40	40	39
60 6	60 7	60 8	60 8	60 6	60 8	60 7	62 6	62 7	63 6	63 6
56 6	56 6	56 7	56 7	60 6	60 6	60 7	54 6	58 6	54 6	56 –
50 6	56 6	74 6	78 6	48 6	56 6	60 6	52 6	52 6	50 6	56 6
17,111	16,426	16,280	17,160	17,553	15,400	16,163	17,935	16,324	17,325	17,248
38⅓	40	60	60	36	48	49	38¼	45	38	42¼
64 6	64 6	65 7	70 8	70 7	72 8	72 7	80 8	75 10	72 9	72 9
58 6	52 6	62 7	54 7	63 7	63 7	64 7	72 8	72 9	66 8	60 8
50 6	52 6	59 7	68 6	58 7	54 6	58 7	68 8	66 8	60 6	54 6
16,296	17,625	15,250	16,830	16,408	16,035	17,142	16,830	13,200	13,200	11,880
40⅓	39	43¼	53⅓	40	44⅓	38¼	40	60	66	60

13 *Teeth in the Balance-wheel.*

Second wheel 48 6 Third wheel pinions ..	48 6	52 6	54 6	54 6	54 6
Third wheel 45 6 Contrate pinion	45 6	52 6	50 6	52 6	52 6
Contrate wheel 66 6 Balance pinion	68 6	52 6	50 6	48 6	50 6
Beats 17,160 in an hour................	17,680	16,925	16,274	16,224	16,900
Seconds 60, in which the 4th wheel revolves	60	46⅓	48	46	46

54 6	54 6	55 6	56 7	56 6	56 6	56 6	56 6	56 6	58 6	58 6
52 6	52 6	51 6	45 6	50 6	50 6	52 6	52 6	54 6	48 6	50 6
51 6	52 6	51 6	66 6	50 6	51 6	48 6	50 6	49 6	52 6	50 6
17,238	17,576	17,219	17,160	16,851	17,188	16,824	17,525	17,836	17,425	17,453
46	46	46¼	60	46½	46⅔	44⅓	44½	42¼	46¼	44⅔
60 6	60 8	60 6	60 6	60 7	60 6	60 7	60 8	60 7	60 8	60 6
48 6	48 6	50 6	50 6	54 6	54 8	56 7	56 7	58 7	58 6	60 7
48 6	66 6	46 6	48 6	52 6	60 6	56 6	66 6	56 6	56 6	48 6
16,640	17,160	16,611	17,333	17,382	17,550	16,640	17,160	17,234	17,593	17,828
45	60	43	43	46⅔	54	52½	60	50⅔	49½	42
60 8	60 6	62 7	63 7	63 7	64 7	64 7	64 8	64 8	65 7	70 8
60 6	60 7	56 7	52 6	60 7	52 6	60 7	60 8	64 8	62 7	60 7
54 6	56 7	56 6	51 6	60 7	50 6	60 7	66 6	72 7	58 7	52 6
17,550	17,828	17,194	17,238	17,191	17,168	17,464	17,160	17,115	17,717	16,900
48	42	50¼	46¼	46⅔	46	45¼	60	56¼	43½	48
70 8	72 8	72 8	74 8	74 8	75 10	75 10	80 10	96 12	96 12	90 10
66 8	52 6	70 8	64 8	68 8	72 9	72 9	60 8	75 10	75 10	90 10
54 7	52 6	68 8	63 7	68 8	70 7	72 9	60 8	80 8	88 8	90 10
17,160	16,673	17,403	17,316	17,400	15,600	12,480	15,600	15,600	17,160	18,954
50	44½	52⅕	48⅔	60	60	60	60	60	60	44⅓

A Table of Trains for Watches *continued.*

15 *Teeth in the Balance-wheel.*

Second wheel 48 6	Third wheel pinion....	48 6	48 6	54 6	54 6	54 6
Third wheel 45 6	Contrate pinion	45 6	45 6	48 6	48 6	48 6
Contrate wheel 54 6	Balance pinion	58 6	60 6	46 6	48 6	64 8
Beats 16,200 in an hour	....	17,400	18,000	16,560	17,280	17,280
Seconds 60, in which the 4th wheel revolves		60	60	50	50	50

54 6	56 7	56 7	56 7	56 6	56 7	58 6	58 6	60 8	60 8	60 8
50 6	45 6	45 6	45 6	48 6	60 8	48 6	50 8	48 6	48 6	56 7
48 6	56 6	58 6	60 6	46 6	60 6	46 6	58 6	58 6	60 6	48 6
18,000	16,800	17,400	18,000	17,173	18,000	17,786	17,520	17,400	18,000	14,400
48	60	60	60	48	60	46¼	59½	60	60	60
60 8	60 7	60 8	60 8	60 8	60 6	60 6	60 6	60 6	60 10	60 6
56 7	56 7	56 7	56 7	56 7	60 8	60 10	60 8	60 10	60 6	60 10
56 7	58 7	58 6	60 6	60 7	48 6	48 6	56 7	58 6	60 6	64 8
14,400	17,044	17,400	18,000	15,386	18,000	14,400	18,000	17,400	18,000	14,400
60	52½	60	60	60	48	60	48	60	60	60
60 8	60 8	62 8	63 7	63 7	64 8	64 8	64 8	64 6	65 7	70 6
64 8	64 8	60 8	54 7	56 7	45 6	60 8	60 8	60 10	56 7	60 10
66 7	70 7	60 6	50 6	56 7	56 6	58 6	60 6	70 8	56 7	48 6
16,971	18,000	17,437	17,356	17,280	16,800	17,400	18,000	16,800	17,828	16,800
60	60	61⅜	51¼	50	60	60	60	56¼	48½	51⅓
70 7	70 8	70 8	70 10	72 6	72 8	72 8	72 8	72 8	75 8	81 9
60 10	64 8	64 8	65 8	60 10	64 8	64 8	64 8	65 8	64 8	72 9
70 7	50 6	58 7	60 6	48 6	50 6	54 7	64 8	64 8	64 8	72 9
18,000	17,500	17,400	17,062	17,280	18,000	16,662	17,280	17,550	18,000	17,280
60	51⅓	51⅓	56¼	50	50	50	50	49	48	50

17 *Teeth in the Balance-wheel.*

Second wheel 48 6	Third wheel pinion....	56 7	60 8	64 8
Third wheel 45 6	Contrate pinion	45 6	56 7	60 8
Contrate wheel 50 6	Balance pinion	53 6	52 6	60 7
Beats 17,000 in an hour	....	18,020	17,828	17,485
Seconds 60, in which the 4th wheel revolves		60	60	60

G.W.	S.W.P.	T.N.S.	G.W.	S.W.P.	T.N.S.	G.W.	S.W.P.	T.N.S.
48	10	6¼	60	10	5	55	12	6 6/11
50	10	6	62	10	4⅚	56	12	6 3/7
52	10	5¾	64	10	4⅝	58	12	6⅕
54	10	5⅝	48	12	7½	60	12	6
55	10	5 5/11	50	12	7⅕	62	12	5⅘
56	10	5⅜	52	12	6 12/13	64	12	5⅝
58	10	5⅙	54	12	6⅔			

If we divide double the product of all the four wheels by the product of all the three pinions, the quotient will be the number of beats, as given in any of the trains contained in this table; also, if we take the second and third wheels, and their pinions respectively, as a compound fraction of an hour, they will give the seconds in which the contrate-wheel, attached to the latter pinion, will revolve; thus, of $\frac{6}{60}$ of $\frac{7}{56}$ of $60^m = 1^m$ or 60^s, which numbers are consequently proper for a watch that indicates the seconds; and if the beats be 18,000, or 14,400, there will be five or four beats respectively in a second, which are the best trains for measuring fractional parts of a second.

CHRONOMETERS.

CHRONOMETERS differ from an ordinary watch principally in the escapement and balance. These machines deserve more than usual attention, as well from their practical utility in navigation, as from the principles on which they are constructed, in which the irregular forces both of impulse and resistance are greatly diminished by the exactness of form and dimension.

In the reign of queen Anne, the British parliament passed an act, offering a reward of 10,000*l.* for any method of determining the longitude within the accuracy of one degree of a great circle; of 15,000*l.* within the limit of forty geographical miles; and of 20,000*l.* within the limit of thirty such miles, or half of a degree; provided such method should extend more than eighty miles from the coast. The hope of obtaining this reward stimulated a watch-maker named Harrison to be indefatigable in his endeavours to effect the required improvement, which eventually led him to apply the principle of the apposite expansions of different metals to a watch to effect a self-regulating curb, for limiting the effective length of the spiral pendulum-spring to correspond to the successive changes of heat and cold, which changes were now known to alter the force of this spring, and the momentum of the balance.

After Harrison had by his industry and perseverance obtained the large reward, the act was repealed, and another substituted, offering separate rewards to any person who should invent a practicable method of determining, within circumscribed limits, the longitude of a ship at sea; for a time-keeper, the reward held forth to the public is 5,000*l.* for determining the longitude to or within one degree; 7,500*l.* for determining the same to forty geographical miles; and 10,000*l.* for a determination at or within half a degree. This act, notwithstanding its abridged limits and diminished reward, has produced several candidates; of whom Mudge, the two Arnolds, and Earnshaw, have had their labours crowned with partial success.

Although, in respect to Mudge's time-keeper, great expectations were at first raised, it has, from the complexity of the machinery, and consequent expense attendant upon making it, gradually fallen into disrepute, and is now seldom or ever made. Such of our readers who wish to see its

manner of construction and performance, we must refer to
" The Description of Mr. Mudge's Time-keeper," published
in 1799, by Thomas Mudge, jun.

The chronometer we purpose to lay before our readers is
that constructed by Mr. Earnshaw, as we are strongly disposed to conclude, from various documents we have seen,
and from the similarity so evident in the construction of the
escapement, that Mr. Arnold derived the knowledge of his
principle from Mr. Earnshaw.

In Mr. Earnshaw's chronometer the escapement is detached,
which is the best for the equal measurement of time, because
the vibrations of the balance are free from the friction of the
wheels, excepting about one-twelfth part of the circle, while
the scape-wheel is acting on the pallet to keep up the motion
of the balance, which is done with considerably more power
and less friction than by any other escapement, as it receives
but one blow from the wheel, whilst other escapements receive
two; it has also an equal advantage of the same quickness of
train, and when the impulse is given to the balance by the
wheel, it is given in a similar direction, and not in opposition,
as most escapements are which produce a recoil.

The pivots of the balance-axis should be the
size of the verge-pivots of a good sized pocket-
watch, and of the annexed shape, which will greatly
add to their strength, the extreme end, or acting
part, only being straight; the jewel-hole should be
as shallow as possible, so as not to endanger cutting
the pivot, and the part of the action of the hole made quite
back, with only a very shallow chamber behind to retain the
oil; deep holes are very bad, for when the oil becomes
glutinous, it will make the pivots stick, so as to prevent the
balance from its usual vibration. The pallet should be half
the diameter of the wheel, or a little larger, for if smaller, or
one-fourth the diameter, as is the case in Arnold's, the
wheel will have too much action on it, which will increase
friction most considerably, and likewise cause the balance to
swing so much farther to clear the wheel; consequently,
a check in the motion of the balance may stop the watch,
and cause time-keepers so constructed to stop. The face of
the pallet should run in a line of equal distance between the
centre of the pallet and its extremity, and not in a right line
to its centre, as this causes an increase of friction, and a loss
of that power which is obtained by the wheel, acting on the
extremity of the pallet. The scape-wheel teeth should form
the same direction as the face of the pallet, under-cut for

the purpose of avoiding friction, and maintaining the power, and for safe unlocking. The points of the wheel-teeth must not be rounded off, but left as sharp as possible. The pivots of the scape-wheel are to be a very little larger than the balance-pivots.

The wheel is locked by a spring, instead of a detent with pivots, as the French have made them; for those pivots must have oil, and when the oil thickens, the spring of the pivot-detents become so affected by it, as to prevent the detent from falling into the wheel quick enough, which causes irregular time, and ultimately a stoppage of the watch.

When the spring is planted on the side of the wheel, the part on which the wheel rests should be a little short of a right angle, so that the wheel may have a tendency to draw the spring into it; for if sloped the other way, or beyond a right angle, it will have a tendency to push the spring out, in which case the wheel will have liberty to run. The wheel should take no more hold on the spring than just sufficient to stop it, otherwise the friction will be increased. The small return-spring should be as thin as possible at the end fastened to the other spring, but at the outer end a little thicker; the spring should be planted down as close to the wheel as to be just free of it: the discharging pallet about one-third, or near one-half the size of the large or main pallet, the face of it in a right line to the centre, the back of it a little rounding off from the centre. Great care must be used, in taking off the edges of this discharging piece to make it round, to prevent cutting the spring, nor can it be made too thin, provided it does not cut; the end of it nearest the balance should be a little more out from the centre of the balance-axis than the lower part of it towards the potence, for counteracting the natural tendency of the spring downwards from the pressure of the scape-wheel; and that part of the spring on which the wheel rests should be sloped a little down, to give the wheel a tendency to force it up, to counteract the natural inclination which the wheel has to draw it down by its pressure on it.

The balance is to be made of the best steel, and turned from its own centre to the proper size, and then put into a crucible with as much of the best brass as will, when melted, cover it. The brass will adhere to the steel, and when set, is to be turned to its proper thickness, and hollowed out, so as to leave the steel rim about the thickness of a repeating-spring to a small sized repeating-watch. The brass is to be turned to near twice or three times the thickness of the steel; cross it out with only one arm straight across the centre, and at each end of the arm fix two screws, opposite to each other, through the rim of the balance, to regulate the watch to time. The diameter of the heads of these screws must be about equal to the thickness of the balance, a little more or less is not material. The compensation-weights should be made of the best brass, and well hammered, and a groove turned to let the rim of the balance into it; this should be cut into fourteen equal parts, which will leave seven pair of pieces of equal size and weight, one of which pair, being screwed on the rim of the balance at equal distances, will produce an equilibrium. In making balances, great care must be taken that they get no bruises or bendings; for if a bruise be made on one side so as to indent the metal, that part will be less affected by the atmospheric agency of heat and cold than those parts whose pores have not been closed by the same violence.

Balances are likewise spoiled by bending the compensation-pieces, as bending cracks and destroys the compact body of the metal. The soldering up those cracks with a metal very different in expansion to the metal cracked is hurtful, as it is not then possible to bend the compensation-pieces into a true circle, in which case they form so many parts of different circles, that nothing regular can be produced.

To adjust the balance in heat and cold, put the watch into about 85 or 90 degrees of heat by the common thermometer, mark down exactly how much it gains or loses in twelve hours, then put it into as severe a cold as you can get for twelve hours; and if it gain one minute more in twelve hours in cold than in heat, move the compensation-weights farther from the arm of the balance about one-eighth of an inch; and if it gain one minute more in twelve hours in heat than in cold, move the weights one-eighth of an inch nearer to the arm of the balance, and so on in like proportion, trying it again and again, till you find the watch go the same in whatever change of heat or cold you put it in.

Mr. Earnshaw has found out a method of obviating the difficulties attendant in making time-keepers go nearly the same in whatever position they might be put. It merely consists in having the balance-spring well and properly made; but if the spring be made as hereafter described, it only requires that the balance should be of equal weight, and it will go, within a few seconds per day, in all positions alike; and if it vibrate not more than $1\frac{1}{4}$ circle, will, by applying a small weight to that part of the balance which is downwards when in the position that it loses most, correct it with great accuracy. If it vibrate more than 1¼ circle, it will require the weight to be above, instead of below; and after the watch has been going a few months, and its vibrations shorten to 1¼ circle, it will go worse and worse by reason of the weight being in the wrong place; therefore, to avoid this evil, it is absolutely necessary to confine the vibrations to 1¼ circle, which will produce the most steady performance.

The greatest difficulty with which Mr. Earnshaw had to contend in the construction of his chronometers was, to find out the invisible properties of that apparent simple part of the machine, called the balance-spring. He found, in reasoning on bodies, that watch-springs, when kept constantly in motion, relax and tire like the human frame. In proof of this, let a watch, that has been going a few months, go down; let it remain down for a week or two, and then set it going, when it will, if it be a good time-keeper, and not affected by the weather, go some few seconds per day faster than it did when it was let down; but it will again lose its quickness in a gradual manner, gaining less and less till it comes to its former rate. Finding, therefore, that isochronal springs would not do, and having made springs of such shape as would render long and short vibrations equal in time, and which constantly lost the longer the watch went, Mr. Earnshaw made them of such shape as to gain in the short vibrations about five or six seconds per day more than the long ones, which quantity could only be found by long experience; and the way he adopted to prove this, was to try the rate of the watch with the balance vibrating about one-third of a circle, then tried its rate vibrating 1¼ circle; and if the short vibrations went slower than the long ones, he found that the watch would lose in its rate; and if equal, it would likewise lose, but that only from relaxation; he found also, if it gain in the short vibrations more than five or six seconds in twenty-four hours, it will in the long run gain on its rate; but if not more than that quantity, and the time-keeper is perfect in heat and cold in every other part, the above properties will render it deserving the name of a perfect time-keeper. Mr. Earnshaw found the common relaxation of balance-springs to be about five or six seconds per day on their rates in the course of a year; therefore, if the short vibrations are made by the shape of the spring to go about that quantity faster than the long ones, and as the spring relaxes in going by time, so the watch accumulates in dirt and thickening of the oil, which shortens the vibrations, the

short ones then being quicker, compensated for the evil of
relaxation of the balance-spring.

Having thus given our readers Mr. Earnshaw's prefatory
observations to the Board of Longitude, we shall, in the next
place, proceed to give a general description of the different
parts of his chronometer.

Fig. 510 represents the time-keeper put together.

Fig. 511, the pillar-plate from which the calliper may be taken; *a*, the
height of the pillars.

Fig. 512, the barrel and main-spring; *b*, side view of the barrel.

Fig. 513, the fusee and great wheel, with ratchet to keep it going whilst
winding up; *c*, side view of fusee.

Fig. 514, second wheel and pinion; *d*, side view of second wheel.

Fig. 515, third wheel and pinion; *e*, side view of it.

Fig. 516, fourth wheel and pinion; *f*, side view of it.

Fig. 517 represents the upper plate, with the escapement on it, from
which the calliper may be taken. In this figure the draftsman has not
placed the pallet near enough the wheel; but this is of no consequence, as
a proper and exact draft of the escapement on a much larger scale is given
in fig. 522; the escapement, therefore, is to be understood from that figure;
this only shows the sizes of the wheels.

Fig. 518 represents a side view of the scape-spring which locks the wheel.

Fig. 519, one of the brass weights to be fixed on the rim of the balance
for the compensation for heat and cold; *g*, the groove cut in it to receive
the rim of the balance. The rim of the balance is cut through in two places
in opposite directions, as in fig. 510, and two of these weights are to be
placed on the balance-rim, at equal distances, as there represented, and
fastened by the screw as at *h*. These weights are to be moved backwards
or forwards on the rim of the balance, to make the watch go faster or
slower in heat or in cold, as by trial may be found necessary.

Fig. 520 is a side view of said brass weights; *k*, the groove to receive
the rim of the balance; its depth shows the breadth for balance-ring.

Fig. 521, the cylindrical balance-spring. The only advantage attending
the cylindrical shape is, that it is rather easier made, being a saving of about
one hour of time; for if the real body or form of the spring be like the shape
of the stem of a feather, or common writing quill, it is of no consequence
whether it be turned into a spiral or cylindrical figure.

The model, from which the four following figures were taken, contains,
besides the parts necessary to explain the nature of the escapement, a box
enclosing a spring, which, when wound up, communicates, by means of
some more wheels, a force to the balance-wheel sufficient, when the balance
is put in motion, to keep it in action for some time. These wheels are
contained between two brass plates, fastened together by four upright pillars. The uppermost of these plates is that which is represented in fig. 522,
where P Q R S are the four screws that take into the heads of the four pillars
above-mentioned, and connect it to the remaining part of the model. The
plate P Q R S contains, however, the whole of the parts necessary for the
present purpose. The side of this plate represented to view, is the undermost when fixed in the model; so that the figure represents this plate as
taken off, with the side next to the balance laid upon a table, and the eye is
supposed to be placed perpendicularly over it.

In the plate P Q R S is an opening, or a piece taken out, represented by

T U W X Y Z. In this opening the balance-wheel A B C D, pallet M S K, and part of the balance U V, are seen. The balance-wheel is supported by two pieces of brass, O N H, O I; the piece O N H is screwed to the side of the plate nearest to view by a strong screw, V, and made firm by small pins, represented by и и и и и; these pins are called steady-pins; they are rivetted fast into the supporting-piece O H, and take into holes in the plate P Q R S, made exactly to fit them. The part O N of this supporting-piece is supposed to be raised above the part O H by a joint or bend at N: the other supporting-piece O I is fastened to the opposite side of the plate; and between these two pieces the balance-wheel turns freely and steadily in the direction of the letters A BC D. The small wheel M S K is called the large pallet; it is a cylindrical piece of steel, having a notch or piece cut out of it at l h i; against the side of this notch is a square, flat piece of ruby, or any hard stone, h l, ground and polished very smooth, and fixed into the pallet. The cylinder is so placed, with respect to the balance-wheel, that it may not be more than just clear of two adjoining teeth. E F is a long, thin spring, which is made fast at one end, by being pinned into a stud G, and made to bear gently against the head of an adjusting screw, m; the other end is bent a little in the form of a hook; to this spring there is fixed another very slender spring at γ, which projects to a small distance beyond it. This small spring lies on the side of the thick spring nearest to the balance-wheel. The adjusting screw m takes into a small brass cock at e p, which is screwed fast to the upper plate by a strong screw. Upon the spring E F there is fixed a semi-cylindrical pin, which stands up perpendicularly upon it, and of a sufficient length to fall between the teeth of the balance-wheel A B C D. This pin is called the locking-pallet, and is placed on the opposite side of the spring represented to view. Through the centre of the cylindrical pallet M S K, a strong steel axis passes, called the verge; the pallet is made fast to this axis, which also passes through the centre of the balance, and is made fast to it; it has two fine pivots at its extremities, upon which it turns very freely, between two firm supporting pieces of brass, screwed firmly, and made as permanent as possible, by steady-pins, to the principal plate. A little above the cylindrical pallet M S K, is fixed a small cylindrical piece of steel, i n, having a small part projecting out at i, through which the verge also passes; this is called the lifting-pallet, and is from one-third to half the diameter of the large pallet; it fixes upon the verge like a collar, and is made fast by a twist, so as to be set in any position with respect to the large pallet M S K. The end E G of the long spring E F being made very slender, if a small force be applied at the point o to press that end out from the wheel A B C D, it yields easily in that direction, turning, as it were, upon a centre at G; it is also made to slide in a groove made in this stud, in such a manner that the end o may be placed at any required distance from the centre of the verge.

Having described the several parts as they appear in the figure, we next come to their situation or connection with respect to each other. Let the long spring E F be supposed to be so placed, that the end of the slender spring γ i may project a little way over the point of the lifting-pallet i n, but not so close but that the point of the pallet may pass by the hooked end of the spring E F without touching it; the head of the adjusting-screw m is also supposed to bear gently on the inner side of the said spring E F, or that nearest to the wheel, and at the same time the locking-pallet is so placed, that one of the teeth, D, of the balance-wheel, may just take hold of it. This pallet is not visible in its proper place in the figure, being covered from sight by the screw m, and part of the spring E F; its position is therefore represented by the dot h, on the opposite side of the wheel,

2 L

off from the pallet, until the force of the pendulum-spring, (which is not represented in the figure,) being continually increased by being wound up, overcomes the momentum of the balance, which for an instant of time is then stationary, but immediately returns by the action of the pendulum-spring, which exerts a considerable force upon it in unwinding itself. As the balance returns, the point i, of the lifting-pallet i n, passes by the ends of two springs, E F and Y O, and, in passing by, pushes the projecting end o of the slender spring in towards the balance-wheel, until it has passed it; after this, the projecting end o again returns, and applies itself close to the hooked end of the spring E F as before. The spring y o is made so slender, that it gives but little resistance to the balance, during the time the point i of the lifting-pallet is passing it, and of course causes but little, if any, decrease in its momentum. During the time the point i of the lifting-pallet is pressing in the small spring y o, the long spring E F remains steadily bearing against the head of the adjusting screw m, as the hooked end at o just lets the end of the lifting-pallet pass by without touching it. As the spring has now been continually acting upon the balance, from the extremity of its vibration in the direction M S K, it has given it the greatest velocity when the point i of the lifting-pallet is passing the end o of the slender spring; for at this instant the spring which was wound up by the contrary direction of the balance, is now unwound again, or in the same state as it was in its quiescent position at first, and of course has no effect at all upon the balance in either direction; but the balance, having now all the velocity it would acquire from the unwinding the spring, goes on in the direction S M K until the force of this spring again stops it, and brings it back again, moving in the same direction as at first, with a considerable velocity. By this return of the balance, the point i of the lifting-pallet comes up again to the projecting end o of the slender spring, pushes back the long spring E F, and unlocks the wheel; and another tooth falling upon the face of the pallet A L, gives fresh energy to the balance; and thus the action is carried on as before.

ESCAPEMENT, OR 'SCAPEMENT.

THE motions of a clock or watch are regulated by a pendulum or balance, which serves as a check, without which, the wheels impelled by the weight in the clock, or spring in the watch, would run round with a rapidly accelerating motion, till this should be rendered uniform by friction, and the resistance of the air; if, however, a pendulum or balance be put in the way of this motion, in such manner that only one tooth of a wheel can pass, the revolutions of the wheel will depend on the vibration of the pendulum or balance.

We know that the motion of a pendulum or balance is alternate, while the pressure of the wheels is constantly exerted in the same direction. Hence it is evident that some means must be employed to accommodate these different motions to each other. Now, when a tooth of the wheel has given the pendulum or balance a motion in one direction, it must quit it, that the pendulum or balance may receive an impulsion in the opposite direction. This *escaping* of the tooth has given rise to the term *escapement*

The ordinary 'scapement is extremely simple, and may be thus illustrated. Let xy, fig. 523, represent a horizontal axis, to which the pendulum is attached by a slender rod. This axis has two leaves, c and d, one near each end, and not in the same plane, but so that when the pendulum hangs perpendicularly at rest, c spreads a few degrees to the right, and d as much to the left. These are called the pallets. Let afb represent a wheel, turning on a perpendicular axis, eo, in the order $afeb$. The teeth of this wheel are in the form of those of a saw, leaning forward in the direction of the rim's motion. This wheel is usually called the *crown-wheel*, or in watches the *balance-wheel*. It in general contains an odd number of teeth. In the figure the pendulum is represented at the extremity of its excursion towards the right, the tooth a having just escaped from the pallet c, and b having just dropped on d.

Now it is evident, that while the pendulum is moving to the left, in the arc pg, the tooth b still presses on the pallet d, and thus accelerates the pendulum, both in its descent along ph, and its ascent up hg, and that when d, by turning round the axis xy, raises its point above the plane of the wheel, the tooth b escapes from it, and i drops on c, now nearly perpendicular. Thus c is pressed to the right, and the motion of the pendulum along gp is accelerated. Again, while the pendulum hangs perpendicularly in the line xh, the tooth b, by pressing on d, will force the pendulum to the left, in proportion to its lightness, and if it be not too heavy, will force it so far from the perpendicular, that b will escape, and i will catch on c, and force the pendulum back to p, when the same motion will be repeated. This effect will be the more remarkable if the rod of the pendulum be continued through xy, and have a ball q, on the other end, to balance p.

When b escapes from d, the balls are moving with a certain velocity and momentum, and in this condition the balance is checked when i catches on c. It is not, however, instantly stopped, but continues to move a little to the left, and i is forced a little backward by the pallet c. It cannot make its escape over the top of the tooth i, as all the momentum of the balance was generated by the force of b, and i is of equal power. Besides, when i catches on c, and the motion of c, to the left, continues, the lower point of c is applied to the face of i, which now acts on the balance by a long lever, and soon stops its motion in that direction, and, continuing to press on c, urges the balance in the opposite direction. In this, it is evident that the motion of the wheel is hobbling and unequal, by which this escapement has received the appellation of the *recoiling 'scapement*.

In considering the utility of the following improved 'scapement for clocks, we must keep in mind the following proposition, which, after the above illustration, scarcely requires any proof. It is, that the natural vibrations of a pendulum are *isochronous*, or are performed in equal times. The great object of the 'scapement is to preserve this isochronous motion of the pendulum.

As the defect of the recoiling 'scapement was long apparent, several ingenious artists attempted to substitute in its place a 'scapement that should produce a more regular and uniform motion. Of these, the 'scapement contrived by Mr. Cumming appears to be one of the most ingenious in its construction, and most perfect in its operation. The following construction is similar to that of Mr. Cumming, but rendered rather less complex for the purpose of shortening the description.

Let A B C, fig. 524, represent a portion of the swing-wheel, of which O is the centre, and A one of the teeth, and Z the centre of the crutch, the pallets, and pendulum. The crutch is represented of the form of the letter A, having in the circular cross-piece a slit *i k*, which is also circular, Z being the centre. The arm Z F forms the first detent, and the tooth A is represented as locked on it at F. D is the first pallet on the end of the arm Z *d*, movable round the same centre with the detents, but independent of them. The arm, *d e*, to which the pallet D is attached, lies wholly behind the arm Z F of the detent, being fixed to a round piece of brass, *e f g*, having pivots turning concentric with the axis of the pendulum. To the same piece of brass is fixed the horizontal arm *e* H, carrying at its extremity the ball H, of such size that the action of the tooth A on the pallet D is just able to raise it up to the position represented. Z P *p* represents the fork, or pendulum-rod, behind both detent and pallet. A pin *p* projects forward, coming through the slit *i k*, without touching either margin of it. Attached to the fork is the arm *m n*, of such length that, when the pendulum-rod is perpendicular, the angular distance of *n q* from the rod *e q* H is just equal to the angular distance of the left side of the pin *p* from the left end *i* of the slit *i k*.

Now the natural position of the pallet D is at *δ*, represented by the dotted lines, resting on the back of the detent F. It is naturally brought into this position by its own weight, and still more by the weight of the ball H. The pallet D, being set on the foreside of the arm at Z, comes into the same plane with the detent F and the swing-wheel, though represented in the figure in a different position. The tooth C of the wheel is supposed to have escaped from the second pallet, on which the tooth A immediately seizes the pallet D, situated at *δ*, forces it out, and then rests on the detent F, the pallet D leaning on the tip of the tooth. After the escape of C, the pendulum, moving down the arch of semi-vibration, is represented as having attained the vertical position. Proceeding still to the left, the pin *p* reaches the extremity *i* of the slit *i k;* and, at the same instant, the arm *n* touches the rod *e* H in *q*. The pendulum proceeding a hair's-breadth further, withdraws the detent F from the tooth, which now pushes off the detent, by acting on the inclining face of it.

The wheel being now unlocked, the tooth, following C on the other side, acts on its pallet, pushes it off, and rests on its detent, which has been rapidly brought into a proper position by the action of A on the inclined face of F. By a similar action of C on its detent at the moment of escape, F was brought into a position proper for the wheels being locked by the tooth A. As the pendulum still goes on, the ball H, and pallet connected with it, are carried by the arm *m n*, and before the pin *p* again reaches the end of the slit, which had been suddenly withdrawn by the action of A on F, the pendulum comes to rest. It now returns towards the right, loaded with the ball H on the left, and thus the motion lost during the last vibration is restored. When the pin *p*, by its motion to the right, reaches the end *k* of *i k*, the wheel on the right side is unlocked, and at the same instant the weight H, being raised from the pendulum by the action of a tooth like B on the pallet D, ceases to act.

In this 'scapement, both pallets and detents are detached from the pendulum, except in the moment of unlocking the wheel, so that, excepting during this short interval, the pen dulum may be said to be free during its whole vibration, and of course its motion must be more equable and undisturbed.

The constructing of a proper 'scapement for watches requires peculiar delicacy, owing to the small size of the machine, from which the error of $\frac{1}{70}$ of an inch has as much effect as the error of a whole inch in a common clock. From the necessary lightness of the balance too, it is extremely difficult to accumulate a sufficient quantity of regulating power. This can only be done by giving the balance a great velocity, which is effected by concentrating as much as possible of its weight in the rim, and making its vibrations very wide. The balance-rim of a tolerable watch should pass through at least ten inches in every second.

In considering the most proper 'scapements for watches, we may assume the following principle; viz. : that the oscillations of a balance urged by its spring, and undisturbed by extraneous forces, are isochronous.

In ordinary pocket-watches, the common recoiling 'scapement of clocks is still employed, and answers the common purposes of a watch tolerably well, so that, if properly executed, a good ordinary watch will keep time within a minute in the day. These watches, however, are subject to great variation in their rate of going, from any change in the power of the wheels.

The following is considered as the best construction of the common watch 'scapement, and is represented by fig. 525, as it appears when looking straight down on the end of the balance arbor. C marks the centre of the balance and verge; C A represents the upper pallet, or that next the balance, and C B the lower pallet; F and D are two teeth of the crown-wheel, moving from left to right; E G are two teeth in the lower part, moving from right to left. The tooth D appears as having just escaped from the point of C A, and the tooth E as having just come in contact with C B. In practice, the 'scapement should not be quite so close, as, by a small inequality of the teeth, D might be kept from escaping at all. In the best proportioned watches, the distance between the front of the teeth, that is, of G F E D, and the axis C of the balance, is $\frac{1}{8}$ of F A, the distance between the points of the teeth. The length C A, C B, of the pallets is $\frac{3}{8}$ of the same degrees, and the front D H, or F K of the teeth makes an angle of 25° with the axis of the crown-wheel. The sloping side of the tooth must be of an epicycloidal form, suited to the relative motion of the tooth and pallet.

It appears from these proportions, that by the action of the tooth D, the pallet A can throw out till it reach a, 120° from C L, the line of the crown-wheel axis. To this if we add B C A = 95°, we shall have L C a = 120°. Again, B will throw out as far on the other side. Now, if from 240°, the sum of the extent of vibration of both pallets, we take 95°, the angle of the pallets, the remainder 145° will express the greatest vibration which the balance can make without striking the front of the teeth. From several causes, however, this measure is too great, and 120° is reckoned a sufficient vibration in the best ordinary 'scapement. *Encyclopædia Britannica.*

the detent D, which detains the wheel in the position first described, the renovating spring being wound up ready to give another impulse to the pendulum.

The pin *b*, fig. 527, is not fixed to the renovating spring itself, but is part of a piece of brass, which is screwed fast to the renovating spring, and is made very slender near the screw which fastens it; this permits the renovating spring to give way, if, by the weight being taken off the clock, or any other accident, the escape-wheel should be wound backwards, so as to catch on the detents improperly.

In this escapement it is necessary to attend to the following observations:

1st. That the renovating and detent springs must spring from one centre, and as similarly as possible.

2dly. That the force applied to the train must be so much more than what will wind up the renovating spring, as will overcome the influence of oil and friction on the pivots of the machine.

3dly. That the renovating spring, when unwound, must rest against the point of the tooth of the wheel, which will be an advantage, as it thereby takes as much force off the tooth of the wheel resting against the detent spring as is equal to the pressure of the renovating spring C, against the face of the tooth of the wheel.

4thly. The detent springs must be made as slender and light as possible, though whatever force they take from the pendulum by their elasticity in removing them to unlock the wheel, so much force they return to the pendulum in following it, to where it removed them from, therefore action and reaction will be equal in contrary directions.

5thly. That it is necessary for the pendulum to remove the detents or renovating springs, much further than it is necessary to free the teeth of the wheel, as it will always vibrate on the same arc; in table clocks it ought to remove them further, so that it can go when not placed exactly level, or what is generally termed out of the beat.

The following description of a clock escapement, contrived by Mr. Reid, about twelve or fifteen years ago, is extracted from the Edinburgh Encyclopædia :

Fig. 528. S W is the swing-wheel, whose diameter may be so large as to be sufficiently free of the arbor of the wheel that runs into its pinion, which in eight-day clocks is the third. The teeth of this swing-wheel are cut thus deep, in order that the wheel may be as light as possible, and the strength of the teeth little more than what is necessary to resist the action or force of a common clock weight through the wheels. They are what may be called the locking-teeth, as will be more readily seen from the use of them afterwards to be explained. Those called the impulse-teeth, consist of very small tempered steel pins, inserted on the surface of the rim of the wheel on one side only. They are nearly two-tenths of an inch in height; and the

tooth at this instant of unlocking, meeting with the flanch of the pallet at the lower edge inside, and pushing forward on the flanch, by this means impels the pendulum, and after having escaped the pallet, the next locking tooth is received by the detent, on the right-hand side, where the wheel is now again locked. In the mean time, while the pendulum is describing that part of its vibration towards the left hand free and detached, as the pallets are now at liberty to move freely and independently of the small pin teeth, on the return of the pendulum to the right-hand side, the detent, by means of the pallet on that side, is pushed out from locking the wheel, and at the instant of unlocking, the wheel gets forward, and the pin tooth is at the same instant ready to get on the flanch of its pallet, and gives new impulse to the pendulum, as is obvious by what is represented in the drawing, fig. 528. After the pin tooth has escaped the pallet, the wheel is again locked on the opposite or left-hand side; the pendulum moves on to the right freely and independently till the next locking on the left takes place, and so on. It may be observed, that the unlocking takes place when the pendulum is near the lowest point, or point of rest, and of course where its force is nearly a maximum.

Without attaching any thing to the merits of this 'scapement, we may remark' that the clock was observed from time to time by a very good transit instrument, and, during a period of eighty-three days, it kept within the second, without any interim apparent deviation. This degree of time-keeping seemed to be as much a matter of accident as otherwise; and cannot reasonably be expected from this or any other clock as a fixed or settled rate.

This 'scapement being a detached or free 'scapement, can at pleasure be converted into a recoiling or dead beat one, without so much as once disturbing or stopping the pendulum a single vibration. To make dead beat of it, put in a peg of wood, or small wire, to each, so as to raise the detents free of the pallets; and these being left so as to keep them in the position, the pin-teeth will now fall on the circular parts of the pallets, and so on to the flanch, and the 'scapement is then, to all intents and purposes, a dead beat one. To make a recoiling one of it, let there be fixed to each arbor of the detents, a wire to project horizontally from them about 3½ or 4 inches long; the outer ends of the wires must be tapped about half an inch in length; provide two small brass balls, half an ounce weight each, having a hole through them, and tapped so as to screw on the wires; the balls can be put more or less home, and be adjusted proportionably to the force of the clock on the pendulum. No recoil will be seen on the seconds' hand; yet these will alternately oppose and assist the motion of the pendulum, as much as any recoiling pallets can possibly do; and as their efforts on the pendulum will be exactly the same, it may be considered as a good recoiling 'scapement. This sort of detached 'scapement, by becoming a dead beat, or a recoiling one, at any time when required, makes it convenient for making various experiments with the different 'scapements.

Another 'scapement, in which a considerable degree of ingenuity is united with comparative simplicity, is that of Mr. De Lafons. The inventor's description, and some of his observations, as presented to the Society of Arts, are as follows:

"Although the giving of an equal impulse to the balance has been already most ingeniously done by Mr. Mudge and Mr. Haley, (from whose great merit I would not wish to detract,) yet the extreme difficulty and expense attending the first, and the very compound locking of the second, render them far from completing the desired object.

"The perfections and advantages arising from my improvements on the remontoire detached 'scapement for chronometers, which gives a perfectly equal impulse to the balance, and not only entirely removes whatever irregularities arise from the different states of fluidity in the oil, from the train of wheels, or from the mainspring, but does it in a simpler way than any with which I am acquainted. I trust it will not be thought improper in me to answer some objections made at the examinations before the committee, as I am fully persuaded the more mathematically and critically the improvements are investigated, the more perfect they will prove to be.

"It was first observed, that my method did not so completely detach the train of wheels from the balance as another 'scapement then referred to. I beg leave to remark, that the train of wheels in mine is prevented from pressing against the locking by the whole power of the remontoire-spring; so that the balance has only to remove the small remaining pressure, which does away that objection, and also that of the disadvantage of detents, as this locking may be compared to a light balance turning on fine pivots, without a pendulum-spring; and has only the advantage of banking safe at two turns of the balance, and of being firmer, and less liable to be out of repair, than any locking where spring-work is used, but likewise of unlocking with much less power. It was then observed, it required more power to make it go than usual. Permit me to say, it requires no more power than any other remontoire-'scapement, as the power is applied in the most mechanical manner possible. And, lastly, it was said, that it set or required the balance to vibrate an unusually large arch before the piece would go. This depends on the accuracy of the execution, the proportionate diameter and weight of the balance, the strength of the remontoire-spring, and the length of the pallets. If these circumstances are well attended to, it will set but little more than the most generally detached 'scapements."

A shows the scape-wheel, fig. 534.

B, the lever-pallet, or an arbor with fine pivots, having at the lower end.

C, the remontoire or spiral spring fixed with a collar and stud, as pendulum-springs are.

D, the pallet of the verge, having a roller turning in small pivots for the lever-pallet to act against.

E, pallets to discharge the locking, with a roller between, as in fig. 535.

F, the arm of the locking-pallet continued at the other end to make it plain, having studs and screws to adjust and bank the quantity of motion.

e and b, the locking-pallets, being portions of circles, fastened on an arbor turning on fine pivots.

G, the triple fork, at the end of the arm of the locking-pallets.

" The centre of the lever-pallet in the draft, is in a right line between the centre of the scape-wheel and the centre of the verge, though in the model it is not : but may be so or not, as best suits the calliper, &c.

" The scape-wheel A, with the tooth 1, is acting on the lever-pallet B, and has wound up the spring C; the verge pallet D (turning the way represented by the arrow) the moment it comes within the reach of the lever-pallet, the discharging pallet E, taking hold of one prong of the fork, removes the arm F, and relieves the tooth 3 from the convex part of the lock a. The wheel goes forward a little, just sufficient to permit the lever-pallet to pass, while the other end gives the impulse to the balance : the tooth 4 of the wheel is then locked on the concave side of the lock b, and the lever-pallet is stopped against the tooth 5, as in fig. 536. So far the operation of giving the impulse, in order again to wind the remontoire-spring, (the other pallet at E, in the return, removing the arm F the contrary direction,) relieves the tooth 3 from the lock b. The wheel again goes forward, almost the whole space, from tooth to tooth, winds the spiral spring again, and comes into the situation of fig 534, and thus the whole performance is completed. The end of the lower pallet B resting on the point of the tooth 1, prevents the wheel exerting its full force on the lock a, as in fig. 534. The same effect is produced by the pallet lying on the tooth 5, by preventing the wheel from pressing on b; and thus the locking becomes the tightest possible. This 'scapement may be much simplified by putting a spring with a pallet made in it, as in fig. 534, instead of the lever-pallet and spiral spring. The operation will be in other respects exactly the same, avoiding the friction of the pivots of the lever-pallet. This method I prefer for a piece to be in a state of rest, as a clock, but the disadvantage, from the weight of the spring in different positions, is obvious. The locking may be on any two teeth of the wheel, as may be found most convenient."

PENDULUMS.

THE pendulum is a simple ponderous body, so suspended, that it may swing backwards and forwards, about some fixed point, by the mere force of gravity.

These alternate ascents and descents of the pendulum are called its *oscillations*, or *vibrations;* each oscillation being the arc which the pendulum describes from the highest point on one side to the highest point on the other side. The point round which the pendulum moves, or vibrates, is called the *axis of suspension*, or *centre of motion;* and a right line drawn through the centre of motion, parallel to the horizon, and perpendicular to the plane in which the pendulum moves, is called the *axis of oscillation.* There is also a certain point within every pendulum, into which, if all the matter that composes the pendulum were collected, or condensed, as into a point, the times in which the vibrations would be performed would not be altered by such condensation ; and this point is called the *centre of oscillation.* The length of the pendulum is always estimated by the distance of this point below the centre of motion, being usually near the bottom of the pendulum; but in a slender cylinder, or any other uniform prism or rod suspended at the top, it is at the distance of one-third from the bottom, or two-thirds below the centre of motion.

The length of a pendulum, so measured to its centre of oscillation that it will perform each vibration in a second of time, thence called the seconds' pendulum, has, in the latitude of London, been generally taken at 39$\frac{1}{7}$, or 39$\frac{1}{7}$ inches; but by some very ingenious and accurate experiments, the late celebrated Mr. George Graham found the true length to be 39$\frac{1}{7}$ inches, or 39$\frac{1}{7}$ inches very nearly.

The length of the pendulum vibrating seconds at Paris was found by Varin, Des Hays, De Glos, and Godin, to be 440$\frac{2}{3}$ lines; by Picard, 440$\frac{1}{2}$ lines; and by Mairan, 440$\frac{1}{2}$ lines.

As all woods and metals are more or less affected by changes of temperature, many ingenious contrivances have been resorted to, to counteract the effects of heat and cold, in lengthening or shortening a pendulum-rod.

The first person who observed that, by change of temperature, metals changed their length, was Godfroi Wendelinus; and he who first endeavoured to take advantage of this knowledge, to counteract the effects of heat and cold upon a pendulum, was Graham, who, in the year 1715, suggested that a combination of rods or wires of different metals would have a tendency to that effect; but being of opinion that this would not be quite adequate to the desired purpose, he did never, we believe, put it in execution. Still continuing his observations, he, a short time afterwards, conceived that mercury, from its great expansion by heat, was more adapted to the end he was pursuing, and accordingly we find, that, by the 9th of June, 1722, he had constructed a clock which had a

must be taken out, to shorten the column. This pendulum, though troublesome to construct, because any filling in or taking out of the mercury from the cylinder or glass jar, to bring about the compensation, will cause a change of place in the index-point on the graduated arch or index-plate, if such a thing be used, is, notwithstanding some defect may arise from the expansion of the mercury commencing sooner than that of the rod, of much practical excellence. The mercurial pendulum has been much improved by Reid; for an account of which we must refer our readers to the article " Horology," written by this gentleman, and inserted in the *Edinburgh Encyclopædia.*

Mr. Harrison, of whom we have already spoken under the article Chronometers, some time previous to 1726, constructed a pendulum in which the compensation was effected by the opposite contraction of different metals. This pendulum, called the *gridiron-pendulum*, from, we suppose, its bearing a near resemblance to the culinary implement of that name, was made of five steel and four brass rods, placed in alternate order, the middle rod, by which the pendulum-ball is suspended, being of steel. These rods are so connected with each other at their ends, that while the expansion of the steel rods has a tendency to lengthen the pendulum, the expansion of the brass rods, acting upwards, tends to shorten it, so that by the combined effect the pendulum is invariably preserved of the same length. This is a very ingenious and simple contrivance, and the only objections we have heard urged against this mode of compensation are, 1st. the difficulty of exactly adjusting the length of the rods; 2dly. of proportioning their thickness, so that they shall all begin to contract or expand at the same instant; 3dly. the connecting bars of a pendulum thus constructed are apt to move by starts; 4thly. this kind of pendulum is more exposed to the air's resistance than a simple pendulum.

Other modes of constructing pendulums on the principle of the opposite contraction of metals have been contrived by other ingenious artists, among whom we may notice Ellicott, Cumming, Troughton, Reid, and Ward.

In Ellicott's pendulum the ball was adjustable by levers, thence called the *lever-pendulum*, which can never be equal to those in which the expansion and contraction act by contact in the direct line of the pendulum-rod; the construction nevertheless evinced great ingenuity. The rod of this pendulum was composed of two bars, one of brass, and the other of steel. It had two levers, each sustaining its half of the

ball or weight, with a spring under the lower part of the ball
to relieve the levers from a considerable part of its weight,
and so to render their motion more smooth and easy. These
levers were placed within the ball, and each had an adjusting
screw to lengthen or shorten the lever, so as to render the
adjustment the more perfect. See the *Philos. Transact.*
vol. xlvii. p. 479; where Mr. Ellicott's methods of construction are described and illustrated by figures.

This pendulum was much improved by Cumming, who
conceived that where there were two bars only, a flexure and
unequal bearing would take place, and consequently an exact
compensation could not be effected. To remedy this, he
constructed a pendulum of one flat bar of brass, and two bars
of steel, and used three levers within the ball of the pendulum,
whereas Mr. Ellicott used only two. Among many other
ingenious contrivances for the more accurate adjusting of this
pendulum to mean time, it is provided with a small ball and
screw below the principal ball or weight, one entire revolution of which on its screw will only alter the rate of the
clock's going one second per day; and its circumference is
divided into 30, one of which divisions will therefore alter its
rate of going one second in a month.

"Troughton's *tubular-pendulum*, which acts on the principle of the gridiron-pendulum, is a very neat and ingenious
contrivance. It is constructed of an exterior tube of brass,
reaching from the bob nearly to the top, within which is
another tube, and five brass wires in its belly, so disposed as
to produce altogether, (like Harrison's gridiron-pendulum,)
three expansions of steel downwards, and two of brass upwards, whose lengths being inversely proportioned to their
dilatation, when properly combined, destroy the whole effect
that either metal would have singly. The small visible part
of the rod, near the top, is a brass tube, whose use is to cover
the upper end of the middle wire, which is single, and otherwise unsupported. Drawings of this pendulum may be seen
in *Nicholson's Journal*, No. 36, N.S.

Reid's pendulum is composed of a zinc tube, and three long
and one short steel rods, connected by means of traverses.
Two of these long rods are inserted at one end in the ball of
the pendulum, and terminate at the other in the upper traverse, which keeps them exactly parallel with respect to each
other. At the lower ends of these rods, not far above the
ball, is another traverse, in the middle of which the short
steel rod is pinned, descending thence through the centre of
the ball. Another traverse is placed a little above this, on

the centre of which the zinc tube rests, extending upwards, and pressing against, or rather pressed by the upper traverse. The third or centre steel rod passes through a hole in the upper traverse, equidistant from each of the other two steel rods, thence down the zinc tube, and finally is pinned to the second traverse, or that traverse on which the zinc tube rests. By this means, the centre steel rod, when lengthened by heat, will make the lower end of the. zinc tube descend with it; but the same cause which lengthens the steel rod downwards will expand the zinc tube upwards, and this will carry up the two outside steel rods with which the ball of the pendulum is connected; their expansion downwards, as well as that of the centre rod, is compensated by the upward expansion of the zinc tube. In constructing a pendulum upon this principle, it would be proper to have a few holes in the tube, for the purpose of admitting air more freely to the centre rod.

Ward's pendulum consists of two flat bars of steel, and one of zinc, connected together by three screws. The description which has been given of it in the Transactions of the Society of Arts, &c. for the year 1807, and the pamphlet which Mr. Ward published at Blandford in 1808, contain sufficient details to enable any common clock-maker to copy it.

Before we conclude this article, we shall briefly notice the sympathy or mutual action of the pendulums of clocks.

It is now nearly a century since it was known that when two clocks are set agoing on the same shelf, they will disturb each other; that the pendulum of the one will stop that of the other; and that the pendulum which was stopped will, after a while, resume its vibrations, and, in its turn, stop that of the other clock, as was observed by the late Mr. John Ellicott. When two clocks are placed near one another, in cases very slightly fixed, or when they stand on the thin boards of a floor, it has been long known that they will affect a little the motions of each other's pendulum. Mr. Ellicott observed, that two clocks resting against the same rail, which agreed to a second for several days, varied 1' 36" in twenty-four hours when separated. The slower having a longer pendulum, set the other in motion in 16½ minutes, and stopped itself in 36½ minutes.

BUILDING.

———

UNDER this general term, which implies the construction of an edifice according to the rules laid down by the different artificers employed, we purpose to treat of the respective business of the Mason, Bricklayer, Carpenter, Joiner, Plasterer, Plumber, Painter, and Glazier; previous to which it will be necessary to consider the sinking of the foundation, the due mixture of the ingredients which compose the mortar, and the art of making bricks; upon the whole of which materially depends the stability of an edifice.

As firmness of foundation is indispensable, wherever it is intended to erect a building, the earth must be pierced by an iron bar, or struck with a rammer, and if found to shake, must be bored with a well-sinker's implement, in order to ascertain whether the shake be local or general. If the soil is in general good, the loose and soft parts, if not very deep, must be excavated until the labourers arrive at a solid bed capable of sustaining the pier or piers to be built. If not very loose, it may be made good by ramming into it very large stones, packed close together, and of a breadth proportionate to the intended weight of the building; but where very bad, it must be piled and planked.

In places where the soil is loose to any great depth, and over which it is intended to place apertures, such as doors, windows, &c. while the parts on which the piers are to stand are firm, the best plan is to turn an inverted arch under each intended aperture, as then the piers in sinking will carry with them the inverted arch, and by compressing the ground compel it to act against the under sides of the arch, which, if closely jointed, so far from yielding, will, with the abutting piers, operate as one solid body; but, on the contrary, if this expedient of the inverted arch is not adopted, the part of the wall under the aperture, being of less height, and consequently of less weight than the piers, will give way to the resistance of the soil acting on its base, and not only injure the brick-work between the apertures, but fracture the window-heads and cills.

2 M

In constructing so essential a part as the arch, great attention must be paid to its curvature, and we strongly recommend the parabolic curve to be adopted, as the most effectual for the purpose ; but if, in consequence of its depth, this cannot conveniently be introduced, the arch should never be made less than a semi-circle. The bed of the piers should be as uniform as possible, for, though the bottom of the trench be very firm, it will in some degree yield to the great weight that is upon it, and if the soil be softer in one part than in another, that part which is the softest, of course will yield more to the pressure, and cause a fracture.

If the solid parts of the trench happen to be under the intended apertures, and the softer parts where piers are wanted, the reverse of the above practice must be resorted to ; that is, the piers must be built on the firm parts, and have an arch that is not inverted between them. In performing this, attention must be paid to ascertain whether the pier will cover the arch ; for if the middle of the pier rest over the middle of the summit of the arch, the narrower the pier is, the greater should be the curvature of the arch at its apex. When suspended arches are used, the intrados ought to be kept clear of the ground, that the arch may have its due effect.

When the ground is in such a state as to require the foundation merely to be rammed, the stones are hammer-dressed, so as to be as little taper as possible, then laid of a breadth proportioned to the weight that is to be rested upon them, and afterwards well rammed together. In general, the lower bed of stones may be allowed to project about a foot from the face of the wall on each side, and on this bed another course may be laid to bring the bed of stones on a level with the top of the trench. The breadth of this upper bed of stones should be four inches less than the lower one ; that is, projecting about eight inches on either side of the wall. In all kinds of walling, each joint of every course must fall as nearly as possible in the centre, between two joints of the course immediately below it ; for in all the various methods of laying stones or bricks, the principal aim is to procure the greatest lap on each other. ,

MORTAR.

In making mortar, particular attention must be paid to the quality of the sand, and if it contain any propor-

n of clay or mud, or is brought from the sea-shore and
ntains saline particles, it must be washed in a stream
clear water till it be divested of its impurities. The
cessity of the first has been clearly proved by Mr. Smeaton,
10, in the course of a long and meritorious attention to
profession of an engineer, has found, that when mortar,
ough otherwise of the best quality, is mixed with a small
portion of unburnt clay, it never acquires that hardness
ich, without it, it would have attained; and, with respect
the second, it is evident, that so long as the sand contains
inc particles it cannot become hard and dry. The sharper
d coarser the sand is the better for the mortar, and the less
e quantity of lime to be used; and sand being the
eapest of the ingredients which compose the mortar,
is more profitable to the maker. The exact proportions
lime and sand are still undetermined; but in general no
are lime is required than is just sufficient to surround
e particles of the sand, or sufficient to preserve the
cessary degree of plasticity.

Mortar, in which sand forms the greater portion requires
is water in its preparation, and consequently is sooner
t. It is also harder and less liable to shrink in drying,
cause the lime, while drying, has a greater tendency to
rink than sand, which retains its original magnitude.
e general proportions given by the London builders is 1¼
rt., or 37 bushels, of lime to 2¼ loads of sand; but if
oper measures be taken to procure the best burnt lime
d the best sand, and in tempering the materials, a greater
rtion of sand may be used. There is scarcely any
ortar that has the lime well calcined, and the com-
sition well beaten, but that will be found to require
o parts of sand to one part of unslacked lime; and it is
rthy of observation, that the more the mortar is beaten
e less proportion of lime suffices.

Many experiments have been made with a view to obtain
most useful proportion of the ingredients, and among
e rest Dr. Higgins has given the following :—

> Lime newly slacked one part,
> Fine sand three parts; and
> Coarse sand four parts.

He also found that one-fourth of the lime of bone-ashes
ently improved the mortar, by giving it tenacity, and ren-
ring it less liable to crack in the drying.

It is best to slack the lime in small quantities as required for use, about a bushel at a time, in order to secure to the mortar such of its qualities as would evaporate were it allowed to remain slacked for a length of time. But if the mortar be slacked for any considerable time previous to being used, it should be kept covered up, and when wanted be re-beaten. If care be taken to secure it from the action of the atmosphere, it may thus remain covered up for a considerable period without its strength being in the least affected ; and, indeed, some advantages are gained, for it sets sooner, is less liable to crack in the drying, and is harder when dry.

Grout, which is a cement containing a larger proportion of water than the common mortar, is used to run into the narrow interstices and irregular courses of rubble-stone walls ; and as it is required to concrete in the course of a day, it is composed of mortar that has been a long time made and thoroughly beaten.

Mortar, composed of pure lime, sand, and water, may be employed in the linings of reservoirs and aqueducts, provided a sufficient time is allowed for it to dry before the water is let in ; but if a sufficient time is not allowed, and the water is admitted while the mortar is wet, it will soon fall to pieces. There are, however, certain ingredients which may be put into the common mortar to make it set immediately under the water ; or, if the quick-lime composing the mortar contain in itself a certain portion of burnt clay, it will possess this property. For further information on this head the reader is referred to the sub-head—*Plastering.*

BRICKS.

The earth best adapted for the manufacture of brick is of a clayey loam, neither containing too much argillaceous matter, which causes it to shrink in the drying, nor too much sand, which has a tendency to render the ware both heavy and brittle. It should be dug two or three years before it is wrought, that it may, by an exposure to the action of the atmosphere, lose the extraneous matter of which it is possessed when first drawn from its bed ; or, at least, should be allowed to remain one winter, that the frost may mellow and pulverize it sufficiently to facilitate the operation of tempering. As the quality of the brick is greatly dependent upon the tempering of the clay, great care should be taken to have this part of the process well done. Formerly the manner of performing it consisted in throwing the clay into

by arching the bricks over so as to leave a space of about a brick in width. The flues run straight through the clamp, and are filled with a mixture of coals, wood, and breeze, which are pressed closely together. If the bricks are required to be burnt off quickly, which can be accomplished in the space of from twenty to thirty days according to the state of the weather, the flues must not exceed six feet distance apart; but if there is no urgent demand, the flues need not be nearer than nine feet, and the clamp may be allowed to burn slowly.

Coke has been recommended as a more suitable fuel for bricks than either coal or wood, as the dimensions of the flues and the stratum of the fuel are not required to be so great, which, since the measurement of the clamp has been restricted to certain limits by the interference of the legislature, is a point of some consideration; besides, the heat arising from the coke is more uniform and more intense than what is produced by the other materials, so that the burning of the bricks is more likely to be perfect throughout. The saving which is thus produced may be calculated at about 32 per cent.

Kilns are also in common use, and are in many respects preferable to the clamp, as less waste arises, less fuel is consumed, and the bricks are sooner burnt. A kiln will burn about 20,000 bricks at a time. The walls of a kiln are about a brick and a half thick, and incline inwards towards the top, so that the area of the upper part is not more than 114 square feet. The bricks are set on flat arches, with holes left between them, resembling lattice-work; and, when the kiln is completed, are they covered with pieces of broken bricks and tiles, and some wood is kindled and put in to dry them gradually. When sufficiently dried, which is known by the smoke changing from a dark to a light transparent colour, the mouths of the kiln are stopped with pieces of brick, called *shinlog*, piled one upon another, and closed over with wet brick-earth. The shinlogs are carried so high as just to leave room for one faggot to be thrust into the kiln at a time, and when the brush-wood, furze, heath, faggots, &c. are put in, the fire is kindled, and the burning of the kiln commences. The fire is kept up till the arches assume a white appearance, and the flames appear through the top of the kiln; upon which the fire is allowed to slacken, and the kiln to cool by degrees. This process of alternately heating and slacking the kiln is continued till the bricks are thoroughly burnt, which, in

which are made of a stronger clay, and are of a red colour. The largest are about twelve inches square, and one inch and a half thick : the next, though called ten-inch tiles, are about nine inches square, and one inch and a quarter thick.

About the year 1795, a patent was obtained by Mr. Cartwright, for an improved system of making bricks, of which the following extract will furnish the reader with all necessary information.

" Imagine a common brick, with a groove or rabate on each side down the middle, rather more than half the width of the side of the brick ; a shoulder will thus be left on either side of the groove, each of which will be nearly equal to one quarter of the width of the side of the brick, or to one half of the groove or rebate. A course of these bricks being laid shoulder to shoulder, they will form an indented line of nearly equal divisions, the grooves or rebates being somewhat wider than the adjoining shoulders, to allow for the mortar or cement. When the course is laid on, the shoulders of the bricks, which compose it, will fall into grooves of the first course, and the shoulders of the first course, will fit into the grooves or rabates of the second, and so with every succeeding course. Buildings constructed with this kind of brick, will require no bond timbers, as an universal bond runs through the whole building, and holds all the parts together ; the walls of which will neither crack nor bilge without breaking through themselves. When bricks of this construction are used for arches, the sides of the grooves should form the radii of the circle, of which the intended arch is a segment ; yet if the circle be very large, the difference of the width at the top and bottom will be so very trifling, as to render a minute attention to this scarcely if at all necessary. In arch-work, the bricks may either be laid in mortar, or dry, and the interstices afterwards filled up by pouring in lime, putty, plaster of Paris, &c. Arches upon this principle, having any lateral pressure, can neither expand at the foot, nor spring at the crown, consequently they want no abutments, requiring only perpendicular walls to be let into, or to rest upon ; neither will they want any superincumbent weight on the crown to prevent their springing up. The centres also may be struck immediately, so that the same centre, which never need be many feet wide, may be regularly shifted as the work proceeds. But the most striking advantage attending this invention is, the security it affords against the ravages of fire ; for, from the peculiar properties of this kind of arch, requiring no abutments, it may be laid upon, or let into common walls, no stronger than what is required for timbers so as to admit of brick floorings."

Having said thus much on the laying of the foundation, the mixing of the mortar, and the manufacture of the brick, we shall next proceed to treat on the principles of the art of masonry, as practised in the present day.

MASONRY,

Is the art of cutting stones, and building them into a mass, so as to form the regular surfaces which are required in the construction of an edifice.

The chief business of the mason is to prepare the stones, make the mortar, raise the wall with the necessary breaks, projections, arches, apertures, &c., and to construct the vaults, &c. as indicated by the design.

A wall built of unhewn stone, whether it be built with mortar or otherwise, is called a *rubble wall*. Rubble work is of two kinds, coursed and uncoursed. In coursed rubble the stones are gauged and dressed by the hammer, and thrown into different heaps, each heap containing stones of equal thickness; and the masonry, which may be of different thicknesses, is laid in horizontal courses. In uncoursed rubble the stones are placed promiscuously in the wall, without any attention being paid to the placing them in courses; and the only preparation the stones undergo, is that of knocking off the sharp angles with the thick end of a tool called a *scabling* hammer. Walls are generally built with an ashlar facing of fine stone, averaging about four or five inches in thickness, and backed with rubble work or brick.

Walls backed with brick or uncoursed rubble, are liable to become convex on the outside, from the great number of joints, and the difficulty of placing the mortar, which shrinks in proportion to the quantity, in equal portions, in each joint; consequently, walls of this description are much inferior to those where the facing and backing are built of the same material, and with equal care, even though both of the sides be uncoursed. When the outside of a wall is faced with ashlar, and the inside is coursed rubble, the courses of the backing should be as high as possible, and set within beds of mortar. Coursed rubble and brick backings are favourable for the insertion of bond timber; but, in good masonry, wooden bonds should never be in continued lengths, as in case of either fire or rot the wood will perish, and the masonry will, by being reduced, be liable to bend at the place where the bond was inserted.

When timber is to be inserted into walls for the purposes of fastening buttons for plastering, or skirting, &c., the pieces of timber ought to be so disposed that the ends of the pieces be in a line with the wall.

In a wall faced with ashlar, the stones are generally about 2 feet or 2½ feet in length, 12 inches in height, and 8 inches in thickness. It is a very good plan to incline the back of each stone, to make all the backs thus inclined run in the same direction, which gives a small degree of lap in the setting of the next course; whereas, if the backs are paral·

lel to the front, there can be no lap where the stones run of
an equal depth in the thickness of the wall.   It is also advantageous to the stability of the wall to select the stones, so
that a thicker and a thinner one may succeed each other
alternately.   In each course of ashlar facing, either with
rubble masonry, or brick backing, thorough-stones should
occasionally be introduced, and their number be in proportion to the length of the course.   In every succeeding
course, the thorough stones should be placed in the middle
of every two thorough-stones in the course below ; and
this disposition of bonds should be punctually attended to
in all cases where the courses are of any great length.
Some masons, in order to prove that they have introduced
sufficient bonds into their work, choose thorough-stones of
a greater length than the thickness of the wall, and afterwards cut off the ends ; but this is far from an eligible plan,
as the wall is not only subject to be shaken, but the stone is
itself apt to split.   In every pier, between windows and
other apertures, every alternate jamb-stone ought to go
through the wall with its bed perfectly level.   When the
jamb-stones are of one entire height, as is frequently the
case when architraves are wrought upon them, upon the
lintel crowning them, and upon the stones at the ends of
the courses of the pier which are adjacent to the architrave-
jamb, every alternate stone ought to be a thorough-stone :
and if the piers between the apertures be very narrow, no
other bond-stone is required ; but where the piers are wide,
the number of bond-stones are proportioned to the space.
Bond-stones must be particularly attended to in all long
courses below and above windows.

All vertical joints, after receding about an inch with a close
joint, should widen gradually to the back, thereby forming hollow spaces of a wedge- like figure, for the reception of mortar,
rubble, &c. The adjoining stones should have their beds and
vertical joints filled, from the face to about three quarters
of an inch inwards, with oil and putty, and the rest of the
beds must be filled with well-tempered mortar.   Putty cement will stand longer than most stones, and will even
remain permanent when the stone itself is mutilated.   All
walls cemented with oil-putty, at first look unsightly ; but
this disagreeable effect ceases in a year or less, when, if
care has been taken to make the colour of the putty suitable
to that of the stone, the joints will hardly be perceptible.

In selecting ashlar, the mason should take care that each
stone invariably lays on its natural bed ; as from careless-

tical, cycloidal, catenarian, parabolical, &c. There are also *pointed, composite,* and *lancet,* or *Gothic arches.*

A *rampant arch* is when the springing lines are of two unequal heights.

When the intrados and extrados of an arch are parallel, it is said to be *extradossed.*

There are, however, other terms much used by masons; for example, the semicircular are called *perfect arches,* and those less than a semicircle, *imperfect, surbused,* or *diminished* arches.

Arches are also called *surmounted,* when they are higher than a semicircle.

A *vault* is an arch used in the interior of a building, overtopping an area of a given boundary, as a passage, or an apartment, and supported by one or more walls, or pillars, placed without the boundary of that area.

Hence an arch in a wall is seldom or never called a vault; and every vault may be called an arch, but every arch cannot be termed a vault.

A *groin vault,* is a complex vault, formed by the intersection of two solids, whose surfaces coincide with the intrados of the arches, and are not confined to the same heights. An arch is said to stand upon splayed jambs, when the springing lines are not at right angles to the face of the wall.

In the art of constructing arches and vaults, it is necessary to build them in a mould, until the whole is closed: the mould used for this purpose is called a *centre.* The intrados of a simple vault is generally formed of a portion of a cylinder, cylindroid, sphere, or spheroid, that is, never greater than the half of the solid: and the springing lines which terminate the walls, or when the vault begins to rise, are generally straight lines, parallel to the axis of the cylinder, or cylindroid.

A circular wall is generally terminated with a spherical vault, which is either hemispherical, or a portion of a sphere less than an hemisphere.

Every vault which has an horizontal straight axis, is called a *straight vault;* and in addition to what we have already said, the concavities which two solids form at an angle, receive likewise the name of arch.

An arch, when a cylinder pierces another of a greater diameter, is called *cylindro-cylindric.* The term cylindro is applied to the cylinder of the greatest diameter, and the term cylindric to the less.

cylinder intersect a sphere of greater diameter than
inder, the arch is called a *sphero-cylindric arch ;* but
other hand, if a sphere pierce a cylinder of greater
er than the sphere, the arch is called a *cylindro-sphe-*
l,
cylinder pierce a cone, so as to make a complete per-
n through the cone, two complete arches will be
, called *cono-cylindric arches;* but, on the contrary,
ne pierce a cylinder, so that the concavity made by
ie is a conic surface, the arch is called *cylindro-conic*

1 a straight wall, there be a cylindric aperture con-
quite through it, two arches will be formed, called
:ylindric arches.
y description of arch is, in a similar manner to
ove, denoted by the two preceding words; the
ending in *o*, signifying the principal vault, or sur-
t through; and the latter in *ic*, signifying the de-
n of the aperture which pierces or intersects the wall
t. .
n groins are introduced merely for use, they may be
ther of brick or stone; but, when introduced by way
ortion or decoration, their beauty will depend on the
ing figures of the sides, the regularity of the sur-
id the acuteness of the angles, which should not be
ed. In the best buildings, when durability and
e are equally required, they may be constructed of
t stone; and, when elegance is wanted, at a trifling
:, of plaster, supported by timber ribs.
one-cutting, a narrow surface formed by a point or
on the surface of a stone, so as to coincide with a
edge, is called a *draught.*

———

formation of stone arches has always been considered
useful and important acquisition to the operative
in order, therefore, to remove any difficulties
night arise in the construction of arches of different
ions, both in straight and circular walls, we shall here
ce a few examples, which, it is hoped, with careful
ation, will greatly facilitate a knowledge of some of
t abstruse parts of the art.

1, No. 1. To find the moulds necessary for the construction of
iular arch, cutting a straight wall obliquely.

Let ABCDEFGH be the plan of the arch ; IKLM the outer line; and NOPQ the inner line on the elevation.

a b c d e, on the elevation, shows the bevel of each joint or bed from the face of the wall ; and *a b c d e* below, gives the mould for the same, where *x y* on the elevation corresponds with *x y* at *e*.

The arch mould, fig. 551, No. 2, is applied on the face of the stone, and on being applied to the parts of the plan, gives, of course, the bevel of each concave side of the stone with the face, that is K to O, on the elevation.

Fig. 552. To find the mould for constructing a semicircular arch in a circular wall.

No. 1 is the elevation of the arch; and No. 2 the plan of the bottom bed from *q* to *r*.

a to *b* is what the arch gains on the circle from the bottom bed *k o* to *l*; and *c* to *d* is the projection of the intrados to *p*, on the joint *l. p.*

Nos. 2, 3, 4, are plans of the three arch-stones, 1, 2, 3, in the elevation ; and Nos. 5 and 6 are moulds to be applied to the beds of stones 1 and 2, in which *s c* equals *s c* in No. 2, and *t w* equals *t w* in No. 3.

In No. 1, *k l p o* is the arch or face mould.

When the reader is thoroughly proficient in the construction of arches, under given datas, as the circumstances of the case may point out, he may proceed to investigate the principles of spherical domes and groins.

Figs. 553 and 554 show the principles of developing the soffits of the arches in the two preceding examples. In each the letters of reference are alike, and the operation is precisely the same.

Let ABDE be the plan of the opening in the wall ; and AFB the elevation of the arch: produce the chord AB to C, divide the semicircle AFB into any number of parts, the more the better, and with the compasses set to any one of these divisions, run it as many times along AC as the semicircle is divided into ; then draw lines, perpendicular to BC, through every division in the semicircle and the line CA, and set the distance 1 *b*, 2 *d*, 3 *f*, &c. respectively equal to *a b*, *c d*, *e f*, &c. and then by tracing a curve through these points, and finding the points in the line GD, in the same manner, the soffit of the arch is complete.

Fig. 555, shows the method of constructing spherical domes.

No. 1 mould is applied on the spherical surface to the vertical joints; and No. 2 mould on the same surface to the other joints ; and in both cases, the mould tends to the centre of the dome.

3, 4, 5, 6, 7, and 8, are moulds which apply on the convex surface to the horizontal joint, the lines *a b*, *c d*, *e f*, &c. being at right angles to the different radii, *b c*, *d c*. *f c*, &c. and produced until they intersect the perpendicular *a c* ; the different intersections are the centres which give the circular leg of the mould, and the straight part gives the horizontal joint.

Fig. 556 exhibits the plan of a groined vault.

Lay down the arch, either at the full or half size, on a floor or piece of floor-cloth, then divide and draw on the plan the number of joints in the semicircular arch, and from the intersections with the diagonals, draw the transverse joints on the plan, and produce them till they touch the intradoes of the elliptical arch, the curve of which may be found by setting the corresponding distances from the line of the base to the curve ; thus *a b* equal to *a b*. This being accomplished, draw the joints of the elliptical arch in the manner of which we give c d, as a specimen. To draw the joint c d, draw the chord e c and bisect it, draw a line from the centre c, through the bisecting point, and produce it till it touches the perpendicular e f; and c d, being at right angles to e f, will be the joint required. In the same manner the others are found.

By examination, it will be seen, that a rectangle circumscribing the mould 3, 3, gives the size of the stone in its square state, and, that if each stone in both arches be thus enclosed, the dimensions for each will be found, as also the position in which the moulds must be placed. The dark lines give the different bevels which must be carefully prepared and applied to the stones in the manner represented in the figure.

Fig. 557. To draw the joints of the stones for an elliptical arch in a wall, &c.

The curve is here described by the intersection of lines, which, certainly, gives the most easy and pleasing curve, as segments of circles apply only under certain data, or in the proportion which the axis major has to the axis minor, while the intersection of lines apply to any description of ellipsis. Find the foci F. In an ellipsis the distance of either focus from one extremity of the axis minor is equal to the semi-axis major; that is, DF is equal to c C. Then to find any joint, a b, draw lines from both foci through the point b, as F e, f d, and bisect the angle d b e by the line a b, which is the joint required.

Having thus given a general outline of the principles of masonry, and accompanied the same with a few examples on the most abstruse parts of the art, we shall conclude this part of our treatise with the methods employed in the mensuration of masons' work.

Rough stone or marble is measured by the foot cube : but in measuring for workmanship, the superficies or surface, for plain work, is measured before it is sunk. In measuring ashlar, one bed and one upright joint are taken and considered plain work. In taking the plain sunk, or circular work, and the straight moulded, or circular moulded work, particular care is required to distinguish the different kinds of work in the progress of preparing the stone. In measuring strings, the weathering is denominated *sunk work*, and the grooving *throatings*.

Stone cills to windows, &c. are, in general, about 4¼ inches thick and 8 inches broad, and are weathered at the top, which reduces the front edge to about 4 inches, and the horizontal surface at the top to about 1¼ inch on the inside ; so that the part taken away is 6¼ inches broad and three quarters of an inch deep. Cills, when placed in the wall, generally project about 2¼ inches. The horizontal part left on the inside of the cill is denominated *plain work;* and the sloping part *sunk work;* and in the dimension book are entered thus,—

$$1\tfrac{1}{4}$$
$$4$$
$$2\tfrac{1}{4}$$

8 inches the breadth of the plain work in the cill according to the above dimensions,—then

¼			
8	2 8	Plain work.	
4			
6½	2 2	Sunk work.	
8			
2 4	6	Plain to ends.	
¼ 0		of throating.	

No account is taken of the sawing.

Cornices are measured by girthing round the moulded parts, that is, the whole of the vertical and under parts, called moulded work :—for example, suppose a cornice project one foot, girth two feet, and is 40 feet in length, then the dimensions will be entered as under,—

40		
2	80	Moulded work.
40		
1	40	Sunk work at top.

All the vertical joints must be added to the above.

Cylindrical work is measured in the girth ; and the surface is calculated to be equivalent to plain work twice taken.

For example, suppose it be required to measure the plain work or a cylinder, 10 feet long, and 5 feet in circumference, the dimensions would then be entered

$$10 \quad 0$$
$$5 \quad 0 \quad \text{500 Sup'. plain work, double measure.}$$

Paving-slabs and chimney-pieces are found by superficial measure, as also are stones under two inches thick.

The manner in which the dimensions of a house are taken, vary according to the place and the nature of the agreement.

In Scotland, and most parts of England, if the builder engages only for workmanship, the dimensions are taken round the outside of the house for the length, and the height is taken for the width, and the two multiplied together gives the superficial contents. This, however, applies only when the wall is of the same thickness all the way up; and when not, as many separate heights are taken as there are thicknesses. This mode of measuring gives something more than the truth, by the addition of the four quoins, which are pillars of two feet square; but this is not more than considered sufficient to compensate the workmen for the extra labour in plumbing the quoins.

If there be a plinth, string, course-cornice, or blocking course, the height is taken from the bottom of the plinth to the top of the blocking course, including the thickness of the same; that is, the measurer takes a line or tape and begins, we will suppose, at the plinth, then stretching the line to the top, bends it into the offset, or weathering, and, keeping the corner tight at the internal angle, stretches the line vertically upon the face of the wall, from the internal angle to the internal angle of the string; then girths round the string to the internal angle at the top of the string, and keeping the line tight at the upper internal angle, stretches it to meet the cornice; he then bends it round all the mouldings to the internal angle of the blocking course, from which he stretches the string up to the blocking course, to the farther extremity of the breadth of the top of the same; so that the extent of the line is the same as the vertical section stretched out: this dimension is accounted the height of the building.

With respect to the length, when there are any pilasters, breaks, or recesses, the girth of the whole is taken at the length. This method is, perhaps, the most absurd of any admitted in the art of measuring; since this addition in height and length, is not sufficient to compensate for the value of the workmanship on the ornamental parts.

The value of a rood of workmanship must be first obtained by estimation, that is, by finding the cost of each kind of work, such as plinth, strings, cornices, and architraves, &c. and adding to them the plain ashlar work, and the value of the materials, the amount of which, divided by the number of roods contained in the whole, give the mean price of a single rood. When the apertures or openings in a building are small, it is not customary to make deductions either for the materials or workmanship which are there deficient, as the trouble of plumbing and returning the quoins, is considered equivalent to the deficiency of materials occasioned by such aperture.

2 N

Elsam's Gentleman's and Builder's Assistant, gives the following information on the practice of measuring. rough stone work.

To find the number of perches contained in a piece of rough stone-work.

If the wall be at the standard thickness, that is, 12 inches high, 18 inches thick, and 21 feet long, divide the area by 21, and the quotient, if any, will be the answer in perches, and the remainder, if any, is feet. If the wall be more or less than 18 inches thick, multiply the area of the wall by the number of inches in thickness, which product, divided by 18, and that quotient by 21, will give the perches contained.

Example. A piece of stone-work is 40 feet long, 20 feet high, and 24 inches thick, how many perches are contained in it?

```
                40 length.
                20 height.
                ————
                800
                 24
                ————
               3200
               1600
               ————— 21)       P.  F.  In.
           18) 19200  (1066  ( 50  16   8
               18      105
               ————    ————
               120      16
               108     ————
               ————
               120
               108
               ————
         12 equal to 8 inches.
```

The method last described, of finding the value of mason's work, is usually adopted, the perch being the standard of the country; but the most expeditious way of ascertaining the value, is to cube the contents of the wall, and to charge the work at per foot. To ascertain the value of common stone-work, a calculation should be made of the prime cost of all the component parts, consisting of the stones in the quarry, the expense of quarrying, land-carriage to the place where it is to be used, with the extra trouble and consequent expense in carrying the stone one, two, three, or more stories higher. Also the price of the lime when delivered, together with the extra expense of wages to workmen, if in the country; all these circumstances must be taken into consideration in finding the value of a perch of common stone-work, the expense of which will be found to vary according to

ircumstances, in degrees scarcely credible ; wherefore
uite price cannot, with propriety, be fixed.

BRICKLAYING

uilding upon an inclined plane, or rising ground, the
ition must be made to rise in a series of level steps,
ling to the general line of the ground, to insure a firm
r the courses, and prevent them from sliding ; for if
ode be not adopted, the moisture in the foundations in
eather, will induce the inclined parts to descend, to
nifest danger of fracturing the walls and destroying
ilding.

ralling, in dry weather, when the work is required to
n, the best mortar must be used ; and the bricks must
tted, or dipped in water, as they are laid, to cause
to adhere to the mortar, which they would not do if
y ; for the dry sandy nature of the brick absorbs the
ure of the mortar and prevents adhesion.

arrying up the wall, not more than four or five feet of
rt should be built at a time ; for, as all walls shrink im-
tely after building, the part which is first carried up will
before the adjacent part is carried up to it, and, con-
itly, the shrinking of the latter will cause the two
to separate ; therefore, no part of a wall should be
l higher than one scaffold, without having its contin-
urts added to it. In carrying up any particular part,
ds should be regularly sloped off, to receive the bond
adjoining parts on the right and left.

re are two descriptions of bonds ; *English bond*, and
h bond. In the *English bond*, a row of bricks is laid
wise on the length of the wall, and is crossed by ano-
w, which has its length in the breadth of the wall,
an alternately. Those courses in which the lengths
bricks are disposed through the length of the wall,
med *stretching courses*, and the bricks *stretchers* ; and
urses in which the bricks run in the thickness of
ths of the walls, *heading courses*, and the bricks
L

other description of bond, called *Flemish bond*, com-
placing a header and a stretcher alternately in the
urse. The latter is deemed the neatest, and most
; but in the execution is attended with great incom-
e, and, in most cases, does not unite the parts of a

wall with the same degree of firmness as the English bond. In general, it may be observed, that, whatever advantages are gained by the English bond in tying a wall together in its thickness, they are lost in the longitudinal bond ; and *vice-versa.* To remove this inconvenience, in thick walls, some builders place the bricks in the cone at an angle of forty-five degrees, parallel to each other, throughout the length of every course, but reversed in the alternate courses ; so that the bricks cross each other at right angles. But even here, though the bricks in the cone have sufficient bond, the sides are very imperfectly tied, on account of the triangular interstices formed by the oblique direction of the internal bricks against the flat edges of those in the outside.

Concerning the English bond, it may be observed, that, as the longitudinal extent of a brick is nine inches, and its breadth four and a half, to prevent two vertical joints from running over each other at the end of the first stretcher from the corner, it is usual, after placing the return corner stretcher, which occupies half of the length of this stretcher, and becomes a header in the face, as the stretcher is below, to place a quarter brick on the side, so that the two together extend six inches and three-quarters, being a lap of two inches and a half for the next header. The bat thus introduced is called a *closer.* A similar effect may be obtained by introducing a three-quarter bat at the corner of the stretching course, so that the corner header being laid over it, a lap of two inches and a quarter will be left, at the end of the stretchers below, for the next header, which being laid on the joint below the stretchers, will coincide with its middle.

In the winter, it is very essential to keep the unfinished wall from the alternate effects of rain and frost ; for if it is exposed, the rain will penetrate into the bricks and mortar, and, by being converted into ice, expand, and burst or crumble the materials in which it is contained.

The decay of buildings, so commonly attributed to the effects of time, is, in fact, attributable to this source ; but as finished edifices have only a vertical surface, the action and counter-action of the rain and frost extend not so rapidly as in an unfinished wall, where the horizontal surface permits the rain and frost to have easy access into the body of the work. Great care, therefore, must be taken as soon as the frost or stormy weather sets in, to cover the un-

product by 9; and the quotient will give the area of the wall at t
standard: divide this standard area by 272, and the quotient will gi
the number of rods; the remainder the reduced feet.

Example. The length of a wall is O feet, the height 20 feet, and t
thickness equal to the length of three bricks; it is therefore requir
to know how many rods of brick-work is contained in the said wall?

By Case I.   60
      20
      ——
      1200
      6
      ——
3) 7200
      ——
272) 2400 (8 rods 224 feet the answer.
      2176
      ——
      224

Case II   60
      20
      ——
      1200
      2.3   thickness of wall
      ——
      2400
      300
      ——
306) 2700 (8 rods 252 feet the answer.
      2448
      ——
      252

Case III.   60
      20
      ——
      1200
      2.3
      ——
      2400
      300
      ——
      2700
      8
      ——
9) 21600
      ——
272) 2400 (8 rods 224 feet, as in Case I
      2176
      ——
      224

In the calculation of brick-work, where there are several walls of different thicknesses, it will be quite unnecessary to use the divisors 3 and 272, as will be hereafter shown.

In taking dimensions for workmanship, it is usual to allow the length of each wall on the external side, to compensate for plumbing the angles; but this practice must not be resorted to for labour and materials, as it gives too much quantity in the height of the building or story by two pillars of brick; and in the horizontal dimensions by the thickness of the walls.

In measuring walls, faced with bricks of a superior quality, most surveyors measure the whole as common work, and allow an additional price per rod for the facing, as the superior excellence of the work, and quality of the bricks may deserve.

Every recess or aperture made in any of the faces must be deducted; but an allowance per foot lineal should be made upon every right angle, whether external or internal, excepting when two external angles may be formed by a brick in breadth, and then only one of them must be allowed.

Gauged arches are sometimes deducted and charged separate; but as the extra price must be allowed in the former case, it will amount to the same thing.

In measuring walls containing chimneys, it is not customary to deduct the flues; but this practice, so far as regards the materials, is unjust, though, perhaps, by taking the labour and materials together, the overcharge, with respect to the quantity of bricks and mortar, may, in some degree, compensate for the loss of time: on the other hand, if the proprietor finds the materials, it is not customary to allow for the trouble of forming the flues, which, consequently, is a loss to the contractor who has engaged by task-work or measure.

If the breast of a chimney project from the face of the wall, and is parallel to it, the best method is, to take the horizontal and vertical dimensions of the face, multiply them together, and multiply the product by the thickness, taken in the thinnest part, without noticing the breast of the chimney; then find the solidity of the breast itself, add these solidities together, and the sum will give the solidity of the wall, including the vacuities, which must be deducted for the real solidity. Nothing more is necessary to be said of the shaft, than to take its dimensions in height, breadth, and thickness, in order to ascertain its solidity.

If a chimney be placed at an angle, with the face of the breast intersecting the two sides of the wall, the breast of the chimney must be considered a triangular prism. To take the dimensions :—from the intersections of the front of the breast into the two adjacent walls, draw two lines on the floor, parallel to each adjacent wall ; then the triangle on the floor, included between the front and these lines, will be equal to the triangle on which the chimney stands, and, consequently, equal to the area of the base. To attain the area of the triangular base, the dimensions may be taken in three various ways, almost equally easy ; one of which is, to take the extent of the base, which is the horizontal dimension of the breast, and multiply it by half of the perpendicular ; or multiply the whole perpendicular by half the base : but, as this calculation would, in cases of odd numbers, run somewhat long, a more preferable method is, to multiply the whole base by the whole perpendicular, and take half of the product, which will give the area on which the chimney stands ; and which, multiplied by the height, gives the solid contents of the chimney. From this contents is to be deducted the vacuity for the fire-place.

A row of plain tiles, laid edge to edge, with their broad surfaces parallel to the termination of a wall, so as to project over the wall at right angles to the vertical surface, is called *single plain tile creasing* ; and two rows, laid one above the other, the one row breaking the joints of the other, are called *double plain tile creasing*.

Over the plain tile creasing a row of bricks is placed on edge, with their length in the thickness of the wall, and are called a *barge course,* or *cope.*

The bricks in gables, which terminate with plain tile creasing coped with bricks, in order to form the sloping bed for the plain tile creasing, must be cut, and the sloping of the bricks thus, is called *cut splay.*

Plain tile creasing and cut splay are charged by the foot run ; and the latter is sometimes charged by the superficial foot.

A brick wall built in pannels between timber quarters is called *brick nogging* ; and is generally measured by the yard square, the quarters and nogging pieces being included in the measure.

Pointing is the filling up the joints of the bricks after the walls are built. It consists in raking out some of the mortar from the joints, and filling them again with blue mortar, and in one kind of pointing, the courses are simply rked with the end of a trowel, called *flat-joint pointing*; if, in addition to flat-joint pointing, plaster be inserted the joint with a regular projection, and neatly paved to parallel breadth, it is termed *kick pointing*, or *kick-joint pointing*, or formerly, *kick, and patt*. Pointing is measured the foot superficial, including in the price, mortar, labour, scaffolding.

Rubbed and gauged work is set in putty or mortar; and measured either by the foot superficial, or the foot run, according to the manner in which it is constructed.

In measuring canted bow windows, the sides are considered as continued straight lines; but the angles on the interior side of the building, whether they be external or internal, are allowed for in addition, and paid for under the nomination of *run of bird's mouth*. All angles within the building, if oblique, from whatever cause they are made, whether by straight or circular bows, or the splays of windows, allowed for, under the head o: *run of cut splay*.

Brick cornices are measured by the lencal foot; but as various kinds of cornices require more or less difficulty in execution, the price must depend on the labour and the use of the material used.

Garden walls are measured the same as other walls, but if interrupted by piers, the thin part may be measured as in common walling, and the piers by themselves, making an allowance, at per foot run, for the right angles The coping measured by itself, according to the kind employed.

Paving is laid either with bricks, or tiles, and is measured the yard square. The price, per yard, is regulated by the manner in which the bricks or tiles are laid, whether flat or edge-ways, or whether any of them be laid in sand or mortar. The circular parts of drains may be reduced either to the standard, or the cubic foot; and the number of rods may, if required, be taken. The mean dimensions of the arch may be found, by taking the half sum of the exterior and interior circumferences; but, perhaps, it were better to make a price of the common measure, whether it be a foot, yard, or rod, greater as the diameter is less; but as the reciprocal ratio would increase the price too much in small diameters, perhaps prices at certain diameters would be a sufficient regulation.

The following tables will be found an acquisition to those persons to whom a saving of time is an object:—

TABLE I.

This Table shews what quantity of bricks are necessary to construct piece of brick-work of any given dimensions, from half a brick to two bricks and a half in thickness; and by which the number for any thickness may be found.

This Table is at the rate of 4500 bricks to the rod of reduced brick work, including waste.

Area of the face of wall.	The number of bricks thick and the quantity required.				
	½ brick.	1 brick.	1½ brick.	2 bricks.	2½ bricks.
1	5	11	16	22	27
2	11	22	33	44	55
3	16	33	49	66	82
4	22	44	66	88	110
5	27	55	82	110	137
6	33	66	99	132	165
7	38	77	115	154	193
8	44	88	132	176	220
9	49	99	148	198	288
10	55	110	165	220	275
20	110	220	330	441	551
30	165	330	496	661	827
40	220	441	661	882	1102
50	275	551	827	1102	1378
60	330	661	992	1323	1655
70	386	772	1158	1544	1930
80	441	882	1323	1764	2205
90	496	992	1488	1985	2480
100	551	1102	1654	2205	2757
200	1102	2205	3303	4411	5514
300	1654	3308	4963	6617	8272
400	2205	4411	6617	8823	11,029
500	2757	5514	8272	11,029	13,786
600	3308	6617	9926	13,235	16,544
700	3860	7720	11,580	15,441	19,301
800	4411	8823	13,235	17,647	22,058
900	4963	9926	14,889	19,852	24,816
1000	5514	11,029	16,544	22,058	27,573
2000	11,029	22,058	33,088	44,117	55,147
3000	16,544	33,088	49,632	66,176	82,720
4000	22,058	44,117	66,176	88,235	110,294
5000	27,573	55,147	82,720	110,294	137,867
6000	33,088	66,176	99,264	132,352	165,441
7000	38,602	77,205	115,808	154,411	193,014
8000	44,117	88,235	132,352	176,470	220,588
9000	49,632	99,264	148,896	198,529	248,161
10,000	55,147	110,294	165,441	220,588	275,735
20,000	110,294	220,588	330,882	441,176	551,470
30,000	165,441	330,882	496,323	661,764	827,205
40,000	220,588	441,176	661,764	882,352	1,102,940
50,000	275,735	551,470	827,205	1,102,940	1,378,675
60,000	330,882	661,764	992,646	1,323,528	1,654,410
70,000	386,029	772,058	1,158,087	1,544,116	1,930,145
80,000	441,176	882,352	1,323,528	1,764,704	2,205,880
90,000	496,323	992,646	1,488,969	1,985,292	2,481,615

The left-hand column contains the number of superficial t contained in the wall to be built : the adjacent columns w the number of bricks required to build a wall of the erent thicknesses of ½, 1, 1½, 2, and 2½ bricks.

xample. Suppose it be required to find the number of bricks neces- to build a wall 1 brick thick, containing an area of 5760 feet? it look for 5000 in the left hand column, and you will find that it takes 147 bricks, add to this quantity, the number necessary for each of the r component parts, and we shall have the following

5000 will require 55147
700 7720
60 661
———
5760   63,528

TABLE II.

ews the number of rods contained in any number of superficial feet, 1 to 10,000, and from ½ a brick to 2½ bricks and thence by addi- to any number, and to any thickness, at the rate of 4500 bricks to od.

Feet sup.	½ brick.				1 brick.				1½ brick.				2 bricks.				2½ bricks.			
	R.	Q.	F.	In.	R.	Q.	F.	In.	R.	Q.	F.	In.	R.	Q.	F.	In.	R.	Q.	F.	In.
1	0	0	0	4	0	0	0	8	0	0	1	0	0	0	1	4	0	0	1	8
2	0	0	0	8	0	0	1	4	0	0	2	0	0	0	2	8	0	0	3	4
3	0	0	1	0	0	0	2	0	0	0	3	0	0	0	4	0	0	0	5	0
4	0	0	1	4	0	0	2	8	0	0	4	0	0	0	5	4	0	0	6	8
5	0	0	1	8	0	0	3	4	0	0	5	0	0	0	6	8	0	0	8	4
6	0	0	2	0	0	0	4	0	0	0	6	0	0	0	8	0	0	0	10	0
7	0	0	2	4	0	0	4	8	0	0	7	0	0	0	9	4	0	0	11	8
8	0	0	2	8	0	0	5	4	0	0	8	0	0	0	10	8	0	0	13	4
9	0	0	3	0	0	0	6	0	0	0	9	0	0	0	12	0	0	0	15	0
10	0	0	3	4	0	0	6	8	0	0	10	0	9	6	13	4	0	0	16	8
11	0	0	3	8	0	0	7	4	0	0	11	0	0	0	14	8	0	0	18	4
12	0	0	4	0	0	0	8	0	0	0	12	0	0	0	16	0	0	0	20	0
13	0	0	4	4	0	0	8	8	9	0	13	0	0	0	17	4	0	0	21	8
14	0	0	4	8	0	0	9	4	0	0	14	0	0	0	18	8	0	0	23	4
15	0	0	5	0	0	0	10	0	0	0	15	0	0	0	20	0	0	0	25	0
16	0	0	5	4	0	0	10	8	0	0	16	0	0	0	21	4	0	0	26	8
17	0	0	5	8	9	0	11	4	0	0	17	0	0	0	22	8	0	0	28	4
18	0	0	6	0	0	0	12	0	0	0	18	0	0	0	24	0	0	0	30	0
19	0	0	6	4	0	0	12	8	0	0	19	0	0	0	25	4	0	0	31	8
20	0	0	6	8	0	0	13	4	0	0	20	0	0	0	26	8	0	0	33	4
21	0	0	7	0	0	0	14	0	0	0	21	0	0	0	28	0	0	0	35	0
22	0	0	7	4	0	0	14	8	0	0	22	0	0	0	29	4	0	0	36	8
23	0	0	7	8	0	0	15	4	0	0	23	0	0	0	30	8	0	0	38	4
24	0	0	8	0	0	0	16	0	0	0	24	0	0	0	32	0	0	0	40	0
25	0	0	8	4	0	0	16	8	0	0	25	0	0	0	33	4	0	0	41	8
26	0	0	8	8	0	0	17	4	0	0	26	0	0	0	34	8	0	0	43	4
27	0	0	9	0	0	0	18	0	0	0	27	0	0	0	36	0	0	0	45	0
28	0	0	9	4	0	0	18	8	0	0	28	0	0	0	37	4	0	0	46	8
29	0	0	9	8	0	0	19	4	0	0	29	0	0	0	38	8	0	0	48	4
30	0	0	10	0	0	0	20	0	0	0	30	0	0	0	40	0	0	0	50	0
31	0	0	10	4	0	0	20	8	0	0	31	0	0	0	41	4	0	0	51	8

Feet sup.	¾ brick.				1 brick.				1½ brick.				2 bricks.				2½ bricks.			
	R.	Q.	F.	In.	R.	Q.	F.	In.	R.	Q.	F.	In.	R.	Q.	F.	In.	R.	Q.	F.	In.
32	0	0	10	8	0	0	21	4	0	0	32	0	0	0	42	8	0	0	53	4
33	0	0	11	0	0	0	22	0	0	0	33	0	0	0	44	0	0	0	55	0
34	0	0	11	4	0	0	22	8	0	0	34	0	0	0	45	4	0	0	56	8
35	0	0	11	8	0	0	23	4	0	0	35	0	0	0	46	8	0	0	58	4
36	0	0	12	0	0	0	24	0	0	0	36	0	0	0	48	0	0	0	60	0
37	0	0	12	4	0	0	24	8	0	0	37	0	0	0	49	4	0	0	61	8
38	0	0	12	8	0	0	25	4	0	0	38	0	0	0	50	8	0	0	63	4
39	0	0	13	0	0	0	26	0	0	0	39	0	0	0	52	0	0	0	65	0
40	0	0	13	4	0	0	26	8	0	0	40	0	0	0	53	4	0	0	66	8
41	0	0	13	8	0	0	27	4	0	0	41	0	0	0	54	8	0	1	0	4
42	0	0	14	0	0	0	28	0	0	0	42	0	0	0	56	0	0	1	2	0
43	0	0	14	4	0	0	28	8	0	0	43	0	0	0	57	4	0	1	3	8
44	0	0	14	8	0	0	29	4	0	0	44	0	0	0	58	8	0	1	5	4
45	0	0	15	0	0	0	30	0	0	0	45	0	0	0	60	0	0	1	7	0
46	0	0	15	4	0	0	30	8	0	0	46	0	0	0	61	4	0	1	8	8
47	0	0	15	8	0	0	31	4	0	0	47	0	0	0	62	8	0	1	10	4
48	0	0	16	0	0	0	32	0	0	0	48	0	0	0	64	0	0	1	12	0
49	0	0	16	4	0	0	32	8	0	0	49	0	0	0	65	4	0	1	13	8
50	0	0	16	8	0	0	33	4	0	0	50	0	0	0	66	8	0	1	15	4
60	0	0	20	0	0	0	40	0	0	0	60	0	0	1	12	0	0	1	32	0
70	0	0	23	4	0	0	46	8	0	1	2	0	0	1	25	4	0	1	48	8
80	0	0	26	8	0	0	53	4	0	1	12	0	0	1	38	8	0	1	65	4
90	0	0	30	0	0	0	60	0	0	1	22	0	0	1	52	0	0	2	14	0
100	0	0	33	4	0	0	66	8	0	1	32	0	0	1	65	4	0	2	30	8
200	0	0	66	8	0	1	65	4	0	2	64	0	0	3	62	8	1	0	61	4
300	0	1	32	0	0	2	64	0	1	0	28	0	1	1	60	0	1	3	24	0
400	0	1	65	4	0	3	62	8	1	1	60	0	1	3	57	4	2	1	54	8
500	0	2	30	8	1	0	61	4	1	3	24	0	2	1	54	8	3	0	17	4
600	0	2	64	0	1	1	60	0	2	0	56	0	2	3	52	0	3	2	48	0
700	0	3	29	4	1	2	58	8	2	2	20	0	3	1	49	4	4	1	10	8
800	0	3	62	8	1	3	57	4	2	3	52	0	3	3	46	8	4	3	41	4
900	1	0	28	0	2	0	56	0	3	1	16	0	4	1	44	0	5	2	4	0
1000	1	0	61	4	2	1	54	8	3	2	48	0	4	3	41	4	6	0	34	8
2000	2	1	54	8	4	3	41	4	7	1	28	0	9	3	14	8	12	1	1	4
3000	3	2	48	0	7	1	28	0	11	0	8	0	14	2	56	0	18	1	36	0
4000	4	3	41	4	9	3	14	8	14	2	56	0	19	2	29	4	24	2	2	8
5000	6	0	34	8	12	1	1	4	18	1	36	0	24	2	2	8	30	2	37	4
6000	7	1	28	0	14	2	56	0	22	0	16	0	29	1	44	0	36	3	4	0
7000	8	2	21	4	17	0	42	8	25	2	64	0	34	1	17	4	42	3	38	8
8000	9	3	14	8	19	2	29	4	29	1	44	0	39	0	58	8	49	0	5	4
9000	11	0	8	0	22	0	16	0	33	0	24	0	44	0	32	0	55	0	40	0
10000	12	1	1	4	24	2	2	8	36	3	4	0	49	0	5	4	61	1	6	8

The left-hand column contains the area of the wall in superficial feet the adjacent columns the quantity, reduced to the standard thickness, according to the different thicknesses on the top.

Example. What is the quantity of reduced brick-work in a wall containing 4540 superficial feet, 2 bricks thick?

Divide the number as in the preceding table, into its component part say 4540 = 4000 + 500 + 40, then by the table.

	R.	Q.	F.	In.
4000 contains	19	2	29	4
500 . . .	2	1	54	8
40 . . .	0	0	53	4
	22	1	1	4

The same by rule.

4540
4 number of half bricks.
—
3)18160(R. Q. F. In. as above.
—
272) 6053 + 4(22   1   1   4
544
—
613
54½
—
¼ of a rod 68) 69 (1
68
—
1

——

TABLE III.

Shows the value of reduced brick-work per rod, calculated at the several prices of £3 5s. £3 10s. £3 15s. £4 0s. £4 5s. and £4 10s. per rod in mortar, labour, and scaffolding; and of bricks from £1 10s. to £3 0s. per thousand; allowing 4500 bricks to the rod.

Bricks per thousand.	Mortar and Labour 3l. 5s. per rod.			Mortar and Labour 3l. 10s. per rod.			Mortar and Labour 3l. 15s. per rod.			Mortar and Labour 4l. 0s. per rod.			Mortar and Labour 4l. 5s. per rod.			Mortar and Labour 4l. 10s. per rod.		
£. s. d.	£.	s.	d.	£.	s.	d.	£.	s.	d.	£.	s.	d.	£.	s.	d.	£.	s.	d.
1 10 0	10	0	0	10	5	0	10	10	0	10	15	0	11	0	0	11	5	0
1 12 0	10	9	0	10	14	0	10	19	0	11	4	0	11	9	0	11	14	0
1 14 0	10	18	0	11	3	0	11	8	0	11	13	0	11	18	0	12	3	0
1 16 0	11	7	0	11	12	0	11	17	0	12	2	0	12	7	0	12	12	0
1 18 0	11	16	0	12	1	0	12	6	0	12	11	0	12	16	0	13	1	0
2 0 0	12	5	0	12	10	0	12	15	0	13	0	0	13	5	0	13	10	0
2 2 0	12	14	0	12	19	0	13	4	0	13	9	0	13	14	0	13	19	0
2 4 0	13	3	0	13	8	0	13	13	0	13	18	0	14	3	0	14	8	0
2 6 0	13	12	0	13	17	0	14	2	0	14	7	0	14	12	0	14	17	0
2 8 0	14	1	0	14	6	0	14	11	0	14	16	0	15	1	0	15	6	0
2 10 0	14	10	0	14	15	0	15	0	0	15	5	0	15	10	0	15	15	0
2 12 0	14	19	0	15	4	0	15	9	0	15	14	0	15	19	0	16	4	0
2 14 0	15	8	0	15	13	0	15	18	0	16	3	0	16	8	0	16	13	0
2 16 0	15	17	0	16	2	0	16	7	0	16	12	0	16	17	0	17	2	0
2 18 0	16	6	0	16	11	0	16	16	0	17	1	0	17	6	0	17	11	0
3 0 0	16	15	0	17	0	0	17	5	0	17	10	0	17	15	0	18	0	0

Example. What is the price of a rod of brick-work, when the rate of bricks £2 2s. per thousand, and the price of mortar £4 5s. per rod?
Look from the given column of bricks until you come under £4 5s. the even price of labour and mortar, and you will find £13 14s. the price of a rod.

iis department we shall treat first, of the most approved
ds of lengthening beams, by what is termed scarfing,
iing them in pieces ; secondly, of the strengthening
ns by trussing ; thirdly, of the methods of joining
mbers at angles, in any given direction ; and lastly, of
nde of connecting several timbers in order to com-
the design, and to effect certain powers respectively
cd by each individual piece.

lengthen a piece of timber implies the act of joining
tening two distinct pieces, so that a part of the end of
hall lap upon the end of another, and the surfaces
th, being one continued plane, form a close joint,
by workmen a *scarf*. It is manifest, that two bodies,
together and intended to act as one continued piece,
tate of tension, or compression, cannot, by any possi-
cans, be so strong as either pieces taken separately.
crefore, requires much attention, and careful discri-
ion, in the choice and selection of such methods as
ie most applicable to the peculiar circumstances o.
isc. Every two pieces of timber joined in the manner
described, and, indeed, in most other cases, require
force to compress them equally on each side, and
particularly when the pieces are light ; for this pur-
iron bolts are used, which act as a tie, and possess the
effect as two equal and opposite forces would have in
iressing the beam on each side the joint : and as the co-
c power of iron is very great, the hole, which is made
cive the bolt, may be of such dimensions as will not,
e least degree, tend to diminish the strength of the
r. When wooden pins are used, the bore is larger, and
oints weaker ; consequently the two pieces, thus con-
d, are not held together by any compression of the pin,
ierely by the friction of the individual pieces.

specific distance can be laid down for the length of the
, though, in general, it may be observed, that, a long
has but little effect in diminishing the cohesive strength
compound piece of timber ; on the contrary, it affords
iportunity of increasing the number of bolts.

. 558 shows the method of joining two pieces of timber by means
ingle step on each piece:

this method more than one-half the power is lost ; and
carf is not calculated to resist the force of tension
l to a single piece sawed half through its thickness from
ipposite side, at a distance equal to the length of the

scarf; by the application of straps, however, it may be made
to resist a much greater force.

Fig. 559 represents a scarf with parallel joints, and a single table upon
each piece.

In this the cohesive strength is decreased in a greater
degree than the preceding example, by the projection of the
table; but this affords an opportunity of driving a wedge
through the joint between the ends of the tables, and thereby forcing the abutting parts to a joint.

A scarf of this description to be longer than those which
have no tables, and the transverse parts of the scarf, must
be strapped and bolted.

Fig. 500 presents us with the same opportunity of wedging as before.
In this figure, if the parts LM and NO be compressed together by bolts
as firmly as if they were but one piece, and if the projection of the tables
be equal to the transverse parts of the joints L and O, the loss of strength,
compared with that of a solid piece, will be no more than what it would
be at L and O.

Strapping across the transverse part of the joint is much
the best and most effectual way of preventing the pieces
from being drawn from each other, by the sliding of the
longitudinal parts of the scarf, and, therefore, giving to the
bolts an oblique position.

Fig. 561 is a scarf formed by several steps.

In this, if all the transverse parts of the steps be equal,
and the longitudinal parts strongly compressed by bolts, the
loss of strength will only be a fourth, compared to that of a
solid piece, there being four transverse parts, that is, the
part which the end of the steps is of the whole.

Fig. 562 is a scarf with a bevel joint, and equally as eligible for ordinary purposes as any in use.

Figs. 561 and 563. Scarfs intended for longer bearings than the preceding one.

Fig. 564 represents the method of constructing a compound timber,
when two pieces are not of adequate length to allow them to lap, by
means of a third piece joined to both by a double scarf, formed by several
gradations or steps, the pieces abutting upon each other with the middle of
the connecting piece over their abutment.

That which shall next claim our attention is a consideration of the principles and the best methods of strengthening beams by trussing.

When girders are extended beyond a certain length, they
bend under their own weight, and the degree of curvature
increases in a proportion far greater than that of their
lengths. The best method to obviate this *sagging*, as it is

termed, without the support of posts, &c. is to make the beam in two equal lengths, and insert a truss, so that when the two pieces are confined together by bolts, the truss may be included between them, and cause them to act as a tie. To prevent any unfavourable results from natural tendency of the timbers to shrink, the posts of the truss may be made of iron, and screwed, and nutted at the ends; and to give a still stronger abutment, the braces may be let in with grooves into the side of each flitch, or piece, which form the beam. The ends of the abutments are also made of iron, screwed, or nutted, at each of the ends, and bolted through the thickness of both pieces, with a broad part in the middle, that the braces may abut upon the whole dimension of their section; or, otherwise, the abutments are made in the form of an inverted wedge at the bottom, and rise cylindrically to the top, where they are screwed and nutted.

These methods may be constructed either with one king-bolt in the middle, or with a truss-bolt at one-third of the length from each end. When two bolts are applied, they include a straining place in the middle. The two braces may be constructed of oak, or cast or wrought iron; but the latter material is seldom used: for, as all metals are liable to contract, wood is considered the best material. With respect to the bolts, iron is indispensable.

The higher the girder is, the less are the parts liable to be effected by the stress; and, consequently, the risk of their giving way under heavy weights, or through long bearings, is less.

Figs. 565 and 566 are two examples of girders calculated from their size to sustain very heavy weights. If the tie beam be very strong, the abutments may be wedged; but the wedges ought to be very long, and a little taper, that there may be no inclination to rise. The excess of length may afterwards be taken off.

In joining two timbers together, in any given direction, the joinings, as practised by carpenters, are almost infinitely various; and though some are executed with a view merely to gratify the eye, the majority have decided advantages, and each, in peculiar cases, is to be preferred. In this treatise, our limits will not permit us to enter upon a description of such as yield no substantial benefit, or are employed only in connecting small work; but, even in these, the skill of the workman may at all times be discovered by his selection of materials. It may here be observed, that, as all timber is either more or less, according to the dryness, and the quality of the timber used, subject to shrink, the carpenter should very carefully consider how much the dimensions of his framings will be affected by it, and so arrange the inferior pieces that their shrinkage shall be in the same direction as the shrinkage of the framing, and so conduce to the greater stability of the whole. If this be not attended to, the parts will separate and split asunder.

Two pieces of timber may be connected either by making both planes of contact parallel with or at right angles to the fibres, or by making the joint parallel with the fibres of the one piece, and at right or oblique angles to the other, or at oblique angles to the fibres of both pieces.

If two pieces of timber are connected, so that the joint runs parallel with the fibres of both, it is called a *longitudinal joint ;* but when the place of the joint is at right angles to the fibres of both, an *abutting joint.* Butting and mitre joints are seldom used in carpentry.

When two pieces of timber are joined together at one or more angles, the one piece will meet the other and form one angle, or by crossing it make two angles, or the two pieces will cross each other and form four angles.

In all the following cases of connecting two timbers, it is supposed, that the sides of the pieces are parallel with the fibres, or, when the fibres are crooked, as nearly so as possible ; and that each piece, the four sides being at right angles to each other, has at least one of its surfaces in the same plane with those of the other. The angle or angles so formed will be either right or obtuse.

Fig. 567, is an example of a notched joint, which is the most common and simple form, and, in some cases, the strongest for joining two timbers at one or more angles, particularly when bolted at the joint. The form of the joint may be varied, according to the position of the sides of the pieces, the number of angles, the quantity and direction of the stress on the one or both pieces, or by any combination of their circumstances. Notching admits two pieces to be joined at from one to four angles ; but joining by mortise and tenon admits only from one to two angles.

In joining by mortise and tenon, four sides of the mortise should, if possible, be at right angles to each other, and to the surface whence it is recessed, and two of these sides parallel with each of the sides which forms a right angle with the side from which the mortise is made : the fifth plane, that is, the bottom of the mortise, is parallel with the top or surface from which the mortise is made. Four sides of the tenon should be parallel to the four sides of the piece ; but there are many cases where a digression is unavoidable.

In the application of timbers to buildings, we will here suppose, that all pieces cut for use have a rectangular section, and when laid down, have their sides perpendicular to, and parallel with, the horizon. If two pieces of timber, therefore, are to be joined at four angles, cut a notch in one piece equal to the breadth of the other, so as to leave the remaining part of the thickness sufficiently strong, and insert the other piece in the notch ; or, if the work is required to be very firm, notch each piece reciprocally to each other's breadth, and fasten them together by pins, spikes, or bolts, as the case may require. This form is applicable when the pieces are equally exposed to a strain.

Fig. 568 will fully elucidate this description of joint.

The framing of timber by dove-tail notching is principally applicable to horizontal framing, where the lower timber is sufficiently supported. Where the lower timber is unsupported it is common to use mortise and tenon, which does not materially weaken the timber ; but when the timber is notched from the upper side, the operation reduces its thickness, and consequently impairs its strength, though, if the solid of one piece fill the excavation of the other, and both be lightly driven or forced together, according to Du Hamel, it will, if not cut more than one third through, rather increase than decrease in strength. It may, however, be observed, that in large works, where heavy timbers are employed, it is difficult, and almost impossible, to fit the mortise and tenon with due accuracy ; and even if the joints were closely fitted at first, the shrinking would occasion cavities on the sides, that would render the tenons of no avail, because the axis of fracture would be nearer to the breaking or under-side of the supporting piece. What has been here said with respect to timbers placed horizontally, applies to framing in every position, when the force is to fall on the plane of the sides ; and if a number of pieces thus liable to lateral pressure on either side, are to be framed into two other stiff pieces, the mortise and tenon will prove best for the purpose.

If it be required to connect two pieces of timber so as to form two right angles, and to be immovable, when the transverse is held or fixed fast, and the standing piece pulled in a direction of its length, cut a dove-tail notch across the breadth of the transverse piece, and notch out the vertical sides of the standing piece at the end, so as to form a similar and equal solid. In some kinds of work, besides the dove-tail, an additional notch is cut to receive the shoulder

of the lower piece. If the position of these pieces
zontal, and the upper is of sufficient weight, or i
ed down by any considerable force, when the p
placed together, the dove-tail will be sufficientl
without the assistance of pins, spikes, or bolts. T
struction requires the timbers to be well seaso
otherwise the shrinking will permit the standing pi
drawn out of the transverse, and thus defeat the p
the construction.

In introducing binding joists, which will, as th
support the bridging joists and boarding of the
framed into girders, there will be a considerable
the extremities, so that it is necessary, in orde
the tenons sufficiently strong, to have a shorter b
non attached to the principal tenon, with a slopin
above, called a *tusk*, which term is likewise appl
tenon, called the *tusk tenon.*

When two parallel pieces, which are quite i
are to have another piece framed between them
ciple is, to insert the one end of the tenon of the
framed in a shallow mortise, and make a long mor
opposite side of the other timber; so that when
piece is moved round the shoulder of the other ex
a centre, it may slide home to its situation. Th
framing a transverse piece between two others, is
in trimming in ceiling joists, which joists are
never cut and fitted into the binding joists before
ing is covered over. The binding joists are alway
before they are disposed in the situation to receiv
ing joists.

When a transverse piece of timber is to be
tween two parallel joists, whose vertical surfaces
rallel, turn the upper edge of the transverse piece
upon the upper horizontal surface of the joists, m
terval, or distance between them, upon the sur
transverse piece now under; then placing the ed
place where it is intended to let down, turn the
piece in the way it is intended to be framed, apply
edge to the oblique surface of the joist, and slide
verse piece so as to bring the mark on the upper s
a line with the straight edge, which being don
in the same manner with the other end, and the
drawn on the vertical sides of the intermediate
give the shoulders of the tenons. This act of
transverse joist between two others is termed *t*

joists; and is particularly useful when the timber is warped
or twisted.

In order that the reader may the more fully understand the
preceding description of the joinings of timbers, we have
annexed a plate (to which the subjoined description refers,)
of the best methods now in practice.

Fig. 457. No. 1 and 2, and 3 and 4, exhibit two methods of a simple
joint, where the two pieces are halved upon each other; in both of which
the end of one piece does not pass the outer surface of the other. No. 3
and 4 represent the two pieces before put together.

Fig. 568, is a method of joining timber, when the end of one piece
passes the end of the other at a small distance. No. 1 represents the
pieces before joined.

Fig. 569 shews how two pieces may be joined by what is termed a
niche.—In this case, the two pieces should be fixed to another by a bolt at
right angles to the niche joint.

Fig. 570. How one piece of timber may be joined to another, when one
of the pieces is extended on both sides of the other piece. Nos. 1 and 2
show the pieces before put together.

Fig. 571 shows the manner of joining the binding joists and girders.
No. 1. The binding joist prepared for being joined to the girder.

Fig. 572 is the general and most approved method of framing the rafter
foot into the girder.

Fig. 573 is a section of the beam, shewing the different shoulders of
the rafter foot.

Fig. 574 is another example, preferable to the former, because the
abutment of the inner part is better supported. In this the beam, when
no broader than the rafter is thick, may be weakened, in which case, it
would require a much deeper socket than is here given; and perhaps an
advantage would be gained by introducing a joint like fig. 576.

Fig. 576 is the method of introducing iron straps to confine the foot of
the rafter to the tie-beam.

When it is found necessary to employ iron straps for
strengthening a joint, considerable attention is required to
place them properly. The first thing to be ascertained is
the direction of the strain. We must then endeavour, as
near as we can, to resolve this strain into a strain parallel
to each piece, and another perpendicular to it. Then the
strap which is to be made fast to any of the pieces, must be
so fixed that it shall resist in the direction parallel to the
piece.

The strap which is generally misplaced, is that which
connects the foot of the rafter with the tie-beam. It binds
down the rafter; but does not act against its horizontal thrust. It should be placed farther back on the beam,
and have a bolt through it, to allow it to turn round; and
should embrace the rafter almost horizontally near the foot,
and be notched square with the back of the rafter. The
example given in No. 10 combines these requisites. By

moving round the eye-bolt, it follows the rafter, and can not pinch and cripple it, which it always does in its ordinary form. Straps which have eye-bolts on the very angles, and allow motion round them, are considered the most perfect.

Fig. 577 exhibits two methods of connecting the struts of a roof, or partition, &c. with the king-post.

' If the action of a piece of timber on another does not extend, but compress, the same, there is no difficulty whatever in the joint, indeed joining is unnecessary: it is enough that the pieces abut on each other; and we have only to take care that the mutual pressure be equally borne by all the parts, and that no lateral pressure, which may cause one of the pieces to slide on the butting joint, be produced. At the joggle of a king-post, a very slight mortise and tenon, with a rafter, or straining beam, is sufficient. It is generally best to make the butting plain, bisecting the angle formed by the sides, or else perpendicular to one of the pieces. For instance, the joint *a* is preferable to *b*, and, indeed, to any uneven joints, which never fail to produce very unequal pressures, by which some of the parts are crippled, and others splintered off.

Fig. 578 is the method of securing the tie-beam and principals, when the king-post is made of an iron rod. ﹔

Fig. 579 shows a method of joining the principals with the king-post by means of an iron dove-tail, which is received in a mortise at the head of each principal.

Trusting that the reader will be able, from the above description, to comprehend the best methods of joining timbers, we shall next proceed to describe the modes of connecting several timbers, in order to complete the design, and to effect certain powers respectively required by each individual piece.

In framing centres for groins, the boarding which forms the interior surface is supported by transverse ribs of timber, which are either constructed simply, or with trusses, according to the magnitude of the work; and, as a groin consists generally of two vaults intersecting each other, one of them is always boarded over the same as a plain vault, without any respect to the other, which is afterwards ribbed and boarded so as to make out the regular surface.

Timbers inserted in walls, and at returns, or angles, are joined together where the magnitude of the building or exposure to strain may require. There are three denominations, viz. *bond timber*, *lintels*, and *wall-plates*.

ooring is supported by one or more rows of parallel
ms, called *naked*, or *carcase flooring*, and is denominated
ier single or double. During the construction of the
lding, the flooring, if not supported by walls or parti-
is, must be shored. The framing of flooring, whether
le or double, depends upon the magnitude of the build-
the horizontal dimensions of the apartments, or the
ss with which the surface of the boarding is likely to be
cted. When the flooring is intended to be very stiff and
l, it is necessary to introduce truss girders. Naked
ring, for ball-rooms, should be framed very strong, and
upper part contrived with a spring, to bend with the im-
sion of the force, while the lower part, which sustains
ceiling, remains immovable.

artitions are constructed of a number of pieces of tim-
, called *scantling*, placed vertically, at a specified dis-
ce from each other, dependent on the purposes for which
intended to answer. If to support girders, they should
trussed, and afterwards filled in with parallel pieces,
ed *studs*.

he framing ought to be so contrived, as to supersede the
essity of hanging up the floor, in whatever situation the
rs may be placed. Truss partitions are also of the
atest utility in supporting floors which are above them.

he rafters which support the covering in a roof are sus-
led by one, two, or several pieces of framing, called a
r *of principals*, placed at right angles to the ridge of the
f. In roofing, many ingenious contrivances are resorted
their application depending upon the pitch of the roof,
number of compartments into which it may be divided,
l the introduction of tie-beams. In cases where apart-
nts are required to be within the framing of the roof,
it is inconvenient to introduce tie-beams, the sides of
roof may be prevented from descending, by arching them
h cast-iron, or trussing them with wood in the inclined
nes of their sides. To restrain the pressure of the raf-
s, which would be discharged at the extremities of the
lding, a strong wall-plate, well connected in all its parts,
ist be introduced, to act as a tie, and prevent the lateral
ssure from forcing out the walls.

In this construction, as well as in the former, the rafters
uld have a tendency to become hollow, so that it is
essary, in order to counteract this tendency, to introduce
ining beams at convenient heights; and if it be requisite
occupy very little space by the wood-work, cast-iron

arches, abutting upon each other, and screwed with their
planes upon the upper sides of the rafters, are best adapted
for the purpose. If this and the former principle were adopted, the combined effect would be very great.

We shall now present the reader with a few practical observations.

Timber, except it stand perpendicular to the horizon, is
much weakened by its own weight. The bending of timber
is nearly in proportion to the weight laid on it. No beam
ought to be trusted for any long time, with above one-third
or one-fourth part of the weight it will absolutely carry;
for experiments prove, that a far less weight will break a
piece of timber when hung to it a considerable time, than
is sufficient to break it when first applied.

The strain occasioned by pulling timber in the direction
of its length, is called *tension*. It frequently occurs in roofs,
and is therefore worthy of consideration.

The absolute strength of a fibre, or small thread of timber, is the force by which every part of it is held together,
and is equal to the force that would be required to pull it
asunder. The force required to tear any number of threads
asunder, is proportional to that of their sum ; but the areas
of the sections of two pieces of timber, composed of fibres
of the same kind, are as the number of fibres in each ;
therefore, the strength of the timber is as the areas of the
sections. Hence all prismatic bodies are equally strong ;
that is, they will not break in one part rather than in another.

Bodies which have unequal sections, will break at their
smallest part; therefore if the absolute strength required to
tear a square inch of each kind of timber be known, we
shall be able to determine the strength of any other quantity whatever.

The wood next to the bark, commonly called *white* or
blea, is also weaker than the rest : and the wood gradually
increases in strength as we recede from the centre to the
blea.

The heart of a tree is never in its centre, but always nearer to the north side, and on that side the annual coats of
wood are thinner. In conformity to this, it is a general
opinion among carpenters, that that timber is strongest
whose annual plates are thickest. The *Tracheæ*, or *air-
vessels*, are weaker than the simple ligneous fibres. These
air-vessels make the separations between the annual plates,
and are the same in diameter, and number of rows, in
all trees of the same species ; consequently, when these

are thicker, they contain a greater proportion of the simple ligneous fibre.

The wood is stronger in the middle of the trunk than at the springing of the branches, or at the root; and the wood of the branches is weaker than that of the trunk.

The part of the tree towards the north, in the European climates, is the weakest, and that of the south side the strongest: and the difference is most remarkable in hedge-row trees, and such as grow singly.

All description of wood is more tenacious while green; and loses very considerably by drying, after the tree is felled.

We shall now conclude these remarks with the following useful problem.

Fig. 580. To cut the strongest beam possible out of a round tree whose section is a given circle. Let $a b c d$ be the section of the tree; draw the diameter $c b$, divide it into three equal parts, e and f, and from one of them, as f, draw $f a$ perpendicular to the diameter $c b$; draw $a b$ and $a c$,—$b d$ and $d c$, and $a b c d$ is the strongest piece that can be cut out of the tree. From this it is manifest, that the strongest beam which can be cut out of a round tree, does not contain the most timber, for the greatest rectangle that can be inscribed in a circle is a square, and therefore the square $g h i k$ is greater than the rectangle $a b c d$, and yet is not the strongest.

Fig. 581. Plan of a floor.—1. Girder resting upon the walls.—2. Bridging-joists.—3. Binding-joists.—4. Trimmers.

Nos. 1 and 2, sections of the floor.

Fig. 582. A trussed partition with an opening in the middle for folding doors.—1. Head.—2. Sill.—3. Posts.—4. Braces.—5. Studs.—6. Door-head.—This partition, as may be seen, supports itself.

Fig. 583. A simple trussed roof.

DEFINITIONS.

Wall-plates; pieces of timber laid on the wall, in order to distribute equally the pressure of the roof, and to bind the walls together. They are sometimes called *raising plates*.

Tie-beam; a horizontal piece of timber, connected to two opposite principal rafters; it answers a two-fold purpose, viz. that of preventing the walls from being pushed outwards by the weight of the covering, and of supporting the ceiling of the rooms below. When placed above the bottom of the rafters, it is called a *collar-beam.*

Principal rafters; two pieces of timber in the sides of the truss, supporting a grated frame of timber over them, on which the covering or slating rests.

Purlines; horizontal pieces of timber notched on the principal rafters.

Common rafters; pieces of timber of a small section, placed equidistantly upon the purlines, and parallel to the principal rafters : they support the boarding to which the slating is fixed.

Pole-plates; pieces of timber resting on the ends of the tie-beams, and supporting the lower ends of the common rafters.

King-post; an upright piece of timber in the middle of a truss, framed at the upper end into the principal rafters. and at the lower end into the tie-beam : this prevents the tie-beam from sinking in the middle.

Struts; oblique straining pieces, framed below into the king-posts, or queen-posts, and above into the principal rafters, which are supported by them ; or sometimes they have their ends framed into beams, that are too long to support themselves without bending, they are often called *braces.*

Other pieces of timber are introduced in roofs of a greater span ; which we shall here describe.

Queen-posts; two upright pieces of timber, framed below into the tie-beam, and above into the principal rafters ; placed equidistantly from the middle of the truss, or its extremities.

Puncheons; short transverse pieces of timber, fixed between two others for supporting them equally; so that when any force operates on the one, the other resists it equally ; and if one break the other will also break. These are sometimes called *studs.*

Straining-beam; a piece of timber placed between two others, called *queen-posts,* at their upper ends, in order to withstand the thrust of the principal rafters.

Straining-cill; a piece of timber placed upon the tie-beam at the bottom of two queen-posts, in order to withstand the force of the braces, which are acted upon by the weight of the covering.

Camber-beam; horizontal pieces of timber, made on the upper edge sloping from the middle towards each end in an obtuse angle, for discharging the water. They are placed above the straining-beam in a truncated roof, for fixing the boarding on which the lead is laid : their ends run three or four inches above the sloping plane of the common rafters, in order to form a roll for fixing the lead.

Auxiliary rafters; pieces of timber framed in the same vertical plane with the principal rafters, under, and parallel to them, for giving additional support. They are sometimes called *principal braces,* and sometimes *cushion rafters.*

The inhabitants of cold countries make their roofs very high; and those of warm countries, where it seldom rains or snows, very flat. But even in the same climate the pitch of the roof is greatly varied. Formerly the roofs were made very high, probably with the notion that the snow would slide off easier; but where there are parapets, a high roof is attended with very bad effects, as the snow slips down and stops the gutters, and an overflow of water is the consequence; besides, in heavy rains, the water descends with such velocity, that the pipes cannot convey it away soon enough to prevent the gutters from being overflowed.

The height of roofs at the present time is very rarely above one-third of the span, and should never be less than one-sixth. The most usual pitch for slates is that when the height is one-fourth of the span, or at an angle of 26¼ degrees with the horizon. Taking this as a standard, the following table will show the degree of inclination which may be given for other materials :—

Kind of covering.	Inclination to the horizon in degrees.		Height of roof in parts of Span.	Weight upon a square of roofing.
	Deg.	Min.		
Copper or lead	3	50	$\frac{1}{14}$	copper 100 / lead 700
Slates large	22	0	$\frac{1}{4}$	1120
Ditto ordinary.	26	33	$\frac{1}{4}$	from 900 / to 500
Stone slate	29	41	$\frac{3}{7}$	2380
Plain tiles	29	41	$\frac{3}{7}$	1780
Pan-tiles	24	0	$\frac{2}{9}$	650
Thatch of straw, reeds.	45		$\frac{1}{2}$	

A roof for a span of from 20 to 30 feet may have a truss of the form shown in Fig. 583. Within this limit, the purlines do not become too wide apart, nor the points of support of the tie-beam.

For spans exceeding 30 feet, and not more than 45 feet, the truss shown in Fig. 584 is well adapted. Each purline is supported, consequently, there are no cross strains on the principal rafters; and the points of support divide the tie-beams into three comparatively short bearings. The sagging, which usually takes place from the shrinking of the heads of the queen-posts, may be avoided by letting the end of the principal rafter abut against the end of the straining beam A, and notching pieces and bolting them together in pairs at each joint.

When the span exceeds 45 feet, and is not more than 60 feet, the truss shown in Fig. 585 is sufficiently strong for the purpose, and leaves a considerable degree of free space in the middle. For this span the tie-beam will most likely require to be scarfed, and as the bearing of that portion of the tie-beam between a and b is short, the scarf should be made there. The middle part of the tie-beam may be made stronger by bolting the straining cill c to it.

It often occurs, that the centre aisles or naives of churches are higher than the side aisles ; a similar effect, as when the tie-beam continues through, may be produced by connecting the lower beams to the upper one, by means of braces, so that the whole may be as a single beam. To illustrate this mode of construction, we have given a design for a roof of a church, somewhat similar to St. Martin's in the fields, London.

Fig. 586, the lower ties, AA, are so connected with the principal tie-beam, B, by means of the braces, a, a, that the foot of the principal rafters, c, c, cannot spread without stretching the tie-beam, B. The iron rods, b, b, perform the office of king-posts to the ties, A, A, and are much better than timber, in consequence of the shrinkage, which in this situation would be very objectionable.

Fig. 587 is a design for a roof of a church, or other building, requiring a semicircular arched ceiling.

Domes derive their names according to the plans on which they are built, circular, elliptical, or polygonal : of these, the circular may be spherical, spheroidal, ellipsoidal, hyperboloidal, paraboloidal, &c. Those which rise higher than the radius of the base, are called *surmounted domes* ; those that are of a less height than the radius, *diminished*, or *surbased*; and such as have circular bases, *cupolas*. The most usual form for a dome is the spherical, in which case, the plan is a circle, and the section a segment of a circle.

The top of a large dome is often finished with a lantern, supported by the framing of the dome.

The interior and exterior forms of domes are seldom alike, and in the space between them, a staircase to the lantern is usually made. According to the space left between the external and internal domes, the framing must be designed. Sometimes the framing may be trussed with ties across the opening ; but generally the interior dome rises so high that ties cannot be obtained.

Fig. 588. No. 1, shows the construction of a dome without ties. This is the most simple method, and one which is particularly applied domes of ordinary dimensions. This example consists in placing a

ber of curved ribs, so that the lower ends stand upon and are well framed into the kirb at the base, and the upper ends meet at the top, or are framed into the upper kirb on which the lantern is placed.

When it occurs, as it generally does, that the pieces are so long, and so much curved, that they cannot be cut out of timber without being cut across the grain, so much as will weaken them, they should be put together in thicknesses, with the joints crossed, and well bolted together.

No. 2, shows the ribs fixed, and bolted together, with horizontal rafters to receive the boarding on the exterior, and the laths on the interior. These ribs should be placed about two feet, or two feet six inches apart at the base, and be composed of three or four thicknesses of one and a half inch-deal, about 11 or 12 inches wide, which, when carefully bolted together with the joints judiciously broken, will stand exceedingly firm and well.

To construct the ribs of a spherical dome, with eight axal ribs, and one purline in the middle.

(Fig. 589.) No. 1. Let ABCDE be the plan of half the dome, which divide into four equal parts at BCD and E, these points of division will mark the centre of the back, or convex sides of the ribs. This being done, let B b, C c, D d, be the plans of these ribs, with the points of division in the centre. F, G, H, I, K, are the seats of the upper ends of the ribs; on the upper kirb draw $x y$. No. 2, parallel to AE, then from the different seats of the ribs on the plan draw perpendiculars cutting $x y$. Draw the cill, $x y$, its intended thickness, and complete the elevation of the front and back ribs. The front ribs are quadrants, forming a semicircle on the upper side of the wall-plate, which, of course, is the diameter. The curves of the sides of each of the other ribs are the quadrants of an ellipsis of the same height with the front rib. Place the purlines in their intended situation, and having drawn the elevation and plan, as shown by the dotted line, the construction is complete.

The ribs of an elliptical dome are found precisely on the same principle.

Given the plan of a polygonal dome, and one of the axal ribs, at right angles to one of the sides, to find the curve of the angle rib and the covering.

Fig. 590. Let A, B, C, D, E, F, G, H, be the plan of an octangular polygonal dome, and e a b the given rib; produce c a to d, divide the curve line a BA b into any number of equal parts, the more the better, in this case four, 1, 2, 3, b, which extend on the line a d; the first from a to 1, the second from 1 to 2, &c.: from the points of division, 1, 2, 3, b, draw lines parallel to B e, cutting C c, and from these points draw lines, parallel to c d, or at right angles to B e, and through the points, 1, 2, 3, draw k l, m n, o p, and tracing a curve through the points d, p, n, l C, and making d o m k B similar, then the space comprehended between the curve lines d B e; and the side BC of the plan, will give the form of the whole covering, for each side of the dome.

To find the hip-line of the angle-rib, whose base is C c.

Draw CE, 2 e, 1 f, at right angles to C c, and make CE equal to c b,

2 e equal 2-8, and 1 f equal to 1, 2, &c. and trace the curve through these points, and it will give the angle-rib.

The method of covering spherical domes is, to suppose them polygonal, and the principle the same as the foregoing operation for an octangular dome.

A *niche*, in carpentry, is the wood-work to be lathed over for plastering. The general construction of niches is with cylindrical backs and spherical heads, called *cylindro-spheric niches*; the execution of which depends upon the principles of spheric sections.

As every section in a sphere is a circle, and that section passing through its centre is equal, and the greatest that can be formed by cutting the sphere; it is evident, that if the head of a niche is intended to form a spherical surface, the ribs may be all formed by one mould, whose curvature must be equal to that of the greatest circle of the sphere; viz. one passing through its centre; but the same spherical surface may, though not so eligible, be formed by ribs of wood, moulded from the sections of lesser circles, in a variety of ways.

The reason why these latter spherical surfaces are not so eligible as those of greater circles is, because their disposition for sustaining the lath is not so good, and the trouble of moulding them to different circles, and of forming the edges according to different bevels, in order to range them in the spherical surface, is very great, compared with those made from great circles.

The disposition of the ribs of niches is generally in a vertical plane, parallel to each other, or intersecting each other in a vertical line. When the line of intersection passes through the centre of a sphere, all the ribs are great circles; but if the line of intersection does not pass through the centre of the sphere, the circles which form the spherical surface are all of different radii. When the ribs are fixed in parallel vertical planes, their disposition is either parallel to the face of the wall, or parallel to a vertical plane, passing through the centre of the sphere, perpendicular to the surface of the wall; but this method is not so eligible for the purposes of lathing.

Another method is, by making the planes of the ribs parallel to the horizon: this is not only attended with great labour in workmanship, but is incommodious for lathing. The various positions in which the ribs of a niche may be placed, are very numerous; but the regular positions, already enumerated, ought to be those to which the carpenter should direct his attention.

To get out the ribs for the head of a niche, all of them being in vertical planes passing through the centre of the sphere.

Fig. 591, No. 1. From the centre C draw the ground-plan of the ribs, and set out as many ribs upon the plan as you intend to have in the head of the niche. With the foot of your compasses in C, and from the ends of each rib, at *k* and *l*, draw the small concentric dotted circles round to the centre rib, at *o* and *p*, and draw *o m*, and *p n*, parallel to *a b*, the face of the wall ; then from *r* round to *s* on the plan is the length and sweep of the centre rib to stand over ; and from *n* round to *s* the length and curve of the rib that stands from *b* to *g* ; and from *m* round to *s*, the curve of the shortest rib, that stands from *k* to *h* on the plan.

How to find the bevel of the ends of the back ribs against the front rib.

The back ribs are laid down distinct by themselves, at A B and C from the plan. Take *b* 1, in No. 1, and set it to *b* 1, at B, draw the perpendiculars, and when they intersect the rib, it will show the bevel required. The same operation being done to C, the bevel is found in the same manner.

The places of the back-ribs when fixed upon the front-rib are ascertained by drawing perpendi culars, and completing the elevation of the niche No. 2 from the plan.

To find the radius of curvature of the ribs of a spherical niche, when the ribs all meet in a vertical line, which divides the front rib into two equal parts.

Fig. 592, No. 1. Complete the circle, of which the inside of the plan is an arc ; produce the middle line of the plan of any rib, as of *a b*, to meet the opposite side of the circumference in *b* ; on the whole line *a b*, as a diameter, describe a semicircle, and from the point *c*, when the ribs intersect, draw *a* perpendicular to *c d*, to meet the arc *a d* at *d*, which is the curve of the rib, whose seat is *d*. The other rib, as AD, is found in the same manner. No. 2 is the elevation of the niche.

Pendentive cradling, is a cove bracketing, springing from the rectangular walls of an apartment upwards to the ceiling, so as to form the horizontal part of the ceiling into a complete circle or ellipsis.

The proper criterion for such bracketing, if the walls are cut by horizontal planes through the coved parts, is, that all the sections through such parts will be portions of circles, or of ellipses, and have their arcs proportioned to the sides of the apartment, so that each section will be a compound figure. Besides having four curvilinear parts, it will have four other parts, which are portions of the sides of the rectangular apartment : and the axis of the ellipsis will bisect each side of the rectangle.

Fig. 593. Let ABCD be the plan of a room, or stair-case, to be bracketad, so as to form the surface of a pendentive ceiling; and let A b c D be the section across the diagonal; it is required to find the curvature of the springing ribs?

Draw C d perpendicular to AC, meeting AC, take the distance from C to the line AC, and set it from C on the line CA, and from this point draw a perpendicular to meet the curve A b c D of the diagonal rib; make the versed sine of the segment A d C equal to this perpendicular, and describe the segment A d C, which is the springing line required. If from the centre C an arc be described, with a radius equal to the length of the seat of a rib, to meet the seat of the diagonal rib AD; and, if from this point of meeting a perpendicular be drawn to meet the curve A b, the portion of the line of the diagonal rib, intercepted between A and the perpendicular, will give the length of the rib, corresponding to the seat which was taken.

Fig. 504. The diagonal rib is a semicircle: the operation is exactly the same, and may be described in the same words.

MENSURATION OF CARPENTERS' WORK.

All large and plain articles in which an uniform quantity of materials and workmanship is expended, are generally measured by the square of 100 superficial feet.

Piles used in the foundations are valued at per piece, and driven by the foot run, according to their diameter, and the quality of the ground.

Sleepers and planking are measured by taking the superficial contents in yards or squares.

Plain centreing is measured by the square; but as the ribs and boarding are two different qualities of work, they ought to be measured and valued separately; one dimension of the boarding being taken by girting it round the arch, the other being the length of the vault.

Centreing for groins should be measured and valued as common centreing; but in addition thereto, the angles should be paid for by the foot run, that is, the ribs and boarding ought to be measured and valued separately, according to the exact superficial contents of each; and the angles by the lineal foot for workmanship, in fitting the rib and boards, and for the waste of wood occasioned by the operation.

Wall-plates, lintels, and bond-timbers, are measured by the cubic foot, under the denomination of fir-in-bond.

Naked flooring may either be measured by the square, or by the cubic foot, according to the description of the work, and the quantity of timber employed. In forming an estimate of its value, it should be observed, that in equal cubic quantities of small and large timbers, the small tim-

bers will have more superficies than the large ones, and,
therefore, the saving will not be in a ratio with the solid
contents; consequently the value of the workmanship will
not follow the cubic quantity, or said ratio. The difficulty
of handling timbers of the same length increases with the
weight or solidity, as the greater quantity requires greater
power to handle it, and consequently more time.

In naked flooring, where girders are introduced, the uniformity of the work is interrupted by mortises and tenons,
so that the sum ascertained by the cubic quantity of the
girders, at the same rate per foot as the other parts, is not
sufficient; not only on account of the great difference of
size, but the great disparity in the quantity of workmanship, occasioned by its being cut full of mortises to receive
the tenons of the binding-joists; the best method, therefore, to value the labour and materials is, to measure and
estimate the whole by the cubic quantity, and allow an additional rate upon every solid foot of girders; or, if the binding-joists are not inserted in the girders, at the usual distances, a fixed price for every mortise and tenon, in proportion to their size, which will keep a ratio with the area
of the end of the girder.

Partitions may be measured by the cubic feet; but the
cills, top-pieces, and door-heads, should be measured by
themselves, according to the solid quantity, at an additional
rate; because, both the uniform solidity, and the uniform
quantity of workmanship are interrupted by them.   In
trussed partitions, the braces should be rated by the foot
cube, at a superior price to that of the quarterings, for the
trouble of fitting the ends of the uprights upon their upper
and lower sides, and for forming the abutments at the ends.

The timbers in roofing should be measured by the cubic
foot, classed as the difficulty of execution, or as the waste
occasioned, may require.

Battening to walls is best measured by the square, according to the dimensions and distances in the clear of the
battening.

It would be endless to enumerate the various methods
of measuring each particular species of carpenters' work;
the leading articles only need be noticed.

When the shell of a building is finished, that is, previous
to the floors being laid, or the ceilings lathed, all the timbers
should be measured, that no doubt may arise as to the actual
scantlings of the timbers, or of the description of the workmanship.   In taking dimensions it must be observed that,

which have tenons, must be measured to the extre-
the tenons.

possible to determine on any proper rate, including both materials and workmanship, as the one may be
stationary, while the other is variable. With respect to
materials, the value of any quantity may be easily ascertained, whatever be the price per load ; but the difficulty is
far greater in fixing proper rates of workmanship ; however,
were the time of executing every species of work known,
there would be no difficulty in establishing certain uniform
which would give the real value.

JOINERY.

is the next branch of art which comes under our consideration, and comprises the practice of employing wood in
the external and internal finishings of houses.

In the execution of this branch of building, it is almost
unnecessary to observe that, as joinery is employed principally by way of decoration, and is liable to close inspection,
it is one of the departments which demands the strictest
care and attention in the workmen; and it requires the
greatest ingenuity, skill, and experience, to become fully
master of every subject under the joiner's consideration.

The first and most important thing to be attended to, is
the judicious selection of materials ; as, without a strict ob-
servance of this particular, the care, ingenuity, and exertions of the workman will be wholly frustrated.

As the temperature of the atmosphere has a great influence on wood, and more particularly in the winter season, it would be advisable to put that which is to be
used in fine work over an oven for a day or two. In the
different descriptions of joint used by the joiner, a hot
tenacious liquid, called *glue*, is almost universally used, and
when applied, the two surfaces of the wood, which have been
previously rendered smooth, are rubbed together until the
glue is nearly all forced out. One piece is then set to its
situation with respect to the other.

For outside work, such as gates, doors, &c. white-lead is
used in all the joints.

When a frame, consisting of several pieces, is required,
the mortises and tenons are fitted together, and the joints
glued all at one time, then entered to their places, and forced
together by the assistance of an instrument called a cramp.

The operation of rendering a rough surface smooth, by taking away the superfluous wood, is called *planing*; and the tools used for this purpose are called *planes*.

The planes used by joiners in the primary operation of their work are called *jack-planes, trying-planes, long-planes,* and *smoothing-planes*; the respective uses of which are as follow :—The jack-plane is used for taking away the rough occasioned by the saw, and removing all superfluous and other uneven parts. The trying-plane more particularly to bring the surface perfectly level and true : the long plane succeeds, when the surface is long, and is required to be very straight, as in jointing long boards for the purpose of gluing them together; and the smoothing-plane is used to smooth and clean off the work.

In addition to the above, termed *bench-planes*, others are occasionally used in forming any kind of prismatic surfaces, viz. *rebating-planes, grooving-planes, moulding-planes,* &c.; under which head is included the *fillister* and *plough*.

Rebating-planes are used for cutting out rebates, a kind of half groove, upon the edge of a board, or other piece of wood, formed by taking down or reducing a small part of the breadth of the board to half, more or less, of the general thickness. By this means, if a rebate be cut on the upper side of one board, and the lower side of another, the two may be made to overlap each other, without making them any thicker at the joint.

Rebates are also used for ornamenting mouldings, and for many other purposes in joiners' work. The planes for cutting them are of different kinds, some having the cutting edge at the side of the iron and stock; others at the bottom edge of the iron and the face of the stock; and others cutting in both these directions. The former are used to smooth the side of a rebate, and therefore are called *side*

when used, the edge of the fence is applied against the edge of the piece to be rebated, and thus gauges the breadth its iron should cut away. The cutting-iron of this plane is not situated at right angles to the length of the stock, but has an obliquity of about forty-five degrees; the exposed side of the iron being more forward than the one next to the fence. By this obliquity, the plane has a tendency or drift to run further into the breadth of the wood; but as the fence sliding against the edge prevents this, the drift always keeps the fence in contact with the piece without the attention of the workman: it also causes the iron to cut the bottom of the rebate smoother, particularly in a transverse direction to the fibres, or where the wood is cross-grained, or where the edge is perpendicular to the sides of the plane. It is chiefly used, however, to throw the shaving into a cylindrical form, and thereby make it issue from one side of the plane. Besides this iron, there is another of smaller dimensions, called the *tooth*, which precedes the other, to scratch or cut a deep crack in the width of the rebate, thus making the shaving, which the iron cuts up from the bottom, separate sideways from the rest of the wood. The *sash-fillister* differs in many particulars from the moving fillister: the fence is adapted to be moved to a considerable distance, not being fixed, as in the moving fillister, by screws upon the face, but sustained by two bars, fixed fast to it, passing through the two vertical sides of the stock at right angles to the sides: these bars, when set to their intended places, are tightened by small wedges. This kind of plane is usually employed to rebate narrow pieces of wood, such as are used in sashes; and the fence is applied against the opposite edge to that on which the rebate is to be formed.

The *plough* is a plane with a very narrow face, made of iron, fixed beneath a wooden stock, and projecting down from the wood of the stock; the edge of the cutting-iron being the full width of the groove required: it is guided by a fence with bars like the sash-fillister, and has also a stop to regulate the depth intended for the grooves.

Moulding-planes are those which have their faces and cutting edges curved, to produce all the varieties of ornamental mouldings: they are known by the names of *snipe's-bills, side snipe's-bills, beads, hollows, rounds, ovolos*, and *ogees*. Of these there are a great variety of sizes, with which every good joiner is furnished.

The whole of these planes have their faces straight in the

direction of their length ; but a section across the face is
the impression or reverse of the moulding they are intended
to make.

The tools employed in boring cylindric holes are a stock
with *bits*, *gimlets*, and *brad-awls* of various descriptions and
sizes. The tools used for paring the wood obliquely, or
across the fibres, and for cutting rectangular prismatic cavities, are in general denominated *chisels ;* those for paring
the wood across the fibres being called *firmers*, or *paring-
chisels*, and those for cutting mortises, *mortise-chisels*. The
best paring-chisels are made entirely of cast steel. Chisels
for paring concave surfaces, are called *gouges*.

Wood is generally divided or reduced by means of *saws*,
of which there are several sorts ; as the *ripping-saw*, for
dividing boards into separate pieces, in the direction of the
fibres ; the *hand-saw*, for cross-cutting, or for sawing thin
pieces in the direction of their length ; the *panel-saw*, either
for cross cutting, or cutting very thin boards longitudinally ;
the *tenon-saw*, with a thick iron back, for making an incision of any depth below the surface of the wood, and for
cutting pieces opposed to the length of the fibres ; also a
sash-saw, and a *dovetail-saw*, used much in the same way
as the tenon-saw.

From the thinness of the plates of these three last-mentioned saws, it is necessary to stiffen them by a strong piece
of metal called the back, which is grooved to receive the
upper edge of the plate, fixed to the back, and which is
thereby secured and prevented from crippling.

When it is required to divide boards into curved surfaces,
a very narrow saw without a back, called a *compass-saw*,
is used ; and in cutting a very small hole, a saw of a similar
description is used, called a *key-hole-saw*. Both of these
description of saws are called *turning-saws*, and have their
plates thin and narrow towards their bottoms, and each
succeeding tooth finer.

The external and internal angles of the teeth of all saws
are generally formed at one angle of 60 degrees, and the
front edge teeth slope backwards in a small degree. The
teeth of every description of saw, except turning-saws, are
alternately bent on contrary sides of the plate, so that all
the teeth on the same side are alike bent throughout the
length of the plate, for the purpose of clearing the sides of
the cut made by it in the wood. The foregoing are generally
termed *edge-tools*.

When it is necessary to ascertain if an angle be exactly

c, or inclined to any number of degrees, a tool called
arc is used, and in the latter instance, a bevel is set to
ngle; when any piece is to be reduced to a parallel
th or thickness, an instrument, called a gauge, formed
square piece with a mortise, having a sliding bar, called
m, running through it at right angles, and furnished
a tooth, projecting a little from the surface, is used;
at when the stock of the gauge is applied to the verti-
de or edge of the piece, with the toothed side of the
upon the horizontal surface, and is pushed and drawn
nately backwards and forwards by the workman, the
will make an incision from the surface into the wood,
)arallel distance from the edge to which the stock part
plied.

hen a mortise is to be made in a piece of wood, the
e used has two teeth. The construction of this gauge
same as that before described, except that the tooth
st the stock moves by means of a longitudinal slider
stem, which is to be set at a distance from the other
, as occasion may require.

a piece of wood is to be sawn across the fibres, a flat
of wood, which has two projecting knobs, on opposite
one at each end, called a side-hook, is used, to keep
iece which has to undergo the operation of the saw
y; the knob at one end presses against the piece, while
t the other end is hooked to the bench. Two of
are necessary when the pieces are long.

hen a piece of wood is required to be cut to a mitre,
is, to half a right angle, joiners use a trunk of wood
three sides, like a box that has neither ends nor top, the
and bottom being parallel pieces, and the sides of equal
t. Through each of the opposite sides, in a plane per-
icular to the bottom, and at the oblique angles of 45°
35° with the planes of the sides, a kerf is cut; and
er kerf is made with its plane at right angles to the
ormer. Into this trunk, termed a mitre-box, the piece
cut is put, and the saw, guided by the kerfs, cuts the
to the angle required.

making a straight surface, a strip of wood called a
ht-edge, which has one of its edges perfectly straight,
quently applied, to detect the irregularities, and the
is accordingly planed with the trying plane until the
cc coincides with the straight-edge.

as certain if the surface of a piece of wood be in one
, the joiner takes two slips of wood, each straightened

and their surfaces, on which the last stave is intended to rest, must be all in the same plane, that its back may rest firmly upon them. In closing up the remaining space, the part of the column that is glued together should be kept from spreading by confining it in a kind of cramp, or cradle, while driving the remaining stave to close the joints.

Instead of the foregoing mode, some joiners glue up the columns in halves and then glue them together. When an iron core is necessary to support a floor or roof, the column must necessarily be glued up in halves; in which case the two halves are to be dowelled together, and the joints filled with white-lead. Instead of a cramp, a rope is used, twisted by means of a lever. In bringing the two halves together, the percussive force of the mallet must be applied upon the middle of the surface of one half, while an assistant holds something steady against the middle of the other, that the opposition may be equal, and by this means the surfaces will be brought into contact, and form the joint as desired. In this operation pieces of wood ought to be inserted between the column and the rope.

Boards can be connected together at any given angle, either by pins or nails, mortise and tenon, or by indenting them together.

This last mode, from the sections of the hollows and projecting parts being formed like a dove's tail, is called *dovetailing*.

There are three sorts of *dovetailing*; viz. common, lap, and mitre. Common dovetailing shews the form of the projecting parts, as well as of the excavations made to receive them; lap dovetailing conceals the dovetail, but shews the thickness of the lap on the return side; and mitre dovetailing conceals the dovetail and shews only a mitre on the edges of the planes at the surface of the concourse; that is, the edges in the same plane, the seam or join being in the concourse of the two faces, making the given angle with each other.

Concealed dovetailing is particularly useful where the faces of the boards are intended to form a saliant angle; but when the faces form a re-entrant angle, common dovetailing is preferable.

There is another simple and expeditious manner of connecting the ends of boards together where the faces form a re-entrant, or internal angle, by means of a groove in the one, and a tongue in the other; and if the pieces be pro

door. Doors ought to be made of the best
fectly seasoned, and firmly put together;
scribings should be brought together with tl
actness, and the whole of their surfaces be pe

The mortising, tenoning, ploughing, and
mouldings, ought to be worked correctly to tl
otherwise the door, when put together, v
truth, and occasion the workman a great d
paring the different parts to make it appe
the door will also lose much of its firmness, c
mortises and tenons require to be pared.

In bead and flush doors, make the work
wards put in the panels, and smooth the
gether; then, marking the panels at the part
ing to which they agree, take the door to piece
beads on the stiles, mountings, and rails. H
double margin, that is, representing a pair of
the staff stile, which imitates the meeting-e
inserted into the top and bottom rails of the
ing the ends into notches cut in the top and I

. In the hanging of doors, the chief aim is to
pet or ground; which may be accomplished
the following rules. First, let the floor be
door, according to the intended thickness
secondly, let the knuckles of the top and botto
placed, that the top hinge hang, or proje
eighth of an inch over the lower; that is, if t
equally into the door and into the jamb, pro
yond the surface of the door; but if the ce
surface of the door, it must be placed at the v
is seldom done, except when the door is hung
Thirdly, let the jamb on which the door h
about an eighth of an inch out of the perp
upper part inclining towards the opposit
fourthly, let the inclination of the robate be
door shall, when shut, project at the bottom
room, about an eighth of an inch.

These several methods, practised on so sm
not perceptible; but, nevertheless, will th
when opened, to a square sufficiently out of
is, at least half an inch, when the height
double the width.

Several kinds of rising hinges have been
this purpose: some of the best, constructed
no means objectionable, even to the best do

Before we proceed to the principles of *hanging* doors, we shall submit to the reader some information on the subject of *hinging* in general.

The placing of hinges depends entirely on the form of the joint, and as the motion of the door or closure is angular, and performed round a fixed line as an axis, the hinge must be so fixed that the motion be not interrupted : thus, if the joint contain the surface of two cylinders, the convex one in motion upon the edges of the closure, and sliding upon the concave one which is at rest on the fixed body, the motion of the closure must be performed on the axis of the cylinder, which axis must be the centre of the hinges. In this case, whether the aperture be shut or open, the joint will be close ; but if the joint be a plane surface, it is necessary to consider upon what side of the aperture the motion is to be performed, as the hinge must be placed on the side of the closure where it revolves.

The hinge is made in two parts, movable in any angular direction, the one upon the other.

The knuckle of the hinge is a portion contained under a cylindric surface, and is common both to the moving part and the part which is at rest ; the cylinders are indented into each other, and are made hollow to receive a concentric cylindric pin, which passes through them, and connects the moving parts together.

The axis of the cylindrical pin, is called the *axis of the hinge.*

When two or more hinges are placed upon a closure, the axis of the hinges must be in the same straight line.

The straight line in which the axis of the hinges are placed is called the *line of hinges.*

We shall now proceed to the principle of hanging doors, shutters, or flaps, with hinges.

The centre of the hinge is generally put in the middle of the joint, as A, Fig. 605, but in many cases there is a necessity for throwing back the flap to a certain distance from the joint ; in order to effect this, suppose the flap when folded back, were required to be at a certain distance from the joint, as BA, Fig. 605, divide BA in two equal parts at the point C, and it will give the centre of the hinge. The centre of the hinge must be placed a small degree beyond the surface of the closure, otherwise it will not fall freely back on the jamb, or partition. It must also be observed, that, the centre of the hinge must be on the same side as the rebate, or it will not open without the joint being constructed in a particu-

To hang two flaps, so that when folded back, they shall be at a certain distance from each other.

This is easily accomplished by means of hinges having knees projecting to half that distance, as appears from Fig. 607: this sort of hinge is used in hanging the doors of pews, in order to clear the moulding of the coping. Fig. 607, No. 2, shows the same hinge opened.

To make a rule joint for a window-shutter, or other folding flap.

Fig. 606, No. 1. Let *a* be the place of the joint, draw *a c* at right angles to the flap, shutter, or door, take *c*, in the line *a c*, for the centre of the hinge, and the plain part *a b*, as may be thought necessary ; or *c*, with a radius, *c b*, describe the arc *b d;* then will *a b d* be the true joint. The knuckle of the hinge is always placed in the wood ; because the further it is inserted, the more of the joint will be covered when it is opened to a right angle, as in Fig. 606, No. 2; but if the centre of the hinge were placed the least without the thickness of the wood, it would show an open space, which would be a blemish.

To form the joints of stiles, to be hung together, when the knuckle of the hinge is placed on the contrary side of the rebate.

Fig. 608. Let *c* be the centre of the hinge, *m i* the joint on the same side, *c b* the depth of the rebate in the middle of the thickness of the styles, perpendicular to *i m*, and *l f* the joint on the other side, parallel to *i m;* bisect *i l* at *k*, join *k c*, on *k c* describe a semicircle *c i k*, cutting *i m* at *h*, through the points *h* and *k* draw *h k g*, cutting *f l* at *g;* then will *f g, h m*, be the true joint.

Fig. 609 represents the common method of hanging shutters together, the hinge being let the whole of its thickness into the shutter, and not into the sash-frame. By this mode it is not so firmly hung as when half of it is let into the shutter, and half into the sash-frame ; but the lining may be made thinner.

It may here be proper to observe, that the centre of the hinge must be in the same plane with the face of the shutter, or beyond it, but not within the thickness.

How to construct a joint for hanging doors with centres.

Fig. 614. Let *a d* be the thickness of the door, bisect it in *b*, draw *b e* perpendicular to *a b*, make *b c* equal to *b a*, or *b d*, or *c*, the centre of the hinge, with a radius *c a*, or *c d*, describe an arc, *a e d*, which will give the joint required.

Another plan is represented in Fig. 613. Draw *a b* parallel to the jamb, meeting the other side in *b*, make *b d* equal to *b a*, and join *a d* and *a c*, bisect *a c* by a perpendicular *e f*, meeting *a d* in *f*, then *f* is the centre of the hinge.

Figs. 610, 611, and 612, exhibit different methods of hanging flaps, &c. These are so very simple, that by a little attention the reader will readily perceive their uses and manner of construction.

We shall now detail the construction of sash-frames, sashes, and shutters, and the manner of putting the several parts together.

Fig. 615, No. 1, the elevation ; No. 2, the plan ; and No. 3, the section of the same ; showing the manner in which the different parts are connected.

No. 1. A Back.—B Flush skirting, separated from the back by flush reeds, and showing the same depth of plinth as the blocks of the pilasters.—C C Blocks or plinths to pilasters.—D D Pilasters.—E E Patteras.—d d d Inside bead of sash-frame.—b b b rounded edge of boxing-stile.

No. 2. Plan of sash-frame, shutters, pilasters, and the different parts are explained in the figures.

No. 3. a thickness of the pilaster or architrave; b the rounded edge of the boxing-stile; c the breadth of the shutter; d bead of the sash-frame; e under sash; f top ditto; g parting bead; h outside lining and bead. i the breadth of the reveal or outer brick-work; k k lintels made of strong yellow deal or oak; l the head of the ground; m the architrave or pilaster fixed upon the grounds; n the soffit, tongued into the top of the sash-frame-head; and, on the other edge, into the head of the architrave m; o the sash-frame head; p the elbow; q capping; r sash-frame cill; s sash-cill; t stone-cill.

The face of the pulley-stile of every sash-frame ought to project about three-eighths of an inch beyond the edge of the brick-work; that is, the distance between the face of each pulley-stile ought to be less by three quarters of an inch than in the clear of the reveals on the outside; so that the face of the shutters ought to be in the same plane with the stone or brick-work on the outside.

Fig. 616 shows a plan of a sash-frame and shutter on the same principle as the foregoing, and which may be applied to a similar window.

As the thickness of the wall is here conceived to be less than in the foregoing example, another back-flap is introduced:—a the outside lining; b the pulley-stile; c the inside lining; d the back lining; e f the weights; g parting slip of weights; h parting bead to sashes; i inside bead; k back lining of boxing; l ground, or boxing-stile, grooved to receive the plastering; m front shutter hung to the inside lining, c, of the sash-frame by the hinge n; o p back flaps hinged together at q, and to the shutter at r; s architrave or pilaster.

Fig. 617. Is a vertical section of the cill, &c. of the same sash-frame; a bottom rail of sash; b cill of the sash-frame; c back of recess of window; d capping bead, or capping let into the sash-frame cill; e inside bead, tongued on the top of the cill; h outside lining; f space for the top-sash to run in; g parting bead.

STAIRS.

This is one of the most important subjects connected with a joiner's art, and should be attentively considered, not only with regard to the situation, but as to the design and execution. The convenience of the building depends on the situation; and the elegance, on the design and execution of the workmanship. In contriving a grand edifice, particular attention must be paid to the situation of the space occupied by the stairs, so as to give them the most easy command of the rooms.

With regard to the lighting of a good staircase, a sky-light, or rather lantern, is the most appropriate; for these

places, the steps being fixed to strings, newels, and carriages, and the ends of the steps of the inferior kind terminating only upon the side of the string, without any housing. In taking dimensions and laying down the plan and section of stair-cases, take a rod, and, having ascertained the number of steps, mark the height of the story, by standing the rod on the lower floor: divide the rod into as many equal parts as there are to be risers, then, if you have a level surface to work upon below the stair, try each of the risers as you go on, and this will prevent any excess or defect; for any error, however small, when multiplied, becomes of considerable magnitude, and even the difference of an inch in the last riser, will not only have a bad effect to the eye, but will be apt to confuse persons not thinking of any such irregularity. In order to try the steps properly by the story rod, if you have not a level surface to work from, the better way will be, to lay two rods on boards, and level their top surface to that of the floor: place one of these rods a little within the string, and the other near or close to the wall, so as to be at right angles to the starting line of the first riser, or which is the same thing, parallel to the plan of the string; set off the breadth of the steps upon these rods, and number the risers; you may set not only the breadth of the flyers, but that of the winders also. In order to try the story-rod exactly to its vertical situation, mark the same distances of the risers upon the top edges, as the distances of the plan of string-board, and the rods are from each other.

In bracket-stairs, as the internal angle of the steps is open to the end, and not closed by the string as in common dog-legged stairs, and the neatness of workmanship is as much regarded as in geometrical stairs, the balusters must be neatly dove-tailed into the ends of the steps, two in every step. The face of each front baluster must be in a straight surface with the face of the riser, and, as all the balusters must be equally divided, the face of the middle baluster must stand in the middle of the face of the riser of the preceding step and succeeding one. The risers and heads are all previously blocked and glued together, and when put up, the under side of the step nailed or screwed into the under edge of the riser, and then rough brackets to the rough strings, as in dog-legged stairs, the pitching pieces and rough strings being similar. In glueing up the steps, the best method is to make a templet, so as to fit the external angle of the steps with the nosing.

2 Q 2

The steps of geometrical stairs ought to be constructed so as to have a very light and clean appearance when put up; for this purpose, and to aid the principle of strength, the risers and treads, when planed up, ought not to be less than one eighth of an inch, supposing the going of the stair, or length of the step, to be four feet, and for every six inches in length, another one-eighth may be added. The risers ought to be dove-tailed into the cover, and when the steps are put up, the treads are screwed up from below to the under edge of the risers. The holes for sinking the heads of the screws ought to be bored with a centre-bit, then fitted closely in with wood, well matched, so as entirely to conceal the screws, and appear as one uniform surface. Brackets are mitred to the riser, and the nosings are continued round. In this mode, however, there is an apparent defect; for the brackets instead of giving support, are themselves unsupported, and dependent on the steps, being of no other use, in point of strength, than merely tying the risers and treads of the internal angles of the steps together : and, from the internal angles being hollow, or a re-entrant angle, except at the ends, which terminate by the wall at one extremity, and by the brackets at the other, there is a want of regular finish. The cavetto, or hollow, is carried round the front of the riser, and is returned at the end, and mitred round the bracket, and if an open string, that is, the under side of the stairs open to view, the hollow is continued along the angle of step and riser.

The best plan, however, of constructing geometrical stairs is, to put up the strings, and to mitre the brackets to the risers, as usual, and enclose the soffit with lath and plaster, which will form an inclined plane under each flight, and a winding surface under the winders. In superior staircases, for the best buildings, the soffit may be divided into panels. If the risers are made from two inch planks, it will greatly add to the solidity. The method of drawing and executing the scroll, and other wreath parts of the hand-rail, will be given in a subsequent part of this article.

In constructing a flight of geometrical stairs, where the soffit is inclosed as above, the bearers should all be framed together, so that when put up, they will form a perfect staircase. Each piece of frame-work, which forms a riser, should, in the partition, be well wedged at the ends. This plan is always advisable when strength and firmness are requisite, as the steps and risers are entirely dependent on the framed carriages, which, if carefully put together, will never yield to the greatest weight.

Fig. 619 will show the section of this framing firmly put together, and wedged into the partition, as above described.

In preparing the string for the wreath part, a cylinder should be made of the size of the well-hole of the staircase, which can be done at a trifling expense ; then set the last tread and riser of the flyers on one side, and the first tread and riser of the returning flight on the opposite side, at their respective heights ; then on the centre of the curved surface of this cylinder, mark the middle between the two, and with a thin slip of wood, bent round with the ruling edge, cutting the two nosings of these flyers and passing through the intermediate height marked on the cylinder, draw a line, which will give the wreath line formed by the nosings of the winders ; then draw the whole of the winders on this line, by dividing it into as many parts as you want risers, and each point of division is the nosing of each winder. Having thus far proceeded, and carefully examined your heights and widths, so that no error may have occurred, prepare a veneer of the width intended for your string, and the length given by the cylinder, and after laying it in its place on the cylinder, proceed to glue a number of blocks about an inch wide on the back of the veneer, with their fibres parallel to the axis of the cylinder. When dry, this will form the string for the wreath part of the staircase, to be framed into the straight strings. It is here necessary to observe, that about five or six inches of the straight string should be in the same piece as the circular, so that the joints fall about the middle of the first and last flyers. This precaution always avoids a cripple, to which the work would otherwise be subject.

Fig. 618, No. 1, is a plan of a dog-legged staircase, *a* the seats of the newels, *c* the seat of the upper newel.

No. 2. The elevation of the same.

AB. The newels ; the part AC being turned.—DE the upper newel.—FG the carriage piece.—HI upper string board framed into the newel.—K a joist framed into the trimmer.

To describe the ramps ; produce the horizontal part of the knee to L, and also the under side of the rail until it meets the face of the first baluster, at *c*, make *c d* equal to *c* D, and upon A *d*, and from the point *d*, draw the perpendicular *d* L, and L is the centre for describing the ramps *d* D.

The story-rod *a b* is a very necessary article in fixing the steps ; for, if a common rule be used for this purpose, the workmen will be very liable to err and render the stairs extremely faulty, which cannot take place if the story-rod be applied to every riser, and the successive risers be regulated by it.

In the construction of dog-legged staircases, the first thing is, to take the dimensions of the stair, and the height of the story, and lay down a plan and section upon a floor to the full size, representing all the newels and steps ; then the situation of the carriages, pitching pieces, long and cross bearers, as also the string boards; the strings, rails, and newels, being framed together, must be fixed with temporary supports. The string-board will show the situation of the pitching-pieces, which must be put up in order, wedging one end firmly into the wall, and fixing the other to the string-board ; this being done, pitch up the rough strings, and finish the carriage part of the flyers. Having proceeded thus far, the steps are next applied, beginning at the bottom and working upwards, the risers being all firmly nailed into the treads.

In the best kind of dog-legged stairs, the nosings are returned ; sometimes the risers are mitred to the brackets, and sometimes mitred with quaker strings. In the latter case a hollow is mitred round the internal angle of the under side of the tread, and the face of the riser. Sometimes the string is framed into the newel, and notched to receive the ends of the steps ; the other end having a corresponding notch-board, and the whole flight being put up like a step-ladder.

Fig. 619, No. 1 and 2, is a plan and elevation of a geometrical staircase. The lower part, No. 2, shows the section of the steps and carriages, which are framed together as directed in a former part of this article.

The methods of finding the different moulds necessary in the formation of the wreath part of the hand-rail, will be found in the next plate.

To draw the scroll of a hand-rail.

Fig. 620. First make a circle 3½ inches in diameter, divide the diameter into three equal parts, make a square in the centre of the circle equal to one of those parts, and divide each side of the square into six equal parts.

Fig. 4, shows this square on a larger scale, and laid in the same position as the little square above, with the different centres marked. The centre at 1 draws from a to b, the centre at 2 from b to c, and the centre at 3 from c to d, &c. which will complete the outside revolution at A : set the thickness of the rail from $c f$ and to s, draw the inside the reverse way, and the scroll will be completed.

To draw the curtail-steps.

Set the balusters in their proper places on each quarter of the scroll, Fig. 8, the first baluster showing the return of the nosing round the step, the second placed at the beginning of the twist, and the third a quarter distant, and straight with the front of the last riser ; then set the projection of the nosing without, and draw it round equally distant from the scroll, which will give the form of the curtail.

As the method of getting a scroll out of a solid piece of wood, having the grain of the wood to run in the same direction with the rail, is far preferable to any other method with joints, being much stronger and more beautiful than any other scroll with one or two joints, we shall here give the method of finding a face-mould to apply on the face of the plank.

Place your pitch board *l m n* with *m n* passing through the eye of the scroll, then draw ordinates across the scroll at discretion, and take the length of the line *o n*, with its divisions, and lay it on *o n*, at Fig. 621, then the ordinate being drawn, take the different distances 2 *x*, 3 *x*, 4 *v*, &c. and transfer them to 2 *y*, 3 *z*, 4 *v*, &c. and the rest of the points being taken in the same manner, a curve may be traced which will be the face-mould required.

To find the parallel thickness of the plank.

Fig. 622. Let *l m n* be the pitch board, and let the level of the scroll rise one-sixth, that is, divide *l m* into six equal parts, and the bottom division is the top of the level of the scroll: from the end of the pitch board, set on *n* to *o*, half the thickness of a baluster, to the inside; then set, from *o* to *p*, half the width of the rail, and draw the form of the rail on the end at *p*, the point *n* being where the front of the riser comes, the point *p* will be the projection of the rail before it: draw a dotted line to touch the nose of the scroll, parallel with *l n*, then the distance between this dotted line and the under tip of the scroll, will show the exact thickness of planking; but there is no occasion for the thickness to come quite to the under side, for if it come to the under side of the hollow it will be sufficient, as a little bit glued under the hollow could not be discernible, and can be no hurt to the scroll. In ordinary cases, where the tread is about 11 inches, and rise 6½, a scroll can be got out of a piece, about 4½ inches thick.

To describe a section of a hand-rail, supposing it to be two inches deep, and two and a quarter inches broad, the usual dimensions.

Fig. 622. Let ABCD be a section of the rail, as squared; on AB describe an equilateral triangle AB *a*; from *a*, as a centre, describe an arc to touch AB, and to meet *a* A and *a* B; take the distance between the point of section in *a* A and the point A, and transfer it from the point *o*, section to *k*, upon the same line *a* A, join D *k*; from *k*, with the distance between *k* and the end of the arc, describe another arc, to meet D *k*; with the same distance describe a third arc, of contrary curvature, and draw a vertical line to touch it; which will form one side of the section of the rail, and the counter part may be formed by a similar operation.

The branch of JOINERY that falls under our next and last consideration is that of hand-railing; which calls into action all the ingenuity and skill of the workman. This art consists in constructing hand-rails by moulds, according to the geometrical principles, that if a cylinder be cut in any direction, except parallel to the axis, or base, the section will be an ellipse; if cut parallel to the axis, a rectangle; and if parallel to the base, a circle.

Now, suppose a hollow cylinder be made to the size of the well-hole of the stair-case, the interior concave, and the exterior convex; and the cylinder be cut by any inclined or oblique plane, the section formed will be bounded by two concentric similar ellipses; consequently, the section will be at its greatest breadth at each extremity of the larger axis, and its least breadth at each extremity of the smaller axis. Therefore, in any quarter of the ellipsis there will be a continued increase of breadth from the extremity of the lesser axis to that of the greater. Now it is evident that a cylinder can be cut by a plane through any three points; therefore, supposing we have the height of the rail at any three points in the cylinder, and that we cut the cylinder through these points, the section will be a figure equal and similar to the face-mould of the rail; and if the cylinder be cut by another plane parallel to the section, at such a distance from it as to contain the thickness of the rail, this portion of the cylinder will represent a part of the rail with its vertical surfaces already worked: and, again, if the back and lower surface of this cylindric portion be squared to vertical lines, either on the convex or concave side, through two certain parallel lines drawn by a thin piece of wood which is bent on that side, the portion of the cylinder thus formed, will represent the part of the rail intended to be made.

Though the foregoing only relates to cylindrical well-holes, it is equally applicable to rails erected on any seat whatever.

The *face-mould* applies to the two faces of the plank, and is regulated by a line drawn on its edge, which line is vertical when the plank is elevated to its intended position. This is also called the *raking-mould*.

The *falling-mould*, is a parallel piece of thin wood applied and bent to the side of the rail-piece, for the purpose of drawing the back and lower surface, which should be so formed, that every level straight line, directed to the axis of the well-hole, from every point of the side of the rail formed by the edges of the falling mould, coincide with the surface.

In order to cut the portion of rail required, out of the least possible thickness of stuff, the plank is so turned up on one of its angles, that the upper surface is no where at right angles to a vertical plane passing through the chord of the plane; the plank in this position is said to be *sprung*

The *pitch-board*, is a right-angled triangular board made to the rise and tread of the step, one side forming the right angle of the width of the tread, and the other of the height of the riser. When there are both winders and flyers, two pitch-boards must be made to their respective treads, but, of course, of the same height, as all the steps rise the same.

The bevel by which the edge of the plank is reduced from the right angle when the plank is sprung, is termed the *spring of the plank*, and the edge thus bevelled is called the *sprung edge.*

The bevel by which the face mould is regulated to each side of the plank, is called the *pitch.*

The formation of the upper and lower surfaces of a rail is called the *falling of the rail*; the upper surface of the rail is termed the *back.*

In the construction of hand-rails, it is necessary to spring the plank, and then to cut away the superfluous wood, as directed by the draughts, formed by the face-mould; which may be done by an experienced workman, so exactly, with a saw, as to require no further reduction; and when set in its place, the surface on both sides will be vertical in all parts, and in a surface perpendicular to the plan. In order to form the back and lower surface, the falling mould is applied to one side, generally the convex, in such a manner, that the upper edge of the falling mould at one end, coincides with the face of the plank; and the same in the middle, and leaves so much wood to be taken away at the other end as will not reduce the plank on the concave side;—the piece of wood to be thus formed into the wreath or twist being agreeable to their given heights.

In the following figures, we have given the method o. finding the moulds necessary for constructing a hand-rail on a circular plan.

Fig. 623, is the plan, showing part of the winders, which in this case are eight, as also the seat of the joint.

Fig. 624. Let AAA, &c. be the outside, and *a a a*, &c. the inside of the plan. BCD a line passing through the middle of the breadth, BC being straight, and CD one-fourth of the circumference of the circle, the point E in the middle of the arc CD, B at one extremity of the line BCED, and D at the other.

Divide the quadrant CD into any number of equal parts, which in this example are four. Draw the straight line MN, and make MN equal to the developement of the quadrant AAA, &c. on the convex side. Draw MO perpendicular to MN, and make MO equal to the height of a step; draw OP parallel to MN, and make OP equal in length to the width of a step, and join PM.

Draw N *s* perpendicular to MN. In N *s* make N *o* equal to the height of four of the winders, and join *o* M · curve off the angle at M, in the manner shown below, by intersection of lines: Through *o* draw *s y* perpendicular to M *s*, make *o s* and *o y* each equal to half the width of the falling mould, and draw the upper and lower edges of the mould.

Join DE. Fig. 624, and produce DE to F. Draw DG and EL. Make DG equal to one-fourth (or any part of) the height from N to the upper edge of the falling mould, Fig. 625, and EL equal to one-fourth, or the same part, of the height from Q to the upper edge of the falling mould. Join GL and produce it to meet DE in F, join the dotted line BF. Draw IK, through the centre F, perpendicular to BF. Draw *a b*, *a b*, &c. meeting IK. At any convenient distance from KI draw *c d* parallel to IK. Make the perpendicular of the face-mould equal to its corresponding height on the falling mould, and draw the straight line *c e*; then draw ordinates A *b*, A *b*, &c. continue them until they meet *c e*, and from the points of intersection draw perpendiculars to *c e*, and set off the distances as shown by corresponding letters. Then by tracing a curve through these points the face mould will be completed.

The top line *r r r*, &c. is left on the falling mould, to regulate its position when bent upon the convex surface, as the line *r r r*, and will fall into the plane surface of the top of the plank. This line is obtained by making the perpendiculars *f r*, 2 *r*, *f r*, &c. equal to the corresponding perpendiculars *f b*, *f b*, &c. Fig. 624. To find the face-mould of a staircase, so that when set to its proper rake it will be perpendicular to the plan whereon it stands for a level landing.

Fig. 626. Draw the central line, *a b*, parallel to the sides of the rail, on the right line *a b* apply the pitch-board of a flyer, from *b* to *c* draw ordinates *n m*, *o p*, *q r*, *s t*, *u v*, at discretion, observing to draw one from the point *r*, so that you may obtain the same point exactly in the face-mould; then take the parts which the ordinates give on the line *a b*, and apply them at Fig. 627, and take the distances *m n*, *p o*, &c. and transfer them to Fig. 627, and a curve through these points will be the face-mould required.

To find the falling mould.

Fig. 626. Divide the radius of the circle into four equal parts, and set three of these parts from 4 to *a*; through *x y*, the extremities of the diameter of the rail, draw *a x* and *a y*, producing them till they touch the tangent AB; then will AB be the circumference of the semicircle *x b y*, which is applied from A to B, Fig. 628, as a base line. Make A *a* the height of a step; draw the hypotenuse *a* B, apply the pitch board of a flyer at *a b c*, and B *d e*, then curve off the angle by intersection of lines, and draw a line parallel to it, for the upper edge of the mould.

MEASURES CUSTOMARY IN JOINERS' WORK.

Prepared boarding is measured by the foot superficial; the following being the different distinctions:—edges shot; edge shot, ploughed, and tongued; wrought on one side, and edges shot; wrought on both sides, and edges shot; wrought on both sides, ploughed, and tongued; boards keyed and clamped, mortise-clamped, and mortise and mitre-clamped. The prices are regulated according to the thickness. If the boards be glued, an additional price per foot is allowed; if tongued, still more, according to the description of tongue. In boarded flooring, the dimensions are taken to the extreme parts, from which the squares are to be computed. Deductions for chimneys, stair-cases

&c. are taken from this. The price depends on the surface, whether wrought or plain, the manner of the longitudinal and heading-joints, the thickness of stuff, whether
the boards be laid one after the other, or folded, or whether
the floor be laid with boards, battens, or wainscot.[1]

Skirting, when wide, is also measured by the foot superficial; the price depending upon the position, whether level,
raking, or ramping, or upon the manner of finishing, whether plain, torus, or rebated, or scribed to the floor, or to the
steps, or upon the plan, whether straight or circular.

Weather-boarding, is measured by the square of 100 superficial feet.

Boarded partitions are measured by the square, from
which must be deducted the doors and windows, except an
agreement be made to the contrary.

The price of all kinds of framing depends on the thickness, or whether the framing be plain or moulded; and if
moulded, the description of moulding, whether struck on the
solid, or laid in, mitred, or scribed; as also upon the number of panels in a given height and breadth, and upon the
nature of the plan.

The different kinds of wainscotting, as window linings,
door linings, back linings, partitions, doors, shutters, &c.
are all measured by the superficial foot.

Windows are in general valued by the foot superficial;
though sometimes by the window. When measured, the
dimensions are taken for height, from the top of the cill to
the under side of the head, allowing seven inches for the
head and cill; and for width in clear of pulley-stiles, allowing eight inches. The sash and frame are either measured
together or separately.

Skylights are measured by the foot superficial, their price
depending on the plan and elevation. Framed grounds at
per foot run.

Ledged doors by the foot superficial, dado by the superficial foot; the price depending whether the plan be straight
or circular, or the elevation level or inclined.

In measuring stair-cases, the risers, treads, and carriages,
are generally classed together, and measured by the foot superficial: the price varying as the steps are flyers or winders, as the risers are mitred into the string-board, the treads
dove-tailed for balusters, and the nosings returned, or whether the bottom of the risers be tongued into the treads. The
curtail step is generally valued as a whole. Returned nosings
at so much each; and if circular, double the price of straight

ones. The brackets at so much each, according to the pattern, and whether straight or circular.

Hand-railing is measured by the foot run, the price depending on the materials, the diameter of the well-hole, or whether ramped, swan-necked, level, circular, or wreathed, or whether made out of the solid, or in thicknesses. The scroll is paid at per piece. The joints at so much each, and three inches of the straight part at each end of the wreath are included in the measurement. Deal balusters are prepared and fixed at per piece; as also iron balusters, iron column to curtail, housings to steps, &c. An extra allowance is made for the additional labour in fixing the iron balusters.

The price of string-board is regulated by the foot superficial, according to the manner in which it is moulded, whether straight, circular, or wreathed, and the manner in which such string is backed. The shafts of columns are measured by the foot superficial; the price depending upon the diameter, and whether it be straight or curved, or properly glued and blocked. If the column be fluted or reeded, the flutes or reeds are measured by the foot run, their price depending upon the size of the flute or reed The beadings of flutes and reeds are at so much each. Pilasters, straight or curved, in the height, are measured in the same way, and the price taken per foot superficial in the caps and bases if pilasters; besides the mouldings, the mitres must be so much each, according to the size.

Mouldings are valued by the foot run, as double-faced architraves, base and surbase. The head of an architrave in a circular wall, is four times the price of the perpendicular parts, not only on account of the time required to form the mouldings to the circular plan, but on account of the greater difficulty of forming the mitres.

All horizontal mouldings, circular upon plan, are three or four times the price of those on a straight plan; being charged more, as the radius of the circle is less: housings to mouldings are valued at so much each, according to the size.

The price per superficial foot of mouldings is regulated by the number of quirks, for each of which an addition is made to the foot.

The price of mouldings depends also upon the materials of which they are made, and upon their running figure, whether curved or raking.

In grooving, the stops are paid over and above, and so much more must be allowed for all grooves wrought by hand, particularly in the parts adjoining the concourse of

lc : circular grooving must be paid still more. Water
arc measured by the foot run ; the rate depending
he side of their square : the hopper-heads and shoes
ued at so much each, as also are the moulded weather
ind the joints. Scaffolding, &c. used in fixing, is
d extra.

ring-boards are prepared, that is, planed, gauged, and
l to a thickness at so much each, the price depending
he length of each board ; if more than nine inches
the rate is increased according to the additional
each board listing at so much per list.

following is a classification of such articles in joinery
isually rated at so much each.

ers.
its.
kets for shelves.
s in general.
s to steps.
to standards.
ippings.
l moulded nosings to steps.
iand-rails.
hand-rails.
and fixing joists of hand-
ith joint-screws.
on columns in curtails.
ron baluster, and prepar-
uld.
g and fixing deal balusters.

Brackets to stairs.
Curtail step.
Clamp-mitres.
Mitres of pilasters according to
their size.
Mitres of cornices.
Headings to flutes and reeds.
Hopper-heads and shoes to water-
trunks.
Joints to water-trunks.
Preparing flooring-boards and bat-
tens.
Fixing locks and fastenings, per
article.
Hole in seat of water-closet.
Patteras.

Articles at per foot running, or lineal.

to shelves.
raisings of panels.
d panels in the extremity
raising to be charged ex-

to wainscot.
ircular string-boards to

ls.
o stairs.
planiers in stairs.
in rail for iron rail or ba-
.
unks and spouts.
and door-grounds.
fillets.

Fillets mitred on panels,
Square or beaded angle-staff, re-
bated.
Mouldings.
Single cornice.
Single faced architrave.
Pilasters under four inches wide.
Boxings to windows.
Ornamental grooving.
Narrow linings.
Legs, rails, and runners of dres-
sers.
Border to hearth.
Base-moulding.
Surbase-moulding.
Narrow skirting.

Articles at per foot superficial.

ned, ploughed, tongued,
, glued, and clamped.

Skirting.
Sash-frames and sashes.

Skylights.

Back, elbow, soffits.

Shutters.

Framed or plain back-linings.

Door-linings, jambs.

Wainscotting.

Dado.

Partitions.

Steps and rises to stairs, including carriages,

Cradling.

Double-faced architraves.

Mouldings wrought by hand, if large.

Shafts of columns.

PLASTERING.

The Plasterer is a workman to whom the decorative part of architecture owes a considerable portion of its effect, and whose art is requisite in every kind of building.

The tools of the plasterer consist of a *spade* or shovel of the usual description; a *rake,* with two or three prongs, bent downwards from the line of the handle, for mixing the hair and mortar together; *trowels* of various kinds and sizes; stopping and picking-out tools; rules called *straight-edges;* and wood *models.*

The trowels used by plasterers are more neatly made than tools of the same name used by other artificers. The *laying and smoothing tool* consists of a flat piece of hardened iron, about ten inches in length, and two inches and a half wide, very thin, and ground to a semicircular shape at one end, but left square at the other; and at the back of the plate, near the square end, is rivetted a small iron rod with two legs, one of which is fixed to the plate, and the other to a round wooden handle. With this tool all the first coats of plaster is laid on, as are also the last, or, as it is technically termed, the *setting.* The other kinds of trowels are made of three or four sizes, for *gauging* the fine stuff and plaster, used in forming cornices, mouldings, &c. The longest size of these is about seven inches on the plate, which is of polished steel, about two inches and three quarters broad at the heel, diverging gradually to a point. To the heel or broad end a handle is adapted.

The *stopping and picking-out tools* are made of polished steel, of different sizes, though most generally about seven or eight inches in length, and half an inch in breadth, flattened at both ends, and ground somewhat round. These tools are used in modelling and finishing mitres and returns to cornices; as likewise in filling-up and perfecting the ornaments at the joinings.

The *straight-edges* are for keeping the work in an even, or perpendicular line; and the *models* or *moulds* are for run-

seem, that it might be safely inferred, that the moderns in general rather err in giving too little, than in giving too much sand. It deserves, however, to be noticed, that the sand, when naturally in the lime-stone, is more intimately blended with the lime, than can possibly be ever effected by any mechanical operation ; so that it would be in vain, to hope to make equally good mortar artificially from pure lime, with so small a proportion of caustic calcareous matter, as may sometimes be effected when the lime naturally contains a very large proportion of sand. Still, however, there seems to be no doubt, that if a much larger proportion of sand than is common were employed, and that more carefully and expeditiously blended and worked, the mortar would be made much more perfect, as has been proved by actual experiments.

Another circumstance, which greatly tends to vary the quality of cement, and to make a greater or smaller proportion of sand necessary, is, the mode of preparing the lime before it is beaten up into mortar. When for plaster, it is of great importance to have every particle of the lime-stone slaked before worked-up, for, as smoothness of surface is the most material point, if any particles of lime be beaten-up before sufficiently slaked, the water still continuing to act on them, will cause them to expand, which will produce those excrescences on the surface of the plaster, termed blisters. Consequently, in order to obtain a perfect kind of plaster, it is absolutely necessary that the lime, before being worked, be allowed to remain a considerable time macerating or *souring* in water : the same sort of process, though not absolutely required, would considerably improve the lime intended for mortar. Great care is required in the management ; the principal thing being the procuring of well-burnt lime, and allowing no more lime, before worked, than is just sufficient to macerate or *sour* it with the water: the best burnt lime will require the maceration of some days.

It has been almost universally admitted, that the hardest lime-stone affords the lime which will consolidate into the firmest cement ; hence, it is generally concluded, that lime made of chalk produces a much weaker cement than that made of marble, or lime-stone. It would seem, however, that, if ever this be the case, it is only incidentally, and not necessarily. In the making of mortar, other substances are occasionally mixed with lime, which we shall here proceed to notice, and endeavour to point out their excellencies and defects. Those commonly used, besides sand of various denominations, are powdered sand-stone, brick-dust, and sea-shells: and for forming plaster, where closeness rather than hardness is required, lime which has been slaked and kept in a dry place till it has become nearly effete, and powdered chalk, or whiting, and gypsum, in various proportions, besides hair and other materials of a similar nature. Other ingredients have been more lately recommended, such as earthy balls, slightly burnt and pounded, old mortar rubbish, powdered and sifted, and various things of the like kind, the whole of which are, in some respect or other, objectionable.

Plaster of Paris is employed by the plasterer to give the requisite form and finish to all the superior parts of his work. It is made of a fossile stone, called gypsum, which is excavated in several parts of the neighbourhood of Paris, whence it derives its name, and is calcined to a powder, to deprive it of its water of crystallization. The best is Montmartre.

The stones are burnt in kilns, which are generally of very simple construction, being not unfrequently built of the gypsum itself. The pieces to be calcined are loosely put together in a parallelopiped heap, below which are vaulted pipes or flues, for the application of a moderate heat.

The calcination must not be carried to excess; as otherwise the plaster will not form a solid mass when mixed with a certain portion of water. During the process of calcination, the water of crystallization rises as white vapour, which, if the atmosphere be dry, is quickly dissolved in air.

The pounding of the calcined fragments is performed sometimes in mills constructed for the purpose, and sometimes by men, whose health is much impaired by the particles of dust settling upon their lungs.

On the river Wolga, in Russia, where the burning of gypsum constitutes one of the chief occupations of the peasantry, all kinds of gypsum are burnt promiscuously on grates made of wood; afterwards the plaster is reduced to powder, passed through a sieve, and finally formed into small round cakes, which are sold at so much per thousand.

These balls are reduced into an impalpable powder by the plasterer, and then mixed with mortar. The less the gypsum is mixed with other substances, the better it is qualified for the purpose of making casts, stucco, &c. The sparry gypsum, or selenite, which is the purer kind, is employed for taking impressions from coins and medals, and

for making those beautiful imitations of marble, ~~~~
and porphyry, known by the name of ~~scagliola~~, which is
derived from the Italian word, ~~scagli~~.

Finely powdered alabaster, or plaster of Paris, when
heated in a crucible, assumes the appearance of a fluid,
rolling in waves, yielding to the touch, steaming, &c. all
which properties it again loses on the departure of the
heat: if taken from the crucible and thrown upon paper, it
will not wet it; but immediately be as motionless as it was
before exposed to the heat.

Two or three spoonfuls of burnt alabaster mixed up
with water, will, at the bottom of a vessel filled with water,
coagulate into a hard lump, notwithstanding the water
that surrounds it. The coagulating or setting property
of burnt alabaster will be very much impaired, or lost,
if the powder be kept for any considerable time, and
more especially in the open air. When it has been once
tempered with water, and suffered to grow hard, it cannot
be rendered of any further use.

Plaster of Paris, diluted with water into the consistence of
a soft or thin paste, quickly sets, or grows firm, and at the
instant of its setting, has its bulk increased. This expansive
property, in passing from a soft to a firm state, is one of its
valuable properties; rendering it an excellent matter for
filling cavities in sundry works, where other earthy mixtures would shrink and leave vacuities, or entirely separate
from the adjoining parts. It is also probable that this expansion of the plaster might be made to contribute to the
elegance of the impressions it receives from medals, &c. by
properly confining it when soft, so that, at its expansion it
would be forced into the minutest traces of the figure.

A plaster of a coarser description, made of a fossil
stone, much like that of which Dutch terras are made,
is sometimes used in this country, for floors in gentlemen's
houses, and for corn-granaries. This stone, when burnt
after the manner of lime, assumes a white appearance, but
does not ferment on being mixed with water: when cold, it
is reduced to a fine powder. About a bushel of this powder
is put into a tub, and water is applied till it becomes fluid.
In this state it is well stirred with a stick, and used immediately; for in less than a quarter of an hour it becomes
hard and useless, as it will not allow of being mixed a
second time.

Other cements are used by plasterers for inside work.
The first is called *lime and hair*, or *coarse stuff*, and is

pared as common mortar, with the addition of hair from the tan-yards. The mortar is first mixed with a requisite quantity of sand, and the hair is afterwards worked in by the application of a rake.

Next to this is *fine stuff*, which is merely pure lime, slaked first with a small quantity of water, and afterwards, without any extraneous addition, supersaturated with water, and put into a tub in a half fluid state, where it is allowed to remain till the water is evaporated. In some particular cases, a small portion of hair is incorporated. When this fine stuff is used for inside walls, it is mixed with very fine washed sand, in the proportion of one part sand to three parts of fine stuff, and is then called *trowelled* or *bastard stucco*, with which all walls intended to be painted are finished.

The cement called *gauge stuff*, consists of three-fifths of fine stuff, and one-fifth plaster of Paris, mixed together with water, in small quantities at a time, to render it more ready to set. This composition is mostly used in forming cornices and mouldings run with a wooden mould. When great expedition is required, plasterers gauge all their mortars with plaster of Paris, which sets immediately.

The technical divisions of plasterer's work shall now claim our attention.

Lathing, the first operation, consists in nailing laths on the ceiling, or partition. If the laths be of oak, they will require wrought iron nails; but if of deal, nails made of cast iron may be used. Those mostly used in London are of fir, imported from America and the Baltic, in pieces called staves. Laths are made in three foot and four foot lengths: and with respect to their thickness and strength, are either single, lath and half, or double. The single are the thinnest and cheapest; those called *lath and half*, are supposed to be one third thicker than the single; and the double laths are twice that thickness. In lathing ceilings, the plasterer should use both the lengths alluded to, and in nailing them up, should so dispose them, that the joints be as much broken as possible, that they may have the stronger key or tie, and thereby strengthen the plastering with which they are to be covered. The thinnest laths are used in partitions, and the strongest for ceilings.

Laths are also distinguished into heart and sap laths: the former should always be used in plain tiling; the latter, which are of inferior quality, are most frequently used by the plasterer.

Laths should be as evenly split as possible. Those that are very crooked should not be used, or the crooked part should be cut out; and such as have a short concavity on the one side, and a convexity on the other, not very prominent, should be placed with the concave sides outward.

The following is the method of rending or splitting laths. The lath-cleavers having cut their timber into the proper lengths, cleave each piece with wedges, into eight, twelve, or sixteen pieces, according to the scantling of the timber, called *bolts*; and then, with dowl-axes, in the direction of the felt-grain, termed *felting*, into sizes for the breadth of the laths; and, lastly, with the *chit*, cleave them into thicknesses by the *quarter grain*.

Having nailed the laths in their appropriate order, the plasterer's next business is to cover them with plaster, the most simple and common operation upon which is *laying*; that is, spreading a single coat of lime and hair over the whole ceiling, or partition; carefully observing to keep it smooth and even in every direction. This is the cheapest kind of plastering.

Pricking up is performed in the same manner as the foregoing; but is only a preliminary to a more perfect kind of work. After the plaster is laid on, it is crossed all over with the end of a lath, to give it a tie or key to the coat which is afterwards to be laid upon it.

Lathing, laying, and set, or what is termed *lath and plaster, one coat and set,* is, when the work, after being lathed, is covered with one coat of lime and hair, and afterwards, when sufficiently dry, a thin and smooth coat spread over it, consisting of lime only, or, as the workmen call it, putty, or *set*. This coat is spread with a smoothing-trowel, used by the workman with his right hand, while his left hand moves a large flat brush of hog's bristles, dipped in water, backwards and forwards over it, and thus produces a surface tolerably even for cheap work.

Lathing, floating, and set, or *lath and plaster, one coat, floated and set,* differs from the foregoing, in having the first coat pricked up to receive the set, which is here called the *floating*. In doing this, the plasterer is provided with a substantial straight edge, frequently from ten to twelve feet in length, which must be used by two workmen. All the parts to be floated are tried by a plumb-line, to ascertain whether they be perfectly flat and level, and wherever any deficiency appears, the hollow is filled up with a trowel full or more of lime and hair only, which is termed filling up and when these preliminaries are settled, the *screeds* are next formed. The term *screed* signifies a style of lime and hair, about seven or eight inches in width, gauged quite true, by drawing the straight edge over it until it be so. These screeds are made at the distance of about three or four feet from each other, in a vertical direction, all round the partitions and walls of a room. When all are formed, the intervals are filled up with lime and hair, called by the workmen, *stuff*, till flush with the face of the screeds. The straight edge is then worked horizontally on the screeds, by which all the superfluous stuff, projecting beyond them in the intervals is removed, and a plain surface produced. This operation is termed *floating*, and may be applied to ceilings as well as to partitions, or upright walls, by first forming the screeds in the direction of the breadth of the apartment, and filling up the intervals as above described. As great care is requisite to render the plaster sound and even, none but skilful workmen should be employed.

The *set* to floated-work is performed in a mode similar to that already prescribed for *laying ;* but being employed only for best rooms, is done with more care. About one-sixth of plaster of Paris is added to it, to make it set more expeditiously, to give it a closer and more compact appearance, and to render it more firm and better calculated to receive the white-wash or colour when dry. For floated stucco-work the pricking up coat cannot be too dry ; but, if the floating which is to receive the setting coat be too dry, before the set is laid on, there will be danger of its peeling off, or of assuming the appearance of little cracks, or shells, which would disfigure the work. Particular care and attention therefore must be paid to have the under coats in a proper state of dryness. It may here be observed, that cracks, and other unpleasant appearances in ceilings, are more frequently the effect of weak laths being covered with too much plaster, or too little plaster upon strong laths, rather than of any sagging or other inadequacy in the timbers, or the building. If the laths be properly attended to, and the plaster laid on by a careful and judicious workman, no cracks or other blemishes are likely to appear.

The next operation combines both the foregoing processes, but requires no lathing; it is called *rendering and set, or rendering, floated, and set.* What is understood by *rendering*, is the covering of a brick or stone wall with a coat of lime and hair, and by *set* is denoted a superficial coat of fine stuff or putty upon the rendering. These styles

rations are similar to those described for setting of ceilings and partitions; and the *floated and set* is laid on the rendering in the same manner as on the partitions, &c. already explained, for the best kind of work.

Trowelled stucco, which is a very neat kind of work, used in dining-rooms, halls, &c. where the walls are prepared to be painted, must be worked upon a floated ground, and the floating be quite dry before the stucco is applied. In this process the plasterer is provided with a wooden tool, called *a float*, consisting of a piece of half inch deal, about nine inches long and three wide, planed smooth, with its lower edges a little rounded off, and having a handle on the upper surface. The stucco is prepared as above described, and afterwards well beaten and tempered with clear water. The ground intended to be stuccoed is first prepared with the large trowel, and is made as smooth and level as possible; when the stucco has been spread upon it to the extent of four or five feet square, the workman, with a float in his right hand and a brush in his left, sprinkles with water, and rubs alternately the face of the stucco, till the whole is reduced to a fine even surface. He then prepares another square of the ground, and proceeds as before, till the whole is completed. The water has the effect of hardening the face of the stucco. When the floating is well performed, it will feel as smooth as glass.

Rough casting, or rough walling, is an exterior finishing, much cheaper than stucco, and, therefore, more frequently employed on cottages, farm-houses, &c. than on buildings of a higher class. The wall intended to be rough-cast, is first pricked-up with a coat of lime and hair; and when this is tolerably dry, a second coat is laid on, of the same materials as the first, as smooth as it can possibly be spread. As fast as the workman finishes this surface, he is followed by another with a pail-full of rough-cast, with which he bespatters the new plastering, and the whole dries together. The rough-cast is composed of fine gravel, washed from all earthy particles, and mixed with pure lime and water till the whole is of a semi-fluid consistency. This is thrown from the pail upon the wall with a wooden float, about five or six inches long, and as many wide, made of half-inch deal, and fitted with a round deal handle. While, with this tool, the plasterer throws on the rough-cast with his right hand, he holds in his left a common whitewashers' brush, dipped in the rough-cast also, with which he brushes and colours the mortar and the rough-cast he has already spread, to give them, when finished, a regular uniform colour and appearance.

Cornices, are either plain or ornamented, and sometimes embrace a portion of both classes. The first point to be attended to is, to examine the drawings, and measure the projections of the principal members, which, if projecting more than seven or eight inches, must be bracketted. This consists in fixing up pieces of wood, at the distance of about ten or twelve inches from each other, all round the place proposed for the cornice, and nailing laths to them, covering the whole with a coat of plaster. In the brackets, the stuff necessary to form the cornices must be allowed, which in general is about one inch and a quarter. A beech mould is next made by the carpenter, of the profile of the intended cornice, about a quarter of an inch in thickness, with the quirks, or small sinkings, of brass or copper. All the sharp edges are carefully removed by the plasterer, who opens with his knife all the points which he finds incompetent to receive the plaster freely.

These preliminaries being adjusted, two workmen, provided with a tub of putty and a quantity of plaster of Paris, proceed to run the cornice. Before using the mould, they gauge a screed of putty and plaster upon the wall and ceiling, covering so much of each as will correspond with the top and bottom of the intended cornice. On this screed one or two slight deal straight-edges, adapted to as many notches or chases made in the mould for it to work upon, are nailed. The putty is then mixed with about one-third of plaster of Paris, and brought to a semi-fluid state by the addition of clean water. One of the workmen, with two or three trowels-full of this composition upon his *hawk*, which he holds in his left hand, begins to plaster over the surface intended for the cornice, with his trowel, while his partner applies the mould to ascertain when more or less is wanted. When a sufficient quantity of plaster is laid on, the workmen holds his mould firmly against both the ceiling and the wall, and moves it backwards and forwards, which removes the superfluous stuff, and leaves an exact impression of the mould upon the plaster. This is not effected at once; for while he works the mould backwards and forwards, the other workman takes notice of any deficiences, and fills them up by adding fresh supplies of plaster. In this manner a cornice from ten to twelve feet in length may be formed in a very short time; indeed, expedition is essentially requisite, as the plaster of Paris occasions a very great tendency in the putty to set, to prevent which, it is necessary to sprinkle the composition frequently with water, as plasterers, in order to secure the truth and correctness of the cornice, generally endeavour to finish all the lengths, or pieces, between any two breaks or projections, at one time. In cornices which have very large proportions, and in cases where any of the orders of architecture are to be introduced, three or four moulds are required, and are similarly applied, till all the parts are formed. Internal and external mitres, and small returns, or breaks, are afterwards modelled and filled up by hand.

Cornices to be enriched with ornaments, have certain indentations, or sinkings, left in the mould in which the casts are laid. These ornaments were formerly made by hand; but now are cast in plaster of Paris, from clay models. When the clay model is finished, and has, by exposure to the action of the atmosphere, acquired some degree of firmness, it is let into a wooden frame, and when it has been retouched and finished, the frame is filled with melted wax, which, when cold, is, by turning the frame upside down, allowed to fall off, being an exact cameo, or counterpart, of the model. By these means, the most enriched and curiously wrought mouldings may be cast by the common plasterer. These wax models are contrived to cast about a foot in length of the ornament at once; such lengths being most easily got out from the cameo. The casts are made of the finest and purest plaster of Paris, saturated with water; and the wax mould is oiled previously to its being put in. When the casts, or intaglios, are first taken from the mould, they are not very firm; but being suffered to dry a little, either in the open air or an oven, they acquire sufficient hardness to allow of being scraped and cleaned.

Basso-relievos and friezes are executed in a similar manner, only the wax mould is so made, that the cast can have a back-ground at least half an inch thick of plaster-cast to the ornament or figure, in order to strengthen and secure the proportions, at the same time that it promotes the general effect.

The process for capitals to columns is also the same, except that numerous moulds are required to complete them. In the Corinthian capital a shaft or belt is first made, on which is afterwards fixed the foliage and volutes; the whole of which require distinct cameos.

In running cornices which are to be enriched, the plasterer takes care to have proper projections in the running-mould, so as to make a groove in the cornice, for the reception of the cast ornament, which is laid in and secured by spreading a small quantity of liquid plaster of Paris on its back. Detached ornaments intended for ceilings or other parts, and where no running mould has been employed, are cast in pieces corresponding with the design, and fixed upon the ceiling, &c. with white-lead, or with the composition known by the name of *iron-cement*.

The manufacture of stucco has, for a long time past, attracted the attention of all connected with this branch of building, as well as chemists and other individuals; but the only benefit resulting from such investigation is, a more extensive knowledge of the materials used. It would seem, that the great moisture of our climate prevents its being brought to any high degree of perfection; though, among the various compositions which have been tried and proposed, some, comparatively speaking, are excellent.

Common stucco, used for external work, consists of clean washed Thames sand and ground Dorking lime, which are mixed dry, in the proportion of three of the latter to one of the former: when well incorporated together, these should be secured from the air in casks till required for use. Walls to be covered with this composition, must first be prepared, by raking the mortar from the joints, and picking the bricks or stones, till the whole is indented: the dust and other extraneous matter must then be brushed off, and the wall well saturated with clean water. The stucco is supersaturated with water, till it has the appearance and consistence of ordinary white-wash, in which state it is rubbed over the wall with a flat brush of hogs' bristles. When this process, called *roughing in*, has been performed, and the work has become tolerably dry and hard, which may be known by its being more white and transparent, the screeds are to be formed upon the wall with fresh stucco from the cask, tempered with water to a proper consistency, and spread on the upper-part of the wall, about eight or nine inches wide; as also against the two ends, beginning at the top and proceeding downwards to the bottom. In this operation, two workmen are required; one to supply the stucco, the other to apply the plumb-rule and straight-edge. When these are truly formed, other screeds must be made in a vertical direction, about four or five feet apart, unless apertures in the wall prevent it, in which case, they must be formed as near together as possible. When the screeding is finished, compo is prepared in larger quantities, and both the workmen spread it with their trowels over the wall in the space left between each pair of screeds. When this operation is complete, the straight-edge is applied, and dragged from the top to the bottom of each pair, to remove whatever superfluous stones may project above the screeds. If there be any hollow places, fresh stucco is applied, and the straight-edge is again drawn over the spot, till the compo is brought even to the face of the screeds, and the whole is level with the edge of the rule. Another interval is then filled up, and the workmen thus proceed till the whole of the wall is covered. The wall is finished by floating, that is, hardening the surface, by sprinkling it with water, and rubbing it with the common wood-float, which is performed similarly to trowelling stucco.

This description of compo. is frequently used by plasterers for cornices and mouldings, in the same manner as described in common plastering; but if the workman finds it necessary, he may add a small quantity of plaster of Paris, to make it fix the better while running or working the mould. Such addition is not, however, calculated to give strength to the stucco, and is only made through the necessity of having a quick set.

In the year 1796, Mr. Parker obtained a patent for a cement that is impervious to water, and which may be successfully employed in ice-houses, cisterns, tanks, &c. In his specification Mr. Parker states, that " nodules of clay, or argillaceous stone, generally contain water in their centre, surrounded by calcareous crystals, having veins of calcareous matter. They are formed in clay, and are of a brown colour like the clay." These nodules he directs should, after being broken into small pieces and burnt in a kiln, with a heat that is nearly sufficient to vitrify them, be reduced to powder: when two measures of water added to five of this powder, will produce *tarras.* Lime and other matters may be added or withheld at pleasure; and the proportion of water may be varied.

The term of the patent being now expired, many other manufactories of this cement have been established, which produce it of equal goodness, and some of them of rather better colour, which is of importance, since the fresco-painting or white-wash, laid on Mr. Parker's composition, is soon taken off by the rain, and leaves the walls of a dingy and unpleasant appearance.

The fresco-painting, or staining, is laid on the walls covered with this cement, to give them the appearance of

... buildings; and is performed by diluting sulphuric acid, (... of vitriol,) with water, and adding fluid-ochres, &c. of ... required tint.

When stucco is washed over with this mixture, the affinity ... in the iron of the cement ceases; and the acid and ... suspended in and upon the stucco are fixed. When ... managed, the surface assumes the appearance ... ashlar bond of masonry.

Scagliola is a distinct branch of plastering, discovered or ... and much used in Italy, and thence introduced ... France, where it obtained its name: the late Mr. H. ... who introduced it into England engaged artists ... Paris, some of whom, finding a demand for their labours, ... in this country, and instructed the natives in ... .

Columns and pilasters are executed in this branch of ... in the following manner: A wooden cradle, ... of thin strips of deal, or other wood, is made to ... the column designed; but about two inches and a ... in diameter than the shaft is intended to be when ... . This cradle is lathed round, as for common plas... , and then covered with a pricking up coat of lime ... hair. When this is quite dry, the artists in scagliola ... operations, by imitations of the most rare and ... marbles, with astonishing and delusive effect; in... , as the imitation takes as high a polish, and feels ... and hard as the most compact and solid marble, ... short of actual fracture can possibly discover the ... counterfeit.

In preparing the scagliola, the workman selects, breaks, ... calcines the purest gypsum, and as soon as the largest ... in the process of calcination, lose their brilliancy, ... the fire, and passes the calcined powder through ... fine sieve, and mixes it, as required for use, with a ... of glue, isinglass, &c. In this solution the colours ... in the marble to be imitated are diffused; but ... the work is to be of various colours, each colour is ... separately, and afterwards mingled and combined, ... in the same manner as a painter mixes on his palette ... primitive colours to compose his different tints.

When the powdered gypsum is prepared, it is laid on the, ... of the intended column, over the pricked-up coat of ... and hair, and is then floated with moulds of wood, ... to the requisite size: the artist uses the colours neces... ry for the imitation during the floating, by which means they mingle and incorporate with the surface. To obtain the glossy lustre, so much admired in works of marble, the workman rubs the work with one hand with a pumice-stone, while with the other he cleans it with a wet sponge; he next polishes it with tripoli, charcoal, and a piece of fine linen; afterwards with a piece of felt dipped in a mixture of oil and tripoli, and finally completes the work by the application of pure oil. This imitation is, certainly, the most complete that can be conceived; and when the bases and capitals are made of real marble, as is the common practice, the deception is beyond discovery. If not exposed to the weather, it is, in point of durability, little inferior to real marble, retains its lustre full as long, and is not one-eighth of the expense of the cheapest kind.

There is another species of plastering, used in the decorative parts of architecture, and for the frames of pictures, looking-glasses, &c. which is a perfectly distinct branch of the art. This composition, which is very strong, and, when quite dry, of a brownish colour, consists of the proportion of two pounds of powdered whiting, one pound of glue in solution, and half a pound of linseed oil, mixed together, and heated in a copper, and stirred with a spatula, till the whole is incorporated. When cool, it is laid upon a stone, covered with powdered whiting, and beaten till it assumes a tough and firm consistence; after which it is covered with wet cloths, to keep it fresh, till required for use.

The ornaments to be cast in this composition, are modelled in clay, as in common plastering, and afterwards a cameo, or mould, is carved in box-wood. This carving requires to be done with the utmost care, otherwise the symmetry of the ornament which is to be cast from it will be spoiled. The composition, when required for use, is cut with a knife into pieces of the requisite size, and forced into the mould; after which it is put into a press, worked by an iron screw, and still further compressed. When the mould is taken from the press, the composition, which is generally cast about a foot in length, is dislodged from the mould, and the superfluous parts pared off with a knife, and cast into the copper for the next supply.

The ornaments thus formed, are glued upon wooden, or other grounds, or fixed by means of white lead, &c.; after which they are painted or gilt, according to the purposes for which they are intended. This composition is at least 80 per cent. cheaper than carving, and, in most cases, equally calculated to answer all the purposes of the art.

It is much to be wished, that the art of plastering could be restored to its ancient perfection; for the Romans possessed an art of rendering works of this kind much more firm and durable than can be accomplished at the present time.

The specimens of ancient Roman plastering still visible, which have not been injured by force, are found to be firm and solid, free from cracks or crevices, and as smooth and polished on the surface as when first applied. The sides and bottoms of the Roman aqueducts were lined with this plastering, and endured many ages.

At Venice, some of the roofs of houses, and the floors of rooms, are covered with a sort of plaster of later date, and yet strong enough to endure the sun and weather for several ages, without either cracking or spoiling.

The method of making the Venetian composition is not known in England; but such might probably be made by heating the powder of gypsum over a fire, and when boiling, which it will do without the aid of water, or other fluid, mixing it with resin, or pitch, or both together, with common sulphur, and the powder of sea-shells. If these be mixed together, water added to it, and the composition kept on the fire till the instant of its being used, it is not improbable that the secret may be discovered. Oil of turpentine and wax, which are the common ingredients in such cements as are accounted firmest, may also be tried as additions; as also may strong ale wort, which is by some directed to be used instead of water, to make mortar of lime-stone of more than ordinary strength.

SLATING.

This branch of building, which is principally employed in the covering of roofs, is not unfrequently combined with that of plastering. The slates chiefly used in London are brought from the quarries at Bangor, in Caernarvonshire, which supply all parts of the United Kingdom. Another kind of slate, of a pale blue-green colour, is used, and most esteemed, being brought from Kendal, in Westmoreland, called *Westmoreland slates.* These slates are not large; but of good substance, and well calculated to give a neat appearance to a roof. The Scottish slate, which assimilates in size and quality to a slate from Wales, called *ladies,* is in little repute.

be remarked, that this kind of slate owes its lightness, not so much to any diversity in the component parts of the stone, as to the thinness to which it is reduced by the workmen; consequently, it is not so well calculated to resist violent winds as those which are heavier.

Slates, when brought from the quarry, are not sufficiently square for the slater's use; he therefore picks up and examines the slates separately, and observes which is the strongest and squarest end; then, seating himself, he holds the slate a little slanting upon, and projecting about an inch over, the edge of a small block of wood, which is of the same height as his seat, and cuts away and makes straight one of its edges; then, with a slip of wood, he gauges, and cuts off the other edge parallel to it, and squares the end. The slate is now considered prepared for use, with the exception of perforating through its opposite ends two small holes, for the reception of the nails which are to confine it to the roof. Copper and zinc nails, or iron nails tinned, are considered the best, being less susceptible of oxidation than nails made of bar iron.

Before we proceed further with the operations necessary in the slating of building, we shall give some account of the tools used by this class of artificers.

Slaters' tools are very few, which sometimes are found by the masters, and sometimes by the men. The tool called the *saire*, is made of tempered iron, about sixteen inches in length, and two inches in width, somewhat bent at one end, with a handle of wood at the other. This tool is not unlike a large knife, except that it has on its back a projecting piece of iron, about three inches in length, drawn to a sharp point. This tool is used to chip or cut all the slates to the required sizes.

The *ripper* is also of iron, about the same length as the *saire*; it has a very thin blade, about an inch and three-quarters wide, tapered somewhat towards the top, where a round head projects over the blade about half an inch on each side: it has also two little round notches in the two internal angles at their intersections. The handle of this tool is raised above the blade by a shoulder, which enables the workman to hold it firm. This instrument is used in repairing old slating, and the application consists in thrusting the blade under the slates, so that the head, which projects, may catch the nail in the little notch at its intersection, and enable the workman to draw it out. During this operation the slate is sufficiently loosened to allow of its being removed, and another inserted in its place.

. The *hammer*, which is somewhat different in shape to the ordinary tool of that name, is about five inches in height on the hammer, or driving part, and the top is bent back, and ground to a tolerably sharp point, its lower or flat end, which is quite round, being about three-quarters of an inch in diameter. On this side of the driving part is a small projection, with a notch in the centre, which is used as a claw to extract such nails as do not drive satisfactorily.

The *shaving-tool* is used for getting the slates to a smooth face for skirtings, floors of balconies, &c. It consists of an iron blade, sharpened at one of its ends like a chisel, and mortised through the centre of two round wooden handles, one fixed at one end, and the other about the middle of the blade. The blade is about eleven inches long, and two inches wide, and the handle is about ten inches long, so that they project about four inches on each side of the blade. In using this tool, the workman places one hand on each side of the handle that is in the middle of the blade, and allows the other to press against both his wrists. In this manner he removes all the uneven parts from off the face of the slate, and gets it to a smooth surface.

The other tools used by the slater consist of chisels, gouges, and files of all sizes; by means of which he finishes the slates into mouldings and other required forms.

In slating roofs, it is necessary to form a base or floor for the slates to lay compactly and safely upon; for *doubles* and *ladies*, boarding is required, which must be laid very even, with the joints close, and properly secured by nails to the rafters. This being completed, the slater provides himself with several slips of wood, called *tilting fillets*, about ten inches and a half wide, and three-quarters of an inch thick on one edge, and chamfered to an arris on the other, which he nails down all round the extreme edges of the roof, beginning with the hips, if any, and if not, with the sides, eaves, and ridge. He next selects the largest of the slates, and arranges them regularly along the eaves with their lower edges to a line, and nails them to the boarding. This part of the work being completed, he takes other slates to form the bond to the under sides of the eaves, and places them under those previously laid, so as to cross and cover all their joints. Such slates are pushed up lightly under those which are above them, and are seldom nailed, but left dependent for support on the weight of those above them, and their own weight on the boarding. The *countesses* and all other description of slates, when intended to be laid in a good manner, are also laid on boards.

When the slater has finished the eaves, he strains a line on the face of the upper slates, parallel to its outer edge, and as far from it as he deems sufficient for the lap of those he intends shall form the next course, which is laid and nailed even with the line, crossing the joints of the upper slates of the eaves. This lining and laying is continued close to the ridge of the roof, observing throughout to cross the different joints, by laying the slates one above another. The same system is uniformly followed in laying all the different sorts of slates, with the exception of those called *patent slates*, as are hereafter explained.

The largest kinds of slate, are found to lay firm on *battens*, which are, consequently, much employed, and produce a very considerable saving of expense in large buildings. A batten is a narrow portion of deal, about two inches and a half, or three inches wide; four of them being commonly procured from an eleven inch board.

For countess slates, battens three-quarters of an inch thick, will be of adequate substance; but for the larger and heavier kinds, inch battens will be necessary. In battening a roof for slates, the battens are not placed at an uniform distance from each other, but so as to suit the length of the slates; and as these vary as they approach the apex, or ridge of the roof, it follows that the slater himself is the best judge where to fix them, so as best to support the slates.

A roof, to be covered with patent slates, requires that the common rafters be left loose upon their purlines, as they must be so arranged that a rafter shall lie under every one of the meeting-joints. Neither battening nor boarding is required for these slates. The number of rafters will depend on the width of the slates; hence if they be of a large size, very few will suffice. This kind of slating is likewise commenced at the eaves; but no crossing or bonding is required, as the slates are laid uniformly, with each end reaching to the centre of the rafter, and butted up to each other throughout the length of the roof. When the eaves-course is laid, the slates which compose it are screwed down to the rafters by two or three strong inch and half screws at each of their ends. A line is then strained about two inches below the upper edge, in order to guide the laying of the next course, which is laid with its lower edge touching the line. This lining, laying with a lap, and screwed

2 s

down, is continued till the roof is completely covered. The joints are then secured by filletting, which consists in covering all the meeting-joints with fillets of slate, bedded in glazier's putty, and screwed down through the whole into the rafters. The fillets are usually about three inches wide, and of a length proportionate to that of the slates, whose joints they have to cover. These fillets are solidly bedded in the putty, and their intersecting joints are lapped similar to those of the slates. The fillets being so laid, and secured by one in the middle of the fillet and one in each lap, are next neatly pointed all round their edges with more putty, and then painted over with the colour of the slate. The hips and ridges of such slating are frequently covered by fillets, which produces a very neat effect; but lead, which is not much dearer, is by far the best kind of covering for all hips and ridges. The patent slating may be laid so as to be perfectly water-tight, with an elevation of the rafters considerably less than for any other slate or tile covering. The rise in each foot of length in the rafter is not required to be more than two inches, which, in a rafter of fifteen feet, will amount to only two feet six inches : a rise scarcely perceptible from the ground.

Slating is performed in several other ways, but the principles already explained, embrace the most of them. Some workmen shape and lay their slates in a lozenge form. This kind of work consists in getting all the slates to an uniform size, of the shape of a geometrical square. When laid on the roof, which must be boarded, they are bonded and lapped as in common slating, observing only to let the elbow, or half of the square, appear above each slate that is next beneath it, and be regular in the courses all over the roof. One nail or screw only can be used for such slating ; hence it soon becomes dilapitated. It is commonly employed in places near to the eye, or where particular neatness is required.

It has been ascertained, that a slate one inch thick will, in an horizontal position, support as much, in weight, as five inches of Portland stone similarly suspended. Hence slates are now wrought and used in galleries, and other purposes, where it is essential to have strength and lightness combined.

Slates are also fashioned into chimney-pieces ; but are incapable of receiving a polish like marble. It makes excellent skirtings of all descriptions, as well as casings to walls, where dilapidations, or great wear and tear are to be ex-

cted. For these purposes, it is capable of being fixed with
ints, equally as neat as wood : and may, if required, be
inted over so as to appear like it. Stair-cases may also
executed in slate, which will produce a resemblance of

MENSURATION OF PLASTERERS' AND SLATERS' WORK.

Plasterers' work is executed by the yard square ; and the
mensions are taken in feet and inches.

If a room consists of more than four quoins, the addi-
nal corners must be allowed at per foot run.

In measuring ceilings with ribs, the superficies must be
ken for plain work ; then an allowance must be made for
ch mitre, and the ribs must be valued at so much per foot
n, according to the girth ; or by the foot superficial, al-
wing moulding work.

In measuring common work the principal things to be
served are as follow :—first, to make deductions for
imneys, windows, and doors ; secondly, to make deduc-
ns for rendering upon brick work, for doors and windows;
irdly, if the workman find materials for rendering between
arters, one-fifth must be added for quarters ; but if work-
anship only is found, the whole must be measured as whole
rk, because the workman could have performed the
lole much sooner if there had been no quarters ; fourthly,
mouldings in plaster work are measured by the foot su-
rficial, the same as joiners, by girting over the mouldings
th a line.

Slaters' work is measured and reduced into squares, con-
ining 100 feet superficial. If in measuring the slating on
roof, it be hipped on all sides with a flat at top, and the
an of the building be rectangular, add the length and breadth
two adjoining sides of the eaves, and the length and
cadth of two adjoining sides at the flat together, multiply
e sum by the breadth of the slope, and the product will
re the area of the space that is covered. Add the number
square feet produced, by multiplying the girts of the roof
the length of the slates at the eaves ; to the area also, for
c trouble of putting on the double row of slates, add the
mber of square feet produced by multiplying the length
the hips by one foot in breadth, and the sum will be the
ole contents, and yield a compensation for the trouble
l waste of materials. If there be no flats, add the two
oining sides and twice the length of the ridge for the
gth ; multiply the sum by the breadth of the slips, for

2 s 2

the area of the space covered, and add the allowances as before.

Another plan is to allow in addition to the nett dimensions of the work, six inches for all the eaves, and four inches for the hips.

All faced work in slate skirting, stair-cases, galleries, &c. is charged by the foot superficial, without any addition.

PLUMBING,

Is the art of casting and working in lead, and using the same in the covering and for other purposes in building.

To the plumber is also confided the pump-work, as well as the making and forming of cisterns and reservoirs, large or small closets, &c. for the purposes of domestic œconomy. The plumber does not use a great variety of tools, because the ductility of the metal upon which he operates does not require it.

The tools used, consist of an iron hammer, rather heavier than a carpenter's, with a short thick handle; two or three wooden mallets of different sizes; and a dressing and flatting tool.

This last is of beech, about eighteen inches long, and two inches square, planed smooth and flat on the under surface, rounded on the upper, and one of its ends tapered off round as a handle. With this tool he stretches out and flattens the sheet-lead, or dresses it to the shape required, using first the flat side, then the round one, as occasion may require.

The plumber has also occasion for a jack and trying plane, similar to that of the carpenter.

With this he reduces the edges of sheet-lead to a straight line, when the purposes to which it is to be applied require it.

Also a chalk line, wound upon a roller, for marking out the lead into such breadths as he may want.

His cutting tools consist of a variety of chisels and gouges as well as knives.

The latter of these are used for cutting the sheet lead into slips and pieces after it has been marked out by the chalk line.

Files of different sizes; ladles of three or four sizes, for melting the solder; and an iron instrument called *grosing-irons*.

These grozing-irons are of several sizes, generally about twelve inches in length, tapered at both ends, the handle end being turned quite round, to allow of its being firmly held while in use: the other end is a bulb of a spindle, or spherical shape, of a size proportioned to the soldering intended to be executed. They are, when required for use, heated to redness.

The plumber's measuring rule is two feet in length, divided into three equal parts of eight inches each; two of its legs are of box-wood, duodecimally divided; and the third consists of a piece of slow tempered steel, attached to one of the box legs by a pivot on which it turns, and falls, when not in use, into a groove cut in such leg for its reception. This steel leg can be passed into places where the others cannot enter; and it is also useful for occasionally removing the oxide or any other extraneous matters from the surface of the heated metal.

Scales and weights are also necessary; and he must be supplied with centre-bits of all sizes; and a stock to work them, for the purpose of making perforations in lead or wood, through which he may want to insert pipes, &c. Compasses, to strike circular pieces, to line or cover figures of that shape, are occasionally required.

Lead is obtained from ore, and, from its being generally combined with sulphur, it has been denominated "*sulphuret.*" After the ore has been taken from its bed it is smelted, first being picked, in order to separate the unctuous and rich, or genuine ore from the stony matrix, and other impurities; the picked ore is then pounded under stampers worked by machinery, and afterwards washed to carry off the remainder of the matrix, which could not be separated in picking. It is next put into a reverberatory furnace, to be *roasted;* during which operation, it is repeatedly stirred, to facilitate the evaporation of the sulphur. When the surface begins to assume the appearance of a paste, it is covered with charcoal, and well shaken together: the fire is then increased, and the purified lead flows down on all sides into the basin of the furnace, whence it runs off into moulds prepared for its reception. The moulds are capable of receiving 154lbs. of lead each, and their contents, when cool, are, in the commercial world, called *pigs.*

Lead is of a bluish-white colour, and when newly melted, or cut, is quite bright; but it soon becomes tarnished on exposure to the atmosphere; assuming first a dirty grey colour, and afterwards becomes white. It is capable of

being hammered into very thin plates, and may be drawn
into wire ; but its tenacity is very inferior to that of other
metals ; for a leaden wire, the hundred and twentieth part
of an inch in diameter, is only capable of supporting about
18lb. without breaking. Lead, next to tin, is the most fusible of all metals ; and if a stronger heat be applied,
it boils and evaporates. If cooled slowly, it crystallizes.
The change of its external colour is owing to its gradual
combination with oxygen, which converts its exterior surface into an oxyd. This outward crust, however, preserves
the rest of the metal for a long time, as the air can penetrate but very slowly.

Lead is not acted upon immediately by water, though that
element greatly facilitates the action of the air upon it : for
it is known that, when lead is exposed to the atmosphere,
and kept constantly wet, the process of oxidation takes
place much more rapidly than it does under other circumstances : hence the white crust that is to be observed on
the sides of leaden vessels containing water, just at the
place where the surface of the water terminates.

Lead is purchased by plumbers, in *pigs*, and they reduce
it into sheets or pipes, as they have occasion. Of sheet-lead
they have two kinds, cast and milled. The former is used
for covering flat roofs of buildings, laying of terraces, forming gutters, lining reservoirs, &c. ; and the latter, which is
very thin, for covering the hips and ridges of roofs. This
last they do not manufacture themselves, but purchase it
of the lead merchants, ready prepared.

For the casting of sheet lead, a copper is provided, and well
fixed in masonry, at the upper end of the workshop, near
the mould or casting table, which consists of strong deal
boards, well jointed together, and bound with bars of iron
at the ends. The sides of this table, of which the shape is
a parallelogram, vary in size from four to six feet in width,
and from 16 to 18 feet and upwards in length, and are guarded
by a frame or edging of wood, 3 inches thick, and 4 or 5
inches higher than the interior surface, called the *shafts*.
This table is fixed upon firm legs, strongly framed together,
about 6 or 7 inches lower than the top of the copper. At
the upper end of the mould, nearest the copper, is a box,
called the *pan*, which is adapted in its length to the breadth
of the table, having at its bottom a long horizontal slit,
from which the heated metal is to issue, when it has been
poured in from the copper. This box moves upon rollers
along the surface of the rim of the table, and is put in mo-

tion by means of ropes and pulleys, fixed to beams above.
While the metal is melting, the surface of the mould, or
table, is prepared by covering it with a stratum of dry and
clean sand, regularly smoothed over with a kind of rake,
called a *strike*, which consists of a board about 5 inches
broad, and rather longer than the inside of the mould, so
that its ends, which are notched about two inches deep, may
ride upon the shafts. This being passed down the whole
length of the table, reduces the sand to an uniform surface. The pan is now brought to the head of the table, close
to the copper, its sides having previously been guarded by a
coat of moistened sand, to prevent its firing from the heat of
the metal, which is now put in by ladles from the copper.

These pans, or boxes, it must be observed, are made to
contain the quantity of melted lead which is required to cast
a whole sheet at one time; and the slit in the bottom is so
adjusted as to let out, during its progress along the table,
just as much as will completely cover it of the thickness and
weight per foot required. Every thing being thus prepared,
the slit is opened, and the box moved along the table, dis-
pensing its contents from the top to the bottom, and leaving
in its progress a sheet of lead of the desired thickness.
When cool, the sheet is rolled up and removed from the
table, and other sheets are cast, till all the metal in the copper is exhausted. The sheets thus formed are then rolled
up and kept for use.

In some places, instead of having a square box upon
wheels, with a slit in the bottom, the pan consists of a kind
of trough, being composed of two planks nailed together at
right angles, with two triangular pieces fitted in between
them, at their ends. The length of this pan, as well as that
of the box, is equal to the whole breadth of the mould. It
is placed with its bottom on a bench at the head of the table
leaning with one side against it: to the opposite side is fixed
a handle, by which it may be lifted up in order to pour out
the liquid metal. On the side of the pan next the mould
are two iron hooks, to hold it to the table, and prevent it
from slipping while the metal is being poured into the
mould.

The mould, as well as the pan, is spread over, about two
inches thick, with sand, sifted and moistened, and rendered
perfectly level by moving over it the strike, and smoothing it
down with a plane of polished brass, about a quarter of an
inch thick, and nine inches square, turned up on the edges.
Before they proceed to casting the lead, the strike is made

prevent its falling into the mould, when the metal is poured out. It generally happens, that the lead, when first taken from the copper, is too hot for casting; it is therefore suffered to cool in the pan, till it begins to stand with a shell or wall on the sand with which the pan is lined. Two men then take the pan by the handle, or one of them takes it by means of a bar and chain fixed to a beam in the ceiling, and turn it down, so that the metal runs into the mould: while another man stands ready with the strike, and, as soon as all the metal is poured in. sweeps it forward and draws the residue into a trough at the bottom, which has been prepared to receive it. The sheet is then rolled up, as before.

In this mode of operation, the table inclines in its length about an inch, or an inch and a half, in the length of sixteen or seventeen feet, or more, according to the required thickness of the sheets; the thinner the sheet the greater the declivity; *and vice versâ.* The lower end of the mould is also left open, to admit of the superfluous metal being thrown off.

When a cistern is to be cast, the size of the four sides is measured out; and the dimensions of the front having been taken, slips of wood, on which the mouldings are carved, are pressed upon the sand. Figures of birds, beasts, &c. are likewise stamped in the internal area, by means of leaden moulds. If any part of the sand has been disturbed in doing this, it is made smooth, and the process of casting goes on as for plain sheets; except that, instead of rolling up the lead when cast, it is bent into four sides, so that the two ends, when they are soldered together, may be joined at the back; the bottom is afterwards soldered up.

The lead which lines the Chinese tea-boxes is reduced

a pot, containing the melted metal, and has two large stones, the lower one fixed and the upper one movable, having their surfaces of contact ground to each other, directly before him. He raises the upper stone by pressing his foot upon its side, and with an iron ladle pours into the opening a sufficient quantity of the fluid metal. He then lets fall the upper stone, and thus forms the lead into an extremely thin and irregular plate, which is afterwards cut into its required form."

Cast sheet lead, used for architectural purposes, is technically divided into 5lb. 5¼lb. 6lb. 6¼lb. 7lb. 7¼lb. 8lb. and 8¼lb.; by which is understood, that every superficial foot is to contain those respective weights, according to the price agreed upon.

The milled lead used by plumbers is very thin, seldom containing more than 5lb. to the foot. It is by no means adapted to gutters or terraces, nor, indeed, to any part of a building that is much exposed either to great wear or to the effects of the sun's rays: in the former case, it soon wears away; in the latter, it expands and cracks. It is laminated in sheets of about the same size as those of cast lead, by means of a roller, or flatting-mill.

Lead-pipes, besides the various ways of manufacture described in page 362, are sometimes made of sheet lead, by beating it on round wooden cylinders of the length and dimensions required, and then soldering up the edges.

Solder is used to secure the joints of work in lead, which by other means would be impossible. It should be easier of fusion than the metal intended to be soldered, and should be as nearly as possible of the same colour. The plumber therefore uses, what is technically called, *soft solder*, which is a compound of equal parts of tin and lead, melted together and run in to moulds. In this state it is sold by the manufacturer by the pound.

In the operation of soldering, the surfaces or edges intended to be united are scraped very clean, and brought close up to each other, in which state they are held by an assistant, while the plumber applies a little resin on the joints, in order to prevent the oxidation of the metal. The heated solder is then brought in a ladle and poured on the joint; after which it is smoothed and finished by rubbing it about with a red-hot soldering iron, and when completed is made sn.ooth by filing.

In the covering of roofs or terraces with lead, (the sheets never exceeding six feet in breadth,) it becomes necessary in large surfaces, to have joints; which are managed several ways, but in all, the chief object is to have them water-tight. The best plan of effecting this, is to form lapt or roll joints, which is done by having a roll, or strip of wood, about two inches square, but rounded on its upper side, nailed under the joints of the sheets, where the edges lap over each other; one of these edges is to be dressed up over the roll on the inside, and the other is to be dressed over them both on the outside, by which means the water is prevented from penetrating. No other fastening is requisite than what is required from the hammering of the sheets together down upon the flat; nor should any other be resorted to, when sheet lead is exposed to the vicissitudes of the weather; because it expands and shrinks, which, if prevented by too much fastening, would cause it to crack and become useless. It sometimes, however, occurs, that rolls cannot be used, and then the method of joining by seams is resorted to. This consists in simply bending the approximate edges of the lead up and over each other, and then dressing them down close to the flat, throughout their length. But this is not equal to the roll, either for neatness or security.

Lead flats and gutters should always be laid with a current, to keep them dry. About a quarter of an inch to the foot run is a sufficient inclination.

In laying gutters, &c. pieces of milled-lead, called flashings, about eight or nine inches wide, are fixed in the walls all round the edges of the sheet-lead, with which the flat is covered, and are suffered to hang down over them, so as to prevent the passage of rain through the interstice between the raised edge and the wall. If the walls have been previously built, the mortar is raked out of the joint of the bricks next above the edge of the sheet, and the flashings are not only inserted into the crack at the upper side, but their lower edges are likewise dressed over those of the lead in the flat, or gutter. When neither of these modes can be resorted to, the flushings are fastened by wall-hooks, and their lower edges dressed down as before.

Drips in flats, or gutters, are formed by raising one part above another, and dressing the lead, as already described, for covering the rolls. They are resorted to when the gutter or flat, exceeds the length of the sheet; or sometimes for convenience. They are also an useful expedient to avoid soldering the joints.

Sheet lead is also used in the lining of reservoirs, which

The tools requisite for the performance of the first of these departments are, a iamond, a ranging lath, a short lath, a square, a rule, a g. hung-knife, a cutting-chisel, a beading-hammer, a duster, an sash-tool; and in addition, for stopping in squares, a cking-knife and hammer.

The diamond is r speck of that precious stone, polished to a cutting point, and set in brass on an iron socket, to receive a wooden han'le, which is so set as to be held in the hand in the cutting direction. The top of the handle goes between the root o. the fore-finger and the middle finger, and the hinder part, between the point of the fore-finger and thumb; there is, in general, a notch in the side of the socket, which shou'd be held next to the lath. Some diamonds have more cuts than one. Plough diamonds have a square nut on the end of the socket, next the glass, which, on running the nut square on the side of the lath, keeps it in the cutting direction.

Glass binders have these plough diamonds without long handles, as, in cutting their curious productions, they cannot apply a lath, but direct them by the point of their middle finger, gliding along the edge of the glass.

The ranging lath must be long enough to extend rather beyond the boundary of the table of glass.

Ranging of glass is the cutting it in breadths as the work may require, and is best done by one uninterrupted cut from one end to the other.

The square is used in cutting the squares from the range, that they may with greater certainty be cut at right angles. The glazing knife is used for laying in the putty in the rebates of the sash, for binding in the glass, and for finishing the front putty.

Of the glass used in building, three qualities are in common use, denominated *best*, *second*, and *third*.

The best is that which is the purest metal and free of blemishes, as blisters, specks, streaks, &c.; the second is inferior, from its not being so free from these blemishes; and the third are still inferior, both in regard to quality and colour, being of greener hue.

They are all sold at the same price per crate; but the number of tables varies according to the quality. Best twelve, second fifteen, and third eighteen tables.

These tables are ci cular when manufactured, and about four feet in diameter, having in the centre a knot, to which, in the course of the process, the flashing rod was fixed; but for the safety of carriage, and convenience of handling, as well as utility in practice, a segment is cut off about four inches from the knot. The large piece with the knot, stil retains the name of *table;* the smaller piece is technically called *a slab.* From these tables being of a given size, it is reasonable to suppose that, when the dimensions of squares are such as cut the glass to waste, the price should be advanced.

A superior kind of glass may be obtained at some of the first houses in London, which is very flat, and of large dimensions; some of it being 2 feet 8 inches by 2 feet 1 inch; these are sold only in squares.

Rough glass is well adapted to baths, and other places of privacy; one side is ground with emery or sand, so that no objects can be seen through it, though the light be still transmitted.

The glass, called *German-sheet,* is of a superior kind, as it can be had of much larger dimensions than common glass; it is also of a purer substance, and for these reasons, is frequently appropriated to picture frames. Squares may be had at the astonishing size of 3 feet 8 inches, by 3 feet 1 inch, and 3 feet 10 inches by 2 feet 8 inches, and under.

The glass is first blown in the form of a globe, and afterwards flatted in a furnace, in consequence of which it has a very forbidding appearance from the outside, the surface being uneven.

Plate-glass is the most superior in quality, substance, and flatness, being cast in plates, and polished. The quantity of metal it contains, must be almost, if not altogether, colourless; that sort which is tinged being of an inferior quality. Plate-glass when used in sashes, is peculiarly magnificent; and it can be had of larger dimensions than any other kind of glass.

Stained-glass is of different colours, as red, orange, yellow, green, blue, and purple.

These colours are fixed by burning, and are as durable as the glass.

Glass can be bent to circular sweeps, which is much used in London for shop windows, and is carried to great perfection in covers, for small pieces of statuary, &c.

The application of stained glass to the purposes of glazing is called *fret-work.* This description of work consists of working ground and stained glass, in fine lead, into different patterns. In many cases family arms and other devices are worked in it. It is a branch capable of great improvement; but at present is much neglected. Old pieces are very much esteemed, though the same expense would furnish elegant modern productions. They are placed in halls and staircase windows, or in some particular church windows. In many instances they are introduced where there is an unpleasant aspect, in a place of particular or genteel resort.

Lead-work is used in inferior offices, and is in general practice all through the country. Frames intended to receive these lights are made with bars across, to which the lights are fastened by leaden bars, called saddle bars ; and where openings are wanted, a casement is introduced either of wood or iron. Sometimes a sliding frame answers the same purposes. Church windows are generally made in this manner, in quarries or in squares.

The tools with which this work is performed are, in addition to the foregoing, as follow :—

A *vice*, with different cheeks and cutters, to turn out the different kinds of lead as the magnitude of the window or the squares may require.

The German vices, which are esteemed the best, are furnished with moulds, and turn out lead in a variety of sizes. The bars of lead cast in these vices are received by the mill, which turns them out with two sides parallel to each other, and about $\frac{1}{8}$ of an inch broad, with a partition connecting the two sides together, about $\frac{1}{8}$ of an inch wide, forming on each side a groove, nearly $\frac{1}{16}$ by $\frac{1}{8}$ of an inch, and about 6 feet long.

Besides a vice and moulds there are a *setting-board, latterkin, setting-knife, resin-box tin, glazing-irons,* and *clips.*

The *setting-board* is that in which the ridge of the light is marked and divided into squares, struck out with a chalk line, or drawn with a lath, which serves to guide the workmen. One side and end is squared with a projecting bead or fillet.

The *latterkin* is a piece of hard wood pointed, to run in the groove of the lead, and widen it for the easier reception of the glass.

The *setting-knife* consists of a blade with a round point, loaded with lead at the bottom and terminating in a long square handle. The square end of the handle serves to force the square of glass tight in the lead. All the intersections are soldered on both sides, except the outside joints of the outer sides, that is, where they come to the outer edge. These lights should be cemented by pouring thin paint along the lead bars, and filling up the chasms with dry whiting, to which, after the oil in the paint has secreted a little, a little more dry whiting, or white lead, must be added. This will dry hard, and resist the action of the atmosphere.

MENSURATION OF GLAZIERS' WORK.

Glaziers' work is measured by superficial feet, and the dimensions are taken in feet, tenths, &c. For this purpose, their rules are generally divided into decimal parts, and their dimensions squared according to decimals. Circular, or oval windows are measured as if they were rectangular; because in cutting squares of glass there is a very great waste, and more time is expended than if the window had been of a rectangular form.

PAINTING,

As applied to purposes of building, is the application of artificial colours, compounded either with oil or water, in embellishing and preserving wood, &c.

This branch of painting is termed *economical*, and applies more immediately to the power which oil and varnishes possess, of preventing the action of the atmosphere upon wood, iron, and stucco, by interposing an artificial surface; but it is here intended to use the term more generally, in allusion to the decorative part, and as it is employed by the architect, throughout every part of his work, both externally and internally.

In every branch of painting in oil, the general processes are very similar, or with such variations only, as readily occur to the workman.

The first coatings, or layers, if on wood or iron, ought always to be of ceruse or white lead, of the best quality, previously ground very fine in nut or linseed oil, either over a stone with a muller, or, as that mode is too tedious for large quantities, passed through a mill. If used on shutters, doors, or wainscottings, made of fir or deal, it is very requisite to destroy the effects of the knots; which are generally so completely saturated with turpentine, as to render it, perhaps, one of the most difficult processes in this business. The best mode, in common cases, is, to pass a brush over the knots, with ceruse ground in water, bound by a size made of parchment or glue; when that is dry, paint the knots with white lead ground in oil, to which add some powerful siccative, or dryer, as red lead, or litharge of lead; about one-fourth part of the latter. These must be laid very smoothly in the direction of the grain of the wood.

When the last coat is dry, smooth it with pumice-stone, or give it the first coat of paint, prepared or diluted with nut or linseed oil; after which, when sufficiently dry, all the nail-holes or other irregularities on the surface, must be carefully stopped with a composition of oil and Spanish white, commonly known by the name of putty. The work must then be again painted with white lead and oil, somewhat diluted with the essence of oil of turpentine, which process should, if the work be intended to be left of a plain white, or stone colour, be repeated not less than three or four times; and if of the latter colour, a small quantity of ivory or lamp-black should be added. But if the work is to be finished of any other colour, either grey, green, &c. it will be requisite to provide for such colour, after the third operation, particularly if it is to be finished flat, or, as the painters style it, dead white, grey, fawn, &c. In order to finish the work flatted or dead, which is a mode much to be preferred for all superior works, not only for its appearance, but also for preserving the colour and purity of the tint, one coat of the flatted colour, or colour mixed up with a considerable quantity of turpentine, will be found sufficient; although in large surfaces it will frequently be requisite to give two coats of the flatting colour, to make it quite complete. Indeed, on stucco it will be almost a general rule.

In all the foregoing operations, it must be observed that, some sort of dryer is absolutely requisite; a very general and useful one is made by grinding in linseed, or, perhaps, prepared oils boiled are better, about two parts of the best white copperas, which must be well dried with one part of litharge of lead: the quantity to be added, will much depend on the dryness or humidity of the atmosphere, at the time of painting, as well as the local situation of the building. It may here be noticed, that there is a sort of copperas made in England, and said to be used for some purposes in medicine, that not only does *not* assist the operation of drying in the colours, but absolutely prevents those colours drying, which would otherwise have done so in the absence of this copperas.

The best dryer for all fine whites, and other delicate tints, is sugar of lead, ground in nut oil, but being very active, a small quantity, about the size of a walnut, will be sufficient for twenty pounds of colour, when the basis is white lead.

It will be always necessary to caution painters to keep their uteusils, brushes, &c. very clean, as the colour would otherwise soon become very foul, so as to destroy the surface of the work. If this should happen, the colour must be passed through a fine sieve, or canvass, and the surface of the work be carefully rubbed down with sand-paper, or pumice-stone : the latter should be ground in water, if the paint be tender, or recently laid on. 'The above may suffice as to painting on wood, either on inside or outside work, the former being seldom finished otherwise than in oil : four or five coats are generally sufficient.

It does not appear that painting in oil can be serviceable in stucco, unless the walls have been erected a sufficient time to permit the mass of brick-work to have acquired a sufficient degree of dryness. When stucco is on battened work, it may be painted over much sooner than when prepared on brick. Indeed, the greatest part of the art of painting stucco, so as to stand or wear well, consists in attending to these observations, for whoever has observed the expansive power of water, not only in congelation, but also in evaporation, must be well aware that when it meets with any foreign body, obstructing its escape, as oil painting, for instance, it immediately resists it, forming a number of vesicles or particles, containing an acrid lime-water, which forces off the layers of plaster, and frequently causes large defective patches, not easily to be eradicated.

Perhaps, in general cases, where persons are building on their own estates, or for themselves, two or three years are not too long to suffer the stucco to remain unpainted, though frequently, in speculative works, as many weeks are scarcely allowed to pass.

The foregoing precautions being attended to, there can be no better mode adopted for priming, or laying on the first coat on stucco, than by linseed or nut-oil, boiled with dryers, as before mentioned ; taking care, in all cases, not to lay on too much, so as to render the surface rough and irregular, and not more than the stucco will absorb. It should then be covered with three or four coats of white-lead, prepared as described for painting on wainscotting, allowing each one a sufficient time to dry hard. If time will permit, two or three days between each layer, will be advantageous. When the stucco is intended to be finished in any given tint, as grey, light green, &c. it will then be proper, about the third coat of painting, to prepare the ground or such tint, by a slight advance towards it. Grey is made

2 T

with white-lead, Prussian-blue, ivory-black, and lake; sage-green, pea, and sea-greens, with white, Prussian-blue, and fine yellows; apricot and peach, with lake, white, and Chinese vermilion; fine yellow fawn colour with burnt terra sienna, or umber and white; and olive-greens with fined Prussian-blues, and Oxfordshire ochre.

Distemper, or painting in water colours, mixed with size, stucco, or plaster, which is intended to be painted in oil when finished, but not being sufficiently dry to receive the oil, may have a coating in water colours, of any given tint required, in order to give a more finished appearance to that part of the building. Straw colours may be made with French whites and ceruse, or white lead and masticot, or Dutch pink. Greys, full, with some whites and refiner's verditer. An inferior grey may be made with blue-black, or bone-black and indigo. Pea-greens with French green, Olympian green, &c. Fawn-colour with burnt terra de sienna, or burnt umber and white, and so of any intermediate tint. The colours should all be ground very fine, and mixed with whiting and a size made with parchment, or some similar substance. Less than two coats will not be sufficient to cover the plaster, and bear out with an uniform appearance. It must be recollected, that when the stucco is sufficiently dry, and it is desirable to have it painted in oil, the whole of the water-colours ought to be removed, which may easily be done by washing, and when quite dry, proceed with it after the direction given on oil-painting in stucco.

If old plastering has become disfigured by stains, or other blemishes, and it be desired to have it painted in distemper, it is, in this case, advisable to give the old plastering, when properly cleaned off and prepared, one coat, at least, of white-lead ground in oil, and used with spirits of turpentine, which will generally fix old stains; and, when quite dry, take water-colours very kindly.

MENSURATION OF PAINTERS' WORK.

Painters' work is measured by the yard square, and the dimensions are taken in feet, inches, and tenths. Every part which the brush has passed over is measured, consequently the dimensions must be taken with a line, that girts over the mouldings, breaks, &c. All kinds of ornamental work produces an extra price, according to the nature of the imitations, &c. Carved work is also valued according to the time taken in painting it

RAIL-ROADS

AND

LOCOMOTIVE ENGINES.

———

AMIDST the various speculations of the day, perhaps none have more deservedly excited the public interest than that of the numerous projected lines of rail-road for diminishing the friction of carriages, and for propelling carriages on them by either gas or steam power.

The lessening the friction, produces a consequent diminution in the power which otherwise would be required to propel a given weight; and therefore, is, in a commercial nation, like that of the united kingdom, a subject worthy of the highest consideration.

Railways were originally made of wood, and appear to have been first introduced between the river Tyne and some of the principal coal-pits, as early as the year 1680. The scarcity of this material, and the expense of frequent repairs, soon suggested an idea that iron might be more advantageously employed. At first, flat rods of bar-iron were nailed upon the original wooden rails, or, as they were technically called, *sleepers*; which, though an expensive process, was found to be a great improvement. But as the wood on which these rested was liable to rot and give way, these railings were soon after superseded by others made entirely of iron.

These *tram* or *rail-roads* have, for a considerable length of time, been much used in the colliery and mining districts; and some few have been carried from one town or manufacturing district to another. The principal of these latter in England and Wales are, the Cardiff and Merthyr, 26¼ miles long, running near the Glamorganshire canal; the Caermarthen; the Lexhowry, 28 miles, in the counties of Monmouth and Brecknock; the Surrey 26 miles; the Swansea, 5 miles; one between Gloucester and Cheltenham; besides several in the north of England.

Railways are of two kinds, arising from the disposition of

2 T 2

the flanch that is to guide the wheels of the carriage, and prevent it from running off the rail. In the one, the flanch is at right angles, and of one piece with the flat surface of the rail : in the other, the flat surface of the rail is raised above the level of the ground, and the flanch is fixed on the wheel of the carriage, at right angles to the tyre, or iron placed on the circumference of the wheel, to strengthen it. Beside these, another kind of railway has lately been introduced by Mr. Palmer, which consists of a single rail, supported some height from the surface of the ground : on this, two wheels confined in sufficient frame-work, are placed, suspending the load equally balanced on either side. This arrangement certainly seems to ensure the grand principle of lessening friction, and doubtless will, in many situations, be found a great improvement.

Previously to entering upon the probable advantages likely to result from a general introduction of railways, we shall give the substance of the specification of a patent, obtained in Sept. 1816, by Messrs. Losh and Stephenson, both of whom are well known to those interested in the subject.

These gentlemen preface a description of their method of facilitating carriages along tram and railways, with an observation, that there are two kinds of railways in general use ; the one consisting of bars of cast iron, generally of the shape of that described by *a*, fig.631 , the other of the shape of that described by figs. 630 and 631. That shewn at *a*, fig. 629, is known in different situations by the denomination of the edge rail, round-top rail, fish-backed rail, &c. That shewn at figs. 632 and 633, by the denomination of the plate-rail, tram-way plate, barrow-way plate, &c. The first we shall distinguish by the name of the edge railway ; the second, by that of the plate railway.

In the construction of edge railways, Messrs. Losh and Stephenson's objects are, first, to fix both the ends of the rails, or separate pieces, of which the ways are formed, immovable, in or upon the chairs or props by which they are supported ; secondly, to place them in such a manner, that the end of any one rail shall not project above or fall below the correspondent end of that with which it is in contact, or with which it is joined ; thirdly, to form the joinings of the rails, with the pedestals or props which support them, in such a manner, that if these props should vary from their perpendicular position in the line of the way, (which in other railways is often the case) the joinings of the rails with each other would remain as before such variation, and so that the rails shall bear upon the props as firmly as before. The formation of the rails or plates of which a plate railway consists, being different from the rails of which the edge railways are composed, they are obliged to adopt a different manner of joining them, both with each other, and with the props and sleepers on which they rest. But in the joining these rails or plates upon their chairs and sleepers, they fix them down immovably, and in such a manner that the end of one rail or plate does not project above, or fall below the end of the adjoining plate, so as to present an obstacle, or cause a shock to the wheels of the carriages which pass over them, and they also form the joinings of these rails or plates in such a manner as to prevent the possibility of the nails, which are employed in fixing them in their chairs, from starting out of their places from the vibration of the plates, or from other causes.

In what relates to the locomotive engines and their carriages, which may be employed for conveying goods or materials along edge railways or plate-railways, or for propelling or drawing after them the carriages or waggons employed for that purpose, their invention consists in sustaining the weight, or a proportion of the weight, of the engine, upon pistons, movable within cylinders, into which the steam or the water of the boiler is allowed to enter, in order to press upon such pistons; and which pistons are, by the intervention of certain levers and connecting rods, or by any other effective contrivance, made to bear upon the axles of the wheels of the carriage upon which the engine rests. In the formation of the wheels it is their object to construct them in such a manner, and to form them of such materials, as shall make them more durable and less expensive in the repairs than those hitherto in use. This is accomplished by forming the wheels either with spokes of malleable iron, and with cast iron rims, or by making the wheels and spokes of cast iron, with hoops, tyres, or trods, of malleable iron, and in some instances, particularly for wheels of very small diameters, instead of spokes of malleable iron, employing plates of malleable iron, to form the junction between the naves and the cast iron rims of the wheels.

The advantages gained by this method of constructing railways are, first, that the separate pieces of which they consist are, *cæteris paribus*, rendered by this mode of joining them, capable of sustaining a much heavier pressure th those which are joined in the usual way. Secondly, by

mode of joining the rails, they remove the liability to which
rails joined in the usual plan, (where the end of one rail is
seldom in the same plane with the correspondent end of the
next) are exposed, of receiving blows and shocks from the
carriages which move over them, and to which blows and
shocks the great breakage which often occurs in railways,
when not made of enormous weight, may generally be referred; and as action and re-action are mutual and contrary, if they prevent the communication of shocks to the
rails, they at the same time preserve the wheels, the carriages, and engines which move over them, from the re-
action which is often destructive to them. As the centre of
gravity in a loaded coal-waggon is, from its shape, much
elevated, there is generally a great waste of coal from the
shaking of the waggons, to which that circumstance (the
position of the centre of gravity) makes them more liable
when they encounter obstacles, as they do at the junction
of almost every two rails on the common railways. On
Losh and Stephenson's railways, the loss thus arising is,
if not entirely prevented, at least considerably diminished,
by the steady and regular motion of the waggons. . The
usual method of fixing down the plates, of which the plate
railways employed in coal-mines, and there called tram
and rolley-ways, are formed, is by a single nail, nearly at
each end of each plate; which nail passes through a hole in
the plate, and fixes it to a sleeper of wood. These nails,
from the vibration of the plate, or the motion of the sleeper,
or some other cause, generally very soon start up, and consequently the plates work loose, and very frequently the
nails come entirely out. The delay of work, the breakage
of plates, wheels, &c. and the injury which the horses receive from the loose nails which result from the mode of
fixing the plate railways, are generally complained of, and
therefore the advantages of a plan which will remove these
inconveniences must be apparent.

When locomotive steam-engines are employed as the
moving or propelling power on railways, these gentlemen
have, from much practice, found it of the utmost importance, that they should move steadily, and as free as possible from shocks or vibrations, which have the effect of
deranging the working parts of the machinery, and lessening
their power. It is therefore to produce that steadiness of
motion, and to prevent the engines from receiving shocks,
and to preserve their equilibrium, that they employ the
floating pistons, which, acting on an elastic fluid, produce

the desired effect with much more accuracy than could be obtained by employing the finest springs of steel to suspend the engine. The wheels which are constructed on this plan will be found, when compared with those already in use (the weights of both being equal) to be more durable; for the arms, when made of malleable iron, being infinitely less liable to be broken by shocks or concussions, than those of cast iron, may be of less weight, and in fewer numbers, so that the excess of weight of the extra arms of the cast iron wheels may be applied on the rims of these wheels, and thus add to the substance of that part which alone suffers from the friction of the rails. The rims of wheels thus constructed, can also be case-hardened without risk of breaking, either in cooling or afterwards, which is not the case when wheels are cast in one piece. The advantage of hooping cast iron wheels with malleable iron tyres or trods, is, that when such tyres or trods are worn through, they can very easily be replaced at a small expense, and that the tyre, which is not liable to break, receiving the shocks from the re-action of the rails, preserves the cast-iron wheel, by considerably lessening the effect of such shocks on the cast metal.

As it is perhaps impossible to cast the bars or plates of metal of which railways and plate-ways are composed perfectly straight, and correctly even and smooth on their surfaces, and equally difficult to fit the joints with mathematical accuracy, the wheels of the engines and waggons will always have some inequalities and obstacles to encounter. From these circumstances, therefore, Messrs. Losh and Stephenson are induced to employ the improvements which they have made in the construction of the locomotive engine, and in the wheels of carriages upon edge railways and plate railways, constructed according to their own plans; but it is apparent that their adoption on the rail and plate-ways on the usual construction, is of still more importance.

They therefore claim as a method of facilitating the conveyance of goods, and all manner of materials along edge railways or plate railways, the use of any of the plans they have described singly, as well as the whole of them collectively. They have no hesitation in saying, that on a railway constructed on their plan, and with a locomotive engine and carriage-wheels on their principle, the expedition with which goods can be conveyed with safety, be increased to nearly double the rate with which the

at present usually taken along railways, and with less interruption from the breakage of wheels, rails, &c. than at present occurs, and with much less injury to the working parts of the engine.

In order that their specification may be more clearly understood, we have annexed a schedule of drawings.

Fig. 629 represents a longitudinal view of the locomotive engine on the edge railway. *a a a*, are the cylinders containing the floating pistons *b b*, which are more fully described in the next figure.

Fig. 630 represents a cross section of Fig. 629, at the middle cylinders *a, a ; b b* are the floating pistons, connected with the wrought iron rods *c c*, the ends of which rest upon the bearing brasses of the axles of the wheels *d d*. These pistons press equally on all the axles, and cause each of the wheels to press with an equal stress upon the rails, and to act upon them with an equal degree of friction, although the rails should not all be in the same plane, for the bearing brasses have the liberty of moving in a perpendicular direction in a groove or slide, and, carrying the axles and wheels along with them, force the wheels to accommodate themselves to the inequalities of the rail-way.

Fig. 634, is a view of the wheel, with wrought iron arms. *a a a a a* show how the arms are cast in the nave *b b*, and dropped into the mortise holes *c c c c c c* in the rim, which are dovetailed, to suit the dovetailed ends of the arms *d d d d d d*. The arms are heated red hot previously to dropping them into the holes, in order to cause them to extend sufficiently for that purpose, for when cold they are too short. In doing this they take advantage of that quality which iron possesses of expanding on the application of heat, and of contracting again to its former dimensions on cooling down to the same temperature from which it was raised; the arms, therefore, on cooling are drawn with a force sufficient to produce a degree of combination between their dovetailed ends and the mortises of the rim, which prevents the possibility of their working loose ; they are afterwards keyed up ; the mortise holes are also dovetailed, from the tail side of the wheel (*a a*, fig. 635) to the crease side (*b b*, on the same figure).

Fig 635, is a cross section through the centre of the wheel, with wrought iron arms.

Fig. 636 is an end view of Fig. 635.

Fig. 637 represents a view of their edge railway ; shewing a rail *a*, connected with the two adjoining rails, the ends of which are shewn by *b b*, and also with the props or pedestals on which they rest. *d d* show the metal chairs, and *c c* the stone supports. The joints *e e* are made by the ends of the rails being applied to each other by what is denominated a half lap, and the pin or bolt *g*, which fixes them to each other, and to the chair in which they are inserted, is made to fit exactly a hole which is drilled through the chair, and both ends of the rails at such a height as to allow both ends of the rails to bear on the chair, and the bearance being the apex of a curve, they both bear at the same point. Thus the end of one rail cannot rise above that of the adjoining one; for although the chair may move on the pin in the direction of the line of the road, yet the rails will still rest upon the curved surface of their bearance without moving.

Fig. 638 is a cross section of their edge railway through the middle of one of the chairs *a*, and across the ends of the two adjoining rails *c d* and the pin *e* ; *f* is the stone support or sleeper.

Fig. 639 is a cross section of a rail a, at the centre, and shows the carriages c behind.

Fig. 640 is a plan of the rail-way described at fig. 637, shewing the butt lap joinings of the rails e e, placed in their carriages d d.

Fig. 641 is a view of the cast iron wheel with the malleable iron tyre. This wheel is made with curved spokes, as shewn at a a a a a a a a a, in the figure, and with a slit or aperture in the rim, shewn at b, into which a key is inserted. The reason of this is, that on the application of the hot tyre the cast metal expands unequally, and the rim is liable to be cracked, and the arms drawn off, unless the first is previously slit or opened, and the latter curved, which allows them to accommodate themselves to the increased diameter of the wheel; by this formation of the wheel the tyre might be forced on when cold, and keyed up afterwards.

Fig. 642 is a cross section of fig. 641, through the centre. a a shew the tyre, b b b b shew the metal rim. This cast metal rim is dovetailed; so that when the tyre, which is dovetailed to suit it, is put on hot, it contracts and applies itself to the rim with a degree of adhesion which prevents its coming off from the motion of the wheel on the rail-way. This wheel is of the form to suit an edge railway, and to make it answer for a plate rail it only requires the rim to be round or flat.

Fig. 643 is an end view of fig. 641, without the malleable iron tyre.

Fig. 644 represents a view of a rolley or tram-wheel, calculated to move upon a plate railway. a a a a show the malleable iron arms, fastened to the projections b b b b, on the inside of the rim c c c, by the bolts d d d d.

Fig. 645 is a cross section of fig. 644, through the centre of the wheel. a a show the arms, c c the rim, d d the bolts.

Fig. 646 represents a view of a rolley or tram-wheel, with a plate of malleable iron a a a a, to form the junction between the nave b b and the cast metal rim c c c c.

Fig. 647 is a cross section of fig. 646. a a show the plate upon which the nave b b is cast. c c show the cast iron rim which is cast upon the plate, the edges of which plate are previously covered with a thin coating of loam and charcoal dust, or other fit substance, to prevent the too intimate adhesion between the iron plate and metal rim, so that if the rim should break, it can easily be taken off and replaced by casting another on the plate.

Fig. 648 represents the plate railway on their plan. At the end of each plate are projections a a a a, to fit into the dovetail carriage b b, and at each end of each plate are projections or tenons c c c c, which fall into the mortise hole (d, in Figs. 649 and 650) in the carriage b b, and secure the rail from an end motion; and when the pin or key e is driven into its place, it secures the plates from rising, thus they are fixed immovable in their carriages.

Fig. 649 is a front view of fig. 648.

Fig. 650 is a plan of the carriage, in which a a show the nail holes through which the nails are driven, to secure it to the sleeper. When the nails are laid in this carriage, and secured by the pin or key, they keep these nails from starting up by resting upon them.

Fig. 651 is a cross section of the carriage, and the end of one of the plate rails.

Fig. 652 shews a rail of the common way, inclining out of the horizontal position, as they very often do from the yielding of the prop or pedestals, and of course a shock is sustained by the waggons in passing the joining to the next rail.

The ease with which cast-iron can be made into any required shape has till very recently given to rails of that material a decided superiority over those of malleable-iron. But the brittleness of the former renders such rails very liable to be broken, unless, indeed, they be of such substance as will resist the effects of the blows or shocks to which they are exposed, and which will require them to be of considerably greater weight than otherwise would be necessary. To obviate this, numerous experiments have been made with a view to substitute malleable-iron for cast-ron rails.

Rails of malleable-iron appear to have been first used at Lord Carlisle's works, at Tindal Fell, in Cumberland, about the year 1808; and though found there, and also at two or three other places at which they were tried, to be a saving in the first cost, and much less liable to accident, they have not till very lately been much used. In fact, it was not till some time after Mr. Birkinshaw, of the Bedlington Iron Works, had obtained a patent for malleable-iron rails of a new and improved construction, that rails of, this material came into competition with the cast-iron rails.

The form of the malleable-iron rails previously to this was that of a parallelopipedon; which was liable to two objections, either that the narrowness of the surfaces, when compared to the breadth of the rim of the carriage wheel, was so considerable as to expose both the wheel and the rail to great injury from wear; or, if the breadth of the rail was increased to remove this objection, the weight of the rail would make the cost amount to almost a prohibition of its use.

Mr. Birkinshaw obtained his patent in October 1820; and the improvement consisted in making the rails in the form of prisms, though their sides need not of necessity be flat. The upper surface, on which the wheel of the carriage is to run, is slightly convex, in order to reduce the friction; and the under part, which rests on the supporting blocks, chains, rests, standards, or pedestals, is mounted upon the sleeper. The wedge form is proposed, because the strength of the rail is always in proportion to the square of its breadth and depth. Hence this form possesses all the strength of a cube equal to its square, with only half the quantity of metal, and consequently half the cost of the former rail. Sufficient strength, however, may be still retained, and the weight of metal further reduced, by forming the bars with concave sides, which is the form of rail the patentee decidedly prefers, although the prism or wedge form, in all its varieties, is the principle upon which his patent-right is founded.

The mode of making these wedge-formed rails of malleable-iron is, by passing bars of iron, when heated, through rollers, having grooves or indentations cut upon their peripheries, agreeably to the intended shape of the bar to be produced. But, though the patentee recommends, and adopts this mode as the most eligible means of producing these rails, he claims the exclusive right of manufacturing and rendering the wedge-formed bars or rails of any length, for the purpose of forming or constructing rail-roads.

The advantages derived from this method of constructing railways may be as follows :—

1st. The original cost of a malleable-iron railway is less than a cast-iron railway of equal strength.

2dly. As the rails can be made in the lengths of 9,12,15, or 18 feet each, and even longer when required, the number of joints is hereby reduced ; and thus is removed, in a great measure, the liability to which the short rails now in use are exposed, of receiving blows and shocks from the carriages which move over them.

3dly. In order to remedy the evil arising from the rails being imperfectly joined, the plan of welding the ends together has been adopted ; by this means making one continued rail the whole length of the road without any joint whatever.

4thly. It hence follows, that on iron railways, the loss of coals, occasioned by the jolting of the waggons at the joints of the rails, and the injury done to the wheels, the carriages, and engines from the same cause, are, if not entirely prevented, at least considerably diminished.

In September, 1821, Mr. Losh took out another patent for further improvements in the construction of railways. These improvements consist, first, in fixing bars of malleable iron on the upper surface of a line of cast or malleable iron rails, of whatever form such rails may be, in the longitudinal direction of the rails when laid, so as to form an uninterrupted line the whole length of the bar, which may be as long as it shall be found convenient, and of the same breadth, or a little broader or narrower than the upper surface of the rails to which it is fixed. Secondly, in fixing, in some cases, a band or strap of malleable iron to the under surface of cast-iron rails, in order that such strap or band may, by its power of tension, give support to the cohesion of the parts of cast iron rails, and admit of its being made lighter, of less expense, and less liable to breakage. Thirdly, in forming a rail, by fixing two bars of malleable iron on their sides or edges, and fixing them in that position by bolts and studs, or any other convenient method; and in placing and fixing on their upper edges a flat bar of malleable iron, or one which is slightly curved or rounded at the edges to diminish friction, so that the bar or plate, placed and fixed on the upper edges of the two malleable iron bars, shall form the surface upon which the wheels of the carriage are to revolve.

Mr. Losh states, in the specification of his patent, that rail-roads are now become so general, that for the information of mechanical men, or those who have the direction of constructing and laying them, drawings would be quite superfluous; he therefore proceeds to state the methods which he has found the most convenient, for forming the junction of the plate or flat bar, which he applies upon the surface of the body of the rail; and also the mode by which he attaches the band or strap to the lower edge of the cast iron rail.

He recommends the dimensions of the bars meant to form the upper surface of a railway, calculated to carry locomotive engines of seven or eight tons, and waggons of three or four tons weight each, to be fifteen or sixteen feet long, two and a quarter inches broad, and half to five-eighths of an inch thick. At every eighteen inches or two feet of the length of this surface-plate, a tenon is firmly welded or riveted; or otherwise attached to the under side, taking care in this operation to leave the upper surface of the plate even as before. These tenons have holes through them in the transverse direction of the bars, to take a pin or rivet of from about a quarter to half an inch in diameter; and at each extremity of the plate, a tenon is fixed on by welding, having previously cut off a piece of about two inches long, and of half the breadth of the bar, from the opposite ends of the bar or plate, and at the opposite angles, so that when two bars, so prepared, are brought to join at the ends, the joint is what is denominated a half-lap, or scarfed joint.

If it be required to place malleable iron plates or bars on cast iron rails, nothing more is necessary than to make the rails with mortise holes, to receive the tenons with transverse holes, to correspond with those in the tenons fixed on the plates; and, after placing the rails in their chairs or carriages, to apply the plate to the surface of the rails, and to drop the tenons into the mortise holes, and to secure them there by a pin driven tightly into and through the transverse holes of the tenons and mortise holes. The mortise holes are made in the rails by placing a core in the mould previously to running in the metal, and lest this core should weaken the rail, it is advisable to add as much metal on the outside of the rail, in the form of a boss, where the hole is, as will make up the deficiency. A chair is then placed on a pedestal at every three or four feet distance, less or more, according to the length of the cast iron rails ; and each of these must be supported at its ends : these rails are generally made with half-lap joints, and to rest on a curb bearance. Care is taken that, where the ends of the surface-plates meet to form a joint, they shall be sustained by a chair ; and the reason for making the joints half-lapped, or scarfed, with tenons welded to these half laps is, that one pin or bolt will secure both the adjoining ends of the surface-plates, and of the bars of cast iron, more perfectly in the chair, than any other known contrivance, when the bearance is the apex of a curve. Surface-plates thus prepared with tenons, as described above, may be attached and fixed to the upper surface of a series of malleable iron rails placed in chairs, which rails consist of flat bars (generally three or four feet long, more or less, but sometimes also as long as the surface plate), fixed on their thin edges, so as to present the greatest resistance to a weight bearing upon them. For this purpose, pins or rivets may be driven through the transverse holes in the tenons on the surface-plate, and the corresponding transverse holes made in the supporting bars ; and thus may be formed a cheap and very serviceable railway. In this case, the supporting bars should not be less than two and a half inches deep, by half an inch thick, if meant to carry locomotive engines. For smaller carriages, the bars may be of less dimensions, in proportion to the decreased weight of the carriages.

In forming the rail, consisting of a plate of malleable iron, supported by two flat bars of the same material, Mr. Losh prepares the surface-plate as above with tenons, and having fixed the two bars intended to support it on their edges, parallel to each other, in a series of chairs, and secured them in that position by bolts passing through them, and by intervening studs, to keep them at a proper distance, which is such, that the sides or edges of the surface-plate, which may be a little curved or rounded, to diminish the friction from the wheels passing over it, shall project above a quarter of an inch beyond them. By these intervening studs, the surface-plate is laid upon them, and the tenons are dropped in between them, and fixed by pins or bolts passing in a transverse direction through holes in the bars, which are made to correspond with holes in the tenons, and thus securing them as if they were in mortise-holes. The strap or band of malleable iron is fixed by Mr. Losh to the under edge of the cast-iron rail, by perforating both ends of the strap, near the extremities, with a long hole, calculated to pass over studs of malleable iron which are fixed at each end of the rail, by being run at the time of casting the rail or otherwise. The studs should be about one and a half inches broad, by three-eighths of an inch thick, and placed so, that when the strap has been put over them in a heated state, it cannot, in contracting, slip its hold; but will, on the contrary, fix itself the closer. These straps are made of malleable iron bars, about one and a half inches broad, three-eighths to half an inch thick, and of such length as to draw strongly against the studs and bottom of the rail, when in its position. The under edge of the cast iron rail to which this strap is applied being curved, it will, when the strap is fixed upon the studs, by an extension of its length by heat, apply itself firmly to, and support every part of the lower edge of the rail, in contracting, by parting with its heat; and till the power of tension of this strap is overcome, and it extends in length, or the studs break, the rail cannot give way.

Many other methods, perhaps equally secure, may be made use of to place and fix surface-plates on the surface of rails; but Mr. Losh prefers the plan pointed out, by tenons and mortise-holes, and by rivets passed through holes in such tenons, and through corresponding holes in the supporting bars; because, when worn or damaged, these plates can easily be taken off and replaced, without injury to that part of the rail which supports them.

The principal patents obtained before the above described, are those by Blenkinsop, Brunton, and Chapman; specifications and drawings of which may be seen in the Repertory of Arts.

Mr. Blenkinsop's patent was obtained the 10th of April, 1811, and is for a method of fixing into the ground a toothed rack, or longitudinal piece of cast iron, or other fit material, having teeth, or protuberances, into which a toothed or cogged wheel, connected with a locomotive carriage, plays.

1799 stated, that a horse of the value of 20*l.* drew down the declivity of an iron road, $\frac{4}{5}$ths of an inch in a yard, 21 carriages or waggons, laden with coals and timber, weighing 30 tons, overcoming the *vis inertiæ* repeatedly with ease. The same horse, up this declivity, drew five tons with ease. On a different railway, one horse, value 30*l.* drew 21 waggons of five cwt. each, which, with their loading of coals, amounted to 43 tons eight cwt., down the declivity of 1-3d of an inch in a yard ; and up the same place he afterwards drew seven tons ; the cwt. in all these experiments by Mr. W. being 120lbs.

Though in the preceding statements there is an apparent variance, the authors are not the less entitled to credit ; because the variations may have arisen from difference in the physical strength of the animals, or in the method of constructing the railways. To make the case, however, as clear as possible, we shall here present our readers with some observations and calculations deduced from known data, which have lately appeared in a very able pamphlet, entitled "A Report on Rail-Roads and Locomotive Engines," by Mr. Charles Sylvester, civil engineer.

Mr. Sylvester, having made some judicious observations on the principles of railways, and the nature of the friction to be overcome, states, that, " agreeably to the principles laid down in the commencement, when a force is applied equal to the friction, the smallest force above that would, if continued, generate any required velocity. But it will be desirable to have such a force at command, as will generate the necessary velocity in a short time, and when that has been accomplished, to reduce this force, but still to leave it fully equal to the friction. If any part of the route has an inclination, there ought to be an extra force at command, above what would be required for a dead level. The plane on which this experiment was made, inclined, in the direction of the load, about $\frac{1}{9}$ of an inch to a yard. This is as great, or perhaps a greater, inclination than any rail-road ought to have, where loaded carriages go up and down. The moving force ought, therefore, to be always greater than the friction added to the force which is required to overcome the inclination of the plane. The latter force assists the body to go down, and equally resists it in moving upwards.

" On this account" says he, " I have used, or supposed, a moving force, which will give the velocity of 5 miles an hour, or $7\frac{1}{3}$ feet per second, in the space of one minute. This will be performed down the above plane by the engine making

45 strokes per minute, (the circumference of the wheel being nine feet), with a pressure of 9·7lbs. upon an inch, of each of the two cylinders, the area of each being 63·6 square inches. The weight of the engine and 16 waggons is equal to 154,560lbs, or nearly 70 tons. The velocity of five miles an hour being acquired after one minute, the only force to keep the whole in motion, at the same rate, will be the difference between the gravity of the weight down the plane and the friction. The friction is 900lbs; the gravitating force of the weights down the plane 540lbs; therefore 900 − 540 = 360lbs.

" If the same weight, at that speed, had to move on a dead level, and acquired the same velocity in one minute as before, the moving force would require to be 1781lbs. which would require a pressure of 13·7lbs. upon one inch. But after the speed is obtained, it will require only 7lbs. to keep it moving at the same rate. If the same load were required to move up the plane, it would require a moving force of 2328lbs. or a pressure upon every square inch of 18·3lbs. And this velocity would be kept up by a constant pressure of 1447lbs. which will be 11·3lbs. upon every inch of the piston.

" In starting the engine, in the first instance, and giving the required velocity, it is probable the effects will agree very nearly with these calculations; namely, 154,560lbs. moved at the rate of five miles an hour, with a pressure of 9·7lbs. upon every inch of the piston. Whether the pressure were reduced to the difference between the friction and the force upon the plane, which is calculated at 2·8lbs. it is difficult to say, as there was no steam-gauge to indicate the pressure when the engine was going."

In table 1, at a more advanced part of the work, Mr. Sylvester states, that, when the engine is required to travel at the rate of nine miles per hour, the force necessary to overcome the weight, 154,560lbs. will be for the first minute, when the engine is travelling on a level 2890·81lbs; when moving down the plane 2461·61lbs; and when moving up the plane 3320·01lbs. But that, when the velocity is attained, a force that will balance the friction is sufficient to keep up the required velocity. This force is, for travelling on a level, 900lbs; for moving down the plane, 471lbs; and for moving up the plane 1329lbs.

By this, therefore, it is evident that, when the lading is expected to be considerably more in one direction of the line of rail-road than it is in the other, the advantage which

2 U

will arise from making the road with a gentle slope, is very great. This kind of railing is also preferable when the lading is only equal at certain periods. For then the expense of extra horses, to draw the additional weights up the plane during these periods, will fall infinitely short of the expense saved by making the plane with a gentle inclination.

The necessary preliminaries being settled, the engineer will obtain much greater facility, as also a diminution of expense, by beginning to lay down the rails on any part of the intended line of road where stone, gravel, and other materials that are wanted, are to be most conveniently had ; as, by that means, he will evade the slow and expensive mode of common cartage.

The immense sums that have been invested in the hands of certain companies, for the purpose of establishing general lines of rail-road throughout the country, have excited much interest and elicited many able papers from practical men, in several of the publications of the day. Amongst these, perhaps those inserted in the Scotsman, an Edinburgh newspaper, and in the Manchester Guardian, are the most deserving of our notice.

The Scotsman first commences with some theoretical statements, and then continues :

Having developed the theory of the motion of carriages on horizontal railways, we shall have little more to do with mathematical discussions, and shall now turn our attention to points of a practical nature, better adapted to the taste of ordinary readers. But first, we shall bring under the eye again, the effect of a given quantity of power on a railway, and on a canal, in a calm atmosphere—for it is only in a calm atmosphere that the results can be properly compared.

We have found that a *boat* weighing with its load 15 *tons*, *and a waggon of the same weight*, the one on a canal, and the other on a rail-way, would be impelled at the following rates, by the following quantities of power—which we have stated both in pounds and in horse power—reckoning one horse power equal to 180 pounds.

Miles per hour.	Boat on a Canal.		Waggon on a Rail-way.	
	power in pounds.	Horse power.	power in in pounds.	Horse power.
2	33	1-5th	100	
4	133	2-3ds.	102	
6	300	1½	105	
8	533	3	109	
12	1200	7	120	2-3ds.
16	2133	12	137	
20	3335	18	158	1

We have not taken into account the time lost in over-coming the inertia of the waggon where a small power is applied, because, in point of fact, the casual resistance of the wind would render it necessary to provide double or triple the power above stated. But if necessary, the time lost by the slow motion at first might be saved. Suppose there are a certain number of places where the steam-coach or waggon was to stop, to take in or put out passengers or goods; and farther, that the waggon, by travelling a few miles, has acquired an uniform velocity of 20 miles an hour. Then, if it is made to ascend an inclined plane of 10 feet perpendicular height, this velocity will be extinguished, and the vehicle will stop at the head of the plane. When it is to proceed again on its journey its descent along an in-clined plane of the same height on the other side, will enable it to recommence its career in a few seconds with the full velocity of 20 miles an hour. By raised platforms of this kind, at the two extremities of the journey, and at the in-termediate stages, the velocity thus generated, might be treasured up for permanent use. The platforms should be of different heights, corresponding to the various velocities of the vehicles plying on the railway. But, in point of fact, the terminal velocity is attained so soon from a state of rest, that this contrivance would probably be found un-necessary.

Where locks or *lifts* occur, the stationary steam-engine should drag up the vehicle (supposing it to be along an in-clined plane), not simply from the one level to the other, but to a platform some feet above the higher level, that the vehicle, by its descent, might recover the lost velocity. It is plain, however, that when the difference of level did not exceed eight or ten feet, the momentum of the vehicle would carry it up without any assistance from a stationary engine, and with merely a small temporary loss of velocity.

Some persons imagine erroneously that teethed wheels and rackwork would be necessary where the railway was not perfectly level. But the friction of iron on iron being 25 per cent. of the weight, if the whole load was upon the wheels to which the moving power was applied, and if the quantity of power was sufficient, the waggon would ascend without slipping though the plane rose one foot in four—while even cart roads scarcely ever rise more than one foot in 19 or 20. If four-fifths of the load, however, were placed on separate cars, and only one-tenth of the whole pressure, for instance, was upon the axle to which the moving force was applied, the power of ascent by friction would only be one-tenth of one foot in four, or one foot in forty.

The steam engine, as we commonly see it, is so bulky, and with the addition of its fuel and supply of water, so ponderous, as to create an impression on a first view, that its whole power would scarcely, under the most favourable circumstances, transport its own weight. The steam-boat, however, which cuts its way through the ocean, in defiance of tide and tempest, shews that this is a mistake. For all velocities above four miles an hour, the locomotive engine will be found superior to the steam-boat; that is to say, it will afford a greater amount of *free* power, above what is required to move its own weight.

We have seen various statements respecting the locomotive engine, few of them so detailed as could be desired—from which we subjoin the following particulars:

Trevithick and Vivian's high pressure locomotive engine, with a cylinder of eight inches diameter, and a pressure of 65 pounds per square inch (apparently about eight horse power), drew carriages containing ten and a half tons of iron, at five and a half miles per hour, for a distance of nine miles. (Stuart's History of Steam Engine, p. 164.) Whether on a road or railway is not mentioned.

We find it stated in a Liverpool paper, as the result of inquiries made respecting the locomotive engines, that one of these, of ten horse power, conveys fifty tons of goods at the rate of six miles an hour on a level railway. But was the road an edge or tram road?

Mr. Blenkinsop states, in replies to queries put by Sir John Sinclair, that his patent locomotive engine, with two eight-inch cylinders, weighs five tons, consumes 2-3d cwt. of coal, and fifty gallons of water per hour, draws 27 waggons weighing 94 tons on a dead level, at three and a half miles per hour, or 15 tons up an ascent of two inches in the yard; when ' lightly loaded' travels 10 miles an hour, does the work of 16 horses in 12 hours, and costs 400l. Another person says, that the weight of this engine with its water and coals is six tons, and that it draws 40 or 50 tons (waggons included) at four miles an hour on a level railway. (Repertory of Arts, 1818, p. 19-21 This seems to have been a high pressure engine of about eight or ten horse power. But we are not informed what sort of railway it worked on, how long its journies were, or what is meant by ' lightly loaded.'

We shall take for granted then that an eight-horse

power high pressure engine, with its charge of water and coal, and with the car which bears it, weighs six tons, and that it requires an additional supply of 100 weight of coal, and 400 weight of water for each hour it works. This is very consistent with other ascertained facts. We find, for instance, in the parliamentary report on steam navigation, that the low pressure engines used in vessels, which are made twice as strong as stationary engines, weigh about one ton and one-fifth for each horse power, including their charge of water and coal. Now the high-pressure engines want the condensing apparatus which must diminish the weight probably by one-fourth part. The estimate for coal we have increased one-half, because we think it rather below the truth. It is only about nine pounds per hour for each horse power, while Mr. Watt allows twelve pounds for his low pressure engines.

It follows, therefore, that an eight-horse power locomotive engine, with coal and water for eight hours, would weigh eight tons. Hence, bulky and ponderous as the steam-engine appears, we find that a locomotive engine, weighing eight tons, moves 50 tons beside itself, (taking the more moderate estimate,) that is, it consumes only one-seventh part of the power it creates, when travelling at four miles an hour; or *the free power applicable to other purposes, is seven-eighths of the whole.* This is the result of an early experiment, made probably upon a rail-road not of the best kind, and with vehicles much less perfect than they may yet be rendered. Though it falls much under the effect calculated theoretically, it does not strike us as being inconsistent with the truth of the principles on which the calculation was founded.

The high pressure engine, on account of its smaller weight and bulk, is evidently best adapted for railways; and it can be used with perfect safety, because it may be easily placed in a car by itself, a few feet before the vehicle in which the passengers are. The vehicle itself, by its regular and steady motion on the railway, would answer the purpose of a *fly-wheel* in the most perfect manner. The engine might run upon six wheels, which should be locked together by teeth pinions, that the tendency to slip might be resisted by the friction of the whole mass of eight tons.

The best form of a steam coach for the conveyance of passengers would probably be the following:—A gallery seven feet high, eight wide, and 100 feet in length, formed into 10 separate galleries 10 feet long each, connected with each other by joints working horizontally, to allow the train to bend where the road turned. A narrow covered footway, suspended on the outside over the wheels on one side, would serve as a common means of communication for the whole. On the other side might be outside seats, to be used in fine weather. The top, surrounded with a rail, might also be a sitting place of promenade, like the deck of a track boat. Two of the 10 rooms might be set apart for cooking, stores, and various accommodations; the other eight would lodge 100 passengers, whose weight, with that of their luggage, might be 12 tons. The coach itself might be 12 tons more; and that of the locomotive machine, eight tons, added to these, would make the whole 32 tons. Each of the short galleries might have four wheels; but to lessen the friction, the two first wheels only should be grooved, the two last cylindrical, and three or four times as broad as the thickness of the rail. The conveyance of goods would be effected by a train of small waggons loosely attached to each other.

It will be observed from the table we have given above, that it would require seven horse power to impel a steamboat weighing 15 tons at 12 miles an hour. This gives a load of two tons so moved; however, the engine, if a low pressure one, with water and eight hours' coals, would weigh nearly 10 tons, and the vessel would weigh at least five; so that the whole power of the engine would be expended in impelling itself and the ship containing it, at the supposed rate, and no *free* power would remain for freight. Facts show that the resistance is actually rather greater in water than theory in this case represents it. We have calculated from data furnished by the Parliamentary Report on steam navigation, that the entire burden on the engine in vessels going only eight or nine miles an hour in calm weather, rarely exceeds three tons for each horse power, while, according to the table, it should be five tons. Indeed, in our common steam-vessels for passengers, going eight or nine miles an hour, the ship and engine may be considered as constituting the whole burden. For 50 passengers, weighing perhaps with their luggage six or eight tons, placed on board a ship weighing, with her engine of 60 or 70 horse power, a hundred and fifty or hundred and eighty tons, form but an addition of one-twentieth or one-thirtieth to the mass—a quantity of no importance in a practical point of view. If we convert the steam-engine power into real horse power, and figure to ourselves 100 horses employed to draw 50 persons, we see what an enormous waste of power there is in the mode of conveyance. We may remark further, that the tenor of the evidence given before the Parliamentary Committee renders it extremely doubtful, whether any vessel could be constructed, that would bear an engine capable of impelling her at the rate of two miles an hour, without the help of wind or tide.

When the steam coach is brought fully into use, practice will teach us many things respecting it, of which theory leaves us ignorant. With the facilities of rapid motion which it will afford, however, we think we are not too sanguine, in expecting to see the present extreme rate of travelling doubled."

This practicability of conveying individuals or merchandize at the speed required in the present improved state of our internal intercourse with the different parts of the kingdom, has created much doubt and discussion with many able and practical mechanics. The question seems to resolve itself thus, Do the friction incurred by any moving body, laying aside the resistance of the atmosphere, increase in proportion to its velocity?

Without going into any diffuse or theoretical argument on this point, we shall merely cite that by the results of actual experiments instituted by Vince and Coulomb, it appears *that friction does not increase in proportion to the velocity.*

By experiments made also by Stephenson and Wood, it appears that the force required to keep a given weight in motion does not vary with the velocity: thus, a force of 14lbs. was found to overcome friction, and keep in motion an empty coal waggon, weighing 23·25 cwt. on a rail-road; and that on doubling the velocity, no more force was required. Further also it appears, that on increasing the weight, or load, the power required to overcome the friction, and keep the waggon in motion, did not increase in similar proportion, but up to 76·25 cwt. was about one-fourteenth less.

Notwithstanding the simple and satisfactory manner by which the experiments that led to these results were conducted, the fact has been still much doubted. We cannot therefore do better than to extract from the Manchester Guardian the following article, which contains an account of experiments, with most conclusive results, made by that able mechanic, Mr. Roberts of Manchester :—

" The object of the papers on rail-roads which appeared in the Scotsman, was, in a great measure, to shew the practicability of transporting commodities upon rail-roads at a very considerable speed; and (with some fallacies, which we shall endeavour to point out) they contain a great deal of valuable information, on the relative merits of highways, canals, and rail-roads. The principal point, however, and the one to which we shall confine our observations, is an enunciation of the laws which regulate the friction of rolling and sliding bodies, as deduced from the experiments of Vince and Coulomb. With a view to the illustration of this part of the subject, some very important and conclusive experiments have recently been made in this town, to which we shall by and by have occasion to refer at some length; but before doing so, we must make a few observations on the rule laid down by the Scotsman, and the misconceptions which appear to have prevailed respecting it, both in that journal and in other quarters.

After comparing the resistance experienced by a boat moving through the water, with the friction which retards the progress of a waggon on a rail-road, and stating that they are governed by different laws, the Scotsman notices the conclusions established by the experiments of Vince and Coulomb; the most important of which is, that the *friction of rolling and sliding bodies is the same for all velocities.* The writer then observes :—

' It is with this last law only that we have to do at present; and it is remarkable that the extraordinary results to which it leads, have been, as far as we know, entirely overlooked by writers on roads and railways. These results, indeed, have an appearance so paradoxical, that they will shock the faith of practical men, though the principle from which they flow is admitted without question by all scientific mechanicians.

' First. It flows from this law, that (abstracting the resistance of the air,) if a car were set in motion on a level railway, with a constant force greater in any degree than is required to overcome its friction, the car would proceed with a motion continually accelerated, like a falling body acted upon by the force of gravitation; and however small the original velocity might be, it would in time increase beyond any assignable limit. It is only the resistance of the air (increasing as the space of the velocity) that prevents this indefinite acceleration, and ultimately renders the motion uniform.

' Secondly. Setting aside again, the resistance of the air (the effects of which we shall estimate by and by,) the very same amount of constant force which impels a car on a railway at two miles an hour, would impel it at ten or twenty miles an hour, if an extra force were employed at first to overcome the *inertia* of the car, and generate the required velocity. Startling as this proposition may appear, it is an indisputable and necessary consequence of the laws of friction.

' Now it would at all times be easy, as we shall afterwards show, to convert this accelerated motion into a uniform of any determinate velocity; and from the nature of the resistance, a high velocity would cost almost as little, and be as readily obtained as a low one. For all velocities, therefore, above four or five miles an hour, rail-ways will afford facilities for communication prodigiously superior to canals, or arms of the sea.'

Now we are perfectly satisfied, both by the experiments of Vince and Coulomb, and those more recent and conclusive experiments, to which we have already alluded, that the rule laid down here is correct; but the writer. ought to have guarded against the misconception to which his last paragraph is liable. When he says that a high velocity would cost almost as little as a low one, he should have said that it would cost as little per mile, or as little over any given space: for it cannot be his meaning, that a carriage can be kept moving for an hour, or for any given time, at a high velocity, with as little expenditure of power, as at a low velocity. Yet this he has been generally understood to mean, and a great deal has been written and said with a view to prove that he was mistaken; when in fact he was only misunderstood. In a subsequent article, however, the author appears, in some degree, to have fallen into the same error into which he has led other persons. He says:

' Every body knows that the rate of stage coach traveling in this country has increased within the last twenty-five years, from six or seven miles an hour to eight or nine, and this, too, before roads were M'Adamized, and with much less injury to the horses than was anticipated. Supposing that a coach-horse could run fourteen miles unloaded, with the same muscular exertion which carries forward the stage-coach at eight or nine miles, then professor Leslie's formula becomes 3-4ths $(14 \ v)2$. Each horse would, of course, draw with a force of 48lbs. at six miles, and of 27lbs. at eight miles an hour. But if the friction increased in the ratio of the velocity, the load upon each horse would increase from 48lbs. to 60lbs., when the speed increased

from six to eight miles an hour : and as the horse exerting
the same strength, would only pull with a force of 27lbs.,
he would thus have more than double work to do, which is
plainly impossible. But admit that the friction is equal in
equal times ; then, since the time is diminished 1-4th by
increasing the speed from six to eight miles an hour, the
horses have actually 4-5ths less to do ; the load upon each is
reduced from 48lbs. to 36, and the horse would have to increase its exertion only 1-3rd, that is, from 27lbs. to 36.
The facts, we believe, will be found strictly consistent with
this hypothesis, and decidedly at variance with the other.
However strange it may sound, then, to common observers,
it is practically true, that a smaller absolute amount of force
will drag a coach over the same space in three hours than in
four, and in one than in two.'

This paragraph seems to us to contain a very obvious
fallacy. If the speed be increased from six miles an hour to
eight, the horses have by no means 1-4th less work to do,
supposing the friction a constant quantity, and the traction
consequently the same. It is true that they exert this
power for a shorter time, but it is over the same distance.
Supposing the power of traction necessary to overcome the
friction is 1000lbs., then that power must be extended over
every yard of the distance, whether the carriage moves at
six or eight miles an hour : and it is by the distance, not
the time, that the power must be measured. That this
must be the case, will be obvious if the experiment be put
in another shape. Suppose a perfectly horizontal railway,
a mile long, with a perpendicular descent of a mile at one
end of it, as represented in fig. 652.

Suppose a waggon placed on this railway at A, attached to a rope passing over a pulley at B, and loaded at
that point with a weight exactly sufficient to overcome the
friction, then, if the resistance of the air is nothing, and
the rope be without weight, it follows, from the rule laid
down, that if the waggon is set in motion at any given
speed, it will continue to move at that rate, until it reaches
the point B and the weight falls to C. But whether the
waggon passes over the railway in an hour or in three
minutes, it is obvious that the same weight will descend
through the same space, and that consequently, the same
amount of power will be expended. It is, perhaps, necessary to observe here, that if the weight is only just sufficient to overcome the friction, there will (as is proved by

the experiments of Mr. Vince) be no acceleration of motion on the principle of falling bodies.

However, though a carriage cannot, as we think we have shewn, be moved ten miles in one hour, with a similar expenditure of power than in two, it is very interesting to know that it can be moved with the same expenditure, (excepting the resistance of the air.) In many cases distance is of so much consequence, that the elucidation and application of this rule will probably lead to very important results. Many persons, however, are very sceptical on this subject, and contend that the experiments of Vince and Coulomb do not authorise any such conclusions as have been drawn from them. It has been asked, if the same constant force will move a carriage as well at a high as at a low velocity, why we do not see something like this in practice; why a carriage moved by a steam-engine instead of acquiring, as it proceeds, a high degree of velocity, moves on at one uniform rate after it has overcome the *vis inertiæ* at the commencement of its journey? We think the reason is very obvious. A locomotive steam-engine does not exert the same constant force on the peripheries of the wheels of the carriage, when it moves at different velocities. For instance, suppose the piston of an engine to move 200 feet in a minute, and to impel the peripheries of the travelling wheels at a velocity of two miles, and with a force just sufficient to overcome the friction, how can the speed be augmented without increasing the power of the engine? If the diameter of the wheels be increased with the view of increasing the speed, the force with which they are impelled will be diminished in the same proportion; and the engine will stop, unless the pressure is increased. To increase that, of course, will be to augment the power. As it is obvious, therefore, that a steam-engine cannot exert the same force at different velocities, some other means must be devised for putting to the test of experiment the rule laid down in the *Scotsman*.

We now come to the most important and interesting part of this article. As none of the experiments of Vince or Coulomb (so far as we have seen or heard them detailed) were made with bodies resembling railway waggons, either in form, or in the nature of their motion, the correctness of the conclusions deduced from them with respect to such carriages, was doubted by many persons of considerable scientific attainments. It became desirable, therefore, that other experiments should be tried, with carriages

upon railways, which, of course, would be much more
satisfactory. This, however, it did not, at first sight, appear very easy to accomplish in a satisfactory manner : but
Mr. Roberts, of this town, recently devised a mode of determining the point, which appears to us wholly unobjec-
tionable, and which exhibits, in a high degree, the simplicity and facility of execution, by which that gentleman's
inventions are so eminently distinguished. It was very
difficult to devise means for measuring accurately the friction of a carriage moving over a railway; but it occurred
to Mr. Roberts, that the difficulty would be obviated if the
railway were made to move under the carriage. When this
idea once presented itself, it was easy to reduce it to practice. Mr. Roberts therefore constructed an apparatus, of
which fig. 654 will give a pretty correct notion.

A is a small waggon with four cast iron wheels, placed on the periphery of a cast iron drum B, three feet in diameter, and six inches broad,
(which acts as the rail-road.) This drum is fastened on the same shaft as
the pulley C, which is driven at different speeds by a strap from another
pulley. The waggon is attached by a wire to one of Marriot's patent
weighing machines D, for the purpose of measuring the friction, and the
board G, prevents the current of air, occasioned by the motion of the
drum, from acting upon the carriage. Now if the drum be driven with
any given velocity, say four miles an hour, in the direction indicated by
the mark E, and the waggon held in its place by the wire which attaches
it to the index, it is perfectly obvious that the wheels will revolve on the
drum in precisely the same manner as if the waggon moved forward on a
horizontal road; and the friction will also be the same, except, perhaps, a
small addition occasioned by the curvature of the drum, but which will
not affect the *relative* frictions of different speeds. As the waggon is stationary, the resistance of the air will be entirely got rid of ; and the index
of the machine will indicate the precise amount of traction necessary to
overcome the friction. Of course, in making the experiment, it will be
necessary to keep the centre of the waggon *exactly over the axis of the
drum;* for if it were permitted to go beyond the centre, a part of the
weight would be added to the friction ; if, on the contrary, it was brought
nearer the index, a part of the weight would act against the friction, and
diminish the apparent quantity. The tempering screw F, is therefore
added to keep the waggon in its proper situation, in whatever way the
spring of the weighing machine may be acted upon by the friction.

This simple apparatus having been constructed, a number of experiments were made, chiefly with a view to determine whether the friction
were the same at different velocities. The waggon was loaded with fifty
pounds, (including its own weight) and the drum was driven at different
velocities, varying from two to twenty-four miles an hour on the periphery: but in every case, the friction, as indicated by the weighing machine,
was precisely the same. No increase of speed affected the index at all,
but on increasing the weight, it immediately shewed a corresponding increase of friction.

We consider these experiments as perfectly conclusive

of the fact, that the friction on a railway is the same for all velocities; and that a carriage may be propelled twenty miles in one hour, with the same amount of force which would be necessary to drive it twenty miles in ten hours, provided the resistance of the atmosphere was out of the question: and, if the carriage was properly constructed, that would not amount to much. In other words, goods may be conveyed from Manchester to Liverpool, on a rail-road, with very nearly the same expenditure of steam, whether they are carried two miles, or four miles, or twenty miles an hour. A steam engine, which will propel twenty tons at four miles an hour, will, with the same expense of coals, propel ten tons at eight miles an hour; so that, with the smaller load, it might make a journey to Liverpool and back, in the same time which would be occupied in going thither with the larger load Or, to put the matter in another shape: suppose a four-horse engine will convey forty tons to Liverpool in eight hours, an eight horse engine will convey the same weight thither in four hours. There will be the same expenditure of steam in both cases, but, in the latter, a saving of half the time; a saving which, we need not add, will frequently be of immense importance."

These practical results are very satisfactory, as the hope of propelling carriages at a suitable speed, for the more rapid dispatch of business, and conveyance of passengers, is thereby placed almost beyond a doubt.

We ought to notice here, the striking difference in the force requisite to give rapid motion on a rail-road to that on a canal or navigable river. These latter are governed by a totally different law, as the resistance, or head of water on the bows of the boat, increase as the squares of its velocity; consequently it will require four times the power to double the speed. But, on the other hand, it must be admitted, that in all speeds under three miles per hour, the canal has a decided advantage, as the force increases as the speed diminishes.

With respect to the horse, it is well known, that his power decreases as his speed increases; and that when he is travelling at his greatest speed, which, with a weight, seldom exceeds 13 miles per hour, he is able to exert little or no strength. We, therefore, take it for granted, that in the present improved state of our manufactures, artificial power of some description must be resorted to, and whatever experience may prove to be the most economical, the application of that power is the most important part of the

to the paper alluded to by Mr. Thompson, in the Repertory of Arts, for March, 1822, to a work which will shortly issue from the Press, by Mr. N. Wood of the Killingworth Colliery, of whose experiments, in conjunction with Mr. Sylvester, we have already had occasion to speak, and to Observations on a General Iron Railway, by Mr. Gray.

APPENDIX.

GEOMETRY.

GEOMETRY is that branch of mathematics which treats of the description and properties of magnitudes in general.

Definitions or Explanation of Terms.

1. A *point* has neither length, breadth, nor thickness. From this definition it may easily be understood that a mathematical point cannot be seen nor felt; it can only be imagined. What is commonly called a point, as a small dot made with a pencil or pen, or the point of a needle, is not in reality a mathematical point; for however small such a dot may be, yet if it be examined with a magnifying glass, it will be found to be an irregular spot, of a very sensible length and breadth; and our not being able to measure its dimensions with the naked eye, arises only from its smallness. The same reasoning may be applied to every thing that is usually called a point; even the point of the finest needle appears like that of a poker when examined with the microscope.

2. A *line* is length, without breadth or thickness. What was said above of a point, is also applicable to the definition of a line. What is drawn upon paper with a pencil or pen, is not in fact a line, but the representation of a line. For however fine you may make these representations, they will still have some breadth. But by the definition, a line has no breadth whatever, yet it is impossible to draw any thing so fine as to have no breadth. A line therefore, can only be imagined. The ends of a line are points.

3. A *right line* is what is commonly called a *straight line*, or that tends every where the same way.

4. A *curve* is a line which continually changes its direction between its extreme points.

5. *Parallel* lines are such as always keep at the same distance from each other, and which, if prolonged ever so far, would never meet. Fig. 1.

6. An *angle* is the inclination or opening of two lines meeting in a point, Fig. 2.

7. The lines AB, and BC, which form the angle, are called the legs or sides; and the point B where they meet, is called the *vertex* of the angle, or the *angular point*. An angle is sometimes expressed by a letter placed at the vertex, as the angle B, Fig. 2; but most commonly by three letters, observing to place in the middle the letter at the vertex, and the other two at the end of each leg, as the angle ABC.

3 X

8. When one line stands upon another, so as not to lean more to one side than to another, both the angles which it makes with the other are called *right angles*, as the angles ABC and ABD, *Fig.* 3, and all right-angles are equal to each other, being all equal to 90°; and the line AB is said to be *perpendicular* to CD.

Beginners are very apt to confound the terms *perpendicular*, and *plumb* or *vertical line*. A line is vertical when it is at right-angles to the plane of the horizon, or level surface of the earth, or to the surface of water, which is always level. The sides of a house are vertical. But a line may be perpendicular to another, whether it stands upright or inclines to the ground, or even if it lies flat upon it, provided only that it makes the two angles formed by meeting with the other line equal to each other ; as for instance, if the angles ABC and ABD be equal, the line AB is perpendicular to CD, whatever may be its position in other respects.

9. When one line, BE (*Fig.* 3.) stands upon another, CD, so as to incline, the angle EBC, which is greater than a right-angle, is called an *obtuse* angle ; and that which is less than a right-angle, is called an *acute* angle, as the angle EBD.

10. Two angles which have one leg in common, as the angles ABC, and ABE, are called *contiguous* angles, or *adjoining* angles ; those which are produced by the crossing of two lines, as the angles EBD and CBF, formed by CD and EF, crossing each other, are called *opposite* or *vertical* angles.

11. A *figure* is a bounded space, and is either a *surface* or a *solid*.

12. A *superficies*, or *surface*, has length and breadth only. The extremities of a superficies are lines.

13. A *plane*, or *plane surface*, is that which is every where perfectly flat and even, or which will touch every part of a straight line, in whatever direction it may be laid upon it. The top of a marble slab, for instance, is an example of this, which a strait edge will touch in every point, so that you cannot see light any where between.

14. A *curved surface* is that which will not coincide with a straight line in any part. Curved surfaces may be either convex or concave.

15. A *convex surface* is when the surface rises up in the middle, as, for instance, a part of the outside of a globe.

16. A *concave surface* is when it sinks in the middle, or is hollow, and is the contrary to convex.

A surface may be bounded either by straight lines, curved lines, or both these.

17. Every surface, bounded by straight lines only, is called a *polygon*. If the sides are all equal, it is called a *regular polygon*. If they are unequal, it is called an *irregular polygon*. Every polygon, whether equal or unequal, has the same number of sides as angles, and they are denominated sometimes according to the number of sides, and sometimes from the number of angles they contain. Thus a figure of three sides is called a *triangle*, and a figure of four sides a *quadrangle*.

A *pentagon* is a polygon of five sides.,

A *hexagon* has six sides.

A *heptagon* seven sides.

An *octagon* eight sides.

A *nonagon* nine sides.

A *decagon* ten sides.

An *undecagon* eleven sides.

A *duodecagon* twelve sides.

When they have a greater number of sides, it is usual to call them polygons of 13 sides, of 14 sides, and so on.

Triangles are of different kinds, according to the lengths of their sides.

18. An *equilateral triangle* has all its sides equal, as ABC, *Fig.* 4.

19. An *isosceles triangle* has two equal sides, as DEF, *Fig.* 5.

20. A *scalene triangle* has all its sides unequal, as GHI, *Fig.* 6.

Triangles are also denominated according to the angles they contain.

21. A *right-angled* triangle is one that has in it a right angle, as ABC, *Fig.* 7.

22. A triangle cannot have more than one right-angle. The side opposite to the right-angle B, as AC, is called the *hypothenuse*, and is always the longest side.

23. An *obtuse-angled* triangle has one obtuse-angle, as *Fig.* 8.

24. An *acute-angled* triangle has all its angles acute, as *Fig.* 4.

25. An isosceles, or a scalene triangle, may be either right-angled, obtuse, or acute.

26. Any side of a triangle is said to *subtend* the angle opposite to it: thus AB *(Fig.* 7), subtends the angle ACB.

27. If the side of a triangle be drawn out beyond the figure, as AD *(Fig.* 8), the angle A, or CAB, is called an *internal* angle, and the angle CAD, or that without the figure, an *external* angle.

28. A *quadrangle* is also called a *quadrilateral* figure. They are of various denominations, as their sides are equal or unequal, or as all their angles are right-angles or not.

29. Every four-sided figure whose opposite sides are parallel, is called a *parallelogram*. Provided that the sides opposite to each other be parallel, it is immaterial whether the angles are right or not. *Fig.* 9, 10, 11, and 12, are all parallelograms.

30. When the angles of a parallelogram are all right-angles, it is called a *rectangular parallelogram* or a *rectangle*, as *Fig.* 11 and 12.

31. A rectangle may have all its sides equal, or only the opposite sides equal. When all its sides are equal, it is called a *square*, as *Fig.* 12.

32. When the opposite sides are parallel, and all the sides equal to each other, but the angles not right-angles, the parallelogram is called a *rhombus*, as *Fig.* 10.

33. A parallelogram having all its angles oblique, and only its opposite equal, is called a *rhomboid*, as *Fig.* 9.

34. When a quadrilateral or four-sided figure has none of its sides parallel, it is called a *trapezium*, as *Fig.* 13; consequently every quadrangle, or quadrilateral which is not a parallelogram, is a trapezium.

2 x 2

the complements.

38. *Base* of a figure is the side on which it is supposed to stand erect, as AB, and CD, *Fig.* 16.

39. *Altitude* of a figure is its perpendicular height from the base to the highest part, as EF, *Fig.* 16.

40. *Area* of a plane figure, or other surface, means the quantity of space contained within its boundaries, expressed in square feet, yards, or any other superficial measure.

41. *Similar figures* are such as have the same angles, and whose sides are in the same proportion, as *Fig.* 17.

42. *Equal figures* are such as have the same area or contents.

43. A *circle* is a plane figure, bounded by a curve line returning into itself, called its *circumference*, ABCD *(Fig.* 18*)*, every where equally distant from a point E within the circle, which is called the *centre.*

44. The *radius* of a circle is a straight line drawn from the centre to the circumference, as EF *(Fig.* 18*)*. The radius is the opening of the compass when a circle is described; and consequently all the radii of a circle must be equal to each other.

45. A *diameter* of a circle is a straight line drawn from one side of the circumference to the other through the centre, as CB *(Fig.* 18*)*. Every diameter divides the circle into two equal parts.

46. A *segment* of a circle is a part of a circle cut off by a straight line drawn across it. This straight line is called the *chord*. A segment may be either equal to, greater, or less than a *semi-circle*, which is a segment formed by the diameter of the circle, as CEB, and is equal to half the circle.

47. A *tangent* is a straight line, drawn so as just to touch a circle without cutting it, as GH (*Fig.* 18). The point A, where it touches the circle, is called the *point of contact*. And a tangent cannot touch a circle in more points than one.

48. A *sector* of a circle is a space comprehended between two radii and an arc, as BIK *(Fig.* 19).

49. The circumference of every circle, whether great or small, is supposed to be divided into 360 equal parts, called *degrees*; and every degree into 60 parts, called minutes; and every minute into 60 seconds. To measure the inclination of lines to each other, or angles, a circle is described round the angular point, as a centre, as IK, *Fig.* 19; and according to the number of degrees, minutes, and

seconds, cut off by the sides of the angle, so many degrees, minutes, and seconds, it is said to contain. Degrees are marked by °, minutes by ', and seconds by "; thus an angle of 48 degrees, 15 minutes, and 7 seconds, is written in this manner, 48° 15' 7".

50. A *solid* is any body that has length, breadth, and thickness : a book, for instance, is solid, so is a sheet of paper ; for though its thickness is very small, yet it has some thickness. The boundaries of a solid are *surfaces*.

51. *Similar solids* are such as are bounded by an equal number of similar planes.

52. A *prism* is a solid, of which the sides are parallelograms, and the two ends or bases are similar polygons, parallel to each other. Prisms are denominated according to the number of angles in the base, *triangular* prisms, *quadrangular, heptangular,* and so on, as Fig. 20, 21, 22, 23. If the sides are perpendicular to the plane of the base, it is called an *upright* prism ; if they are inclined, it is called an *oblique* prism.

53. When the base of a prism is a parallelogram, it is called a *parallelopipedon,* as Fig. 22 and 23. Hence, a parallelopipedon is a solid, terminated by six parallelograms.

54. When all the sides of a parallelopipedon are squares, the solid called a *cube,* as Fig. 23.

55. A *rhomboid* is an oblique prism, whose bases are parallelograms. (*Fig.* 24.)

56. A *pyramid* AB (*Fig.* 25 and 26) is a solid, bounded by, or contained within, a number of planes, whose base may be any polygon, and whose faces are triangles terminated in one point, B, commonly called the *summit,* or *vertex* of the pyramid.

57. When the figure of the base is a triangle, it is called a *triangular pyramid ;* when the figure of the base is a quadrilateral, it is called a *quadrilateral pyramid,* &c.

58. A pyramid is either *regular* or *irregular,* according as the base is regular or irregular.

59. A pyramid is also *right* or *upright,* or it is *oblique.* It is right, when a line drawn from the vertex to the centre of the base, is perpendicular to it, as *Fig.* 25 ; and oblique, when this line inclines, as *Fig.* 26.

60. A *cylinder* is a solid (*Fig.* 27 and 28) generated or formed by the rotation of a rectangle about one of its sides, supposed to be at rest ; this quiescent side is called the *axis* of the cylinder. Or it may be conceived to be generated by the motion of a circle, in a direction perpendicular to its surface, and always parallel to itself.

61. A cylinder is either *right* or *oblique,* as the axis is perpendicular to the base or inclined.

62. Every *section* of a right cylinder taken at right-angles to its axis, is a *circle ;* and every section taken across the cylinder, but oblique to the axis, is an *ellipsis.*

63. A circle being a polygon of an infinite number of sides, it fol-

lows, that the cylinder may be conceived as a prism, having such polygons for bases.

64. A *cone* is a solid (*Fig.* 29 and 30) having a circle for its base, and its sides a convex surface, terminating in a point A, called the *vertex* or *apex* of the cone. It may be conceived to be generated by the revolution of a right-angled triangle about its perpendicular.

65. A line drawn from the vertex to the centre of the base is the *axis* of the cone.

66. When this line is perpendicular to the base, the cone is called an upright or *right* cone; but when it is inclined, it is called an *oblique* cone.

67. If it be cut through the axis, from the vertex to the base, the section will be a *triangle*.

68. If a right cone be cut by a plane at right-angles to the axis, the section will be a *circle*.

69. If it be cut oblique to the axis, and quite across from one side to the other, the section will be an *ellipsis*, as *Fig.* 31. A section of a cylinder, made in the same manner, is also an ellipsis; and this is easily conceived; but it does not appear so readily to most people, that the oblique section of a cone is an ellipsis: they frequently imagine that it will be wider at one end than the other, or what is called an *oval*, which is of the shape of an egg. But that this is a mistake, any one may convince himself, by making a cone, and cutting it across obliquely: it will be then seen, that the section, in whatever direction it is taken, is a regular ellipsis; and this is the case, whether the cone be right or oblique, except only in one case, in the oblique cone, which is, when the section is taken in a particular direction, which is called *sub-contrary* to its base.

70. When the section is made parallel to one of the sides of the cone, as *Fig.* 32, the curve ABC, which bounds the section, is called a *parabola*.

71. When the section is taken parallel to the axis, as *Fig.* 33, the curve is called an *hyperbola*.

These curves, which are formed by cutting a cone in different directions, have various properties, which are of great importance in astronomy, gunnery, perspective, and many other sciences.

72. A *sphere* is a solid, terminated by a convex surface, every point of which is at an equal distance from a point within, called the *centre*, *Fig.* 34.

73. It may be conceived to be formed by making a semicircle revolve round its diameter. This may be illustrated by the process of forming a ball of clay by the potter's wheel, a semicircular mould being used for the purpose. The diameter of the semicircle, round which it revolves, is called the *axis* of the sphere.

74. The ends of the axis are called *poles*.

75. Any line passing through the centre of the sphere, and terminated by the circumference, is a *diameter* of the sphere.

76. Every section of a sphere is a circle; every section taken

through the centre of the sphere, is called a *great circle*, as AB, Fig. 34 ; every other is a *lesser circle*, as CD.

77. Any portion of a sphere cut off by a plane, is called a *segment* ; and when the plane passes through the centre, it divides the sphere into two equal parts, each of which is called a *hemisphere*.

78. A *conoid* is a solid, produced by the circumvolution of a section of the cone, about its axis, and, consequently, may be either an *elliptical conoid*, a *hyperbolical conoid*, or a *parabolical conoid*. When it is elliptical, it is generally called a *spheroid*. These solids are also called *ellipsoid*, *hyperboloid*, and *paraboloid*.

79. A *spheroid* is a solid (*Fig.* 35), generated by the rotation of a semi-ellipsis about the transverse or conjugate axis ; and the centre of the ellipsis is the centre of the spheroid.

80. The line about which the ellipsis revolves, is called the *axis*. If the spheroid be generated about the conjugate axis of the semi-ellipsis, it is called a *prolate spheroid*.

81. If the spheroid be generated by the semi-ellipsis, by revolving about the transverse axis, it is called an *oblong spheroid*.

82. Every *section* of a spheroid is an *ellipsis*, except when it is perpendicular to that axis about which it is generated ; in which case, it is a circle.

83. All sections of a spheroid parallel to each other, are similar figures.

A *frustum* of a solid, means a piece cut off from the solid, by a plane passed through it, usually parallel to the base of the solid, as the frustum of a cone, a pyramid, &c.

There is a *lower* and an *upper* frustum, according as the piece spoken of does or does not contain the base of the solid.

84. A regular body is a solid, contained under a certain number of equal and regular plane figures of the same sort.

85. The *faces of the solid* are the plane figures under which it is contained ; and the *linear sides*, or *edges of the solid*, are the sides of the plane faces.

86. There are only five regular bodies :—viz. 1st. the *tetraedon*, which is a regular pyramid, having four triangular faces ; 2d. the *hexaedron*, or cube, which has six equal square faces ; 3d. the *octaedron*, which has eight triangular faces ; 4th. the *dodecaedron*, which has twelve pentagonal faces ; 5th. the *icosaedron*, which has twenty triangular faces.

NOTE.—If the figures marked A, B, C, D, E, be exactly drawn on pasteboard, and the lines cut half through, so that the parts be turned up, and glued together, they will represent the five regular bodies, viz.—*Fig.* A, the tetraedon ; B, the hexaedron ; C, the octaedron ; D, the dodecaedron ; and E, the icosaedron.

87. *Ratio* is the proportion which one magnitude bears to another of the same kind, with respect to quantity, and is usually marked thus, A : B.

Of these, the first is called the *antecedent*, and the second the *consequent*.

88. The *measure* or *quantity* of a ratio, is conceived by considering what part of the consequent is the antecedent; consequently, it is obtained by dividing the consequent by the antecedent.

89. *Three* magnitudes or quantities, A, B, C, are said to be *proportional*, when the ratio of the first to the second is the same as that of the second to the third. Thus, 2, 4, 8, are proportional, because 4 is contained in 8 as many times as 2 is in 4.

90. *Four* quantities, A, B, C, D, are said to be proportional, when the ratio of the first, A, to the second, B, is the same as the ratio of the third, C, to the fourth, D. It is usually written, A : B :: C : D, or, if expressed in numbers, 2 : 4 :: 8 : 16.

91. Of *three* proportional quantities, the middle one is said to be a *mean proportional* between the other two; and the last a *third proportional* to the first and second.

92. Of *four* proportional quantities, the last is said to be a *fourth proportional* to the other three, taken in order.

93. *Ratio of equality* is that which equal numbers bear to each other.

94. *Inverse ratio* is when the antecedent is made the consequent, and the consequent the antecedent. Thus, if 1 : 2 :: 3 : 6; then, *inversely*, 2 : 1 :: 6 : 3.

95. *Alternate proportion* is when the antecedent is compared with antecedent, and consequent with consequent. Thus, if 2 : 1 :: 6 : 3; then, by *alternation*, 2 : 6 :: 1 : 3.

96. Proportion by *composition* is when the antecedent and consequent, taken as one quantity, are compared either with the consequent or with the antecedent. Thus, if 2 : 1 :: 6 : 3; then, by *composition*, 2+1 : 1 :: 6+3 : 3, and 2+1 : 2 :: 6+3 : 6.

97. *Divided proportion* is when the difference of the antecedent and consequent is compared either with the consequent or with the antecedent. Thus, if 3 : 1 :: 12 : 4; then, by *division*, 3—1 : 1 :: 12—4 : 4, and 3—1 : 3 :: 12—4 : 12.

98. *Continued proportion* is when the first is to the second as the second to the third; as the third to the fourth; as the fourth to the fifth; and so on.

99. *Compound ratio* is formed by the multiplication of several antecedents and the several consequents of ratios together, in the following manner:

If A be to B as 3 to 5, B to C as 5 to 8, and C to D as 8 to 6; then A will be to D, as $\frac{3 \times 5 \times 8}{5 \times 8 \times 6} = \frac{120}{240} = \frac{1}{2}$; that is, A : D :: 1 : 2.

100. *Bisect*, means to divide any thing into two equal parts.

101. *Trisect*, is to divide any thing into three equal parts.

102. *Inscribe*, to draw one figure within another, so that all the angles of the inner figure touch either the angles, sides, or planes of the external figure.

103. *Circumscribe*, to draw a figure round another, so that either the angles, sides, or planes of the circumscribed figure, touch all the angles of the figure within it.

104. *Rectangle under any two lines*, means a rectangle which has two of its sides equal to one of the lines, and two of them equal to the other. Also, the rectangle under AB, CD, means AB × CD.

105. *Scales of equal parts*. A scale of equal parts is only a straight line, divided into any number of equal parts, at pleasure. Each part may represent any measure you please, as an inch, a foot, a yard, &c. One of these is generally subdivided into parts of the next denomination, or into tenths or hundredths. Scales may be constructed in a variety of ways. The most usual manner is, to make an inch, or some aliquot part of an inch, to represent a foot ; and then they are called inch scales, three-quarter inch scales, half-inch scales, quarter-inch scales, &c. They are usually drawn upon ivory or box-wood.

106. An *axiom* is a manifest truth, not requiring any demonstration.

107. *Postulates* are things required to be granted true, before we proceed to demonstrate a proposition.

108. A *proposition* is when something is either proposed to be done, or to be demonstrated, and is either a *problem* or a *theorem*.

109. A *problem* is when something is proposed to be done, as some figure to be drawn.

110. A *theorem* is when something is proposed to be demonstrated or proved.

111. A *lemma* is when a premise is demonstrated, in order to render the thing in hand the more easy.

112. A *corollary* is an inference drawn from the demonstration of some proposition.

113. A *scholium* is when some remark or observation is made upon something mentioned before.

114. The sign = denotes that the quantities betwixt which it stands, are equal.

115. The sign + denotes that the quantity after it, is to be added to that immediately before it.

116. The sign — denotes, that the quantity after it is to be taken away or subtracted from the quantity preceding it.

Geometrical Problems.

Prob. 1. To divide a given line AB into two equal parts.

From the points A and B, as centres, and with any opening of the compasses greater than half AB, describe arches, cutting each other in c and d. Draw the line c d ; and the point E, where it cuts A B, will be the middle required.

Prob. 2. To raise a perpendicular to a given line A B, from a point given at C.

Case 1. When the given point is near the middle of the line, on each side of the point C. Take any two equal distances, C d and C e; from d and e, with any radius or opening of the compasses greater than C d or C e, describe two arcs cutting each other in f. Lastly, through the points f, C, draw the line f C, and it will be the perpendicular required.

Case 2. When the point is at, or near the end of the line. Take any point d, above the line, and with the radius or distance d C, describe the arc e C f, cutting AB in e and C. Through the centre d, and the point e, draw the line e d f, cutting the arc e C f in f. Through the points f C, draw the line fC, and it will be the perpendicular required.

Prob. 3. From a given point f, to let fall a perpendicular upon a given line AB.

From the point f, with any radius, describe the arc d e, cutting AB in e and d. From the points e d, with the same or any other radius, describe two arcs, cutting each other in g. Through the points f and g, draw the line f g, and f C will be the perpendicular required.

Prob. 4. To make an angle equal to another angle which is given, as a B b.

From the point B, with any radius, describe the arc a b, cutting the legs B a, B b, in the points a and b. Draw the line D e, and from the point D, with the same radius as before, describe the arc e f, cutting D e in e. Take the distance B a, and apply it to the arc e f, from e to f. Lastly, through the points D, f, draw the line D f, and the angle e D f will be equal to the angle b B a, as was required.

Prob. 5: To divide a given angle, ABC, into two equal angles.

From the point B, with any radius, describe the arc AC. From A and C, with the same, or any other radius, describe arcs cutting each in d. Draw the line B d, and it will bisect the angle ABC, as was required.

Prob. 6. To lay down an angle of any number of degrees.

There are various methods of doing this. One is by the use of an instrument called a *protractor*, with a semicircle of brass, having its circumference divided into degrees. Let AB be a given line, and let it be required to draw from the angular point A, a line making, with AB, any number of degrees, suppose 20. Lay the straight side of the protractor along the line AB, and count 20° from the end B of the semicircle; at C, which is 20° from B, mark; then, removing the protractor, draw the line AC, which makes, with AB, the angle required. Or, it may be done by a divided line, usually drawn upon scales, called a *line of chords*. Take 60° from the line of chords, in the compasses, and setting one at the angular point B, *Prob.* 4, with that opening as a radius, describe an arch, as a b : then take the number of degrees of which you intend the angle to be; and set it from b to a, then is a B b the angle required.

Prob. 7. Through a given point C, to draw a line parallel to a given line AB.

Case 1. Take any point d, in AB ; upon d and C, with the distance C d, describe two arcs, e C, and d f, cutting the line AB in e and d. Make d f equal to e C ; through C and f draw C f, and it will be the line required.

Case 2. When the parallel is to be at a given distance from AB: From any two points, c and d, in the line AB, with a radius equal to the given distance, describe the arcs e and f : draw the line CB to

.ouch those arcs without cutting them, and it will be parallel to AB, as was required.

Prob. 8. To divide a given line AB, into any proposed number of equal parts.

From A, one end of the line, draw A c, making any angle with AB; and from B, the other end, draw B d, making the angle A B d equal to B A c. In each of these lines, A c, B d, beginning at A and B, set off as many equal parts, of any length, as AB is to be divided into. Join the points C 5, 46, 57, and AB will be divided as required.

Prob. 9. To find the centre of a given circle, or of any one already described. Draw any chord AB, and bisect it with the perpendicular CD. Bisect CD with the diameter EF, and the intersection O will be the centre required.

Prob. 10. To draw a tangent to a given circle that shall pass through a given point, A.

From the centre O, draw the radius OA. Through the point A, draw DE perpendicular to OA, and it will be the tangent required.

Prob. 11. To draw a tangent to a circle, or any segment of a circle ABC, through a given point B, without making use of the centre of the circle.

Take any two equal divisions upon the circle, from the given point B, towards d and e, and draw the chord e B. Upon B, as a centre, with the distance B d, describe the arc f d g, cutting the chord e B in f. Make d g equal to d f; through g draw g B, and it will be the tangent required.

Prob. 12. Given three points, A, B, C, not in a straight line, to describe a circle that shall pass through them.

Bisect the lines AB, BC, by the perpendiculars a b, b d, meeting at d. Upon d, with the distance d A, d B, or d C, describe ABC, and it will be the required circle.

Prob. 13. To describe the segment of a circle to any length AB, and height CD.

Bisect AB by the perpendicular D g, cutting AB in c. From c make c D, on the perpendicular, equal to CD. Draw AD, and bisect it by a perpendicular e f, cutting D g in g. Upon g the centre, describe ADB, and it will be the required segment.

Prob. 14. To describe the segment of a circle by means of two rules, to any length AB, and perpendicular height CD in the middle of AB, without making use of the centre.

Place the rules to the height at C ; bring the edges close to A and B ; fix them together at C, and put another piece across them to keep them fast. Put in pins at A and B, then move the rulers round these pins, holding a pencil at the angular point C, which will describe the segment.

Prob. 15. In any given triangle to inscribe a circle.

Bisect any two angles A and C, with the lines AD and DB. From D, the point of intersection, let fall the perpendicular DE; it will be the radius of the circle required.

Prob. 16. In a given square, to describe a regular octagon.

Draw the diagonals AC and BD, intersecting at e. Upon the points A, B, C, D, as centres, with a radius e C, describe the arcs h e l, k e n, m e g, f e i. Join f n, m h, k i, l g, and it will be the required octagon,

Prob. 17. In a given circle, to describe any regular polygon.

Divide the circumference into as many parts as there are sides in the polygon to be drawn, and join the points of division.

Prob. 18. Upon a given line AB, to construct an equilateral triangle.

Upon the points A and B, with a radius equal to AB, describe arches cutting each other at C. Draw AC and BC, and ABC will be the triangle required.

Prob. 19. To make a triangle, whose sides shall be equal to three given lines D, E, F, any two of them being greater than the third.

Draw AB equal to the line D. Upon A, with the radius F, describe an arc CD. Upon B, with the radius E, describe another arc intersecting the former at C. Draw AC and CB, and ABC will be the triangle required.

Prob. 20. To make a trapezium equal and similar to a given trapezium ABCD.

Divide the given trapezium ABCD into two triangles, by the diagonal DB. Make EF equal to AB; upon EF construct the triangle EFH, whose sides shall be respectively equal to those of the triangle ABD, by the last problem. Upon HF, which is equal to DB, construct the triangle HFG, whose sides are respectively equal to DBC; then EFGH will be the trapezium required.

By the help of this problem, any plan may be copied; as every figure, however irregular, may be divided into triangles. Upon this the practice of land-surveying and making plans of estates, is founded.

Prob. 21. To make a square equal to two given squares. Make the sides DE and DF of the two given squares A and B, form the sides of a right-angled triangle FDE; draw the hypothenuse FE; on it describe the square EFGH; and it will be the square required.

Prob. 22. Two right lines AB, CD, being given, to find a third proportional. Make an angle HEI at pleasure; from E make EF equal to AB, and EG equal to CD: join FG. Make EI equal to EF, and draw HI parallel to FG; then EH will be the third proportional required; that is, EF : EG : : EH : EI, or AB : CD : : CD : EI.

Prob. 23. Three lines being given, to find a fourth proportional. Make the angle HGI at pleasure; from G make GH equal to AB, GI equal to CD, and join HI. Make GK equal to EF; draw KL through K, parallel to HI; then GL will be the fourth proportional required, that is, GH : GI : : GK : GL, or AB : CD : : EF : GL.

Prob. 24. To divide a given line AB in the same proportion as another CD is divided.

Make any angle KHI, and make HI equal to AB; then apply the

several divisions of CD, from H to K, and join KI. Draw the lines e, i f, k g. parallel to IK ; and the line HI will be divided in e, f, g, as was required.

Prob. 25. Between two given lines AB and CD to find a mean proportional.

Draw the right line EG, in which make EF equal to AB, and FG equal to CD. Bisect EG in H, and with HE or HG, as radius, describe the semicircle EIG. From F draw FI perpendicular to EG, cutting the circle in I ; and IF will be the mean proportional required.

Prob. 26. To describe an ellipsis.

If two pins are fixed at the points E and F, a string being put about them, and the ends tied together at C ; the point C being moved round, keeping the string stretched will describe an ellipsis.

The points E and F, where the pins were fixed, are called the *foci*.

The line AB passing through the foci, is called the *transverse axis*.

The point G bisecting the transverse axis, is the *centre* of the ellipsis.

The line CD crossing this centre at right-angles to the transverse axis, is the *conjugate axis*.

The *latus rectum* is a right line passing through the focus at F, at right-angles to the transverse axis terminated by the curve : this is also called the *parameter*.

A *diameter* is any line passing through the centre, and terminated by the curve.

A *conjugate diameter* to another diameter, is a line drawn through the centre, parallel to a tangent at the extreme of the other diameter, and terminated by the curve.

A *double ordinate* is a line drawn through any diameter parallel to a tangent, at the extreme of that diameter terminated by the curve.

Prob. 26. The transverse axis AB, and conjugate axis CD, of any ellipsis, being given, to find the two foci, and from thence to describe the ellipsis.

Take the semi-transverse AE, or EB, and from C as a centre, describe an arc, cutting AB at F and G, which are the foci. Fix pins in these points ; a string being stretched about the joints FCG, the ellipsis is described as above.

Prob. 27. The same being given, to describe an ellipsis by a *trammel*.

The trammel is an instrument consisting of two rulers fixed at right-angles to each other, with a groove in each. A rod with two moveable nuts works in this groove, and, by means of a pencil fixed in the end of the rod, describes the curve. The operation is as follows:

Let the distance of the first pin at B, from the pencil at A, be equal to half the shortest axis, and the distance of the second pin at C, from A, to half the longest axis ; the pins being put in the grooves, move the pencil at A, which will describe the ellipsis.

Prob. 28. To draw the representation of an ellipsis with a compass to any length AB, and width CD.

Draw BP parallel and equal to EC, and bisect it at 1 ; then draw

1 C and PD, cutting each other at K.; bisect KC by a perpendicular meeting CD in O; and on O, with the radius OC, describe the quadrant CGQ.

Through Q and A, draw QG, cutting the quadrant at G; then draw GO, cutting AB at M; make EL equal to EM, also EN equal to EO. From N, through M and L draw NH and NI; then M, L, N, O, are the four centres by which the four quarters of the ellipsis are drawn.

It must be observed, that this is not a true ellipsis, but only an approximation to it; for it is impossible to draw a perfect ellipsis by means of compasses, which can only describe parts of circles. But the curve of an ellipsis differs essentially from that of a circle in every part; and no portions of circles put together, can ever form an ellipsis. But by this means, a figure may be drawn, which approaches nearly to an ellipsis, and therefore may be often substituted for it when a trammel cannot be had, or when the ellipsis is too small to be drawn by it. At the joining of the portions of circles in this operation, the defect is not perceivable; and the best way is not to join them quite, and to help the curve by hand.

Prob. 29. An ellipsis, ACDB, being given, to find the transverse and conjugate axis.

Draw any two parallel lines, AB and CD, cutting the ellipsis at the points A, B, C, D; bisect them in e and f. Through e and f, draw GH, cutting the ellipsis at G and H; bisect GH at I; and it will give the centre.

Upon I, with any radius, describe a circle, cutting the ellipsis in the four points k, l, m, n; join k, l, and m, n; bisect k l, or m n, at o or p. Through the points o, I, or I, p, draw QR, cutting the ellipsis at Q and R; then QR will be the transverse axis. Through I draw TS, parallel to k l, cutting the ellipsis at T and S; and TS will be the conjugate axis.

Prob. 30. To describe an ellipsis similar to a given one ADBC, to any given length IK, or to a given width ML.

Let AB and CD be the two axes of the given ellipsis. Through the points of contact A,D,B,C, complete the rectangle GEHF; draw the diagonals EF and GH: they will pass through the centre at R. Through I and K draw PN and OQ parallel to CD, cutting the diagonals EF and GH, at P,N,Q,O. Join PO and NQ, cutting CD at L and M; then IK is the transverse, and ML the conjugate axis of an ellipsis, that will be similar to the given ellipsis ADBC, which may be described by some of the foregoing methods.

Prob. 31. To describe a parabola. If a thread equal in length to BC, be fixt at C, the end of a square ABC, and the other end be fixt at F.; and if the side AB of the square be moved along the line AD, and if the point E be always kept close to the edge BC of the square, keeping the string tight, the point or pin E will describe a curve EGIH, called a *parabola*.

The *focus* of the parabola is the fixed point F, about which the string revolves.

The *directrix* is the line AD, which the side of the square moves along.

The *axis* is the line LK, drawn through the focus F, perpendicular to the directrix.

The *vertex* is the point I, where the line LK cuts the curve.

The *latus rectum*, or *parameter*, is the line GH passing through the focus F, at right-angles to the axis IK, and terminated by the curve.

The *diameter* is any line MN, drawn parallel to the axis IK.

A *double ordinate* is a right line RS, drawn parallel to a tangent at M, the extreme of the diameter MN, terminated by the curve.

The *abscissa* is that part of a diameter contained between the curve and its ordinate, as MN.

Prob. 32. To describe a *parabola*, by finding points in the curve ; the axis AB, or any diameter being given, and a double ordinate CD.

Through A draw EF parallel to CD ; through C and D draw DF and CE parallel to AB, cutting EF at E and F. Divide BC and BD, each into any number of equal parts, as four ; likewise divide CE and DF into the same number of equal parts. Through the points 1, 2, 3, &c. in CD, draw the lines 1 a, 2 b, 3 c, &c. parallel to CD ; also through the points 1, 2, 3, in CE and DF, draw the lines 1 A, 2 A, 3 A, cutting the parallel lines at the points a, b, c · then the points a, b, c, are in the curve of the parabola.

Prob. 33. To describe an *hyperbola*.

If B and C are two fixed points, and a rule AB be made moveable about the point B, a string ADC being tied to the other end of the rule, and to the point C ; and if the point A be moved round the centre B, towards G, the angle D of the string ADC, by keeping it always tight and close to the edge of the rule AB, will describe a curve DHG, called an hyperbola.

If the end of the rule at B were made moveable about the point C, the string being tied from the end of the rule A to B, and a curve being described after the same manner, is called an *opposite hyperbola*.

The *foci* are the two points B and C, about which the rule and string revolves.

The *transverse axis* is the line IH terminated by the two curves passing 'through the foci, if continued.

The *centre* is the point M, in the middle of the transverse axis IH.

The *conjugate axis* is the line NO, passing through the centre M, and terminated by a circle from H, whose radius is MC, at N and O.

A *diameter* is any line VW, drawn through the centre M, and terminated by the opposite curves.

Conjugate diameter to another, is a line drawn through the centre, parallel to a tangent with either of the curves, at the extreme of the other diameter terminated by the curves.

Abscissa is when any diameter is continued within the curve, terminated by a double ordinate and the curve ; then the part within is called the abscissa.

2. To find the area of a rectangle, whose length is 9, and breadth 4 inches, or feet, &c.

$$9$$
$$4$$
$$\overline{\quad}$$

Ansr. 36
$$\overline{\quad}$$

3. To find the area of a rhombus, whose length is 6 chains, and perpendicular height 5.

$$6$$
$$5$$
$$\overline{\quad}$$

Ansr. 30
$$\overline{\quad}$$

Prob. 2. To find the Area of a Triangle.
Rule 1. Multiply the base by the perpendicular height, and half the product will be the area.
Rule 2. When the three sides only are given : Add the three sides together, and take half the sum ; from the half sum subtract each side separately ; multiply the half sum and the three remainders continually together ; and the square root of the last product will be the area of the triangle.
Ex. Required the area of the triangle whose base is 6 feet, and perpendicular height 5 feet.

$$6$$
$$5$$
$$\overline{\quad}$$

2) 30 (15 Ansr.

Prob. 3. To find one Side of a right-angled Triangle, having the other two Sides given.
The square of the hypotenuse is equal to both the squares of the two legs. Therefore,
1. To find the hypotenuse ; add the squares of the two legs together, and extract the square root of the sum.
2. To find one leg ; subtract the square of the other leg from the square of the hypotenuse, and extract the root of the difference.
Ex. 1. Required the hypotenuse of a right-angled triangle, whose base A B is 40, and perpendicular BC 30.

4	3
4	3
16	9
9	
25	(5 the square root of the sum of the two squares, being
25	the hypotenuse AC.

2 Y

2. What is the perpendicular of a right-angled triangle, whose base AB is 56, and hypotenuse, AC 65 ?

$$
\begin{array}{cc}
56 & 65 \\
56 & 65 \\
\hline
336 & 325 \\
280 & 390 \\
\hline
3136 & 4225 \\
 & 3136 \\
\end{array}
$$

1089 (33 The perpendicular, which is the root
9   of the remainder of the square of the
         hypotenuse AC, when the square
63 | 189   of AB has been subtracted.
3 | 189

Prob. 4. To find the Area of a Trapezoid.

Multiply the sum of the two parallel sides by the perpendicular distance between them, and half the product will be the area.

Ex. In a trapezoid, the parallel sides are AB 7, and CD 12, and the perpendicular distance AP or CN is 9 : required the area.

$$
\begin{array}{c}
7 \\
12 \\
\hline
19 \\
9 \\
\hline
171 \\
\end{array}
$$

85½ the area.

Prob. 5. To find the Area of a Trapezium.

Case for any trapezium.—Divide it into two triangles by a diagonal ; then find the areas of these triangles, and add them together.

Note. If two perpendiculars be let fall on the diagonal, from the other two opposite angles, the sum of these perpendiculars being multiplied by the diagonal, half the product will be the area of the trapezium.

Ex. To find the area of the trapezium ABCD, the diagonal AC being 42 , the perpendicular BF 18 , and the perpendicular DE 16.

$$
\begin{array}{c}
18 \\
16 \\
\hline
34 \\
42 \\
\hline
68 \\
136 \\
\end{array}
$$

2 | 1428

714 the answer.

Prob. 6. To find the Area of an Irregular Polygon.

Draw diagonals dividing the figure into trapeziums and triangles. Then find the areas of all these separately, and their sum will be the content of the whole irregular figure.

Es. To find the content of the irregular figure ABCDEF, in which are given the following diagonals and perpendiculars : namely,

$$
\begin{aligned}
c.a &= 10 \\
d.f &= 6 \\
c.i &= 4 \\
k.e &= 2 \\
m.f &= 3 \\
n.b &= 4
\end{aligned}
$$

For trapez. d c f e.		For trapez. c f a b	
ci.	4	n.b.	4
ke.	2	m.f.	3
	—		—
	6		7
df.	6	c.a.	10
2)	36	2)	70
	—		—
	18 contents.		35 contents.

18 contents d. c f e
35 ——— c. f. a b
—
53 contents of the irregular
— polygon.

Prob. 7. To find the Area of a Regular Polygon.

Rule. Multiply the perimeter of the figure, or sum of its sides, by the perpendicular falling from its centre upon one of its sides, and half the product will be the area.

Prob. 8. In a Circular Arc, having any two of the following lines, viz. the chord AB, the versed sine DP, the chord of half the arc AD, and the diameter, or the radius AC or CD given, to find the others.

If any two of these lines be given, two sides of one of the right-angled triangles, APC or APD, will be known, and from them the remaining side, and other lines in the arc, may be found by Prob. 3.

Suppose AB and PD be given, then, by Prob. 3., the half of AB, or AP is a mean proportional between DP and PC + CD; for PC + CD + PD is the diameter of the circle, half of which is the radius or CA, and by Prob. 3, $AC^2 - AP^2 = CP^2$, and $AP^2 + PD^2 = AD^2$.

Suppose CD and AB be given, then half of AB = AP, and CD = AC; therefore $\sqrt{CD^2 - AP^2} = CP$, and CD − CP = PD. $\sqrt{PD^2 + AP^2} = AD$.

Prob. 9. To find the Diameter and Circumference of a Circle, the one from the other.

Rule 1. As 7 is to 22, so is the diameter to the circumference.

As 22 is to 7, so is the circumference to the diameter.

Rule 2. As 113 is to 355, so is the diameter to the circumference.
As 355 is to 113, so is the circumference to the diameter

Rule 3. As 1 is to 3·1416, so is the diameter to the circumference.
As 3·1416 is to 1, so is the circumference to the diameter.

Ex. 1. To find the circumference of a circle, whose diameter AB is 10.

By *Rule* 1.

7 : 22 : : 10 : 31·42857
    10,
    ——
7) 220
   31$\frac{1}{7}$
or 31·42857 ans.

—————

By *Rule* 2.

113 : 355 : : 10 : 31$\frac{47}{113}$
    10
    ——
113) 3550 (31·41593
    ——   the ans.
    160
    ——
    470
    ——
    180
    ——
    670
    ——
    1050
    ——
    330
    ——

By *Rule* 3.

1 : 3·1416 : : 10 : 31·416
the circumference nearly,
the true circumference
being
31·415926535**8**979, &c.

So that the 2d rule is
nearest the truth.

2. To find the diameter when the circumference is 100.

By *Rule* 1.

$$22 : 7 : : 50 : \frac{7 \times 25}{11} = \frac{175·}{11} = 15 \tfrac{10}{11} = 15·90\overset{..}{90} \text{ ans}$$

By *Rule* 2.	By *Rule* 3

355 : 113 : : 50 : 15 4/5   3·1416 : 1 : : 50 : 15·9156
         50   50

355 | 5650   3·1416)50·000 (15·9156

71 | 1130 (15·9155   . ....) 18584

       420   2876

       650   49

       110   18

       390   2

       350

Prob. 10. To find the Length of any Arc of a Circle.

Rule 1. As 180 is to the number of degrees in the arc,
            So is 3·1416 times the radius, to its length.
            Or as 3 is to the number of degrees in the arc,
            So is ·05236 times the radius, to its length.

Ex. 1. To find the length of an arc ADB (Prob. 8,) of 30 degrees, the radius being 9 feet.,

              3·1416
                 9
As 180 : 30   ———
Or 6 : 1 : : 282744 : 4·7124
Or 3 : 30 : : ·05236 × 9 : 4·7124
             90

     4·7124 the answer.

Rule 2. From 8 times the chord of half the arc subtract the chord of the whole arc, and ⅓ of the remainder will be the length of the arc nearly.

Ex. 2. The chord AB (*Prob.* 8.) of the whole arc being 4·65874, and the chord AD of the half arc 2·34947 ; required the length of the arc.

          2·34947
             8

        18·79576
         4·65874

    3) 14·13702

     4·71234 answer.

Prob. 11. To find the Area of a Circle, the diameter or circumference being given.

Rule 1. Multiply half the circumference by half the diameter. Or, take ¼ of the product of the whole circumference and diameter.

Rule 2. Multiply the square of the diameter by ·7854.

Rule 3. Multiply the square of the circumference by ·07958.

Rule 4. As 14 is to 11, so is the square of the diameter to the area.

Rule 5. As 88 is to 7, so is the square of the circumference to the area.

Ex. To find the area of a circle whose diameter is 10, and circumference 314·159265

By Rule 1.	*By Rule* 2.	*By Rule* 4.
31·4159265		14 : 11 : : 100
10	·7854	11 area
————	100	———
4)314·159265	————	14 | 1100 | 78·57
	area 78·54	98
area 78·539816		———

```
                          120
                          112
                          ———
                           80
                           70
                          ———
                          100
                           98
                          ———
                            2
```

By Rule 3.
```
sq. circ. 986·96044
   invert.    85970
          ————————
            6908723
             888264
              49348
               7896
          ————————
          78·54231  area.
```

By Rule 5.
```
31·4159265 circum.
562951413 invert.
————————
 94247779
  3141593
  1256637
    31416
    15708
     2827
       63
       19
        2
————————
88 : 7 : : 986·96044
                  7
        ————————
     8 | 6908·72308
       ————————
    11 | 863·59038
       ————————
          78·50821
```

Prob. 12. To find the Area of the Sector of a Circle.

Rule 1. Multiply the radius, or half the diameter, by half the arc of the sector, for the area. Or take ¼ of the product of the diameter and arc of the sector.

Note. The arc may be found by problem 10.

Rule 2. As 360 is to the degrees in the arc of the sector, so is the whole area of the circle, to the area of the sector.

Ex. What is the area of the sector CAB, the radius being 10, and the chord AB 16.

$$100 = AC^2$$
$$64 = AE^2$$

$$36 \quad (6 = CE$$
$$10 = CD$$

$$4 = DE$$

$$16 = DE^2$$
$$64 = AE^2$$

$$80 \quad (8{\cdot}9442719 = AD.$$
$$8$$

$$71{\cdot}5541752$$
$$16$$

$$3) \quad 55{\cdot}5541752$$

$$2) \quad 18{\cdot}5180584 \text{ arc ADB}$$

$$9{\cdot}2590297 = \text{half arc}$$
$$10 = \text{radius}$$

$$92{\cdot}590297 \text{ answer.}$$

Prob. 13. To find the Area of a Segment of a Circle.

Rule. Find the area of the sector having the same arc with the segment, by the last problem.

Find the area of the triangle, formed by the chord of the segment and the two radii of the sector.

Then the sum of these two will be the answer when the segment is greater than a semicircle: but the difference will be the answer when it is less than a semicircle.

Ex. Required the area of the segment ACBD, its chord AB being 12, and the radius EA or CE 10.

$$100 \text{ AE}^2$$
$$36 \text{ AD}^2$$
$$\overline{}$$
$$64 \text{ DE}^2$$

its root   8 DE
from   10 CE
$$\overline{}$$
$$2 \text{ CD}$$

$$4 \text{ CD}^2$$
$$36 \text{ AD}^2$$
$$\overline{}$$
$$40 \text{ chord AC}^2$$

its root   6·324555 chord AC
                    8
$$\overline{}$$
$$50·596440$$
$$12·$$
$$\overline{}$$
$$3)38·59644$$
$$\overline{}$$
$$2)12·86548 \text{ arc ACB}$$
$$\overline{}$$
6·43274 ½ arc
10 radius
$$\overline{}$$
64·3274 area of sect. EACB
48·0000 area of triangle EAB
$$\overline{}$$
ans. 16·3274 area of segm. ACBA

6   AD
8   DE
—
48 area of △ EAB

Prob. 14.   To find the Area of a Circular Zone ADCBA.

Rule 1. Find the areas of the two segments AEB, DEC, and their difference will be the zone ADCB.

Rule 2.   To the area of the trapezoid DQP add the area of the small segment ADR; and double the sum for the area of the zone ADCB.

Prob. 15. To find the Area of a Circular Ring, or Space included between two Concentric Circles.

The difference between the two circles will be the ring. Or, multiply the sum of the diameters by their difference, and multiply the product by ·7854 for the answer.

Ex. The diameters of the two concentric circles being AB 10 and DG 6, required the area of the ring contained between their circumferences AEBA, and BFGD.

```
        10            ·7854
         6              64
       ----          -------
  sum  16            31416
  dif.  4            47124
       ----          -------
        64           50·2656 Ansr.
```

Prob. 16. To measure long Irregular Figures.

Take the breadth in several places at equal distances. Add all the breadths together, and divide the sum by the number of them, for the mean breadth; which multiply by the length for the area.

Ex. The breadths of an irregular figure, at five equi-distant places being AD 8·1, mP 7·4, nq 9·2, or 10·1, BC 8·6; and the length AB 39; required the area.

```
             8·1
             7·4
             9·2
            10·1
             8·6
           ------
     5)    43·4
           ------
             8·68
            39
           ------
          7812
          2604
           ------
          338·52 Ansr.
```

MENSURATION OF SOLIDS.

Prob. 1. To find the Solidity of a Cube.

Cube one of its sides for the contents; that is, multiply the side by itself, and that product by the side again.

Ex. If the side of a cube be 24 inches, what is its solidity or contents?

$$
\begin{array}{r}
24 \\
24 \\
\hline
96 \\
48 \\
\hline
576 \\
24 \\
\hline
2304 \\
1152 \\
\hline
13824 \text{ Ansr.}
\end{array}
$$

Prob. 2. To find the Solidity of a Parallelopipedon.

Multiply the length by the breadth, and the products by the depth or altitude.

Ex. Required the contents of the parallelopipedon whose length AB is 6, its breadth AC 2, and altitude BD 3.

$$
\begin{array}{r}
6 \\
2 \\
\hline
12 \\
3 \\
\hline
36 \text{ Ansr.}
\end{array}
$$

Prob. 3. To find the Solidity of any Prism.

Multiply the area of the base, or end, by the height, and it will give the contents.

Which rule will do, whether the prism be triangular or square, or pentagonal, &c. or round, as a cylinder.

Ex. What is the content of a triangular prism, whose length is 12, and each side of its equilateral base 8?

Area of base, $28 \times 12 = 336$ contents.

Prob. 4. To find the Convex Surface of a Cylinder.

Multiply the circumference by the height of the cylinder.

Prob. 5. To find the Convex Surface of a Right Cone.

Multiply the circumference of the base by the slant height, or length of the side, and half the product will be the surface.

Ex. If the diameter of the base be 5 feet, and the side of the cone 18, required the convex surface.

$$
\begin{array}{r}
3\cdot1416 \\
5 \\
\hline
15\cdot7080 \text{ circumf.} \\
18 \\
\hline
125664 \\
15708 \\
\hline
2\,)\,\overline{282\cdot744} \\
\hline
141\cdot372 \text{ Ansr.}
\end{array}
$$

Prob. 6. To find the Convex Surface of the Frustum of a Right Cone.
Multiply the sum of the perimeters of the two ends by the slant
height or side of the frustum, and half the product will be the surface.

Ex. If the circumferences of the two ends be 12·5 and 10·3, and
the slant height 14, required the convex surface of the frustum.

$$
\begin{array}{r}
12\text{·}5 \\
10\text{·}3 \\
\hline
22\text{·}8 \\
14 \\
\hline
912 \\
228 \\
\hline
2\ )\ 319\text{·}2 \\
\hline
159\text{·}6 \ \text{Ansr.}
\end{array}
$$

Prob. 7. To find the Solidity of a Cone, or any Pyramid.
Multiply the area of the base by the perpendicular height of the
area, and one-third of the product will be the contents.

Prob. 8. To find the Solidity of any Frustum of a Cone or Pyramid.
Rule. Add together the area of the base, the area of the upper
surface, and the mean proportional between those areas; take one-
third of this sum for the mean area, which multiplied by the height
will give the contents.—Or, for a cone, take the square of each diameter of the base and upper surface, and the product of these two diameters multiplied together; add these three sums together, and
multiply by ·2618 for the mean area, which multiply as before.

Or, if the circumferences be used in like manner, instead of their
diameters, the multiplier will be ·02654.

Ex. What is the content of the frustum of a cone, whose height
is 20 inches, and the diameters of its two ends 28 and 20 inches?

Area of base	615·79	28	28	20
Area of upper surface	314·16	28	20	20
Mean proportional	439·84			
		224	560	40C
	3) 1369·79	56	784	
			400	
	456·59	784		
	20		1744	
			2618	
	9131·80			
			13952	
			1744	
			10464	
			3488	
			456·5792	
			20	
			9131·5840 Ansr.	

Prob. 9. To find the Solidity of a Wedge. ·

To the length of the edge add twice the length of the back or base, and reserve the sum ; multiply the height of the wedge by the breadth of the base ; then multiply this product by the reserved sum, and one-sixth of the last product will be the contents.

Ex. What is the contents of a wedge, whose altitude AP is 14 inches, its edge AB 21 inches, and the length of its base DE 32 inches, and its breadth CD 4½ inches ?

```
    21              14
    32              4¼
    32              —
    —              56
    85              7
    —              —
                   63
                   85
                  ———
                   315
                   504
                  ———
        6 | 5355
          |  892·5 Ansr.
```

Prob. 10. To find the Solidity of a Prismoid.

Definition.—A prismoid differs only from the frustum of a pyramid, in not having its opposite ends similar planes.

Rule. Add into one sum, the areas of the two ends and four times the middle section parallel to them, and one-sixth of that sum will be a mean area ; and being multiplied by the height, will give the contents.

Note.—The length of the middle section is equal to half the sum of the lengths of the two ends ; and its breadth is equal to half the sum of the breadths of the two ends.

Ex. What are the contents of a prismoid whose ends are rectangles, the length and breadth of the one being 14 and 12 ; and the corresponding sides of the other 6 and 4, the perpendicular height being 30½ ?

```
    14          10          6
    12          8           4
   ———         ——          —
   168          80          24
   ———          4           —
               ———
                320
                168
                24
               ———
        6 ) 512
```

85⅓. mean area.
30½ height.

2560
42⅓

2602·6 Ansr.

Prob. 11. To find the Convex Surface of a Sphere or Globe.
Multiply its diameter by its circumference.

Note.—In like manner the convex surface of any zone or segment is found, by multiplying its height by the whole circumference of the sphere.

Ex. Required the convex superficies of a globe, whose diameter or axis is 24.

3·1416
24 diam.

125664
62832

75·3984 circumf.
24

3015936
1507968

1809·5616 Ansr.

Prob. 12. To find the Solidity of a Sphere or Globe.
Multiply the cube of the axis by ·5236·

Ex. What is the solidity of the sphere, whose axis is 12 ?

12
12

144
12

1728
·5236

10368
5184
3456
8640

904·7808 Ansr.

Prob. 13. To find the Solidity of a Spherical Segment.

To 3 times the square of the radius of its base add the square of its height; then multiply the sum by the height, and the product again by ·5236.

Ex. Required the contents of a spherical segment, its height AB being 4, and the radius of its base CD 8.

```
    8           4              ·5236
    8           4               832
   ──          ──           ─────────
   64          ¡6             10472
    3          192            15708
  ────        ────           41888
  192          208         ─────────
  ────           4          435·6352 Ansr.
                ───
                832
```

Prob. 14. To find the Solidity of a Spherical Zone or Frustum.

Add together the square of the radius of each end and ⅓ of the square of their distance, or the height; then multiply the sum by the said height, and the product again by 1·5708.

Ex. What is the solid contents of a zone, whose greater diameter is 12, the less 8, and the height 10 inches?

```
    6           4              10
    6           4              10
   ──          ──            ────
   36          16         3 ) 100
   ──          36            ────
               33⅓            33⅓
              ────
               85⅓
              1·5708
             ────────
              78540
             125664
               5236
             ────────
             134·0416
                 10
             ────────
```

1340·416 Ansr.

Prob. 15. To find the Surface of a Circular Spindle.

Multiply the length AB of the spindle by the radius OC of the revolving arc. Multiply also the said arc ACB by the central distance OE, or distance between the centre of the spindle and centre of the revolving arc. Subtract the latter product from the former, and multiply the remainder by 6·2832, for the surface.

Note. The same rule will serve for any segment or zone cut off perpendicular to the chord of the revolving arc, only using the particular length of the part, and the part of the arc which describes it, instead of the whole length and whole arc.

Ex. Required the surface of a circular spindle, whose length AB is 40, and its thickness CD 30 inches.

Here, by the remarks at pa. 688.

The chord $AC = \sqrt{AE^2 + CE^2} = \sqrt{20^2 + 15^2} = 25$,

and $2\ CE : AC :: AC : CO = \frac{25^2}{30} = 20\frac{5}{6}$,

hence $OE = OC - CE = 20\frac{5}{6} - 15 = 5\frac{5}{6}$.

Also, by problem 10, rule 2, pa. 693

25 AC
8
———
200
40 AB
———
3) 160
$53\frac{1}{3}$ arc ACB
———

Then, by our rule,

$20\frac{5}{6}$	$53\frac{1}{3}$
40	$5\frac{5}{6}$
800	$266\frac{2}{3}$
$33\frac{1}{3}$	$44\frac{4}{9}$
$833\frac{1}{3}$	$311\frac{1}{9}$
$311\frac{1}{9}$	

$522\frac{2}{9}$ or $522\dot{2}$ or $\frac{4700}{9}$

6·2832
———
10444
156666
4177777
10444444
313333333
———
3281·22666
———

Or thus,
6·2832
4700
———
439824
251328
———
9) 29531·04
———
3281·226 ans. nearly
———

Prob. 16. To find the Solidity of a Circular Spindle.

Multiply the central distance OE by half the area of the revolving segment ACBEA. Subtract the product from $\frac{1}{3}$ of the cube of EA, half the length of the spindle. Then multiply the remainder by 12·5664, or 4 times 3·1416, for the whole contents

Ex. Required the contents of the circular spindle, whose length AB is 40, and middle diameter CD 30.

By the work of the last problem,

we have OE =	$6\frac{5}{6}$	20 half length
and arc AC =	$26\frac{2}{3}$	20
and rad. OC =	$20\frac{5}{6}$	——
	——	400
	$533\frac{1}{3}$	20
	$22\frac{2}{9}$	——
	——	3) 8000
Sector OACB	$555\frac{5}{9}$	——
AE × OE=OAB	$116\frac{2}{3}$	$2666\frac{2}{3}$
	——	$1280\frac{2}{9}$
	2) $438\frac{8}{9}$	——
	——	$1386\frac{4}{9}$
$\frac{1}{2}$ seg. ACE	$219\frac{4}{9}$	——
OE	$5\frac{5}{6}$	or 1386·44
	——	4665·21 mult. inver.
	$1097\frac{2}{9}$	——
	183 nearly	138644
	——	27739
	$1280\frac{2}{9}$	6932
		832
		83
		5
		——
		17423·5 Ansr.

Prob. 17. To find the Solidity of the Middle Frustum or Zone of a Circular Spindle.

From the square of half the length of the whole spindle, take $\frac{1}{3}$ of the square of half the length of the middle frustum, and multiply the remainder by the said half length of the frustum.——Multiply the central distance by the revolving area, which generates the middle frustum.—Subtract this latter product from the former; and the remainder multiplied by 6·2832, or twice 3·1416, will give the contents.

Ex. Required the solidity of the frustum, whose length m n is 40 inches, also its greatest diameter EF is 32, and least diameter AD or BC 24.

Draw DG parallel to m n, then we have DG = $\frac{1}{2}$m n = 20, and $\quad EG = \frac{1}{2}EF - \frac{1}{2}AD = 4$,

chord $DE^2 = DG^2 + GE^2 = 416$,

and $\quad DE^2 \div EG = \dfrac{416}{4} = 104$ the diameter of the generating
circle.

or the radius OE = 52,

hence OI = 52 − 16 = 36 the central distance,

and $HI^2 = OH^2 - OI^2 = 52^2 - 36^2 = 1408$,

$\quad \frac{1}{3}DG^2 = \frac{1}{3}$ of 400 = $\quad$ .. $\quad$ .. $\quad$ $133\frac{1}{3}$,

$$\frac{\qquad\qquad}{1274\frac{2}{3}}$$

$$\text{DG} \quad .. \quad .. \quad 20$$

$$\frac{\qquad\qquad}{25493\frac{1}{3}}$$
$$\text{———— 1st. prod.}$$

$\quad GE \div 2\,OE = \dfrac{4}{104} = \dfrac{1}{26} = \cdot03846$ a ver. sine

Its tab. segment $\quad$ .. $\quad$ ·00994

but 104^2 $\quad$ is $\quad$. $\quad$ 10816

$$\frac{\qquad\qquad}{}$$

area of seg. DECGD $\quad$ 107·51104

m D × mn = 12 × 40 $\quad$ 480·

$$\frac{\qquad\qquad}{}$$

gener. area m DEC n $\quad$ 587·51104

$\qquad\qquad$ OI $\qquad$ 36

$$\frac{\qquad\qquad}{}$$

21150·39744 2d product

25493·33333 1st product

$$\frac{\qquad\qquad}{}$$

4342·93589

2382·6 $\quad$ mult. inv.

$$\frac{\qquad\qquad}{}$$

260576

8686

3474

130

9

$$\frac{\qquad\qquad}{}$$

27287·5 Ansr.

$$\frac{\qquad\qquad}{}$$

Prob. 18. To find the Superficies or Solidity of any Regular Body.

1. Multiply the tabular area (taken from the following table) by
he square of the linear edge of the body for the superficies.

2. Multiply the tabular solidity by the cube of the linear edge,
r the solid contents.

No. of Sides	Names	Surfaces	Solidities
4	Tetraedron	1·73205	0·11785
6	Hexaedron	6·00000	1·00000
8	Octaedron	3·46410	0·47140
12	Dodecaedron	20·64573	7·66312
20	Icosaedron	8·66025	2·18169

Ex. If the linear edge or side of a tetraedron be 3, required its surface and solidity.

The square of 3 is 9, and the cube 27,. Then,

tab. sur.	1·73205		0·11785 tab. sol.
	9		27
superf.	15·58845		82495
			23570
		solidity	3·18195

Prob. 19. To find the Surface of a Cylindrical Ring.

This figure being only a cylinder bent round into a ring, its surface and solidity may be found as in the cylinder, namely, by multiplying the axis, or length of the cylinder, by the circumference of the ring or section, for the surface; and by the area of a section, for the solidity. Or use the following rules :

For the surface.——To the thickness of the ring add the inner diameter ; multiply this sum by the thickness, and the product again by 9·8696, or the square of 3·1416.

Ex. Required the superficies of a ring, whose thickness AB is 2 inches, and inner diameter BC is 12 inches.

12	9·8696
2	28
14	789568
2	197392
28	276·3488 Ansr.

Prob. 20. To find the Solidity of a Cylindrical Ring.

To the thickness of the ring add the inner diameter ; then multiply the sum by the square of the thickness, and the product again by 2·4674, or ¼ of the square of 3·1416, for the solidity.

Ex. Required the solidity of the ring whose thickness is 2 inches, and its inner diameter 12.

12	2·4674
2	56
14	148044
4	123370
56	138·1744 Ansr.

USEFUL RECEIPTS.

———

Compounds of Metals.

Fusible Metal.—No. 1.

> 4 oz. of bismuth,
> 2½ oz. of lead, and
> 1½ oz. of tin.

Put the bismuth into a crucible, and, when it is melted, add the lead and tin. This will form an alloy fusible at the temperature of boiling water.

> No. 2.——1 oz. of zinc,
> 1 oz. of bismuth, and
> 1 oz. of lead.

This alloy is so very fusible, that it will remain in a state of fusion if put on a sheet of paper, and held over the flame of a candle or lamp.

> No. 3.——3 parts of lead,
> 2 parts of tin, and
> 5 parts of bismuth,

will form an alloy fusible at 197° Fahrenheit, peculiarly applicable to casting, or the taking of impressions from gems, seals, &c. In making casts with this and similar alloys, it is necessary to use the metal at as low a temperature as possible; otherwise, the water adhering to the things from which the casts are to be taken, forms vapour, and produces bubbles. The fused metal should be poured into a tea-cup, and allowed to cool, till just ready to set at the edges, when it must be poured into the mould. In taking impressions from gems, seals, &c. the fused alloy should be placed on paper or pasteboard, and stirred about till it has, by cooling, attained the consistence of paste, at which moment the die, gem, or seal should be stamped on it, and a very sharp impression will then be obtained.

Bath Metal, is a mixture of

> 4½ oz. of zinc, with
> 1 lb. of brass.

Brass is composed of

> 4½ lb. of copper, and
> 1½ lb. of zinc.

But brass that is to be cast into plates, from which pans and kettles are to be made, and wire is to be drawn, must, instead of using the zinc in a pure state, be composed of

> 56 lb. of the finest calamine, or ore of zinc, and
> 34 lb. of copper.

2 Z 2

Old brass, which has been frequently exposed to the action of fire, when mixed with the copper and calamine, renders the brass far more ductile, and fitter for the making of fine wire, than it would be without it ; but the German brass, particularly that of Nuremburgh, is, when drawn into wire, said to be far preferable to any made in England, for the strings of musical instruments.

Pinchbeck.—No. 1.
> 5 oz. of pure copper, and
> 1 oz. of zinc.

The zinc must not be added till the copper is in a state of fusion. Some use only half this quantity of zinc, in which proportion the alloy is more easily worked, especially in the making of jewellery.

> No. 2.——1 oz. of brass.
> 2 oz. of copper,

Fused together, under a coat of charcoal dust.

Prince's Metal.—No. 1.
> 3 oz. of copper, and
> 1 oz. of zinc.

> Or, 8 oz. of brass, and
> 1 oz. of zinc.

> No. 2.——4 oz. of copper, and
> 2 oz. of zinc.

In this last, the copper must be fused before the zinc is added. When they have combined, a very beautiful and useful alloy is formed, called Prince Rupert's metal.

Bell Metal.—No. 1.
> 6 parts of copper, and
> 2 parts of tin.

These proportions are the most approved, for bells, throughout Europe, and in China. In the union of the two metals, the combination is so complete, that the specific gravity of the alloy is greater than that of the two metals in an uncombined state.

> No. 2.——10 parts of copper, and
> 2 parts of tin.

It may, in general, be observed, that a less proportion of tin is used for making church-bells than clock bells ; and that a little zinc is added for the bells of repeating watches, and other small bells.

Tutania, or Britannia Metal.—No. 1.
> 4 oz. of plate brass, and
> 4 oz. of tin ; when in fusion, add
> 4 oz. of bismuth, and
> 4 oz. of regulus of antimony.

This is the composition, or hardening, that is to be added, at discretion, to melted tin, until it has acquired the requisite degree of colour and hardness.

No. 2.—Melt together, 2 lb. of plate brass,
> 2 lb. of tin,

       2 lb. of bismuth,
       2 lb. regulus of antimony,
       2 lb. of a mixture of copper and arsenic,
          either by cementation or melting.

This composition is to be added, at discretion, to melted tin.

    No. 3.——1 lb. of copper,
         1 lb. of tin, and
         2 lb. of regulus of antimony, with or without a
          little bismuth.

    No. 4.——8 oz. of shruff brass,
         2 lb. regulus of antimony, and
         10 lb. of tin.

German Tutania.

       2 drachms of copper,
       1 oz. regulus of antimony, and
       12 oz. of tin.

Spanish Tutania.—No. 1.

       8 oz. of scrap iron or steel,
       1 lb. of antimony, and
       3 oz. of nitre.

The iron or steel must be heated to a white heat, and the antimony and nitre must be added in small portions. Melt and harden 1 lb. of tin with 2 oz. of this compound.

No. 2.—Melt together, 4 oz. of antimony,
              1 oz. of arsenic, and
              2 lb. of tin.

The first of these Spanish alloys would be a beautiful metal if arsenic were added.

Engestroom Tutania.

       4 parts copper,
       8 parts regulus of antimony, and
       1 part bismuth,

When added to 100 parts of tin, this compound will be ready for use.

Queen's Metal.—No. 1.

       4½ lb. of tin,
       ½ lb. bismuth,
       ¼ lb. antimony, and
       ¼ lb. lead.

This alloy is used for the making of tea-pots, and other vessels, which are required to imitate silver. It retains its lustre to the last.

    No. 2.——100 lb. of tin,
         8 lb. regulus of antimony,
         1 lb. bismuth, and
         4 lb. copper.

White Metal.—No. 1.
>10 oz. of lead,
>6 oz. of bismuth, and
>4 drachms regulus of antimony.

No. 2.——2 lb. of regulus of antimony,
>8 oz. of brass, and
>10 oz. of tin.

Common Hard White Metal.
>1 lb. of brass,
>1¼ oz. of zinc, and
>½ oz. of tin.

Tombac.
>16 lb. of copper,
>1 lb. of tin, and
>1 lb. of zinc.

Red Tombac.
>5½ lb. of copper, and
>½ lb. of zinc.

The copper must be fused in the crucible before the zinc is added. This alloy is of a reddish colour, and possesses more lustre, and is of greater durability, than copper.

White Tombac. Copper and
Arsenic,
Put together in a crucible, and melted, covering the surface with muriate of soda, to prevent oxidation, will form a white brittle alloy.

Gun Metal.—No. 1. 112 lb. of Bristol brass,
>14 lb. zinc, and
>7 lb. block tin.

No. 2.——9 parts copper, and
>1 part tin.

The above compounds are those used in the manufacture of small and great brass guns, swivels, &c.

Blanched Copper. 8 oz. of copper, and
>½ oz. of neutral arsenical salt,
fused together, under a flux composed of calcined borax, charcoal dust, and fine powder glass.

Specula of Telescopes.
>7 lb. of copper, and when fused, add
>3 lb. of zinc, and
>4 lb. of tin.

These metals will combine and form a beautiful alloy of great lustre, and of a light yellow colour, fitted to be made into specula for telescopes. Mr. Mudge used only copper and grain tin, in the proportion of two pounds to fourteen and a half ounces.

Kustitien's Metal for Tinning.
>To 1 lb. of malleable iron, at a white heat, add

5 oz. of regulus of antimony, and
24 lb. of the purest Molucca tin.

This alloy polishes without the blue tint, and is free from lead or arsenic.

Metal for Flute-key Valves.

4 oz. lead, and
2 oz. antimony,

fused in a crucible, and cast into a bar, forms an alloy of considerable hardness and lustre. It is used by flute manufacturers (when turned into small buttons in a lathe,) for making valves to stop the key-holes of flutes.

Printers' Types.   10 lb. of lead, and
2 lb. of antimony.

The antimony must be thrown into the crucible when the lead is in a state of fusion. The antimony gives a hardness to the lead, without which, the type would speedily be rendered useless, in a printing press. Different proportions of lead, copper, brass, and antimony, frequently constitute this metal. Every artist has his own proportions, so that the same composition cannot be obtained from different foundries; each boasts of the superiority of his own mixture.

Small Types and Stereotype Plates.—No. 1.

9 lb. of lead, and when melted, add
2 lb. of antimony, and
1 lb. of bismuth.

This alloy expands as it cools, and is, therefore, well suited for the formation of small printing types, (particularly when many are cast together, to form stereotype plates,) as the whole of the mould is accurately filled with the alloy; consequently, there can be no blemish in the letters.

No. 2.——8 parts lead,
2 parts antimony, and
½ part tin.

For the manufacture of stereotype plates, plaster of Paris, of the consistence of a batter pudding before baking, is poured over the letter-press page, and worked into the interstices of the types, with a brush. It is then collected from the sides, by a slip of iron or wood, so as to lie smooth and compact. In about two minutes, the whole mass is hardened into a solid cake. This cake, which is to serve as the matrix of the stereotype plate, is now put upon a rack in an oven, where it undergoes great heat, so as to drive off superfluous moisture. When ready for use, these moulds, according to their size, are placed in flat cast-iron pots, and are covered over with another piece of cast-iron, perforated at each end, to admit the metallic composition intended for the preparation of the stereotype plates. The flat cast-iron pots are now fastened in a crane, which carries them steadily to the metallic-bath, or melting-pot, where they are immersed, and kept for a considerable time, until all the pores and crevices of the mould, are completely and accurately filled. When this has taken place, the pots are elevated from the bath, by working the crane, and are placed over a water-trough, to cool gradually. When cold, the whole is turned out of the pots, and the plaster being separated, by hammering and washing, the plates are ready for use, having received the most exact and perfect impression.

Metallic Casts from Engravings on Copper.

A most important discovery has lately been made, which promises to be of considerable utility in the fine arts : some very beautiful specimens of metallic plates, of a peculiar composition, have lately appeared, under the name of "cast engravings." This invention consists in taking moulds from every kind of engravings, with line, mezzotinto, or aquatinta, and pouring on this mould an alloy, in a state of fusion, capable of taking the finest impression. The obvious utility of this invention, as applicable to engravings which meet with a ready sale, and of which great numbers are required, will be incalculable ; as it will wholly prevent the expense of retracing, which forms so prominent a charge in all works of an extended sale. No sooner is one cast worn out, than another may be immediately procured from the original plate, so that every impression will be a proof. Thus, the works of our most celebrated artists may be handed down, *ad infinitum*, for the improvement and delight of future ages, and will afford, at the same time, the greatest satisfaction to every lover of the fine arts.

Common Pewter.   7 lb. of tin,
                     1 lb. of lead,
                     6 oz. of copper, and
                     2 oz. of zinc.

The copper must be fused before the other ingredients are added. This combination of metals will form an alloy of great durability and tenacity ; also, of considerable lustre.

Best Pewter.   100 parts tin, and
               17 parts regulus of antimony.

Hard Pewter.   12 lb. of tin,
               1 lb. regulus of antimony, and
               4 oz. of copper.

Common Solder.   2 lb. of lead, and
                 1 lb. of tin.

The lead must be melted before the tin is added. This alloy, when heated by a hot iron, and applied to the tinned iron with powdered rosin, acts as a cement or solder ; it is also used to join lead pipes, &c. &c.

Soft Solder.   2 lb. of tin, and
              1 lb. of lead.

Solder for Steel Joints.
               19 dwts. of fine silver,
               1 dwt. copper, and
               2 dwts. brass,

melted together under a coat of charcoal dust. This solder possesses several advantages over the usual zinc soda, or brass, when employed in soldering cast steel, &c. as it fuses with less heat, and its whiteness has a better appearance than brass.

Silver Solder for Jewellers.

19 dwts. of fine silver,
1 dwt. copper, and
10 dwts. brass.

Silver Solder for Plating.

10 dwts. brass, and
1 oz. pure silver.

Gold Solder.   12 dwts. pure gold,
2 dwts. pure silver, and
4 dwts. copper.

Brass Solder for Iron.—Thin plates of brass are to be melted between the pieces that are to be joined. If the work be very fine, as when two leaves of a broken saw are to be brazed together, cover it with pulverized borax, melted with water, that it may incorporate with the brass powder, which is added to it: the piece must be then exposed to the fire, without touching the coals, and heated till the brass is seen to run.

Bronze.——7 lbs. pure copper,
3 lbs. zinc, and
2 lbs. tin.

The copper must be fused before the other ingredients are added. These metals, when combined, form the bronze so much used, both in ancient and modern times, in the formation of busts, medals, and statues.

Composition of ancient Statues.

According to Pliny, the metal used by the Romans for their statues, and for the plates on which they engraved inscriptions, was composed in the following manner. They first melted a quantity of copper, into which they put 1-3d of its weight of old copper, which had been long in use ; to every 100lbs. weight of this mixture, they added 12½ lbs. of an alloy composed of equal parts of lead and tin.

Mock Platina.—Melt together
8 oz. of brass, and
5 oz. of zinc.

Useful alloy of Gold with Platinum.

$7\frac{1}{4}$ dr. pure gold, and
⅛ dr. platinum.

The platinum must be added when the gold is perfectly melted. The two metals will combine intimately, forming an alloy rather whiter than pure gold, but remarkably ductile and elastic ; it is also less perishable than pure gold, or jeweller's gold : but more readily fusible than that metal.

These excellent qualities must render this alloy an object of great interest to workers in metals. For *springs*, where steel cannot be used, it will prove exceedingly advantageous.

It is a curious circumstance, that the alloy of gold and platinum is soluble in nitric acid, which does not act on either of the metals in a separate state. It is remarkable, too, that the alloy has very nearly the colour of platinum, even when composed of eleven parts of gold to one of the former metal.

Ring-gold.   6 dwts. 12 grs. Spanish copper,
                        3 dwts. 16 grs. fine silver, and
                        1 oz. 5 dwts. gold coin.

Gold from 35s. to 40s. per ounce.
                        8 oz. 8 dwts. Spanish copper,
                        10 dwts. fine silver, and
                        1 oz. gold coin.

Manheim-gold, or Similor.
                        3½ oz. of copper,
                        1¼ oz. of brass, and
                        15 gr. of pure tin.

Gilding-Metal.   4 parts of copper,
                        1 part of Bristol old brass, and
                        14 oz. of tin to every pound of copper.

For common jewellery. 3 parts of copper,
                        1 part of Bristol old brass, and
                        4 oz. of tin to every pound of copper.

If this alloy is for fine polishing, the tin may be omitted, and a mixture of lead and antimony substituted. Paler polishing metal is made by reducing the copper to two or to one part.

Yellow Dipping Metal.—No 1.
                        2 parts of Cheadle brass,
                        1 part of copper, with a little
                        Bristol old brass, and
                        ¼ oz. of tin to every pound of copper.

This alloy is almost of the colour of gold coin. Cheadle brass is the darkest, and gives the metal a greenish hue. Old Bristol brass is pale and yellow.

            No. 2.—1 lb. of copper, and
                        5 oz. of zinc.

The copper should be tough cake, and not tile.

When antimony is used instead of tin, it should be in smaller quantity, or the metal will be brittle.

Imitation of Silver. ¼ oz. of tin, and
                        1 lb. of copper,

Will make a pale bell-metal, which will roll and ring very near to sterling silver.

PREPARATION OF FOILS.

Foils are thin plates or leaves of metal that are put under stones, or compositions in imitation of stones, when they are set.

The intention of foils is either to increase the lustre or play of the stones, or more generally improve the colour, by giving an additional force to the tinge, whether it be natural or artificial, by that of a ground of the same hue, which the foil is in this case made to be.

There are consequently two kinds of foils ; the one is colourless, where the effect of giving lustre or play to the stone is produced by the polish of the surface, which makes it act as a mirror, and, by reflecting the light, prevents that deadness which attends the having a duller ground under the stone, and brings it, by the double refraction of the light that is caused, nearer to the effect of the diamond. The other is coloured with some pigment or stain of the same hue as the stone, or of some other which is intended to modify and change the hue of the stone in some degree ; as, where a yellow foil may be put under green, which is too much inclining to the blue, or under crimson, where it is desired to have the appearance more orange or scarlet.

Foils may be made of copper or tin ; and silver has been sometimes used, with which it has been advised, for some purposes, to mix gold, but the expense of either is needless, as copper may be made to answer the same end.

To prepare Copper for Foils.—Where coloured foils are wanted, copper may therefore be best used, and may be prepared for the purpose by the following means.

Take copper plates beaten to a proper thickness, and pass them betwixt a pair of fine steel rollers very close set, and draw them as thin as is possible to retain a proper tenacity. Polish them with very fine whiting, or rotten-stone, till they shine, and have as much brightness as can be given them, and they will then be fit to receive the colour.

To whiten Foils.—Where the yellow, or rather orange-colour of the ground would be injurious to the effect, as in the case of purples, or crimson red, the foils should be whitened, which may be done by the following manner.

Take a small quantity of silver, and dissolve it in *aqua-fortis*, and then put bits of copper into the solution, and precipitate the silver ; which being done, the fluid must be poured off, and fresh water added to it, to wash away all the remainder of the first fluid ; after which the silver must be dried, an equal weight of cream of tartar and common salt must then be ground with it, till the whole is reduced to a very fine powder ; and with this mixture the foils, being first slightly moistened, must be rubbed by the finger, or a bit of linen rag, till they be of the degree of whiteness desired ; after which, if it appear to be wanted, the polish must be refreshed.

The tin-foils are only used in the case of colourless stones, where quicksilver is employed; and they may be drawn out by the same rollers, but need not be further polished, so that effect is produced by other means in this case.

Foils for crystals, pebbles, or paste, to give the lustre and play of diamonds.—The manner of preparing foils, so as to give colourless stones the greatest degree of play and lustre, is by raising so high a polish or smoothness on the surface, as to give them the effect of a mirror, which can only be done, in a perfect manner, by the use of quicksilver, applied in the same general way as in the case of looking-glass. The method by which it may be best performed is as follows.

Take leaves of tin, prepared in the same manner as for silvering looking-glasses, and cut them into small pieces of such size as to cover the surface of the sockets of the stones that are to be set. Lay three of these then, one upon another, and having moistened the inside of the socket with thin gum-water, and suffered it to become again so dry, that only a slight stickiness remains, put the three pieces of leaves, lying on each other, into it, and adapt them to the surface in as even a manner as possible. When this is done, heat the socket, and fill it with warm quicksilver, which must be suffered to continue in it three or four minutes, and then gently poured out. The stone must then be thrust into the socket, and closed with it, care having been taken to give such room for it that it may enter without stripping off the tin and quicksilver from any part of the surface. The work should be well closed round the stone, to prevent the tin and quicksilver contained in the socket from being shaken out by any violence.

The lustre of stones set in this manner, will continue longer than when they are set in the common way, as the cavity round them being filled, there will be no passage found for moisture, which is so injurious to the wear of stones treated in any other way.

This kind of foil likewise gives some lustre to glass or other transparent matter, which has little of itself; but to stones or pastes, that have some share of play, it gives a most beautiful brilliance.

To colour Foils.—Two methods have been invented for colouring foils: the one by tinging the surface of the copper of the colour required by means of smoke, the other by staining or painting it with some pigment or other colouring substance.

The colours used for painting foils may be tempered with either oil, water rendered duly viscid by gum-arabic, size, or varnish. Where deep colours are wanted, oil is most proper, because some pigments become wholly transparent in it, as lake, or Prussian blue; the yellow and green may be better laid on in varnish, as these colours may be had in perfection from a tinge wholly dissolved in spirit of wine, in the same manner as in the case of lacquers; and the most beautiful green is to be produced by distilled verdigris, which is apt to lose its colour and turn black with oil. In common cases, however, any of the colours may be, with the least trouble, laid on with isinglass size, in the same manner as the glazing colours used in miniature painting.

Ruby-Colours.—For red, where the ruby is to be imitated, a little lake used in isinglass size, carmine, or shell-lac varnish, is to be employed, if the glass or paste be of a full crimson, verging towards the purple; but if the glass incline to the scarlet, or orange, very bright lake (that is, not purple) may be used alone in oil.

Garnet Red.—For the garnet red, dragon's blood dissolved in seed-lac varnish may be used; and for the vinegar garnet, the orange-lake tempered with shell-lac varnish, will be found excellent.

Amethyst.—For the amethyst, lake, with a little Prussian blue, used with oil, and very thinly spread on the foil, will completely answer the end.

Blue.—For blue, where a deep colour, or the effect of the sapphire is wanted, Prussian blue, that is not too deep, should be used in oil, and it should be spread more or less thinly on the foil according to the lightness or deepness of the colour required.

Eagle Marine.—For the eagle-marine, common verdigris, with a little Prussian blue, tempered in shell-lac varnish.

Yellow.—Where a full yellow is desired, the foil may be coloured with a yellow lacquer, laid on as for other purposes; and for the slighter colour of topazes, the burnish and foil itself will be sufficiently strong without any addition.

Green.—For green, where a deep hue is required, the crystals of verdigris, tempered in shell-lac varnish, should be used, but where the emerald is to be imitated, a little yellow lacquer should be added, to bring the colour to a truer green, and less verging to the blue.

Other Colours.—The stones of more diluted colour, such as the amethyst, topaz, vinegar-garnet, and eagle-marine, may be very cheaply imitated by transparent white glass or paste, even without foils. This is to be done, by tempering the colours above enumerated with turpentine and mastic, and painting the socket in which the counterfeit stone is to be set with the mixture, the socket and stone itself being previously heated. In this case, however, the stone should be immediately set, and the socket closed upon it before the mixture cools and grows hard. The orange-lake above-mentioned was invented for this purpose, in which it has a beautiful effect, and was used with great success by a considerable manufacturer. The colour it produces is that of the vinegar-garnet, which it affords with great brightness. The colours before directed to be used in oil should be extremely well ground in oil of turpentine, and tempered with nut or poppy-oil; or, if time can be given for their drying, with strong fat oil; diluted with spirit of turpentine, which will gain a fine polish of itself.

The colours used in varnish should be likewise thoroughly well ground and mixt; and in the case of the dragon's blood in the seed-lac varnish and the lacquer, the foils should be warmed before they are laid out. All the mixtures should be laid on the foils with a broad soft brush, which must be passed from one end to the other, and no part should be crossed, or twice gone over, or, at least, not till the first coat can be dry; when, if the colour do not lie strong enough, a second coat may be given.

GILDING, SILVERING, AND TINNING.

Gold powder for Gilding.—Gold powder may be prepared in 3 different ways :—1st, put into an earthen mortar some gold-leaf, with a little honey, or thick gum-water, and grind the mixture till the gold is reduced to extremely minute particles. When this is done, a little warm water will wash out the honey or gum, leaving the gold behind in a pulverulent state.

2nd.—Dissolve pure gold (or the leaf), in nitro-muriatic acid, and then to precipitate it by a piece of copper, or by a solution of sulphate of iron. The precipitate (if by copper,) must be digested in distilled vinegar, and then washed, (by pouring water over it repeatedly,) and dried. This precipitate will be in the form of a very fine powder : it works better, and is more easily burnished than gold leaf ground with honey as above.

And 3d, or the best method of preparing gold powder, is by heating a prepared amalgam of gold, in an open clean crucible, and continuing the strong heat until the whole of the mercury is evaporated; at the same time constantly stirring the amalgam with a glass rod. When the mercury has completely left the gold, the remaining powder is to be ground in a Wedgewood's mortar, with a little water, and afterwards dried. It is then fit for use.

Although the last mode of operating has been here given, the operator cannot be too much reminded of the danger attending the sublimation of mercury. In the small way here described, it is impossible to operate without danger ; it is therefore better to prepare it according to the former directions, than to risk the health by the latter.

To cover Bars of Copper, &c. with Gold, so as to be rolled out into Sheets.—This method of *gilding* was invented by Mr. Turner, of Birmingham. Mr. Turner first prepares ingots or pieces of copper or brass, in convenient lengths and sizes. He then cleans them from impurity, and makes their surfaces level, and prepares plates of pure gold, or gold mixed with a portion of alloy, of the same size as the ingots of metal, and of suitable thickness. Having placed a piece of gold upon an ingot intended to be plated, he hammers and compresses them both together, so that they may have their surfaces as nearly equal to each other as possible ; and then binds them together with wire, in order to keep them in the same position during the process required to attach them. Afterwards he takes silver filings, which he mixes with borax, to assist the fusion of the silver. This mixture he lays upon the edge of the plate of gold, and next to the ingot of metal. Having thus prepared the two bodies, he places them on a fire in a stove or furnace, where they remain until the silver and borax placed along the edges of the metals melt, and until the adhesion of the gold with the metal is perfect. He then takes the ingot carefully out of the stove. By this process the ingot is plated with gold, and prepared ready for rolling into sheets.

To Gild in Colours.—The principal colours of gold for gilding are red, green, and yellow. These should be kept in different amalgams. The part which is to remain of the first colour, is to be stopped off with a composition of chalk and glue; the variety required is produced by gilding the unstopped parts with the proper amalgam, according to the usual mode of gilding.

Sometimes the amalgam is applied to the surface to be gilt, without any quicking, by spreading it with aqua-fortis; but this depends on the same principle as a previous quicking.

Grecian Gilding.—Equal parts of sal-ammoniac and corrosive sublimate, are dissolved in spirit of nitre, and a solution of gold made with this menstruum. The silver is brushed over with it, which is turned black, but on exposure to a red heat it assumes the colour of gold.

To dissolve Gold in Aqua-Regia.—Take an aqua-regia, composed of two parts of nitrous acid, and one of marine acid; or of one part of sal-ammoniac, and four parts of aqua-fortis; let the gold be granulated, put into a sufficient quantity of this menstruum, and exposed to a moderate degree of heat. During the solution, an effervescence takes place, and it acquires a beautiful yellow colour, which becomes more and more intense, till it has a golden or even orange colour. When the menstruum is saturated, it is very clear and transparent.

To gild Iron or Steel with a solution of Gold.—Make a solution of 8 ounces of nitre and common salt, with 5 ounces of crude alum in a sufficient quantity of water; dissolve half an ounce of gold thinly plated and cut; and afterwards evaporate to dryness. Digest the residuum in rectified spirit of wine or æther, which will perfectly abstract the gold. The iron is brushed over with this solution and becomes immediately gilt.

To Gild, by dissolving Gold in Aqua-Regia.—Fine linen rags are soaked in a saturated solution of gold in aqua-regia, gently dried, and afterwards burnt to tinder. The substance to be gilt must be well polished; a piece of cork is first dipped into a solution of common salt in water, and afterwards into the tinder, which is well rubbed on the surface of the metal to be gilt, and the gold appears in all its metallic lustre.

Amalgam of Gold in the large way.—A quantity of quicksilver is put into a crucible or iron ladle, which is lined with clay, and exposed to heat till it begins to smoke. The gold to be mixed should be previously granulated, and heated red hot, when it should be added to the quicksilver, and stirred about with an iron rod till it is perfectly dissolved. If there should be any superfluous mercury, it may be separated by passing it through clean soft leather; and the

remaining amalgam will have the consistence of butter, and contain
about 3 parts of mercury to 1 of gold.

To Gild by Amalgamation.—The metal to be gilt is previously
well cleaned on its surface, by boiling in a weak pickle, which is a
very dilute nitrous acid. A quantity of aqua-fortis is poured into an
earthen vessel, and quicksilver put therein ; when a sufficient quantity of mercury is dissolved, the articles to be gilt are put into the
solution, and stirred about with a brush till they become white. This
is called quicking. But, as during quicking by this mode, a noxious
vapour continually arises, which proves very injurious to the health
of the workmen, they have adopted another method, by which they,
in a great measure, avoid that danger. They now dissolve the quicksilver in a bottle containing aqua-fortis, and leave it in the open air
during the solution, so that the noxious vapour escapes into the air.
Then a little of this solution is poured into a bason, and with a brush
dipped therein, they stroke over the surface of the metal to be gilt,
which immediately becomes quicked. The amalgam is now applied by
one of the following methods :—

1st. By proportioning it to the quantity of articles to be gilt, and
putting them into a white hat together, working them about with a
soft brush, till the amalgam is uniformly spread.

Or, 2dly. By applying a portion of the amalgam upon one part,
and spreading it on the surface, if flat, by working it about with a
harder brush.

The work thus managed is put into a pan, and exposed to a gentle
degree of heat ; when it becomes hot, it is frequently put into a hat,
and worked about with a painter's large brush, to prevent an irregular
dissipation of the mercury, till, at last, the quicksilver is ˙entirely
dissipated by a repetition of the heat, and the gold is attached to the
surface of the metal. This gilt surface is well cleaned by a wire
brush, and then artists heighten the colour of the gold by the application of various compositions ; this part of the process is called
COLOURING.

To gild Glass and Porcelain. No. 1.—Drinking, and other glasses
are sometimes gilt on their edges. This is done, either by an adhesive varnish or by heat. The varnish is prepared by dissolving in
boiled linseed oil an equal weight either of copal or amber. This is
to be diluted by a proper quantity of oil of turpentine, so as to be applied as thin as possible to the parts of the glass intended to be gilt.
When this is done, which will be in about twenty-four hours, the
glass is to be placed in a stove, till it is so warm as almost to burn
the fingers when handled. At this temperature, the varnish will
become adhesive, and a piece of leaf gold, applied in the usual way,
will immediately stick. Sweep off the superfluous portions of the leaf,
and when quite cold, it may be burnished, taking care to interpose
a piece of very thin paper (India paper) between the gold and the
burnisher. If the varnish is very good, this is the best method of
gilding glass, as the gold is thus fixed on more evenly than in any
other way.

No. 2.—It often happens, when the varnish is but indifferent, that by repeated washing the gold wears off; on this account the practice of burning it in is sometimes had recourse to.

For this purpose, some gold powder is ground with borax, and in this state applied to the clean surface of the glass, by a camel's hair pencil; when quite dry, the glass is put into a stove heated to about the temperature of an annealing oven; the gum burns off, and the borax, by vitrifying, cements the gold with great firmness to the glass; after which it may be burnished. The gilding upon porcelain is in like manner fixed by heat and the use of borax; and this kind of ware being neither transparent nor liable to soften, and thus to be injured in its form in a low red heat, is free from the risk and injury which the finer and more fusible kinds of glass are apt to sustain from such treatment. Porcelain and other wares may be platinised, silvered, tinned, and bronzed, in a similar manner.

To Gild Leather.—In order to impress gilt figures, letters, and other marks upon leather, as on the covers of books, edgings for doors, &c. the leather must first be dusted over with very finely powdered yellow resin, or mastich gum. The iron tools or stamps are now arranged on a rack before a clear fire, so as to be well heated, without becoming red hot. If the tools are *letters*, they have an alphabetical arrangement on the rack. Each letter or stamp must be tried as to its heat, by imprinting its mark on the raw side of a piece of waste leather. A little practice will enable the workman to judge of the heat. The tool is now to be pressed downwards on the gold leaf; which will of course be indented, and shew the figure imprinted on it. The next letter or stamp is now to be taken and stamped in like manner and so on with the others; taking care to keep the letters in an even line with each other, like those in a book. By this operation the resin is melted; consequently the gold adheres to the leather: the superfluous gold may then be rubbed off by a cloth; the gilded impressions remaining on the leather. In this, as in every other operation, adroitness is acquired by practice.

The cloth alluded to should be slightly greasy, to retain the gold wiped off; (otherwise there will be a great waste in a few months,) the cloth will thus be soon completely saturated or loaded with the gold. When this is the case, these cloths are generally sold to the refiners, who burn them and recover the gold. Some of these afford so much gold by burning, as to be worth from a guinea to a guinea and a half.

To Gild Writings, Drawings, &c. on Paper or Parchment.—Letters written on vellum or paper are gilded in three ways: in the first, a little size is mixed with the ink, and the letters are written as usual; when they are dry, a slight degree of stickiness is produced by breathing on them, upon which the gold leaf is immediately applied, and by a little pressure may be made to adhere with sufficient firmness. In the second method, some white lead or chalk is ground up with strong size, and the letters are made with this by means of a brush: when the mixture is almost dry, the gold leaf may be laid on, and afterwards burnished. The last method is to mix up some

3 A

gold powder with size, and to form the letters of this by means of a brush. It is supposed that this latter method was that used by the monks in illuminating their missals, psalters, and rubrics.

To Gild the edges of Paper.—The edges of the leaves of books and letter paper are gilded whilst in a horizontal position in the book-binder's press, by first applying a composition formed of four parts of Armenian bole, and one of candied sugar, ground together with water to a proper consistence, and laid on by a brush with the white of an egg. This coating, when nearly dry, is smoothed by the bur-nisher ; which is generally a crooked piece of agate, very smooth, and fixed in a handle. It is then slightly moistened by a sponge dip-ped in clean water, and squeezed in the hand. The gold leaf is now taken up on a piece of cotton, from the leathern cushion, and applied on the moistened surface. When dry, it is to be burnished by rubbing the agate over it repeatedly from end to end, taking care not to wound the surface by the point of the burnisher. A piece of silk or India paper is usually interposed between the gold and the burnisher.

Cotton wool is generally used by bookbinders to take the leaf up from the cushion ; being the best adapted for the purpose on ac-count of its pliability, smoothness, softness, and slight moistness.

To gild Silk, Satin, Ivory, &c. by Hydrogen Gas. No. 1.—Im-merse a piece of white satin, silk, or ivory in a solution of nitro-mu-riate of gold, in the proportion of one part of the nitro-muriate to three of distilled water. Whilst the substance to be gilded is still wet, immerse it in a jar of hydrogen gas : it will soon be covered by a complete coat of gold.

No. 2.—The foregoing experiment may be very prettily and ad-vantageously varied as follows :—Paint flowers or other ornaments with a very fine camel hair pencil, dipped in the above-mentioned solution of gold, on pieces of silk, satin, &c. &c. &c. and hold them over a Florence flask, from which hydrogen gas is evolved, during the decomposition of the water by sulphuric acid and iron filings. The painted flowers, &c. in a few minutes, will shine with all the splen-dour of the purest gold. A coating of this kind will not tarnish on exposure to the air, or in washing.

Oil gilding on Wood.—The wood must first be covered, or primed, by two or three coatings of boiled linseed oil and carbonate of lead, in order to fill up the pores, and conceal the irregularities of the sur-face, occasioned by the veins in the wood. When the priming is quite dry, a thin coat of gold-size must be laid on. This is prepared by grinding together some red oxide of lead with the thickest drying oil that can be procured, and the older the better, that it may work freely : it is to be mixed, previously to being used, with a little oil of turpentine, till it is brought to a proper consistence. If the gold-size is good, it will be sufficiently dry in twelve hours, more or less, to allow the artist to proceed to the last part of the process, which is the application of the gold. For this purpose, a leaf of gold is spread on a cushion (formed by a few folds of flannel secured on a piece of wood, about eight inches square, by a tight covering of leather), and is cut into strips of a proper size by a blunt pallet knife; each strip being then taken upon the point of a fine brush, is applied to the part intended to be gilded, and is then gently pressed down by a ball of soft cotton; the gold immediately adheres to the sticky surface of the size, and after a few minutes, the dexterous application of a large camel's hair brush sweeps away the loose particles of the gold leaf without disturbing the rest. In a day or two the size will be completely dried, and the operation will be finished.

The advantages of this method of gilding are, that it is very simple, very durable, and not readily injured by changes of weather, even when exposed to the open air; and when soiled it may be cleaned by a little warm water and a soft brush; its chief employment is in out-door work. Its disadvantage is, that it cannot be burnished, and therefore wants the high lustre produced by the following method.

To Gild by burnishing.—This operation is chiefly performed on picture frames, mouldings, beadings, and fine stucco work. The surface to be gilt must be carefully covered with a strong size, made by boiling down pieces of white leather, or clippings of parchment, till they are reduced to a stiff jelly; this coating being dried, eight or ten more must be applied, consisting of the same size, mixed with fine Paris plaster or washed chalk; when a sufficient number of layers have been put on, varying according to the nature of the work, and the whole is become quite dry, a moderately thick layer must be applied, composed of size and Armenian bole, or yellow oxide of lead: while this last is yet moist, the gold leaf is to be put on in the usual manner; it will immediately adhere on being pressed by the cotton ball, and before the size is become perfectly dry, those parts which are intended to be the most brilliant are to be carefully burnished by an agate or dog's tooth fixed in a handle.

In order to save the labour of burnishing, it is a common, but bad practice, slightly to burnish the brilliant parts, and to deaden the rest by drawing a brush over them dipped in size; the required contrast between the polished and the unpolished gold is indeed thus obtained; but the general effect is much inferior to that produced in the regular way, and the smallest drop of water falling on the sized part occasions a stain. This kind of gilding can only be applied on in-door work; as rain, and even a considerable degree of dampness, will occasion the gold to peel off. When dirty, it may be cleaned by a soft brush, with hot spirit of wine, or oil of turpentine.

To Gild Copper, &c. by Amalgam.—Immerse a very clean bright piece of copper in a diluted solution of nitrate of mercury. By the affinity of copper for nitric acid, the mercury will be precipitated: now spread the amalgam of gold rather thinly over the coat of mercury just given to the copper. This coat unites with the amalgam, but of course will remain on the copper. Now place the piece or pieces so operated on, in a clean oven or furnace, where there is no smoke. If the heat is a little greater than 66°, the mercury of the amalgam will be volatilised, and the copper will be beautifully gilt.

In the large way of gilding, the furnaces are so contrived that the

3 A 2

· with aqua fortis, and heating
⁻: when the gold may be

' pure silver in
ch add ¼lb.
.f sublimate.
fortis ; preci-
 J ounces of coin-
riol, and ¼ ounce

.ne stone with a mul-
.obed over with a suffi-
a proper degree of heat.
nre, and dipped into weak

nation—Silver will not attach
unless it be first gilt. The pro-
s, only no acid should be used.

.ion salt,
alum, and
of silver, precipitated from the nitrous
.d by copper.
.aste with a little water. This is to be rub-
. be silvered with a cork, &c.
ة pure silver in aqua fortis, and precipitate the
.ion salt; make this precipitate into a paste, by
.ore salt and cream of tartar. It is applied as in the

Copper Ingots.—The principal difficulties in plating cop-
are, to bring the surfaces of the copper and silver into
the same time, and to prevent the copper from scaling ; for
urposes fluxes are used. The surface of the copper on which
ver is to be fixed must be made flat by filing, and should be left
ι. The silver is first annealed, and afterwards pickled in weak
it of salt ; it is planished, and then scraped on the surface to be
.ed on the copper. These prepared surfaces are annointed with
. solution of borax, or strewed with fine powdered borax itself, and
then confined in contact with each other, by binding wire. When
they are exposed to a sufficient degree of heat, the flux causes the
surfaces to fuse at the same time, and after they become cold, they
are found firmly united.

Copper may likewise be plated by heating it, and burnishing leaf-
silver upon it ; so may iron and brass. This process is called FRENCH
PLATING.

To separate the Silver from Plated Copper.—This process is applied to recover the silver from the plated metal, which has been
rolled down for buttons, toys, &c. without destroying any large por-

volatilised mercury is again condensed, and preserved for further use, so that there is no loss in the operation. There is also a contrivance by which the volatile particles of mercury are prevented from injuring the gilders.

To Gild Steel.—Pour some of the ethereal solution of gold into a wine glass, and dip therein the blade of a new pen-knife, lancet, or razor; withdraw the instrument, and allow the ether to evaporate. The blade will be found to be covered by a very beautiful coat of gold. A clean rag, or small piece of very dry sponge, may be dipped in the ether, and used to moisten the blade, with the same result.

In this case there is no occasion to pour the liquid into a glass, which must undoubtedly lose by evaporation; but the rag or sponge may be moistened by it, by applying either to the mouth of the phial. This coating of gold will remain on the steel for a great length of time, and will preserve it from rusting.

This is the way in which swords and other cutlery are ornamented. Lancets too are in this way gilded with great advantage, to secure them from rust.

To heighten the colour of Yellow Gold.

> 6 oz. saltpetre,
> 2 oz. copperas,
> 1 oz. white vitriol, and
> 1 oz. alum.

If it be wanted redder, a small portion of blue vitriol must be added. These are to be well-mixed, and dissolved in water as the colour is wanted.

To heighten the colour of Green Gold.

> 1 oz. 10 dwts. saltpetre,
> 1 oz. 4 dwts sal ammoniac,

paste, made of powdered sal ammoniac, with aqua fortis, and heating it till the matter smokes, and is nearly dry; when the gold may be separated by rubbing it with a scratch-brush.

To Silver by Heat. No. 1.—Dissolve an ounce of pure silver in aqua fortis, and precipitate it with common salt; to which add ½lb. of sal ammoniac, sandiver, and white vitriol, and ½ oz. of sublimate.

No. 2.—Dissolve an ounce of pure silver in aqua fortis; precipitate it with common salt, and add, after washing, 6 ounces of common salt, 3 ounces each of sandiver and white vitriol, and ¼ ounce of sublimate.

These are to be ground into a paste upon a fine stone with a muller; the substance to be silvered must be rubbed over with a sufficient quantity of the paste, and exposed to a proper degree of heat. Where the silver runs, it is taken from the fire, and dipped into weak spirit of salt to clean it.

Silvering on Gilt Work, by Amalgamation—Silver will not attach itself to any metal by amalgamation, unless it be first gilt. The process is the same as gilding in colours, only no acid should be used.

To Silver in the Cold Way.

No. 1.—2 dr. tartar,
2 dr. common salt,
½ dr. of alum, and
20 grs. of silver, precipitated from the nitrous acid by copper.

Make them into a paste with a little water. This is to be rubbed on the surface to be silvered with a cork, &c.

No. 2.—Dissolve pure silver in aqua fortis, and precipitate the silver with common salt; make this precipitate into a paste, by adding a little more salt and cream of tartar. It is applied as in the former method.

To Silver Copper Ingots.—The principal difficulties in plating copper ingots are, to bring the surfaces of the copper and silver into fusion at the same time, and to prevent the copper from scaling; for which purposes fluxes are used. The surface of the copper on which the silver is to be fixed must be made flat by filing, and should be left rough. The silver is first annealed, and afterwards pickled in weak spirit of salt; it is planished, and then scraped on the surface to be fitted on the copper. These prepared surfaces are annointed with a solution of borax, or strewed with fine powdered borax itself, and then confined in contact with each other, by binding wire. When they are exposed to a sufficient degree of heat, the flux causes the surfaces to fuse at the same time, and after they become cold, they are found firmly united.

Copper may likewise be plated by heating it, and burnishing leaf-silver upon it; so may iron and brass. This process is called FRENCH

To separate the Silver from Plated Copper.—This process is applied to recover the silver from the plated metal, which has been rolled down for buttons, toys, &c. without destroying any large por-

tinned is put therein, and the boiling continued, when the tin is precipitated in its metallic form.

To tin Iron and Copper Vessels.—Iron which is to be tinned, must be previously steeped in acid materials, such as sour whey, distillers' wash, &c.; then scoured and dipped in melted tin, having been first rubbed over with a solution of sal ammoniac. The surface of the tin is prevented from calcining, by covering it with a coat of fat. Copper vessels must be well cleansed; and then a sufficient quantity of tin with sal ammoniac is put therein, and brought into fusion, and the copper vessel moved about. A little resin is sometimes added. The sal ammoniac prevents the copper from scaling, and causes the tin to be fixed wherever it touches. Lately, zinc has been proposed for lining vessels instead of tin, to avoid the ill consequences which have been unjustly apprehended.

To prepare the Silver Tree.—Pour into a glass globe or decanter, 4 drachms of nitrate of silver, dissolved in a pound or more of distilled water, and lay the vessel on the chimney piece, or in some place where it may not be disturbed. Now pour in 4 drachms of mercury. In a short time the silver will be precipitated in the most beautiful arborescent form, resembling real vegetation. This has been generally termed the Arbor Dianæ.

To prepare the Tin Tree.—Into the same or a similar vessel to that used in the last experiment, pour distilled water as before, and put in 3 drachms of muriate of tin, adding 10 drops of nitric acid, and shake the vessel until the salt be completely dissolved. Replace the zinc (which must be cleared from the effects of the former experiment,) as before, and set the whole aside to precipitate without disturbance. In a few hours, the effect will be similar to the last, only that the tree of tin will have more lustre. In these experiments, it is surprising to observe the laminæ shoot out as it were from nothing; but this phenomenon seems to proceed from a galvanic action of the metals and the water.

To prepare the Lead Tree.—Put ½ an ounce of the super-acetate of lead in powder, into a clear glass globe or wine decanter, filled to the bottom of the neck with distilled water, and 10 drops of nitric acid, and shake the mixture well. Prepare a rod of zinc with a hammer and file, so that it may be a quarter of an inch thick and 1 inch long; at the same time form notches in each side for a thread, by which it is to be suspended, and tie the thread so that the knot shall be uppermost, when the metal hangs quite perpendicular. When it is tied, pass the two ends of the thread through a perforation in the cork, and let them be again tied over a small splinter of wood which may pass between them and the cork. When the string is tried, let the length between the cork and the zinc be such that the precipitant (the zinc) may be at equal distances from the side, bottom, and top, of the vessel, when immersed in it. When all things are thus prepared, place the vessel in a place where it may not be disturbed, and introduce the zinc, at the same time fitting in the cork. The metal will very soon be covered with the lead, which it precipitates from the solution, and this will continue to take place until the whole be precipitated upon the zinc, which will assume the form of a tree or bush, whose leaves and branches are laminal, or plates of a metallic lustre.

Metallic Watering, or for Blanc Moire.—This article of Parisian invention, which is much employed to cover ornamental cabinet work, dressing-boxes, telescopes, opera glasses, &c. &c. is prepared in the following manner.

Sulphuric acid is to be diluted with from seven to nine parts of water; then dip a sponge or rag into it, and wash with it the surface of a sheet of tin. This will speedily exhibit an appearance of crystallization, which is the moiré.

This effect, however, cannot be easily produced upon every sort of sheet tin, for if the sheet has been much hardened by hammering or rolling, then the moiré cannot be effected until the sheet has been heated so as to produce an incipient fusion on the surface, after which the acid will act upon it, and produce the moiré. Almost any acid will do as well as the sulphuric, and it is said, that the citric acid, dissolved in a sufficient quantity of water, answers better than any other.

The moiré may be much improved by employing the blow-pipe, to form small and beautiful specks on the surface of the tin, previous to the application of the acid.

When the moiré has been formed, the plate is to be varnished and polished, the varnish being tinted with any glazing colour, and thus the red, green, yellow, and pearl coloured moirés are manufactured.

Chinese Sheet Lead.—The operation is carried on by two men; one is seated on the floor with a large flat stone before him, and with a moveable flat stone-stand at his side. His fellow workman stands beside him with a crucible filled with melted lead; and having poured

a certain quantity upon the stone, the other lifts the moveable stone,
and dashing it on the fluid lead, presses it out into a flat and thin
plate, which he instantly removes from the stone. A second quantity
of lead is poured in a similar manner, and a similar plate formed, the
process being carried on with singular rapidity. The rough edges of
the plates are then cut off, and they are soldered together for use.

Mr. Waddell has applied this method, with great success, to the
formation of thin plates of zinc, for galvanic purposes.

To plate Looking-Glasses.—This art is erroneously termed silvering, for, as will be presently seen, there is not a particle of silver
present in the whole composition.

On tin-foil, fitly deposed on a flat table, mercury is to be poured,
and gently rubbed with a hare's foot ; it soon unites itself with the
tin, which then becomes very splendid, or, as the workmen say,
is *quickened.* A plate of glass is then cautiously to be slid upon the
tin-leaf, in such a manner as to sweep off the redundant mercury,
which is not incorporated with the tin ; lead weights are then to be
placed on the glass, and, in a little time, the quicksilvered tin-foil
adheres so firmly to the glass, that the weights may be removed without any danger of its falling off. The glass thus coated is a common
looking-glass. About two ounces of mercury are sufficient for covering three square feet of glass.

The success of this operation depends much on the cleanness of
the glass ; and the least dirt or dust on its surface, will prevent the
adhesion of the amalgam or alloy.

Liquid Foil for silvering Glass Globes.—No. 1.

> 1 oz clean lead,
> 1 oz. fine tin,
> 1 oz. bismuth, and
> 10 oz. quicksilver.

The lead and tin must be put into the ladle first, and so soon as
melted the bismuth must be added. Skim off the dross, remove the
ladle from the fire, and before it sets, add the quicksilver : stir the
whole carefully together, taking care not to breathe over it, as the
fumes of the mercury are very pernicious. Pour this through an
earthen pipe, into the glass globe, which turn repeatedly round.

> No. 2.——2 parts mercury,
> 1 part tin,
> 1 part lead, and
> 1 part bismuth.

> No. 3.——4 oz. quicksilver, and
> tin-toil.

The quantity of tin-foil to be added, is so much as will become
barely fluid when mixed. Let the globe be clean and warm, and inject
the quicksilver by means of a pipe at the aperture, turning it about
till it is silvered all over. Let the remainder run out, and hang the
globe up.

LACQUERING.

Lacquer for Brass. 6 oz. seed lac,
   2 oz. amber or copal, ground on porphyry,
   40 gr. of dragon's blood,
   30 gr. extract of red sandal wood, obtained by
    water,
   36 gr. of Oriental saffron,
    4 oz. of pounded glass, and
   40 oz. very pure alcohol.

To apply this varnish to articles or ornaments of brass, expose them to a gentle heat, and dip them into varnish. Two or three coatings may be applied in this manner, if necessary. The varnish is durable, and has a beautiful colour. Articles varnished in this manner, may be cleaned with water and a bit of dry rag.

Lacquer for Philosophical Instruments.—This lacquer or varnish is destined to change, or to modify the colour of those bodies to which it is applied.

   ½ oz. of gum guttæ,
   2 oz. of gum sandarie,
   2 oz. of gum elemi,
   1 oz. of dragon's blood, of the best quality,
   1 oz. of seed lac,
   ½ oz. of terra merita,
   2 grains of Oriental saffron,
   3 oz. of pounded glass, and
   20 oz. of pure alcohol.

The tincture of saffron and of terra merita, is first obtained by infusing them in alcohol for twenty-four hours, or exposing them to the heat of the sun in summer. The tincture must be strained through a piece of clean linen cloth, and ought to be strongly squeezed. This tincture is poured over the dragon's blood, the gum elemi, the seed lac, and the gum guttæ, all pounded and mixed with the glass. The varnish is then made according to the directions before given.

It may be applied with great advantage to philosophical instruments: the use of it might be extended, also, to various cast or moulded articles with which furniture is ornamented.

If the dragon's blood be of the first quality, it may give too high a colour; in this case, the dose may be lessened at pleasure, as well as that of the other colouring matters.

It is with a similar kind of varnish that the artists of Geneva give a golden orange colour to the small nails employed to ornament watch-cases; but they keep the process very secret. A beautiful bright colour might be easily communicated to this mixture; but they prefer the orange colour, produced by certain compositions, the preparation of which has no relation to that of varnish, and which has been successfully imitated with saline mixtures, in which orpiment is a principal ingredient. The nails are heated before they are immersed in the varnish, and they are then spread out on sheets of dry paper.

Gold-coloured Lacquer, for brass Watch-cases, Watch-keys, &c.—

    6 oz. of seed lac,
    2 oz. of amber,
    2 oz. of gum guttæ,
    24 gr. of extract of red sandal wood in water,
    60 gr. of dragon's blood,
    36 gr. of Oriental saffron,
    4 oz. of pounded glass, and
    36 oz. of pure alcohol.

Grind the amber, the seed lac, gum guttæ, and dragon's blood on a piece of porphyry; then mix them with the pounded glass, and add the alcohol, after forming with it an infusion of the saffron and an extract of the sandal wood. The varnish must then be completed as before. The metal articles destined to be covered by this varnish, are heated, and those which will admit of it, are immersed in packets. The tint of the varnish may be varied, by modifying the doses of the colouring substances.

Lacquer of a less drying quality.

    4 oz. seed lac,
    4 oz. sandarac, or mastic,
    ¼ oz. dragon's blood,
    36 gr. terra merita,
    36 gr. gum guttæ,
    5 oz. pounded glass,
    2 oz. clear turpentine,
    32 oz. essence of turpentine,

Extract, by infusion, the tincture of the colouring substances, and then add the resinous bodies according to the directions for compound mastic varnish.

Lacquer or varnishes of this kind are called changing, because, when applied to metals, such as copper, brass, or hammered tin, or to wooden boxes and other furniture, they communicate to them a more agreeable colour. Besides, by their contact with the common metals, they acquire a lustre which approaches that of the precious metals, and to which, in consequence of peculiar intrinsic qualities or certain laws of convention, a much greater value is attached. It is by means of these changing varnishes, that artists are able to communicate to their leaves of silver and copper, those shining colours observed in foils. This product of industry becomes a source of prosperity to the manufacturers of buttons and works formed with foil, which, in the hands of the jeweller, contributes with so much success to produce that reflection of the rays of light which doubles the lustre and sparkling quality of precious stones.

It is to varnish of this kind that we are indebted for the manufacture of gilt leather, which, taking refuge in England, has given place to that of papier maché, which is employed for the decoration of palaces, theatres, &c.

the last place, it is by the effect of a foreign tint obtained
the colouring part of saffron, that the scales of silver dissemi-
in *confection d' hyacinthe* reflect a beautiful gold colour.

o colours transmitted by different colouring substances, require
suited to the objects for which they are destined. The artist
in his own power to vary them at pleasure. The addition of
to to the mixture of dragon's blood, saffron, &c. or some changes
· doses of the mode intended to be made in colours. It is,
ore, impossible to give limited formulæ.

make Lacquer of various Tints.

Infuse separately

> 4 oz. gum guttæ in
> 32 oz. of essence of turpentine,
> 1 oz. annatto, and
> 4 oz. dragon's blood, also in separate doses of es-
> sence.

se infusions may be easily made in the sun. After fifteen
xposure, pour a certain quantity of these liquors into a flask,
· varying the doses different shades of colour will be obtained.
se infusions may be employed also for changing alcoholic var-
: but in this case, the use of saffron, as well as that of red
wood, which does not succeed with essence, will soon give the
cessary for imitating, with other tinctures, the colour of gold.

Bronze Plaster Figures.—For the ground, after it has been
nd rubbed down, take Prussian blue, verditer, and spruce
Grind them separately in water, turpentine, or oil, according
work, and mix them in such proportions as will produce the
desired. Then grind Dutch metal in a part of this composi-
laying it with judgment on the prominent parts of the figure,
produces a grand effect.

rown Gun Barrels.—After the barrel is finished rub it over
quafortis, or spirit of salt, diluted with water. Then lay it
a week, till a complete coat of oil is formed. A little oil is
be applied, and after rubbing the surface dry, polish it with
brush and a little bees' wax.

VARNISHES.

ake White Copal Varnish.—No. 1.—White oxide of lead, ceruse,
h white, white clay. Such of these substances as are preferred
o be carefully dried. Ceruse and clays obstinately retain a
cal of humidity, which would oppose their adhesion to drying
varnish. The cement then crumbles under the fingers, and
t assume a body.

.—On 16 ounces of melted copal, pour 4, 6, or 8 ounces of
oil, boiled and quite free from grease. When well mixed by
d stirrings, and after they are pretty cool, pour in 16 ounces

of the essence of Venice turpentine. Pass the varnish throu
cloth. Amber varnish is made the same way.

Black.—Lamp-black, made of burnt vine twigs, and black of p
stones. The lamp-black must be carefully washed, and after
dried. Washing carries off a great many of its impurities.

Yellow.—Yellow oxide of lead of Naples and Montpellier, bo
duced to impalpable powder. These yellows are hurt by the co
of iron and steel ; in mixing them up, therefore, a horn sp
with a glass mortar and pestle must be employed.

Gum guttæ, yellow ochre, or Dutch pink, according to the n
and tone of the colour to be imitated.

Blue.—Indigo, prussiate of iron, (Prussian blue) blue ver
and ultra-marine. All these substances must be very much divi

Green.—Verdigris, crystallized verdigris, compound green, (a
ture of yellow and blue.) The first two require a mixture of
in proper proportions, from a fourth to two-thirds, according t
tint intended to be given. The white used for this purpose is ei
or the white oxide of lead, or Spanish white, which is less sol
white of Moudon.

Red.—Red sulphurated oxide of mercury, (cinnabar vermil
Red oxide of lead (minium,) different red ochres, or Pru
reds, &c.

Purple.—Cochineal, carmine, and carminated lakes, with o
and boiled oil.

Brick Red.—Dragon's blood.

Chamois Colour.—Dragon's blood with a paste compose
flowers of zinc, or, what is still better, a little red vermilion.

Violet.—Red sulphurated oxide of mercury, mixed with lamp-b
washed very dry, or with the black of burnt vine-twigs ; and to
der it mellower, a proper mixture of red, blue, and white.

Pearl Grey.—White and black ; white and blue ; for exa
ceruse and lamp-black ; ceruse and indigo.

Flaxen Grey.—Ceruse, which forms the ground of the p
mixed with a small quantity of Cologne earth, as much Englis
or carminated lake, which is not so durable, and a partic
prussiate of iron, (Prussian blue.)

To make Varnishes for Violins, &c.—To a gallon of rectified
rit of wine, add six ounces of gum sandarac, three ounces of
mastich, and half a pint of turpentine varnish. Put the whole
a tin can, which keep in a warm place, frequently shaking it
twelve days, until it is dissolved. Then strain, and keep it for

To dissolve Elastic Gum, &c.—M. Grossart, by an ingenious
thod, succeeded in forming India rubber into elastic tubes. C
bottle of the gum circularly, in a spiral slip, of a few lines in brea
then plunge the whole of the slip into vitriolic ether, till it beco
softened ; half an hour is generally sufficient for this purpose.
slip is then taken out of the liquid, and one of the extremities ap
to the end of a mould, first rolling it on itself, and pressing it,
mounting spirally along the cylinder, taking care to lay over

The invention of air balloons led to the idea of applying caout-chouc to the composition of varnish. It was necessary to have a varnish which should unite great pliability and consistence. No var-nish seemed capable of corresponding to these views, except that of caoutchouc, but the desiccation of it is exceedingly tedious.

To make Varnish for Silks, &c.—To one quart of cold-drawn lin-seed oil, poured off from the lees, (produced on the addition of un-slacked lime, on which the oil has stood eight or ten days at the least, in order to communicate a drying quality,—or brown umber, burnt and powdered, which will have the like effect,) and half an ounce of litharge ; boil them for half an hour, then add half an ounce of the copal varnish. While the ingredients are on the fire, in a copper vessel, put in one ounce of chios turpentine, or common resin, and a few drops of neatsfoot oil, and stir the whole with a knife ; when cool, it is ready for use. The neatsfoot oil prevents the varnish from being sticky or adhesive, and may be put into the linseed oil at the same time with the lime, or burnt umber. Resin or chios turpentine may be added, till the varnish has attained the desired thickness.

The longer the raw linseed oil remains on the unslacked lime or umber, the sooner will the oil dry after it is used ; if some months, so much the better : such varnish will set, that is to say, not run, but keep its place on the silk in four hours ; the silk may then be turned, and varnished on the other side.

To make pliable Varnish for Umbrellas.—Take any quantity of caoutchouc, as ten or twelve ounces, cut into small bits with a pair of scissors, and put a strong iron ladle, (such as that in which painters, plumbers, or glaziers melt their lead) over a common pit-coal or other fire ; which must be gentle, glowing, and without smoke. When the la-dle is hot, put a single bit into it : if black smoke issues, it will pre-sently flame and disappear, or it will evaporate without flame ; the ladle is then too hot. When the ladle is less hot, put in a second bit, which will produce a white smoke ; this white smoke will continue during the operation, and evaporate the caoutchouc ; therefore no time is to be lost, but little bits are to be put in, a few at a time, till the whole are melted ; it should be continually and gently stirred with an iron or brass spoon. The instant the smoke changes from white to black, take off the ladle, or the whole will break out into a violent flame, or be spoiled, or lost. Care must be taken that no water be added, a few drops only of which would, on account of its expansi-bility, make it boil over furiously and with great noise ; at this period of the process, 2 pounds or 1 quart of the best drying oil is to be put into the melted caoutchouc, and stirred till hot, and the whole poured into a glazed vessel through a coarse gauze, or wire sieve. When settled and clear, which will be in a few minutes, it is fit for use, either hot or cold.

The silk should be always stretched horizontally by pins or tenter-hooks on frames : (the greater they are in length the better,) and the varnish poured on *cold in hot weather,* and *hot in cold weather.* It is perhaps best, always to lay it on when cold. The art of laying it on properly, consists in making no intestine motion in the varnish, which would create minute bubbles, therefore brushes of every kind are improper, as each bubble breaks in drying, and forms a small hole, through which the air will transpire.

This varnish is pliant, unadhesive, and unalterable by weather.

Varnish used for Indian Shields.—Shields made at Silhet, in Bengal, are noted throughout India, for the *lustre and durability of the black varnish* with which they are covered ; Silhet shields constitute, therefore, no inconsiderable article of traffic, being in request among natives who carry arms, and retain the ancient predilection for the scimitar and buckler. The varnish is composed of the expressed juice of the marking nut, *Semecarpus Anacardium*, and that of another kindred fruit, *Holigarna Longifolia*.

The shell of the *Semecarpus Anacardium* contains between its integuments numerous cells, filled with a black, acrid, resinous juice ; which likewise is found, though less abundantly, in the wood of the tree. It is commonly employed as an indelible ink, to mark all sorts of cotton cloth. The colour is fixed with quick lime. The cortical part of the fruit of *Holigarna Longifolia* likewise contains between its laminæ numerous cells, filled with a black, thick, acrid fluid. The natives of Malabar, extract by incision, with which they varnish targets.

To prepare the varnish according to the method practised in Silhet, the nuts of the *Semecarpus Anacardium*, and the berries of the *Holigarna Longifolia*, having been steeped for a month in clear water, are cut transversely, and pressed in a mill. The expressed juice of each is kept for several months, taking off the scum from time to time. Afterwards the liquor is decanted, and two parts of the one are added to one part of the other, to be used as varnish. Other proportions of ingredients are sometimes employed : but in all the resinous juice of the *Semocarpus* predominates. The varnish is laid on like paint, and when dry, is polished by rubbing it with an agate, or smooth pebble. This varnish also prevents destruction of wood, &c. by the *white ant.*

To give a drying quality to Poppy Oil.

    3lb. of pure water,

    1 oz. of sulphate of zinc, (white vitriol), and

    2 lb. of oil of pinks, or poppy oil.

Expose this mixture in an earthen vessel capable of standing the fire, to a degree of heat sufficient to maintain it in a slight state of ebullition. When one half or two-thirds of the water has evaporated, pour the whole into a large glass bottle or jar, and leave it at rest till the oil becomes clear. Decant the clearest part by means of a glass funnel, the beak of which is stopped with a piece of cork : when the separation of the oil from the water is completely effected, remove the cork stopper, and supply its place by the fore-finger, which must be applied in such a manner as to suffer the water to escape, and to retain only the oil.

Poppy-oil when prepared in this manner becomes, after some weeks, exceedingly limpid and colourless.

To give a drying quality to fat Oils.

No. 1.—8 lb. nut-oil, or linseed-oil,
    1 oz. white lead, slightly calcined,
    1 oz. yellow acetate of lead, (sal saturni) also calcined,
    1 oz. sulphate of zinc, (white vitriol,)
    12 oz. vitreous oxide of lead, (litharge) and
    a head of garlic, or a small onion.

When the dry substances are pulverized, mix them with the garlic and oil, over a fire capable of maintaining the oil in a slight state of ebullition : continue it till the oil ceases to throw up scum, till it assumes a reddish colour, and till the head of garlic becomes brown. A pellicle will then be soon formed on the oil, which indicates that the operation is completed. Take the vessel from the fire, and the pellicle, being precipitated by rest, will carry with it all the unctuous parts which rendered the oil fat. When the oil becomes clear, separate it from the deposit, and put it into wide-mouthed bottles, where it will completely clarify itself in time, and improve in quality.

No. 2.—1½ oz. of vitreous oxide of lead, (litharge)
    ¼ oz. sulphate of zinc, (white vitriol) and
    16 oz. linseed, or nut-oil,

The operation must be conducted as in the preceding case.

The choice of the oil is not a matter of indifference. If it be destined for painting articles exposed to the impression of the external air, or for delicate painting, nut-oil or poppy-oil will be requisite. Linseed oil is used for coarse painting, and that sheltered from the effects of the rain and of the sun.

A little negligence in the management of the fire, has often an influence on the colour of the oil, to which a drying quality is communicated ; in this case it is not proper for delicate painting. This inconvenience may be avoided by tying up the drying matters in a small bag ; but the dose of the litharge must then be doubled. The bag must be suspended by a piece of packthread fastened to a stick, which is made to rest on the edge of the vessel in such a manner as to keep the bag at the distance of an inch from the bottom of the vessel. A pellicle will be formed as in the first operation, but it will be slower in making its appearance.

No. 3.—A drying quality may be communicated to oil by treating, in a heat capable of maintaining a slight ebullition, linseed, or nut oil, to each pound of which is added 3 oz. of vitreous oxide of lead, (litharge) reduced to fine powder.

The preparation of floor-cloths, and all paintings of large figures or ornaments, in which argillaceous colours, such as yellow and red boles, Dutch pink, &c. are employed, require this kind of preparation, that the desiccation may not be too slow ; but painting for which metallic oxides are used, such as preparations of lead, copper, &c. require only the doses before indicated, because these oxides contain a great deal of oxygen, and the oil, by their contact, acquires more of a drying quality.

No. 4.—2 lbs. of nut-oil,
3 lbs. of common water, and
2 oz. of sulphate of zinc, (white vitriol).

Mix these matters, and subject them to a slight ebullition, till little water remains. Decant the oil, which will pass over with a small quantity of water, and separate the latter by means of a funnel. The oil remains nebulous for some time ; after which it becomes clear, and seems to be very little coloured.

No. 5.—6 lbs. of nut-oil, or linseed-oil,
4 lbs. of common water,
1 oz. of sulphate of zinc, and
1 head of garlic.

Mix these matters in a large iron or copper pan ; then place them over the fire, and maintain the mixture in a state of ebullition during the whole day : boiling water must from time to time be added to make up for the loss of that by evaporation. The garlic will assume a brown appearance. Take the pan from the fire, and having suffered a deposit to be formed, decant the oil, which will clarify itself in the vessels. By this process the drying oil is rendered somewhat more coloured : it is reserved for delicate colours.

Resinous Drying Oil.—Take 10 lbs. of drying nut oil, if the paint is destined for external, or 10 lbs. of drying linseed oil, if for internal articles.

3 lbs. of resin, and
6 oz. of turpentine.

Cause the resin to dissolve the oil by means of a gentle heat. When dissolved and incorporated with the oil, add the turpentine : leave the varnish at rest, by which means it will often deposit portions of resin and other impurities ; and then preserve it in wide-mouthed bottles. It must be used fresh ; when suffered to grow old it abandons some of its resin. If this resinous oil assumes too much consistence, dilute it with a little essence, if intended for articles sheltered from the sun, or with oil of poppies.

In Switzerland, where the principal part of the mason's work consists of stones subject to crumble to pieces, it is often found necessary to give them a coating of oil paint, to stop the effects of this decomposition. This painting has a great deal of lustre, and when the last coating is applied with resinous oil, it has the effect of a varnish. To give it more durability, the first ought to be applied exceedingly warm and with plain oil, or oil very little charged, with the grey colour, which is added to the two following.—

Fat Copal Varnish.

16 oz. of picked copal,
8 oz. of prepared linseed oil, or oil of poppies, and
16 oz. of essence of turpentine.

Liquefy the copal in a matrass over a common fire, and then add the linseed oil, or oil of poppies, in a state of ebullition ; when these

3 B

matters are incorporated, take the matrass from the fire, stir the matter till the greatest heat is subsided, and then add the essence of turpentine warm. Strain the whole, while still warm, through a piece of linen, and put the varnish into a wide-mouthed bottle. Time contributes towards its clarification ; and in this manner it acquires a better quality.

Varnish for Watch-Cases, in imitation of Tortoise-shell.

> 6 oz. of copal of an amber colour,
> 1½ oz. Venice turpentine,
> 24 oz. prepared linseed oil, and
> 6 oz. essence of turpentine.

It is customary to place the turpentine over the copal, reduced to small fragments, in the bottom of an earthen or metal vessel, or in a matrass exposed to such a heat as to liquefy the copal : but it is more advantageous to liquefy the latter alone, to add the oil in a state of ebullition, then the turpentine liquefied, and in the last place, the essence. If the varnish is too thick, some essence may be added. The latter liquor is a regulator for the consistence in the hands of an artist.

To make a Colourless Copal Varnish.—As all copal is not fit for this purpose, in order to ascertain such pieces as are good, each must be taken separately, and a single drop of pure essential oil of rosemary, not altered by keeping, must be let fall on it. Those pieces which soften at the part that imbibes the oil, are good ; reduce them to powder, which sift through a very fine hair-sieve, and put it into a glass, on the bottom of which it must not lie more than a finger's breadth thick. Pour upon it essence of rosemary to a similar height ; stir the whole for a few minutes, when the copal will dissolve into a viscous fluid. Let it stand for two hours, and then pour gently on it two or three drops of very pure alcohol, which distribute over the oily mass, by inclining the bottle in different directions with a very gentle motion. Repeat this operation by little and little, till the incorporation is effected, and the varnish reduced to a proper degree of fluidity. It must then be left to stand a few days, and, when very clear, be decanted off. This varnish, thus made without heat, may be applied with equal success, to pasteboard, wood, and metals, and takes a better polish than any other. It may be used on paintings, the beauty of which it greatly heightens.

Gold-coloured Copal Varnish.

> 1 oz. Copal in powder,
> 2 oz. essential oil of lavender, and
> 6 oz. essence of turpentine.

Put the essential oil of lavender into a matrass of a proper size, placed on a sand-bath heated by an Argand's lamp, or over a moderate coal-fire. Add to the oil while very warm, and at several times, the copal powder, and stir the mixture with a stick of white wood, rounded at the end. When the copal has entirely disappeared, add at three different times the essence almost in a state of ebullition; and keep continually stirring the mixture. When the solution is completed, the result will be a varnish of a gold colour, exceedingly durable and brilliant, but less drying than the preceding.

No. 2. To obtain this varnish colourless, it will be proper to rectify the essence of the shops, which is often highly coloured, and to give it the necessary density by exposure to the sun in bottles closed with cork stoppers, leaving an interval of some inches between the stopper and the surface of the liquid. A few months are thus sufficient to communicate to it the required qualities. Besides, the essence of the shops is rarely possessed of that state of consistence, without having at the same time a strong amber colour.

The varnish resulting from the solution of copal in oil of turpentine, brought to such a state as to produce the maximum of solution, is exceedingly durable and brilliant. It resists the shock of hard bodies much better than the enamel of toys, which often becomes scratched and whitened by the impression of repeated friction ; it is applied with the greatest success to philosophical instruments, and the paintings with which vessels and other utensils of metal are decorated.

No. 3.—4 oz. copal, and
1 oz. clear turpentine.

Put the copal, coarsely pulverized, into a varnish pot, and give it the form of a pyramid, which must be covered with turpentine. Shut the vessel closely, and placing it over a gentle fire, increase the heat gradually that it may not attack the copal ; as soon as the matter is well liquefied, pour it upon a plate of copper, and when it has resumed its consistence reduce it to powder.

Put half an ounce of this powder into a matrass with four ounces of the essence of turpentine, and stir the mixture till the solid matter is entirely dissolved.

Camphorated Copal Varnish.—This varnish is destined for articles which require durability, pliableness, and transparency, such as the varnished wire-gauze, used in ships instead of glass.

2 oz. of pulverized copal,
6 oz. of essential oil of lavender,
$\frac{1}{2}$ of an oz. of camphor, and
essence of turpentine, a sufficient quantity, according to the consistence required to be given to the varnish.

Put into a phial of thin glass, or into a small matrass, the essential oil of lavender and the camphor, and place the mixture on a moderately open fire, to bring the oil and the camphor to a slight state of ebullition ; then add the copal powder in small portions, which must be renewed as they disappear in the liquid. Favour the solution, by continually stirring it with a stick of white wood ; and when the copal is incorporated with the oil, add the essence of turpentine boiling ; but care must be taken to pour in, at first, only a small portion.

This varnish is little coloured, and by rest it acquires a transparency which, united to the solidity observed in almost every kind of copal varnishes, renders it fit to be applied with great success in

many cases, and particularly in the ingenious invention of substituting varnished metallic gauze in the room of Muscovy talc, a kind
of mica, in large laminæ, used for the cabin windows of ships, as
presenting more resistance to the concussion of the air during the
firing of the guns. Varnished metallic gauze, of this kind, is manufactured at Rouen.

Ethereal Copal Varnish.—½ oz. of amberry copal, and
　　　　　　　2 oz. of ether.

Reduce the copal to a very fine powder, and introduce it by small
portions into the flask which contains the ether; close the flask with
a glass or cork stopper, and having shaken the mixture for half an
hour, leave it at rest till the next morning. In shaking the flask, if
the sides become covered with small undulations, and if the liquor be
not exceedingly clear, the solution is not complete. In this case, add
a little ether, and leave the mixture at rest. The varnish is of a
light lemon colour. The largest quantity of copal united to ether
may be a fourth, and the least a fifth. The use of copal varnish made
with ether seems, by the expense attending it, to be confined to repairing those accidents which frequently happen to the enamel of
toys, as it will supply the place of glass to the coloured varnishes
employed for mending fractures, or to restoring the smooth surface
of paintings which have been cracked and shattered.

The great volatility of ether, and in particular its high price, do
not allow the application of this varnish to be recommended, but for
the purpose here indicated. It has been applied to wood with complete success, and the glazing it produced, united lustre to solidity.
In consequence of the too speedy evaporation of the liquid, it often
boils under the brush. Its evaporation, however, may be retarded,
by spreading over the wood a slight stratum of essential oil of rosemary, or lavender, or even of turpentine, which may afterwards be
removed by a piece of linen rag; what remains is sufficient to retard
the evaporation of the ether.

Turpentine Copal Varnish.
　　　1½ oz. of copal, of an amber colour, and in powder, and
　　　8 oz. of best oil of turpentine.

Expose the essence to a balneum mariæ, in a wide-mouthed matrass
with a short neck; as soon as the water of the bath begins to boil,
throw into the essence a large pinch of copal powder, and keep the
matrass in a state of circular motion. When the powder is incorporated with the essence, add new doses of it; and continue in this
manner till you observe that there is formed an insoluble deposit.
Then take the matrass from the bath, and leave it at rest for some
days. Draw off the clear varnish, and filter it through cotton.

At the moment when the first portion of the copal is thrown into
the essence, if the powder precipitate itself under the form of lumps,
it is needless to proceed any further. This effect arises from two
causes: either the essence does not possess the proper degree of
concentration, or it has not been sufficiently deprived of water.
Exposure to the sun, employing the same matrass, to which a cork

... ought to be added, will give it the qualities requisite for the ... of the copal. This effect will be announced by the disap-... of the portion of copal already put into it.

Another Copal Varnish.

3 oz. of copal, liquefied, and
20 oz. of essence of turpentine.

Place the matrass containing the oil in a balneum mariæ, and when ... water boils add the pulverized copal in small doses. Keep stir-... the mixture, and add no more copal till the former be incorpo-... with the oil. If the oil, in consequence of its particular dispo-..., can take up 3 ounces of it, add a little more; but stop, if the ... becomes nebulous, then leave the varnish at rest. If it be too ..., dilute it with a little warm essence, after having heated it in ... balneum mariæ. When cold, filter it through cotton, and pre-... it in a clean bottle.

This varnish has a good consistence, and is as free from colour as ... best alcoholic varnish. When extended in one stratum over ... wood, which has undergone no preparation, it forms a very ... glazing, which, in the coarse of two days, in summer, ac-... all the solidity that may be required.

The facility which attends the preparation of this varnish by the new method here indicated, will admit of its being applied to all coloured grounds which require solidity, pure whites alone excepted; painted boxes, therefore, and all small articles, coloured or not coloured, whenever it is required to make the veins appear in all the richness of their tones, call for the application of this varnish, which produces the most beautiful effect, and which is more durable than ... varnishes composed with other resinous substances.

... Amber Varnish.

16 oz. of amber, coarsely powdered,
2 oz. of Venice turpentine, or gum lac,
10 oz. of prepared linseed oil, and
16 oz. of essence of turpentine.

The circumstances of the process are the same as those prescribed ... preparation of the camphorated copal varnish.

... varnish was formerly much used; but it has given place, in ... to that of copal, which is preferred on account of its being less ... Watin introduces more essence and less linseed oil; ... and long practice are the only authority on which I recom-... the adoption of the present formula.

Amber Varnish with Essence of Turpentine.

6 or 7 oz. of liquefied amber, and separated from
the oily portions which alter its consistence.

... the amber to powder, and if the operation of pounding forms ... a paste, break it with your fingers: then mix it with the ..., and heat the whole in a balneum mariæ. It will speedily ... and the essence will take up, at the least, a fourth part of ... of the prepared amber.

... one coating of it is applied to white smooth wood, but

without any preparation, it forms a very pure and very durable glazing, which speedily dries, but slower than copal varnish.

Fat Amber, or Copal Varnish.

> 4 oz. of amber or copal of one fusion,
> 10 oz. of essence of turpentine, and
> 10 oz. of drying linseed oil.

Put the whole into a pretty large matrass, and expose it to the heat of a balneum mariæ, or move it over the surface of an uncovered chaffing-dish, but without flame, and at the distance from it of two or three inches. When the solution is completed, add still a little copal or amber to saturate the liquid; then pour the whole on a filter prepared with cotton, and leave it to clarify by rest. If the varnish is too thick, add a little warm essence to prevent the separation of any of the amber.

This varnish is coloured, but far less so than those composed by the usual methods. When spread over white wood, without any preparation, it forms a solid glazing, and communicates a slight tint to the wood.

If it be required to charge this varnish with more copal, or prepared amber, the liquid must be composed of two parts of essence for one of oil.

Compound Mastic Varnish.

> 32 oz. of pure alcohol,
> 6 oz. of purified mastic,
> 3 oz. of gum sandarac,
> 3 oz. of very clear Venice turpentine, and
> 4 oz. of glass, coarsely pounded.

Reduce the mastic and sandarac to fine powder; mix this powder with white glass, from which the finest parts have been separated by means of a hair sieve; put all the ingredients, with alcohol, into a short-necked matrass, and adapt to it a stick of white wood, rounded at the end, and of a length proportioned to the height of the matrass, that it may be put in motion. Expose the matrass in a vessel filled with water, made at first a little warm, and which must afterwards be maintained in a state of ebullition for one or two hours. The matrass may be made fast to a ring of straw.

When the solution seems to be sufficiently extended, add the turpentine, which must be kept separately in a phial or a pot, and which must be melted, by immersing it for a moment in a balneum mariæ. The matrass must be still left in the water for half an hour, at the end of which it is taken off; and the varnish is continually stirred till it is somewhat cool. Next day it is to be drawn off, and filtered through cotton. By these means it will become exceedingly limpid.

The addition of glass may appear extraordinary; but this substance divides the parts of the mixture, which have been made with the dry ingredients, and it retains the same quality when placed over the fire. It therefore obviates with success two inconveniences, which are exceedingly troublesome to those who compose varnishes. In the first place, by dividing the matters, it facilitates the action of the alcohol and in the second its weight, which surpasses that of resins, prevent these resins from adhering to the bottom of the matrass, and also the coloration acquired by the varnish when a sand-bath is employed, as is commonly the case.

The application of this varnish is suited to articles belonging to the toilette, such as dressing-boxes, cut paper-works, &c. The following possess the same brilliancy and lustre ; but they have more solidity, and are exceedingly drying.

Camphorated Mastic Varnish for Paintings.

> 12 oz. of mastic, cleaned and washed,
> 1¼ oz. of pure turpentine,
> ⅜ oz. of camphor,
> 5 oz. of white glass, pounded, and
> 36 oz. of ethereous essence of turpentine.

Make the varnish according to the method indicated for compound mastic varnish of the first genus. The camphor is employed in pieces, and the turpentine is added when the solution of the resin is completed. But if the varnish is to be applied to old paintings, or paintings which have been already varnished, the turpentine may be suppressed, as this ingredient is here recommended only in cases of a first application to new paintings, and just freed from white of egg varnish.

The ethereous essence recommended for varnish, is that distilled slowly, without any intermediate substance, according to the second process already given for its rectification.

The question by able masters, respecting the kind of varnish proper to be employed for paintings, has never yet been determined.

Some artists, who have paid particular attention to this object, make a mystery of the means they employ to obtain the desired effect. The real end may be accomplished by giving to the varnish, destined for painting, pliability and softness, without being too solicitous in regard to what may add to its consistence or its solidity. The latter quality is particularly requisite in varnishes which are to be applied to articles much exposed to friction, such as boxes, furniture, &c.

To make Painter's Cream.—Painters, who have long intervals between their periods of labour, are accustomed to cover the parts they have painted with a preparation which preserves the freshness of the colours, and which they can remove when they resume their work. This preparation is as follows :

> 3 oz. very clear nut oil,
> ¼ oz. mastic in tears, pulverized, and
> ¼ oz. sal saturni, in powder, (acetate of lead.)

Dissolve the mastic oil over a gentle fire, and pour the mixture into a marble mortar, over the pounded salt of lead ; stir it with a wooden pestle, and add water in small quantities, till the matter assume the appearance and consistence of cream, and refuse to admit more water.

Sandarac Varnish.

> 8 oz. gum sandarac,

        2 oz. pounded mastic,
        4 oz. clear turpentine,
        4 oz. pounded glass, and
        32 oz. alcohol,
Mix, and dissolve as before.

Compound Sandarac Varnish.
        3 oz. pounded copal of an amber colour, once liquefied,
        6 oz. gum sandarac,
        3 oz. mastic, cleaned,
        2½ oz. clear turpentine,
        4 oz. pounded glass, and
        32 oz. pure alcohol.
Mix these ingredients, and pursue the same method as above.

This varnish is destined for articles subject to friction, such as furniture, chairs, fan-sticks, mouldings, &c. and even metals, to which it may be applied with success. The sandarac gives it great durability.

Camphorated Sandarac Varnish for cut Paper Works, Dressing Boxes, &c.
    No. 1.—6 oz. gum sandarac,
        4 oz. gum elemi,
        1 oz. gum anima,
        ½ oz. camphor,
        4 oz. pounded glass, and
        32 oz. pure alcohol.
Make the varnish according to the directions already given. The soft resins must be pounded with the dry bodies. The camphor is to be added in pieces.
    No 2.—6 oz. gallipot, or white incense,
        2 oz. gum animi,
        2 oz. gum elemi,
        4 oz. pounded glass, and
        32 oz. alcohol.
Make the varnish with the precautions indicated for the compound mastic varnish.

The two last varnishes are to be used for ceilings and wainscots, coloured or not coloured : they may even be employed as a covering to parts painted with strong colours.

Spirituous Sandarac Varnish for Wainscotting, small Articles of Furniture, Balustrades, and inside Railing.
    No. 1.—6 oz. of gum sandarac,
        2 oz. of shell lac,
        4 oz. of colophonium, or resin,
        4 oz. of white glass powdered,
        4 oz. of clear turpentine, and
        32 oz. of pure alcohol.
Dissolve the varnish according to the directions given for compound mastic varnish.

This varnish is sufficiently durable to be applied to articles destined

daily and continual use. Varnishes composed with copal ought,
however, in these cases, to be preferred.

No. 2—There is another composition which, without forming part
of the compound varnishes, is employed with success for giving a
polish and lustre to furniture made of wood: wax forms the basis
of it.

Many cabinet-makers are contented with waxing common furniture,
such as tables, chests of drawers, &c. This covering, by means of
heated friction, soon acquires a polish and transparency which resembles those of varnish. Waxing seems to possess qualities peculiar to itself; but, like varnish, it is attended with inconveniences as
well as advantages.

Varnish supplies better the part of glazing; it gives a lustre to
the wood which it covers, and heightens the colours of that destined,
in particular, for delicate articles. These real and valuable advantages are counterbalanced by its want of consistence; it yields too
easily to the shrinking or swelling of the wood, and rises in scales,
or flies, on being exposed to the slightest shock. These accidents can
be repaired only by new strata of varnish, which render application
of the varnisher necessary, and occasion trouble and expense.

Waxing stands shocks; but it does not possess, in the same degree
as varnish, the property of giving lustre to the bodies on which it is
applied, and of heightening their tints. The lustre it communicates
is dull, but this inconvenience is compensated by the facility with
which any accident that may have altered its polish can be repaired,
by rubbing it with a piece of fine cork. There are some circumstances, therefore, under which the application of wax ought to be
preferred to that of varnish. This seems to be the case in particular
with tables of walnut-tree wood, exposed to daily use, chairs, mouldings, and for all small articles subject to constant employment.

But as it is of importance to make the stratum of wax as thin as
possible in order that the veins of the wood may be more apparent,
the following process will be acceptable to the reader.

Melt over a moderate fire, in a very clean vessel, two ounces of
white or yellow wax; and, when liquefied, add four ounces of
good essence of turpentine. Stir the whole until it is entirely cool,
and the result will be a kind of pomade fit for waxing furniture, and
which must be rubbed over them according to the usual method.
The essence of turpentine is soon dissipated; but the wax, which by
this mixture is reduced to a state of very great division, may be extended with more ease, and in a more uniform manner. The essence soon penetrates the pores of the wood, calls forth the colour of
it, causes the wax to adhere better, and the lustre which thence results is equal to that of varnish, without having any of its inconveniences.

*Coloured Varnish for Violins, and other stringed Instruments, also
for Plum-tree, Mahogany, and Rose-wood.*

            4 oz. gum sandarac,
            2 oz. seed lac,

oz. mastic,

1 oz. Benjamin in tears,

4 oz. pounded glass,

2 oz. Venice turpentine, and

32 oz. pure alcohol.

The gum sandarac and lac render this varnish durable : it may be coloured with a little saffron or dragon's blood.

Fat Varnish of a gold colour.

8 oz. amber,

2 oz. gum lac,

8 oz drying linseed oil, and

16 oz. essence of turpentine.

Dissolve separately the gum lac, and then add the amber, prepared and pulverized, with the linseed oil and essence very warm. When the whole has lost part of its heat, mix in relative proportions, tinctures of annatto, of terra merita, gum guttæ, and dragon's blood. This varnish, when applied to white metals, gives them a gold colour.

Fat Turpentine or Golden Varnish, being a mordant to gold and dark colours.

16 oz. boiled linseed oil,

8 oz. Venice turpentine, and

5 oz. Naples yellow.

Heat the oil with the turpentine ; and mix the Naples yellow pulverized.

Naples yellow is an oxide of lead, the composition of which will be given when we come to treat of colouring substances.   It is substituted here for resins, on account of its drying qualities, and in particular of its colour, which resembles that of gold; great use is made of the varnish in applying gold leaf.

The yellow, however, may be omitted when this species of varnish is to be solid and coloured coverings.   In this case an ounce of litharge to each pound of composition may be substituted in its stead, without this mixture doing any injury to the colour which is to constitute the ground, *(la teinte dure.)*

Turners' Varnish for Box Wood.

5 oz. seed lac,

2 oz. gum sandarac,

1½ oz. gum elemi,

2 oz. Venice turpentine,

5 oz. pounded glass, and

24 oz. pure alcohol.

The artists of St. Claude do not all employ this formula, which required to be corrected on account of its too great dryness, which is here lessened by the turpentine and gum elemi.   This composition is secured from cracking, which disfigures these boxes after they had been used for some months.

No. 2.—Other turners employ the gum lac united to a little elemi and turpentine digested some months in pure alcohol exposed to the sun.   If this method be followed, it will be proper to substitute for

Mastic Gallipot Varnish, for grinding Colours.

4 oz. new gallipot, or white incense,
2 oz. mastic,
6 oz. Venice turpentine,
4 oz. pounded glass, and
32 oz. essence of turpentine.

When the varnish is made with the precautions already indicated, add prepared nut oil or linseed oil two ounces.

The matters ground with this varnish dry more slowly; they are then mixed up with the following varnish, if it be for common painting, or with particular varnishes destined for colours and for grounds

Mordant Varnish for Gilding.

1 oz. of mastic,
1 oz. of gum sandarac,
½ oz. of gum gattæ,
½ oz. of turpentine. and
6 oz. of essence of turpentine.

Some artists who make use of mordant, substitute for the turpentine an' ounce of the essence of lavender, which renders this composition still less drying.

In general, the composition of mordants admits of modifications, according to the kind of work for which they are destined. The application of them, however, is confined chiefly to gold. When it is required to fill up a design with gold leaf on any ground whatever, the composition which is to serve as the means of union between the metal and the ground, ought to be neither too thick nor too fluid; because both these circumstances are equally injurious to delicacy in the strokes; it will be requisite also that the composition should not dry till the artist has completed his design.

Other Mordants. No. 1.—Some prepare their mordants with Jew's pitch and drying oil diluted with essence of turpentine. They employ it for gilding pale gold, or for bronzing.

Other artists imitate the Chinese, and mix with their mordants colours proper for assisting the tone which they are desirous of giving to the gold, such as yellow, red, &c.

Others employ merely fat varnish, to which they add a little red oxide of lead (minium).

Others make use of thick glue, in which they dissolve a little honey. This is what they call *batture.* When they are desirous of heightening the colour of the gold, they employ this glue, to which the gold leaf adheres exceedingly well.

No. 2.—The qualities of the following are fit for every kind of application, and particularly to metals. Expose boiled oil to a strong heat in a pan : when a black smoke is disengaged from it, set it on fire, and extinguish it a few moments after by putting on the cover of the pan. Then pour the matter still warm, into a heated bottle, and add to it a little essence of turpentine. This mordant dries very speedily ; it has body and adheres to, and strongly retains, gold leaf, when applied to wood, metals, and other substances.

Varnish for Pales and coarse Wood-work.—Take any quantity of tar, and grind it with as much Spanish brown as it will bear, without rendering it too thick to be used as a paint or varnish, and then spread it on the pales, or other wood, as soon as convenient, for it quickly hardens by keeping.

This mixture must be laid on the wood to be varnished by a large brush, or house painter's tool ; and the work should then be kept as free from dust and insects as possible, till the varnish be thoroughly dry. It will, if laid on smooth wood, have a very good gloss, and is an excellent preservation of it against moisture ; on which account, as well as its being cheaper, it is far preferable to painting, not only for pales, but for weather-boarding, and all other kinds of wood-work for grosser purposes. Where the glossy brown colour is not liked, the work may be made of a greyish brown, by mixing a small proportion of white lead, or whiting and ivory black, with the Spanish brown.

A Black Varnish for Old Straw or Chip Hats.

½ oz. of best black sealing wax, and
2 oz. of rectified spirit of wine.

Powder the sealing-wax, and put it with the spirit of wine, into a four-ounce phial ; digest them in a sand heat, or near a fire, till the wax is dissolved ; lay it on warm with a fine soft hair-brush, before a fire or in the sun. It gives a good stiffness to old straw hats, and a beautiful gloss, equal to new, and resists wet.

To make Varnish for coloured Drawings.

1 oz. of Canada balsam, and
2 oz. of spirit of turpentine.—Mix them together.

Before this composition is applied, the drawing or print should be sized with a solution of isinglass in water ; and when dry, apply the varnish with a camel's-hair brush.

To make Varnish for Wood, which resists the action of Boiling Water.
—Take a pound and a half of linseed-oil, and boil it in a red copper vessel, not tinned, holding suspended over it, in a small linen bag, 5 oz. of litharge, and 3 oz. of pulverized minium ; taking care that the bag does not touch the bottom of the vessel. Continue the ebullition until the oil acquires a deep brown colour ; then take away the bag, and substitute another in its place, containing a clove of garlic ; continue the ebullition, and renew the clove of garlic seven or eight times, or rather put them all in at once.

Then throw into the vessel a pound of yellow amber, after having melted it in the following manner :—Add to the pound of amber, well pulverized, two ounces of linseed oil, and place the whole on a strong fire. When the fusion is complete, pour it boiling into the prepared linseed-oil, and continue to leave it boiling for two or three minutes, stirring the whole up well. It is then left to settle ; the composition is decantered and preserved, when it becomes cold, in well-corked bottles.

After polishing the wood on which this varnish is to be applied, give to the wood the colour required ; for instance, for walnut wood, a slight coat of a mixture of soot with the essence of turpentine.

When this colour is perfectly dry, give it a coat of varnish with a fine sponge, in order to spread it very equal; repeat these coats four times, taking care always to let the preceding coat be dried.

To Varnish Drawings and Card-work.—Boil some clear parchment cuttings in water, in a glazed pipkin, till they produce a very clear size. Strain it, and keep it for use.

Give the work two coats of the size, passing the brush quickly over the work, not to disturb the colours.

To prepare a Composition for making coloured Drawings and Prints resemble Paintings in Oil.

　　　　1 oz. of Canada balsam, and
　　　　2 oz. of spirit of turpentine.—Mixt together.

Before this composition is applied, the drawing or print should be sized with a solution of isinglass in water, and when dry, apply the varnish with a camel-hair brush.

To Varnish Harps and Dulcimers.—Prepare the work with size and red ochre, then take ochre, burnt umber, and red lead, well ground, and mix up a dark brown colour in turpentine varnish, adding as much oil of turpentine that the brush may just be able to pass over the work fair and even. While yet wet, take a muslin sieve, and sift as much Dutch metal, previously powdered, upon it, as is requisite to produce the effect, after which, varnish and polish it.

To Varnish Glass.—Pulverize a quantity of gum adragant, and let it dissolve for twenty-four hours in the white of eggs well beat up; then rub it gently on the glass with a brush.

To Varnish Balloons. No. 1.—The compositions for varnishing balloons have been variously modified; but, upon the whole, the most approved appears to be the bird-lime varnish of M. Faujas St. Fond, prepared after M. Cavallo's method as follows : " In order to render linseed oil drying, boil it with 2 ounces of sugar of lead, and 3 ounces of litharge, for every pint of oil, till they are dissolved, which may be in half an hour. Then put a pound of bird-lime, and half a pint of the drying oil, into an iron or copper vessel, whose capacity should equal about a gallon, and let it boil very gently over a slow charcoal fire, till the bird-lime ceases to crackle, which will be in about half, or three-quarters, of an hour; then pour upon it two and a half pints more of the drying oil, and let it boil about an hour longer; stirring it frequently with an iron or wooden spatula. As the varnish, whilst boiling, and especially when nearly ready, swells very much, care should be taken to remove, in those cases, the pot from the fire, and to replace it when the varnish subsides ; otherwise it will boil over. Whilst the stuff is boiling, the operator should occasionally examine whether it has boiled enough, which may be known by observing whether, when rubbed between two knives, which are then to be separated from one another, the varnish forms threads between them, as it must then be removed from the fire. When nearly cool, add about an equal quantity of oil of turpentine. In using the varnish, the stuff must be stretched, and the varnish applied lukewarm. In 24 hours it will dry."

No. 2. As the elastic resin, known by the name of Indian rubber, has been much extolled for a varnish, the following method of making it, as practised by M. Blanchard, may not prove unacceptable.— Dissolve elastic gum, cut small, in five times its weight of rectified essential oil of turpentine, by keeping them some days together : then boil 1 ounce of this solution in 8 ounces of drying linseed oil for a few minutes ; strain the solution, and use it warm.

To Varnish Rarefied Air Balloons.—With regard to the rarefied-air machines, M. Cavallo recommends, first, to soak the cloth in a solution of sal-ammoniac and common size, using one pound of each to every gallon of water ; and when the cloth is quite dry, to paint it over on the inside with some earthy colour, and strong size or glue. When this paint has dried perfectly, it will then be proper to cover it with oily varnish, which might dry before it could penetrate quite through the cloth. Simple drying linseed oil will answer the purpose as well as any, provided it be not very fluid.

To Paint Sail-Cloth, &c. so as to be pliant, durable, and impervious to Water.—This process, which is extracted from the *Transactions of the Society of Arts,* is now universally practised in the public dock-yards.

The paint usually laid upon canvas hardens to such a degree as to crack, and eventually to break the canvas, which renders it unserviceable in a short time : but the canvas painted in the new manner is so superior, that all canvas used in the navy is thus prepared ; and a saving of a guinea is made in every one hundred square yards of canvas so painted.

The old mode of painting canvas, was to wet the canvas, and prime it with Spanish brown ; then to give it a second coat of a chocolate colour, made by mixing Spanish brown and black paint : and, lastly, to finish it with black.

The new method is to grind 96 lbs. of English ochre with boiled oil, and to add 16 lbs. of black paint, which mixture forms an indifferent black. A pound of yellow soap dissolved in six pints of water over the fire, is mixed, while hot, with the paint. This composition is then laid upon the canvas, (without being wetted, as in the usual way,) as stiff as can conveniently be done with the brush, so as to form a smooth surface ; the next day, or still better, on the second day, a second coat of ochre and black (without any, or but a very small portion of soap) is laid on, and allowing this coat an intermediate day for drying, the canvas is then finished with black paint as usual. Three days being then allowed for it to dry and harden, it does not stick together when taken down, and folded in cloths containing 60 or 70 yards each ; and canvas finished entirely with the composition, leaving it to dry one day between each coat, will not stick together, if laid in quantities.

It has been ascertained from actual trials, that the solution of yellow soap is a preservative to red, yellow, and black paints, when ground in oil and put into casks, as they acquire no improper hardness, and dry in a remarkable manner when laid on with the brush, without the use of the usual drying articles.

It is surprising that the adoption of soap, which is so well known to be miscible with oily substances, or, at least, the alkali of which it is composed, has not already been brought into use in the composition of oil colours.

Coloured Compositions for rendering Linen and Cloth impenetrable to Water.—Begin by washing the stuff with hot water; then dry and rub it between the hands until such time as it becomes perfectly supple; afterwards spread it out by drawing it into a frame, and give it, with the aid of a brush, a first coat, composed of a mixture of eight quarts of boiling linseed oil, 15 *grammes* of calcined amber and acetate of lead, (of each 7½ *grammes*) to which add 90 *grammes* of lamp-black. For the second coat use the same ingredients as above, except the calx of lead. This coat will give a few hours, according to the season; afterwards take a dry plasterer's brush, and rub the stuff strongly with it, when the hair by this operation will become very smooth. The third and last coat will give a perfect and durable jet black.

Or rather take 12 quarts of boiling linseed oil, 30 *grammes* of amber, 15 *grammes* of acetate of lead, 7½ sulphate of zinc, 15 Prussian blue, and 7½ verdigris; mix them very fine with a little oil, and add 120 *grammes* of lamp-black. These coats are used at discretion, as is done with painting.

To Thicken Linen Cloth for Screens and Bed Testers.—Grind whiting with zinc, and to prevent cracking, add a little honey to it; then take a soft brush, and lay it upon the cloth, and so do two or three times, suffering it the mean while to dry between layings on, and for the last laying, smooth it over with Spanish white, ground with linseed oil, the oil being first heated, and mixed with a small quantity of the litharge of gold, the better to endure the weather, and so it will be lasting.

Common Wax, or Varnished Cloth.—The manufacture of this kind of cloth is very simple. The cloth and linseed oil are the principal articles required for the establishment. Common canvas, of an open and coarse texture, is extended on large frames, placed under sheds, the sides of which are open, so as to afford a free passage to the external air. The manner in which the cloth is fastened to these frames is as follows : it is fixed to each side of the frame by hooks which catch the edge of the cloth, and by pieces of strong packthread passing through holes at the other extremity of the hooks, which are tied round moveable pegs placed in the lower edge of the frame. The mechanism by which the strings of a violin are stretched or unstretched, will give some idea of the arrangement of the pegs employed for extending the cloth in this apparatus. By these means the cloth can be easily stretched or relaxed, when the oily varnish has exercised an action on its texture in the course of the operation. The whole being thus arranged, a liquid paste made with drying oil, which may be varied at pleasure, is applied to the cloth.

To make Liquid Paste with Drying Oil.—Mix Spanish white or tobacco-pipe clay, or any other argillaceous matter, with water, and leave it at rest some hours, which will be sufficient to separate the argillaceous parts, and to produce a sediment. Stir the sediment with a broom, to complete the division of the earth ; and after it has rested some seconds, decant the turbid water into an earthen or wooden vessel. By this process the earth will be separated from the sand and other foreign bodies, which are precipitated, and which must be thrown away. If the earth has been washed by the same process, on a large scale, it is divided by kneading it. The supernatant water is thrown aside, and the sediment placed, in sieves, on pieces of cloth, where it is suffered to drain: it is then mixed up with oil rendered drying by a large dose of litharge, that is about a fourth of the weight of the oil. The consistence of thin paste being given to the mixture, it is spread over the cloth by means of an iron spatula, the length of which is equal to that of the breadth of the cloth. This spatula performs the part of a knife, and pushes forward the excess of matter above the quantity sufficient to cover the cloth. When the first stratum is dry, a second is applied. The inequalities produced by the coarseness of the cloth, or by an unequal extension of the paste, are smoothed down with pumice-stone. The pumice-stone is reduced to powder, and rubbed over the cloth with a piece of soft serge or cork dipped in water. The cloth must then be well washed in water to clean it ; and after it is dried, a varnish of gum lac dissolved in linseed oil boiled with turpentine, is to be applied to it.

This preparation produces yellowish varnished cloth. When wanted black, mix lamp-black with the Spanish white, or tobacco-pipe clay, which forms the basis of the liquid paste. Various shades of grey may be obtained, according to the quantity of lamp-black which is added. Umber, Cologne earth, and different ochry argillaceous earths, may be used to vary the tints, without causing any addition to the expense.

To prepare fine Printed Varnished Cloths.—The process just described for manufacturing common varnished and polished cloths, may serve to give some idea of that employed for making fine cloths of the same kind, decorated with a coloured impression. The manufactories of Germany have varnished cloths embellished with large and small subjects, figures, and landscapes, well executed, and which are destined for covering furniture subjected to daily use.

This process, which is only an improvement of the former, requires a finer paste, and cloth of a more delicate texture. The stratum of paste is applied in the same manner, and when dry and polished, the cloth is taken from the frame and removed to the painter's table, where the art of the colourist and designer is displayed under a thousand forms ; and, as in that of printed cottons, exhibits a richness of tints, and a distribution of subjects, which discover taste, and ensure a ready sale for the articles manufactured.

The processes, however, employed in these two arts to extract the

3 C

colouring parts are not the same. In the art of cotton printing the colours are extracted by the bath, as in that of dyeing. In printing varnished cloths, the colouring parts are the result of the union of drying oil mixed with varnish; and the different colours employed in oil painting or painting in varnish.

The varnish applied to common oil cloth is composed of gum lac and drying linseed oil; but that destined for printed varnished cloths requires some choice, both in regard to the oil and the resinous matter which gives it consistence. Prepared oil of pinks and copal form a varnish very little coloured, pliable, and solid.

To prepare Varnished Silk. No. 1.—Varnished silk, for making umbrellas, capots, coverings for hats, &c. is prepared in the same manner as the varnished and polished cloths already described, but with some variation in the liquid paste or varnish.

If the surface of the silk be pretty large, it is made fast to a wooden frame furnished with hooks and moveable pegs, such as that used in the manufacture of common varnished cloths. A soft paste, composed of linseed oil, boiled with a fourth part of litharge; tobacco-pipe clay, dried and sifted through a silk sieve, 16 parts; litharge ground on porphyry with water, dried and sifted in the same manner, 3 parts; and lamp-black, 1 part. This paste is then spread in an uniform manner over the surface of the silk, by means of a long knife, having a handle at each extremity. In summer, twenty-four hours are sufficient for its desiccation. When dry, the knots produced by the inequalities of the silk are smoothed with pumice-stone. This operation is performed with water, and when finished, the surface of the silk is washed. It is then suffered to dry, and flat copal varnish is applied.

If it be intended to polish this varnish, apply a second stratum; after which polish it with a ball of cloth and very fine tripoli. The varnished silk thus made, is very black, exceedingly pliable, and has a fine polish. It may be rumpled a thousand ways without retaining any fold, or even the mark of one. It is light, and thereby proper for coverings to hats, and for making cloaks and caps so useful to travellers in wet weather.

No. 2.—A kind of varnished silk, which has only a yellowish colour, and which suffers the texture of the stuff to appear, is prepared with a mixture of 3 parts boiled oil of pinks, and 1 part of fat copal varnish, which is extended with a coarse brush or a knife. Two strata are sufficient when oil has been freed from its greasy particles over a slow fire, or when boiled with a fourth part of its weight of litharge.

The inequalities are removed by pumice-stone and water; after which the copal varnish is applied. This simple operation gives to white silk a yellow colour, which arises from the boiled oil and the varnish.

This varnished silk possesses all those qualities ascribed to certain preparations of silk which are recommended to be worn as jackets by persons subject to rheumatism.

To recover Varnish.—Clear off the filth with a ley made of potash, and the ashes of the lees of wine; then take 48 ounces of potash, and 16 of the above-mentioned ashes, and put them into 6 quarts of water, and this completes the ley.

To polish Varnish.—This is effected with pumice-stone and Tripoli earth. The pumice-stone must be reduced to an impalpable powder, and put upon a piece of serge moistened with water; with this rub lightly and equally the varnished substance. The tripoli must also be reduced to a very fine powder, and put upon a clean woollen cloth, moistened with olive oil, with which the polishing is to be performed. The varnish is then to be wiped off with soft linen, and when quite dry, cleaned with starch or Spanish white, and rubbed with the palm of the hand.

ANTI-ATTRITION.

To prepare Anti-Attrition.—According to the specification of the patent, this mixture consists of one hundred weight of plumbago, to four hundred weight of hog's-lard, or other grease; the two to be well incorporated. The application is to prevent the effects of friction in all descriptions of engines or machines; and a sufficient quantity must be rubbed over the surface of the axle, spindle, or other part where the bearing is.

ASSAYING OF METALLIC ORES.

Before metallic ores are worked upon in the large way, it will be necessary to enquire what sort of metal, and what portion of it, is to be found in a determined quantity of the ore; to discover whether it will be worth while to extract it largely, and in what manner the process is to be conducted, so as to answer that purpose. The knowledge requisite for this, is called the art of assaying.

Assay of Ores in the Dry Way.—The assaying of ores may be performed either in the dry or moist way; the first is the most ancient, and, in many respects, the most advantageous, and consequently still continues to be mostly used.

Assays are made either in crucibles with the blast of the bellows, or in tests under a muffle.

Assay Weights.—The assay weights are always imaginary, sometimes an ounce represents an hundred weight on the large scale, and is subdivided into the same number of parts, as that hundred weight is in the great; so that the contents of the ore obtained by the assay, shall accurately determine by such relative proportion, the quantity to be expected from any weight of the ore on a larger scale.

Roasting the Ore.—In the lotting of the ores, care should be taken to have small portions from different specimens, which should be pulverized, and well mixed in an iron or brass mortar. The proper quantity of the ore is now taken, and if it contain either sulphur or arsenic, it is put into a crucible or test, and exposed to a moderate degree of heat, till no vapour arises from it; to assist this volatilization, some add a small quantity of powdered charcoal.

Fluxes. To assist the fusion of the ores, and to convert the extraneous matters connected with them into scoria, assayers use different kinds of fluxes. The most usual and efficacious materials for the composition of these are, borax, tartar, nitre, sal ammoniac, common salt, glass, fluor-spar, charcoal powder, pitch, lime, litharge, &c. in different proportions.

Crude of White Flux.—This consists of 1 part of nitre, and 2 of tartar, well mixed together.

Black Flux.—The above crude flux detonates by means of kindled charcoal, and if the detonation be effected in a mortar slightly covered, the smoke that rises unites with the alkalised nitre and the tartar, and renders it black.

Cornish Reducing Flux.

> 10 oz. of tartar,
> 3 oz. and 6 drachms of nitre, and
> 3 oz. and 1 drachm of borax.—Mixt well together.

Cornish Refining Flux.—Deflagrate, and afterwards pulverize, 2 parts of nitre, and 1 part of tartar.

The above fluxes answer the purpose very well, provided the ores be deprived of all their sulphur; or, if they contain much earthy matters, because, in the latter case, they unite with them, and convert them into a thin glass: but if any quantity of sulphur remain, these fluxes unite with it, and form a liver of sulphur, which has the power of destroying a portion of all the metals; consequently, the assay under such circumstances must be very inaccurate. The principal difficulty in assaying appears to be in the appropriation of the proper fluxes to each particular ore, and it likewise appears, that such a discriminating knowledge can only be acquired from an extensive practice, or from a knowledge of the chemical affinities and actions of different bodies upon each other.

In assaying, we are at liberty to use the most expensive materials to effect our purpose, hence, the use of different saline fluxes; but in the working at large, such expensive means cannot be applied; as by such processes the inferior metals would be too much enhanced in value, especially in working very poor ores. In consequence of which, in smelting works, where the object is the production of metals in the great way, cheaper additions are used; such as lime-stone, feldt-spar, fluor-spar, quartz, sand, slate, and slags. These are to be chosen according to the different views of the operator, and the nature of the ores, Thus iron ores, on account of the argillaceous earth they contain, require calcareous additions, and the copper ores, rather slags or vitrescent stones, than calcareous earth.

Humid Assay of Metallic Ores.—The mode of assaying ores for their particular metals by the dry way, is deficient so far as relates to pointing out the different substances connected with them, because they are always destroyed by the process for obtaining the assay metal. The assay by the moist way is more correct, because the different substances can be accurately ascertained. The late celebrated Bergman first communicated this method. It depends upon a knowledge of the chemical affinities of different bodies for each other ; and must be varied according to the nature of the ore ; it is very extensive in its application, and requires great patience and address in its execution. To describe the treatment of each variety of metallic ores, would take up too much of our room ; but to give a general idea, we shall describe the procedure, both in the dry and the humid way, on one species of all the different ores.

To Assay Iron Ores. No. 1.—The ore must be roasted till the vapour ceases to arise. Take 2 assay quintals of it, and triturate them with one of fluor-spar, ¼ of a quintal of powdered charcoal, and 4 quintals of decrepitated sea salt ; this mixture is to be put into a crucible, lined on the inside with clay and powdered charcoal ; a cover must be luted upon the crucible, and the crucible itself exposed to a violent fire for an hour, and when it is cool, broken. When, if the operation has been well conducted, the iron will be found at the bottom of the crucible ; to which must be added those metallic particles, which may adhere to the scoria. The metallic particles so adhering may be separated, by pulverising it in paper, and then attracting them with a magnet.

No. 2.—If the ore should be in a calciform state, mixed with earths, the roasting of it previous to assaying, if not detrimental, is at least superfluous ; if the earths should be of the argillaceous and siliceous kind, to half a quintal of them, add of dry quick lime and fluor-spar of each 1 quintal and ½, reduced to powder, and mix them with ¼ of a quintal of powdered charcoal, covering the whole with one ounce of decrepitated common salt ; and expose the luted crucible to a strong forge fire for an hour and a quarter, then let it gradually cool, and let the regulus be struck off and weighed.

No. 3.—If the ore contain calcareous earth, there will be no occasion to add quick lime ; the proportion of the ingredients may be as follows :—viz. 1 assay quintal of the ore ; 1 of decrepitated sea-salt ; ¼ of powdered charcoal ; and 1 of fluor-spar, and the process conducted as above.

There is a great difference in the reguli of iron ; when the cold regulus is struck with a hammer and breaks, the iron is called cold short : if it break on being struck red-hot, it is called red short ; but if it resist the hammer, both in its cold and ignited state, it is good iron.

Humid Assay of Iron Ore.—To assay the calciform ores, which do not contain much earthy or stony matter, they must be reduced to a fine powder, and dissolved in the marine acid, and precipitated by the Prussian alkali. A determinate quantity of the Prussian alkali must be tried previously, to ascertain the portion of iron which it will precipitate, and the estimate made accordingly. If the iron contain any considerable portion of zinc or manganese, the precipitate must be calcined to redness, and the calx treated with dephlogisticated nitrous acid, which will then take up only the calx of zinc; when this is separated, the calx should again be treated either with nitrous acid, with the addition of sugar, or with the acetous acid, which will dissolve the manganese, if any; the remaining calx of iron may then be dissolved by the marine acid, and precipitated by the mineral alkali; or it may be farther calcined, and then weighed.

Zinc Ores.—Take the assay weight of roasted ore, and mix it well with ⅛ part of charcoal dust, put it into a strong luted earthen retort, to which must be fitted a receiver; place the retort in a furnace, and raise the fire, and continue it in a violent heat for two hours, suffer it then to cool gradually, and the zinc will be found adhering to the neck of the retort in its metallic form.

In the humid way.—Distil vitriolic acid over calamine to dryness; the residuum must be lixiviated in hot water; what remains undissolved is siliceous earth; to the solution add caustic volatile alkali, which precipitates the iron and argil, but keeps the zinc in solution. The precipitate must be re-dissolved in vitriolic acid, and the iron and argil separated.

Tin Ores.—Mix a quintal of tin ore, previously washed, pulverized, and roasted, till no arsenical vapour arises, with half a quintal of calcined borax, and the same quantity of pulverized pitch; these are to be put into a crucible moistened with charcoal-dust and water, and the crucible placed in an air furnace. After the pitch is burnt, give a violent heat for a quarter of an hour, and on withdrawing the crucible, the regulus will be found at the bottom. If the ore be not well washed from earthy matters, a larger quantity of borax will be requisite, with some powdered glass; and if the ore contain iron, some alkaline salt may be added.

In the humid way.—The assay of tin ores in the liquid way, was looked upon as impracticable, till Bergman devised the following method, which is generally successful. Let the tin ore be well separated from its stony matrix, by well washing, and then reduced to the most subtle powder; digest it in concentrated oil of vitriol, in a strong heat for several hours, then, when cooled, add a small portion of concentrated marine acid, and suffer it to stand for an hour or two; then add water, and when the solution is clear, pour it off, and precipitate it by fixed alkali—131 grains of this precipitate, well washed and dried, are equivalent to 100 of tin in its reguline state, if the precipitate consist of pure tin; but if it contain copper or iron, it must be calcined in a red heat for an hour, and then digested in nitrous acid, which will take up the copper; and afterwards in marine acid, which will separate the iron.

Lead Ores.—As most of the lead ores contain either sulphur or arsenic, they require to be well roasted. Take a quintal of roasted ore, with the same quantity of calcined borax, half a quintal of fine powdered glass, a quarter of a quintal of pitch, and as much clean iron filings. Line the crucible with wetted charcoal dust, and put the mixture into the crucible, and place it before the bellows of a forge-fire. When it is red hot, raise the fire for 15 or 20 minutes, then withdraw the crucible, and break it when cold.

In the humid way.—Dissolve the ore by boiling it in dilute nitrous acid; the sulphur, insoluble stony parts, and calx of iron will remain. The iron may be separated by digestion, in the marine acid, and the sulphur by digestion, in caustic fixed alkali. The nitrous solution contains the lead and silver, which should be precipitated by the mineral fixed alkali, and the precipitate well washed in cold water, dried, and weighed. Digest it in caustic volatile alkali, which will take up the calx of silver, the residuum being again dried and weighed, gives the proportion of the calx of lead, 132 grains of which are equal to 100 of lead in its metallic state. The difference of weight of the precipitate before and after the application of the volatile alkali, gives the quantity of silver, 129 grains of which are equal to 100 of silver in its metallic state.

Copper Ores.—Take an exact troy ounce of the ore, previously pulverized, and calcine it well; stir it all the time with an iron rod, without removing it from the crucible; after the calcination add an equal quantity of borax. half the quantity of fusible glass, one-fourth the quantity of pitch, and a little charcoal-dust; rub the inner surface of the crucible, with a paste composed of charcoal-dust, a little fine powdered clay and water. Cover the mass with common salt, and put a lid on the crucible, which is to be placed in a furnace: the fire is to be raised gradually, till it burns briskly, and the crucible continued in it for half an hour, stirring the metal frequently with an iron rod, and when the scoria which adheres to the rod appears clear, then the crucible must be taken out, and suffered to cool; after which it must be broken, and the regulus separated and weighed; this is called black copper, to refine which, equal parts of common salt and nitre are to be well mixed together. The black copper is brought into fusion, and a tea-spoonful of the flux is thrown upon it, which is repeated three or four times, when the metal is poured into an ingot mould, and the button is found to be fine copper.

In the humid way.—Make a solution of vitreous copper ore, in 5 times its weight of concentrated vitriolic acid, and boil it to dryness; add as much water as will dissolve the vitriol thus formed; to this solution add a clean bar of iron which will precipitate the whole of the copper in its metallic form. If the solution be contaminated with iron, the copper must be re-dissolved in the same manner, and precipitated again. The sulphur may be separated by filtration.

Bismuth Ores.—If the ore be mineralized by sulphur, or sulphur and iron, a previous roasting will be necessary. The strong ores require no roasting, but only to be reduced to a fine powder. Take a troy ounce weight and mix it with half the quantity of calcined tartar, and the same of powdered glass; line the crucible with the same, push it as quickly as possible; and when well done, take out the crucible, and let it cool gradually. The regulus will be found at the bottom.

In the Humid way.—Bismuth is easily soluble in nitrous acid or aqua-regia. Its solution is colourless, and is precipitable by the addition of pure water; 118 grains of the precipitate from nitrous acid well washed and dried, are equal to 100 of bismuth in its metallic form.

Antimonial Ores.—Take a common crucible, bore a number of small holes in the bottom, and place it in another crucible a size smaller, luting them well together, then put the proper quantity of ore in small lumps into the upper crucible, and lute thereon a cover; place these vessels on a hearth, and surround them with stones about six inches distant from them; the intermediate space must be filled with ashes, so that the undermost crucible may be covered with them; but upon the upper charcoal must be laid, and the whole made red hot by the assistance of hand-bellows. The antimony being of easy fusion is separated, and runs through the holes of the upper vessel into the inferior one, where it is collected.

Humid Assay of Arseniated Antimony.—Dissolve the ore in aqua-regia, both the regulus and arsenic remain in the solution, the sulphur is separated by filtration. If the solution be boiled with twice its weight of strong nitrous acid, the regulus of antimony will be precipitated, and the arsenic converted into an acid, which may be obtained by evaporation to dryness.

Manganese Ore.—The regulus is obtained by mixing the calx or ore of manganese with pitch, making it into a ball, and putting it into a crucible, lined with powdered charcoal, 1-10th of an inch on the sides, and ⅛ of an inch at bottom, then filling the empty space with charcoal dust, covering the crucible with another inverted and luted on, and exposing it to the strongest heat of a forge for an hour or more.

In the Humid Way.—The ores should be first well roasted to dephlogisticate the calx of manganese and iron, if any, and then treated with nitrous acid to dissolve the earths. The residuum should now be treated with nitrous acid and sugar, by which means a colourless solution of manganese will be obtained, and likewise of the iron, if any. Precipitate with the Prussian alkali, and digest the precipitate in pure water; the prussiate of manganese will be dissolved, whilst the prussiate of iron will remain undissolved.

Arsenical Ores.—This assay is made by sublimation in close vessels. Beat the ore into small pieces, and put them into a matrass, which place in a sand-pot, with a proper degree of heat; the arsenic sublimes in this operation, and adheres to the upper part of the vessel; when it must be carefully collected with a view to ascertain its weight. Sometimes a single sublimation will not be sufficient, for the arsenic in many cases will melt with the ore, and prevent its total volatilization; in which case it is better to perform the first sublimation with a moderate heat, and afterwards bruise the remainder again, and expose it to a stronger heat.

In the Humid Way.—Digest the ore in marine acid, adding the nitrous by degrees to help the solution. The sulphur will be found on the filter; the arsenic will remain in the solution, and may be precipitated in its metallic form by zinc, adding spirit of wine to the solution.

Nickel Ore.—The ores must be well roasted to expel the sulphur and arsenic; the greener the calx proves during this torrefaction, the more it abounds in the nickel; but the redder it is, the more iron it contains. The proper quantity of this roasted ore is fused in an open crucible, with twice or thrice its weight of black flux, and the whole covered with common salt. By exposing the crucible to the strongest heat of a forge-fire, and making the fusion complete, a regulus will be produced. This regulus is not pure, but contains a portion of arsenic, cobalt, and iron. Of the first it may be deprived by a fresh calcination, with the addition of powdered charcoal; and of the second by scorification; but it is with difficulty that it is entirely freed from the iron.

In the Humid Way.—By solution in nitrous acid, it is freed from its sulphur; and by adding water to the solution, bismuth, if any, may be precipitated: as may silver, if contained in it, by the marine acid; and copper, when any, by iron.

To separate cobalt from nickel, when the cobalt is in considerable quantity, drop a saturated solution of the roasted ore in nitrous acid into liquid volatile alkali; the cobaltic part is instantly re-dissolved and assumes a garnet colour; when filtered, a grey powder remains on the filter, which is the nickel. The cobalt may be precipitated from the volatile alkali by any acid.

Cobalt Ores.—Free them as much as possible from earthy matters by well washing, and from sulphur and arsenic by roasting. The ore thus prepared is to be mixed with three parts of black flux, and a little decrepitated sea-salt: put the mixture in a lined crucible, cover it, and place it in a forge-fire, or in a hot furnace, for this ore is very difficult of fusion.

When well fused, a metallic regulus will be found at the bottom, covered with a scoria of a deep blue colour: as almost all cobalt ores contain bismuth, this is reduced by the same operation as the regulus of cobalt; but as they are incapable of chemically uniting together, they are always found distinct from each other in the crucible. The regulus of bismuth having a greater specific gravity, is always at the bottom, and may be separated by a blow with a hammer.

In the Humid Way.—Make a solution of the ore in nitrous acid, or aqua-regia, and evaporate to dryness; the residuum, treated with the acetous acid will yield to it the cobaltic part; the arsenic should be first precipitated by the addition of water.

Mercurial Ores.—The calciform ores of mercury are easily reduced without any addition. A quintal of the ore is put into a retort, and a receiver luted on, containing some water; the retort is placed in a sand-bath, and a sufficient degree of heat given it, to force over the mercury which is condensed in the water of the receiver.

Sulphurated Mercurial Ores.—The sulphureous ores are assayed by distillation in the manner above, only these ores require an equal weight of clean iron-filings to be mixed with them, to disengage the sulphur, while the heat volatilises the mercury, and forces it into the receiver. These ores should likewise be tried for cinnabar, to know whether it will answer the purpose of extracting it from them: for this a determinate quantity of the ore is finely powdered and put into a glass vessel, which is exposed to a gentle heat at first, and gradually increased till nothing more is sublimed. By the quantity thus acquired a judgment may be formed whether the process will answer. Sometimes this cinnabar is not of so lively a colour as that which is used in trade; in this case it may be refined by a second sublimation, and if it be still of too dark a colour, it may be brightened by the addition of a quantity of mercury, and subliming it again.

Humid Assay of Cinnabar.—The stony matrix should be dissolved in nitrous acid, and the cinnabar, being disengaged, should be boiled in 8 or 10 times its weight of aqua regia, composed of 3 parts nitrous, and 1 of marine acid. The mercury may be precipitated in its running form by zinc.

Silver Ores.—Take the assay quantity of the ore finely powdered, and roast it well in a proper degree of heat, frequently stirring it with an iron rod; then add to it about double the quantity of granulated lead, put it in a covered crucible, and place it in a furnace; raise the fire gently at first, and continue to increase it gradually, till the metal begins to work; if it should appear too thick, make it thinner by the addition of a little more lead; if the metal should boil too rapidly, the fire should be diminished. The surface will be covered by degrees with a mass of scoria, at which time the metal should be carefully stirred with an iron hook heated, especially towards the border, lest any of the ore should remain undissolved; and if what is adherent to the hook when raised from the crucible, melts quickly again, and the extremity of the hook, after it is grown cold, is covered with a thin, shining, smooth crust, the scorification is perfect; but, on the contrary, if while stirring it, any considerable clamminess is perceived in the scoria, and when it adheres to the hook, though red hot, and appears unequally tinged, and seems dusty or rough, with grains interspersed here and there, the scorification is incomplete; in consequence of which the fire should be increased a little, and what adheres to the hook should be gently beaten off, and returned with a small ladle into the crucible again. When the scorification is perfect, the metal should be poured into a cone, previously rubbed with a little tallow, and when it becomes cold, the scoria may be separated by a few strokes of a hammer. The button is the produce of the assay.

By Cupellation.—Take the assay quantity of ore, roast and grind it with an equal portion of litharge, divide it into 2 or 3 parts, and wrap each up in a small piece of paper; put a cupel previously seasoned under a muffle, with about 6 times the quantity of lead upon it. When the lead begins to work, carefully put one of the papers upon and after this is absorbed, put on a second, and so on till the whole quantity is introduced; then raise the fire, and as the scoria is formed it will be taken up by the cupel, and at last the silver will remain alone. This will be the produce of the assay, unless the lead contains a small portion of silver, which may be discovered by putting an equal quantity of the same lead on another cupel, and working it at the same time; if any silver be produced it must be deducted from the assay. This is called the witness.

In the humid way.—Boil vitreous silver ore in dilute nitrous acid, using about 25 times its weight, until the sulphur is quite exhausted. The silver may be precipitated from the solution by marine acid, or common salt; 100 grains of this precipitate contain 75 of real silver; if it contain any gold it will remain undissolved. Fixed alkalies precipitate the earthy matters, and the Prussian alkali will show if any other metal be contained in the solution.

To Assay the value of Silver.—The general method of examining the purity of silver is by mixing it with a quantity of lead proportionate to the supposed portion of alloy: by testing this mixture, and afterwards weighing the remaining button of silver. This is the same process as refining silver by cupellation.

It is supposed that the mass of silver to be examined, consists of 12 equal parts, called penny-weights; so that if an ingot weighs an ounce, each of the parts will be 1-12th of an ounce. Hence, if the mass of silver be pure, it is called silver of 12 penny-weights; if it contain 1-12th of its weight of alloy, it is called silver of 11 penny-weights: if 2-12ths of its weight be alloy, it is called silver of 10 penny-weights; which parts of pure silver are called 5 penny-weights. It must be observed here, that assayers give the name penny-weight, to a weight equal to 24 real grains, which must not be confounded with their ideal weights. The assayers' grains are called fine grains. An ingot of fine silver, or silver of 12 penny-weights, contains, then, 288 fine grains; if this ingot contain 1-288th alloy, it is said to be silver of 11 penny-weights and 23 grains; if it contain 4-288ths of alloy, it is said to be 11 penny-weights, 20 grains, &c. Now a certain real weight must be taken to represent the assay-weights: for instance, 36 real grains represent 12 fine penny-weights; this is subdivided into a sufficient number of other smaller weights, which also represent fractions of fine penny-weights and grains. Thus, 18 real grains represent 6 fine penny-weights; 3 real grains represent 1 fine penny-weight, or 24 grains; a real grain and a half represents 12 fine grains: 1-32d of a real grain represents a quarter of a fine grain, which is only 1-752nd part of a mass of 12 penny-weights.

Double Assay of Silver.—It is customary to make a double assay. The silver for the assay should be taken from opposite sides of the ingot, and tried on a touch-stone. Assayers know pretty nearly the value of silver merely by the look of the ingot, and still better by the test of the touch-stone. The quantity of lead to be added is regulated by the portion of alloy, which being in general copper, will be nearly as follows :

Silver of dwts.	grs.	dwts. grs:	Requires from	
11	6 — — —		5 to 6	
0	12 — — —		8 — 9	
From 19	18 to 9 0		12 — 13	Times its weight of lead.
8	6 — 7 12		13 — 14	
6	18 — 6 0		14 — 15	
3	0 — 1 12		0 — 16	
1	12 — 0 18		0 — 20	

The cupel must be heated red hot for half an hour before any metal is put upon it, by which all moisture is expelled. When the cupel is almost white by heat, the lead is put into it, and the fire increased till the lead becomes red hot, smoking, and agitated by a motion of all its parts, called its circulation. Then the silver is to be put on the cupel, and the fire continued till the silver has entered the lead ; and when the mass circulates well, the heat must be diminished by closing more or less the door of the assay furnace. The heat should be so regulated, that the metal on its surface may appear convex and ardent, while the cupel is less red ; that the smoke shall rise to the roof of the muffle ; that undulations shall be made in all directions ; and that the middle of the metal shall appear smooth, with a small circle of litharge, which is continually imbibed by the cupel. By this treatment, the lead and alloy will entirely be absorbed by the cupel, and the silver become bright and shining, when it is said to lighten ; after which, if the operation has been well performed, the silver will be covered with rainbow colours, which quickly undulate and cross each other, and then the button becomes fixed and solid.

The diminution of weight shews the quantity of alloy. As all lead contains a small portion of silver, an equal weight with that used in the assay, is tested off, and the product deducted from the assay weight. This portion is called the witness.

To Assay Plated Metals.—Take a determinate quantity of the plated metal ; put it into an earthen vessel, with a sufficient quantity of the above menstruum, and place it in a gentle heat. When the silver is stripped, it must be collected with common salt ; the calx must be tested with lead, and the estimate made according to the product of silver.

Ores and Earths containing Gold.—No. 1.—That which is now most generally used is by amalgamation, the proper quantity is taken and reduced to a powder ; about 1-10th of its weight of pure quicksilver is added, and the whole triturated in an iron mortar. The attraction subsisting between the gold and quicksilver, quickly unites them in the form of an amalgam, which is pressed through shamois leather ; the gold is easily separated from this amalgam, by exposure to a proper degree of heat, which evaporates the quicksilver, and leaves the gold. This evaporation should be made with luted vessels.

This is the foundation of all the operations by which gold is obtained from the rich mines of Peru, in Spanish America.

No. 2.—Take a quantity of the gold-sand, and heat it red-hot, quench it in water; repeat this two or three times, and the colour of the sand will become a reddish brown. Then mix it with twice its weight of litharge, and revive the litharge into lead, by adding a small portion of charcoal-dust, and exposing it to a proper degree of heat; when the lead revives, it separates the gold from the sand; and the freeing of the gold from the lead must be afterwards performed by cupellation.

No. 3.—Bergman assayed metallic ores containing gold, by mixing two parts of the ore, well pounded and washed, with 1½ of litharge, and 3 of glass; covering the whole with common salt, and melting it in a smith's forge, in a covered crucible; he then opened the crucible, put a nail into it, and continued to do so, till the iron was no longer attacked. The lead was thus precipitated which contained the gold, and was afterwards separated by cupellation.

Humid Assay of Gold mixed with Martial Pyrites.—Dissolve the ore in twelve times its weight of dilute nitrous acid, gradually added; place it in a proper degree of heat; this takes up the soluble parts, and leaves the gold untouched, with the insoluble matrix, from which it may be separated by aqua regia. The gold may be again separated from the aqua regia by pouring ether upon it; the ether takes up the gold, and by being burnt off leaves it in its metallic state. The solution may contain iron, copper, manganese, calcareous earth, or argil; if it be evaporated to dryness, and the residuum heated to redness for half an hour, volatile alkali will extract the copper; dephlogisticated nitrous acid, the earths; the acetous acid, the manganese; and the marine acid, the calx of iron. The sulphur floats on the first solution, from which it should be separated by filtration.

PARTING.

By this process gold and silver are separated from each other. These two metals equally resisting the action of fire and lead, must therefore be separated by other means. This is effected by different menstrua. Nitrous acid, marine acid and sulphur, which cannot attack gold, operate upon silver; and these are the principal agents employed in this process.

Parting by nitrous acid is most convenient, consequently most used,—indeed, it is the only one employed by goldsmiths. This is called simply parting.

That made by the marine acid is by cementation, and is called centrated parting; and parting by sulphur is made by fusion, and called DRY PARTING.

Parting by Aqua-Fortis. This process cannot succeed unless we attend to some essential circumstances: 1st. The gold and silver must be in a proper portion, viz. the silver ought to be three parts to one of gold; though a mass containing two parts of silver to one of gold may be parted. To judge of the quality of this metal to be parted, assayers make a comparison upon a touch-stone, between it and certain needles composed of gold and silver, in graduated proportions, and properly marked; which are called PROOF NEEDLES. If this trial shews that the silver is not to the gold as three to one, the mass is improper for the operation, unless more silver be added; and 2dly, that the parting may be exact, the aqua-fortis must be very pure, especially free from any mixture of the vitriolic or marine acid. For if this were not attended to, a quantity of silver proportionable to these two foreign acids would be separated during the solution; and this quantity of sulphate of silver would remain mingled with the gold, which consequently would not be entirely purified by the operation.

The gold and silver to be parted ought previously to be granulated, by melting it in a crucible, and pouring it into a vessel of water, giving the water at the same time a rapid circular motion, by quickly stirring it round with a stick. The vessels generally used in this operation are called parting glasses, which ought to be very well annealed, and chosen free from flaws; as one of the chief inconveniences attending the operation is, that the glasses are apt to crack by exposure to cold, or even when touched by the hand. Some operators secure the bottom of the glasses by a coating composed of a mixture of new-slaked lime, with beer and whites of eggs spread on a cloth, and wrapped round the glasses at the bottom; over which they apply a composition of clay and hair. The parting glasses should be placed in vessels containing water supported by trivets, with a fire under them; because if a glass should break, the contents are caught in the vessel of water. If the heat communicated to the water be too great, it may be properly regulated by pouring cold water gradually and carefully down the side of the vessel into a parting glass 15 inches high, and 10 or 12 inches wide at the bottom; placed in a copper pan 12 inches wide at bottom, 15 inches wide at top, and 10 inches high; there is usually put about 80 oz. of metal, with twice as much of aqua-fortis.

The aqua-fortis ought to be so strong as to act sensibly on silver when cold, but not so strong as to act violently. Little heat should be applied at first, as the liquor is apt to swell and rise over the vessel; but when the acid is nearly saturated, the heat may safely be increased. When the solution ceases, which is known by the effervescence discontinuing, the liquor is to be poured off, if any grains appear entire, more aqua-fortis must be added, till the silver is all dissolved. If the operation has been performed slowly, the remaining gold will have the form of distinct masses. The gold appears black after parting; its parts have no adhesion together, because the silver dissolved from it has left many interstices. To give them more solidity, and improve their colour, they are put into a test under a muffle, and made red hot, after which they contract and become more solid, and the gold resumes its colour and lustre. It is then called GRAIN GOLD. If the operation has been performed hastily, the gold will have the appearance of black mud or powder, which, after well washing, must be melted.

The silver is usually recovered by precipitating it from the aqua-fortis by means of pure copper. If the solution be perfectly saturated, no precipitation can take place, till a few drops of aqua-fortis are added to the liquor. The precipitate of silver must be well washed with boiling water, and may be fused with nitre, or tested off with lead.

Parting by Cementation.—A cement is prepared, composed of 4 parts of bricks powdered and sifted ; of one part of green vitriol calcined till it becomes red ; and of one part of common salt : this is to be made into a firm paste with a little water. It is called the CEMENT ROYAL.

The gold to be cemented is reduced into plates as thin as money. At the bottom of the crucible or cementing pot, a stratum of cement, of the thickness of a finger, is put, which is covered with plates of gold ; and so the strata are placed alternately. The whole is covered with a lid, which is luted with a mixture of clay and sand. This pot must be placed in a furnace or oven, heated gradually till it becomes red hot, in which it must be continued during 24 hours. The heat must not melt the gold. The pot or crucible is then suffered to cool ; and the gold carefully separated from the cement, and boiled at different times in a large quantity of pure water. It is then assayed upon a touch-stone, or otherwise ; and if it be not sufficiently pure, it is cemented a second time. In this process the vitriolic acid of the bricks, and of the calcined vitriol, decomposes the common salt during the cementation, by uniting to its alkaline base, while the marine acid becomes concentrated by the heat, and dissolves the silver alloyed with the gold. This is a very troublesome process, though it succeeds when the portion of silver is so small that it would be defended from the action of aqua-fortis by the super-abundant gold ; but is little used, except to extract silver, or base metals, from the surface of gold, and thus giving to an alloyed metal, the colour and appearance of pure gold.

Dry Parting.—This process is performed by sulphur, which will easily unite with silver, but does not attack gold. As this dry parting is even troublesome as well as expensive, it ought not to be undertaken but on a considerable quantity of silver alloyed with gold. The general procedure is as follows.—The metal must be granulated; from 1·8 to 1·5 of it (according as it is richer or poorer in the gold,) is reserved, and the rest well mingled with an eighth of powdered sulphur ; and put into a crucible, keeping a gentle fire, that the silver, before melting, may be thoroughly penetrated by the sulphur ; if the fire be hastily urged, the sulphur will be dissipated. If to sulphurated silver infusion, pure silver be added, the latter falls to the bottom, and forms there a distinct fluid, not miscible with the other. The particles of gold having no affinity with the sulphurated silver, join themselves to the pure silver wherever they come in contact, and are thus transferred from the former into the latter, more or less perfectly, according as the pure silver was more or less thoroughly diffused through the mixture. It is for this use that a part of the granulated silver was reserved. The sulphurated mass being brought into fusion, and kept melting for nearly an hour in a covered crucible, one third of the reserved grains is thrown in, which, when melted, the whole is well stirred, that the fresh silver may be distributed through the mixed to collect the gold from it; this is performed with a wooden rod. This is repeated till the whole reserved metal be introduced. The sulphurated silver appears, in fusion, of a dark brown colour; after it has been kept in fusion for a certain time, a part of the sulphur having escaped from the top, the surface becomes white, and some bright drops of silver, about the size of a pea, are perceived on it. When this happens the fire must be immediately discontinued, for otherwise more and more of the silver thus losing its sulphur, would subside and mingle with the part at the bottom, in which the gold is collected. The whole is poured into an iron mortar greased and duly heated. The gold diffused at first through the whole mass, is now found collected in a part of it at the bottom, (amounting only to about as much as was reserved unsulphurated from the mass) by a chisel or hammer, or more perfectly by placing the whole mass with its bottom upwards in a crucible, the sulphurated part quickly melts, leaving, unmelted, that which contains the gold. The sulphurated silver is assayed, by keeping a portion of it in fusion in an open crucible, till the sulphur is dissipated, and then by dissolving it in aqua-fortis. If it should still be found to contain gold, it must be subjected to the same treatment as before. The gold thus collected may be concentrated into a smaller part by repeating the whole process, so that at last it may be parted by aqua-fortis without too much expense.

IRON AND STEEL:

Expeditious mode of reducing Iron Ore into Malleable Iron.—The way of proceeding is by stamping, washing, &c. the calcine and materials, to separate the ore from extraneous matter; then fusing the prepared ore in an open furnace, and instead of casting it, to suffer it to remain at the bottom of the furnace till it becomes cold.

New Method of Shingling and Manufacturing Iron.—The ore being fused in a reverberating furnace, is conveyed, whilst fluid, into an air-furnace, where it is exposed to a strong heat, till a bluish flame is observed on the surface; it is then agitated on the surface, till it loses its fusibility, and is collected into lumps called *loops*. These *loops* are then put into another air-furnace, brought to a white or welding heat, and then *shingled* into *half-blooms* or *slabes*. They are again exposed to the air-furnace, and the half-blooms taken out and forged into *anconies, bars, half-flats,* and *rods* for *wire;* while the *slabes* are passed, when of a welding heat, through the grooved rollers. In this way of proceeding, it matters not whether the iron is prepared from *cold* or *hot-short* metal, nor is there any occasion for the use of finery, charcoal, coke, chafery, or hollow fire ; or any blast by bellows ; or otherwise : or the use of fluxes, in any part of the process.

Approved Method of Welding Iron.—This consists in the skilful *bundling* of the iron to be welded ; in the use of an extraordinary large forge-hammer, in employing a *balling-furnace*, instead of a *hollow fire* or *chafery*; and in passing the iron, reduced to a melting heat, through grooved mill-rollers of different shapes and sizes, as required.

Common Hardening.—Iron, by being heated red hot, and plunged into cold water, acquires a great degree of hardness. This proceeds from the coldness of the water which contracts the particles of the iron into less space.

Case-Hardening.—Case-hardening is a superficial conversion of iron into steel by cementation. It is performed on small pieces of iron, by enclosing them in an iron box, containing burnt leather, bone-dust, or any other carbonic material, and exposing them for some hours to a red heat. The surface of the iron thus becomes perfectly metallized. Iron thus treated is susceptible of the finest polish.

To convert Iron into Steel by Cementation.—The iron is formed into bars of a convenient size, and then placed in a cementing furnace, with sufficient quantity of cement, which is composed of coals of animal or vegetable substances, mixed with calcined bones, &c. The following are very excellent cement :—1st, one part of powdered charcoal, and half a part of wood-ashes well mixed together ; or, 2dly, two parts of charcoal, moderately powdered, one part of bones, horn, hair, or skins of animals, burnt in close vessels to blackness and powdered ; and half a part of wood-ashes ; mix them well together. The bars of iron to be converted into steel, are placed upon a stratum of cement, and covered all over with the same ; and the vessel which contains them, closely luted, must be exposed to a red heat for 8 or 10 hours, when the iron will be converted into steel.

Steel is prepared from bar iron by fusion ; which consists of plunging a bar into melted iron, and keeping it there for some time, by which process it is converted into good steel.

All iron which becomes harder by suddenly quenching in cold water is called steel ; and that steel which in quenching acquires the greatest degree of hardness in the lowest degree of heat, and retains the greatest strength in and after induration, ought to be considered as the best.

Improved process of hardening Steel.—Articles manufactured of steel for the purposes of cutting, are, almost without an exception, hardened from the anvil ; in other words, they are taken from the forger to the hardener without undergoing any intermediate process ; and such is the accustomed routine, that the mischief arising has escaped observation. The act of forging produces a strong scale or coating, which is spread over the whole of the blade ; and to make the evil still more formidable, this scale or coating is unequal in substance varying in proportion to the degree of heat communicated to the steel in forging ; it is, partially, almost impenetrable to the action of

3 D

water when immersed for the purpose of hardening. Hence it is
that different degrees of hardness prevail in nearly every razor manufactured : this is evidently a positive defect ; and so long as it
continues to exist, great difference of temperature must exist likewise. Razor-blades not unfrequently exhibit the fact here stated in
a very striking manner : what are termed clouds, or parts of unequal
polish, derive their origin from this cause ; and clearly and distinctly,
or rather *distinctly* though not *clearly*, show how far this partial
coating has extended, and where the action of the water has been
yielded to, and where resisted. It certainly cannot be matter of asto-
nishment, that so few improvements have been made in the hardening of steel, when the evil here complained of so universally obtains,
as almost to warrant the supposition that no attempt has ever been
made to remove it. The remedy, however, is easy and simple in the
extreme, and so evidently efficient in its application, that it cannot
but excite surprise, that, in the present highly improved state of our
manufactures, such a communication should be made as a discovery
entirely new.

Instead, therefore, of the customary mode of hardening the blade
from the anvil, let it be passed immediately from the hands of the
forger to the grinder ; a slight application of the stone will remove
the whole of the scale or coating, and the razor will then be properly
prepared to undergo the operation of hardening with advantage. It
will be easily ascertained, that steel in this state heats in the fire with
greater regularity, and that when immersed, the obstacles being removed to the immediate action of the water on the body of the steel,
the latter becomes equally hard from one extremity to the other. To
this may be added, that, as *the lowest possible heat at which steel becomes hard is indubitably the best*, the mode here recommended will
be found the only one by which the process of hardening can be
effected with a less portion of fire than is, or can be required in any
other way. These observations are decisive, and will, in all probability, tend to establish in general use what cannot but be regarded
as a very important improvement in the manufacturing of edged steel
instruments.

English cast Steel.—The finest kind of steel, called *English cast
steel*, is prepared by breaking to pieces blistered steel, and then melting
it in a crucible with a flux composed of carbonaceous and vitrifiable ingredients. The vitrifiable ingredient is used only inasmuch as a fusible body, which flows over the surface of the metal in the crucibles,
and prevents the access of the oxygen of the atmosphere. Broken glass
is sometimes used for this purpose.

When thoroughly fused it is cast into ingots, which by gentle heating and careful hammering, are tilted into bars. By this process the
steel becomes more highly carbonized in proportion to the quantity of
flux, and in consequence is more brittle and fusible than before.
Hence, it surpasses all other steel in uniformity of texture, hardness,
and closeness of grain, and is the material employed in all the finest
articles of English cutlery.

To make Edge-tools from cast Steel and Iron.—This method consists in fixing a clean piece of wrought iron, brought to a welding heat, in the centre of a mould, and then pouring in melted steel, so as entirely to envelope the iron; and then forging the mass into the shape required.

To colour Steel Blue.—The steel must be finely polished on its surface, and then exposed to an uniform degree of heat. Accordingly, there are three ways of colouring: first, by a flame producing no soot, as spirit of wine; secondly, by a hot plate of iron; and thirdly, by wood-ashes. As a very regular degree of heat is necessary, wood-ashes for fine work bears the preference. The work must be covered over with them, and carefully watched; when the colour is sufficiently heightened, the work is perfect. This colour is occasionally taken off with a very dilute marine acid.

To distinguish Steel from Iron.—The principal characters by which steel may be distinguished from iron, are as follow :—

1. After being polished, steel appears of a whiter, light grey hue, without the blue cast exhibited by iron. It also takes a higher polish.

2. The hardest steel when not annealed, appears granulated, but dull, and without shining fibres.

3. When steeped in acids the harder the steel is, of a darker hue is its surface.

4. Steel is not so much inclined to rust as iron.

5. In general, steel has a greater specific gravity.

6. By being hardened and wrought, it may be rendered much more elastic than iron.

7. It is not attracted so strongly by the magnet as soft iron. It likewise acquires magnetic properties more slowly, but retains them longer; for which reason, steel is used in making needles for compasses, and artificial magnets.

8. Steel is ignited sooner, and fuses with less degree of heat, than malleable iron, which can scarcely be made to fuse without the addition of powdered charcoal; by which it is converted into steel, and afterwards into crude iron.

9. Polished steel is sooner tinged by heat, and that with higher colours, than iron.

10. In a calcining heat, it suffers less loss by burning, than soft iron does in the same heat, and the same time. In calcination a light blue flame hovers over the steel, either with or without a sulphureous odour.

11. The scales of steel are harder and sharper than those of iron; and consequently more fit for polishing with.

12. In a white heat, when exposed to the blast of the bellows among the coals, it begins to sweat, wet, or melt, partly with light-coloured and bright, and partly with red sparkles, but less crackling than those of iron. In a melting heat too, it consumes faster.

13. In the vitriolic, nitrous, and other acids, steel is violently attacked, but is longer in dissolving than iron. After maceration, according as it is softer or harder, it appears of a lighter, or darker grey colour; while iron on the other hand is white.

GLOSSARY.

Æolopile .... A hollow metallic ball, with a small orifice, to shew the power of steam.

Anneal. .... To expose iron or other metals to the action of fire, in order to reduce them to a greater degree of tenacity.

Anvil. .... A block or mass of iron, with a hardened steel surface, on which smiths and other artificers hammer and fashion their work.

Arbor. .... The principal spindle or axis which communicates motion to the other parts of a machine.

Arm. .... The length of the sail of a windmill measured from the axis.

Arms (Axle) .... The two ends of an axle-tree: projecting supports in machinery.

Ash-hole. .... A receptacle for the ashes which fall from the hearth of a furnace.

Attraction of Cohesion. The attraction which holds the particles of matter to each other.

——— of Gravitation. The force which causes all ponderous bodies to fall towards the earth's centre.

Augur .... The wimble or tool used in the boring of woods.

Automaton .... A machine which, by an internal arrangement, seems to move of itself.

Axis. .... The spindle or centre of any rotatory motion.

—— of oscillation.... The shaft upon which any body vibrates.

—— in peritrochio. .. One of the six mechanical powers; usually called the wheel and axle.

—— of rotation...... The shaft round which any body revolves.

Backboards. .... Boards attached to the rims of the water-wheel, to prevent the water running off the floats into the interior of the wheel.

Backlash .... The hobbling movement of a wheel not fixed firm on its axis.

Back-water. .... The water which impedes the motion of a water-wheel during floods, or from other causes.

Balance .... An instrument which, by the application of the lever, exhibits the weights of bodies.

Batten .... The movable lath or bar of a loom which serves to strike in or close, more or less, the threads of a woof : a long narrow slip of wood in carpentry.

Batter .... A machine used early in the process of the cotton manufacture.

Bayonet. .... A piece of wood or metal with two legs to disengage and re-engage machinery : *vide* Mill-Geering.

Beats .... The strokes made by the pallets or fangs of a spindle in clock or watch movements.

Beetle. .... An implement for flattening the texture of linen or woollen cloth : a heavy mallet.

Bevel-geer .... Wheels in which the teeth are set at angles of various degrees from the radius.

Bitts. .... Small tools used in boring.

Bloom .... A bar of iron to be passed through the rollers of an iron-mill to be elongated into a bar, rod, or hoop.

Blunging .... The act of mixing or kneading clay for the potter's use.

Bobbins. .... Little circular pieces of wood on which the thread of cotton, silk, &c. is wound.

Bolter .... A machine for sifting meal.

Bolting-cloth .... A cloth through which the sifted meal runs.

Brace .... A curved instrument of iron or wood for moving small boring tools called bitts.

Bracket. .... A support fixed to a wall.

Brake. .... A machine for separating the cuticle or outer skin from the flax plant.

Brazing .... The soldering or joining two pieces of metal by melting of brass between the pieces to be joined.

Breast. .... The first part of a revolver carding-engine.

Breasting .... The circular sweep of masonry, &c. which surrounds the shuttle side of a breast-wheel.

Breast-plate. .... A small piece of steel with holes to receive the ends of a drill.

Breast-wheel. .... A water-wheel on which water is admitted at or nearly level with the axis.

Buff-stick. .... A piece of wood covered with buff leather, used for polishing.

Bullet. .... To alter the wards of a lock in such manner that they may be passable by more than one key.

Bush. .... A hole in the nave of a wheel.

Cæteris paribus. ...... Other things being equal.

Calibre. .... The diameter of a hole.

Calk. .... To force oakum, tow, or other material in the joints of vessels, to make them steam, air, or water-tight.

Camb ....,..... An eccentric.

Capstan.............. A vertical post resting on a pivot and turned by powerful arms or levers to raise heavy weights by crane work; a windlass.

Carbon. .... Charcoal.

Card............... Piece of leather containing numerous iron-wire teeth, forming a species of comb; *vide* Cotton-Manufacture.

Case-harden .... The process of converting the surface of iron into steel.

Casting.............. The act of forming metal or other matter into any required shape, by pouring it into moulds while in a fluid state.

Catch............... Various contrivances in mechanics, to act on the principle of a latch.

Cement.............. A composition for joining hard bodies.

Centre-bit............ A boring tool in carpentry.

Centrifugal. .... Flying from the centre.

Centripetal........... Flying to the centre.

Chafery............. A kind of forge in the iron manufacture, where the metal is exposed to a welding heat.

Chaliometer.......... An instrument to measure heat.

Chamfer............. A groove to receive the tenon in carpentry.

Cheeks.............. A term generally applied to those pieces of timber in machinery, which are double, and correspond with each other.

Chord............... Perpendicular let fall from any radius of a circle.

Chuck............... That part of a lath which revolves with the arbor : to this is affixed the article to be turned.

Circumference. .... The measure round any circle.

Clack.............. A bell so contrived that it shall ring when more corn is required to be put in the mill.

Clamp. .... A pile of unburnt bricks raised for burning.

Clip. .... An arrangement to impede velocity by friction.

Clutch .... *Vide* Bayonet.

Cockling............. To entangle.

Cocoon .... A small ball of silk spun by a silk-worm.

Cog .... This word, correctly speaking, implies teeth formed of a different material to the body of the wheel; but is generally used to express all kinds of toothed wheels.

Concentric .... Having the same centre.

Conspiring forces. .... Various forces combined into one.

Constant forces. ...... Force without interruption.

Contractile forces.	Forces which decrease.
Core.	The internal mould which forms a hollow in foundry: as the hollow of a tub or pipe.
Countersink.	To take off the edge round a hole to let in a screw-head, that it may be even with the surface.
Couplings.	To connect two shafts or spindles longitudinally.
Coupling-box.	A strong piece of hollow iron to connect shafting and throw machinery in and out of geer.
Crank.	A bent part of a shaft, by means of which a rectilinear motion is gained.
Crow-bar.	A strong bar of iron used as a temporary lever.
Crown-wheel	A wheel which has teeth at right angles to its radii.
Cycloid.	A geometric curve.
Cylinder.	A long round body; a roller.
Dam.	The bank or wall which pens back the water in a mill-head.
Data.	Facts from which we may deduce results.
Decimetre.	To measure by tenths.
Dent.	The wire staple which constitutes the tooth of a card.
Devil	A machine for dividing rags or cotton in the first process of the manufacture of paper or cotton.
Diameter.	The line which passes through the centre of a circle.
Die.	Pieces of steel for cutting screws, having the threads countersunk on them: a stamp.
Doffer	That part of a carding machine which takes the cotton from the cylinder.
Doffing-plate	The plate which receives the cotton from the doffer.
Dog.	A piece in small machinery which acts as a pall.
Draw-plate	A steel plate, having a gradation of conical holes, through which metals are drawn to be reduced and elongated.
Drench.	To wet or inundate.
Drill-bow.	A small bow moved by hand to impart motion to a drill.
Drum.	A hollow cylinder.
Ductile.	Malleable and soft.
Eccentric	Deviating from the centre; as cambs, attached to the rim or circumference of a shaft for lifting forge hammers, stampers, &c.

Effective-head .... The real head, or that which can be applied to practice.

Effluent............... Flowing from ; running out.

Efflux........... ..... The act of flowing out.

Epicycloid............ The curve described in the air by a point on the circumference of a circle, when this circle rolls on another circle as its base.

Equilibrium........... That peculiar state of rest in which a body is maintained by the force of gravitation, when the quantity of matter in it is exactly equal on each side of the bar or point on which it is supported.

Escapement ....,..... The part of a clock or watch movement which receives the force of the spring or weight, to give motion to the pendulum or balance.

Face of the tooth..... The curved part of a tooth which imparts impulse to another wheel.

Faggot .... Pieces of iron bound together for re-manufacture.

Fan................. Small vanes or sails to receive the impulse of the wind, and, by a connexion with machinery, to keep the large sails of a smock wind-mill always in the direction of the wind : an instrument to winnow corn ; also to decrease speed by its action on the air.

Female-screw......... The spiral threaded cavity in which a screw operates.

File .... A tool used by smiths for the abrasion of metals ; denominated, according to its fineness, rough, bastard, or smooth.

First-mover .... Power, either natural or artificial.

Flanch .... An edge or projection for the better connexion of piping or castings of any description.

Flank of the tooth..... The straight part of a tooth which receives impulse from another wheel.

Float............... The board which receives the impulse of the water either in breast or undershot-wheels.

Floodgate .... A strong framing of timber to pen back or let out water.

Flux .... Ingredients put into a smelting furnace to fuse the ore of metals.

Fly-wheel. .... A heavy wheel to maintain equable motion.

Foot-brake .. .. A machine used in the flax manufacture.

Forge .... A manufactory in which metals are made malleable ; a furnace.

Forge .... To form by the hammer.

Friction............ Inequality of surface ; act of rubbing together.

Frisket .... An iron frame used in printing to keep the sheet of paper on the tympan, and to prevent the margin from being blacked.

Fulcrum .... The point or bar on which a lever rests.

Geering .... Part of mill-work.

Gibbet .... That part of a crane which sustains the weight of goods.

Gig-mill .... A mill in which the nap of woollen cloth is raised by the application of teasels.

Girder .... The largest timber in a floor.

Girt .... Vide Gripe.

Gravity .... Tendency towards the centre of the earth : weight

Gripe .... A pliable lever which can be pressed against a wheel to retard or stop its motion by friction.

Governor .... A pair of heavy balls connected with machinery to regulate the speed on the principle of central force.

Gudgeon .... The centres or pivots of a water-wheel.

Half-stuff .... This term, in general, implies any thing half-formed in the process of the manufacture.

Heald or Heddle ...... Vide Heddle.

Heckle .... A metal comb for the manufacture of flax.

Heddle .... That portion of a loom which imparts motion to the warp of a web during the process of manufacture.

Helve .... The shaft of a forge or tilt-hammer.

Hopper .... A funnel in which grain is deposited, whence it runs between the stones of a flour-mill.

Horology .... The art of constructing machines for measuring time.

Hydraulics .... The science which treats of the motion of fluids, of the resistance which they oppose to moving bodies, and of the various machines in which fluids are the principal agent.

Hydrodynamics .... The science which embraces the phenomena exhibited by water and other fluids, whether they be at rest or in motion : it is generally divided into two heads, hydrostatics and hydraulics.

Hydrostatics .... The science which considers the pressure, equilibrium, and cohesion of fluids.

Impact .... Transmission of force.

Impinge .... To dash against.

Inertia .... That tendency which every piece of matter has, when at rest, to remain at rest; and when in motion, to continue that motion.

In Vacuo...............	Empty space, void.
Isochronal.............	Of equal duration.
Isochronous..........	The vibrations of a pendulum.
Jenney................	A machine used in the process of the cotton manufacture.
Jib...................	*Vide* Gibbet.
Kiln..................	A place where bricks are burnt.
Kink or Kinkle........	The entangling of cordage from overtwisting.
Lateral...............	A horizontal or lengthwise movement.
Lathe.................	Machine used by turners.
Lantern...............	A wheel with staff-teeth; the trundle or wallower.
Leaves................	The teeth of a pinion.
Lever.................	One of the mechanical powers.
Line of centres.......	A line drawn from the centre of one wheel to the centre of another when their circumferences touch each other.
Locomotive...........	The power of changing place.
Loom.................	A machine used by weavers in the making of cloth.
Machinist............	One who makes machines.
Mandrel..............	Part of a lathe; Cone used by smiths; a cylindrical piece of polished iron or steal put down the core or hole of a pipe during the process of elongation.
Mastering............	Preparation of lime used by tanners.
Matrice..............	The concave form of a letter in which the types are cast.
Maximum............	Is the utmost extent of any movement or power.
Mechanist...........	One acquainted with the laws of mechanics.
Mill-head...........	The head of water which is to turn a mill.
Mill-tail...........	The water which has passed through the wheel-race; or is below the mill.
Minimum............	The reverse of maximum.
Momentum...........	The force possessed by matter in motion.
Monkey.............	A weight or mass of iron let fall from a height to drive piles into the earth.
Mortise.............	A joint.
Movement...........	The working part of a watch or clock.
Nave...............	The centre, or that part, of a wheel in which the spokes or arms are fixed.
Nealing............	*Vide* Annealing.
Nippers............	Pincers with cutting edges for dividing metals.
Nitric acid........	A corrosive acid extracted from nitre.
Ouse...............	Preparation of bark used by tanners.

Overshot-wheel. ...... A wheel which receives the water in buckets at not more than 45 degrees from the apex.

Oxyd. .... A combination of oxygen with a metallic or other base.

Oxygen. .... A gas which supports combustion

Paddle .... A kind of oar ; floats to a wheel.

Pall. .... A small piece of metal which falls between the teeth of a ratchet-wheel, to prevent a load which has been raised from descending when the operative power is removed.

Pallet.. .... That part of a watch or clock escapement on which the crown-wheel strikes.

Pendulum .... A weight suspended by a flexible cord to an axis, so as to swing backwards and forwards, when once raised, by the force of gravitation.

Periphery. .... The circumference of a wheel.

Perpendicular. .... At right angles to a given base.

Pick. .... A chisel for dressing the stones of a flour-mill.

Pile .... A large piece of timber, pointed at one end, to drive into the earth to sustain the piers of bridges, &c.

Pin .... To strike a piece of metal with the narrow end of a hammer to form dents and produce elongation.

Pincers. .... A tool formed by placing two levers on one fulcrum, regulated by a screw-movement, for holding bodies firmly.

Pinion .... A small toothed wheel.

Pirn. .... The wound yarn that is on a weaver's shuttle.

Piston.. .... A plug made to fit tight and work up and down a cylinder in hydraulic engines.

Pitch-lines .... The touching circumferences of two wheels which are to act on each other.

Pitch of the wheel..... The distance from the centres of two teeth, measured upon their pitch line.

Pivot. .... A short shaft on which a body turns or vibrates.

Platina. .... A white metal capable of withstanding great heats.

Pliers .... A small tool constructed similarly to pincers.

Plumb .... A leaden weight suspended by a cord to ascertain the perpendicular.

Plunger .... A body that is forced into a fluid in hydraulic engines, to displace its own weight.

Portable steam-engine. A steam-engine built in a compact form, and not attached to the wall of the building in which it works.

Proportional circles. ... *Vide* Pitch-lines.

Proportional radii. .... The radii of two circles whose circumferences are in contact.

Puddling. .... The act of ramming with clay to arrest the progress of water.

Puddling-furnace. ..... A furnace used in the iron manufactures.

Pulley. .... A small wheel over which a strap is passed.

Quintal. .... A French or Spanish weight equivalent to 100lbs. of those respective nations.

Rabbit or Rap-it. ...... The strong wooden spring against which the forge hammer strikes on its ascent.

Race. .... The canal along which the water is conveyed to and from a water-wheel.

Rack. .... A straight bar which has teeth similar to those on a toothed wheel.

Radii. .... The plural of radius.

Radius .... The semi-diameter of a circle; the arm or spoke of a wheel.

Rasp. ... . A species of file, on which the cutting prominences are distinct, being raised by a point instead of an edge.

Rasure .... The act of scraping.

Ratch. .... A bar containing teeth into which the pall drops to prevent machines running back.

Ratchet-wheel. .. .. A wheel having teeth similar to those of a ratch.

Reciprocating. .... Acting alternately.

Rectilinear or Rectilineal Consisting of right lines.

Reed. .... Part of a loom resembling a comb for dividing the warp.

Regulator. .... A small lever in watch-work, which, by being moved, increases or decreases the amount of the balance spring that is allowed to act.

Reel. .... A frame on which yarn may be wound.

Reeling. .... The act of winding yarn on a reel.

Resolution of Forces. .. *Vide* "Of the Action of Forces," page 5.

Reservoir .... A large basin or conservatory of water.

Reverberatory .... Beating back.

Reverberatory-furnace. A furnace used in the iron and copper manufactures.

Rivet. .... To form a head by the percussion of a hammer, to prevent a piece of metal which has been passed through an orifice, to connect things together, from returning.

Roller-gin. ... . A machine to divest cotton of the husk and other superfluous parts, previous to the commencement of the manufacture.

Rotatory.............. Revolving.

Rowans.............. Cotton in that part of the manufacture before it goes to the roving frame.

Rubber.............. A heavy file used for coarse work.

Rubble.............. A mode of building ; *vide* Masonry, page 537.

Rynd.............. The piece of iron that goes across the hole in an upper mill-stone.

Safety-valve......... A valve which fits on the boiler of a steam-engine to guard against accidents by the steam obtaining too high a pressure.

Saw-gin.............. A machine on the principle of the roller-gin.

Scantling .... The length, breadth, and thickness of any solid body taken lineally.

Scapement .... *Vide* Escapement.

Scotching .... The operation of packing hemp before it goes to the market.

Scoria.............. Slag from a smelting furnace.

Scowering Barrel...... An octagonal, or other shaped barrel, in which scrap-iron, &c. is cleansed from rust by friction as it revolves.

Scrap-iron.......... Various pieces of old iron to be re-manufactured.

Screw.............. One of the mechanical powers.

Scribbler-Engine...... An engine used in the process of the cotton manufacture.

Shaft .... A long piece of wood or metal, on which large wheels are fixed in mill-work.

Sheeve .... A small kind of pulley.

Shoulder............ A support by means of a projection from a surface.

Shrouding .... The boards, &c. which form buckets of water-wheels.

Shuttle .... An arrangement to allow or shut off water from a water-wheel ; a small piece of wood which carries the thread in weaving.

Size .... Gelatinous matter made from animal or vegetable substances, and applied to fibrous materials to impart stiffness.

Slag .... Scoria, or refuse from an iron furnace.

Sledge-hammer ...... A heavy hammer, used by a smith with both hands.

Slip.............. Potter's clay of the requisite consistency.

Sluices.............. Vent for water ; a kind of flood-gate

Snail-movement ...... An eccentric.

Solder.............. Various compounds of metals for conjoining other metals that are less fusible than such compound.

Sparables .... From sparrow-bill, small nails to drive into shoes.

Spatula .... A thin knife, used mostly to extend superficially some semi-fluid matter.

Spindle .... A thin piece of wood or steel on which yarn is wound after it has been twisted : a small kind of shaft.

Spokes .... The radial pieces which connect the periphery of a wheel with its centre-piece or nave : this term is only applied to carriages.

Spring .... An elastic body formed of metal or wood.

Spring-arbor .... The arbor or spring round which the main spring of a watch is wound.

Spring-box .... The box which contains the main spring.

Spur-geer .... Wheels whose axes are parallel to each other.

Splice .... To conjoin lengthwise two flexible pieces : by the interposition of their respective parts, so as to maintain them in conjunction by friction.

Staff .... The teeth of a trundle, lantern, or wallower.

Staking-on .... To drive wedges in the bush of a wheel or pulley, to fix it firm on a shaft or spindle.

Start or Strut .... The partitions which determine the form of a bucket in an over-shot wheel ; the shoulder or wrest.

Staves .... The plural of staff.

Steam-boat .... A boat moved by steam power.

Steam-engine .... A machine for applying the force of steam to create motion.

Steel-yard .... A machine which denotes the weight of bodies by placing them at different distances from its fulcrum.

Stereotype .... The art of casting solid plates from movable types, to print from.

Strike .... A thing used to strike any thing level in a measure : the strickle.

Strata .... The plural of stratum.

Stratum .... A single layer or bed of any one thing.

Stuff .... This term is applied to an infinite variety of things ; wood is, by the carpenter, called stuff, so is lime and hair by the bricklayer, and plaster by the plasterer, &c.

Swag .... An unequal or hobbling motion.

Swifts .... The rapid movement in a carding machine.

Swingling .... *Vide* Scotching.

Swing-tree .... Any beam that vibrates.

Swivel .... A thing fixed in another body to turn round upon.

Syphon .... A bent tube with unequal legs through which a fluid will flow by the force of gravity.

Tail-water .... Water which impedes the water-wheel in mill-work.

Tank .... Reservoir for water, &c.

Teasels .... Thistles used to raise the nap of cloth in the gig-mill.

Tenon.............. That part which fills up the mortise.

Tilt-hammer .... A hammer lifted by machinery, to forge iron or steel.

Treadle .... A lever affixed to a crank which communicates motion to machinery by a foot movement.

Throwsting.......... Spinning.

Triblet .... *Vide* Mandrel.

Truckles .... Small rollers for diminishing friction.

Trundle .... A small wheel with staff teeth; the lantern or wallower.

Tuyere or Tue-iron. ... An orifice through which a blast or strong current of air is passed into forges

Tympan .... That part of a printing-press on which the paper is laid to receive the impression.

Undershot-wheel ...... A wheel acted on by water below its centre.

Vacuum............. Void of air.

Valve .... A cover to an aperture, in hydraulic machines, to prevent fluids taking a wrong course.

Vane .... A flat surface capable of being moved by the current of a fluid; as, for instance, the vanes of a windmill, moved by the wind.

Tappets .... Projections on the plug-tree of a steam-engine which open and shut the valves at proper intervals.

Varnish .... A solution of certain resinous bodies in spirits or oils, which assumes a solid form on dissication.

Velocity .... The measure of quickness with which a body moves.

Vertical .... Perpendicular to the horizon.

Vibration .... Rapid alternating motion.

Virtual head.......... The real or effective head.

Vis-inertia........... *Vide* Inertia.

Wabble .... A hobbling unequal motion.

Wallower .... Small wheel with staff teeth; the trundle or lantern.

Warp .... The layer of threads which extends the length of the piece to be woven.

Washers .... Small pieces of metal placed under a nut to reduce friction.

Water-wheel .... A wheel which receives its impulse from water.

Weathering .... The angle at which the sails of a windmill are set, to receive the impulse of the wind.

Wedge .... An angularly shaped piece of wood or metal; one of the mechanical powers.

Weft................. *Vide* Woof.

Weight .... The measure of the amount of the attraction of gravitation in any body compared with that of other bodies.

Welding............. The property of conjunction possessed by some metals at high temperatures.

Wheel and Axis ...... One of the mechanical powers.

Wheel-race.......... The place in which a water-wheel is fixed.

Whip .... To bind two rods together with small twine: the length of the sail of a windmill measured from the axis.

Whirl............... A rotatory motion with a decreasing speed.

Winch .... The lever or handle to which force is applied in machines turned by manual labour.

Wiper.. .... An eccentric.

Wire-draw.......... To reduce any longitudinal body exceedingly in the transverse section: rapid passage of a fluid through a conical orifice.

Woof.. .... Those portions of thread or yarn in cloth, which lie across the length of the warp.

Wrest or Wrist ...... The partitions which determine the form of the bucket in an overshot wheel; the start or shoulder.

Yarn .... The combination of fibrous materials into a linear form by torsion.

INDEX.

3 E

SD - #0010 - 280920 - C0 - 229/152/43 - PB - 9780282481056